HANDBOOK OF PSYCHIATRIC GENETICS

Edited by

Kenneth Blum, Ph.D.
Ernest P. Noble, M.D., Ph.D.

Associate Editors
Robert S. Sparkes, M.D., Ph.D.
Thomas H. J. Chen, Ph.D.
John G. Cull, Ph.D.

CRC Press
Boca Raton   New York   London   Tokyo

Senior Acquisitions Editor:	Paul Petralia
Editorial Assistant:	Cindy Carelli
Project Editor:	Carrie L. Unger
Marketing Manager:	Susie Carlisle
Direct Marketing Manager:	Becky McEldowney
Cover design:	Denise Craig
PrePress:	Kevin Luong
Manufacturing:	Sheri Schwartz

Library of Congress Cataloging-in-Publication Data

Handbook of psychiatric genetics / edited by Kenneth Blum, Ernest P. Noble: associate editors, Robert S. Sparkes, John G. Cull, Thomas H. J. Chen
  p. cm.
  Includes bibliographical references and index.
  ISBN 0-8493-4486-7
  1. Mental illness—Genetic aspects. 2. Behavior genetics. 3. Psychophysiology—Genetic aspects.
I. Blum, Kenneth. II. Noble, Ernest P.
  [DNLM: 1. Mental Disorders—genetics. WM 140 H236 1996]
  RC455.4.G4H36 1996
  616'.89'042—dc20
  DNLM/DLC
  for Library of Congress   96-19053
                                                                                                                        CIP

This book contains information obtained from authentic and highly regarded sources. Reprinted material is quoted with permission, and sources are indicated. A wide variety of references are listed. Reasonable efforts have been made to publish reliable data and information, but the author and the publisher cannot assume responsibility for the validity of all materials or for the consequences of their use.

Neither this book nor any part may be reproduced or transmitted in any form or by any means, electronic or mechanical, including photocopying, microfilming, and recording, or by any information storage or retrieval system, without prior permission in writing from the publisher.

All rights reserved. Authorization to photocopy items for internal or personal use, or the personal or internal use of specific clients, may be granted by CRC Press, Inc., provided that $.50 per page photocopied is paid directly to Copyright Clearance Center, 27 Congress Street, Salem, MA 01970 USA. The fee code for users of the Transactional Reporting Service is ISBN 0-8493-4486-7/97/$0.00+$.50. The fee is subject to change without notice. For organizations that have been granted a photocopy license by the CCC, a separate system of payment has been arranged.

The consent of CRC Press does not extend to copying for general distribution, for promotion, for creating new works, or for resale. Specific permission must be obtained in writing from CRC Press for such copying.

Direct all inquiries to CRC Press, Inc., 2000 Corporate Blvd., N.W., Boca Raton, Florida 33431.

© 1997 by CRC Press, Inc.

No claim to original U.S. Government works
International Standard Book Number 0-8493-4486-7
Library of Congress Card Number 96-19053
Printed in the United States of America 1 2 3 4 5 6 7 8 9 0
Printed on acid-free paper

Preface

The study of psychiatric genetics has become increasingly important as a specialty area within medical genetics. This domain, originally restricted to a few researchers, has now become a vast (although somewhat uncharted) common ground for scientists from very diverse fields including psychiatry, psychology, medical and population genetics, anthropology, molecular biology, biochemistry, pharmacology, neurology, and medical ethics. The increased interest stems principally from advances in molecular genetic techniques, the genome project, the neurosciences, enhanced public awareness of the role of genes in somatic diseases, and more recently, the finding of genes for complex mental disorders. The announcements of genes associated with such devastating genetically based single-gene disorders such as Huntington's disease, cystic fibrosis, muscular dystrophy, lung cancer, breast cancer, colon cancer, diabetes, and even aging has profoundly aroused the interest of people all over the world.

In this regard, although the flow of information identifying genes for a number of somatic diseases continues at an almost breathtaking pace, many gaps in the field of genes and behavior exist due to the controversial nature of the subject. To date, while finding genes for such important mental illnesses as schizophrenia, bipolar disorder, Alzheimer's disease, substance use disorder, obesity, and violent and aggressive behavior continues to capture the interest of humans everywhere, because of the potential effect of environment on net expression, more data are needed to make any definitive conclusions.

As researchers caught up on the field themselves, the more they worked the more they began to appreciate the complexity of the field we refer to as *Psychiatric Genetics*. To put things in perspective, the idea for compiling this *Handbook* came from Paul Petralia of CRC Press.

While feeling privileged to take on the task of editing, coordinating, and attracting world class scientists to contribute to this compendium, we were cognizant of the controversies that existed in the field, the politics related to the genome project, and the polarization of established scientific institutions and organizations as well as the scientists themselves.

Prior to the report by Janice Egeland and colleagues of linkage of a genetic marker on chromosome 11 to manic depressive disorder in Old Order Amish families, little existed in terms of linking specific genetic markers in psychiatric medicine. Following this report, we published in 1990 a paper in the *Journal of the American Medical Association* suggesting that a specific genetic variant was associated to a severe form of alcoholism. Unfortunately, it was often erroneously reported that we had found the "alcoholism gene", implying that there is a one-to-one relation between a gene and a specific behavior. Such misinterpretations are common—readers may recall accounts of an "obesity" or a "personality" gene. Needless to say, there is no such thing as a specific gene for alcoholism, obesity or a particular type of personality. However, it would be naive to assert the opposite, that these aspects of human behavior are not associated with any particular genes, rather the issue at hand is to understand how certain genes and behavioral traits are connected.

Expression of complex human behavior involves both gene and environmental effects. Plomin and colleagues assert that the success of molecular genetics in elucidating the genetic bases of behavioral disorders has largely relied on a reductionistic one gene, one disorder (OGOD) approach. In contrast, a quantitative trait loci (QTL) approach involves the search for multiple genes, each of which is neither necessary nor sufficient for the development of the trait. The OGOD and QTL approaches are highlighted in this volume. Other concepts explored involve the appropriate use of linkage vs. allelic association studies, in terms of complex diseases having polygenic inheritable components as well as the appropriate use of "super" normal controls and the characterization of the true disease phenotype.

This volume provides an outline of the basic principles and statistical approaches to dissect the genes involved in a variety of complex mental disorders. The book's general approach is "clinically oriented", i.e., attention is given primarily to research geared to gene searches as they relate to psychiatric genetics in humans and to other currently relevant human problems, either for epidemiological reasons or because of their health and social implications.

This handbook consists of six units. The first covers current concepts and analytic approaches to unravel genetic mechanisms in psychiatry. This includes association vs. linkage analysis in mental disorders and the value of narrow psychiatric phenotypes and appropriate "super" normal controls.

The second section surveys novel methods in DNA analyses involving isolation of coding sequences and identification of microsatellite repeat markers and transcribed sequences in cloned genomic DNA, pointing out important implications for neurological diseases.

The third section reviews molecular biology of receptors and associated proteins. It includes receptor cloning methodologies, opioid, muscarinic, dopaminergic, serotonergic, and adrenergic receptors, and the characterzation of the GABA transporter.

The fourth section includes articles on the genetics of psychiatric disorder including common childhood psychopathology, Alzheimer's disease, polygenic inheritance in psychiatric disorders, bipolar disorders, personality, attention-deficit/hyperactivity and, finally, reward deficiency syndrome.

Section five deals with impulsive-addictive and compulsive disorders. It includes polymorphisms of the D2 dopamine receptor gene in alcoholism, cocaine, and nicotinic dependence and obesity, as well as polysubstance abusers and association vs. linkage studies. It also includes studies involving alleles of the D2 dopamine receptor gene in severe Japanese alcoholics. Another chapter discusses the behavioral genetics of cigarette smoking in twins. The final chapter reviews evidence for and against D2 dopamine receptor gene locus in reward behaviors.

The final section includes articles reviewing animal genetic research and how genetic discoveries impact society and includes the role of QTL in animal genetics of behavior, genetic determinants of alcohol preference and aversion, ethical issues, genetic screening, gene therapy, and scientific conduct.

In a review volume of this size, it is not possible to cover every aspect of the subject; however, the editors have attempted to compile an outline that is comprehensive that could serve as a "state of the art" framework for a rather new discipline. Every effort has been made to provide an informative, basic text that presents as wide a view as possible of the current status of psychiatric genetics and illustrates the direction in which this emerging discipline is rapidly moving.

It is hoped that the contents of this compendium will be useful in providing geneticists and other relevant scientists with a greater appreciation of the general principles of their discipline and appropriate approaches in the study of the central nervous system. The volume also serves as an important reference source and review to students interested in the field of psychiatric genetics. The volume also is addressed to basic scientists, clinicians, and other professionals who have a specialized interest in the molecular genetics of the central nervous system.

The editors are greatly indebted to all who have contributed to this manuscript given their busy schedules. We wish to thank Boris Tabakoff for encouraging CRC Press to go forward with the project. We thank both Sadie Phillips and Anne Jaeger for secretarial assistance and also Cindy Carelli for important liaison work with all the contributors.

Finally, the editors acknowledge the wisdom and clear vision of Paul Petralia for inviting us to edit such an important first of its kind compendium.

The Editors

Kenneth Blum, Ph.D., is an internationally recognized authority in psychopharmacology and is a distinguished academician. Dr. Blum has been a Full Professor of Pharmacology at the University of Texas Health Science Center in San Antonio, Texas for the past 23 years. He now serves as a Research Professor in the Department of Behavioral Sciences, School of Public Health, University of Texas Health Science Center/Houston in San Antonio. Dr. Blum is Chairman and President of NeuGen, Inc. in San Antonio, Texas and is a Co-manager of Synergen Neutraceuticals, a Limited Liability Corporation based in San Antonio.

Dr. Blum has published over 300 scientific articles relating to Neuropsychopharmacogenetics on the aspects of addictive/impulsive/compulsive behaviors. In addition to his *Handbook of Psychiatric Genetics*, published by CRC Press, Inc., he has edited seven scientific volumes. Dr. Blum has chaired two distinguished Gordon Research Conferences and is currently the Chairman of the first Gordon Research Conference on Psychoneurogenetics. He is the recipient of the National Institute on Drug Abuse (NIDA) Career Teacher Award.

Dr. Blum was the co-founder and former editor-in-chief of the *Journal of Substance Use and Alcohol: Actions and Misuse* (Pergamon). He is a member of a number of editorial boards and scientific societies. He has been named Chairman of the International Scientific Advisory Board of the Secular Organization of Sobriety (SOS). Among other honors, Dr. Blum is the recipient of "Mind Science Imagineer Award" (Finalist), as well as the National Council on Alcoholism's John B. Sullivan Presidential Award for Excellence in Research (San Antonio). He discovered, along with his colleagues, the first molecular genetic variant of the dopamine D2 receptor gene (DRD2) in severe alcoholism and other addictive behaviors and is credited with defining the "Reward Deficiency Syndrome" (RDS). He is also the holder of many domestic and international patents. He is the developer of the concept of "Amino Acid Therapy and Enkephalinase" inhibition to treat Reward Deficiency Syndrome behaviors.

Dr. Blum has authored the *Handbook of Abusable Drugs* (1984, 1997), *Alcohol and the Addictive Brain* (1991), and *Overload: ADHD and the Addictive Brain* (1996). Dr. Blum received his Bachelor's degree in Pharmacy from Columbia University in New York City, his Master's degree of Medical Sciences from the New Jersey College of Medicine, Jersey City, and he received his Doctorate in Neuropharmacology from the New York Medical College, New York City.

Dr. Ernest P. Noble, Ph.D., M.D., is a distinguished educator, biochemist, and clinical psychiatrist. He is considered by many of his colleagues to be one of the world's foremost leaders in the field of alcohol research and is widely recognized for his pioneering research on the effects of alcohol on the brain.

Dr. Noble came to UCLA in 1981 with a distinguished academic and service career. After many years as a researcher at both Stanford University and the University of California at Irvine and as a former Director of the National Institute of Alcohol Abuse and Alcoholism (NIAAA) in Washington, D.C., he was appointed the Pike Professor of Alcohol Studies at UCLA and Director of the UCLA Alcohol Research Center. There he is directing fundamental studies aimed at understanding the biochemical, physiological, and genetic causes of alcoholism. Dr. Noble was also the former Vice President of the National Council on Alcoholism and former President of the International Commission for the Prevention of Alcoholism and Drug Dependency.

Dr. Noble received his B.S. degree in Chemistry from the University of California at Berkeley. He obtained a Ph.D. in Biochemistry from Oregon State University and an M.D. from Case Western Reserve University, followed by a medical internship and psychiatric residency at Stanford Univeristy. He has been a Fulbright Scholar at the Sorbonne in Paris, a Guggenheim Fellow at the Centre de Neurochimie in Strasbourg, France, and a Senior Fulbright Scholar at the Max-Planck Institute for Psychiatry in Munich, Germany.

Dr. Noble has over 350 scientific and medical publications to his credit and is a member of numerous editorial boards and scientific and professional societies. He has also received many honors for his achievements. In 1990, he and his colleagues were the first to discover the association of the D2 dopamine receptor (DRD2) with alcoholism. This study received wide national and international recognition. Subsequently, he has also found the same gene to be involved in other substance abuse disorders including cocaine and nicotine addiction and obesity. It is believed that the finding of this reward or pleasure gene opens a window of hope and an opportunity for the prevention and treatment of alcoholism, other drug addictions, and obesity.

Contributors

Roberta Agabio, M.D.
Department of Neuroscience
University of Cagliari
Cagliari, Italy

Tadao Arinami, M.D.
Department of Medical Genetics
Institute of Basic Medical Sciences
University of Tsukuba
Ibaraki, Japan

John K. Belknap, Ph.D.
Department of Behavioral Neuroscience
Oregon Health Sciences University and
  Veterans Affairs Medical Center
Portland, Oregon

Wade Berrettini, M.D., Ph.D.
Department of Psychiatry
Thomas Jefferson University
Philadelphia, Pennsylvania

Kenneth Blum, Ph.D.
Department of Behavioral Sciences
University of Texas School of Public Health
Houston, Texas and
Consumatory Behaviorial Associates
San Antonio, Texas

Thomas J. Bouchard, Jr., Ph.D.
Twin Study
University of Minnesota Health Center
Minneapolis, Minnesota

Mark R. Brann, Ph.D.
Department of Psychiatry and Pharmacology
University of Vermont
Burlington, Vermont

Eric R. Braverman, Ph.D.
Department of Psychiatry
New York University School of Medicine
New York, New York

Kari J. Buck, Ph.D.
Department of Behavioral Neuroscience
Oregon Health Sciences University
Portland, Oregon

Dorit Carmelli, Ph.D.
Center for Health Sciences
Stanford Research Institute
Menlo Park, California

Thomas H. J. Chen, Ph.D.
Chang Jung University
Taiwan, Republic of China

C. Robert Cloninger, M.D.
Department of Psychiatry and Genetics
Washington University School of Medicine
St. Louis, Missouri

Giancarlo Colombo, Ph.D.
C.N.R. Center for Neuropharmacology
University of Cagliari
Cagliari, Italy

David E. Comings, M.D.
Department of Medical Genetics
City of Hope National Medical Center
Duarte, California

P. Michael Conneally, Ph.D.
Department of Medical and Molecular Genetics
Indiana University School of Medicine
Indianapolis, Indiana

Edwin H. Cook, Jr., M.D.
Department of Psychiatry
University of Chicago
Chicago, Illinois

John C. Crabbe, Ph.D.
Department of Research
Veterans Affairs Medical Center
Portland, Oregon

John G. Cull, Ph.D.
NeuRecovery International, Inc.
San Antonio, Texas

Christopher Dubay, Ph.D.
Biomedical Information and Communications Center and Department of Medical and Molecular Genetics
Oregon Health Sciences University
Portland, Oregon

Fabio Fadda, Ph.D.
Institute of Human Physiology
University of Cagliari
Cagliari, Italy

Joel Gelernter, M.D.
Department of Psychiatry
Yale University School of Medicine and West Haven Veterans Affairs Medical Center
New Haven, Connecticut

Susan R. George, M.D., F.R.C.P., F.A.C.P.
Departments of Pharmacology and Medicine
University of Toronto and the Addiction Research Foundation
Toronto, Ontario, Canada

Gian Luigi Gessa, Ph.D.
Department of Neuroscience
University of Cagliari
Cagliari, Italy

Elizabeth Gettig, M.S.
Department of Human Genetics
University of Pittsburgh
Pittsburgh, Pennsylvania

Julie G. Hensler, Ph.D.
Department of Pharmacology
University of Texas Health Science Center
San Antonio, Texas

Shirley Y. Hill, Ph.D.
Department of Psychiatry
University of Pittsburgh
Pittsburgh, Pennsylvania

James J. Hudziak, M.D.
Center for Children
Department of Psychiatry
University of Vermont School of Medicine
Burlington, Vermont

Masanari Itokawa, M.D.
Department of Neuropsychiatry
Tokyo Medical and Dental University
School of Medicine
Tokyo, Japan

Pudur Jagadeeswaran, Ph.D.
Department of Cellular and Structural Biology
University of Texas Health Science Center
San Antonio, Texas

Rajendra P. Kandpal, Ph.D.
Fels Institute for Cancer Research
Temple University
Philadelphia, Pennsylvania

Yasuhiro Kimura, M.D., Ph.D.
Department of Psychiatry and Pharmacology
University of Vermont
Burlington, Vermont

Steven C. King, Ph.D.
Department of Physiology and Biophysics
University of Texas Medical Branch
Galveston, Texas

Tokutaro Komiyama, M.D.
National Hospital for Mental, Nervous, and Muscular Disorders, N.C.N.P.
Tokyo, Japan

Bennett L. Leventhal, M.D.
Department of Psychiatry
University of Chicago
Chicago, Illinois

Yuan C. Liu, Ph.D.
Department of Cellular and Structural Biology
University of Texas Health Science Center
San Antonio, Texas

Carla Lobina, Ph.D.
C.N.R. Center for Neuropharmacology
University of Cagliari
Cagliari, Italy

Horace H. Loh, Ph.D.
Department of Pharmacology
University of Minnesota
Minneapolis, Minnesota

Hideo Mifune, M.D.
Hatsuishi Hospital
Chiba, Japan

Hiroshi Mitsushio, M.D.
National Hospital for Mental, Nervous, and Muscular Disorders, N.C.N.P.
Tokyo, Japan

Hiroshi Mori, M.D.
Asai Hospital
Chiba, Japan

David A. Morilak, Ph.D.
Department of Pharmacology
University of Texas Health Science Center
San Antonio, Texas

Katherine Neiswanger, Ph.D.
University of Pittsburgh
Western Psychiatric Institute and Clinic
Pittsburgh, Pennsylvania

Gordon Y.K. Ng, B.S. Pharm., Ph.D.
Departments of Pharmacology and Psychiatry
University of Toronto
Toronto, Ontario, Canada

Ernest P. Noble, Ph.D., M.D.
UCLA Neuropsychiatric Institute
Los Angeles, California

Brian F. O'Dowd, Ph.D.
Department of Pharmacology
University of Toronto and the Addiction Research Foundation
Toronto, Ontario, Canada

Michael W. Odom, Ph.D.
Department of Cellular and Structural Biology
University of Texas Health Science Center
San Antonio, Texas

Lisa S. Parker, Ph.D.
Department of Human Genetics
University of Pittsburgh
Pittsburgh, Pennsylvania

Abbas Parsian, Ph.D.
Departments of Psychiatry and Genetics
Washington University School of Medicine
St. Louis, Missouri

Sankhavaram R. Patanjali, Ph.D.
Department of Genetics
Boyer Center for Molecular Medicine
Yale University School of Medicine
New Haven, Connecticut

Antonio M. Persico, M.D.
Laboratory of Neuroscience
Libero Instituto Universitario
Campus Bio-Medico
Rome, Italy

Roberta Reali, Ph.D.
Department of Neuroscience
University of Cagliari
Cagliari, Italy

Joanne M. Scalzitti, Ph.D.
Department of Pharmacology
University of Texas Health Science Center
San Antonio, Texas

Gerard D. Schellenberg, Ph.D.
Veterans Affairs Puget Sound Health Care System
Seattle, Washington

Peter Sheridan, Ph.D.
Department of Cellular and Structural Biology
University of Texas Health Sciences Center
San Antonio, Texas

Mark A. Stein, Ph.D.
Department of Psychiatry
University of Chicago
Chicago, Illinois

Gary E. Swan, Ph.D.
Center for Health Sciences
Stanford Research Institute
Menlo Park, California

R.D. Todd, Ph.D., M.D.
Departments of Psychiatry and Genetics
Washington University School of Medicine
St. Louis, Missouri

Michio Toru, M.D.
Department of Neuropsychiatry
Tokyo Medical and Dental University
School of Medicine
Tokyo, Japan

George R. Uhl, M.D., Ph.D.
Laboratory of Neurobiology
NIDA Addiction Research Center
Baltimore, Maryland

Li-Na Wei, Ph.D.
Department of Pharmacology
University of Minnesota
Minneapolis, Minnesota

Sherman M. Weissman, M.D.
Department of Genetics
Boyer Center for Molecular Medicine
Yale University School of Medicine
New Haven, Connecticut

Ellen M. Wijsman, Ph.D.
Division of Medical Genetics
Department of Biostatistics
University of Washington
Seattle, Washington

Robert C. Wood, Ph.D.
Computing Resources
University of Texas Health Science Center
San Antonio, Texas

Hongxia Xu, Ph.D.
Department of Genetics
Boyer Center for Molecular Medicine
Yale University School of Medicine
New Haven, Connecticut

Table of Contents

Introductory Remarks: Historical Overview of Molecular
Genetic Advances in Psychiatry 1
P. Michael Conneally

SECTION I
Genetic Mechanisms in Psychiatry: Analytic Approaches

Chapter 1
Association Vs. Linkage Analysis in Mental Disorders 7
Ellen M. Wijsman

Chapter 2
Genetic Association Studies in Psychiatry: Recent History 25
Joel Gelernter

Chapter 3
The Value of Narrow Psychiatric Phenotypes and "Super" Normal Controls 37
Shirley Y. Hill and Katherine Neiswanger

SECTION II
DNA Analysis

Chapter 4
Isolation of Coding Sequences from the Human Genome: Implication for
Neurological Functions and Diseases 49
Pudus Jagadeeswaran, Michael W. Odom, and Yuan C. Liu

Chapter 5
Identification of Microsatellite Repeat Markers and Transcribed
Sequences in Cloned Genomic DNA 63
Sankhavaram R. Pantanjali, Rajendra P. Kandpal, Hongxia Xu, and
Sherman M. Weissman

SECTION III
Molecular Biology of Receptors and Associated Proteins

Chapter 6
Molecular Biology of Opiod Receptors and Associated Proteins 77
Li-Na Wei and Horace H. Loh

Chapter 7
Phenotypic Analysis of Receptor Genes: Role of Muscarinic
Receptor Genes in Neuropsychiatric Disease 89
Mark R. Brann and Yasuhiro Kimura

Chapter 8
Studies on Dopamine Receptors and Role in Drug Addiction 95
Gordon Y. K. Ng, Susan R. George, and Brian F. O'Dowd

Chapter 9
Serotonergic Receptors: Role in Psychiatry 113
Joanne M. Scalzitti and Julie G. Hensler

Chapter 10
Brain Adrenergic Receptors 147
David A. Morilak

Chapter 11
Emerging Bacterial Models for GABA Receptors and Transporters 163
Steven C. King

SECTION IV
Psychiatric Genetics

Chapter 12
Identification of Phenotypes for Molecular Genetic Studies of Common
Childhood Psychopathology 201
James J. Hudziak

Chapter 13
Molecular Genetics of Alzheimer's Disease 219
Gerard D. Schellenberg

Chapter 14
Polygenic Inheritance in Psychiatric Disorders 235
David E. Comings

Chapter 15
Molecular Linkage Studies of Manic-Depressive Illness 261
Wade Berrettini

Chapter 16
The Genetics of Personality 273
Thomas J. Bouchard, Jr.

Chapter 17
Family-Based Association of Attention-Deficit/Hyperactivity Disorder
and the Dopamine Transporter 297
Edwin H. Cook, Jr., Mark A. Stein, and Bennett L. Leventhal

Chapter 18
Reward Deficiency Syndrome: Neurobiological and Genetic Aspects..311
*Kenneth Blum, John G. Cull, Eric R. Braverman, Thomas H. J. Chen, and
David E. Comings*

SECTION V
Substance Use Disorders

Chapter 19
Polymorphisms of the D2 Dopamine Receptor Gene in Alcoholism,
Cocaine and Nicotine Dependence, and Obesity 331
Ernest P. Noble

Chapter 20
Polymorphisms of the D2 Dopamine Receptor Gene in Polysubstance Abusers 353
Antonio M. Persico and George R. Uhl

Chapter 21
Association Vs. Linkage Analysis in Compulsive Disorder 367
Abbas Parsian, R. D. Todd, and C. Robert Cloninger

Chapter 22
Association of the A1 and B1 Alleles of the Dopamine D2 Receptor
Gene in Severe Japanese Alcoholics 375
*Tadao Arinami, Masanari Itokawa, Tokutaro Komiyama, Hiroshi Mitsushio,
Hiroshi Mori, Hideo Mifune, and Michio Toru*

Chapter 23
Behavior Genetic Investigations of Cigarette Smoking and
Related Issues in Twins 387
Gary E. Swan and Dorit Carmelli

Chapter 24
The Dopamine D2 Receptor Gene Locus in Reward Deficiency
Syndrome: Meta-Analyses 407
*Kenneth Blum, Peter J. Sheridan, Thomas H. J. Chen, Robert C. Wood,
Eric R. Braverman, John G. Cull, and David E. Comings*

SECTION VI
From Animal Research to Society: Genetic Impact on Behavior

Chapter 25
Mapping Quantitative Trait Loci for Behaviorial Traits in the Mouse 435
John K. Belknap, Christopher Dubay, John C. Crabbe, and Kari J. Buck

Chapter 26
Drug Discrimination: A Tool to Unravel the Genetic Determinants
of Alcohol Preference and Aversion 455
*Giancarlo Colombo, Roberta Agabio, Carla Lobina, Roberta Reali, Fabio Fadda, and
Gian Luigi Gessa*

Chapter 27
Ethical Issues: Genetic Screening, Gene Therapy, and Scientific Conduct 469
Lisa S. Parker and Elizabeth Gettig

Index 481

Introductory Remarks: Historical Overview of Molecular Genetic Advances in Psychiatry

P. Michael Conneally

A. INTRODUCTION

The first milestone in gene mapping occurred in 1971 with the publication by Elston and Stewart[1] of their efficient algorithm to determine the likelihood of a pedigree to map human disease genes. This algorithm became the basis of the well-known computer program LIPED, written by Ott,[2] which was utilized almost exclusively by all gene mappers. The next major landmark in the use of molecular genetic techniques for mapping human disease genes was the 1980 publication by Botstein and colleagues,[3] advocating the uses of newly discovered Restriction Fragment Length Polymorphisms (RFLPs) as markers for the construction of a genetic linkage map. Previous to this the available genetic markers consisted of blood group and protein polymorphisms. Although 30 to 40 such markers were available, many required tedious techniques and, furthermore, a significant number were not highly polymorphic. Thus, the advent of recombinant DNA ushered in a new era for the localization and study of human disease genes.

B. DISEASE MAPPING

The first inherited disorder to be mapped using molecular genetic approaches was Huntington's disease in 1983.[4] Though the number of probes defining marker loci were relatively scant at that time, Gusella and colleagues,[4] using 12 probes, had the good fortune to have chosen a set that contained one that was linked to the HD gene. The pace of gene mapping increased exponentially during the 1980s so that by the end of the decade the majority of the common Mendelian genetic disorders had been localized to a short region of a chromosome and many of the genes involved had been cloned. McKusick[5] provides a historical perspective of the localization and cloning of these genes.

C. GENETIC MARKERS

The next major breakthrough came with the discovery of dinucleotide (CA) repeats, independently by Weber and May[6] and by Litt and Luty.[7] This discovery increased the number of known genetic markers significantly, sufficient to cover the human genome at an average distance of at least 5 cM.

Polymorphic Chain Reaction (PCR) technology and automated genotyping drastically reduced the average time and effort required to locate a disease gene. Tetranucleotide repeats, though less polymorphic than dinucleotide repeats, are easier to genotype and are now supplanting the latter in many laboratories. It is now a routine matter to map a disease gene locus to a small region of a chromosome if the inheritance pattern is Mendelian. There are over 6000 Mendelian disorders in humans; however, many are very rare and sufficient family data for an optimal linkage study are often difficult to obtain. Notwithstanding this, many genetic disorders have already been mapped in humans.

D. COMPLEX DISEASES

There are, however, a number of human disorders with a strong genetic component not inherited in simple Mendelian fashion. Terms such as "polygenic" or "multifactorial" have been used to describe these disorders, but the term "complex diseases" is now generally used. These diseases tend to be common and thus, from a public health perspective, have a much greater impact than Mendelian disorders. While molecular genetic techniques have been successful in mapping and cloning Mendelian traits, this is not the case for complex disorders.

The majority of psychiatric disorders have a genetic component most likely involving the interaction among different genes and between the expression of these genes and the environment. Since the mode of inheritance is therefore not known, classical linkage methodology is usually not appropriate. For this reason early genetic studies of psychiatric disorders avoided such methods and turned instead to association studies. The first study of an association, by Shapiro et al.,[8] reported an increased frequency of the HLA-BW16 allele in patients with affective disorder when compared to controls. A number of association studies between the HLA locus and both affective disorder and schizophrenia have been published. As is common in association studies, no consistent findings have emerged. Linkage studies also were inconsistent for both disorders. In summary, there is no compelling evidence that genes at the HLA complex are involved in the etiology of major psychiatric illness.

E. GENE LOCALIZATION OF BIPOLAR AFFECTIVE DISORDER AND SCHIZOPHRENIA

Historically, the first positive linkage study to have a major impact in psychiatry was the linkage of bipolar affective disorder to chromosome 11 in a large Amish pedigree by Egeland and colleagues.[9] This was an "ideal" linkage study. The family was not only large but also homogeneous since the Amish are a genetic isolate. The disorder segregated in the family as an autosomal dominant with approximately 63% penetrance. The maximum LOD score was 4.3. The fact that others had excluded linkage to the same markers did not mitigate the importance of the original linkage. Unfortunately, however, later evaluation of the family by Kelsoe et al.[10] failed to confirm the original result.

A similar situation existed with an early linkage study in schizophrenia. In 1988 Sherrington et al.[11] reported linkage between schizophrenia in seven Icelandic and British families to markers in the region of 5q11-q13. However, many other investigators (as documented by Hovatta et al.[12]) were unable to confirm the finding. The euphoria engendered by the localization of the two major psychiatric illnesses rapidly abated. In retrospect, it was unfortunate that both findings caused so much attention from the media. Scientists began to question the wisdom of molecular approaches to psychiatric disease. It became clear that positional cloning, or as it was then termed "reverse genetics", while still envisioned as the best approach, would require much more rigorous methodology both in terms of clinical phenotyping and the availability of more markers in order to define a much finer genome search and, probably most important of all, to develop new analytic techniques in linkage analysis. In the area of clinical phenotyping, it was not unusual to broaden or sometimes narrow the definition of the phenotype simply to drive the LOD score higher. This, and the arbitrary selection of densely affected families, could easily lead to Type I errors. A multifactorial disorder such as bipolar illness or schizophrenia can mimic an autosomal dominant with reduced penetrance. This is especially so if only families with multigenerations of affecteds are selected; in the extreme, such families could fit an autosomal dominant model with nearly complete penetrance.

F. LINKAGE ANALYSIS METHODOLOGY

The classical likelihood (LOD score) method of linkage analysis is a parametric test in the sense that the mode of inheritance of a phenotype and the marker loci are specified. For psychiatric and other complex disorders, a nonparametric approach is necessary.

Nonparametric linkage analysis tests the sharing of marker alleles among pairs of relatives. When the same allele is inherited by a pair of relatives from the same ancestral source, the allele is considered to be identical by descent (IBD). However, when the same allele may or may not be inherited from a common ancestor, then it is considered to be identical by state (IBS). One of the earliest methods for linkage analysis was the sib-pair method developed by Penrose[13] in 1935. The basic premise is that regardless of the type of inheritance, affected sibs would be expected to share marker alleles identical by descent in the region of a genetic locus for the trait being mapped. Thus, the mode of inheritance need not be specified. It is of interest that this method was used to map characteristics such as hair and eye color, which are not inherited in simple Mendelian fashion. Since the number of markers available at that time were minuscule the method was rarely utilized. However, with the advent of the very polymorphic di- and tetranucleotide markers, this method has become very powerful, especially for complex diseases in which the mode of inheritance is unknown, and therefore parametric methods are inadequate.

More recently, Lange[14,15] published a newer method known as the affected-sib-pair (ASP) method using IBS relations. The method was later generalized by Weeks and Lange[16] to the affected-pedigree-member (APM) method. The method compares the IBS sharing of marker alleles. Since it is more striking for distant-affected relatives to share a rare rather than a common marker allele, the statistic includes a weighting factor based on allele frequencies. One problem with the APM approach is that it is highly sensitive to errors in gene frequency estimates. A major successful use of the method was in locating a gene for late-onset familial Alzheimer's disease by Pericak-Vance et al.,[17] which later led to the discovery of the importance of the APOE-4 allele as a risk factor in Alzheimer's disease.

Current linkage analysis methodology has been refined, and more precise and uniform diagnostic criteria have been promulgated and are discussed in detail in other chapters.

G. CONCLUSIONS

The title of these introductory remarks includes the word "advances". Some may argue that this is an overstatement considering the fact that the two major genetic linkage advances in psychiatric illness could not be replicated. However, it is our belief that the euphoria experienced 10 years ago has been supplanted by quite a bit of soul searching in the whole area of complex disease mapping. This has led to much more conservative approaches. It is now clear that regardless of the magnitude of significance of a mapping statistic, independent replication of the findings are critical. The ultimate aim, of course, is to clone and identify the gene or genes involved in a disorder. This in itself is a tedious process. Thus, it is imperative that the localization of a disease gene or genes be defined before embarking on the new venture of identifying the protein produced as a product of this gene.

REFERENCES

1. Elston, R. C. and Stewart, J., A general model for the analysis of pedigree data, *Hum. Hered.*, 21, 523, 1971.
2. Ott, J., Estimation of the recombination fraction in human pedigrees: efficient computation of the likelihood for human linkage studies, *Am. J. Hum. Genet.*, 26, 588, 1974.
3. Botstein, D., White, R. L., Skolnick, M., and Davis, R. W., Construction a of genetic linkage map using restriction fragment length polymorphisms, *Am. J. Hum. Genet.*, 32, 373, 1980.
4. Gusella, J. F., Wexler, N. S., Conneally, P. M., Naylor, S. L., Anderson, M. A., Tanzi, R. E., Watkins, P. C., Ottina, K., Wallace, M. R., Sakaguchi, A. Y., Young, A. B., Shoulson, I., Bonilla, E., and Martin, J. B., A polymorphic DNA marker genetically linked to Huntington's disease, *Nature*, 306, 234, 1983.
5. McKusick, V. A., *Mendelian Inheritance in Man*, 10th ed., Johns Hopkins University Press, Baltimore, 1992.
6. Weber, J. L. and May, P. E., Abundant class of human polymorphisms which can be typed using the polymerase chain reaction, *Am. J. Hum. Genet.*, 44, 388, 1989.
7. Litt, M. and Luty, J. A., A hypervariable microsatellite revealed by in vitro amplification of a dinucleotide repeat with the cardiac muscle actin gene, *Am. J. Hum. Genet.*, 44, 397, 1989.
8. Shapiro, R. W., Bock, E., Rafaelson, O. J., Ryder, L. P., and Svejgaard, A., Histcompatability antigens and manic-depressive disorders, *Arch. Gen. Psychiatry*, 33, 823, 1976.
9. Egeland, J. A., Gerhard, D. S., Pauls, D. L., Sussex, J. N., Kidd, K. K., Allen, C. R., Hostetter, A. M., and Housman, D. E., Bipolar affective disorders linked to DNA markers on chromosome 11, *Nature*, 325, 783, 1987.
10. Kelsoe, J. R., Gings, E. I., Egeland, J. A., Gerhard, D. S., Goldstein, A. M., Bale, S. J., Pauls, D. L., Long, R. T., Kidd, K. K., Conte, G., Housman, D. E., and Paul, S. M., Re-evaluation of the linkage relationship between chromosome 11p loci and the gene for bipolar affective disorder in the Old Order Amish, *Nature*, 342, 238, 1989.
11. Sherrington, R., Brynjolfsson, J., Petursson, H., Potter, M., Dudleston, K., Barraclough, B., Wasmuth, J., Dobbs, M., and Gurling, H., Localization of a susceptibility locus for schizophrenia on chromosome 5, *Nature*, 336, 164, 1988.
12. Hovatta, I., Seppala, J., Pekkarinen, P., Tanskanen, A., Lonnquist, J., and Peltonen, L., Linkage analysis in two schizophrenic families originating from a restricted subpopulation of Finland, *Psychiatr. Genet.*, 4, 143, 1994.
13. Penrose, L. S., The detection of autosomal linkage in data which consists of pairs of brothers and sisters of unspecified parentage, *Ann. Eugen.*, 133, 1935.
14. Lange, K., The affected sib-pair method using identity by state relations, *Am. J. Hum. Genet.*, 39, 148, 1986.
15. Lange, K., A test statistic for the affected-sib-set method, *Ann. Hum. Genet.*, 50, 283, 1986.
16. Weeks, D. E. and Lange, K., The affected-pedigree-member method of linkage analysis, *Am. J. Hum. Genet.*, 42, 315, 1988.
17. Pericak-Vance, M. A., Bebout, J. L., Gaskell, P. C., Jr., Yamaoka, L. H., Hung, W. Y., Alberts, J. J., Walker, A. P., Bartlett, R. C., Haynes, C. A., Welsh, K. A., Earl, N. L., Heyman, A., Clark, C. M., and Roses, A. D., Linkage studies in familial Alzheimer disease: evidence for chromosome 19 linkage, *Am. J. Hum. Genet.*, 48, 1034, 1991.

Section I

Genetic Mechanisms in Psychiatry: Analytic Approaches

1 Association Vs. Linkage Analysis in Mental Disorders

Ellen M. Wijsman

CONTENTS

A. Introduction 7
B. Description of Analysis Components 9
    1. Data Sets 9
        a. Pedigrees 9
        b. Unrelated Individuals 10
    2. Strategies 10
        a. Linkage Analysis 10
        b. Linkage Disequilibrium 11
C. Causes of Disequilibrium/Association 11
    1. Population Stratification 12
    2. Population Admixture 13
    3. Sample Stratification 13
    4. Other Causes 13
D. Persistence of Disequilibrium/Association 14
E. Statistical Interpretation of Tests 14
    1. Statistical Significance 14
    2. Repeatability 16
    3. Power 16
F. Linkage Disequilibrium in Human Populations 16
    1. Detectable Linkage Disequilibrium 17
    2. Absence of Statistically Detectable Disequilibrium 18
G. Conclusions 19
References 21

A. INTRODUCTION

Probably less is understood about the underlying physiology and biochemistry of mental disorders than about any other category of human disease. This has presented tremendous problems in combatting one of the major modern public health problems. One of the barriers in making headway against these diseases stems from the phenotype, which thus limits most studies to humans and increases their difficulty. However, strong evidence from twin studies and family studies suggests that at least some of these disorders have a genetic component. This, in turn, suggests that one fruitful approach to the study of these diseases is to identify pathways involved in the physiology and biochemistry of these disorders by identifying genes which are involved in hereditary predisposition to psychiatric diseases.

Mental disorders rarely exhibit a simple mode of inheritance, and for most such disorders, there are likely to be both genetic and environmental components. Presence of a genetic component does not, however, preclude the role of the environment in the development of the disease, and variability among environments is also likely to play a role in such diseases. The possibility that yet-unidentified environmental factors may also play a role in the development of mental disorders adds a complication to the genetic analysis of such traits since it is no longer possible to be sure that the difference between an affected an unaffected individual must be found only in differences at the genetic level. Any method of genetic analysis must, at some level, accommodate this uncertainty.

Disorders such as psychiatric diseases, where both genetic and environmental variation have important effects on the resulting phenotype, are examples of complex genetic traits. Because of their relatively high frequency in the population relative to diseases with known single-gene modes of inheritance and because of the variation in the observed phenotype, it is likely that variation in a number of genes may contribute to these disorders. These genes may act either independently and/or epistatically and may consist of relatively few loci with large effects on susceptibility or as many loci that have small effects on susceptibility. For complex diseases we also cannot assume that one model for the mode of inheritance will be correct in all families. This complicates the identification of the underlying genetic causes behind the disease since there may be no single identifiable cause, and it complicates the design and analysis of such studies because a clear choice for the approach of the study design is not available, as it would be for single-gene disorders.

Nevertheless, because of the evidence that genes may play a role in these disorders, because the genetic approach has been enormously successful in recent years for other disorders, and because biochemical and epidemiological studies have failed to identify causes of mental illness, the use of genetic analyses has become increasingly popular in attempts to track down and identify genes leading to such disorders. This similar approach has been of enormous help in our understanding of the underlying biology behind other complex diseases; identification of even a single genetic defect involved in a complex disorder can lead to the identification of physiological pathways involved in a number of related diseases. An excellent example of the issues involved in using genetics to try to dissect the biology behind complex disorders can be found in coronary artery disease (CAD). The recognition of a phenotypically distinguishable group of patients,[1] followed by the eventual identification of the defect in the low density lipoprotein receptor gene, or LDLR gene,[2] was the key which led investigators into much of what we currently understand about lipid metabolism and the development and treatment of cardiovascular disease.

It is relatively straightforward to map genes for single-locus diseases with linkage analysis methods. However, unlike linkage analyses of more straightforward diseases, reports of localization of genes through linkage analyses of psychiatric disorders have generally been difficult to confirm. For example, the initially positive results reported for bipolar affective disorder[3] disappeared when the pedigrees were extended.[4] The evidence for linkage of schizophrenia to chromosome 5 was immediately questioned by failure to find comparable results in a second sample,[5] and recent reports of another location still need confirmation.[6]

It is, however, important to note three things in terms of designing and evaluating strategies for mapping and ultimately identifying genes contributing to psychiatric diseases. First, apparently false positive results in linkage analyses of complex diseases are not confined to the area of psychiatric illness, but occur in attempts to map genes for a variety of complex diseases. For example, a recent report of linkage of familial combined hyperlipidemia to chromosome 11[7] was not confirmed in an independent set of families.[8] Second, an initial false positive result does not preclude the eventual successful identification of the location of a disease gene. For example, the initial report of linkage of early-onset Alzheimer's disease to chromosome 21[9] was not confirmed in a second set of families.[10] However, this locus was more recently successfully mapped to chromosome 14,[11] with rapid confirmation in three other studies,[15–17] including the four families which were originally used in the initial report.[9] Third, any strategy based on statistical methods will have some rate of false positive and negative results; the key in choosing an efficient strategy is to choose a design which minimizes these false positive and negative results overall.

Because of difficulties surrounding linkage analyses of complex disorders, in particular with respect to parameterization of the mode of inheritance of the traits, some investigators have chosen to use approaches that require fewer assumptions about parameter values than do likelihood-based linkage analysis methods. In particular, these methods generally do not require assumptions about the mode of inheritance and the values of the associated parameter values. These approaches include a variety of nonparametric methods which are based on data in family members, including relative-pair methods suitable for both qualitative[15,16] and quantitative[17,18] traits.

These approaches, which make few assumptions about the mode of inheritance, also include methods that are based on association studies at the population level between the disease and marker alleles in unrelated individuals and/or independent chromosomes. Such studies have recently been reintroduced as a possible way of identifying genes involved in psychiatric disorders.[19] The results of this study, the results of follow-up studies,[20–22] and the general approach have been hotly debated. There is skepticism among some investigators that association tests are useful for identifying genes involved in diseases with a genetic component,[23] partly because similar approaches in the 1970s and earlier were so unsuccessful with diseases other than autoimmune diseases and HLA associations. However, the changes in technology and available data in the past two decades may have improved the prospects of eventually obtaining positive results from association studies. Some of the issues pertinent for evaluating the prospects of using association studies in addition to or in lieu of linkage analysis studies for mapping and identifying genes involved in psychiatric disorders are discussed in this chapter.

B. DESCRIPTION OF ANALYSIS COMPONENTS

1. DATA SETS

The two types of data that can be collected from human populations for use in genetic analyses include unrelated individuals and pedigrees of related individuals. Both types have been used to search for genes which contribute to diseases. These two types have certain advantages and disadvantages, with data sets consisting of unrelated individuals being cheaper to collect than those of pedigrees, and analyses based on pedigree data being more robust than association analyses based on unrelated individuals if the assumptions behind the analyses are appropriate.

a. Pedigrees

Data sets of pedigrees are difficult and expensive to collect and to define the rate-limiting step of most mapping studies. The pedigree of a proband identified through a phenotype of interest may be ineligible for further use because of a number of possible events, including small family size, inaccessibility of family members to further study, absence of the phenotype of interest in a sufficient number of family members, or because of nonpaternity which is exposed during initial studies with genetic markers, which occurs according to some studies in approximately 20% of families even in the absence of psychiatric illness.[24] Pedigrees may also become ineligible for further study if there are too many individuals in the pedigree who refuse to participate, a problem which increases with the study of mild traits and those which carry social stigma. Family studies of mental disorders often suffer loss of potential pedigrees because of several of these issues in addition to the usual fraction of small or otherwise ineligible pedigrees: the typically adult onset increases the chance that important individuals will be deceased or geographically unavailable; affected family member may be more likely to be missing because the nature of the diseases may cause loss of contact; and the negative social implications of the diseases decrease the willingness of family members to participate in genetic studies. For rare diseases, late-onset diseases, or diseases with high mortality or loss of contact, the prospects of collecting sufficient pedigree material to carry out a linkage study are often bleak.

b. Unrelated Individuals

It is much easier and much less expensive to collect a data set of unrelated individuals than it is to collect a data set of pedigrees. There is no loss of data because of unavailability or ineligibility of family members. Finally, the expenses of sampling ineligible individuals before knowing their eligibility status need not occur, as happens when family members in pedigrees are excluded because of knowledge which appears through, e.g., initial marker typing. Association studies which are based on independent individuals therefore have the initial appeal of lower cost of sample collection, as compared to pedigree studies. Also, because the additional constraints imposed by the need for available pedigree data are not present, it is easier to collect a data set of appropriate size for diseases for which pedigree data are difficult to collect.

2. STRATEGIES

Linkage and association analyses use very different strategies for mapping or identifying genes. Linkage analysis methods rely on the knowledge that close proximity of two loci on a chromosome will cause specific alleles at the two loci to be transmitted together in meiosis more often than is expected by chance. The alleles involved are likely to differ among families, but the degree of joint transmission is highest when the two loci are very closely linked. Association analysis, on the other hand, relies on identification of specific alleles at both loci which are found together or are in linkage disequilibrium, more often than expected by chance in the population.

a. Linkage Analysis

When data have been collected on several individuals in a pedigree, linkage analysis methods are frequently used. There are a wide variety of methods available which vary the strength of assumptions and the amount of the total data actually used. All are based on statistical tests of the degree of joint transmission of alleles at two or more loci during meiosis.

The assumptions needed for different methods of linkage analysis vary. The LOD score method[25] is the most widely used, can be used with pedigrees of virtually any structure, and is the most powerful method when assumptions are approximated. The assumptions, however, include specification of the mode of inheritance of the disease trait and parameterization thereof, including gene frequency and penetrance functions which are rarely known in the case of complex diseases. Recent results also suggest that when there are multiple unknown parameter values, false positive results may be much more common than originally believed.[26] In addition, because this method is based on maximum likelihood methods, it requires computation of *all* possible outcomes of the data in the pedigrees, and the computational burden can be enormous, especially when there are a lot of missing data, as is typical for adult-onset diseases.

As a result of these issues, various methods of analysis which are based on, for example, only affected individuals, also have been devised. Unfortunately, these methods currently suffer from low power,[27] partly because of inefficient use of the data. In addition, recent results indicate that false positive results are also possible with these methods under some of the same situations which may produce false positive results under likelihood-based methods.[26] Despite these potential problems, linkage analysis methods, primarily based on likelihood-based methods, have remained the methods of choice for mapping genes, even for complex disorders, partly because they have worked so well for so many diseases. In addition, while the assumed parameter values can affect the analysis, the sensitivity of the results to the assumptions can be evaluated,[28] so that careful use of these methods for complex disorders is unlikely to produce irreconcilable conclusions.

b. **Linkage Disequilibrium**

Studies which are based on the use of unrelated individuals rely on the search for associations in populations between the disease phenotype of interest and the presence/absence of particular genotypes and/or alleles at one or more marker loci. An association may exist under circumstances which are discussed in the next section. These include but are not restricted to the situation where the genetic marker is itself involved in the trait. In the case of complex diseases for which some understanding of the underlying physiology is available, for example in the study of traits associated with CAD,[29] this approach has met with some successes in identifying additional loci which contribute to phenotypic variability. Past attempts to use this approach to map genes involved in genetic disorders in the absence of strong candidate genes was also very common during the 1970s but led to only a minuscule number of results which have stood the test of time.

If there is a genetic component to the disease under study, an association between the disease phenotype and allele(s) at a marker locus may result from linkage disequilibrium between the marker locus and one or more of the loci contributing to the disease, whether the disease is measured on a quantitative or discrete scale. The details of the techniques used to look for evidence of association or linkage disequilibrium will vary according to the details of the type of association and the available data, but the general principles will hold regardless of whether the trait is qualitative or quantitative, and whether disequilibrium can be estimated at the gametic level or whether association can only be measured at the phenotypic level.

A complete description of different statistical tests which can be used in the search for linkage disequilibrium is beyond the scope of this chapter. Detailed descriptions of such tests based on gametic or genotypic data can be found elsewhere.[30–33] The design on which these tests are based, however, suffers somewhat from potential problems of poorly matched cases and controls, as is discussed below. However, association tests need not be restricted to unrelated individuals. One approach to achieving well-matched case and control samples for disequilibrium testing is to use as control chromosomes those which were *not* transmitted from parents to affected children for discrete diseases.[34] This approach and the related transmission disequilibrium test has been proposed as a method to circumvent some of the difficulties inherent in population-based association studies. It is important to note that this method require availability of parents of affected individuals for typing; however, extended pedigrees and/or siblings need not be available.

C. **CAUSES OF DISEQUILIBRIUM/ASSOCIATION**

Unlike physical linkage of two genes, which is the only probable reason for observing statistically significant joint transmission of alleles at two loci in a linkage analysis, there are a number of possible causes of association between loci at the population level. Appropriate interpretation of a finding of positive association (or lack thereof) depends on an understanding of the possible mechanisms leading to the association. We will therefore consider here some causes of linkage disequilibrium between loci which could be used to explain observed disequilibrium in human populations.

1. **POPULATION STRATIFICATION**

Population stratification occurs if there are multiple subpopulations living in one geographical location, but random mating between subpopulations is not achieved even though there may be random mating within each of the subpopulations. If a sample of gametes is randomly selected from all individuals residing in the geographical area and if allele frequencies at two or more loci differ among the populations, then this sample may exhibit evidence of linkage disequilibrium between the loci. It is important to note that the loci need not be on the same chromosome.

To understand how population stratification can lead to linkage disequilibrium, consider the following scenario. Suppose we draw a random sample of gametes from a large population. We sample genotypes from locus 1, with alleles A and a, and from locus 2, with alleles B and b. Assume, also, that the population consists of two separate subpopulations which mate randomly within themselves, but do not mate between subpopulations. A fraction, f, of the total population is composed of subpopulation 1, and a fraction, $(1 - f)$, is composed of subpopulation 2. Suppose, also, that the frequencies of the alleles at both loci differ in the two populations. Denote the frequency of allele A at the first locus in the full population, as p_A, and denote the frequency of allele B at the second locus as p_B in the full population. In the subpopulations, the frequencies of alleles A and B are q_A and q_B, respectively, in subpopulation 1, and r_A and r_B in subpopulation 2. In the full population, therefore, $p_A = fq_A + (1 - f)r_A$, and $p_B = fq_B + (1 - f)r_B$. Finally, assume that *within* each of the subpopulations there is no linkage disequilibrium between the two loci. Therefore, in subpopulation 1 the frequency of the gamete containing both A and B is $Q_{AB} = q_A q_B$. In subpopulation 2, the gamete containing both A and B has frequency $R_{AB} = r_A r_B$.

In the sample of gametes taken from the larger population, the frequency, P_{AB}, of gamete AB is

$$\begin{aligned} P_{AB} &= fq_A q_B + (1 - f)r_A r_B, \\ &= fQ_{AB} + (1 - f)R_{AB} \end{aligned} \quad (1\text{-}1)$$

with similar expressions for frequencies of the three other gametes.

In this case, the linkage disequilibrium coefficient, D, which measures the association between the two loci, will be

$$\begin{aligned} D &= f(1 - f)[p_A - q_A][p_B - q_B] \\ &= P_{AB}P_{ab} - P_{Ab}P_{aB} \end{aligned} \quad (1\text{-}2)$$

D is zero only if either or both of the loci have the same allele frequencies in the two populations, or if there is only one population (either $f = 0$ or $f = 1$). If D is non-zero, there is an association between the two loci in the full population. Note that the above derivation does not include any measure of the location of the two loci with respect to one another. Thus if the two loci are on different chromosomes and if the frequencies of alleles at both loci differ between the two subpopulations, the combined population will be in linkage disequilibrium with respect to the two loci. This also illustrates the confusion of the commonly used term "linkage disequilibrium", which does not, in fact, imply anything about the presence or absence of linkage, but only refers to statistical association at the population level.

2. **POPULATION ADMIXTURE**

Another cause of linkage disequilibrium is admixture between two or more populations. Admixture can occur in a relatively brief period of time through large-scale migration of one or more populations into the territory of another, or through the slower trickle of gene flow from one population into another over an extended period of time. In this discussion no distinction will be made between these two types of admixture since the way in which admixture occurs is not important in understanding the effect on the resulting linkage disequilibrium.

Immediately after two populations have fused, the situation will be as in the case of population stratification. However, the newly fused population may now mate at random within the larger population. In the case of each individual locus, under random mating the population will attain Hardy-Weinberg equilibrium in a single generation of random mating. However, linkage equilibrium between the two loci will persist for some generations even if the loci under consideration are on different chromosomes. This is discussed further below.

An example of a situation in which an originally unrecognized admixture was responsible for an apparent association between a marker locus and a complex disease can be found in a study of diabetes mellitus and Gm polymorphisms in a Native American population.[36] In the full population there was a strong negative association between one haplotype at the Gm locus and the presence of diabetes mellitus. However, this association disappeared when the population was stratified in the analysis by the amount of Caucasian admixture, thus demonstrating that the association was simply a function of combining data from populations with different amounts of admixture.

3. **SAMPLE STRATIFICATION**

Sample stratification may also cause an association between two loci in a sample. This kind of stratification may occur when the *sampling* process is not random, or if individuals in the sample are derived from different geographical locations or from different ethnic groups. For example, if a group of individuals with a particular disease is drawn from a clinic-based population and a group of controls is drawn from a population of volunteers, the two populations may be drawn from somewhat different geographical regions and/or from different genetic backgrounds. If the two populations are not well matched, the resulting sample may be a mixed population with the same effect on estimated linkage disequilibrium as population stratification.

4. **OTHER CAUSES**

There are a number of other situations which also may result in linkage disequilibrium between loci. Most, including mutation and selection on joint genotypes, are likely to have such small effects that they cannot be detected in human samples. Founder effects and genetic drift in small populations, however, may cause detectable associations which also take some time to decay.

D. PERSISTENCE OF DISEQUILIBRIUM/ASSOCIATION

While linkage disequilibrium can be established through any of a number of causes, the decay of this disequilibrium is a function of time, the recombination fraction between the two loci, and whether the forces which established the disequilibrium continue. For example, in the case of disequilibrium which is created by admixture between two populations, if there is random mating in the new population the disequilibrium will decay as described below. Even if loci are on different chromosomes, several generations of random mating will be needed to eliminate the disequilibrium. If mating between the subpopulations occurs less often than expected, this decay in disequilibrium will be retarded.

Under random mating in a population with discrete generations in which the disequilibrium is D_t at generation t, the disequilibrium in successive generations can be related to the disequilibrium in previous generations as $D_t = (1 - r)D_{t-1}$, where r is the recombination rate between the two loci. Rewriting this in terms of the initial disequilibrium present at the time of fusion of the populations, D_0, gives $D_t = (1 - r)^t D_0$. The amount of linkage disequilibrium in a randomly mating population is thus a function of three quantities: the initial disequilibrium level, time (in generations), and the recombination rate between the two loci. For two unlinked loci ($r = 1/2$), the initial disequilibrium decays by a factor of 1/2 each generation, and will eliminate all but the strongest disequilibrium in several generations; for loci which are more closely linked, the decay is slower. Figure 1-1 shows the decay in disequilibrium as a function of time for three different recombination fractions. The recombination fractions, r, of $r = 0.067$ and $r = 0.0069$ were chosen to illustrate the eventual decay of disequilibrium for even closely linked markers. These two recombination rates will produce a curve identical to that shown for $r = 0.5$ if the units on the x-axis are, respectively, 10 and 100 times those shown for $r = 0.5$. Thus there are two important things to note about these curves. The first is that disequilibrium can exist in the presence of random mating and in the absence of linkage ($r = 0.5$) if the cause of the disequilibrium is relatively recent. The second is that disequilibrium between even closely linked loci decays over time so that there is no guarantee that closely linked loci will exhibit disequilibrium. Even for fairly closely linked markers ($r = 0.067$), very little remaining linkage disequilibrium will remain after 40 to 50 generations of random mating.

E. STATISTICAL INTERPRETATION OF TESTS

Interpretation of the results of a test must include consideration of the various ways in which the data might give rise to the observed results. For any statistical test, of course, there is statistical noise to consider. It is also important to consider alternative tests needed to verify or exclude the initial results. Similarly, if a test fails to indicate that there is linkage or association between two loci, it is necessary to consider mechanisms by which linkage may fail to be detected, or by which loci may be found in linkage equilibrium, as well as whether the sample sizes used were sufficient to detect the desired association, if present.

1. Statistical Significance

When a single locus disorder has been mapped to a marker locus by the LOD score method of linkage analysis, the interpretation of a statistically significant positive result is straightforward when the disorder is monogenic. The prior probability that the disorder is linked to the marker can be combined with the error rate under the null hypothesis of free recombination in an individual test of linkage (the type I error) to produce a Bayesian estimate of the posterior probability of linkage.[25] The low prior probability of linkage means that a type I error, or *p* value, of 0.0001 to 0.001 for an individual test of linkage corresponds to a

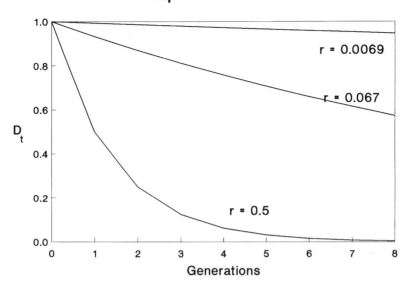

FIGURE 1-1 Linkage disequilibrium as a function of time, shown as a proportion of the disequilibrium initially present. Curves for r = 0.067 and r = 0.0069 superimpose upon that for r = 0.5 if the number of generations is 10 times or 100 times that shown, respectively.

posterior probability of linkage of approximately 5%, if the traditional LOD score of 3 is used as the cutoff of statistical significance.[25,37]

For diseases with unknown and possibly complex etiologies there are two complicating factors in interpretation of analyses. First, the posterior probability of linkage given a statistically significant outcome of an individual test is unlikely to be greater than an equivalent result for a single locus disorder since the prior probability is likely to be no greater than for simple disorders. This affects any test which is used to try to map genes, including association tests, and thus such tests need a significance level of less than 0.0001 to 0.001 before they even begin to approach the significance level normally desired in linkage analyses. For diseases where individual major-gene influences are unlikely, in fact, the lower significance level of 0.0001 is usually desired to partially accommodate the uncertainties in mode of inheritance. It is also worth noting here that if each of two independent tests is significant at the 5% level, the joint significance level of the two studies is 0.0025 (or 0.05^2), or somewhat larger than this maximum desired level of 0.001. Two *independent* studies, each with a significance level of 0.01, would be needed to achieve a joint significance level of 0.0001.

The second complication for complex diseases derives from the assumptions inherent in use of the LOD score method of linkage analysis. Two important assumptions are (1) that the mode of inheritance of the disease locus is known, and (2) that there is no epistasis. This latter assumption can be investigated with methods suggested by Risch.[38] The former assumption can be addressed for complex diseases by using both a carefully chosen dominant and recessive model in the analysis, with a corresponding increase in the critical positive LOD score to adjust for the additional tests performed. This may avoid false rejection of linkage.[39] It is not critical that the mode of inheritance of the complex disease is monogenic. However, misspecification of marker gene frequencies can cause inflation of LOD scores[40] in any linkage analysis in which there is missing marker data on founders in the pedigrees, including nonparametric analyses. Recent results indicate that misspecification of the parameters of the disease model along with misspecification of marker gene frequencies can result in substantial inflation of the LOD score,[26] which is likely to provide problems in interpretation of the statistical significance of linkage analyses of complex disorders.

2. REPEATABILITY

It is sometimes argued that positive results from multiple studies strengthen the argument that the locus under investigation is involved in the disease. However, although independent confirmation is important to assure that the result is not a statistical aberration, this does not imply that a causal relationship between the marker and the disease is the only explanation. Since parameter values used in the analysis can affect the results, independent confirmations are most convincing if they are performed by independent investigators, who are less likely to use identical parameters in the analysis. In the case of linkage disequilibrium, if there truly is disequilibrium in a sample for reasons such as admixture or population stratification, then an independent sample drawn from the same population is likely to show the same statistical relationship between loci. In the absence of a linkage analysis in pedigrees, what is needed to strengthen the case in favor of involvement of the locus in question is repeatability of the observed association in additional ethnic groups, since unrelated populations are unlikely to exhibit the same pattern of associations, e.g., because of admixture or some other force which may give rise to associations between unlinked loci. Alternatively, the case in favor of a biologically meaningful association can be strengthened by identifying a restricted set of disease-associated haplotypes, as is discussed later.

3. POWER

The interpretation of a negative result is just as important as the interpretation of a positive result. The failure to find statistically significant evidence of disequilibrium is sometimes taken as evidence that disequilibrium is not present. However, failure to reject the null hypothesis that there is no disequilibrium does not imply that disequilibrium is not present. Disequilibrium may be present, but may not be detectable because of low statistical power due to small sample size, weak disequilibrium, or phase of the alleles for the two loci. A discussion of how to compute power to detect disequilibrium under case-control sampling can be found elsewhere,[41] because in most situations where a disease locus is involved, the study design does not involve a random sample of gametes, as assumed in this discussion. Negative results in linkage analyses must also be interpreted cautiously, with attention to the expected power of the tests used. Nonparametric methods of analysis appear attractive in view of the possibility that a negative result from a LOD score analysis may only mean that the disease model is incorrect. However, such analyses have low power, requiring increased sample sizes. Also, many can only be applied to restricted pedigree structures, although methodological improvements are slowly removing such restrictions.

F. LINKAGE DISEQUILIBRIUM IN HUMAN POPULATIONS

As efforts to map and clone genes involved in genetic disorders begin to yield results, tests of linkage disequilibrium with marker loci in the vicinity of these genes are beginning to provide information on the expected existence and extent of linkage disequilibrium. These data are important for evaluating the extent to which this disequilibrium is the result of genetic linkage, and for using the extent of disequilibrium in the vicinity of genes to design better strategies for using association analyses in future studies to localize genes involved in mental disorders. It is also useful to consider data on the extent of linkage disequilibrium between unlinked markers. If such disequilibrium between unlinked loci is common, this will undermine the potential usefulness of using linkage disequilibrium to localize genes of unknown location, since given the low prior probability of linkage, a large fraction of positive results will turn out to be false positives.

1. **DETECTABLE LINKAGE DISEQUILIBRIUM**

There is a growing body of literature containing information on linkage disequilibrium between disease genes and closely linked markers. It is becoming increasingly clear that there are frequently a number of markers in the close vicinity of such genes which exhibit statistically detectable linkage disequilibrium with the disease locus. Once a disease locus has been mapped to a small chromosomal region, the existence of disequilibrium is now recognized as a key piece of evidence in potentially defining an even smaller region containing the gene of interest. In the case of Huntington's disease, the identification of a region of disequilibrium which was subtelomeric to 4p[42,43] was one of the main reasons the hunt for the gene moved away from the telomeric region of chromosome 4, leading to the recent identification of the gene.[44]

Tests of disequilibrium with a large number of marker loci which cover extensive regions surrounding genes have been done for only a small number of genes. However, information on disequilibrium testing is available for a small number of linked loci for a larger set of genes. The region defined by markers which show statistically detectable disequilibrium with the locus of interest is generally fairly small: on the order of 1 to 2 cM in nonisolated populations (assuming 1 cM is 1000 kb of DNA). Some examples for which a region of linkage disequilibrium can be defined are listed in Table 1-1, although this is not an exhaustive list of all such tests which have been performed. It is important to note that for most of the loci listed in Table 1-1 there have been relatively few markers tested. In the absence of tests done with large numbers of very closely spaced markers, as was done in the case of cystic fibrosis, it is not possible to define precisely the region in which disequilibrium can be identified. For most of these loci, therefore, the region in which there is some disequilibrium may be somewhat larger than that indicated in Table 1-1.

While the regions which exhibit statistically detectable linkage disequilibrium may be relatively large from the point of view of a molecular biologist who is trying to clone a gene from the region, there are two points which must be made about these regions. The first is

TABLE 1-1
Sizes of Regions Exhibiting Linkage Disequilibrium Around Some Loci

Gene/Disease	Region
Huntington's disease[45]	2.0 cM
Familial Mediterranean fever[46]	2.0 cM
Wilson's disease[47]	1.6 cM
Batten disease[48,49]	1.3–1.7 cM
Cystic fibrosis[50]	1.6 cM
Ig heavy-chain[51]	1.5 cM
Myotonic dystrophy[52]	<1 cM
Hematochromatosis[53]	<1 cM
Spinal muscular atrophy[54,55]	<0.7 cM
Freidrich's ataxia[56]	0.45 cM
G6PD[57]	0.35 cM

that they are very small compared to the size of the genome: in order to successfully use disequilibrium mapping to localize a disease locus one *must* choose a marker which is already extremely close to the gene of interest. This may eventually be possible when the regional location of the disease locus is already known, but is improbable when the location of the disease locus is unknown. Linkage analysis methods are likely to remain the method of choice to localize genes responsible for disorders for some time to come. The second point is that there are often subregions within these larger regions which contain markers which are in very strong disequilibrium with the disease loci. For example, there is a probable founder haplotype covering about 500 kb of DNA in the Huntington's disease region,[58] and a region of very strong disequilibrium covering about 300 kb of DNA in the cystic fibrosis gene region.[50] Thus, in general, if disequilibrium between a marker and disease is the result of genetic linkage it should be possible to find additional markers in the region which are also in disequilibrium with the disease.

2. ABSENCE OF STATISTICALLY DETECTABLE DISEQUILIBRIUM

Although disequilibrium can often be found in the close vicinity of genes of interest, close proximity does not guarantee the existence of statistically detectable linkage disequilibrium between loci (Table 1-2). Among the tests done which define the regions of disequilibrium in Table 1-1 there were many pairs of loci within the same regions which failed to give significant evidence of disequilibrium. For example, in a study of linkage disequilibrium in a 1500-kb region containing the immunoglobulin heavy-chain complex, only 9 of 36 comparisons gave nominally significant evidence of disequilibrium.[51] In the region of disequilibrium surrounding the Huntington's locus, only 8 of 153 tests in a 250-kb region remained significant after adjusting for multiple tests.[45]

TABLE 1-2
Number of Statistically Significant and Nonsignificant Tests of Linkage Disequilibrium Between Closely Linked Polymorphic Markers

Gene/Region	Size of Region	Signif.	Non-signif.	Total
D2S3[59]	20 kb	7	15	22
LDL receptor[60,61]	100 kb	34	77	111
TCRβ[62]	600 kb	6	22	28
LPL[63]	30 kb	5	4	9
Huntington's disease[45]	2500 kb	8	145	153

Failure to detect disequilibrium may occur because of lack of statistical power. Alternatively, there may be little or no linkage disequilibrium because the population has had time to come to equilibrium. This is probably the reason that disequilibrium between pairs of loci which are more than 2 to 3 cM apart is almost never detected among the enormous number of tests which have been performed with disease loci and markers which are linked to such loci at distance of 1 to 10 cM. The absence of significant disequilibrium between unlinked markers also seems to be the rule rather than the exception: in a study of 24 unlinked markers, after correction for multiple tests there was no evidence of linkage disequilibrium among the unlinked markers.[64]

A final cause of difficulty in detecting disequilibrium may arise when there are many independent disease locus mutations, each occurring on a different haplotypic background. Even in the case of markers which are very closely linked to a disease locus, it is possible to find examples where disequilibrium is difficult to detect under such circumstances. For example, no significant association can be detected between individual markers and mutations which cause phenylketonuria, or PKU.[65] This is undoubtedly because in the case of PKU, unlike cystic fibrosis, there is no single mutation which accounts for a high frequency in the population of the chromosomes bearing a mutation at the disease locus. However, it is important to note that when multiple markers in the vicinity of the PKU locus are jointly considered, it is possible to identify haplotypes defined by these markers which occur predominantly on PKU chromosomes.[65]

The high probability that a random pair of genes will fail to demonstrate evidence of linkage disequilibrium is illustrated by the high frequency of negative results obtained even for closely linked polymorphisms. A large number of tests have been performed in the search for disequilibrium between polymorphic sites found within genes. While linkage disequilibrium is common among such sites, many of these pairs show no evidence of being in linkage disequilibrium. Some examples of the number of significant vs. non-significant tests found within small genomic regions are shown in Table 1-2. While the tests within each locus are not independent, it is still clear that only about one third of the tests in this table were nominally significant at the 5% level.

G. CONCLUSIONS

This chapter identifies four general issues which must be kept in mind when designing or interpreting studies to map genes involved in mental disorders. These are (1) the conditions under which a false positive result may occur, (2) the conditions under which a true positive result may be detected, (3) the marker spacing needed to obtain a positive result from a test, if the marker is linked to the disease locus, and (4) the strength of a conclusion which can be made on the basis of finding a statistically significant result. The use of association vs. linkage studies for mapping genes responsible for psychiatric disorders must be evaluated in terms of the total cost and effort needed to correctly localize relevant genes, keeping these four issues in mind. This will include evaluation of the cost needed to follow up false positive leads with additional and/or alternative methods of analysis, as well as the cost of performing tests on a sufficiently large number of markers within each region to avoid missing a true association. These overall costs for disequilibrium studies, while not yet known in comparison to those of linkage analyses, will determine if and when association studies should be used in lieu of pedigree studies.

It is important when designing studies to avoid excessive numbers of false positive results. In the case of association studies, careful choice of cases and controls is therefore necessary. Unfortunately, since historical events can also introduce associations between unlinked loci, confirmation of associations through use of alternative data and analyses is critical. In the case of linkage analyses, careful attention must be paid to the choice of models and methods of analyses, and sensitivity analyses should be done when results appear positive. In addition, the very low prior probability that the disease and marker locus are linked requires stringent statistical significance levels in the absence of a strong candidate locus.

Positive results from an association analysis which takes place *after* a gene has been mapped to the region through linkage analysis is likely to indicate that the marker is very close to the disease locus. Positive results from an association analysis which is conducted ***prior to*** a linkage analysis will require additional data and/or analyses to verify that the association implies the presence of a disease locus in the vicinity of the marker locus. Significant positive results from a linkage analysis in a data set of pedigrees is one convincing way to confirm an association. An alternative, which itself is an association study, is to demonstrate that there is a ***restricted*** set of disease-associated haplotypes defined by multiple markers in the close vicinity of the marker which gave the initial positive result. Because of data which suggest that the existence of disequilibrium with multiple markers 0.5 to 2 cM around cloned genes may be the rule rather than the exception, identification of multiple markers in disequilibrium with the disease provides strong evidence that there may be a disease gene in the region. It is important to remember, however, that a reasonably large number of markers in a very small region may need to be typed; many markers will fail to show evidence of disequilibrium even in a region where some disequilibrium is present.

This restricted region expected to show detectable disequilibrium should also be kept in mind when designing studies. A disequilibrium study is likely to require testing of multiple, closely spaced markers. A linkage study, on the other hand, is likely to be able to detect evidence of linkage with a much looser marker spacing. It is not unreasonable to expect reasonable power to detect linkage with 20 cM spacing between highly polymorphic markers, and even at 30 to 40 cM spacing there is likely to be some positive signal which can be followed up with a few additional markers in the region. It therefore seems unlikely that disequilibrium testing can replace linkage analysis for genomic screening unless the cost of collecting individuals for a study becomes the most expensive part of the study with technological advances in molecular biology.

Most mental disorders and other complex diseases are far more common than even the most common Mendelian disorders. This is likely to be a big barrier to the success of association studies since such studies are based on the premise that linkage disequilibrium between a disease locus and marker(s) in the close vicinity of the locus is likely to exist. For a common disease many independent mutations most likely have occurred, diluting any linkage disequilibrium to undetectable levels. This would not affect a linkage analysis. Furthermore, it is likely that there are mutations in multiple loci, adding statistical noise to attempts to find evidence of an association between the disease in question and any arbitrary marker locus. One strategy which may be necessary to eventually identify loci involved in mental disorders, as well as other complex disorders, is to focus studies on rare and unusual forms of these diseases. This strategy parallels current, similar practices which have already led to successful genetic studies of a variety of common diseases with phenotypic and genetic heterogeneity, including CAD[2] and Alzheimer's disease.[11-14]

The probability of successful identification of a susceptibility locus through either a linkage or an association study increases through use of a *strong* candidate locus. Association tests have identified loci involved in a few complex diseases such as autoimmune diseases and cardiovascular disease, involving candidate loci for which there was already substantial circumstantial evidence that the gene might be involved in the disease. Once a biochemical pathway leading to a mental disorder has been identified, proteins which interact in this pathway are likely to provide additional strong candidate genes which may yield results in association tests. However, in the absence of strong biological evidence to indicate involvement of a gene in a particular disease, the probability that any given locus is involved in a particular disease is very small, and must be factored into interpretation of the meaning of "significant" evidence of an association.

The data on linkage disequilibrium around genes indicate that it should be possible to use the presence of linkage disequilibrium to more carefully localize genes which have already been mapped to a region by linkage analysis. Whether it is possible to extend association studies to genomic searches is more open to question, although this, too, is possible in principle. Whether this would be practical or cost effective is not currently known, given problems with sampling, possible artifacts, need for confirmatory tests, and the density of markers which would be needed in order to avoid missing true associations. Linkage analyses, while more difficult for complex disorders than for monogenic disorders, are likely to remain the method of choice for mapping genes contributing to mental disorders for some time. However, it is possible that once one or more of such a gene has been convincingly identified, it may be possible to evaluate the role on the disorder of other interacting loci with association studies. Many statistical issues also remain unresolved with respect to optimal design and sample size requirements of association studies in general. In particular, the total cost of a study, including the costs of replicating initially positive results and the total number of markers which need to be tested, still needs to be determined. It is not known whether association studies will ever be more cost effective than family studies for eventually locating genes contributing to complex disorders, but the issues and data raised here identify the considerations which must be taken into account in determining which approach is most effective.

REFERENCES

1. Goldstein, J. L., Hazzard, W. R., Schrott, H. G., Bierman, E. L., and Motulsky, A. G., Hyperlipidemia in coronary heart disease. II. Genetic analysis of lipid levels in 176 families and delineation of a new inherited disorder, *J. Clin. Invest.*, 51, 1544, 1973.
2. Goldstein, J. L. and Brown, M. S., Familial hypercholesterolemia, in *The Metabolic Basis of Inherited Diseases*, 6th ed., Scriver, C. R., Ed., McGraw-Hill, New York, 1989, 1215.
3. Egeland, J. A., Gerhard, D. S., Pauls, D. L., Sussex, J. N., Kidd, K. K., Allen, C. R., Hostetter, A. M., and Housman, D. E., Bipolar affective disorders linked to DNA markers on chromosome 11, *Nature*, 325, 783, 1987.
4. Kelsoe, J. R., Ginns, E. I., Egeland, J. A., Gerhard, D. S., Goldstein, A. M., Bale, S. J., Pauls, D. L., Long, R. T., Kidd, K. K., Conte, G., Housman, D. E., and Paul, S. M., Re-evaluation of the linkage relationship between chromosome 11p loci and the gene for bipolar affective disorder in the old order Amish, *Nature*, 342, 238, 1989.
5. Kennedy, J. L., Guiffra, L. A., Moises, H. W., Cavalli-Sforza, L. L., Pakstis, A. J., Kidd, J. R., Castiglione, C.M., Sjogren, B., Wetterberg, L., and Kidd, K. K., Evidence against linkage of schizophrenia to markers on chromosome 5 in a northern Swedish pedigree, *Nature*, 336, 167, 1988.
6. Straub, R. E., MacLean, C. J., O'Neill, A. D., Burke, J., Murphy, B., Duke, F., Shinkwin, R., Webb, B. T., Zhang, J., Walsh, D., and Kendler, K. S., A potential vulnerability locus for schizophrenia on chromosome 6p24-22: evidence for genetic heterogeneity, *Nat. Genet.*, 11, 287, 1995.
7. Wojciechowski, A. P., Farrall, M., Cullen, P., Wilson, T. M. E., Bayliss, J. D., Farren, B., Griffin, B. A., Caslake, M. J., Packard, C. J., Shepherd, J., Thakker, R., and Scott, J., Familial combined hyperlipidaemia linked to the apolipoprotein AI-CIII-AIV gene cluster on chromosome 11q23-q24, *Nature*, 349, 161-164, 1991.
8. Wijsman, E. M., Motulsky, A. G., Guo, S. W., Yang, M., Austin, M., Brunzell, J., and Deeb, S., Evidence against linkage of familial combined hyperlipidemia to the AI-CIII-AIV gene complex, *Circulation*, 86, I-420, 1992.
9. St. George-Hyslop, P. H., Tanzi, R. E., Polinsky, R. J., Haines, J. L., Nee, L., Watkins, P. C., Myers, R. H., Feldman, R. G., Pollen, D., Drachman, D., Growdon, J., Bruni, A., Foncin, J.-F., Salmon, D., Frommelt, P., Amaducci, L., Sorbi, S., Piacentini, S., Stewart, G. D., Hobbs, W. J., Conneally, P. M., and Gusella, J. F., The genetic defect causing familial Alzheimer's disease maps on chromosome 21, *Science*, 235, 885, 1987.
10. Schellenberg, G. D., Bird, T. D., Wijsman, E. M., Moore, D. K., Boehnke, M., Bryant, E. M., Lampe, T. H., Nochlin, D., Sumi, S. M., Deeb, S. S., Beyreuther, K., and Martin, G. M., Absence of linkage of chromosome 21q21 markers to familial Alzheimer's disease, *Science*, 241, 1507, 1988.
11. Schellenberg, G. D., Bird, T. D., Wijsman, E. M., Orr, H. T., Anderson, L., Nemens, E., White, J. A., Bonnycastle, L., Weber, J. L., Alonso, M. E., Potter, H., Heston, L. L., and Martin, G. M., Genetic linkage evidence for a familial Alzheimer's disease locus on chromosome 14, *Science*, 258, 668, 1992.
12. St. George-Hyslop, P., Haines, J., Rogaev, E., Mortilla, M., Vaula, G., Pericak-Vance, M., Foncin, J.-F., Montesi, M., Bruni, A., Sorbi, S., Rainero, I., Piness, I. L., Pollen, D., Polinsky, R., Nee, L., Kennedy, J., Macciardi, F., Rogaeva, E., Liang, Y., Aleandrova, N., Lukiw, W., Schlumpf, K., Tanzi, R., Tsuda, T., Farrere, L., Cantu, J.-M., Duara, R., Amaducci, L., Bergamini, L., Gusella, J., Roses, A., and Crapper McLachlan, D. R., Genetic evidence for a novel familial Alzheimer's disease locus on chromosome 14, *Nat. Genet.*, 2, 330, 1992.
13. Mullan, J., Houlden, H., Windelspect, M., Fidani, L., Lombardi, C., Diaz, P., Rossor, M., Crook, R., Hardy, J., Duff, K., and Crawford, F., A locus for familial early-onset Alzheimer's disease on the long arm of chromosome 14, proximal to the alpha1-antichymotrypsin gene, *Nat. Genet.*, 2, 340, 1992.
14. Von Broeckhoven, C., Backhovens, H., Cruts, M., De Winter, G., Bruyland, M., Cras, P., and Martin, J.-J., Mapping of a gene predisposing to early-onset Alzheimer's disease to chromosome 14q24.3, *Nat. Genet.*, 2, 335, 1992.
15. Blackwelder, W. C. and Elston, R. C., A comparison of sib-pair linkage tests for disease susceptibility loci, *Genet. Epidemiol.*, 2, 85, 1985.
16. Weeks, D. and Lange, K., The affected-pedigree-member method of linkage analysis, *Am. J. Hum. Genet.*, 42, 315, 1988.

17. Haseman, J. K. and Elston, R. C., The investigation of linkage between a quantitative trait and a marker locus, *Behav. Genet.*, 2, 3, 1972.
18. Olson, J. M. and Wijsman, E. M., Linkage between quantitative trait and marker loci: methods using all relative pairs, *Genet. Epidemiol.*, 10, 87, 1993.
19. Blum, K., Noble, E. P., Sheridan, P. J., Montgomery, A., Ritchie, T., Jagadeeswaran, P., Nogami, H., Briggs, A. H., and Cohn, J. B., Allelic association of human dopamine D2 receptor gene in alcoholism, *JAMA*, 263, 2055, 1990.
20. Gelernter, J., Goldman, D., and Risch, N., The A1 allele at the D2 dopamine receptor gene and alcoholism, *JAMA,* 269, 1673, 1993.
21. Arinami, T., Masanari, I., Komiyama, T., Mitsushio, H., Mori, H., Mifune, H., Hamaguchi, H., and Toru, M., Association between severity of alcoholism and the A1 allele of the dopamine D2 receptor gene *Taq I* A RFLP in Japanese, *Biol. Psychiatry*, 33, 108, 1993.
22. Goldman, D., Brown, G. L., Albaugh, B., Robin, R., Goodson, S., Trunzo, M., Akhtar, L., Lucas-Derse, S., Long, J., Linnoila, M., and Dean, M., DRD2 dopamine receptor genotype, linkage disequilibrium, and alcoholism in American Indians and other populations, *Alcohol Clin. Exp. Res.*, 17, 199, 1993.
23. Kidd, K. K., Associations of disease with genetic markers. Deja vu all over again, *Am. J. Med. Genet.*, 48, 71, 1993.
24. Jequier, A. M., Non-therapy related pregnancies in the consorts of a group of men with obstructive azoospermia, *Andrologia*, 17, 6, 1985.
25. Morton, N. E., Sequential tests for the detection of linkage, *Am. J. Hum. Genet.*, 7, 277, 1955.
26. Wijsman, E. M., Genetic analysis of Alzheimer's disease: a summary of GAW8, *Genet. Epidemiol.*, 10, 349, 1993.
27. Bishop, D. T. and Williamson, J. A., The power of identity-by-descent methods for linkage analysis, *Am. J. Hum. Genet.*, 46, 254, 1990.
28. Hodge, S. E. and Greenberg, D. A., Sensitivity of LOD scores to changes in diagnostic status, *Am. J. Hum. Genet.*, 50, 1053, 1992.
29. Davignon, J., Gregg, R. E., and Sing, C. F., Apolipoprotein E polymorphism and atherosclerosis, *Arteriosclerosis*, 8, 1, 1988.
30. Weir, B. S., Inferences about linkage disequilibrium, *Biometrics*, 35, 235, 1979.
31. Weir, B. S., *Genetic Data Analysis*, Sinauer Associates, Sunderland, MA, 1990, chap. 3.
32. Hernandez, J. and Weir, B., A disequilibrium coefficient approach to Hardy-Weinberg testing, *Biometrics*, 45, 53, 1989.
33. Weir, B. S. and Cockerham, C. C., Testing hypotheses about linkage disequilibrium with multiple alleles, *Heredity*, 42, 105, 1978.
34. Rubinstein, P., Walker, M., Carpenter, C., Carrier, C., Krassner, J., Falk, C. T., and Ginsburg, F., The use of the haplotype relative risk (HRR) and the 'haplo-delta' (Dh) estimates in juvenile diabetes from three racial groups, *Hum. Immunol.*, 3, 384, 1981.
35. Ewens, W. J. and Spielman, R. S., The transmission disequilibrium test: history, subdivision, and admixture, *Am. J. Hum. Genet.*, 57, 455, 1995.
36. Knowler, W. C., Williams, R. C., Pettitt, D. J., and Steinberg, A. G., Gm$^{3;5,13,14}$ and type 2 diabetes mellitus: an association in American Indians with genetic admixture, *Am. J. Hum. Genet.*, 43, 520, 1988.
37. Ott, J., *Analysis of Human Genetic Linkage, Revised Edition*, The Johns Hopkins University Press, Baltimore, 1991, chap. 4.
38. Risch, N., Linkage strategies for genetically complex traits. I. Multilocus models, *Am. J. Hum. Genet.*, 46, 219, 1990.
39. Greenberg, D. A. and Hodge, S. E., Linkage analysis under "random" and "genetic" reduced penetrance, *Genet. Epidemiol.*, 6, 259, 1989.
40. Ott, J., Strategies for characterizing highly polymorphic markers in human gene mapping, *Am. J. Hum. Genet.*, 51, 283, 1992.
41. Olson, J. M. and Wijsman, E. M., Design and sample-size considerations in the detection of linkage disequilibrium with a disease locus, *Am. J. Hum. Genet.*, 55, 574, 1994.
42. Snell, R. G., Lazarou, L. P., Youngman, S., Quarrell, O. W. J., Wasmuth, J. J., Shaw, D. J., and Harper, P. S., Linkage disequilibrium in Huntington's disease: an improved localisation for the gene, *J. Med. Genet.*, 26, 673, 1989.
43. Theilmann, J., Kanani, S., Shiang, R., Robbins, C., Quarrell, O., Huggins, M., Hedrick, A., Weber, B., Collins, C., Wasmuth, J. J., Buetow, K. H., Murray, J. C., and Hayden, M. R., Non-random association between alleles detected at D4S95 and D4S98 and the Huntington's disease gene, *J. Med. Genet.*, 26, 676, 1989.
44. Huntington's Disease Collaborative Research Group, A novel gene containing a trinucleotide repeat that is expanded and unstable on Huntington's disease chromosomes, *Cell*, 72, 971, 1993.
45. MacDonald, M. E., Lin, C., Srinidhi, L., Bates, G., Altherr, M., Whaley, W. L., Lehrach, H., Wasmuth, J., and Gusella, J. F., Complex patterns of linkage disequilibrium in the Huntington disease region, *Am. J. Hum. Genet.*, 49, 723, 1991.

46. Aksentijevich, I., Pras, E., Gruberg, L., Shen, Y., Holman, K., Helling, S., Prosen, L., Sutherland, G. R., Richards, R. I., Dean, M., Pras, M., and Kastner, D. L., Familial Mediterranean fever (FMF) in Moroccan Jews: demonstration of a founder effect by extended haplotype analysis, *Am. J. Hum. Genet.*, 53, 644, 1993.
47. Petrukhin, K., Fischer, S. G., Pirastu, M., Tanzi, R. E., Chernov, I., Devoto, M., Brzustowicz, L. M., Cayanis, E., Vitale, E., Russo, J. J., Matseonane, D., Boukhgalter, B., Wasco, W., Figus, A. L., Loudianos, J., Cao, A., Sternlieb, I., Evgrafov, O., Parano, E., Pavone, L., Warburton, D., Ott, J., Penchaszadeh, G. K., Scheinberg, I. H., and Gilliam, T. C., Mapping, cloning and genetic characterization of the region containing the Wilson disease gene, *Nat. New Genet.*, 5, 338, 1993.
48. Mitchison, H. M., Thompson, A. D., Mulley, J. C., Kozman, H. M., Richards, R. I., Callen, D. F., Stallings, R. L., Doggett, N. A., Attwood, J., McKay, T. R., Sutherland, G. R., and Gardiner, R. M., Fine genetic mapping of the batten disease locus (CLN3) by haplotype analysis and demonstration of allelic association with chromosome 16p microsatellite loci, *Genomics*, 16, 455, 1993.
49. Lerner, T., Boustany, R.-M., Schultz, E., D'Arigo, K., Schlumpf, K., Gusella, J., and Haines, J., Linkage disequilibrium between the juvenile NCL gene (CLN3) and marker loci on chromosome 16p12.1, *Am. J. Hum. Genet.*, 53s,1032, 1993.
50. Kerem, B.-S., Rommens, J. M., Buchanan, J. A., Markiewicz, D., Cox, T. K., Chakravarti, A., Buchwald, M., and Tsui, L.-C., Identification of the cystic fibrosis gene: genetic analysis, *Science*, 245, 1073, 1989.
51. Walter, M. A. and Cox, D. W., Nonuniform linkage disequilibrium within a 1,500-kb region of the human immunoglobulin heavy-chain complex, *Am. J. Hum. Genet.*, 49, 917, 1991.
52. Cobo, A., Grinberg, D., Balcells, S., Vilageliu, L., Gonzalez-Duarte, R., and Baiget, M., Linkage disequilibrium detected between myotonic dystrophy and the anonymous marker D19S63 in the Spanish population, *Hum. Genet.*, 89, 287, 1992.
53. Jazwinska, E. C., Lee, S. C., Pyper, W. R., Webb, S. I., Burt, M. J., Halliday, J. W., and Powell, L. N., Haplotype analysis in hemochromatosis pedigrees: evidence for a common haplotype and a gene location telomeric of HLA-A, *Am. J. Hum. Genet.*, 53s,1019, 1993.
54. Brzustowicz, L. M., Matseoane, D., Wang, C. H., Kleyn, P. W., Vitale, E., Penchaszadeh, G. K., Hausmanowa-Petrusewicz, I., and Gilliam, T. C., Linkage disequilibrium and haplotype analysis among Polish families with spinal muscular atrophy, *Am. J. Hum. Genet.*, 53s, 982, 1993.
55. Simard, L. R., Prescott, G., Rochette, C., Morgan, K., Melancon, S. B., and Vanasse, M., Linkage disequilibrium between certain chromosome 5q markers and childhood-onset spinal muscular atrophy (SMA) in the French Canadian population, *Am. J. Hum. Genet.*, 53s, 1078, 1993.
56. Pandolfo, M., Sirugo, G., Antonelli, A., Weitnauer, L., Ferretti, L., Leone, M., Dones, I., Cerino, A., Fujta, R., Hanauer, A., Mandel, J.-L., and Di Donato, S., Friedreich's ataxia in Italian families: genetic homogeneity and linkage disequilibrium with the marker loci D9S5 and D9S15, *Am. J. Hum. Genet.*, 47, 228, 1990.
57. Filosa, S., Calabro, V., Lania, G., Vulliamy, T. J., Brancati, C., Tagarelli, A., Luzzatto, L., and Martini, G., G6PD haplotypes spanning Xq28 from F8C to red/green color vision, *Genomics*, 17, 6, 1993.
58. MacDonald, M. E., Novelletto, A., Lin, C., Tagle, D., Barnes, G., Bates, G., Taylor, S., Allitto, B., Altherr, M., Myers, R., Lehrach, H., Collins, F. S., Wasmuth, J. J., Frontali, M., and Gusella, J. F., The Huntington's disease candidate region exhibits many different haplotypes, *Nat. Genet.*, 1, 99, 1992.
59. Litt, M. and Jorde, L. B., Linkage disequilibria between pairs of loci within a highly polymorphic region of chromosome 2Q, *Am. J. Hum. Genet.*, 39, 166, 1986.
60. Hegele, R. A., Plaetke, R., and Lalouel, J.-M., Linkage disequilibrium between DNA markers at the low-density lipoprotein receptor gene, *Genet. Epidemiol.*, 7, 69, 1990.
61. Leitersdorf, E., Chakravarti, A., and Hobbs, H. H., Polymorphic DNA haplotypes at the LDL receptor locus, *Am. J. Hum. Genet.*, 44, 409, 1989.
62. Charmley, P., Chao, A., Concannon, P., Hood, L., and Gatti, R. A., Haplotyping the human T-cell receptor b-chain gene complex by use of restriction fragment length polymorphisms, *Proc. Natl. Acad. Sci. U.S.A.*, 87, 4823, 1990.
63. Heizmann, C., Kirchgessner, T., Kwiterovich, P. O., Ladias, J. A., Derby, C., Antonarakis, S. E., and Lusis, A. J., DNA polymorphism haplotypes of the human lipoprotein lipase gene: possible association with high density lipoprotein levels, *Hum. Genet.*, 86, 578, 1991.
64. Hernandez, J. L., Elston, R. C., and Ward, L. J., Gametic equilibrium between 24 polymorphic markers, *Hum. Genet.*, 85, 343, 1990.
65. Chakraborty, R., Lidsky, A. S., Daiger, S. P., Guttler, F., Sullivan, S., Dilella, A. G., and Woo, S. L. C., Polymorphic DNA haplotypes at the human phenylalanine hydroxylase locus and their relationship with phenylketonuria, *Hum. Genet.*, 76, 40, 1987.
66. Goate, A. M., Chartier-Harlin, M.-C., Mullan, M. J., Brown, J., Crawford, F., Fidani, L., Giuffra, L., Haynes, A., Irving, N., James, L., Mant, R., Newton, P., Rooke, K., Roques, P., Talbot, C., Pericak-Vance, M., Roses, A., Williamson, R., Rosser, M. N., Owen, M. J., and Hardy, J. A., Segregation of a missence mutation in the amyloid precursor protein gene with familial Alzheimer's disease, *Nature*, 349, 704, 1991.

2 Genetic Association Studies in Psychiatry: Recent History

Joel Gelernter

CONTENTS

A. Association Studies and Study Design 25
B. Successes in Neuropsychiatry in Relating Genes and Diseases or Phenotypes 27
C. Problems With Association Studies and Some Potential Solutions 28
D. Association Studies With the D2 Dopamine Receptor (DRD2) 29
E. DRD3 and Excess Homozygosity 31
F. Future Directions 32
Acknowledgments 32
References 32

A. ASSOCIATION STUDIES AND STUDY DESIGN

Genetic association studies have seen increasing application in psychiatric genetics in recent years despite many well-known limitations. We will discuss some advantages and disadvantages of genetic association studies in the context of psychiatry, cite examples where theoretical difficulties have led to important ambiguities in interpretation of results, and discuss solutions that have been proposed to address some specific obstacles.

Study design is a fundamental issue in psychiatric genetics, as for all science. In choosing a study design, we try to identify a method that will lead us quickly and efficiently to a positive result, with minimal risk of false positive or false negative findings. We then want to use statistical tests that give us an accurate reflection of the meaning of our findings; i.e., if we say that a certain finding is significant at the $p = 0.05$ level, there should really be about a 19 in 20 chance that the finding will prove to be correct in the long run. Understanding the subtleties of study design is particularly important for psychiatric genetics; history has demonstrated that miscalculations can be devastating.

Genetic linkage analysis has been considered the method of choice for gene detection in psychiatry over the past decade, but as of this writing replicable results for purely psychiatric (as opposed to neuropsychiatric) illness have been sparse. Several features of psychiatric illness might be responsible for this difficulty including, for example, the genetic complexity of these disorders (Risch, 1990) and difficulties with establishing diagnosis (Tsuang et al., 1993). This has led to innovation in linkage analysis and also to the application of other techniques. For example, a recent study employed innovative linkage techniques for analysis of a complex trait, type 1 diabetes (Davies et al., 1994), perhaps analogous to some psychiatric disorders such as schizophrenia.

Quantitative trait locus (QTL) mapping (Lander and Botstein, 1989) is a special case of linkage analysis where an attempt is made to map multiple genes influencing quantifiable aspects of a phenotype. QTL mapping using animal models has recently grown in popularity and may provide a method to identify individual genes important for genetically complex disorders. For example, this method was used to map loci important for regulating growth and fatness in pigs (Andersson et al., 1994). For application to human disease, however, it is necessary to cope with the possibility that similar phenotypes in different species may arise by different mechanisms (e.g., Hubner et al. [1994] reported that blood pressure in a hypertensive rat strain was not linked to the angiotensinogen locus, whereas such linkage does exist in some human families). The very complexity of some phenotypes may reflect the fact that they may be reached through many pathways; animal models, even with phenotypes resembling those observed in humans, may have arrived at those phenotypes by entirely different mechanisms from those usually seen in humans. The problem is especially great where the phenotypes are not exactly analogous, as must be the case for psychiatric disorders. Nonetheless, initial applications for behaviors have given highly encouraging results, e.g., with the mapping of a quantitative trait locus for reading disability (Cardon et al., 1994).

Considering these factors, association studies provide an alternative that is attractive under some circumstances. Genetic association studies may help identify genes that either increase liability for psychiatric illness or affect certain phenotypes within these illnesses. Samples for association studies are comprised of unrelated individuals. Any alteration in DNA sequence expressed preferentially by a series of unrelated individuals who all have an illness, but not by a series of unrelateds without the illness, may be shown with appropriate statistical testing to be associated with the illness (Gejman and Gelernter, 1993). The populations must meet certain criteria, such as genetic homogeneity. The association has physiological meaning if the ill individuals share the DNA sequence because they have the disease (and the DNA sequence either causes the disease or influences its development). The same basic reasoning can be used to study particular phenotypic traits of ill individuals, such as people with schizophrenia. If everyone (or a higher percentage) with a certain trait (such as paranoia) carries a certain DNA sequence and if everyone without the trait carries a different sequence, this supports a relationship of trait with sequence (although there are potential complicating factors one must take into account as well).

While many aspects of genetic analysis are obviously complicated — linkage analysis, for example — other aspects that are also complicated may not seem so. Genetic association studies fall into the second category. Initially, the idea and application of a genetic association study may seem straightforward, but close inspection shows that many pitfalls await the investigator. These problems are most likely to occur under the following circumstances: (1) when marker polymorphisms, rather than coding region polymorphisms affecting structure, are used — this is especially problematic if the polymorphisms used are far enough away to be in linkage equilibrium with the candidate gene and therefore alleles at the marker locus do not predict alleles at the candidate gene locus; and (2) when selection of affected and unaffected groups does not control sufficiently for race or ethnicity (Gelernter et al., 1993).

The basic physical reality underlying an association study is as follows, if we make the simplifying assumption of genetic homogeneity of the illness for the moment. All genetic effects can ultimately be traced to a variation in DNA sequence. A group of people who are unrelated vary in most genetic characteristics (because they are unrelated) — including not only such easily observable genetic characteristics as eye color, HLA type, and blood groups, but also other measurable genetic polymorphism such as other protein variants and DNA-level polymorphisms — unless they are selected to share one of these traits in common. However, a group of unrelated people ill with a genetic disease *is* selected for homogeneity for a DNA variant: the DNA variant that causes the disease. In the absence of population stratification effects (e.g., many ill people deriving from an ethnic group sharing other DNA variants for reasons related to their ethnic group and unrelated to illness), the DNA variant responsible for causing the illness should be the only such polymorphism shared at a greater rate among ill individuals than among unrelated individuals without the illness.

Therefore, if a DNA variant is found to be more common in a group of ill people than in a group of well people who have no other reason to share alleles at the locus being studied, the reason must be either that the DNA variant has something to do with the illness, or it is in linkage disequilibrium with such a variant. The same general reasoning applies for a genetically heterogeneous disorder, except that some but not all ill individuals would be expected to carry the variant (so the effect would be harder to detect) (Gejman and Gelernter, 1993).

In several branches of medicine, demonstration of genetic associations between gene and disease have, in some cases, resulted in progression to an improved understanding of disease mechanisms, e.g., for ApoE alleles and Alzheimer's disease (discussed below) (Ma et al., 1994). Candidate gene investigations have also led to improved understanding of the pathophysiology of many other disease conditions and appreciation of the importance of critical metabolic pathways; among many other possible examples there is an allelic association between angiotensin-converting enzyme (ACE) and hypertrophic cardiomyopathy (Marian et al., 1993). So far at least, similarly clear advances have not been made for common psychiatric illness. Indeed, until recently, almost all association studies in psychiatry have used noncoding variants as genetic markers rather than variants related to changes in protein sequence, which adds an extra level of complexity to the interpretation of results.

B. SUCCESSES IN NEUROPSYCHIATRY IN RELATING GENES AND DISEASES OR PHENOTYPES

There have been several recent successes relating specific gene variants to specific behaviors or outcomes. Considering dementia in this category, Alzheimer's disease has been the subject of notable advances in genetic understanding in recent years, with two key genetic association results. The most common form of familial early-onset Alzheimer's disease is the chromosome 14-linked variety (Schellenberg et al., 1992). A different form (later onset) linked to markers on chromosome 19 has also been known for some time (Pericak-Vance et al., 1991). The apolipoprotein E (ApoE) locus maps to this region. Dramatic work reported over the past two years demonstrated that the three common alleles at ApoE exert a major effect on risk of Alzheimer's disease. Allele 4 predisposes to Alzheimer's disease (Corder et al., 1993; Saunders et al., 1993) whereas allele 2 has a protective effect (Corder et al., 1994). (This work has been replicated multiple times and extended; e.g., Poirier et al., 1993.) ApoE4 directly promotes increased formation of filamentous amyloid deposits (Ma et al., 1994), suggesting a possible mechanism for the effect. In this case, therefore, there is an allelic association with polymorphic variants of the ApoE gene and Alzheimer's disease. The different alleles correspond directly with different protein products (Hansen et al., 1994) and these different protein products appear directly responsible for the effect on illness.

A previous candidate gene study led to a mutation in amyloid precursor protein responsible for a small percentage of early-onset familial Alzheimer's disease (Goate et al., 1991). It was demonstrated later that another mutation in the same gene can cause a similar but not identical phenotype, i.e., presenile dementia with cerebral hemorrhage (Hendriks et al., 1992). That is, once this gene was identified as one for which mutations lead to one form of dementia, it led to clarification of the pathophysiology of another phenotypically related but not identical disorder as well, and demonstrated a clear relationship between the two disorders.

These studies with Alzheimer's disease do not provide as clear an analogy with candidate gene studies in psychiatry as may appear at first, because in both cases a gene location was known before a candidate gene was studied. This greatly increased the *a priori* chance of success for any gene mapping to the implicated genomic region.

A clear-cut and striking recent presentation of a genotype-phenotype relationship for a behavioral disorder was the work by Brunner et al., demonstrating first, by linkage, the genomic location of a syndrome characterized by violence and abnormal behavior (Brunner et al., 1993a), then identifying the exact lesion involved: a mutation in the monoamine oxidase A (MAOA) gene introducing a stop codon and resulting in inactive protein (Brunner et al., 1993b). In the single family described, all males with the abnormal gene had the abnormal phenotype, and heterozygous females could transmit the illness. This was a particularly satisfying demonstration that some of what we thought we knew about the effects of monoamines on behavior was true. It is a demonstration of the genetic basis for one form of violent behavior, now known as "Brunner's syndrome".

Small steps have already been taken in demonstrating genotype/phenotype relationships in other genetically complex psychiatric disorders. For example, Pelchat and Danowski (1992) reported an association between PROP tasting and alcoholism; there was a higher rate of nontasting of PROP, which is a bitter substance similar to PTC, among children of alcoholics than children of controls. PROP tasting is inherited as a Mendelian dominant. The genetic trait corresponds directly to a protein difference and a phenotype (i.e., ability to taste certain bitter substances). This trait clearly could directly affect the disease phenotype: it could affect the perception of the taste of alcohol and therefore influence its aversive or reinforcing properties. Abnormalities in alcohol-metabolizing enzymes provide another similar example (Thomasson et al., 1991); they protect the affected individuals from alcoholism.

C. PROBLEMS WITH ASSOCIATION STUDIES AND SOME POTENTIAL SOLUTIONS

Population stratification, that is, the presence of subpopulations within the larger population used with different allele frequencies at the locus studied, is a major potential confounder for genetic association studies. It is often the case that members of different genetic populations, e.g., racial or ethnic groups or population isolates, differ in their baseline allele frequencies at many genetic loci (Bowcock et al., 1991; Devlin and Risch, 1992). If the illness studied occurs with greater frequency in one ethnic group than in other groups, that group also has an increased frequency of a certain allele at a certain polymorphic system, and if a stratified sample is studied, a genetic association will be observed between allele and disease. However, the association does not have meaning for the physiology of the disease. Although it is "real" (in that it exists and may be replicated), it does not imply a gene-disease relationship. Methods addressing this potential problem have been devised. The Haplotype Relative Risk (HRR) method is a method applied to control for the possibility of population stratification in association studies by composing a control sample from nontransmitted parental alleles (Falk and Rubinstein, 1987; Terwilliger and Ott, 1992). This method controls for variation in allele frequency by race and ethnicity by constructing a control group of nontransmitted parental alleles. For an HRR study, DNA is collected for probands (who must receive a research diagnosis) and from the proband's parents (who do not need to be diagnosed). All three individuals are genotyped. The two alleles of the proband are added to the "ill" group. The nontransmitted parental alleles are determined by subtracting the set of the offspring's two alleles from the set of the parent's four alleles. Since the two parents each donate one allele to the "ill" group and one allele to the comparison group, it is evident that they must be matched for ethnicity and race, as contributions to the two sets of alleles are completely balanced. This method and other related methods have been surveyed by Schaid and Sommer (1994).

Another potential problem with candidate gene association studies is the potential confusion between genetic marker loci, i.e., polymorphisms that are *near* the candidate gene, and polymorphisms corresponding to sequence variation within the gene itself. An example of the former is the DRD2 "A" *Taq I* RFLP system, used for many association studies, sometimes misidentified as representing variation in the DRD2 gene, and actually located about 10 kb beyond the coding region (Hauge et al., 1992). Because it is outside the coding region this polymorphism is very unlikely to show any direct correspondence to DRD2 protein variation, although it could be in linkage disequilibrium with a variant that did show such a correspondence. An example of the latter is a polymorphism in the D4 dopamine receptor gene (DRD4) that directly results in protein variation and possibly even in substrate binding (Van Tol et al., 1992). (We demonstrated that DRD4 alleles do not influence response to clozapine in schizophrenic patients; Rao et al., 1994.) Association study results using coding region polymorphisms affecting protein sequence are inherently more informative than studies using marker polymorphisms.

D. ASSOCIATION STUDIES WITH THE D2 DOPAMINE RECEPTOR (DRD2)

Dopamine receptors have been enduringly popular as candidate genes for causation of psychiatric illness due to the wealth of data supporting a relationship between dopaminergic abnormalities and psychiatric illness. The first of these receptors to be cloned was DRD2 (Grandy et al., 1989a), followed in quick succession by DRD1, DRD3, DRD4, and DRD5. A *Taq I* restriction fragment length polymorphism (RFLP) was identified by Grandy et al. (1989b). Using this polymorphism, we studied DRD2 as a candidate gene for causation of Tourette's syndrome (Gelernter et al., 1990) and schizophrenia (Moises et al., 1991) using a linkage strategy, and concluded that for the kindreds and genetic models chosen, linkage could be excluded between DRD2 and those illnesses. Despite its value as a marker for linkage analysis, the nature of this polymorphism presented a clear barrier to its use for genetic association studies. The principal problem was that, as for most RFLPs, no direct correlation could be made between the alleles at this locus and variation in the coding sequence of the protein.

Blum et al. (1990) reported an association between DRD2 alleles and alcoholism. The polymorphism studied was the same RFLP used for the linkage studies described above, known as the *Taq I* "A" system. Although the next paper published on the subject reported a failure to replicate the finding (Bolos et al., 1990), several other papers were soon published in rapid succession that appeared to support the positive finding. There have now been multiple positive publications (Blum et al., 1991; Parsian et al., 1991; Amadeo et al., 1993; Comings et al., 1991) and negative publications (Bolos et al., 1990; Gelernter et al., 1991; Schwab et al., 1991; Cook et al., 1992; Goldman et al., 1992; Turner et al., 1992; Goldman et al., 1993; Arinami et al., 1993). A close examination of the positive papers published subsequent to Blum et al. (1990) shows that rather than being clear replications, the hypothesis was altered, sometimes subtly, and ideas about the phenotype associated to DRD2 alleles changed. For example, Blum et al. (1990) described their sample as severely alcoholic on the basis of the observation that many individuals in their autopsy sample had died of complications related to alcoholism. Parsian et al. (1991) then defined severity on the basis of medical complications. Other authors (Bolos et al., 1990; Gelernter et al., 1991) used definitions of severity based on such things as the amount of alcohol consumed, presence of physical withdrawal symptoms, and MAST (Michigan Alcohol Screening Test) scores. Blum et al. (1991) subsequently defined lack of severity on the basis of absence of medical complications, regardless of other symptoms. Studies using wholly corresponding definitions of severity are distinctly rare.

Another area of controversy was the use of screened control groups vs. unscreened "random population" controls. Noble and Blum (1991) and Smith et al. (1991) both contended that the use of control groups screened to eliminate alcoholics was an extremely important factor. Cloninger (1991) hypothesized that Bolos et al. (1990) and Gelernter et al. (1991) did not replicate Blum et al. because of high control A1 allele frequencies, possibly because of failure to exclude alcoholics. We demonstrated that removing known alcoholics from control groups could have only a modest effect (Gelernter et al., 1991), because the frequency of severe alcoholism in the population is not great enough for the "contaminating" ill individuals to have a substantial effect on the allele frequency observed in a large control group, even if there were a relationship between alleles and disease. Also, in order to be able to judge the investigator's possible success at excluding ill individuals, the age of the subjects making up the control group would become an important issue because it would be necessary to know that all individuals in the control group who were judged free of illness (alcohol dependence in this case) had already passed through the age of risk.

Finally, reliability of diagnosis would become an issue not only for the ill group but for the well group also; people who say they are free of mental illness may have an incentive to withhold relevant information if being paid for research participation that is contingent upon their being "normal controls". These factors would tend to decrease the "purity" of the screened control group and decrease the already small difference that might exist between screened and unscreened groups. Some of the largest differences seen in allele frequencies so far for the DRD2 A system are between screened and unscreened control groups (Gelernter et al., 1993; Noble and Blum, 1993). Noble and Blum (1993) argued that this shows that control groups must be screened; we (Gelernter et al., 1993b) argued that, since DRD2 A1 allele frequencies were identical for alcoholics and unscreened controls, but screened controls were different from both, the most likely explanation was artifactual or, alternatively, an association between the other DRD2 allele (A2) and lack of psychiatric diagnosis. Screening the affected individuals out of a control group, besides being of questionable value except for a common illness, could introduce new artifacts relating, for example, to the motivation of volunteers to submit to a detailed diagnostic interview.

Parsian et al. (1991) reported support for association between DRD2 alleles and alcoholism in the absence of linkage. Hodge (1993) however, suggested that what was described as a linkage test in that study was actually better considered a family-based association test than a linkage test. When the same group increased their sample size they were no longer able to support the hypothesis of an association (Suarez et al., 1994).

Population stratification, discussed above, can be a confounding factor for association studies and may easily lead to false positive results. It has been shown specifically for the DRD2 A1 system that the potential for this effect to occur actually does exist, due to great variation in allele frequency between populations of different ethnicity (Goldman et al., 1992; Goldman et al., 1993; Barr and Kidd, 1993). All single-ethnicity DRD2 association studies (Schwab et al., 1991; Goldman et al., 1992; Goldman et al., 1993; Arinami et al., 1993), with one exception (Amadeo et al., 1993), have failed to show elevated DRD2 A1 frequency in alcoholics.

When we considered all of the data published by 1993, some interesting patterns emerged (Gelernter et al., 1993). The most striking was that when the first hypothesis-generating study was omitted from the meta-analysis, allele frequencies for alcoholics and nonalcoholics, considering data from all investigators, were identical. There was significant heterogeneity between studies both for alcoholics and for controls. We argued in conclusion that while a genetic association between DRD2 alleles and alcoholism was possible, a more conservative explanation of the data (i.e., no association, with positive results explained by many factors including sampling error and population stratification) should be accepted until some stronger form of evidence could be produced, such as a mutation in the gene itself associated to alcoholism.

The most rigorous molecular biological test of the proposed relationship between DRD2 alleles and alcoholism so far was a mutational analysis by Gejman et al. (1994). Gejman et al. studied DNA from alcoholics previously studied by several other groups including Blum et al. (1990), Bolos et al. (1990), and Gelernter et al. (1991). Using denaturing gradient gel electrophoresis, they studied all translated exons of DRD2 and demonstrated that there was no common coding region variant that could convey an effect. The only way to account for such an effect now is by hypothesizing a mutation outside of the coding region which might act by an effect on regulation of the gene (resulting, e.g., in altered numbers of D2 receptors in brain, as supported by Noble et al.'s post-mortem study [1991] or stability of the mRNA). If an effect of DRD2 alleles on D2 receptor B_{max} could be demonstrated *in vivo*, this could be explained by a noncoding region polymorphism and would provide a physiological explanation for the positive association findings.

The reports regarding DRD2 and alcoholism led investigators first towards substance abuse, and then away from alcoholism. Motivated by possible etiological relationships between alcoholism and substance abuse, Smith et al. (1992) reported an association between substance abuse and the "B" RFLP system at DRD2. Noble et al. (1993) reported an association between both "A" and "B" system DRD2 alleles and cocaine dependence. Comings et al. (1994), studying the "A" system, supported an association between DRD2 alleles and substance abuse — but not with alcoholism. We could not find an association between DRD2 alleles and cocaine dependence in a preliminary study (Gelernter et al. 1993c).

An interesting possible generalization of the range of effects of DRD2 alleles on psychiatric illness was proposed by Comings (1991), who suggested that DRD2 alleles have an effect on the severity of Tourette's syndrome (TS). Using a within-family design in families where genetic linkage with Tourette's syndrome had already been excluded, we did not find any effect of DRD2 alleles on TS severity (Gelernter et al., 1994).

More recently, an effect of DRD2 alleles on obesity has been reported (Comings et al., 1993). If all of these reports were correct, there would be some interesting and easily testable consequences. For example, with the same DRD2 alleles associated with both obesity and alcoholism, alcohol-dependent subjects should have greater average weight than nonalcoholics.

DRD2 now stands as the most studied gene in the history of psychiatric genetics; it has been the subject of linkage studies, association studies, mutational analysis, and direct sequencing studies (Sarkar et al., 1991). Several rare coding sequence variants are known, but none that correlate with any physiologic effect. The most studied polymorphic site (the *Taq I* "A" system) is well outside the coding region of the gene. It would be impossible to prove that DRD2 alleles do not have an effect on illness, and as such it would be inappropriate to consider such a relationship "ruled out", but one must consider the weight of the various probabilistic arguments offered.

Given the exhaustive study and lack of a clearly replicated positive result not explicable by some common artifact, a straightforward association with a particular diagnostic group now appears unlikely. If there is an association it appears most likely that it is with some phenotype more complex than a single disease trait. This phenotype could be overrepresented in certain substance-abusing populations, for example.

F. DRD3 AND EXCESS HOMOZYGOSITY

The case of DRD3 alleles and a possible relationship of excess homozygosity at that locus to either schizophrenia or Tourette's syndrome is also an interesting one, and the meaning of any such relationship is still not clear. It stands on fundamentally firmer ground that the claims for association with DRD2 and various psychiatric traits, because the DRD3 polymorphism in question occurs in the coding region of the gene and the polymorphic variants differ in amino acid sequence at one point (glycine or serine) (Crocq et al., 1992).

Crocq et al. (1992) reported an association between homozygosity at this locus and schizophrenia. It is of interest that excess homozygosity had already been postulated to be a general feature of the genome in schizophrenia on the basis of several reports showing increased fluctuating dermatoglyphic asymmetry in schizophrenic patients (e.g., Markow and Wandler, 1986). Subsequently, there have been reports tending to be consistent with the original report (Mant et al., 1994) and failures to replicate it (Nöthen et al., 1993; Nanko et al., 1993; Jönsson et al., 1993; Nimgaonkar et al., 1993). There has also been a report of excess homozygosity at DRD3 for Tourette's syndrome (Comings et al., 1993) which was not replicated (Hebebrand et al., 1993; Brett et al., 1993 [using different methodology]).

Although much has been made of deviations from homozygosity predicted by the Hardy-Weinberg squares law at the DRD3 locus in some schizophrenic populations, it is worth recalling that this law holds only when certain conditions are met for the population. It must be large, there must be random mating, the population must be homogeneous, and there must be no migration in or out of the populations (Spiess, 1991). When a population is composed of several smaller populations, the expected observation is an increase in homozygosity, a circumstance known as Wahlund's principle (Spiess, 1991). A population of schizophrenic subjects may well violate other of these assumptions also, such as random mating.

Animal studies have shown that the D3 dopamine receptor may be important for the reinforcing effects of cocaine (Caine and Koob, 1993). We investigated DRD3 alleles in a cocaine-dependent population and found that there was no association (Freimer et al., in press).

G. FUTURE DIRECTIONS

If association studies are conducted with a heightened awareness of their potential problems, results will be more reliable. The problem of unclear meaning of associations with unexpressed polymorphisms, or polymorphisms far from the gene studied, can be addressed by preferentially studying polymorphisms in or near the coding region of the gene studied. Ideally, the polymorphism would affect protein structure; an example of such a polymorphism is the VNTR located in the region of the gene corresponding to the third cytoplasmic loop of the D4 dopamine receptor (Van Tol et al., 1992). The problem of population stratification can be addressed by making allele frequency comparisons only between individuals of similar ethnicity (unless it has been demonstrated that racial or ethnic groups do not differ in allele frequency for a particular marker), and by applying the haplotype relative risk (HRR) method of Falk and Rubenstein (1987) or related methods (Schaid and Sommer, 1994) using non-transmitted parental alleles as controls. These methods may result in the discovery of more reliable and stable genetic associations between candidate genes and psychiatric illness.

ACKNOWLEDGMENTS

Susan Kruger, M.D., Ph.D. and Joseph Cubells, M.D., Ph.D. kindly provided useful comments on the manuscript. This work was supported in part by funds from the U.S. Department of Veterans Affairs (USVA National Center for Schizophrenia Research, the VA-Yale Alcoholism Research Center, and the VA Medical Research Program [Merit Review grant to J.G.]) and NIMH grant MH00931.

REFERENCES

Amadeo, S., Abbar, M., Fourcade, M.L., Waksman, G., Leroux, M.G., Madec, A., Selin, M., Champiat, J.-C., Brethome, A., Leclaire, Y., Castelnau, D., Venisse, J.-L., and Mallet, J., D2 dopamine receptor gene and alcoholism, *J. Psychiatr. Res.*, 1993; 27:173-179.

Andersson, L., Haley, C.S., Ellegren, H., Knott, S.A., Johansson, M., Andersson, K., Andersson-Eklund, L., Edfors-Lilja, I., Fredholm, M., Hansson, I., et al., Genetic mapping of growth and fatness in pigs, *Science*, 1994; 263:1771-1774.

Arinami, T., Itokawa, M., Komiyama, T., Mitsushio, H., Mori, H., Mifune, H., Hamaguchi, H., and Toru, M., Association between severity of alcoholism and the A1 allele of the dopamine D2 receptor gene *Taq I* A RFLP in Japanese, *Biol. Psychiatry*, 1993; 33:108-114.

Barr, C.L. and Kidd, K.K., Population frequencies of the A1 allele at the dopamine D2 receptor locus, *Biol. Psychiatry*, 1993; 34:204-209.

Blum, K., Noble, E.P., Sheridan, P.J., Montgomery, A., Ritchie, T., Jagadeeswaran, P., Nogami, H., Briggs, A.H., and Cohn, J.B., Allelic association of human dopamine D2 receptor gene in alcoholism. *JAMA*, 1990; 263:2055-2060.

Blum, K., Noble, E.P., Sheridan, P.J., Finely, O., Montgomery, A., Ritchie, T., Ozkaragoz, T., Fitch, R.J., Seedlike, F., Sheffield, D., Dahlmann, T., Halbardier, S., and Nogami, H., Association of the A1 allele of the D2 dopamine receptor gene with severe alcoholism, *Alcohol*, 1991; 8:409-416.

Bolos, A.M., Dean, M., Lucas-Derse, S., Ramsburg, M., Brown, G.L., and Goldman, D., Population and pedigree studies reveal a lack of association between the dopamine D2 receptor gene and alcoholism. *JAMA*, 1990; 264:3156-3160.

Bowcock, A.M., Kidd, J.R., Mountain, J.L., et al., Drift, admixture, and selection in human evolution: a study with DNA polymorphisms, *Proc. Natl. Acad. Sci. U.S.A.*, 1991; 88:839-843.

Brett, P., Robertson, M., Gurling, H., and Curtis, D., Failure to find linkage and increased homozygosity for the dopamine D3 receptor gene in Tourette syndrome, *Lancet*, 1993; 341:1225.

Brunner, H.G., Nelen, M.R., van Zandvoort, P., Abeling, N.G.G.M., van Gennip, A.H., Wolters, E.C., Kuiper, M.A., Ropers, H.H., and van Oost, B.A., X-linked borderline mental retardation with prominent behavioral: phenotype, genetic localization, and evidence for disturbed monoamine metabolism, *Am. J. Hum. Genet.*, 1993a; 52:1032-1039.

Brunner, H.G., Nelen, M., Breakefield, X.O., Ropers, H.H., and van Oost, B.A., Abnormal behavior associated with a point mutation in the structural gene for monoamine oxidase A, *Science*, 1993b; 262:578-580.

Caine, S.B. and Koob, G.F., Modulation of cocaine self administration in the rat through D3 dopamine receptors, *Science*, 1993; 260:1814-1816.

Cardon, L.R., Smith, S.D., Fulker, D.W., Kimberling, W.J., Pennington, B.F., and DeFries, J., Quantitative trait locus for reading disability on chromosome 6, *Science*, 1994; 266:276-279.

Cloninger, C.R., D2 dopamine receptor gene is associated but not linked with alcoholism (Editorial), *JAMA*, 1991; 266:1833-1834.

Comings, D.E., Muhleman, D., Dietz, G., Dino, M., Legro, R., and Gade, R., Tourette's syndrome and homozygosity for the dopamine D3 receptor gene — reply, *Lancet*, 1993; 341:1483-1484.

Comings, D.E., Muhleman, D., Dietz, G., Dino, M., Legro, R., and Gade, R., Association between Tourette's syndrome and homozygosity at the dopamine-D3 receptor gene, *Lancet*, 1993; 341:906.

Comings, D.E., Comings, B.G., Muhleman, D., Dietz, G., Shahbahrami, B., Tast, D., Knell, E., Kocsis, P., Baumgarten, R., Kovacs, B.W., Levy, D.L., Smith, M., Borison, R.L., Evans, D., Klein, D.N., MacMurray, J., Tosk, J.M., Sverd, J., Gysin, R., and Flanagan, S.D., The dopamine D2 receptor locus as a modifying gene in neuropsychiatric disorders, *JAMA*, 1991; 266:1793-1800.

Comings, D., In reply [letter], *JAMA*, 1992; 267:652.

Comings, D.E., Muhleman, D., Ahn, C., Gysin, R., Flanagan, S.D., The dopamine D2 receptor gene. A genetic risk factor in substance abuse, *Drug Alcohol Depend.*, 1994; 34:175-180.

Comings, D.E., Flanagan, S.D., Dietz, G., Muhleman, D., Knell, E., and Gysin, R., The dopamine D2 receptor (DRD2) as a major gene in obesity and height, *Biochem. Med. Metab. Biol.*, 1993; 50:176-185.

Cook, B.L., Wang, Z.W., Crowe, R.R., Hauser, R., and Freimer, M., Alcoholism and the D2 receptor gene, *Alcoholism Clin. Exp. Res.*, 1992; 4:806-809.

Corder, E.H., Saunders, A.M., Strittmatter, W.J., Schmechel, D.E., Gaskell. P.C., Small, G.W., Roses, A.D., Haines J.L., and Pericak-Vance, M.A., Gene dose of apolipoprotein E type 4 allele and the risk of Alzheimer's disease in late onset families, *Science*, 1993; 261:921-923.

Corder, E.H., Saunders, A.M., Risch, N.J., Strittmatter, W.J., Schmechel, D.E., Gaskell, P.C., JR., Rimmler, J.B., Locke, P.A., Conneally, P.M., Schmader, K.E., Small, G.W., Roses, A.D., Haines, J.L., and Pericak-Vance, M.A., Protective effect of apolipoprotein E type 2 allele for late onset Alzheimer's disease, *Nat. Genet.*, 1994; 7:180-184.

Crocq, M.-A., Mant, R., Asherson, P., Williams, J., Hode, Y., Mayerova, D., Collier, D., Lannfelt, L., Sokoloff, P., Schwartz, J.-C., Gill, M., Macher, J.-P., McGuffin, P., and Owen, M.J., Association between schizophrenia and homozygosity at the dopamine D3 receptor gene, *J. Med. Genet.*, 1992; 29:858-60.

Davies, J.L., Kawaguchi, Y., Bennett, S.T., Copeman, J.B., Cordell, H.J., Pritchard, L.E., Reed, P.W., Gough, S.C.L., Jenkins, S.C., Palmer, S.M., Balfour, K.M., Rowe, B.R., Farrall, M., Barnett, A.H., Bain, S.C., and Todd, J.A., A genome-wide search for human type 1 diabetes susceptibility genes, *Nature*, 1994; 371:130-136.

Devlin, B. and Risch, N., Ethnic differentiation at VNTR loci, with special reference to forensic applications, *Am. J. Hum. Genet.*, 1992; 51:534-548.

Devor, E.J., The D2 dopamine receptor and Tourette's syndrome [letter], *JAMA*, 1992; 267:651.

Falk, C.T. and Rubinstein, P., Haplotype relative risks: an easy reliable way to construct a proper control sample for risk calculations, *Ann. Hum. Genet.*, 1987; 51:227-233.

Freimer, M., Kranzler, H.R., Satel, S., Lacobelle, J., Skipsey, K., Charney, D.S., and Gelernter, J., No association between D3 dopamine receptor (DRD3) alleles and cocaine dependence. In press, Addiction Biology.

Gejman, P.V., Ram, A., Gelernter, J., Friedman, E., Cao, Q., Pickar, D., Blum, K., Noble, E.P., Kranzler, H., O'Malley, S., Hamer, D.H., Whitsitt, F., Rao, P., DeLisi, L.E., Virkkunen, M., Linnoila, M., Goldman, D., and Gershon, E.S., No structural mutation in the dopamine D2 receptor gene in alcoholism or schizophrenia, *JAMA*, 1993; 271:204-208.

Gejman, P.V. and Gelernter, J., Mutational analysis of candidate genes in psychiatric disorders, *Am. J. Med. Genet.*, 1993; 48:184-191.

Gelernter, J., Goldman, D., and Risch, N., Alcoholism and the D2 dopamine receptor gene — reply (letter), *JAMA*, 1993b; 270:1547-1548.

Gelernter, J., Goldman, D., and Risch, N., The A1 allele at the D2 dopamine receptor gene and alcoholism: a reappraisal, *JAMA*, 1993; 269:1673-1677.

Gelernter, J., Pakstis, A.J., Pauls, D.L., Kurlan, R., Gancher, S., Civelli, O., Grandy, D., and Kidd, K.K., Gilles de la Tourette syndrome is not linked to D2 dopamine receptor, *Arch. Gen. Psychiatry*, 1990; 47:1073-1077.

Gelernter, J., Pauls, D.L., Leckman, J., Kidd, K.K., and Kurlan, R., D2 dopamine receptor (DRD2) alleles do not influence severity of Tourette's syndrome: results from four large kindreds, *Arch. Neurol.*, 1994, 51:397-400.

Gelernter, J., O'Malley, S., Risch, N., Kranzler, H., Krystal, J., Merikangas, K., Kennedy, J., and Kidd, K.K., No association between an allele at the D2 dopamine receptor gene (DRD2) and alcoholism, *JAMA*, 1991; 266:1801-1807.

Gelernter, J., Kranzler, H., and Satel, S., No association between DRD2 alleles and cocaine abuse. College on Problems of Drug Dependence, Toronto, Canada, annual meeting, June 16, 1993c.

Goate, A., Chartier-Harlin, M.C., Mullan, M., Brown, J., Crawford, F., Fidani, L., Giuffra, L., Haynes, A., Irving, N., James, L., Mant, R., Newton, P., Rooke, K., Roques, P., Talbot, C., Pericak-Vance, M., Roses, A., Williamson, R., Rosser, M., Owen, M., and Hardy, J., Segregation of a missense mutation in the amyloid precursor protein gene with familial Alzheimer's disease, *Nature*, 1991; 349:704-706.

Goldman, D., Dean, M., Brown, G.L., Bolos, A.M., Tokola, R., Virkkunen, M., and Linnoila, M., D2 dopamine receptor genotype and cerebrospinal fluid homovanillic acid, 5-hydroxyindoleacetic acid and 3-methoxy-4-hydroxyphenylglycol in Finland and the United States, *Acta Psychiatr. Scand.*, 1992; 86:351-357.

Goldman, D., Brown, G.L., Albaugh, B., Robin, R., Goodson, S., Trunzo, M., Akhtar, L., Lucas-Derse, S., Long, J., Linnoila, M., and Dean, M., DRD2 dopamine receptor genotype, linkage disequilibrium, and alcoholism in American Indians and other populations, *Alcoholism*, 1993; 17:199-204.

Grandy, D.K., Marchionni, M.A., Makam, H., et al., Cloning of the cDNA and gene for a human D2 dopamine receptor, *Proc. Natl. Acad. Sci. U.S.A.*, 1989a; 86:9762-9766.

Grandy, D.K., Litt, M., Allen, L., Bunzow, J.R., Marchionni, M., Makam, H., Reed, L., Magenis, R.E., and Civelli, O., The human dopamine D2 receptor gene is located in chromosome 11 at q22-q23 and identifies a *Taq I* polymorphism, *Am. J. Hum. Genet.*, 1989b; 45:778-785.

Hansen, P.S., Gerdes, L.U., Klausen, I.C., Gregersen, N., and Faergeman, O., Genotyping compared with protein phenotyping of the common apolipoprotein E polymorphism, *Clin. Chim. Acta*, 1994; 224:131-137.

Hauge, X.Y., Grandy, D.K., Eubanks, J.H., Evans, G.A., Civelli, O., and Litt, M., Detection and characterization of additional DNA polymorphisms in the dopamine D2 receptor gene, *Genomics*, 1991; 10:527-530.

Hebebrand, J., Nothen, M.M., Lehmkuhl, G., Poustka, F., Schmidt, M., Propping, P., and Remschmidt, H., Tourette's syndrome and homozygosity for the dopamine D3 receptor gene, *Lancet*, 1993; 341:1483.

Hendriks, L., van Duijn, C.M., Cras, P., Cruts, M., Van Hul, W., van Harskamp, F., Warren, A., McInnis, M.G., Antonarakis, S.E., Martin, J.-J., Hofman, A., and Van Broeckhoven, C., Presenile dementia and cerebral haemorrhage linked to a mutation at codon 692 of the ß-amyloid precursor protein gene, *Nat. Genet.*, 1992; 1:218-221.

Hodge, S.E., Linkage analysis versus association analysis: distinguishing between two models that explain disease-marker associations, *Am. J. Hum. Genet.*, 1993; 53:367-384.

Hubner, N., Kreutz, R., Takahashi, S., Ganten, D., and Lindpainter, K., Unlike human hypertension, blood pressure in a hereditary hypertensive rat strain shows no linkage to angiotensinogen locus, *Hypertension*, 1994; 23:797-801.

Jönsson, E., Lannfelt, L., Sokoloff, P., Schwartz, J.-C., and Sedvall, G., Lack of association between schizophrenia and alleles in the dopamine D3 receptor gene, *Acta Psychiatr. Scand.*, 1993; 87:345-349.

Lander, E.S. and Botstein, D., Mapping Mendelian factors underlying quantitative traits using RFLP linkage maps. *Genetics*, 1989; 121:185-199.

Ma, J., Yee, A., Brewer, H.B., Jr., Das, D., and Potter, H., Amyloid-associated proteins α_1-antichymotrypsin and apolipoprotein E promote assembly of Alzheimer β-protein into filaments, *Nature*, 1994; 372:92-94.

Mant, R., Williams, J., Asherson, P., Parfitt, E., McGuffin, P., and Owen, M.J., Relationship between homozygosity at the dopamine D3 receptor and schizophrenia, *Am. J. Med. Genet.*, 1994; 54:21-26.

Marian, A.J., Yu, Q., Workman, R., Greve, G., and Roberts, R., Angiotensin-converting enzyme polymorphism in hypertrophic cardiomyopathy and sudden cardiac death, *Lancet*, 1993; 342:1085-86.

Markow, T.A. and Wandler, K., Fluctuating dermatoglyphic asymmetry and the genetics of liability to schizophrenia, *Psychiatry Research*, 1986; 19:323-328.

Moises, H.W., Gelernter, J., Giuffra, L.A., Zarcone, V., Wetterberg, L., Civelli, O., Kidd, K.K., and Cavalli-Sforza, L., No linkage between D2-dopamine receptor gene region and schizophrenia, *Arch. Gen. Psychiatry*, 1991; 48:643-647.

Nanko, S., Sasaki, T., Fukuda, R., Hattori, M., Dai, X.Y., Kazamatsuri, H., Kuwata, S., Juji, T., and Gill, M., A study of the association between schizophrenia and the dopamine D3 receptor gene, *Hum. Genet.*, 1993; 92:336-338.

Nimgaonkar, V.L., Zhang, X,R., Caldwell, J.G., Ganguli, R., and Chakravarti, A., Association study of schizophrenia with dopamine D3 receptor gene polymorphisms: probable effects of family history of schizophrenia?, *Am. J. Med. Genet.*, 1993; 48:214-217.

Noble, E.P., Blum, K., Ritchie, T., Montgomery, A., and Sheridan, P.J., Allelic association of the D2 dopamine receptor gene with receptor-binding characteristics in alcoholism, *Arch. Gen. Psychiatry*, 1991; 48:648-654.

Noble, E.P. and Blum, K., The dopamine D2 receptor gene and alcoholism (letter), *JAMA*, 1991; 268:2667.

Noble, E.P. and Blum, K., Alcoholism and the D2 dopamine receptor gene (letter), *JAMA*, 1993; 270:1547.

Noble, E.P., Blum, K., Khalsa, M.E., Ritchie, T., Montgomery, A., Wood, R.C., Fitch, R.J., Ozkaragoz, T., Sheridan, P.J., Anglin, M.D., Paredes, A., Treiman, L.J., Sparkes, R.S., Allelic association of the D2 dopamine receptor gene with cocaine dependence, *Drug Alcohol Depend.*, 1993; 33:271-285.

Nöthen, M.M., Chichon, S., Propping, P., Fimmers, R., Schwab, S.G., and Wildenauer, D.B., Excess homozygosity at the dopamine D3 receptor gene in schizophrenia not confirmed (letter), *J. Med. Genet.*, 1993; 30:708.

O'Hara, B.F., Smith, S.S., Bird, G., Persico, A.M., Suarez, B., Cutting, G.R., and Uhl, G.R., Dopamine D2 receptor RFLPs, haplotypes and their association with substance use in black and Caucasian research volunteers, *Hum. Genet.*, 1993; 43:209-218.

Parsian, A., Todd, R.D., Devor, E.J., O'Malley, K.L., Suarez, B.K., Reich, T., and Cloninger, C.R., Alcoholism and alleles of the human dopamine D2 receptor locus. Studies of association and linkage, *Arch. Gen. Psychiatry*, 1991; 48:655-663.

Pelchat, M.L. and Danowski, S., A possible genetic association between PROP-tasting and alcoholism, *Physiol. Behav.*, 1992; 51:1261-1266.

Pericak-Vance, M.A., Bebout, J.L., Gaskell, P.C., Jr., Yamaoka, L.A., Hung, W.-Y., Alberts, M.J., Walker, A.P., Bartlett, R.J., Haynes, C.A., Welsh, K.A., Earl, N.L., Heyman, A., Clark, C.M., and Roses, A.D., Linkage studies in familial Alzheimer disease: evidence for chromosome 19 linkage, *Am. J. Hum. Genet.*, 1991; 48:1034-1050.

Poirier, J., Davignon, J., Bouthillier, D., Kogan, S., Bertrand, P., and Gauthier, S., Apolipoprotein E polymorphism and Alzheimer's disease, *Lancet*, 1993; 342:697-699.

Rao, P.A., Pickar, D., Gejman, P.V., Ram, A., Gershon, E.S., and Gelernter, J., Allelic variation in the D4 dopamine receptor (DRD4) gene does not predict response to clozapine, *Arch. Gen. Psychiatry*, 1994; 51:912-917.

Risch, N., Genetic linkage and complex diseases, with special reference to psychiatric disorders, *Genet. Epidemiol.*, 1990; 7:3-16.

Sarkar, G., Kapelner, S., Grandy, D.K., Marchionni, M., Civelli, O., Sobell, J., Heston, L., and Sommer, S.S., Direct sequencing of the dopamine D2 receptor (DRD2) in schizophrenics reveals three polymorphisms but no structural change in the receptor, *Genomics*, 1991; 11:8-14.

Saunders, A.M., Strittmatter, W.J., Schmechel, D., St. George-Hyslop, P.J., Pericak-Vance, M.A., Joo, S.H., Rosi, B.L., Gusella, J.F., Crapper-MacLachlan, D.R., Alberts, M.J., Hulette, C., Crain, B., Goldgaber, D., and Roses, A.D., Association of apolipoprotein E allele E4 with late-onset familial and sporadic Alzheimer's disease, *Neurology*, 1993; 43:1467-1472.

Schaid, D.J. and Sommer, S.S., Comparison of statistics for candidate-gene association studies using cases and parents, *Am. J. Hum. Genet.*, 1994; 55:402-409.

Schellenberg, G.D., Bird, T.D., Wijsman, E.M., Orr, H.T., Anderson, L., Nemens, E., White, J.A., Bonnycastle, L., Weber, J.L., Alonso, M.E., Potter, H, Heston, L.L., and Martin, G.M., Genetic linkage evidence for a familial Alzheimer's disease locus on chromosome 14, *Science*, 1992; 258:668-671.

Schwab, S., Soyka, M., Niederecker, M., Ackenheil, M., Scherer, J., and Wildenauer, D.B., Allelic association of human D2-receptor DNA polymorphism ruled out in 45 alcoholics (A1094), *Am. J. Hum. Genet.*, 1991; 49:4(Supl.):203.

Smith, S.S., Gorelick, D.A., O'Hara, B.F., and Uhl, G.R., The dopamine D2 receptor gene and alcoholism (letter), *JAMA*, 1991; 268:2667-2668.

Smith, S.S., O'Hara, B.F., Persico, A.M., Gorelick, D.A., Newlin, D.B., Vlahov, D., Solomon, L., Pickins, R., and Uhl, G.R., Genetic vulnerability to drug abuse. The D2 dopamine receptor *Taq I* B1 restriction fragment length polymorphism appears more frequently in polysubstance abusers, *Arch. Gen. Psychiatry*, 1992; 49:723-727.

Spiess, E.B., *Genes in Populations*, 2nd ed., John Wiley & Sons, New York, 1989.

Suarez, B.K., Parsian, A., Hampe, CL., Todd, R.D., Reich, T., and Cloninger, C.R., Linkage disequilibria at the D2 dopamine receptor locus (DRD2) in alcoholics and controls, *Genomics*, 1994; 19:12-20.

Terwilliger, J.D. and Ott, J., A haplotype-based 'haplotype relative risk' approach to detecting allelic associations, *Hum. Hered.*, 1992; 42:337-346.

Thomasson, H.R., Edenberg, H.J., Crabb, D.W., Mai, X.L., Jerome, R.E., Li, T.K., Wang, S.P., Lin, Y.T., Lu, R.B., and Yin, S.J., Alcohol and aldehyde dehydrogenase genotypes and alcoholism in Chinese men, *Am. J. Hum. Genet.*, 1991; 48:677-681.

Tsuang, M.T., Faraone, S.V., and Lyons, M.J., Identification of the phenotype in psychiatric genetics, *Eur. Arch. Psychiatry Clin. Neurosci.*, 1993; 243:131-142.

Turner, E., Ewing, J., Shilling, P., Smith, T.L., Irwin, M., Schuckit, M., and Kelsoe, J.R., Lack of association between an RFLP near the D2 dopamine receptor gene and severe alcoholism, *Biol. Psychiatry*, 1992; 31:285-290.

Van Tol, H.H.M., Caren, M.W., Guan, H.-C., O'Hara, K., Bunzow, J.R., Civelli, O., Kennedy, J., Seeman, P., Niznik, H.B., and Jovanovic, V., Multiple dopamine D4 receptor variants in the human population, *Nature*, 1992; 358:149-152.

3 The Value of Narrow Psychiatric Phenotypes and "Super" Normal Controls*

Shirley Y. Hill and Katherine Neiswanger

CONTENTS

A. Introduction 37
B. In Search of Perfect Research Designs 38
    1. Sample Selection 39
        a. Affected Samples 39
        b. Control Samples 40
    2. Linkage Vs. Association 40
C. The DRD2/Alcoholism Story 41
    1. Association Studies and Prior Probability 42
    2. Association Studies and Population Stratification 42
    3. Association in the Absence of Linkage in Families 44
    4. The A1 Allele as an Example of a "Common Path" for a Behavioral Tendency 44
D. Conclusions 45
References 45

A. INTRODUCTION

With the recognition that complex phenotypes present special problems for understanding genetic mechanisms, consideration has been recently given to issues of research design. Research design alternatives include choosing the most feasible and powerful strategy to find evidence for specific genes (e.g., linkage vs. association). While this is an important issue, an equally important consideration is sample selection. The first question we raise is what constitutes an adequate sample of persons with a given psychiatric diagnosis and what should we demand of our control groups if we are interested in testing population-based associations? If our aim is to uncover significant linkages between a psychiatric disorder and a marker locus, how might we maximize our chances of finding important loci? These are important questions for which there remains considerable controversy. Among the trade-offs in linkage designs is finding a phenotype that is extreme enough to maximize finding significant loci while at the same time preserving what our epidemiological colleagues term "representativeness". For population-based studies in which the investigator requires a representative control

---

* Portions of this chapter are taken from *What Can the DRD2/Alcoholism Study Teach Us About Association Studies in Psychiatric Genetics?*, an editorial published by John Wiley & Sons. Copyright 1995.

sample, the obvious limitation of this strategy is that this sample may contain individuals with the same disease as that being studied in the experimental group, potentially reducing the power to detect differences between the groups. Thus, removing these confounding cases from the control group may improve chances of finding significant differences between experimental and control groups but risks lack of representativeness in the control group. How do we resolve these problems? This chapter will address these issues and offer possible solutions.

B. IN SEARCH OF PERFECT RESEARCH DESIGNS

A review of the fundamental assumptions of research design across disparate disciplines may reveal why our notions of what is "good design" vary so much. The experimental psychologist who utilizes animal strains to study brain/behavior relationships first learns in graduate school that not only must one do an experiment with the same strain of rats, but they must be from the same supplier, of the same sex and age, and matched as nearly as possible in every dimension. The assumption is that if there is an important contributing variable (brain factor X) which predicts behavior X, then we must search for this relationship in a homogeneous group (Strain X). Once this relationship is adequately replicated by other laboratories, an attempt is made to extend it to other strains such that brain factor X predicts behavior X in Strain Y of rats, and so on, perhaps extending to other mammals. The goal, initially, is not to be "representative" for to do so would mean conducting the experiment with a few Sprague-Dawley rats, a few Long-Evans rats, a few Wistar rats, and so on, but rather to increase the "signal to noise ratio" by making the groups as homogeneous as possible. Extending this analogy to psychiatric samples makes it quite obvious that removal of comorbidity in the initial stages of a search for a gene linked to a given psychiatric disorder is essential.

Heterogeneity creates innumerable problems for genetic studies. While methods have long been developed for handling heterogeneity in the analysis of Mendelian traits (see, for example, Reference 1) the situation is far more difficult for common conditions that do not follow Mendelian inheritance. Given these considerations, would we not want to reduce heterogeneity to a minimum by studying disorder X persons with disorders Y and Z removed? As logical as this seems from the experimental laboratory perspective, psychiatric geneticists, trained usually in human genetics, psychiatry, and epidemiology, are quick to ask, "But is the sample representative?"

The tradition of the epidemiologist is a perspective that is often more familiar to the population geneticist. With parametric linkage analyses and segregation analyses so highly dependent on estimating both the "correct" population prevalence of the disorder in question and its gene frequency in the population, it is no accident that epidemiology and human genetics are "kindred spirits." From an epidemiological perspective, the sample should ideally be the entire population (if it were practical). Without that possibility, epidemiologists have come up with rather clever means for weighting samples (e.g., probability samples, stratified samples) all of which are designed to accurately reflect characteristics of the population. These are clearly useful methods for estimating prevalence within populations.

However, it should be remembered that not all stratified samples are stratified for the purpose of making the sample "representative." Stratification is also important for studying the association of two variables in a population that could not otherwise be done without stratification. If we wish to study the relationship between some variable X and some psychiatric outcome, but that variable is frequently "tied" within the population to another variable, Y, we cannot attain sufficient power to test our variable X / outcome hypothesis without some degree of stratification to remove Y. For example, should we wish to test the notion that single-parent homes are associated with juvenile delinquency, we may have to deal with important variables such as poverty that are highly associated with single-parent status. With a stratified sample, which is deliberately designed to be nonrepresentative of the population, one can pull a quota of single parents with and without poverty status, providing sufficient power to test the original hypothesis.

In the same way that epidemiological methods can help us to "untie" naturally co-occurring variables, it can also help in linking variables that may or may not be tied in the population. For example, if a particular chemical has low carcinogenic potential in the general population but has an especially robust effect in persons with a particular familial form of cancer, then we may miss that relationship altogether without sample stratification. Because cancer-prone families are relatively rare, adequately testing the carcinogenicity of a particular agent cannot be done without stratifying the sample to include a large pool of individuals who are cancer prone by family history who have or have not been exposed to that agent. Again, the cancer-prone families have been chosen intentionally to not be representative of the entire population.

In short, all of these methods are useful depending on the question one is asking. For particular research questions (e.g., to detect a gene), it may be a mistake to demand that our psychiatric phenotype be so broad that we miss the defining relationship between the behavior and locus in the name of "representativeness." In the case of psychiatric disorders, we really do not know if God carved out the psychiatric disorders in the same fashion as is seen in DSM-IV. Genes for behavioral tendencies (anxiety, impulsivity, compulsivity, harm avoidance) may be the ultimate source for our understanding of the psychiatric disorders. Each disorder may be represented by some combination of these behavioral tendencies. The first step would appear to be to define a relatively homogeneous psychiatric entity so that the critical behavioral tendencies defining the disorder along with their genetic underpinnings can be identified.

1. Sample Selection

a. Affected Samples

The need for homogeneity in the affected phenotype is important not only for population-based association studies but also for linkage analysis. In order to achieve homogeneity in a linkage analysis, it is necessary to screen both probands and family members for comorbid conditions. This strategy suggested here for linkage analyses is far more controversial than the relatively more recognized need to screen affected samples for comorbidity when performing population-based association analyses. Traditionally, linkage analyses have been based on a random selection of families from the "universe" of families with affected probands. We have challenged this tradition by choosing a nonrandom set of families based on both the proband's disorder as well as disorders occurring in family members. For example, an alcoholic proband must be free of recurrent depression, schizophrenia, and drug dependence while all efforts are made to keep these conditions absent from the first-degree relatives as well. Thus, this strategy will result in a nonrandom set of families from the universe of alcoholism families.

One can object that this strategy tips the balance of disorders within families so that the co-aggregation of psychiatric illnesses within the family, studied collectively, is not representative of the co-aggregation of disorders in affected families — the epidemiologists' nightmare! However, if one's purpose is not to quantify the co-aggregation of disorders in families but rather to find a gene, then this linkage design is ideal and may be necessary to detect a specific gene possibly unique to the disorder under investigation.

b. Control Samples

Similarly, a screened control group is essential for uncovering population-based associations where the disease in question may be very common. Why is this so? We know that approximately one-third of the population meet lifetime criteria for common psychiatric disorders according to the results of the Epidemiological Catchment Area (ECA) survey.[2] Because persons without any diagnosis can nevertheless carry a genotype for psychiatric illness (presumably, in part because a first-degree relative has a disorder), we see that the likelihood of either having a disorder or carrying a vulnerability gene for any psychiatric disorder is quite common. Therefore, if we hypothesize that a particular gene is present that is a necessary condition for developing "any" psychiatric disorder, which we know is extremely common, it is absolutely necessary that our control groups be "clean." Otherwise, an important "signal" in the experimental group will be washed out in comparison to the unscreened, albeit representative, control group.

The use of a control group that is "squeaky clean" has been criticized by some[3] on the grounds that "supernormal controls" and their relatives will have rates of comorbid disorders lower than that in the general population. As a result, Kendler argues that in comparison with affected samples, these supernormal controls may produce spurious co-aggregation of disorders within families. Note this argument is valid only if the same psychopathology which is removed from the control group is *not* excluded from among the probands and their relatives. If a narrow psychiatric phenotype is chosen to test population-based associations, the resulting removal of comorbid conditions to a similar extent in both disease and control groups will raise the study's potential to detect a genetic signal.

2. LINKAGE VS. ASSOCIATION

The availability of molecular genetic techniques, which currently include CA-repeat genetic markers, makes it possible to perform genome scans with markers spaced at 10-cM intervals. With appropriate family collections of DNA, it is possible to look for linkages to any DNA marker chosen among a given commercially available set. With the discovery of a suggestive linkage, the region can be saturated and linkages calculated. This makes it possible to do positional cloning of the gene whose function may or may not already be known.

With the availability of saturated linkage maps, the need for continuing to search for candidate genes utilizing population-based associations may seem unnecessary. We would argue, however, that there is a place for both linkage and association strategies. First, extended families of psychiatric patients are not readily available for linkage analysis. Nonparametric methods often require 100 to 200 affected sib pairs. Thus, population-based associations may offer an economical alternative as a first stage, pointing the way to future work. This is particularly true where the investigator may have a group of patients who are phenotypically quite homogeneous and who exhibit a particularly severe form of the disorder. An example of the utility of this strategy is the now well-known search for diabetes genes. A combination of genotyping and clinical data utilizing association techniques first suggested that juvenile-onset diabetes and adult-onset diabetes are separate entities.[4]

The second important reason to not abandon the association strategy altogether is the opportunity it presents for finding modifier genes.[5] Linkage techniques can detect genes of sufficient influence that having the gene qualifies as a necessary condition for developing a particular disorder. However, if a given gene modifies the probability of disease but is not a necessary condition, then association studies may be the method of choice. Recent work has shown that families simulated with an association between a marker and a disease may not always reveal close linkage between that marker and disease. This would occur if the marker is neither necessary nor sufficient for the expression of the disease, but merely increases its risk.[5] Thus some form of association study is likely to be a crucial adjunct to linkage analysis in finding the genes involved in the psychiatric illnesses, since it is very possible that modifier genes play an important role in many of these disorders. As noted by Hodge,[6] both association and linkage strategies will be needed to understand the genetics of a complex phenotype. Practically speaking, population-based association tests are an inexpensive and sensitive way to find correlations between biologic/genetic traits and diseases of unknown etiology. Once the samples are collected, the genotyping of a new marker is a trivial endeavor, and it becomes relatively cost effective to replicate an initial finding.

C. THE DRD2/ALCOHOLISM STORY

As noted in Section I, the choice of an ideal sample of persons with a given psychiatric disorder (with or without comorbidity) and the most useful type of control group (screened vs. unscreened) remains controversial in the field. However, the importance of appropriately screening both the control group and the psychiatrically affected group for other psychopathology is well illustrated by the search for the DRD2 population-based association with alcoholism. We would argue that the DRD2 story confirms our belief that a sound first step in any population-based association or proposed linkage analysis where complex phenotypes are concerned should start with affected samples without comorbidity and "screened" controls. Once an adequate signal has been detected, the phenotype can be broadened to include other psychopathology, enabling us to test the generality of the finding.

Disagreement over the validity of reported associations between alcoholism and the A1 allele of the *Taq I* A RFLP from the dopamine D2 receptor gene appears to have cast doubt on population-based association methodologies in psychiatric genetics. Some have suggested that any population-based association is likely to be an artifact, either of low prior probability or due to population stratification. Almost any study could be accused of having ethnic variation in allelic frequencies, restricting investigators from testing potential markers for associations. Thus, this argument seems tantamount to "throwing the baby out with the bath water". A review of the controversy surrounding the DRD2 association makes it evident that ruling out alternative explanations to having found a valid association is a necessary first step in their acceptance. However, once we have ruled out the major sources of contamination (low prior probability and population stratification), acceptance of an association would appear to be reasonable. Moreover, depending on the way in which samples were selected (e.g., those with more severe form of the disorder) the findings may point to important leads for follow-up.

By way of review, the A1 allele has been mapped to a region 12 kb downstream from the DRD2 gene on chromosome 11. The A1 allele is relatively infrequent in the general U.S. population, with approximately 20% carrying A1, and less than 1% being homozygous for A1. Based partly upon the DRD2 experience (for review and meta-analyses, see References 7-9) many psychiatric geneticists are suggesting caution in interpreting population-based association studies.

As discussed by Neiswanger and colleagues,[10] the original positive association of the *Taq I* A1 allele to alcoholism[11] has been replicated several times (see References 9 and 12–15), but not by all investigators in the field (see References 16–22). In two of the three studies employing alcoholism families, the association was present in population-based tests, but not confirmed by either family-based association or linkage analyses,[9,13] while the third family study was completely negative.[16] Other markers in the DRD2 gene region have not shown within-family association with alcoholism,[22] nor have molecular studies of the DRD2 gene detected mutations in the coding regions which could alter DRD2 function in Caucasian alcoholics.[23] However, it has been suggested that the *Taq I* A alleles are in linkage disequilibrium with a functional allelic variant in the promoter or other regulator regions of the DRD2 that affects receptor expression.[14] Also, associations have been seen in phenotypes other than alcoholism — most notably in substance abuse (for the *Taq I* B RFLP[24]). Recently, an association between the A1 allele and cocaine abuse has been reported.[25]

1. ASSOCIATION STUDIES AND PRIOR PROBABILITY

The question then arises: Is the population-based association for DRD2 and alcoholism real? Given the statistical nature of association studies, many published associations will be observed by chance alone. This has become especially apparent in light of recent work showing that a low prior probability makes it highly unlikely that a positive report of an association between a candidate gene and a psychiatric illness is true.[26-28] In particular, Crowe[27] shows that for $\alpha = 0.05$, less than 1% of positive findings will be true if the prior probability is 0.00025. Carey[28] further demonstrates that for the situation where the prior probability is 0.001, $\alpha = 0.00001$, $N = 300$, and the prevalence of the disorder = 0.01, the vast majority of positive findings will be false for a wide range of candidate genotype prevalences and relative risks.[10]

Based on the prior probability argument, the likelihood of DRD2 being involved with alcoholism is low, probably on the order of 0.01 to 0.001 or even lower. Thus, the arguments of Carey[28] are important in assessing the possibility that the *Taq I* A association has been observed by chance alone. Clearly, the original observation is much more likely to be a false positive than a real finding. The important point to keep in mind, however, is that the prior probability changes with each subsequent replication (see Figure 3-1).

According to Figure 3-1, the probability that a positive report is a false positive drops dramatically after two or three replications of the initial finding for a range of standard significance and power levels. However, it is important to note that the number of negative reports needed to generate six positive reports is not considered in the calculations for Figure 3-1, which is based on Carey's original formulation. Clearly, 6 positive reports accompanied by 6 negative reports has a different interpretation than 6 positive reports would have in the presence of 600 negative reports. Practically speaking, the replications needed to verify the initial positive finding would not likely be performed, if the ratio of **all** positive reports to **all** negative reports were as low as 1:100. But if negative reports are not routinely published, then positive reports may appear to inflate the likelihood that an association is real. Hence, it is very important that all positive and negative associations be published, even if only in a very brief format. Since this will never be the case, we are left with an imperfect estimate of the true prior probability associated with a given finding. However, when multiple studies have been completed and numerous studies are significant, the prior probability argument no longer holds as an explanation of a given positive association.

Note that in the case of DRD2 and alcoholism, in addition to several negative reports, at least four independent positive replications have been published. Moreover, Pato and colleagues[8] have shown that the association remains significant when both positive and negative reports are considered and is not likely to be a statistical artifact. Therefore, the *Taq I*/alcoholism association cannot be dismissed as an example of an error based on prior probability.

2. ASSOCIATION STUDIES AND POPULATION STRATIFICATION

Since the DRD2/alcoholism association may well not be due to chance, investigators have attempted to account for conflicting results by noting such methodological problems as race/ethnicity, differential inclusion and exclusion criteria for both control and alcoholic

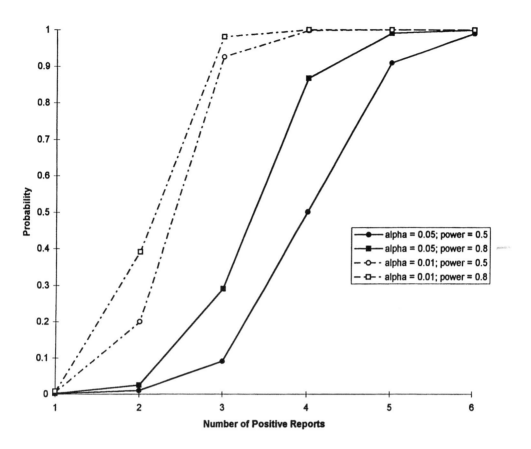

FIGURE 3-1 Probability that a replication is a true positive, given an increasing number of prior positive reports. Calculations were based on the equation given by Carey:[28] $\alpha^* = \alpha(1 - \omega) / [\alpha(1 - \omega) + (1 - \beta)\omega]$, where α^* is the proportion of false positives among all positive reports, α is the significance level, $(1 - \beta)$ the power, and ω the prior probability of the positive report. The initial prior probability (ω_o) was set at 0.0001. Each subsequent prior probability (ω_{n+1}) was set at $(1 - \alpha_n^*)$. Calculations were performed assuming four different values of α and β, which were held constant as the number of prior positive reports was increased from $n = 0$ to 5. Note that this calculation assumes no methodological differences among positive reports.

samples, and different definitions of alcoholism and its severity. Among these methodological differences, population stratification has been singled out as the likely explanation for the positive reports, allowing the association to be dismissed as an artifact of inadequate study design. Population stratification will occur if (1) there are different allelic frequencies in different ethnic groups, and (2) patient and control samples from a given study contain different percentages of those ethnic groups. It will result in an association which has nothing to do with the trait itself, but instead derives from the population structure of the samples being compared. The existence of extensive ethnic variation in A1 allele frequencies has been amply demonstrated.[29,30] The second condition has only been inferred to be present in the studies reporting a positive association, and is based on the assumption that the "U.S. Caucasian" samples in these studies have not been ethnically matched in enough detail.

However, it is extremely unlikely that population stratification between alcoholism and control samples has created the numerous positive *Taq I* A1/alcoholism associations. The important point to note here is that it is the frequency of A1 in control groups which is unexpectedly low in positive reports, ***not*** an elevation in the frequency of A1 in alcoholics (see Figure 3-2). The A1 frequency in screened controls is lower than all A1 frequencies reported by either Goldman and colleagues[30] or by Barr and Kidd,[29] except for the Asian

Yemenite Jews and Druze (frequency of A1 = 0.09 and 0.11, respectively).[29] Moreover, this exceptionally low rate for the A1 allele in "screened" controls has been consistently replicated in three sites — San Antonio, Los Angeles, and Pittsburgh.[9,12,14] It would appear highly unlikely that this low frequency is due to ethnic variation between alcoholics and controls at any particular site. This suggests that the DRD2 associations reported to date are not random, nor likely due to population stratification resulting from the inadvertent selection of persons with different ethnic background for control and alcoholic groups. (See Neiswanger[10] for further discussion of population stratification.)

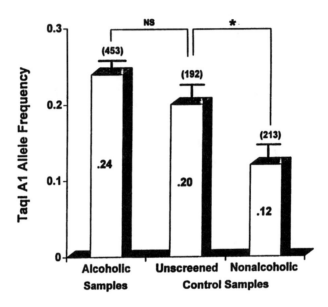

FIGURE 3-2 Frequency of the A1 allele in pooled samples of U.S. Caucasians. All samples were derived from nonoverlapping subsets of the reported data. The alcoholic sample was obtained from References 9, 11, 12, 14, 16, 17, 19, and 22. Unscreened controls came from References 14, 16, 17, and 34. Nonalcoholic controls were obtained from References 9, 11, 12, 14, 19, and 22. The difference between the unscreened and the nonalcoholic controls was significant at $p = .002$. See Reference 9 for details.

3. ASSOCIATION IN THE ABSENCE OF LINKAGE IN FAMILIES

If the association between the *Taq I* A RFLP is due uniquely to alcoholism, one would expect that within-family association or linkage studies in families of alcoholics would be positive. Published studies to date have not found positive linkages. In this regard, it is particularly noteworthy that tests for both within-family association and linkage have been negative.[9,13,16] This leaves open the prospect that *Taq I* A1 is not associated with alcoholism per se, but with a related phenotype that has not yet been accurately defined.

4. THE A1 ALLELE AS AN EXAMPLE OF A "COMMON PATH" FOR A BEHAVIORAL TENDENCY

The idea that A1 may be associated with some trait, common to many psychiatric problems in which a prominent feature is a defect in the "cut-off" mechanism of ongoing behavior (e.g., reward deficiency), has recently been discussed by Blum and colleagues.[31] This formulation is in keeping with our interpretation of the results of our study of alcoholic men and women who differed significantly from controls screened for alcoholism and other common psychiatric illnesses.[9] This contrasts sharply with those studies that failed to find significant population-based associations using unscreened control groups. We have argued[9] that variations in DRD2 alleles represent variations in a very common latent trait associated with dopaminergic function, of which alcoholism is but a single manifestation. By necessity, this trait would be underrepresented in controls screened to exclude alcoholism. Hence, failure to find linkage or within-family association could be due to incomplete understanding of the appropriate phenotype for analysis. Additionally, the presence of assortative mating for alcoholism, which has been commonly observed, could well diminish the power of the within-family association tests. Moreover, the high frequency of the trait in the general population might obscure detecting a population-based association between alcoholics and the A1 allele utilizing an unscreened control sample.

D. CONCLUSIONS

In undertaking psychiatric genetic studies, we have argued that it is critically important to narrow the phenotype for study. For linkage studies this means removing comorbid conditions in both probands and their family members. For association studies both affected and control groups should have comorbid conditions removed, and to equal degree in both groups. The use of a "super-normal" control group can lead to overestimation of comorbid associations in the affected group, if the traits screened from the controls are not screened from the affecteds.[3] However, excluding "other psychopathology" from both controls and affecteds as has been done successfully in one large multiplex alcoholism family study[32] circumvents this problem. Thus, this strategy has been accepted in the field as one way to improve the chance of finding differences between groups without risking distortion of results.[33] We realize that these suggestions may be controversial because the resulting samples are not representative of the general population. However, if the goal of these studies is to detect a genetic signal for a common psychiatric illness, then this strategy appears to offer the most power to successfully achieve that goal.

REFERENCES

1. Elston, R. C. and Stewart, J., A general model for the genetic analyses of pedigree data, *Hum. Hered.*, 21, 523, 1971.
2. Helzer, J. E., Burnam, A., and McEvoy, L. T., Alcohol abuse and dependence: the diagnosis of alcoholism, in *Psychiatric Disorders in America: The Epidemiologic Catchment Area Study*, Robins, N. and Regier, D.A., Eds., Free Press, New York, 1991, pp 81-115.
3. Kendler, K. S., The super-normal control group in psychiatric genetics. Possible artifactual evidence for coaggregation, *Psychiatr. Genet.*, 1, 43, 1990.
4. Nerup, J., Platz, P., Andersen, O. O., Christy, M., Lyngsoe, J., Poulsen, J. E., Ryder, L. P., Nielsen, L. S., Thomsen, M., and Svejggard, A., HLA antigens and diabetes mellitus, *Lancet*, 2, 864, 1974.
5. Greenberg, D. A., Linkage analysis of "necessary" disease loci versus "susceptibility" loci, *Am. J. Hum. Genet.*, 52, 135, 1993.
6. Hodge, S. E., What association analysis can and cannot tell us about the genetics of complex disease, *Am. J. Med. Genet. (Neuropsychiatr. Genet.)*, 54, 318, 1994.
7. Gelernter, J., Goldman, D., and Risch, N., The A1 allele at the D2 dopamine receptor gene and alcoholism: a reappraisal, *JAMA*, 269, 1673, 1993.
8. Pato, C. N., Macciardi, F., Pato, M. T., Verga, M., and Kennedy, J. L., Review of the putative association of dopamine D2 receptor and alcoholism: a meta-analysis, *Am. J. Med. Genet. (Neuropsychiatr. Genet.)*, 48, 78, 1993.
9. Neiswanger, K., Hill, S. Y., and Kaplan, B. B., Association and linkage studies of the *Taq I* A1 allele at the dopamine D2 receptor gene in samples of female and male alcoholics, *Am. J. Med. Genet. (Neuropsychiatr. Genet.)*, 60, 267, 1995.
10. Neiswanger, K., Kaplan, B. B., and Hill, S. Y., What can the DRD2/alcoholism story teach us about association studies in Psychiatric Genetics?, *Am. J. Med. Genet. (Neuropsychiatr. Genet.)*, 60, 272, 1995

11. Blum, K., Noble, E. P., Sheridan, P. J., Montgomery, A., Ritchie, T., Jagadeeswaran, P., Nogami, H., Briggs, A. H., and Cohn, J. B., Allelic association of human dopamine D2 receptor gene in alcoholism, *JAMA*, 263, 2055, 1990.
12. Blum, K., Noble, E. P., Sheridan, P. J., Finley, O., Montgomery, A., Ritchie, T., Ozkaragoz, T., Fitch, R. J., Sadlack, F., Sheffield, D., Dahlmann, T., Halbardier, S., and Nogami, H., Association of the A1 allele of the D2 dopamine receptor gene with severe alcoholism, *Alcohol,* 8, 409, 1991.
13. Parsian, A., Todd, R. D., Devor, E. J., O'Malley, K. L., Suarez, B. K., Reich, T., and Cloninger, C. R., Alcoholism and alleles of the human D2 dopamine receptor locus. Studies of association and linkage, *Arch. Gen. Psychiatry*, 48, 655, 1991.
14. Comings, D. E., Comings, B. G., Muhleman, D., Dietz, G., Shahbahrami, B., Tast, D., Knell, E., Kocsis, P., Baumgarten, R., Kovacs, B. W., Levy, D. L., Smith, M., Borison, R. L., Evans, D. D., Klein, D. N., MacMurray, J., Tosk, J. M., Sverd, J., Gysin, R., and Flanagan, S. D., The dopamine D2 receptor locus as a modifying gene in neuropsychiatric disorders, *JAMA*, 266, 1793, 1991.
15. Amadeo, S., Fourcade, M. L., Abbar, M., Leroux, M. G., Castelnau, D., Venisse, J. L., and Mallet, J., Association between D2 receptor gene polymorphism and alcoholism, *Psychiatr. Genet.*, 3, 130, 1993.
16. Bolos, A. M., Dean, M., Lucas-Derse, S., Ramsburg, M., Brown, G. L., and Goldman, D., Population and pedigree studies reveal a lack of association between the dopamine D2 receptor gene and alcoholism, *JAMA*, 264, 3156, 1990.
17. Gelernter, J., O'Malley, S., Risch, N., Kranzler, H. R., Krystal, J., Merikangas, K., Kennedy, J. L., and Kidd, K. K., No association between an allele at the D2 dopamine receptor gene (DRD2) and alcoholism, *JAMA*, 266, 1801, 1991.
18. Goldman, D., Dean, M., Brown, G. L., Bolos, A.M., Tokola, R., Virkkunen, M., and Linnoila, M., D2 dopamine receptor genotype and cerebrospinal fluid homovanillic acid, 5-hydroxyindoleacetic acid and 3-methoxy-4-hydroxyphenylglycol in alcoholics in Finland and the United States, *Acta Psychiatr. Scand.*, 86, 351, 1992.
19. Cook, B. L., Wang, Z. W., Crowe, R. R., Hauser, R., and Freimer, M., Alcoholism and the D2 receptor gene, *Alcohol Clin. Exp. Res.*, 16, 806, 1992.
20. Turner, E., Ewing, J., Shilling, P., Smith, T. L., Irwin, M., Schuckit, M., and Kelsoe, J. R., Lack of association between an RFLP near the D2 dopamine receptor gene and severe alcoholism, *Biol. Psychiatry*, 31, 285, 1992.
21. Arinami, T., Itokawa, M., Komiyama, T., Mitsushio, H., Mori, H., Mifune, H., Hamaguchi, H., and Toru, M., Association between severity of alcoholism and the A1 allele of the dopamine D2 receptor gene *Taq I* A RFLP in Japanese, *Biol. Psychiatry*, 33, 108, 1993.
22. Suarez, B. K., Parsian, A., Hampe, C. L., Todd, R. D., Reich, T., and Cloninger, C. R., Linkage disequilibria at the D2 dopamine receptor locus (DRD2) in alcoholics and controls, *Genomics*, 19, 12, 1994.
23. Gejman, P. V., Ram, A., Gelernter, J., Friedman, E., Cao, Q., Pickar, D., Blum, K., Noble, E. P., Kranzler, H. R., O'Malley, S., Hamer, D. H., Whitsitt, F., Rao, P., DeLisi, L. E., Virkkunen, M., Linnoila, M., Goldman, D., and Gershon, E. S., No structural mutation in the dopamine D2 receptor gene in alcoholism or schizophrenia, *JAMA*, 271, 204, 1994.
24. Smith, S. S., O'Hara, B. F., Persico, A. M., Gorelick, D. A., Newlin, D. B., Vlahov, D., Solomon, L., Pickens, R., and Uhl, G. R., Genetic vulnerability to drug abuse. The D2 dopamine receptor *Taq I* B1 restriction fragment length polymorphism appears more frequently in polysubstance abusers, *Arch. Gen. Psychiatry*, 49, 723, 1992.
25. Compton, P. A., Anglin, M. D., Khalsa-Denison, E., and Paredes, A., The D2 dopamine receptor gene, addiction and personality: clinical correlates in cocaine abusers, *Biol. Psychiatry*, 39, 302, 1996.
26. Kidd, K. K., Associations of disease with genetic markers. Deja vu all over again, *Am. J. Med. Genet. (Neuropsychiatr. Genet.)*, 48, 71, 1993.
27. Crowe, R. R., Candidate genes in psychiatry: an epidemiological perspective, *Am. J. Med. Genet. (Neuropsychiatr. Genet.)*, 48, 74, 1993.
28. Carey, G., Genetic association study in psychiatry: Analytical evaluation and a recommendation, *Am. J. Med. Genet. (Neuropsychiatr. Genet.)*, 54, 311, 1994.
29. Barr, C. L. and Kidd, K. K., Population frequencies of the A1 allele at the dopamine D2 receptor locus, *Biol Psychiatry*, 34, 204, 1993.
30. Goldman, D., Brown, G. L., Albaugh, B., Robin, R., Goodson, S., Trunzo, M., Akhtar, L., Lucas-Derse, S., Long, J., Linnoila, M., and Dean, M., DRD2 dopamine receptor genotype, linkage disequilibrium, and alcoholism in American Indians and other populations, *Alcohol Clin. Exp. Res.*, 17, 199, 1993.
31. Blum, K., Cull, J. G., Braverman, E. R., and Comings, D.E., Reward deficiency syndrome, *Am. Sci.*, 84, 132, 1996.
32. Hill, S. Y., Etiology, in *Annual Review of Addictions Research and Treatment*, Langenbucher, J., McCrady, B., Frankenstein, W., and Nathan, P., Eds., Elsevier Science, Tarrytown, NY, 1994, pp 127-148.
33. Sayette, M. A., Commentary, *Annu. Rev. Addict Res. Treat.*, 3, 189, 1994.
34. Grandy, D. K., Litt, M., Allen, L., Bunzow, J. R., Marachionni, M., Makam, H., Reed, L., Magenis, R. E., and Civelli, O., The human D2 receptor is located on chromosome 11 at q22-q23 and identifies a *Taq I* RFLP, *Am. J. Hum. Genet.*, 45, 778, 1989.

Section II
DNA Analysis

4 Isolation of Coding Sequences from the Human Genome: Implications for Neurological Functions and Diseases

Pudur Jagadeeswaran, Michael W. Odom, and Yuan C. Liu

CONTENTS

A. Introduction 49
B. Traditional Methods 50
C. Expressed Sequence Tags From Brain cDNA Libraries 51
D. Direct Selection of Coding Sequences by Hybridization 53
    1. Matrix-Based Methods 53
    2. Solution-Based Methods 55
E. Exon Trapping and Exon Amplification 57
F. Conclusion 59
References 59

A. INTRODUCTION

In order to understand basic molecular mechanisms of neurological function and disease, genes which control these processes must be identified and characterized. The human genome is remarkably complex, consisting of large molecules of DNA organized into 23 chromosomal pairs that contain approximately 3×10^9 bp of sequence data. This genetic information is varied and includes an estimated 100,000 genes which code for all the proteins of an organism, repetitive DNA, and arbitrary sequences. Proteins either constitute, or via their enzymatic functions, are involved in the production of all of the molecules of the cell, and thus ultimately determine all cellular properties and functions. However, even though coding sequences are believed to represent the vast majority of the information within the genome, they make up only about 3% of the genomic sequence.

Certain aspects of the human genome and its transcription add further complexity to the identification of coding sequences. First, although coding sequences are organized into genes, even within specific genes DNA sequences are discontinuous, containing exons interrupted by noncoding introns. Second, a significant amount of noncoding sequence, including repetitive sequences and untranslated regions, are present within sequences that are transcribed into mRNA. Moreover, although all cells of the body contain the same genetic information, the expression of genes varies from tissue to tissue and from cell to cell. Some genes code for proteins which are crucial to the integrity of most any cell or tissue and are widely expressed, whereas others encode highly specialized proteins which are only expressed by a single tissue or cell type. Not surprisingly given its high degree of complexity and diversity of functions, a large percentage of genes are expressed by the brain. Of the 100,000 genes present within the human genome, an estimated 30,000, or nearly one third, are expressed by the brain, and as many as two thirds of these genes may encode low-abundance transcripts expressed only by the brain.[1] In addition to tissue-specific differences in gene expression, some genes are only expressed at specific stages of development. Thus, the identification of all the coding sequences from even a single tissue such as the brain is a complex and difficult task.

There are two basic strategies which have been used for the identification of coding sequences. Traditionally, if some information is available about the gene of interest such as a partial protein sequence or an antibody against the protein of interest, a directed search may be undertaken by conventional methods of library screening. Linkage analysis and techniques of positional cloning may also be applied in this directed effort to identify a gene locus. An alternative strategy is an unbiased search for coding sequences whose only defining characteristic is simply that they are transcribed into mRNA. Some of these identified sequences will encode portions of known human proteins, or they will be homologous to known proteins of other species. However, many of these transcribed sequences will encode portions of previously unknown proteins. This second approach will be crucial for a complete understanding of the human genome and the proteins it encodes. Moreover, it will almost certainly provide new insights into basic cellular functions and human development.

Two approaches have recently been applied to this unbiased search for coding sequences. The first is to systematically determine the sequence of clones from a variety of cDNA libraries.[2,3] This straightforward strategy provides ready access to the transcribed sequences from a given tissue or cell type at a specific developmental stage. However, because of the marked diversity of human tissues and cell types, as well as substantial differences in levels of gene expression both within different cell types and at different stages of development, this approach cannot reliably lead to the identification of every mRNA. The second approach begins with a source of genomic DNA and, through one of a variety of methods, transcribed sequences are identified and sequenced. These newer methods may also be combined with the more established methods of positional cloning. This chapter will briefly review traditional methods of gene isolation and then will focus on newer approaches to the isolation of coding sequences, particularly those of relevance to the nervous system.

B. TRADITIONAL METHODS

Traditional strategies to identify coding sequences within genomic DNA have successfully been used to discover genes responsible for several hereditary diseases; however, these methods, in general, are indirect and cumbersome. If some information is available about the gene of interest, a variety of probes may be used in conventional methods to screen a given genomic or cDNA library. Another commonly used starting point for such investigations is linkage analysis of polymorphic DNA markers within affected families. Linkage analysis is based on the principle that markers in close proximity to a disease locus segregate together during meiosis. Restriction fragment length polymorphisms of DNA markers have commonly been used in this type of analysis. These studies can narrow the search for a disease locus to a particular subchromosomal region; however, these areas frequently still encompass several hundred thousand to several million base pairs of DNA. Because important DNA sequences, such as sequences which encode physiologically important proteins, are usually conserved across diverse species, subclones from the region identified by linkage analysis can be hybridized to genomic DNA from a variety of different species to identify candidate gene fragments. Evolutionarily conserved subclones can then serve as probes to screen appropriate human cDNA libraries to identify candidate cDNAs. Monaco et al.[4] used this type of strategy to identify candidate cDNAs for the Duchenne muscular dystrophy gene. Others have used cross-species hybridization to identify the sex-determining region of the human Y chromosome[5] and the gene responsible for Wilms' tumor.[6] Another established method to identify transcribed elements is analysis of genomic sequence data to identify open reading frames[7,8] or consensus genetic control elements.[9,10] Hypomethylated CpG-rich islands frequently mark the 5' ends of genes[11] and can thus indicate the start of a transcribed element. Bione et al.[12] used identification of hypomethylated CpG islands to produce a transcriptional map of Xq28. After linkage analysis revealed that the cystic fibrosis gene was within 7q31, Rommens et al.[13] used cross-species hybridization along with the identification of a hypomethylated CpG island and an open reading frame within cloned DNA fragments to isolate this important gene. A final method to isolate transcribed sequences relies on carefully selected primers to synthesize cDNAs from the heterogeneous nuclear RNA of somatic hybrid cell lines. Primers that have been used for this purpose have included hexamers complementary to 5' consensus splice sites[14] and human *Alu* repetitive sequences.[15]

Recently, these more traditional approaches have been expanded and refined but deficiencies remain. In addition to the identification of open reading frames within short genomic sequences, investigators have developed more extensive algorithms to identify genes and predict their structure. Guigo et al.[16] have reported a hierarchical, rule-based system known as **GeneId** which searches for genetic elements such as promoters, translation initiation and stop codons, splice sites, and poly(A) signals in order to identify probable protein encoding sequences and to predict gene structure. Potential elements are assigned a relative rank and threshold cutoffs and filters are applied. GeneModeler[17] and GRAIL[18] are two other computer-based neural networks designed to identify coding elements within stretches of genomic sequence. Snyder et al.[8] have recently applied the concept of dynamic programming to develop a neural network known as **GeneParser** which can be trained to identify internal exons and introns in genomic DNA sequences. Although these neural networks can recognize potential coding sequences, they are imperfect at best and require confirmatory studies. Despite improvements in many of these traditional approaches, direct identification of coding sequences from large stretches of genomic DNA is not possible by any of these traditional methods. Moreover, the size and complexity of the human genome requires rapid and straightforward methods to identify coding sequences.

C. EXPRESSED SEQUENCE TAGS FROM BRAIN cDNA LIBRARIES

Because up to one third of all human genes are expressed by the brain and an estimated one fourth of all genetic diseases affect neurological function, Venter and colleagues have begun to sequence brain cDNAs. Rather than attempting to determine complete and accurate sequences of full-length cDNAs, these investigators have used automated, single-run sequence analysis to rapidly generate partial sequence data for a large number of brain cDNA clones.[3,19-21] These partial cDNA sequences provide "expressed sequence tags" (ESTs) which can be used to search Genbank and protein databases for matches to known gene and protein sequences and in studies to determine chromosomal and subchromosomal locations of the cDNAs.

For initial experiments to characterize this methodology, clones were selected at random from three oligo(dT), random-primed brain cDNA libraries: one from the hippocampus and one from the temporal cortex of a 2-year-old individual and one from a 17- to 18-week-old fetal brain. Random-primed libraries were selected over full-length cDNA libraries to improve the likelihood of including informative coding sequences in the ESTs rather than only 5' and 3' untranslated end sequences which might result from full-length cDNA clones. To generate ESTs, inserts from the cDNA libraries were cloned *en masse* into a plasmid vector, and double-stranded sequencing was performed on fluorescence-labeled DNA using an automated Applied Biosystems, Inc. (ABI) sequencing system. This automated methodology has recently been adapted for high-throughput sequencing at the Institute for Genomic Research and is currently capable of performing sequence analysis on up to 960 templates per day.[22] All ESTs contained at least 150 bases and <3% sequence ambiguities. Sequence accuracy was estimated at 97%. In these initial pilot experiments more than 600 EST sequences were generated; about one third matched known human sequences based on searches of existing databases. The majority of these known ESTs, however, were from either mitochondrial genes, repetitive sequences, or ribosomal genes. Only about 25% of known ESTs (about 8 to 10% of total ESTs) were from other nuclear genes. About 10% matched nonhuman genes or proteins and 38% had no significant matches. The remaining 22% of ESTs either contained no insert or consisted of only polyA sequences. Seven genes were represented by more than one EST and two ESTs were duplicated by one other EST.

These investigators have recently extended these initial results to provide partial sequence data for a total of nearly 8600 brain cDNAs.[20] Computer-based analyses of these ESTs have identified over 700 known proteins expressed by the brain. These known proteins are quite diverse and include structural proteins of the cytoskeleton, receptors, a variety of enzymes, regulatory proteins important during various developmental stages, and neurotransmitters. Structural proteins of the cytoskeleton were the most abundant of the known ESTs. Many ESTs showed similarities to genes or proteins previously identified from human or other species and may represent new members of known gene families. About half of all ESTs predicted to contain a protein coding region by computer analysis, showed no similarity to any known protein. These unknown ESTs may represent new, previously uncharacterized gene families and, when combined with results from other large-scale sequencing projects that have reported a similar high proportion of predicted coding sequences that do not match known proteins, suggest a much greater degree of diversity to gene families and subfamilies than would be expected from the 1000 basic families of proteins believed to exist.[23] These studies of brain ESTs can provide an overview of gene expression by different regions and different developmental stages of the brain. Moreover, an examination of cDNA libraries from additional tissue sources can provide insights into tissue-specific gene expression.

Polymeropoulos and co-workers have provided further insights into genes expressed by the nervous system by determining the chromosomal locations of 366 brain cDNAs.[24,25] Chromosomal localization was performed by analyzing the segregation of PCR products from human-rodent somatic cell hybrids. For these studies, a computer program identified 80- to 100-bp stretches from a given cDNA sequence unlikely to be interrupted by an intron in genomic DNA. Oligonucleotide primers were then synthesized from brain EST sequences and were used to amplify DNA from human-rodent somatic cell lines. Amplified products were subjected to high-resolution polyacrylamide gel electrophoresis to provide an exact measure of the size of the amplified fragment. About 20% of primer pairs amplified a rodent product, but usually the size of this product was different than that of the human product so chromosomal assignment was still possible. These rodent products were most likely derived from homologous rodent genes conserved across species. Several of these conserved products showed sequence similarities to previously characterized genes. The chromosomal distribution of these brain cDNAs was consistent with results from previously reported mapping data of other human genes in that the distribution of these genes did not correlate closely with chromosomal length. For example, chromosome 13 had many fewer genes than would be anticipated by its length, whereas chromosome 19 had about the same number of genes as chromosome 2, a much larger chromosome.

Other investigators have also reported chromosomal locations of brain ESTs.[26,27] Subchromosomal localizations could be performed in similar studies using DNA from deletion and radiation cell hybrids, and in one such application Durkin et al. reported subchromosomal assignments for 12 of 63 newly mapped brain ESTs located on chromosomes 6, 11, or the X chromosome.[28] ESTs that have been mapped have many potential applications. For example, they may prove useful as starting points for large-scale sequencing projects. Alternatively, primers could be used to assign ESTs to large genomic clones contained in cosmids or YACs and thus facilitate physical mapping of the genome. ESTs are expanding rapidly in number and have now been derived from a wide variety of different sources.[29-32]

Although a great deal of information can be gained by directly sequencing cDNAs, this approach has certain problems. First, this type of an approach is labor intensive and requires significant resources. Second, these studies depend in large part on the quality of the cDNA library to be studied. Even in the best of circumstances a large number of ESTs from conventionally prepared cDNA clones will be uninformative. In pilot experiments described above, Adams et al.[3] found that over 30% of the clones from a commercially obtained library contained inserts consisting of rRNA, mitochondrial cDNAs, or polyA sequences. Moreover, differences in levels of gene expression lead to problems of redundancy, especially for cDNAs from genes that are expressed at high levels. Levels of gene expression vary greatly in a given tissue or cell type, such that genes expressed in high abundance may produce up to 200,000 mRNA molecules per cell, whereas genes expressed in low abundance may yield only 1 to 10 mRNA molecules per cell. On average, in a given cell type, 50% of mRNA species by weight are of high abundance (expressed at >200,000 copies per cell), 20 to 50% are of low abundance (expressed at levels of 1 to 15 copies per cell), and other species are rare (expressed at levels of <1 molecule per cell). Furthermore, out of an average of 12,000 expressed genes per cell type, 11,000 belong to the low-abundance group and the remaining 1000 are either expressed at high or extremely low levels. Thus, genes expressed at very low levels or genes that are only expressed at certain developmental stages may be missed by this approach of random sequencing of clones from a single conventional cDNA library. Indeed, even if as many as 1 million cDNA clones were sequenced, some rare species could be missed. Recently, investigators have developed "equalized" or "normalized" cDNA libraries which use a kinetic approach to develop an approximately equal distribution of different cDNA species within a library.[33] A group known as the IMAGE consortium have also used normalized libraries from brain, breast, and liver to generate a large number of new ESTs. Although these normalization techniques can compensate to some extent for differences in relative abundance of cDNAs, complementary methods are essential to identify the coding sequences within long stretches of genomic DNA. To address these problems, several new techniques have recently been developed to directly isolate coding sequences from a large background of genomic DNA.

D. DIRECT SELECTION OF CODING SEQUENCES BY HYBRIDIZATION

1. Matrix-Based Methods

Two different approaches to identify unknown coding sequences from large genomic regions have been used in these new methodologies. The first approach used by the majority of new methods is the hybridization of genomic DNA to cDNA clones to selectively enrich for or to isolate transcribed sequences encoded by these genomic fragments. The second approach, referred to as exon amplification or exon trapping uses a specially designed plasmid vector to isolate sequences that are transcribed by host cells. Lovett et al. were among the first to apply the principles of hybrid selection in a technique referred to as direct selection.

For this method, DNA from genomic clones is immobilized on filters and then hybridized to amplified inserts from a cDNA library.[34] Before hybridization, genomic DNA and amplified cDNAs were incubated with either genomic DNA from human, yeast, or plasmid sources or Cot1 DNA in order to block repetitive sequences. After hybridization, filters were washed to improve specificity of binding and bound cDNAs were eluted. These selected cDNAs were amplified and cloned or subjected to repeated cycles of hybridization selection to further enrich the cDNA population. In test experiments, cDNAs from two different oligo(dT)-primed cDNA libraries, which were known to be encoded within cosmid and YAC clones, were enriched ~800- to 2000-fold by one cycle of direct selection. Also, an unknown cDNA from a fetal kidney cDNA library was shown to be encoded within a 550-kb YAC from chromosome 7 which contained the human erythropoietin gene. This selected cDNA was subsequently sequenced and shown to encode the β_2 subunit of the guanine nucleotide-binding protein. One problem encountered in direct selection was the frequent selection of false positive cDNAs when YAC clones were used as the genomic source. This artifact was shown to be yeast rRNA or DNA derived from the carrier yeast RNA used during commercial synthesis of cDNA clones and could be overcome by preblocking the amplified cDNAs or the YAC DNA with yeast DNA. Preblocking also helped to reduce, but did not eliminate, the problem of false positives from repetitive sequences in the cDNA population.

A second hybridization enrichment strategy conceptually similar to the direct selection method of Lovett et al. was described by Parimoo et al.[35] For this method, cosmid or YAC genomic DNA was digested by *Eco*RI, denatured, and immobilized on a nylon disc. Several quenching agents including polydIdC, sonicated yeast DNA, plasmid clones containing human genomic DNA from chromosome 15, repetitive DNA, and human rRNA gene sequences were used in a prehybridization reaction to block nonspecific binding. cDNA inserts from a random primer-generated, short-fragment cDNA library were amplified by PCR and hybridized to the nylon discs containing the genomic material. Following several washes, the bound cDNA clones were eluted, amplified, and cloned into λgt10 for immediate analysis, or were subjected to an additional cycle of hybridization selection. This method was capable of isolating an encoded cDNA from as little as 1 pg of target cosmid DNA. A cDNA of moderate abundance and a rare cDNA contained within a YAC genomic source were enriched 70- and 1000-fold over their concentrations in the starting cDNA library after a single cycle of hybridization selection. A second cycle of selection served to enrich the rare species by more than 7000-fold over their original concentration. However, despite preblocking with the quenching agents, selected cDNAs continued to contain a significant percentage of contaminating clones including repetitive DNA and ribosomal cDNAs.

Streptavidin-coated magnetic beads have been incorporated into the direct selection method of Lovett et al. to eliminate the nitrocellulose support used in the original procedure.[36,37] In this significant improvement over the original method, the genomic source is digested by a restriction enzyme and oligonucleotide linkers are ligated to the 5′ and 3′ ends of the digested DNA. Next, the genomic DNA is amplified in a PCR reaction using a 5′ biotinylated primer. Oligo(dT)-primed cDNAs are also digested and a different set of oligonucleotide linkers are ligated to the 5′ and 3′ ends of the cDNAs. These linkers are used for PCR amplification of the cDNAs. Repetitive sequences are blocked by prehybridizing the digested cDNAs to Cot1 DNA. Biotinylated genomic DNA is then hybridized to the blocked cDNA and bound cDNAs are captured on streptavidin-coated magnetic beads. The beads are washed extensively and the selected cDNAs are eluted and cloned into λgt10 or are subjected to a second cycle of hybridization enrichment. In pilot studies, cDNA clones isolated after two cycles of direct selection with a YAC genomic source were examined and two novel cDNAs were identified. None of the clones examined were redundant and all mapped back to the starting YAC clone. Comparison of this group of selected clones to the starting cDNA population showed that all of the selected clones were initially present at very low abundance and were enriched by ~6000-fold to >100,000-fold over their concentrations in the starting library. As in the other forms of hybrid selection described above, a large number of ribosomal clones were present after even a single cycle of direct selection by bead-capture. Ribosomal clones made up 60% of these selected cDNAs. The investigators proposed to address this problem by preblocking with ribosomal DNA or by counterscreening with a ribosomal probe after cloning the selected cDNAs. Direct selection has recently been applied to identify transcribed sequences surrounding the G6PD locus of Xq28.[38]

Isolation of coding sequences by this method of direct selection has certain advantages but also some limitations. A major advantage of direct selection methods is the ability to readily isolate rare cDNAs. For example, Morgan et al.[36] were able to isolate a rare cDNA after two cycles of direct selection, whereas a screen of ~450,000 clones derived from the starting population failed to identify this cDNA. The bead-capture method of direct selection described above is particularly advantageous in this regard, because, since hybridization reactions are carried out in solution, kinetics can be more accurately predicted than for reactions performed on solid supports and thereby optimized. This bead-capture approach should also be faster and less cumbersome than hybridizations on solid matrices. Selection strategies which are based on hybridization also provide a certain degree of normalization of cDNA species so that if the genomic DNA is limiting in the hybridization reaction, more abundant species will saturate their genomic targets more quickly than low-abundance or rare species. The end result is a relative normalization of the abundant cDNAs downwards and of the low-abundance classes upwards.[36]

Although these solid-state methods of direct selection are novel and potentially quite useful, they have certain limitations. First, the hybridization reaction itself can be a limiting factor. Second, as noted above, these methods are only enrichment techniques and require the identification of positive clones from a background of ribosomal clones and repetitive sequences by conventional screening and cloning techniques. One possible reason for this significant remaining background is the nonspecific adsorption of cDNA molecules to the streptavidin beads. Reportedly, this problem can be overcome by using calf thymus or salmon sperm DNA to saturate nonspecific binding sites on the streptavidin beads. Alternatively, the use of iron oxide-coated linkers may decrease nonspecificity.

Another hybridization-based method to isolate the coding sequences from large genomic fragments has been referred to as "cDNA amplification for the identification of genomic expressed sequences" or CAIGES.[39] For this method, a Southern blot of a genomic clone is immobilized on a filter and then hybridized to the amplified, radiolabeled inserts of a cDNA library. To prevent false positives from repetitive sequences, the blot and probes are prehybridized with placental genomic DNA. Positive genomic fragments can then be cloned and sequenced. Guo et al. have applied this method to a cosmid and a human liver cDNA library to clone a 6-kb genomic fragment and cDNAs for the human glycerol kinase gene. The cDNA that hybridized to the genomic fragment was present in low abundance in the initial library (0.0005%) suggesting that the method is relatively sensitive. However, two smaller fragments from the cosmid digest that contained fewer coding sequences than the 6-kb fragment were not identified on the original blot. Also, as in the initial descriptions of direct selection, the use of a solid support for hybridization reactions makes the procedure somewhat cumbersome and time consuming. This method would seem to be particularly useful for the final steps of positional cloning studies, when partial localization of a particular gene has been accomplished. To date it has not been applied to larger genomic fragments such as YACs or to the identification of unknown coding sequences.

2. SOLUTION-BASED METHODS

Our laboratory has recently developed a PCR-based strategy of hybrid selection in order to rapidly isolate expressed sequences from genomic DNA. Fragments from a genomic source are hybridized to cDNA clones in order to form "genomic-cDNA chimeras" that are then amplified by PCR and isolated. This method is summarized schematically in Figure 4-1. The first step in this procedure is to generate a short-fragment genomic library by digesting the genomic DNA with a restriction enzyme and "shotgun subcloning" into a vector, flanked by vector-specific primers P1 and P2. A wide variety of genomic sources can be used including DNA from a microdissected subchromosomal region, or DNA from cosmid or YAC contigs. A random-primed short-fragment cDNA library is also cloned into a second vector which contains different primers, X1 and X2, flanking the inserts. Both the genomic and cDNA libraries must contain relatively short inserts, because the genomic-cDNA chimera must be amplified by PCR in order to be isolated. Thus, the combined size of the genomic and cDNA components, in general, should not exceed 1 to 1.5 kb.

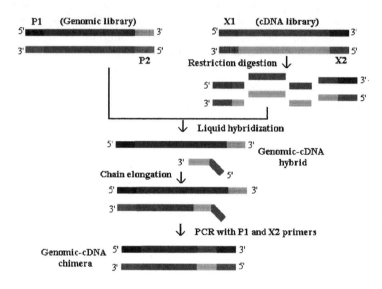

FIGURE 4-1 General scheme for a solution-based method to select genomic-cDNA chimeras. P1, P2, X1, and X2 represent vector-specific primers.

Prior to hybridization, DNA from the cDNA fraction is digested by a restriction enzyme that infrequently cleaves repetitive DNA, that does not cut between primers X1 and X2 and the insertion sites of the cDNAs, and that cleaves most unique cDNA sequences at least once. A recent search of GenBank suggested that *Taq I* would fulfill these criteria. Approximately 800 *Taq I* sites were present in over 650 kb of cDNA sequence so, on average, most cDNAs of 800 bp or greater should contain at least one *Taq I* site. Moreover, *Taq I* sites were present in only about one out of three repetitive sequences. Thus, digestion of the cDNA molecules by *Taq I* should help to minimize the formation of chimeras consisting of repetitive sequences and improve the specificity of this method. Genomic DNA from the short-fragment library and the digested cDNA fragments are then mixed, denatured, and allowed to reanneal to form genomic-cDNA hybrids. Next, the cDNA portion of the hybrid is used to prime a cycle of chain elongation that completes the formation of the genomic strand of the chimera. The restriction enzyme chosen to digest the cDNA fraction must, therefore, also provide free 3'-OH ends to be used by *Taq* polymerase in chain elongation. Following this initial cycle of chain elongation, the hybrid is amplified in a conventional PCR reaction in which one primer anneals to the cDNA (X1 or X2) and the other anneals to the genomic DNA (P1 or P2). Coding sequences are thereby isolated as genomic-cDNA hybrids.

In preliminary experiments to validate this strategy, we used a short-fragment genomic library to isolate a test genomic-cDNA chimera.[40] Known genomic and cDNA clones were added to 500 colonies from a short-fragment genomic library (average insert size 250 to 300 bp) along with increasing numbers of cDNA clones. The test hybrid was able to be isolated in experiments using 10,000, 50,000, and 250,000 cDNA clones, but not in experiments using the entire cDNA library. We are currently testing modifications of the reaction conditions in order to improve the efficiency of this procedure and enhance its applicability. Modifications which may prove useful include converting the supercoiled plasmid DNA that contains the genomic source and drives the hybridization reactions into a linear, double-stranded template. Alternatively, since vector sequences are in large molar excess, the genomic and cDNA inserts could be amplified by PCR and hybridized. Modifications of hybridization conditions may also improve efficiency. For example, the phenol emulsion reassociation technique has been reported to improve hybridization efficiencies from 100- to 1000-fold.[41] Other agents which may improve hybridization efficiencies include dextran sulfate and formamide.[42]

In any event, improvements in the hybridization conditions and kinetics should be possible and will enhance the applicability of this method. Potential advantages of this PCR-based method include ease of use and, as for the bead-capture method of Morgan et al., a certain degree of cDNA normalization is inherent in this method because genomic fragments are rate-limiting to the hybridization reactions. Also, since hybrids are formed from genomic and cDNA sequences, the probability of isolating polymorphisms should be high. Known primer sequences flanking either end of the hybrid could be used to establish single-stranded conformational polymorphisms for some hybrids, which could then serve as genetic markers for certain diseases. Finally, by digesting the genomic cDNA hybrid with *Taq I*, the resulting fragment can be mixed with the cDNA library and used with primers flanking the cDNA inserts to generate longer cDNA fragments by PCR.

E. EXON TRAPPING AND EXON AMPLIFICATION

Recently, several investigators have used a novel strategy to identify coding sequences known as "exon trapping" or "exon amplification" which is based on the eukaryotic splicing mechanism for removal of introns as heterogenous nuclear RNA is processed into mature cytoplasmic RNA. In the first report of this type of a strategy, Duyk et al.[43] used a retroviral shuttle vector to select for 3′ splice sites in random fragments of genomic DNA. However, this method proved to be rather cumbersome and time consuming. Shortly thereafter, Buckler et al.[44] developed a splicing vector known as pSPL1 to efficiently screen genomic fragments *in vivo* for exon sequences. pSPL1 contains the *tat* intron of the human immunodeficiency virus-1 genome with its flanking 5′ and 3′ splice sites and its exon intact. Genomic fragments between 1 and 4 kb are cloned into this intron, and following transfection into COS-7 cells, an SV40 origin of replication, promoter, and polyadenylation signal drive amplification and transcription of the plasmid and its cloned genomic fragment. Two to three days after transfection, cytoplasmic RNA is isolated, reverse transcribed, and amplified by PCR (RT-PCR). If the genomic fragment cloned into the *tat* intron contained an entire exon with its flanking splice sites in the proper orientation, it should be included in the cytoplasmic poly(A) RNA and be amplified by RT-PCR. As an initial test to verify the validity of this method, the investigators cloned a 3.5-kb fragment from a cosmid known to contain murine exon sequences into pSPL1 in both the sense and antisense orientations. Northern analysis showed that cytoplasmic RNA from transfected COS-7 cells contained mRNA from the 3.5-kb fragment cloned in the sense but not the antisense orientation. These exon sequences were able to be amplified and isolated using RT-PCR. To increase the complexity of the genomic DNA, the investigators digested the entire cosmid with restriction endonucleases and "shotgun-cloned" the resulting fragments into pSPL1. Following transfection into COS-7 cells and RT-PCR, the murine exon sequences could still be isolated from this more complex genomic source. At least one artifactual product was present as well. Finally, the authors have used the exon-trapping technique to test previously uncharacterized genomic DNA for transcribed sequences. In initial experiments, 12 previously uncharacterized phage clones that contained 15- to 20-kb human genomic inserts derived from a segment of chromosome 19 were examined for coding sequences by exon trapping; 6 of the 12 phage clones were found to have exon sequences. Southern analysis was used to show that the isolated exons hybridized only to their original genomic fragments and not to other human genomic DNA or lambda DNA. One of the products amplified by RT-PCR in these experiments contained an exon of the DNA excision repair gene ERCC1, which is known to be located in this region of chromosome 19. Approximately 70% of 33 cosmid clones and 45% of 18 phage clones examined by exon trapping contained putative exons.

Since its original descriptions, certain modifications have been applied to the experimental design of the exon amplification strategy to improve its efficiency and broaden its applicability. Hamaguchi et al.[45] attempted to improve the sensitivity and specificity of exon amplification by designing an improved trapping cassette. A portion of the human p53 gene, containing the entire 10th intron and portions of exons 10 and 11, was cloned into the plasmid pEUK-C2 to produce a new vector pMHC2. This portion of the p53 gene was chosen for the trapping cassette because it contained a long pyrimidine tract to prevent exon skipping, consensus sequences at branch and splice sites that closely matched those of small nuclear RNAs involved in splicing, and an intron sufficiently long (>500 bp) to readily distinguish spliced from unspliced fragments. Test experiments suggested that this system was capable of trapping exons from genomic fragments as large as several hundred kilobases in length.

Church et al. recently modified the pSPL1 splicing vector to produce a more sensitive and reliable vector pSPL3.[46] In addition to this new trapping vector, two problems encountered in the original experimental design of Buckler et al. were overcome. First, because of the large excess of introns and noncoding sequences relative to exons in genomic DNA, the majority of fragments amplified by the original method contained only *tat* exon sequences from the pSPL1 vector. Thus, competition among PCR templates favored this smallest and most abundant fragment. In order to overcome this difficulty, the *tat* exon sequences flanking the vector intron were modified so that removal of the intron during RNA processing of heterogenous nuclear RNA would produce a restriction endonuclease recognition site. A *Bst*XI recognition sequence (CCAN$_6$TGG) was chosen so as not to alter the vector exon sequences immediately flanking the splice sites. Digestion by *Bst*XI was performed immediately after RT-PCR. Additional modifications included the introduction of a multiple cloning site within the *tat* intron to increase the flexibility of the vector, and the introduction of a second *Bst*XI recognition sequence at the 3' end of the polylinker so that digestion by this enzyme would eliminate commonly encountered false positives resulting from a cryptic splice site within the *tat* intron. In test experiments using 10 cosmids derived from chromosome 9, these authors were able to isolate a total of 31 unique exons, or about 3 exons per cosmid. Increasing the complexity of the genomic starting material reduced the method's sensitivity. For example, when pools of cosmids were tested the average number of exons isolated per cosmid decreased to 1.4, or an average of 1 exon for every 20 to 25 kb of genomic DNA. Also, 11% of putative exons were found to be false positives and were derived from either repetitive sequences or cryptic splice sites that remained in the *tat* intron of pSPL3. This version of exon amplification was also applied to YAC clones and ~1000 plasmid clones containing a total of ~3 Mb of mouse genomic DNA. The sensitivity of exon amplification was further reduced in these experiments using highly complex genomic sources; 1 exon was isolated for every 75 to 85 kb of genomic DNA tested. However, these investigators estimated that these modifications resulted in about a 60- to 80-fold increase in overall efficiency of the exon amplification procedure. Further refinements of the pSPL3 vector have also been recently reported by Burn et al.[47]

A large number of transcribed sequences from a variety of genomic sources have been identified by exon amplification. These exon-containing sequences have been derived from cosmids from chromosomes 6 and 9,[46,48] human genomic DNA,[46] and human and mouse yeast artificial chromosomes.[46,49] Moreover, this technique has played an important role in the isolation of several important human genes including a copper transporter gene defective in individuals with Menke's disease,[50] the neurofibromatosis type 2 tumor-suppressor gene,[51] and several genes within the region of the Huntington's disease locus, including the causative gene itself.[52-55]

Exon amplification offers significant advantages but also certain disadvantages when compared to other methods for the rapid identification of coding sequences from genomic DNA. Most importantly, exon amplification does not rely on gene expression for the detection of transcribed sequences and it is not subject to many of the problems inherent in technologies that use cDNA libraries to identify expressed sequences. Therefore, it can be used to identify tissue-specific or developmentally regulated genes just as well as genes that are expressed ubiquitously. Also, the level of expression of a particular gene does not affect its likelihood to be identified by exon amplification. Thus, in contrast to some of the methods described previously, exon amplification is robust enough to isolate even very rare cDNA clones. On the other hand, however, exon amplification can not provide insights into tissue-specific or developmentally regulated gene expression. Also, false positives resulting from repetitive DNA and cryptic splice sites within intron sequences persist at a level of about 8 to 10%, even when improved splicing vectors such as pSPL3 are used.[46] Additionally, because exon amplification relies on the presence of functional 5' and 3' splice sites to identify an exon, genes made up of a single exon or those that do not contain introns will be missed by this approach. Moreover, some exons that contain flanking splice sites can be missed by this approach. This false negative rate has been estimated to be as high as 10 to 15%.[48] Also, putative exons isolated by exon amplification require verification by either Northern analysis or screening a cDNA library.

F. CONCLUSION

A number of new approaches and methodologies have been developed recently for the isolation of coding sequences and each has certain advantages and limitations. Thus, a combined approach in which methods of direct selection and exon trapping are combined with positional cloning and characterization of ESTs will almost certainly be the most effective way to identify this crucial portion of the human genome. Several investigators have begun to apply these techniques in combination.[56,57] Further characterization of coding sequences, including elucidation of the functional significance of molecules they encode, will present new challenges but will also provide insights into various neurological and psychological disorders. An understanding of the molecular basis of these disease processes should pave the way for novel methods of testing, treatment, and ultimately prevention. Initial steps towards this end will include the incorporation of transcriptional maps with available physical maps which contain identified disease markers. Progress has been made for such maps of chromosome 21,[58-60] 7q22,[61] Xq13.3,[62] Xq28,[12,63] Xp21, and Xp11.

REFERENCES

1. Sutcliffe, J. G., mRNA in the mammalian central nervous system, *Annu. Rev. Neurosci.*, 11, 157, 1988.
2. Brenner, S., The human genome: the nature of the enterprise, *Ciba Found. Symp.*, 149, 6, 1990.
3. Adams, M. D., Kelley, J. M., Gocayne, J. D., Dubnick, M., Polymeropoulos, M. H., Xiao, H., Merril, C. R., Wu, A., Olde, B., Moreno, R. F., Kerlavage, A. R., McCombie, W. R., and Venter, J. C., Complementary DNA sequencing: expressed sequence tags and Human Genome Project, *Science*, 252, 1651, 1991.
4. Monaco, A. P., Neve, R. L., Colletti-Feener, C., Bertelson, C. J., Kurnit, D. M., and Kunkel, L. M., Isolation of candidate cDNAs for portions of the Duchenne muscular dystrophy gene, *Nature*, 323, 646, 1986.

5. Page, D. C., Moshner, R., Simpson, E. M., Fisher, E. M., Mardon, G., Pollack, J., McGillivray, B., Chapelle, A., and Brown, L. G., The sex-determining region of the human Y chromosome encodes a finger protein, *Cell*, 51, 1091, 1987.
6. Call, K. M., Glaser, T., Ito, C. Y., Buckler, A. J., Pelletier, J., Haber, D. A., Rose, E. A., Kral, A., Yeger, H., Lewis, W. H., Jones, C., and Housman, D. E., Isolation and characterization of a zinc finger polypeptide gene at the human chromosome 11 Wilm's tumor locus, *Cell*, 60, 509, 1990.
7. Weber, F., Villiers, J., and Schaffner, W., An SV40 "enhancer trap" incorporates exogenous enhancers or generates enhancers from its own sequences, *Cell*, 36, 983, 1984.
8. Snyder, E. E. and Stormo, G. D., Identification of coding regions in genomic DNA sequences: an application of dynamic programming and neural networks, *Nucleic Acids Res.*, 21, 307, 1993.
9. Allen, N. D., Cran, D. G., Barton, S. C., Hettle, S., Reik, W., and Surani, M. A., Transgenes as probes for active chromosomal domains in mouse development, *Nature*, 333, 852, 1988.
10. Gossler, A., Joyner, A., Rossant, J., and Skarnes, W., Mouse embryonic stem cells and reporter constructs to detect developmentally regulated genes, *Science*, 244, 463, 1989.
11. Bird, A. P., CpG-rich islands and the function of DNA methylation, *Nature*, 321, 209, 1986.
12. Bione, S., Tamanni, F., Maestrini, E. C. T., Poustka, A. G. T., Rivella, S., and Toniolo, D., Transcriptional organization of a 450-kb region of the human X chromosome in Xq28, *Proc. Natl. Acad. Sci. U.S.A.*, 90, 10977, 1993.
13. Rommens, J. M., Ianuzzi, M. C., Kerem, B.-S., Drumm, M. L., Melmer, G., Dean, M., Rozmahel, R., Cole, J. L., Kennedy, D., Hidaka, N., Zsiga, M., Buchwald, M., Riordan, J. R., Tsui, L.-C., and Collins, F. S., Identification of the cystic fibrosis gene: chromosome walking and jumping, *Science*, 245, 1059, 1989.
14. Liu, P., Legerski, R., and Siciliano, M. J., Isolation of human transcribed sequences from human-rodent somatic cell hybrids, *Science*, 246, 813, 1989.
15. Corbo, L., Maley, J. A., Nelson, D. L., and Caskey, C. T., Direct cloning of human transcripts with hnRNA from hybrid cell lines, *Science*, 249, 652, 1990.
16. Guigo, R., Knudsen, S., Drake, N., and Smith, T., Prediction of gene structure, *J. Mol. Biol.*, 226, 141, 1992.
17. Fields, C. A. and Soderlund, C. A., GM: a practical tool for automating DNA sequence analysis, *Comp. Appl. Biosci.*, 36, 263, 1990.
18. Uberbacher, E. C. and Mural, R. J., Locating protein-coding regions in human DNA sequences by a multiple sensor-neural network approach, *Proc. Natl. Acad. Sci. U.S.A.*, 88, 11261, 1991.
19. Adams, M. D., Dubnick, M., Kerlavage, A. R., Moreno, R., Kelley, J. M., Utterback, T. R., Nagle, J. W., Fields, C., and Venter, J. C., Sequence identification of 2,375 human brain genes, *Nature*, 355, 632, 1992.
20. Adams, M. D., Kerlavage, A. R., Fields, C., and Venter, J. C., The 3,400 new expressed sequence tags identify diversity of transcripts in human brain, *Nat. Genet.*, 4, 256, 1993.
21. Adams, M. D., Soares, B. M., Kerlavage, A. R., Fields, C., and Venter, J. C., Rapid cDNA sequencing (expressed sequence tags) from directional cloned human infant brain cDNA library, *Nat. Genet.*, 4, 373, 1993.
22. Adams, M. D., Kerlavage, A. R., Kelley, J. M., Gocayne, J. D., Fields, C., Fraser, C. M., and Venter, J. C., A model for high-throughput automated DNA sequencing and analysis core facilities, *Nature*, 368, 474, 1994.
23. Clothia, C., One thousand families for the molecular biologist, *Nature*, 367, 543, 1992.
24. Polymeropoulos, M. H., Xiao, H., Glodek, A., Gorski, M., Adams, M. D., Moreno, R. F., Fitzgerald, M. G., Venter, J. C., and Merril, C. R., Chromosomal assignment of 46 brain cDNAs, *Genomics*, 12, 492, 1992.
25. Polymeropoulos, M. H., Xiao, H., Sikela, J. M., Adams, M., Venter, J. C., and Merril, C. R., Chromosomal distribution of 320 genes from a brain cDNA library, *Nat. Genet.*, 4, 381, 1993.
26. Durkin, A. S., Maglott, D. R., and Nierman, W. C., Chromosomal assignment of 39 human brain expressed sequence tags (ESTs) by analyzing fluorescently-labeled PCR products from hybrid cell panels, *Genomics*, 14, 808, 1992.
27. Khan, A. S., Wilcox, A. S., Polymeropoulos, M. H., Hopkins, J. A., Stevens, T. J., Robinson, M., Orpana, A. K., and Sikela, J. M., Single pass sequencing and physical and genetic mapping of human brain cDNAs, *Nat. Genet.*, 2, 180, 1992.
28. Durkin, A. S., Nierman, W. C., Zoghbi, H., Jones, C., Kozak, C. A., and Maglott, D. R., Chromosome assignment of human brain expressed sequence tags (ESTs) by analyzing fluorescently labeled PCR products from hybrid cell panels, *Cytogenet. Cell Genet.*, 65, 86, 1994.
29. Pawlak, A., Toussaint, C., Levy, I., Bulle, F., Poyard, M., Barouki, R., and Guellaen, G., Characterization of a large population of mRNAs from human testis, *Genomics*, 26, 151, 1995.
30. Sudo, K., Chinen, K., and Nakamura, Y., The 2058 expressed sequence tags (ESTs) from a human fetal lung cDNA library, *Genomics*, 24, 276, 1994.
31. Watson, M. A. and Fleming, T. P., Isolation of differentially expressed sequence tags from human breast cancer, *Cancer Res.*, 54, 4598, 1994.
32. Liew, C. C., Hwang, D. M., Fung, Y. W., Laurenssen, C., Cukerman, E., Tsui, S., and Lee, C. Y., A catalogue of genes in the cardiovascular system as identified by expressed sequence tags, *Proc. Natl. Acad. Sci. U.S.A.*, 91, 10645, 1994.

33. Patanjali, S. R., Parimoo, S., and Weissman, S. M., Construction of a uniform-abundance (normalized) cDNA library, *Proc. Natl. Acad. Sci. U.S.A.*, 88, 1943, 1991.
34. Lovett, M., Kere, J., and Hinton, L. M., Direct selection: a method for the isolation of cDNAs encoded by large genomic regions, *Proc. Natl. Acad. Sci. U.S.A.*, 88, 9628, 1991.
35. Parimoo, S., Patanjali, S. R., Shukla, H., Chaplin, D. D., and Weissman, S. M., cDNA selection: efficient PCR approach for the selection of cDNAs encoded in large chromosomal DNA fragments, *Proc. Natl. Acad. Sci. U.S.A.*, 88, 9623, 1991.
36. Morgan, J. G., Dolganov, G. M., Robbins, S. E., Hinton, L. M., and Lovett, M., The selective isolation of novel cDNAs encoded by the regions surrounding the human interleukin 4 and 5 genes, *Nucleic Acids Res.*, 20, 5173, 1992.
37. Tagle, D. A., Swaroop, M., Lovett, M., and Collins, F. S., Magnetic bead capture of expressed sequences encoded within large genomic segments, *Nature*, 361, 751, 1993.
38. Korn, B., Sedlacek, Z., Manca, A., Kioschis, P., Konecki, D., Lehrach, H., and Poustka, A., A strategy for the selection of transcribed sequences in the Xq28 region, *Hum. Mol. Genet.*, 1, 235, 1992.
39. Guo, W., Worley, K., Adams, V., Mason, J., Sylvester-Jackson, D., Zhang, Y.-H., Towbin, J. A., Fogt, D. D., Madu, S., Wheeler, D. A., and McCabe, E. R. B., Genomic scanning for expressed sequences in Xp21 identifies the glycerol kinase gene, *Nat. Genet.*, 4, 367, 1993.
40. Jagadeeswaran, P., Odom, M. W., and Boland, E. J., Novel strategy for isolating unknown coding sequences from genomic DNA by generating genomic-cDNA chimeras, in *Identification of Transcribed Sequences*, Hochgeschwender, U. and Gardiner, K., Eds., Plenum Press, New York, 1994.
41. Kohne, D. E., Levison, S. A., and Byers, M. J., Room temperature method for the rate of DNA reassociation of many thousand-fold: the phenol emulsion reassociation technique, *Biochemistry*, 16, 5329, 1977.
42. Wetmur, R., Acceleration of DNA renaturation rates, *Biopolymers*, 14, 2517, 1975.
43. Duyk, G. M., Kim, S. W., Myers, R. M., and Cox, D. R., Exon trapping: a genetic screen to identify candidate transcribed sequences in cloned mammalian genomic DNA, *Proc. Natl. Acad. Sci. U.S.A.*, 87, 8995, 1990.
44. Buckler, A. J., Chang, D. D., Graw, S. L., Brook, J. D., Haber, D. A., Sharp, P. A., and Housman, D. E., Exon amplification: a strategy to isolate mammalian genes based on RNA splicing, *Proc. Natl. Acad. Sci. U.S.A.*, 88, 4005, 1991.
45. Hamaguchi, M., Sakamoto, H., Tsuruta, H., Sasaki, H., Muto, T., Sugimura, T., and Terada, M., Establishment of a highly sensitive and specific exon-trapping system, *Proc. Natl. Acad. Sci. U.S.A.*, 89, 9779, 1992.
46. Church, D. M., Stotler, C. J., Rutter, J. L., Murrell, J. R., Troffater, J. A., and Buckler, A. J., Isolation of genes from complex sources of mammalian genomic DNA using exon amplification, *Nat. Genet.*, 6, 98, 1994.
47. Burn, T. C., Connors, T. D., Klinger, K. W., and Landes, G. M., Increased exon-trapping efficiency through modifications to the pSPL3 splicing vector, *Gene*, 161, 183, 1995.
48. North, M. A., Sanseau, P., Buckler, A. J., Church, D., Jackson, A., Patel, K., Trowsdale, J., and Lehrach, H., Efficiency and specificity of gene isolation by exon amplification, *Mamm. Genome*, 4, 466, 1993.
49. Gibson, F., Lehrach, H., Buckler, A. J., Brown, S. D. M., and North, M. A., Isolation of conserved sequences from yeast artificial chromosomes by exon amplification, *BioTechniques*, 16, 453, 1994.
50. Vulpe, C., Levinson, B., Whitney, S., Packman, S., and Gitschier, J., Isolation of a candidate gene for Menke's disease and evidence that it encodes a copper transporting ATPase, *Nat. Genet.*, 3, 7, 1993.
51. Trofatter, J. A., A novel moesin-, ezrin-, radixin-like gene is a candidate for the neurofibromatosis 2 tumor suppressor, *Cell*, 72, 791, 1993.
52. Taylor, S. A. M., Cloning of the α-adducin gene from the Huntington's disease candidate region of human chromosome 4 by exon amplification, *Nat. Genet.*, 1, 697, 1992.
53. Ambrose, C., James, M., Barnes, G., Lin, C., Bates, G., Altherr, M., Duyao, M., Groot, N., Church, D., and Wasmuth, J. J., A novel G protein-coupled receptor kinase gene cloned from 4p16.3, *Hum. Mol. Genet.*, 1, 697, 1992.
54. Duyao, M. P., A gene from 4p16.3 with similarity to a superfamily of transporter proteins, *Hum. Mol. Genet.*, 2, 673, 1993.
55. Group, T. H. s. D. C. R., A novel gene containing a trinucleotide repeat that is expanded and unstable on Huntington's disease chromosomes, *Cell*, 72, 971, 1993.
56. Koyama, K., Sudo, K., and Nakamura, Y., Isolation of 115 human chromosome 8-specific expressed sequence tags by exon amplification, *Genomics*, 26, 245, 1995.
57. Brody, L. C., Abel, K. J., Castilla, L. H., Couch, F. J., McKinley, D. R., Yin, G., Ho, P. P., Merajver, S., Chandrasekharappa, S. C., and Xu, J., Construction of a transcription map surrounding the BRCA1 locus of human chromosome 17, *Genomics*, 25, 238, 1995.
58. Tassone, F., Cheng, S., and Gardiner, K., Analysis of chromosome 21 yeast artificial chromosome (YAC) clones, *Am. J. Hum. Genet.*, 51, 1251, 1992.
59. Cheng, S., Lutfalla, G., Uze, G., Chumakov, I. M., and Gardiner, K., GART, SON, IFANR, and CRF2-4 genes cluster on human chromosome 21 and chromosome 16, *Mamm. Genome*, 4, 338, 1993.
60. Patterson, D., Rahmani, Z., Donaldson, D., Gardiner, K., and Jones, C., Physical mapping of chromosome 21, *Prog. Clin. Biol. Res.*, 384, 33, 1993.

61. Scherer, S. W., Rommens, J. M., Soder, S., Wong, E., Plavsic, N., Tompkins, B. J., Beattie, A., Kim, J., and Tsui, L. C., Refined localization and yeast artificial chromosome (YAC) contig-mapping of genes and DNA segment in the 7q21-q32 region, *Hum. Mol. Genet.*, 2, 751, 1993.
62. Gecz, J., Villard, L., Lossi, A. M., Millasseau, P., Djabali, M., and Fontes, M., Physical and transcriptional mapping of DXS56-PGK1 1 Mb region: identification of three new transcripts, *Hum. Mol. Genet.*, 2, 1389, 1993.
63. Sedlacek, Z., Korn, B., Konecki, D. S., Siebenhaar, R., and Coy, J. F., Construction of a transcription map of a 300 kb region around the human G6PD locus by direct cDNA selection, *Hum. Mol. Genet.*, 2, 1865, 1993.

5 Identification of Microsatellite Repeat Markers and Transcribed Sequences in Cloned Genomic DNA

Sankhavaram R. Patanjali, Rajendra P. Kandpal, Hongxia Xu, and Sherman M. Weissman

CONTENTS

- A. Introduction 64
- B. Background 64
    1. Microsatellite Markers 64
    2. Transcribed Sequences of Mapped Genomic Regions 65
        a. Sequence-Based Approaches 66
            - Sequencing of Genomic DNA 66
            - Random Sequencing of cDNAs 66
            - CpG Islands 66
        b. Region-Specific Transcript Identification Approaches 66
            - *In Vivo* Transcription Methods 66
            - *In Vitro* Hybridization Methods 67
- C. Description of Hybridization Selection/cDNA Selection 68
    1. Input 68
        a. For Microsatellite Selection 68
            - Microsatellite-Enriched Genomic DNA Libraries 68
        b. For cDNA Selection 68
    2. Target 69
        a. For Microsatellite Selection 69
        b. For cDNA Selection 69
    3. Quenchers 69
    4. Experimental Details of Selection 69
- D. Results and Discussion 70
    1. Selection Microsatellite Markers 70
        a. Enrichment of CA Repeat Motifs 70
        b. YAC-Specific Markers 71
    2. Selection of cDNAs 72
- E. Conclusion 73
- Acknowledgments 73
- References 73

A. INTRODUCTION

The complex process of molecular analysis of a genome, whether it be as simple as a bacterium's or as complicated as a human's, is greatly facilitated by mapping of various genes or sequence tagged sites (STS) along the length of the genome. Initial mapping efforts using restriction fragment length polymorphisms (RFLPs) were very successful in generating a number of new genetic markers.[1,2] In recent years, however, RFLPs have been largely replaced by microsatellite DNA markers containing multiple tandem copies of simple di-, tri-, and tetranucleotide sequences. These microsatellite markers have two major advantages over RFLPs: (1) they are more highly polymorphic so that a much larger fraction of meiotic events are informative, and (2) they are more easily analyzed on DNA samples, as they do not in general require a hybridization step to be detected. A vast majority of genetic mappings of disease-related genes are being conducted nowadays by the use of over 3000 microsatellite markers that are commercially available. Short of complete genomic sequencing, a major goal of the human genome project is to map the entire genome by specific markers and identify all the transcription units encoded by the genome. It is estimated that at the frequency of 1 in 100,000 bases, at least 30,000 markers will be needed for the high-resolution mapping of the genome.

Recently major successes have been achieved in assembling large YAC contigs covering major portions of the human genome by DNA fingerprinting using Alu, LINE, and THE repeat elements as probes.[3] A rough scaffolding of contigs for almost the entire genome will probably be available in the near future. Availability of such large YAC contigs and the localization of known genes, STSs, and microsatellite markers to those contigs lead to the next step of generating transcription maps of the genomic regions spanned by the respective YAC contigs. A comprehensive map of sequences encoding mRNA would be most useful for rapidly identifying human disease-causing genes. There are a number of approaches available for the identification of transcripts from the cloned genomic DNA material based on screening for sequences conserved between species,[4] identification of HTF islands,[5] probing the cDNA libraries with the genomic DNA in the form of lambda, cosmid, and YAC clones,[6] exon selection,[7,8] and direct cDNA selection approaches.[9,10] Alternatively genomic DNA is sequenced randomly and putative exons are identified by using computational methods that recognize exons.[11] It appears that no single method can identify all the coding sequences encoded by every genomic DNA fragment that is studied. However, cDNA selection has proven to be very sensitive and detects almost all cDNAs present in the accessible tissues.

In view of the abundance of literature on the methods of mapping the genome and identification of transcribed sequences, we have limited our discussion to a brief delineation of the work that has been published on these aspects and explain in more detail the work done in our laboratory in developing hybridization selection methods that facilitate genome mapping and positional cloning studies.

B. BACKGROUND

1. Microsatellite Markers

A major effort has been made to generate an extensive genetic map of man and mouse, using microsatellite repeats. In two noteworthy efforts, Weissenbach and collaborators at Genethon in France have published several hundred CA repeat markers[12] for humans and Lander and colleagues have used semiautomated approaches to prepare a large set of markers for the mouse.[13] In both cases the approach has been to screen short-fragment total genomic libraries with a probe for CA repeats and select and characterize individual clones. An improvement in the random selection approach has been made in which GA or CA repeat motif-containing fragments have been enriched so as to generate repeat-specific libraries.[14] The method used for GA repeats depended on formation of DNA structures that might not be directly generalizable to sequences containing both purines and pyrimidines on the same strand. In the case of CA repeats the approach was to use a single-stranded DNA vector containing deoxyuracil[15] and generate a second strand containing deoxythymidine, with DNA synthesis primed by a CA repeat oligonucleotide. After passage in the appropriate strain of bacteria, the parental vector is largely eliminated and only the DNA strands containing thymidine generate progeny. The latter approach generated a library in which 50% of the inserts contained CA repeats with an average length of 16.4 repeats.

The large-scale automated approaches have been quite successful at generating microsatellite maps defined genetically at a resolution of 1 to 2 cM for most of the genome. In the interim, large parts of the genome have been covered by contigs of megabase YACs, and it appears reasonable that in the future much of the mapping of microsatellites will initially be based on their physical location, with genetic studies following. For future work, particularly in the identification of genes associated with diseases or interesting phenotypes, a dense map of polymorphic markers — probably at the level of 1 marker every 100 kb or less — would be important. This is particularly so because methods to identify the functionally significant mutation in a large block of DNA are imperfect and considerably time consuming. This is reflected in the large fraction of cloned disease genes that were identified because of a chromosomal deletion or translocation, or because an obvious candidate gene in a region was known. Our ability to refine the limits of a suspect DNA region is limited by the availability either of chromosomal breaks or of informative markers.

Once a YAC contig covering a region of genetic interest has been established it may be desirable to identify a large fraction of microsatellite sequences within the contig so as to narrow the region of interest genetically. A common method of doing this is to prepare sublibraries from the YACs using either phage or cosmid vectors. The sublibraries may be screened with Alu sequences and with oligonucleotides corresponding to a microsatellite sequence. Clones that are derived from human DNA and contain microsatellite markers are then taken for further analysis. A variation of this approach is to prime the inserts in individual subclones with a CA oligonucleotide primer and determine the sequences adjacent to any CA microsatellite present in the insert. We have tried to simplify this approach by developing a selection method for identifying microsatellites in cloned DNA. To expedite the construction of a microsatellite map with coverage of all parts of the genome and at an average density of 1 marker per 100 kb (on the average, 0.1 cM) or less, we developed an alternative to the random clone approach.[16] We describe in detail the method and analysis of the selected microsatellite markers in the following sections.

2. Transcribed Sequences of Mapped Genomic Regions

Positional cloning efforts often result in the localization of the gene to a large segment of genomic DNA (500 kb to 1 Mb) that can not be further narrowed down due to the lack of informative markers or any known microdeletions or translocations in that region. Identification of all the transcribed sequences from that region is of utmost importance in identifying any abnormalities in their expression pattern and/or the mutations. Such identification is complicated by nonuniform distribution of genes along the length of the chromosomes. Some of the existing methods to identify transcripts from a genomic DNA region are described briefly in the following sections.

a. **Sequence-Based Approaches**

Sequencing of Genomic DNA

With the advent of automated sequencing machines, large regions (15,000 to 20,000 bases per day per machine) of cloned DNA can be sequenced effectively in a short time. Computational methods (GRAIL 1 and GRAIL 2) have been generated to identify putative exon sites from a given nucleotide sequence. Although the results are encouraging, some of the genes may not be identified due to limitations in the current computational data analysis for the recognition of exons and the experimental error in determining the sequences.

Random Sequencing of cDNAs

cDNA clones from a human brain library were randomly chosen and partially sequenced by Venter et al.[17] The sequences were deemed unique by comparison to the existing sequences in the nonredundant nucleotide databases at GenBank and EMBL. However, these sequences are not mapped on the genome and this approach does not lead to the transcriptional mapping of known genomic regions. In addition, the sequences are often in the range of 200 to 600 bases and do not represent the entire transcript. Thus several cDNAs may represent the same transcript.

CpG Islands

Many genes are known to have a CG-rich sequence towards their 5' ends. It is also known that the CG dinucleotide sequences are usually methylated on the genome except in the areas that contain genes.[5] Based on these observations, a method to selectively identify such regions by restriction digestion with the methylation-sensitive enzyme Hpa II (with the specificity CCGG) has been developed. Novel genes were successfully localized by sequencing the genomic DNA adjacent to Hpa II sites and the exonic regions were identified by consensus sequence homology of splice sites. However, the CpG islands are not uniformly distributed on the mammalian genome. Most of the CpG islands have been localized to early replicating R bands and fewer islands have been discovered on late replicating G bands.[18] In addition, not all genes have CpG islands at their 5' ends, and even when they exist they can be several thousand kilobases upstream of the coding sequences. However, a large number of genes (80%) mapped so far have been localized to the R bands — areas that are rich in CpG islands.[19]

a. **Region-Specific Transcript Identification Approaches**

In Vivo Transcription Methods

This approach, called exon trapping, relies on the eukaryotic transcriptional machinery that mediates the production of mRNA transcripts from larger transcripts of genomic DNA fragments cloned under the regulation of eukaryotic promoter elements. Two methods have been published: (1) to identify the internal exons,[7,8] and (2) to identify specifically the 3' ends of the transcriptional units.[20] The first method, described by Duyk et al.[7] and later substantially modified by Buckler et al.,[8] involves cloning of the genomic DNA into the multiple cloning site of the eukaryotic expression vector, followed by transfection into COS-7 cells. Cells are harvested after 48 h of transfection, then RNA is isolated and reverse transcribed using a vector primer. Subsequently, the insert-specific exons are amplified by the second vector primer and are cloned into one of a variety of vectors. The second method, recently reported by Krizman and Berget,[20] specifically amplifies only the 3' ends of genes by exon trapping. The method also involves cloning of genomic DNA inserts into an expression vector (pTAG4) and transient expression of the exons in COS-7 cells. The vector consists of a promoter and two viral exons followed by a cloning site for the genomic DNA; in the absence of genomic DNA the viral message is not polyadenylated. However, after transient expression, the viral RNA message consists of a polyadenylated 3' end derived from an exon that is transcribed from the genomic DNA. Novel 3' ends are detected by a combination of 3' RACE (rapid amplification of cDNA ends)[20] and RT-PCR approaches.[21]

Both of these elegant methods depend on the *in vivo* transcription of the cloned genomic DNA. Since the size of the genomic DNA insert is limiting, comprehensive exon trapping often requires multiple transfections and analyses of all the resulting clones. It is estimated that at least one third of all the exons of a genomic DNA segment can be detected by exon trapping. Although the "background" is reported to be relatively high, the 3' exon-trapping approach based on the positive selection is said to have approached a theoretical limit in trapping 3' exons from several cosmid clones.[20] The method is attractive in the sense that 3' ends of most of the genes have unique sequences and provide ideal probes to isolate corresponding full-length clones. However, the method may fail to identify 3' exons that are very large, either because large genomic regions are disrupted during cloning or because of PCR bias against large fragments.

In Vitro *Hybridization Methods*

Identification of cDNAs by screening libraries with a small (5 to 10 kb) genomic DNA probe is well tested and is a simplistic approach to identify the genes. In the recent years genomic DNA from lambda, cosmid, and YAC clones have been used to screen cDNA libraries and identify novel transcripts. However, genomic DNA contains repetitive elements and GC-rich sequences that can nonspecifically hybridize with the cDNA clones. Such nonspecific hybrids often are retained in spite of stringent hybridization conditions, due to the high thermal stability of the hybrids with GC-rich regions. In addition, large portions of genomic DNA may be intronic sequences and only a fraction (<5%) constitute exons. A probe made from such a genomic DNA fragment may represent largely intronic and intergenic sequences and only a small fraction of exonic sequences. Hence, the signal intensity from the specific hybridization to cDNA may not be sufficient to permit detection above the background levels.

We and others have developed a polymerase chain reaction-based hybridization selection method to identify cDNAs from genomic DNA cloned in YACs, P1 clones, and cosmids.[9,10] Our method employs short-fragment cDNAs that are amplified from a library as "input" and hybridize with an immobilized genomic DNA target on a nylon membrane. After stringent hybridization, specific cDNAs are recovered by PCR amplification. The process is repeated for a second time using the selected cDNAs as input in the second round of selection. The selected material is cloned in lambda gt10 vector and analyzed by sequencing to identify novel clones. A detailed description of the analysis is described elsewhere.[22,23] In an alternate method,[10] the YAC target is biotinylated and hybridized to the cDNAs and the specific cDNAs are retained on an avidin affinity matrix. After two rounds of selection, both methods have reported up to 10,000-fold amplification of the specific cDNAs. Although initial reports required purification of YAC DNA on pulse field gel electrophoresis, later modifications circumvented the need to purify the YAC DNA.[23,24] This procedure may yield from 1 to 3% of ribosomal clones, a very small fraction of repetitive clones, and almost no nonspecific single-copy sequences. A further simplification can be achieved by using uncloned selected material to screen oligo dT-primed cDNA libraries (Xu and Weissman, manuscript in preparation).

C. DESCRIPTION OF HYBRIDIZATION SELECTION / cDNA SELECTION

The hybridization selection method was initially designed to isolate novel cDNAs and was later modified successfully to identify novel microsatellite markers. In the following sections we describe various aspects of selection of microsatellite markers and transcribed sequences from genomic targets.

1. INPUT

a. For Microsatellite Selection

In order to isolate regions of genomic DNA containing microsatellite markers two different libraries were prepared: (1) a short-fragment genomic DNA library from randomly sheared DNA, and (2) a library of Mbo I-digested genomic DNA. A short-fragment genomic library was prepared by sonicating the genomic DNA to generate 300- to 800-bp-long fragments, ligating an EcoR I adapter, and cloning them in lambda gt10 vector. Alternately, genomic DNA was digested with Mbo I, and ligated to an adaptor with Mbo I-compatible ends and fractionated on the gel. This was used as such with no further cloning into a vector. Hybridization selection of microsatellite repeats from a specific target was achieved in two steps: (1) a genomic DNA library was prepared that was enriched for the microsatellite repeats, and (2) enriched sequences were hybridized to the YAC target under stringent hybridization conditions and specific sequences recovered by PCR amplification. Hence, the microsatellite-enriched genomic DNA serves as input in identifying region-specific microsatellite markers.

Microsatellite-Enriched Genomic DNA Libraries

Short-fragment genomic DNA library inserts or adapter-ligated Mbo I genomic DNA are amplified by primers from vector sequences flanking the EcoR I site of lambda gt10 and the adapter primers, respectively. Then 10 to 15 μg of the amplified DNA is denatured at 98° C and hybridized with 5 μg of biotinylated $(CA)_{22}$, $(CAG)_{15}$, or $(AGAT)_{11}$ primers in 500 μl of hybridization solution containing 0.5 M sodium phosphate buffer, pH 7.4 and 0.5% SDS at 50°C for 16 to 18 h. The hybridized DNA is diluted to 10 ml in 0.1 M Tris-Cl pH 7.5 and 0.15 M NaCl, and the biotinylated hybrids are selected by incubation with a Vectrex-avidin matrix column (Vector Laboratories) as described elsewhere.[25] The genomic DNA hybrids of $(CAG)_{15}$, $(AGAT)_{11}$, and $(CA)_n$ were eluted from the column hybrids with 15 mM NaCl at 50°C, 150 mM NaCl at 65°C, and with water at 65°C, respectively. A second round of affinity capture was performed to further enrich the genomic sequences with the above-mentioned microsatellite repeats. A library of these enriched sequences is constructed in the lambda gt10 vector system.

b. For cDNA Selection

We have constructed cDNA libraries by priming the poly A+ RNA with random hexanucleotides instead of an oligo dT primer. This has two advantages:

1. It is often observed that the 5' ends of messages longer than 5 kb are not represented in oligo dT-primed libraries due to inefficient first strand synthesis by MMLV reverse transcriptase. Random hexanucleotide priming generates cDNA messages from all along the length of the RNA message irrespective of its length.

2. The average length of such messages is between 400 to 2000 bp and is ideally suited for the PCR amplification purposes.

On the other hand, random hexanucleotides can prime 28S and 18S ribosomal RNA that constitutes 5 to 20% of the poly A+ RNA as a contaminant, and thus a significant number of clones in the cDNA library represent ribosomal DNA clones.

Identification of the unknown cDNAs from a given genome target largely relies on the quality of the cDNA library that is used in the selection. Hence it is critical to generate cDNA libraries with at least 3 to 4 million primary recombinant clones, if constructed from any normal tissue. In more complex tissues like brain and liver as many as 10 to 14 million plus may be needed to construct a cDNA library. A comprehensive cDNA library should consist of messages from all the tissues of the organism at all the developmental stages. Since it is impractical to generate such a library, a pool of cDNA libraries generated from several organs may be a viable alternative. In our studies we generated human cDNA libraries containing 5 to 12 million recombinant clones from total fetus (9 weeks), fetal brain (11 weeks), cerebral cortex, spleen, thymus, and testes. About 25 ng of phage DNA from each library is amplified using lambda gt10 vector primers flanking the cDNA insert.[9] Amplified DNA is pooled from all the six libraries and fractionated over 1% agarose gel. Inserts in the size range of 400 to 2000 bp are used in all the subsequent steps.

2. Target

a. For Microsatellite Selection

Several biotinylated oligonucleotides with di-, tri-, and tetranucleotide repeats were employed as targets for hybridization selection. Repeats included CA, CAG, and AGAT nucleotides. A detailed description of these oligonucleotides is given elsewhere.[16]

b. For cDNA Selection

Total yeast genomic DNA containing the YAC of interest is prepared as described by Parimoo et al.[23] and treated with RNase A and T1. Total degradation of RNA from the target DNA is essential to avoid hybridization of ribosomal DNA from the input cDNA library pool. About 200 ng/µl of target DNA is digested by Hind III, denatured by heating, chilled quickly over wet ice, and mixed with an equal volume of 10 × SSC. About 200 ng of DNA is spotted on a 4-mm² nylon membrane (Amersham) in 0.5-µl aliquots, air dried, and baked in a vacuum oven at 80°C for 2 h. These membranes can be stored at 4° pending further use.

3. Quenchers

Since the cDNA libraries contain ribosomal DNA, 3′ Alu sequences, and noncoding GC-rich sequences, it is essential to effectively quench their hybridization to the YAC target. A detailed description of the preparation of quenchers is given elsewhere.[22,23] Table 5-1 describes the nature of various quenchers and the concentrations employed in a typical cDNA selection reaction. In addition, poly d(IC) at 2.0 µg/100 µl is also used to quench the nonspecific GC interactions.

4. Experimental Details of Selection

The general principle of selection, whether to isolate microsatellite markers or cDNAs that are specific for a genomic DNA target, is described below. RNA-free and restriction

TABLE 5-1
Concentrations and the Nature of Various Quenchers Used in cDNA Selection

Quencher[a]	Source	Concentration in 100 μl of hybridization solution
pR 5.8	Human ribosomal	6.25 μg
pR 7.3	Human ribosomal	6.25 μg
pRibH7	Yeast ribosomal	5 μg
pRibH15	Yeast ribosomal	20 μg
pKK3535	*E. coli* ribosomal	4 μg
pL1a	Human LINE element	25 μg
Cot-1	Human Alu-like	2.5 μg

[a] Quenchers are prepared as described by Parimoo et al.[9]

enzyme-digested YAC DNA (200 ng) is immobilized on the nylon membrane and prehybridized at 65°C in 50 to 100 μl of prehybridization buffer containing 5 × SSC/0.5% SDS, and the quencher DNA as described above and in Table 5-1. In addition, selection experiments to isolate microsatellite repeats also included $(CA)_{15}$ nucleotide as a quencher at 0.1 μg/μl. After 24 h of prehybridization the nylon membranes are washed with 5 × SSPE containing 0.5% SDS and hybridized in a solution that is similar to the prehybridization solution, but in addition contains either 10 μg/ml of the enriched microsatellite library or the cDNA library pool. Hybridization is performed for an additional 36 h and the filters are washed with 2 × SSC/0.1% SDS at room temperature three times, 2 × SSC/0.1% SDS at 65°C three times for 20 min, 1 × SSC/0.1%SDS at 65°C once for 20 min, 0.2 × SSC/0.1% SDS at 65°C once for 20 min, 0.1 × SSC/0.1% SDS at 65°C twice for 20 min, and 0.1 × SSC at room temperature twice. The filters are then submerged in 40 μl of water and heated to 98°C for 5 min to release the hybridized sequences; 20 μl of this solution is used in the PCR reaction with appropriate primers. A detailed description of the primers for the selection of microsatellite repeats is given by Kandpal et al.[16] and for cDNA selection by Parimoo et al.[9] The amplified DNA (200 to 600 bp fraction) is used as input DNA in a second cycle of selection. The hybridized DNA at the end of two rounds of selection is recovered by PCR amplification, digested with EcoR I, and cloned into lambda gt10.

D. RESULTS AND DISCUSSION

1. SELECTION OF MICROSATELLITE MARKERS

a. Enrichment of CA Repeat Motifs

We have described in the preceding sections the experimental details for enrichment of CA repeat motifs from a short-fragment genomic DNA library or from Mbo I-digested adapter-ligated genomic DNA by carrying out solution hybridization between denatured genomic DNA and biotinylated $(CA)_n$ oligonucleotides. The biotinylated hybrid DNA was isolated by selective retention on Vectrex-avidin and specific elution from the matrix. The eluted DNA was amplified with the adaptor oligonucleotide, run on agarose gel alongside starting DNA, and probed with ^{32}P-phosphorylated $(CA)_{22}$ oligonucleotide to check the extent of enrichment. This analysis indicated that the affinity-captured DNA was considerably enriched for CA dinucleotide repeats. The PCR-amplified DNA was digested with EcoR I, purified on agarose gel, and ligated to EcoR I digested and dephosphorylated arms of λ gt10. A representative number of random plaques were amplified using vector sequences flanking the EcoR I site, run on gel, and probed with $(CA)_{22}$ oligonucleotide to ascertain the presence of (CA) dinucleotide repeats in the clones.

A significant enrichment of (CA) repeats was demonstrated in the affinity captured DNA by Southern blot analysis of the amplified clones. Of the 18 randomly chosen clones 17 hybridized to a (CA) oligonucleotide, showing that about 95% of the clones contained CA repeats.[16] Sequencing of these clones revealed that the average length of the CA stretch was >18 uninterrupted dinucleotide repeats.

a. **YAC-Specific Markers**

We have used the CA-enriched library described above for isolating YAC-specific CA repeats. We chose an X-chromosome-specific YAC from the factor IX locus, obtained from Dr. Schlessinger (Washington University, St. Louis). The YAC DNA was separated on PFGE and the DNA was electroeluted from the gel piece, digested with Hind III, treated with mung bean nuclease, and precipitated with 2 vol of ethanol in the presence of 0.3 M sodium acetate (pH 5.2). Then 10 ng of the purified YAC was spotted on Hybond (Amersham) membrane as described earlier. The abundant repeats and CA repeat sequences present in the YAC target were quenched by prehybridization. To select for CA repeat clones specific to the YAC sequences, hybridization was carried out with the inserts recovered from the CA-enriched library in the presence of quencher DNAs. Since the CA repeat sequences in the target DNA have been quenched with CA repeat oligonucleotide, the YAC-specific CA repeat clones present in the hybridization mixture will hybridize through the unique sequences flanking the CA repeats.

A representative number of clones from the selected library were PCR amplified, electrophoresed on an agarose gel, and blotted onto a nylon membrane. The membrane filter was probed with ^{32}P-labeled $(CA)_{22}$ oligonucleotide. More than 90% of the selected clones contained CA repeats as revealed by autoradiogram.[16] These results indicate that CA-enriched libraries are well suited for isolating YAC-specific CA clones. Subsequently, we chose ten clones randomly from the selected library for sequence analysis. Sequence data revealed the presence of two clones which recurred two and three times, respectively. PCR primers were synthesized from the sequences flanking the CA repeats in these two clones and used for amplifying corresponding CA repeats from human genomic DNA and the target YACs. The primer sequences appeared to be from repetitive DNA, resulting in the amplification of the DNA smears rather than discrete bands. We therefore resorted to Southern hybridization to ascertain the origin of the selected CA clones. The clones were digested with Mnl I, electrophoresed on agarose, blotted onto a nylon membrane, and probed with a CA oligonucleotide. A DNA fragment which was devoid of CA repeats after Mnl I digestion of the selected clone was used as a probe to hybridize to Southern blots containing DNA from target YACs, unrelated YAC, and human genomic DNA. No hybridization was observed with a YAC unrelated to the target YAC. We assessed the abundance of these clones in the selected library by screening approximately 500 selected clones with unique fragments obtained from the selected CA marker clones. The screening results indicated that these clones account for nearly 60% of the clones in the selected library. These results demonstrate the feasibility and sensitivity of the hybridization selection approach for isolating region-specific markers and indicate that a majority of the clones present in the selected library are specific to the target sequences.

We are currently extending these experiments to total yeast DNA. Successful use of total yeast DNA as target for isolating YAC-specific microsatellites will obviate the need to purify YAC DNA by pulse field gel electrophoresis. However, the yeast genome is also known to have CA dinucleotide repeats. It would therefore require experimentation to confirm that all the CA repeats are blocked in the prehybridization step. Hybridization selection for YAC-specific CA repeats may have two potential risks: (1) cross hybridization of CA sequences may result in the selection of clones that do not correspond to YAC DNA, and (2) the presence of Alu repeat elements flanking the CA repeats may interfere with the specificity of selection. However, our results indicate that these factors did not influence the selection. The success of hybridization selection approach for YAC-specific markers is primarily dependent on complete quenching of microsatellite repeat sequences in the target DNA. As is clear from the preceding discussion, incomplete quenching of repeat sequences would result in the selection of nonspecific clones by cross hybridization. However, the results presented in this article demonstrate that repeat sequences in the target can be efficiently quenched by using appropriate quencher DNAs and oligonucleotides. Given the ease of selection and parallel handling of multiple target YACs simultaneously, this approach may provide a rapid method for saturating any defined region of the genome with highly informative polymorphic markers.

2. SELECTION OF cDNAs

The cDNA selection is a simple but effective method to identify cDNA sequences encoded in genomic DNA fragments. Using this approach, we have performed selections on YAC targets from the MHC region on chromosome 6,[26,27] chromosome 21, and on Kit proto-oncogene region YACs from chromosome 4. The specificity of the selection is routinely validated by probing a Southern blot panel of Hind III- or EcoR I-digested total genomic DNA from (1) the target YAC, (2) corresponding contiguous YACs, (3) human genomic DNA, and (4) wild-type yeast genomic DNA with each of the novel cDNA clones. The specificity is confirmed by the appearance of cognate bands in the target YAC and human genomic DNA.

It is becoming increasingly clear that the human genome consists of areas that are gene dense and may code for a gene at a frequency of 1 or more in 20,000 bases. On the other hand, consecutive stretches of several kilobases of DNA may be transcriptionally silent. cDNA selection experiments in our laboratory clearly reflect this attribute of the genome in the sense that as many as 25 novel cDNAs are identified[26,27] from a YAC contig that spans about 600 kb in size from MHC on chromosome 6, suggesting a gene density of approximately 1 in every 20,000 bases. The frequency of nonspecific clones like ribosomal and Alu repeat clones is less than 5%. By comparing the frequency of some of the selected clones with their relative abundance in the input cDNA library, it is estimated that more than a 10,000-fold enrichment is achieved after two rounds of selection. However, in a few cases ribosomal sequences represent a large fraction of selected material. cDNA selections using a YAC that contained a hematopoietic stem cell-specific marker Kit resulted in a fraction that contained about 20% of Kit inserts and most of the remaining inserts were ribosomal DNA clones. This may be due to the presence of (1) YAC DNA in only a small population of the yeast colonies that are used to prepare total DNA, or (2) the presence of an aberrant ribosomal DNA gene on the YAC from the Kit region, or (3) because of very low levels of Kit cDNA clones in the input cDNA library.

As part of an effort to identify the expressed sequences on chromosome 21, a series of YACs spanning over 5 Mb of the dark band at chromosome 21q2.1 were used for cDNA selection. Selected libraries from all but one of these YACs showed only ribosomal sequences with an occasional repetitive sequence insert or an insert consisting of a multimer of one of the PCR primers used for selections. Repeat selections and sequencing of clones that did not hybridize to ribosomal DNA probes did not detect any unique sequences. To further investigate this, one of the YACs was subcloned as EcoR I fragments in lambda gt10. There were 14 different EcoR I fragments from the YAC identified and sequenced from both ends, giving 28 sequences. The resulting approximately 12,000 bases of sequences were examined by the program GRAIL and no part of the sequence gave even a minimal signal for the presence of an exon. Most, but not all, end sequences contained one or another form of repetitive DNA. A control YAC from a region of human MHC known to contain genes was also examined. In this case 3 of 16 sequences gave a strong signal for an exon and a 4th gave a weak signal. Overall, we concluded that this particular dark band from chromosome 21 contained very few if any expressed sequences, and that interstitial regions of chromosomes that extend over a number of millions of base pairs may essentially lack any genes.

Some of the problems associated with nonspecific recovery of ribosomal DNA clones may be circumvented by the use of oligo dT-primed cDNA as input instead of random hexanucleotide-primed cDNA. Since oligo dT-primed libraries are devoid of ribosomal DNA clones, cDNA selection using total yeast genomic DNA may result in a large fraction of the selected clones that are specific to YAC. However, it is well known that reverse transcriptases fail to synthesize complete cDNA from large mRNA (>5 kb) messages and hence the input is devoid of the 5' ends of such cDNAs. In addition, polymerase chain reactions using *Taq* DNA polymerase tend to preferentially amplify smaller fragments, which may result in little or no amplification of large cDNAs (>3 kb) after cDNA selection. This bias can possibly be circumvented by using novel polymerases (Pfu, KlenTaq, or rTth) and amplification of large fragments as described recently.[28]

E. CONCLUSION

In conclusion, cDNA selection has proven to be a very sensitive and exhaustive method for detecting the presence of transcribed sequences in YACs without requiring the separation of YAC from host yeast DNA. The major effort in identifying transcribed sequences has been to convert the short fragments into full-length cDNAs. In the near future, improved methods should also greatly accelerate this process, making construction of genome-wide expression maps a realistic goal for moderate-sized laboratories. The initial data on extension of selection for the isolation of microsatellite sequences looks promising and, potentially, selection could also be applied to any other type of library prepared from defined subsets of genomic DNA fragments.

ACKNOWLEDGMENTS

This work was supported in part by research funds of the American Otological Society (R.P.K.), National Institutes of Health Grant DC01682 (R.P.K.) and Outstanding Investigator Award of the National Cancer Institute, CA42556 (S.M.W.).

REFERENCES

1. White, R.L. and Lalouel, J.M. (1987), Investigation of genetic linkage in human families, *Advances in Human Genetics*, Vol. 16, Harrs, H. and Hirschhorn, K., Eds., Plenum Press, New York, pp. 121-228.
2. Barker, D., Schafer, M., and White, R.L. (1984), Restriction sites containing CpG show a higher frequency of polymorphism in human DNA, *Cell*, 36: 131-138.
3. Bellanne-Chantelot, C. et al. (1992), Mapping the whole human genome by finger printing yeast artificial chromosomes, *Cell*, 70: 1059-1068.
4. Monaco, A.P., Neve, R.L., Colletti-Feener, C., Bertelson, C.J., Kurnit, D.M., Kunkel, L.M. (1986), Isolation of candidate cDNAs for portions of the Duchenne muscular dystrophy gene, *Nature*, 323: 646-650.
5. Bird, A.P. (1986), CpG-rich islands and the function of DNA methylation, *Nature*, 321: 209-213.
6. Elvin, P., Slynn, G., Black, D., Graham, A., Riley. J., Anand, R., and Markham, A.F. (1990), Isolation of cDNA clones using yeast artificial chromosome probes, *Nucleic Acids Res.*, 18: 3913-3917.

7. Duyk, G.M., Kim, S., Myers, R.M., and Cox, D.R. (1990), Exon trapping: a genetic screen to identify candidate transcribed sequences in cloned mammalian genomic DNA, *Proc. Natl. Acad. Sci. U.S.A.*, 87: 8995-8999.
8. Buckler, A.J., Chang, D.D., Graw, S.L., Brook, J.D., Haber, D.A., Sharp, P.A., and Housman, D.E. (1991), Exon amplification: a strategy to isolate mammalian genes based on RNA splicing, *Proc. Natl. Acad. Sci. U.S.A.*, 88: 4005-4009.
9. Parimoo, S., Patanjali, S.R., Shukla, H., Chaplin, D.D., and Weissman, S.M. (1991), cDNA selection: efficient PCR approach for the selection of cDNAs encoded in large chromosomal DNA fragments, *Proc. Natl. Acad. Sci. U.S.A.*, 88: 9623-9627.
10. Lovett, M., Kere, J., and Hinton, L.M. (1991), Direct selection: a method for the isolation of cDNAs encoded by large genomic regions, *Proc. Natl. Acad. Sci. U.S.A.*, 88: 9628-9632.
11. Xu, Y., Mural, R.J., Shah, J.B., and Uberbacher, J.C. (1994), Recognizing exons in genomic sequences using GRAIL II, in *Genetic Engineering: Principles and Methods*, Vol. 15, Setlow, J., Ed., Plenum Press, New York.
12. Weissenbach, J., Gyapay, G., Dib, C., Vignal, A., Morissette, J., Millasseau, P., Vaysseix, G., and Lathrop, M. (1992), A second generation linkage map of the human genome, *Nature (London)*, 359: 794-801.
13. Arratia, R., Lander, E.S., Tavare, S., and Waterman, M.S. (1991), Genomic mapping by anchoring random clones: a mathematical analysis, *Genomics*, 11(4): 806-27.
14. Ito, T., Smith, C.L., and Cantor, C.R. (1992), Sequence-specific DNA purification by triplex affinity capture, *Proc. Natl. Acad. Sci. U.S.A.*, 89(2): 495-498.
15. Ostrander, E.O., Jong, P.M., Rine, J., and Duyk, G. (1992), Construction of small insert genomic DNA libraries highly enriched for microsatellite repeat sequences, *Proc. Natl. Acad. Sci. U.S.A.*, 89: 3419-3423.
16. Kandpal, R.P., Kandpal, G., and Weissman, S.M. (1994), Construction of libraries enriched for sequence repeats and jumping clones, and hybridization selection for region-specific markers, *Proc. Natl. Acad. Sci. U.S.A.*, 91: 88-92.
17. Adams, M.D., Kerlavage, A.R., Fields, C., and Venter, J.C. (1993), The 3,400 new expressed sequence tags identify diversity of transcripts in human brain, *Nat. Genet.*, 4(3): 256-67.
18. Craig, J. and Brickmore, W.A. (1994), The distribution of CpG islands in mammalian chromosomes. *Nat. Genet.*, 7: 376-382.
19. Sentis, C., Ludena, P., and Piqueras, F. (1993), Non uniform distribution of methylatable CCGG sequences on human chromosomes shown by in situ methylation. *Chromosoma*, 102: 267-271.
20. Krizman, D.B. and Berget, S.M. (1993), Efficient selection of 3'-terminal exons from vertebrate DNA, *Nucleic Acids Res.*, 21: 5198-5202.
21. Frohman, M.A., Dush, M.K., Martin, G.R. (1988), Rapid production of full-length cDNAs from rare transcripts: amplification using a single gene-specific oligonucleotide primer. *Proc. Natl. Acad. Sci. U.S.A.*, 85(23): 8998-9002.
22. Parimoo, S., Patanjali, S.R., and Weissman, S.M. (1993), Normalization and selection with short fragment cDNAs, in *Methods in Molecular Genetics*, Vol. I, Academic Press, New York, pp. 23-50.
23. Parimoo, S., Kolluri, R., and Weissman, S.M. (1993), cDNA selection from total yeast DNA containing YACs, *Nucleic Acids Res.*, 21: 4422-4423.
24. Patanjali, S.R., Xu, H., Parimoo, S., and Weissman, S.M. (1994), Locus specific identification of transcribed sequences using YACs and whole yeast genomic DNA, in *Identification of Transcribed Sequences*, Hochgeschwender, U. and Gardiner, K., Eds., Plenum Press, pp. 29-35.
25. Kandpal, R.P., Ward, D.C., and Weissman, S.M. (1992), Chromosome fishing: an affinity capture method for selective enrichment of large genomic DNA fragments, *Methods Enzymol.*, 216: 39-54.
26. Fan, W., Wei, H., Shukla, H., Parimoo, S., Patanjali, S.R., Li, Z., and Weissman, S.M. (1993), Application of cDNA selection techniques to regions of the human MHC, *Genomics*, 17: 575-581.
27. Wei, H., Fan, W., Xu., H., Parimoo, S., Shukla, H., Chaplin, D., and Weissman, S.M. (1993), Genes on one megabase of the HLA class I region, *Proc. Natl. Acad. Sci. U.S.A.*, 90: 11870-11874.
28. Barnes, W.M. (1994), PCR up to 35 kb DNA with high fidelity and high yield from lambda templates, *Proc. Natl. Acad. Sci. U.S.A.*, 91: 2216-2220.

Section III

Molecular Biology of Receptors and Associated Proteins

6 Molecular Biology of Opioid Receptors and Associated Proteins

Li-Na Wei and Horace H. Loh

CONTENTS

A. Introduction 77
B. Cloning of Opioid Receptors 78
C. Distribution of Opioid Receptors 79
    1. Distribution of δ Opioid Receptor (DOR) 79
    2. Distribution of μ Opioid Receptor (MOR) 79
    3. Distribution of κ Opioid Receptor (KOR) 81
    4. Distribution of Orphan Receptor X (X-OR) 81
D. Genomic Organization of Opioid Receptor Genes 81
    1. The MOR Gene 81
    2. The DOR Gene 82
    3. The KOR Gene 82
    4. The X-OR Gene 83
    5. The Opioid Receptor Gene Family 83
E. Opioid Receptor-Associated Proteins 83
    1. Cloning of Opioid Binding Cell Adhesion Molecules (OBCAM) 83
    2. The 25-kDa Protein 84
    3. The Opioid Downregulated Protein with a Putative Single Transmembrane Domain 85
    4. The Opioid Downregulated Zinc Finger Protein 85
F. Conclusions 85
Acknowledgments 86
References 86

A. INTRODUCTION

Opioid drugs comprise a vast series of compounds including naturally occurring alkaloids such as morphine, synthetic analogs such as heroin, and endogenous peptides such as β-endorphin and enkephalins. They are commonly used clinically as pain killers for postsurgical trauma and for terminal cancer patients. In addition, they have a wide range of other physiological effects including respiratory depression, inhibition of gastrointestinal motility, hypothermia, motor effects, alterations of feeding behavior, and suppression of some immune functions.

Like other drugs and biological agents, opioids act through specific receptors on target cells. Opioid receptors, in fact, were one of the first class of cell surface receptors to be identified, using *in vitro* ligand-binding assays with brain tissue. These receptors are heterogeneous, comprising several classes with selectivity toward particular types of ligands. The major three types of opioid receptors are μ, δ, and κ, defined by their selectivity towards morphine, enkephalins, and ketocyclazocine, respectively. For many years efforts to purify and clone them were unsuccessful, because of their sensitivity to detergents, their heterogeneity, and the lack of a simple biochemical assay for their functions.[1] This situation has been dramatically altered by the recent successful expression cloning of the δ opioid receptor from NG108-15 cells, reported at the end of 1992 by two independent laboratories, Evans et al.[2] and Kieffer et al.[3]

The importance of cloning the first cDNA encoding one of the opioid receptors can not be overestimated, as within months all three types of opioid receptors were cloned by various laboratories, all depending upon use of the first reported cDNA sequence in their cloning procedures such as homologous screening and polymerase chain reaction (PCR) cloning. Several orphan receptors that are highly homologous to the three opioid receptor sequences but do not bind opioids were also cloned. In addition, our knowledge about these receptors, in terms of their genes, anatomical distribution, and relationship with their potential peptide ligands, has advanced dramatically as a result of genomic DNA cloning and detection of expression. The genes for the three opioid receptors were first characterized completely for the mouse, and similar structural features have also been found in human and rat genes. The distribution of the three receptors has been extensively examined in both the mouse and the rat. The dissection of these molecules into functional domains has also begun to unveil the nature of these molecules. In this chapter, we first briefly review cloning of opioid receptors and associated proteins, then address what we know about the genes for these receptors and their distribution. Studies about the function of these molecules will not be reviewed here because results of these types of studies are not as conclusive.

B. CLONING OF OPIOID RECEPTORS

Evans et al.[2] and Kieffer et al.[3] used a similar expression cloning approach without purification of receptor proteins, preparing a cDNA library from NG108-15 cells, transfecting pools of the cDNA library into mammalian cells, then assaying the cells for binding of a radioactive opioid ligand. The cloned, expressed receptor showed typical binding properties expected of a δ opioid receptor including high affinity, stereoselectivity, and preference for a particular class of opioids. In addition, Evans et al.[2] demonstrated that binding to the expressed receptor inhibited adenylyl cyclase, as is the case with native δ opioid receptors on NG108-15 cells.

The availability of δ opioid receptor cDNA allowed the use of nucleic acid probes to screen cDNA libraries for receptors of high homology to the δ opioid receptors. Using this approach, Chen et al.[4] isolated from rat brain a cDNA that encoded a receptor selective for μ ligands such as DAMGO, which was negatively coupled to adenylyl cyclase. Yasuda et al.[5] isolated a clone from mouse brain that appeared to encode a receptor selective for κ opioids. Within a short period of time, cloning of opioid receptors from various animal species were reported, including the rat μ,[6] the rat δ,[7,8] the rat κ,[9-13] the human μ,[14] and the human δ[15] receptor cDNAs. In addition, several putative membrane receptors that are highly homologous to opioid receptors in sequence but do not bind opioids were cloned, including a human clone hORL1,[16] its rat homologue X-OR,[17-21] and mouse homologue MOR-C,[22] and later, a mouse clone isolated from lymphocyte mRNA.[23] The endogenous ligand for ORL1 was found only recently by two groups, both demonstrating a neuropeptide resembling dynorphin A as an agonist of this receptor.[24,25] However, no ligand specificity was assigned to the orphan clone isolated from lymphocytes. To account for various subtypes of each receptor demonstrated pharmacologically, many laboratories tried to look for a splice variant of each receptor, and only two laboratories reported a shorter isoform of the μ receptor, truncated at its C-terminus.[26,27]

Hydropathy analyses of the primary structure of opioid receptors suggested that these receptors belong to the family of G-protein-coupled membrane receptors containing seven transmembrane (TM) domains. Comparison of the amino acid sequences among the three types of opioid receptors and the orphan receptor revealed the most conserved sequence in TM domains 1 to 3. For the same type of receptor, sequence conservation across species is extremely high, especially in the seven TM domains. Figure 6-1 shows amino acid sequence comparison of the four mouse receptors aligned to the optimal homology in the seven TM domains.

C. DISTRIBUTION OF OPIOID RECEPTORS

The cloning of opioid receptors also allowed the detection of their expression with more sensitive molecular techniques such as *in situ* hybridization, immunohistochemistry with peptide antibodies, and reverse transcription-polymerase chain reaction (RT-PCR). By comparing the distribution of specific opioid ligand-binding sites and anatomical maps of opioid receptor mRNA and proteins, it is possible to address issues like transport and trafficking of opioid receptors and their pre- vs. postsynaptic localization. The mRNA expression pattern of three types of opioid receptors has been reviewed.[28] Furthermore, by double and triple labeling, one can also compare the distribution of opioid receptors and their potential endogenous peptide ligands as well as other neurotransmitters. This enables a potential relationship between opioid receptors and their endogenous peptides or neurotransmitters to be addressed. The most significant observation in these studies is the finding of presynaptic localization of the δ receptor and the mostly postsynaptic association of the μ and the κ receptors, as reviewed by Elde et al.[29]

1. Distribution of δ Opioid Receptor (DOR)

Upon cloning of the first opioid receptor δ, its distribution was first examined using *in situ* hybridization, receptor autoradiography, and immunohistochemistry, mainly by two groups.[30,31] Their results showed both common and different distribution patterns between the mRNA and protein of the δ opioid receptor. In the rat, DOR mRNA was detected in many areas of brain such as neocortex, striatum, olfactory tubercle, diagonal band of Broca, and the nucleus of the solitary tract.[31] In mouse forebrain, DOR mRNA expression was primarily detected in the striatum.[32] By immunohistochemistry, DOR protein was detected in several nuclei of the brain stem and spinal cord in the rat.[33] Most strikingly, immunoreactivity (ir) to the DOR protein was primarily in axonal structures. It was suggested that the δ receptor functions in a presynaptic manner and probably modulated afferent input to both serotoninergic and noradrenergic neurons.

2. Distribution of μ Opioid Receptor (MOR)

The expression of MOR mRNA in rat brain was examined by several groups,[34-36] where results in agreement and discrepancies were reported. Agreed among all these studies, brain areas expressing MOR mRNA include the thalamus, striatum, and olfactory bulb. In immunohistochemical studies, ir to MOR protein was found to be mostly associated with neuronal plasma membranes of dendrites and cell bodies, suggesting a major postsynaptic

```
μ         1 MDSSAGPGNI SDCSDPLAPA SCSPAPGSWL NLSHVDGNQS DPCGPNRTGL
δ                      MELVPSARAE LQSSPLVNLS DAFPSAFPSA
κ            M ESPIQIFRGD PGPTCSPSAC LLPNSSSWFP NWAESDSNGS
orphan receptor X       MESLFPAPFW EVLYGSHFQG NLSLLNETVP

                                   ----------- TM1 -------------
       51 GGSHSLCPQT GSPSMVTAIT IMALYSIVCV VGLFGNFLVM YVIVRYTKMK
          ANASGSPGAR SASSLALAIA ITALYSAVCA VGLLGNVLVM FGIVRYTKLK
          VGSEDQQLES AHISPAIPVI ITAVYSVVFV VGLVGNSLVM FVIIRYTKMK
          HHLLLNASHS AFLPLGLKVT IVGLYLAVCI GGLLGNCLVM YVILRHTKMK

            --------- TM2 ----------                    ----------
      101 TATNIYIFNL ALADALATST LPFQSVNYLM GTWPFGNILC KIVISIDYYN
          TATNIYIFNL ALADALATST LPFQSAKYLM ETWPFGELLC KAVLSIDYYN
          TATNIYIFNL ALADALVTTT MPFQSAVYLM NSWPFGDVLC KIVISIDYYN
          TATNIYIFNL ALADTLVLLT LPFQGTDILL GFWPFGNALC KTVIAIDYYN

          -- TM3 ---------                    ----------- TM4 --
      151 MFTSIFTLCT MSVDRYIAVC HPVKALDFRT PRNAKIVNVC NWILSSAIGL
          MFTSIFTLTM MSVDRYIAVC HPVKALDFRT PAKAKLINIC IWVLASGVGV
          MFTSIFTLTM MSVDRYIAVC HPVKALDFRT PLKAKIINIC IWLLASSVGI
          MFTSTFTLTA MSVDRYVAIC HPIRALDVRT SSKAQAVNVA IWALASVVGV

            -------                        ----------- TM5 ---
      201 .PVMFMATTKY RQGSIDCTLTFSHP...TWYWENLLKICVFIFA FIMPVLIITV
          VPIMVMAVTQP RDGAVVCMLQFPSP....WYWDTVTKICVFLFA FVVPILIITV
          .SAIVLGGTKV REDVDVIECSLQFPDDEYSWWDLFMKICVFVFA FVIPVLIIIV
          .PVAIMGSAQV EDEEIECLVEIPAPQD...YWGPVFAICIFLFS FIIPVLIISV

            --------                        ---------- TM6 ---------
      251 CYGLMILRLK SVRMLSGSKE KDRNLPRITR MVLVVVAVFI VCWTPIHIYV
          CYGLMLLRLR SVRLLSGSKE KDRSLRRITR MVLVVVGAFV VCWAPIHIFV
          CYTLMILRLK SVRLLSGSRE KDRNLRRITK LVLVVVAVFI ICWTPIHIFI
          CYSLMIRRLR GVRLLSGSRE KDRNLRRITR LVLVVVAVFV GCWTPVQVFV

            --           ------------ TM7 ----------
      301 IIKALITIPE TTFQTVSW.H FCIALGYTNSC LNPVLYAFLD ENFKRCFREF
          IVWTLVDINR RDPLVVAALH LCIALGYANSS LNPVLYAFLD ENFKRCFRQL
          LVEALGSTSH STAALSSY.Y FCIALGYTNSS LNPVLYAFLD ENFKRCFRDF
          LVQGLGVQPG SETAVAIL.R FCTALGYVNSC LNPILYAFLD ENFKACFRKF

      351 CIPTSSTIEQ QNSARIRQNT REHPSTANTV DRTNHQLENL EAETAPLP
          CRTPCGRQEP GSLRRPRQAT TRERVTACTP SDGPGGGRAA
          CFPIKMRMER QSTNRVRNTV EDPASMRDVG GMNKPV
          CCASALHREM QVSDRVRSIA KDVGLGCKTS ETVPRPA
```

FIGURE 6-1 Amino acid sequence alignment of mouse μ, δ, and κ opioid receptors as well the orphan receptor. The amino acid sequence of each opioid receptor was aligned to its optimal homology in the seven TM domains labeled with TM1-TM7 above the boldface sequences. The position of each intron is underlined at the amino acid residue of each sequence where splicing occurs. Numbering on the left of the sequence indicates the amino acid position of the mouse μ opioid receptor.

function. However, discrete populations of neurons were shown to target MOR to their axons, suggesting a presynaptic role for MOR in certain neurons. Brain areas with positive MOR ir include the cerebral cortex, striatum, hippocampus, locus ceruleus, and the superficial laminae of the dorsal horn. In addition, areas with MOR ir were frequently complementary to areas positive for enkephalin ir, suggesting enkephalins to be physiologically relevant ligands for the μ receptor.[37]

3. DISTRIBUTION OF κ OPIOID RECEPTOR (KOR)

The mRNA expression pattern of KOR was also reported by several groups.[38,39] Common positive areas shown in these studies include the claustrum, striatum, olfactory tubercle, several thalamic and hypothalamic nuclei, and the locus ceruleus. In immunohistochemical studies,[40] κ receptor ir was also seen in both axonal and somatodendritic compartments, but the majority of staining was seen in the somatodendritic compartment, implying a major postsynaptic function. Many brain areas including the ventral forebrain, hypothalamus, thalamus, posterior pituitary, and midbrain appeared to be positive for KOR ir. By comparing the ir patterns of κ receptor and endogenous opioids, it was suggested κ receptor was primarily deployed to postsynaptic membranes where it probably mediates the effects of dynorphin and enkephalin.

4. DISTRIBUTION OF ORPHAN RECEPTOR X (X-OR)

The expression pattern of the X-OR gene was examined, primarily using *in situ* hybridization, by several groups.[17,18,20,21] It appeared that the X-OR mRNA was present widely in brain and spinal cord and its expression pattern was very different from that of MOR, DOR, and KOR. No immunohistochemical studies have been reported for this receptor yet.

Results from these studies presented both common and different expression patterns for each type of opioid receptor, possibly due to variation in the specificity of probes and antibodies as well as experimental conditions. However, it is generally agreed that the four receptors are differentially expressed in the CNS and variation in their expression may also exist in different animal species. More precise information about opioid receptor expression and their physiological relevance to opioid peptides may be obtained from genetic studies such as transgenesis and gene-targeting in the future.

D. GENOMIC ORGANIZATION OF OPIOID RECEPTOR GENES

Upon cloning of cDNAs for the three types of opioid receptors and the orphan receptor, studies of their genomic structures and chromosome locations were quickly conducted. It appeared that the four receptor cDNAs were encoded by four distinct genes at four different chromosomal loci, and each gene was shown to be a single-copy gene. In consistence with their highly conserved amino acid sequences, the genomic structures of the four genes are also very similar and appear to be highly conserved during evolution. The complete genomic structures for the four receptor genes were first determined for the mouse,[9,41-43] which has provided a prototype for this gene family and has allowed subsequent determination of the gene structures of opioid receptors in other animal species in a short time.

1. THE MOR GENE

The murine MOR gene was the first to be characterized.[41] This gene was found to span a distance greater than 53 kb divided by 3 introns. The first exon encodes the N-terminal extracellular domain and TM1 domain; exon 2 covers TM2, TM3, and TM4; exon 3 contains TM5, TM6, TM7, and a major portion of the C-terminal intracellular domain; and the 4th exon encodes only the last 12 amino acid residues, which appear to be unique to MOR as compared to DOR and KOR. Multiple transcription initiation sites were located at approximately 268 to 291 bp upstream of the ATG codon, and no TATA box was found in the highly G/C-rich promoter region. Several putative binding sites for transcription factors were identified by comparing to the Transcription Factors Database, including Sp1, AP1, AP2, GRE, NF-IL6, and CRE within a 1.5-kb upstream region. In a later study,[44] a more upstream promoter located at approximately 800 bp upstream of the ATG codon was reported. The rat MOR gene promoter was also reported, revealing very similar features as observed in the mouse gene.[45] The organization of exons of the human MOR gene appears to be identical to the mouse gene, but no promoter sequence was reported.[46]

Using RT-PCR, one variant was discovered for both the human and the rat MOR, named MOR1A[26] and rMOR1B[27], respectively. Both variants were truncated at the C-terminus of the wild-type MOR at a very similar position. It appeared that the human MOR1A was a result of failure to splice the 3rd intron, thus generating an early termination of translation with a substitution of 4 different amino acid residues for the last 12 amino acid residues in the wild-type MOR. However, this human MOR variant displayed similar ligand specificity and negative coupling to cAMP pathway as the wild-type MOR.[26] The rat isoform differed in agonist-induced desensitization from the wild-type MOR; however, it is not clear how the rat variant is generated from the gene.[27] Among the opioid receptor gene family, this is the only known splice variant with an altered amino acid sequence as compared to its wild-type protein.

2. THE DOR GENE

The intron position of the DOR gene was first determined for the human gene, where two introns were identified in its coding region.[47] The entire genomic structure, including the promoter region, was first completed for the mouse gene.[42] Intron 1 is located at the identical position found in the MOR gene, where an Arg is present, and intron 2 is located at a position very close to intron 2 of the MOR gene. The size of the mouse DOR gene is approximately 32 kb in length and the promoter is also TATA-less, GC rich, and with multiple transcription initiation sites. In addition, several transcription factor binding sites are found within a 1.3-kb upstream region, which appears to be very similar to the panel found in the MOR promoter region.[41,44,45]

3. THE KOR GENE

The intron position of the KOR coding region was first determined for the mouse gene,[22] followed by examination of the entire mouse KOR gene.[43] In the study showing the entire mouse KOR gene, an intron was found in the 5'-untranslated region, which was also observed in the rat KOR gene.[48] It appears that KOR translation starts at an ATG codon in the second exon. The second intron is located to the same Arg residue used for the MOR and the DOR genes where intron 1 is present, and the third intron is also located to a position very close to the second intron of the MOR and the DOR genes. The entire mouse KOR gene spans a distance of approximately 16 kb, and its promoter shares similar features to those observed in the MOR and DOR genes, including TATA-less, multiple transcription initiation sites, and various putative binding sites for transcription factors. A transcription initiation site in the first intron was reported for the rat KOR gene, which resulted in a KOR variant with a possible novel amino acid sequence at its amino terminus.[48] However, the presence of this KOR variant has not been substantiated, and no similar alternative transcript could be detected in mRNA isolated from either mouse brain or a mouse lymphoma cell line R1.1,[49] which has been shown to express a high level of KOR transcript[43] and KOR binding activities.[50] The human KOR gene was found to have a very similar exon/intron organization as used in the mouse and the rat KOR genes; however, its promoter region was not determined.[47]

4. THE X-OR GENE

Only the mouse gene for the orphan receptor identified from various animal species has been briefly determined.[22] The arrangement of its exons appears to be very similar to those of the three opioid receptor genes, with coding region spliced into three exons at very similar locations. More strikingly, the first intron was located at the same Arg residue used in the three opioid receptor genes. However, the promoter region of this gene was not examined.

5. THE OPIOID RECEPTOR GENE FAMILY

In studies of the genomic organization of opioid receptors, all four genes were best characterized in the mouse. Figure 6-2 shows the comparison of the four mouse genes aligned to the three exons (I, II, and III for MOR, DOR, and X-OR, and II, III, and IV for KOR) which contain the major coding sequence of each receptor. It appears that the four genes were evolved from an ancestral gene, being highly conserved in the arrangement of exons, coding sequence, and the promoter region (with the exception of the X-OR which has not been determined). The most striking conservation is the use of the same Arg residue dividing exon I (covering the N-terminal extracellular domain and TM1) and exon II (covering the first intracellular loop, TM2, TM3, and TM4, see Figure 6-1). TM 5, TM6, TM7, and the C-terminus are encoded in exon III, except that a 12-amino-acid sequence in the MOR C-terminus is encoded in an additional exon not found in DOR or KOR. The KOR gene, however, contains an extra exon in the 5'-untranslated region. These variations probably evolved later for functions such as regulation of gene activities (such as the KOR gene) and generation of alternative forms (such as the MOR gene). In the 5'-regulatory region, most striking features include TATA-less, multiple initiation and transcription factor binding sites. However, the biological relevance of these regulatory sequences awaits further studies. For the human genes, exon/intron junctions were found to be identical to those used in the mouse genes. Gene structures of opioid receptors in other animal species are expected to be very similar, based upon limited information available from studies of the rat genes.

E. OPIOID RECEPTOR-ASSOCIATED PROTEINS

1. Cloning of Opioid Binding Cell Adhesion Molecules (OBCAM)

In our initial attempt to purify opioid binding components from animal tissues, we were able to isolate from bovine brain a 58-kDa protein, selective for opioid alkaloid ligands, using a combination of affinity chromatography, lectin chromatography, and gel filtration.[51] This protein required acidic lipids possessing unsaturated fatty acids in order to manifest binding activity; neither the protein nor the lipids alone possessed significant opioid binding.[52] Using oligonucleotide probes deduced from partial amino acid sequence of the purified protein to screen a bovine brain cDNA library, we identified a unique cDNA clone with homology to the immunoglobulin superfamily.[53] This superfamily constitutes a group of proteins characterized by repeating domains flanked by cysteine residues. The sequence of the bovine cDNA was most homologous to a group of cell adhesion molecules, including the neural cell adhesion molecule (N-CAM), myelin-associated glycoprotein (MAG), and the invertebrate proteins amalgam, neuroglian, and fasciclin II. Accordingly, the protein was named OBCAM, for opioid binding cell adhesion molecules.

The physiological role of OBCAM is not completely clear at the present time. However, several lines of indirect evidence indicate that it is involved in opioid receptor function. For

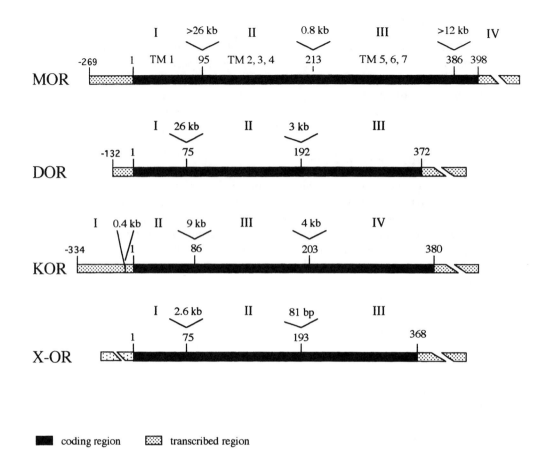

FIGURE 2 Genomic alignment of the mouse MOR, DOR, KOR, and X-OR genes. The transcribed region of each gene was aligned to the initiation codon (number 1 above each gene). Exons are numbered with I to IV above each gene. Introns are indicated at the amino acid residues where splicing events occur and the approximate size of each intron is indicated. In the coding region, numbers indicate the amino acid position. Whereas in the 5′-untranslated region, numbers indicate the nucleotide position relative to the translation initiation codon, representing the 5′-end of each transcribed region. The 5′-end of the orphan receptor gene X-OR and the 3′-tail of each transcribed region have not been reported, thus are presented with broken bars.

example, antibodies to either OBCAM or to peptides representing portions of its predicted amino acid sequence block opioid binding. One of the OBCAM peptide antibodies reacts with a component on the surface of NG108-15 cells and is downregulated by chronic treatment of the cells with opioid agonist in a fashion parallel to that of opioid receptors in these cells. In addition, stable transfection of NG108-15 cells with OBCAM antisense cDNA greatly reduces opioid binding in these cells, alters the coupling of opioid receptors with G-proteins, and consequently blocks the signal transduction of opioid receptor to one of its effector systems, the adenylyl cyclase, as reviewed by Smith et al.[54]

2. THE 25-kDa PROTEIN

Using different cross-linking agents to cross-link β-^{125}I-endorphin to the NG108-15 cell membrane receptor, we have also identified another protein associated with the opioid receptor.[55] Several bands with molecular masses of 55, 35, and 25 kDa were labeled initially. With the use of several criteria to evaluate the relevance of these cross-linked bands to the δ opioid receptor, including selectivity, stereospecificity, affinity, G-protein coupling, downregulation, and correlation with the opioid receptor level in different well-characterized cell lines, only the 25-kDa protein fulfilled all these criteria. The functional role of this 25-kDa protein is currently being investigated.

3. THE OPIOID DOWNREGULATED PROTEIN WITH A PUTATIVE SINGLE TRANSMEMBRANE DOMAIN

Using a subtractive hybridization strategy, we identified two closely related cDNAs that are greatly reduced in NG108-15 cells chronically treated with opioid agonists and which are therefore candidates for coding sequences involved in opioid receptor functions. The two cDNAs are nearly identical in their coding sequences, differing mainly in the 3'-untranslated region. These two clones code for a predicted amino acid sequence that has no significant homology with any protein in the database. An hydropathy plot suggests the existence of a single potential transmembrane region, consistent with the possibility that this protein is a cell surface protein associated with opioid binding.[56]

4. THE OPIOID DOWNREGULATED ZINC FINGER PROTEIN

Using a subtractive hybridization strategy, we also isolated another clone downregulated by chronic opioid treatment of NG108-15 cells, with homology to the zinc finger proteins. This clone codes for a protein of 547 amino acids, containing 14 zinc finger domains,[57] and is possibly involved in cell differentiation. The exact function of this protein is not clear.

F. CONCLUSIONS

Opioid receptors were among the first to be identified by *in vitro* ligand-binding techniques and have since been extensively characterized pharmacologically and biochemically. However, full characterization has lagged behind that of other cell surface receptors due to great difficulty in isolating these proteins. Gene cloning allows the isolation of the DNA fragment encoding the protein of interest without purification of the complete protein molecule and expression cloning allows the isolation of a functional DNA for the protein of interest without any purification procedure. Using an expression cloning strategy, two groups independently isolated the full-length cDNA for the δ opioid receptor, which has changed the progress of research in this field completely. Within two years, all the three opioid receptors were characterized at a molecular level, and an orphan receptor with high sequence homology was identified. To date, all four receptors have been pharmacologically confirmed.

Using various molecular approaches, their expression patterns and gene structures have been thoroughly examined. In multiple-labeling immunohistochemical studies, DOR was localized in a presynaptic fashion, whereas MOR and KOR were localized primarily in postsynaptic compartments, suggesting their physiological and pharmacological roles in the central and peripheral nervous systems. Based upon their sequence homology and the similarity in genomic organization, these four genes appear to have evolved from one ancestral gene during evolution. With this basic genetic information about opioid receptors, it is now possible to conduct molecular genetic studies to understand physiological functions of these receptors, and to address mechanisms underlying many pharmacological problems related to opioids such as tolerance, dependence, and addiction. Molecular dissection to reveal the function of each domain in their effector systems and their ligand-binding specificity is being vigorously conducted in many labs, including ours. It is anticipated that much more information about structure-function relationship of opioid receptors will become available in the very near future. This will allow novel specific analgesics to be developed, and pharmacological application of opioids will be greatly enhanced.

ACKNOWLEDGMENTS

This research was supported in part by NIH grants DA00564, DA01583, DA05695, K05-DA70554, and DK46866, and by the F. & A. Stark Fund of the Minnesota Medical Foundation.

REFERENCES

1. Loh, H. H. and Smith, A. P., Molecular characterization of opioid receptors, *Annu. Rev. Pharmacol.*, 30, 123, 1990.
2. Evans, C. J., Keith, D. E., Jr., Morris, H., Magendzo, K., and Edwards, R. H., Cloning of a delta opioid receptor by functional expression, *Science*, 258, 1952, 1992.
3. Kieffer, B. L., Befort, K., Gaveriaux-Ruff, C., and Hirth, C. G., The δ-opioid receptor: isolation of a cDNA by expression cloning and pharmacological characterization, *Proc. Natl. Acad. Sci. U.S.A.*, 89, 12048, 1992.
4. Chen, Y., Mestek, A., Liu, J., Hurley, J. A., and Yu, L., Molecular cloning and functional expression of a μ-opioid receptor from rat brain, *Mol. Pharmacol.*, 44, 8, 1993.
5. Yasuda, K., Raynor, K., Kong, H., Breder, C. D., Takeda, J., Reisine, T., and Bell, G. I., Cloning and functional comparison of κ and δ opioid receptors from mouse brain, *Proc. Natl. Acad. Sci. U.S.A.*, 90, 6736, 1993.
6. Thompson, R. C., Mansour, A., Akil, H., and Watson, S. J., Cloning and pharmacological characterization of a rat μ opioid receptor, *Neuron*, 11, 903, 1993.
7. Kato, S., Mori, K., Nishi, M., and Takeshima, H., Primary structures and expression from cDNAs of rat opioid receptor delta- and mu-subtypes, *FEBS Lett.*, 327, 311, 1993.
8. Abood, M. E., Noel, M. A., Farnsworth, J. S., and Tao, Q., Molecular cloning and expression of a δ-opioid receptor from rat brain, *J. Neurosci. Res.*, 37, 714, 1994.
9. Nishi, M., Takeshima, H., Fukuda, K., Kato, S., and Mori, K., cDNA cloning and pharmacological characterization of an opioid receptor with high affinities for κ subtype-selective ligands, *FEBS Lett.*, 330, 77, 1993.
10. Meng, F., Xie, G., Thompson, R. C., Mansour, A., Goldstein, A., Watson, S. J., and Akil, H., Cloning and pharmacological characterization of a rat κ opioid receptor, *Proc. Natl. Acad. Sci. U.S.A.*, 90, 9954, 1993.
11. Chen, Y., Mestek, A., Liu, J., and Yu, L., Molecular cloning of a rat kappa opioid receptor reveals sequence similarities to the mu and delta opioid receptors, *Biochem. J.*, 295, 625, 1993.
12. Li, S., Zhu, J., Chen, C., Chen, Y.-W., Dericl, J. K., Ashby, D., and Liu Chen, L. Y., Molecular cloning and expression of a rat κ opioid receptor, *Biochem. J.*, 296, 629, 1993.
13. Minami, M., Toya, T., Katao, Y., Maekawa, K., Nakamura, S., Onogi, T., Kaneko, S., and Satoh, M., Cloning and expression of a cDNA for the rat κ-opioid, receptor. *FEBS Lett.*, 329, 291, 1993.
14. Raynor, K., Kong, H., Mestek, A., Bye, L.S., Tian, M., Liu, J., Yu, L., and Reisine, T., Characterization of the cloned human mu opioid receptor, *J. Pharmacol. Exp. Ther.*, 272, 423, 1995.
15. Knapp, R. J., Malatynska, E., Fang, L., Li, X., Babin, E., Nguyen, M., Santoro, G., Varga, E. V., Hruby, V. J., Roeske, W. R., and Yamamura, H. I., Identification of a human delta opioid receptor: cloning and expression, *Life Sci.*, 54, PL463, 1994.
16. Mollereau, C. Parmentier, M., Mailleux, P., Butour, J. L., Moisand, C., Chalon, P., Caput, D., Vassart, G., and Meunier, J. C., ORL1, a novel member of the opioid receptor family: cloning, functional expression and localization. *FEBS Lett.*, 341, 33, 1994.
17. Bunzow, J. R., Saez, C., Mortrud, M., Bouvier, C. Williams, J. T., Low, M., and Grandy, D. K., Molecular cloning and tissue distribution of a putative member of the rat opioid receptor gene family that is not a μ, δ, or κ opioid receptor type, *FEBS Lett.*, 347, 284, 1994.
18. Chen, Y., Fan, Y., Liu, J., Mestek, A., Tian, M., Kozak, C. A., and Yu, L., Molecular cloning, tissue distribution and chromosomal localization of a novel member of the opioid receptor gene family, *FEBS Lett.*, 347, 279, 1994.
19. Fukuda, K., Kato, S., Mori, K., Nishi, M., Takeshima, H., Iwabe, N., Miyata, T., Houtani, T., and Sugimoto, T., cDNA cloning and regional distribution of a novel member of the opioid receptor family, *FEBS Lett.*, 343, 42, 1994.
20. Wang, J. B., Johnson, P. S., Imai, Y., Persico, A. M., Ozenherger, B. A., Eppler, C. M., and Uhl, G. R., cDNA cloning of an orphan opioiate receptor gene family member and its splice variant, *FEBS Lett.*, 348, 75, 1994.

21. Wick, M. J., Minnerath, S. R., Lin, X., Elde, R. Law, P. Y., and Loh, H. H., Isolation of a novel cDNA encoding a putative membrane receptor with high homology to the cloned µ, δ, and κ opioid receptor, *Mol. Brain Res.*, 27, 37, 1994.
22. Nishi, M., Takeshima, H., Mori, M., Nakagawara, K., and Takeuchi, T., Structure and chromosomal mapping of genes for the mouse κ-opioid receptor and an opioid receptor homologue (MOR-C), *Biochem. Biophys. Res. Commun.*, 205, 1353, 1994.
23. Halford, W. P., Gebhardt, B. M., and Carr, D. J., Functional role and sequence analysis of a lymphocyte orphan opioid receptor, *J. Neuroimmunol.*, 59, 91, 1995.
24. Meunier, J.-C., Mollereau, C., Toll, L., Suaudeau, C., Moisand, C., Alvinerie, P., Butour, J.-L., Guillemot, J.-C., Ferrara, P., Monsarrat, B., Mazarguil, H., Vassart, G., Parmentier, M., and Costentin, J., Isolation and structure of the endogenous agonist of opioid receptor-like ORL$_1$ receptor, *Nature*, 377, 532, 1995.
25. Reinscheid, R., Nothacker, H.-P., Bourson, A., Ardati, A., Henningsen, R. A., Bunzow, J. R., Grandy, D. K., Langen, H., Monsma, F. J., and Civelli, O., Orphan FQ: a neuropeptide that activates an opioidlike G protein-coupled receptor, *Science*, 270, 792, 1995.
26. Bare, L. A., Mansson, E., and Yang, D., Expression of two variants of the human µ opioid receptor mRNA in SK-N-SH cells and human brain, *FEBS Lett.*, 354, 213, 1994.
27. Zimprich, A., Simon, T., and Hollt, V., Cloning and expression of an isoform of the rat µ opioid receptor (rMOR1B) which differs in agonist induced desensitization from rMOR1, *FEBS Lett.*, 359, 142, 1995.
28. Mansour, A., Fox, C. A., Akil, H., and Watson, S. J., Opioid-receptor mRNA expression in the rat CNS: anatomical and functional implications, *Trends Neurosci.*, 18, 22, 1995.
29. Elde, R., Arvidsson, U., Riedl, M., Vulchanova, L., Lee, J. H., Dado, R., Nakano, A., Chakrabarti, S., Zhang, X., Loh, H. H., et al., Distribution of neuropeptide receptors. New views of peptidergic neurotransmission made possible by antibodies to opioid receptors, *Ann. N. Y. Acad. Sci.*, 757, 390, 1995.
30. Dado, R. J., Law, P. Y., Loh, H. H., and Elde, R., Immunofluorescent identification of a delta-opioid receptor on primary afferent nerve terminals, *Neuroreport*, 5, 341, 1993.
31. Mansour, A. Thompson, R. C., Akil, H., and Watson, S. J., Delta opioid receptor mRNA distribution in the brain: comparison to delta receptor binding and proenkephalin mRNA, *J. Chem. Neuroanat.*, 6, 351, 1993.
32. Le Moine, C., Kieffer, B., Gaveriaux-Ruff, C., Befort, K., and Bloch, B., Delta-opioid receptor gene expression in the mouse forebrain: localization in cholinergic neurons of the striatum, *Neurosci.*, 62, 635, 1994.
33. Arvidsson, U., Dado, R. J., Riedl, M., Lee, J.-H., Law, P. Y., Loh, H. H., Elde, R., and Wessendorf, M. W., δ-Opioid receptor immunoreactivity: distribution in brainstem and spinal cord, and relationship to biogenic amines and enkephalin, *J. Neurosci.*, 15, 1215, 1995.
34. Minami, M., Onogi, T., Toya, T., Katao, Y., Hosoi, Y., Maekawa, K., Katsumata, S., Yabuuchi, K., and Satoh, M., Molecular cloning and in situ hybridization histochemistry for rat mu-opioid receptor, *Neurosci. Res.*, 18, 315, 1994.
35. Zastawny, R. L., George, S. R., Cheng, R., Tsatsos, J., Briones-Urbina, R., and O'Dowd, B. F., Cloning, characterization and distribution of a mu-opioid receptor in rat brain, *J. Neurochem.*, 62, 2099, 1994.
36. Delfs, J. M., Kong, H., Mestek, A., Chen, Y., Yu, l., Reisine, T., and Chesselet, M. F., Expression of mu opioid receptor mRNA in rat brain: an in situ hybridization study at the single cell level, *J. Comp. Neurol.*, 345, 46, 1994.
37. Arvidsson, U., Riedl, M., Chakrabarti, S., Lee, J. H., Nakano, A. H., Dado, R. J., Loh, H. H., Law, P. Y., Wessendorf, M. W., and Elde, R., Distribution and targeting of a µ opioid receptor (MOR1) in brain and spinal cord, *J. Neurosci.*, 15, 3328, 1995.
38. Mansour, A., Fox, C. A., Meng, F., Akil, H. and Watson, S. J., κ1 Receptor mRNA distribution in the rat CNS: comparison to κ receptor binding and prodynorphin mRNA, *Mol Cell. Neurosci.*, 5, 124, 1994.
39. Depaoli, A. M., Hurled, K. M., Yasuda, K., Reisine, T., and Bell, G., Distribution of κ opioid receptor mRNA in adult mouse brain: an in situ hybridization histochemistry study, *Mol. Cell. Neurosci.*, 5, 327, 1994.
40. Arvidsson, U., Riedl, M., Chakrabarti, S., Vulchanova, L., Lee, J. H., Nakano, A. H., Lin, X., Loh, H. H., Law, P. Y., Wessendorf, M. W., and Elde, R., The κ opioid receptor is primarily postsynaptic: combined immunohistochemical localization of the receptor and endogenous opioids, *Proc. Natl. Acad. Sci. U.S.A.*, 92, 5062, 1995.
41. Min, B. H., Augustin, L. B., Felsheim, R. F., Fuchs, J. A., and Loh, H. H., Genomic structure and analysis of promoter sequence of a mouse µ opioid receptor gene, *Proc. Natl. Acad. Sci. U.S.A.*, 91, 9081, 1994.
42. Augustin, L. B., Felsheim, R. F., Min, B. H., Fuchs, S. M., Fuchs, J. A., and Loh, H. H., Genomic structure of the mouse δ opioid receptor gene, *Biochem. Biophys. Res. Commun.*, 207, 111, 1995.
43. Liu, H.-C., Lu, S., Augustin, L. B., Felsheim, R. F., Chen, H.-C., Loh, H. H., and Wei, L.-N., Cloning and promoter mapping of mouse κ opioid receptor gene, *Biochem. Biophys. Res. Commun.*, 209, 639, 1995.
44. Liang, Y., Mestek, A., Yu, L. and Carr, L. G., Cloning and characterization of the promoter region of the mouse µ opioid receptor gene, *Brain Res.*, 679, 82, 1995.
45. Kraus, J., Horn, G., Zimprich, A., Simon, T., Mayer, P., and Hollt, V., Molecular cloning and functional analysis of the rat mu opioid receptor gene promoter, *Biochem. Biophys. Res. Commun.*, 215, 591, 1995.

46. Simonin, F., Befort, K., Gaveriaux-Ruff, C., Matthes, H., Nappey, V., Lannes, B., Micheletti, G., and Kieffer, B., The human δ-opioid receptor: genomic organization, cDNA cloning, functional expression and distribution in human brain, *Mol. Pharmacol.*, 46, 1015, 1994.
47. Simonin, F., Gaveriaux-Ruff, C., Befort, K., Matthes, H., Lannes, B., Micheletti, G., Mattei, M.-G., Charron, G., Bloch, B., and Kieffer, B. L., κ-Opioid receptor in humans: cDNA and genomic cloning, chromosomal assignment, functional expression, pharmacology, and expression pattern in the central nervous system, *Proc. Natl. Acad. Sci. U.S.A.*, 92, 7006, 1995.
48. Yakovlev, A. G., Krueger, K. E., and Faden, A. I., Structure and expression of a rat kappa opioid receptor gene, *J. Biol. Chem.*, 270, 6421, 1995.
49. Lu, S., Wei, L.-N., and Loh, H. H., Unpublished observation, 1995.
50. Bidlack, J. M., Saripalli, L. D., and Lawrence, D. M. P., κ-Opioid binding sites on a murine lymphoma cell line, *Eur. J. Pharmacol.*, 227, 257, 1992.
51. Cho, T. M., Hasegawa, J., Ge, B. L., and Loh, H. H., Purification to apparent homogeneity of a mu-type opioid receptor from rat brain, *Proc. Natl. Acad. Sci. U.S.A.*, 83, 4138, 1986.
52. Hasegawa, J., Loh, H. H., and Lee, N. M., Lipid requirement for mu opioid receptor binding, *J. Neurochem.*, 49, 1007, 1987.
53. Schofield, P. R., McFarland, K. C., Hayflick, J. S., Wilcox, J. N., Cho, T. M., Roy, S., Lee, N. M., Loh, H. H., and Seeburg, P. H., Molecular characterization of a new immunoglobin superfamily protein with potential roles in opioid binding and cell contact, *EMBO J.*, 8, 489, 1989.
54. Smith, A. P., Loh, H. H., and Lee, N. M., Characterization of opioid-binding protein and other molecules related to opioid functions. in *Hand Book of Experimental Pharmacology: Opioid I*, Herz, A., Ed., 1993, 37.
55. Ko, J. L., Lee, N. M., and Loh, H. H., Characterization of β-^{125} endorphin cross-linked proteins in NG108-15 cell membranes, *J. Biol. Chem.*, 267, 12722, 1992.
56. Wick, M. J., Ann, D. K., and Loh, H. H., Molecular cloning of a novel protein regulated by opioid treatment of NG108-15 cells, *Mol Brain Res.*, 32, 171, 1995.
57. Wick, M. J., Ann, D. K., Lee, N. M., and Loh, H. H., Isolation of a cDNA encoding a novel zinc-finger protein from neuroblastoma x glioma NG108-15 cells, *Gene*, 152, 227, 1995.

7 Phenotypic Analysis of Receptor Genes: Role of Muscarinic Receptor Genes in Neuropsychiatric Disease

Mark R. Brann and Yasuhiro Kimura

CONTENTS

A. Introduction 89
B. Phenotypic Analysis 90
C. Genetic Analysis 90
    1. Association Studies 90
    2. Linkage Studies 90
    3. Sequence Analysis 91
    4. Phenotypic Testing of Genes 91
D. Conclusions 92
References 93

A. INTRODUCTION

A number of neuropsychiatric diseases are most effectively treated with drugs that affect neurotransmitter receptors. Thus, mutations in genes encoding neurotransmitter receptors are considered among the most likely candidates for disease genes. We have developed procedures that allow the high-throughput screening of patient DNAs for alterations in the function of encoded neurotransmitter receptors. In our initial studies we have examined the functional properties of the m5 muscarinic receptor in schizophrenia and Tourette's syndrome. The m5 receptor is selectively expressed by dopaminergic neurons and is likely to be the muscarinic receptor that controls dopamine release. Alterations in dopaminergic tone have been implicated in these disorders. The studied patients expressed functional m5 receptors, and no functional abnormalities were detected. Using our procedures it is now practical to systematically screen patient DNAs for functional abnormalities in the majority of neurotransmitter receptors. Our approach is compared with traditional candidate gene strategies that do not measure changes in receptor function.

PHENOTYPIC ANALYSIS

Most of the neuropsychiatric diseases are most effectively treated with drugs that affect neurotransmitter receptors. While a complete review is obviously beyond the scope of this chapter, among the most studied examples include dopamine antagonists for treatment of schizophrenia and Tourette's syndrome, inhibitors of serotonin reuptake for depression, benzodiazepines for anxiety, and cholinergic agonists for treatment of Alzheimer's disease.[1] Because of the success of these pharmacological treatments, receptors and associated molecules have been the focus of attempts to understand the molecular basis of neuropsychiatric disease.

The development of radioligand binding assays in the early 1970s made possible the direct examination of the molecular properties of receptors in patient populations. While several hundred papers have been written on the subject, the direct linkage of individual receptors with the pathophysiology of neuropsychiatric disease remains unclear and controversial. One explanation of the uncertainty in the field comes from the molecular genetic analysis of neurotransmitter receptors. It is now known that what used to be considered to be individual receptors are actually families of closely related receptor subtypes.[2] The pharmacological tools (e.g., radioligands) used in most of the studies of disease populations are unable to distinguish among the newly discovered receptor subtypes.[3] Thus, even if an individual receptor subtype were altered in disease, the presence of the other subtypes would obscure the analysis.

C. GENETIC ANALYSIS

With the availability of the sequences of neurotransmitter receptor genes, it is now possible to use techniques of molecular genetics to examine neurotransmitter receptors. While this is a young field, there are already large numbers of papers where molecular genetic techniques have been applied. Unfortunately, controversy and confusion are prominent. Many different approaches are being used, each with its own strengths and weaknesses.

1. Association Studies

Mutations in receptors have been identified in a number of diseases (e.g., cancer[4,5] and endocrine[6,7]). Where functionally relevant mutations are known, screening the association of mutations with disease is extremely powerful and represents a breakthrough in diagnostics. To date, mutations that alter the function of receptors have not been identified in any neuropsychiatric diseases. One exception may be early-onset Alzheimer's disease, where in some families, disease is caused by a mutant gene that may be a receptor.[8] What has been identified are sequence polymorphisms that either do not change the sequences of the encoded proteins[9-11] or cause no detectable change in receptor function.[12,13] In spite of the fact that the sequence changes do not influence receptor function, the prevalence of these polymorphisms has been studied in many neuropsychiatric populations. Unfortunately, these studies do not address the potential of functionally relevant mutations in these receptors being associated with disease.[14-16]

2. Linkage Studies

The availability of polymorphic markers has made possible the study of the inheritance patterns of receptors in families that have disease. Again, once a functionally relevant mutation has been identified, linkage analysis is an extremely powerful tool for showing its causal relationship in a disease (consider the Alzheimer's example[8]). Fortunately, linkage studies can be very powerful even if functionally active mutations have not been identified. In these studies the polymorphic marker is at least in close genetic proximity to mutations that may have functional importance. It should be remembered that unless a given receptor is a major cause of a disease, linkage analysis is unlikely to be successful. Thus, the negative linkage data that are available for several neurotransmitter receptors do not exclude their having a significant role in the studied diseases. On the other hand, this approach is a very cost-effective way of screening for candidate receptors that may play a major role in disease.[16]

3. SEQUENCE ANALYSIS

The most powerful technique for determining the role of receptor mutations in disease is to directly search for them. Because this is extremely labor-intensive using traditional technologies, very few receptors have been examined. All that have turned up so far are a number of apparently silent sequence polymorphisms. Several elegant methodological improvements have made it possible to more efficiently search for sequence polymorphisms.[11,12] While sequence analysis is a very rigorous approach to finding functionally relevant mutations in the coding regions of receptors, this approach has much less applicability to identifying mutations in the noncoding regions of receptors. The latter could influence receptor expression and processing. The problem with the latter is the limited amount of information that is available concerning the relevant regulatory mechanisms, and the large amount of genetic information that may have to be searched. For example, the regulatory elements of some receptors are several kilobases away from the coding region.[18]

4. PHENOTYPIC TESTING OF GENES

In our ongoing work with the pharmacology of cloned receptor subtypes we have developed technology for efficiently evaluating the functional properties of these cloned genes.[19,20] Our approach, called Receptor Selection and Amplification Technology (R-SAT, patents pending), is based on the ability of a wide range of receptors to mediate the proliferation/transformation of mammalian cells. We have used this technology to search libraries of mutant receptors for molecules with a wide range of functional properties.[21,22] One functional property that is particularly easy to identify is constitutive activity. Constitutively active receptors have activity in the absence of added ligand. Others have found that constitutively active receptors are responsible for a variety of diseases.[23,24]

We have recently combined R-SAT with the PCR amplification of receptors from patient DNAs as a method to phenotypically analyze the functional properties of patient receptors (manuscripts in preparation). Our strategy is very similar to that of sequence analysis, except that we directly examine the functional properties of the encoded receptors. Sequences are only determined in cases where receptor function is altered. The disadvantage of our approach is that we will not pick up the functionally silent mutations. This is also our greatest advantage because mass sequencing efforts devote a considerable amount of time demonstrating that most of the identified sequence changes have no functional significance. Thus, it is far more cost effective to screen patient populations using our approach.

In Figure 7-1, the functional properties of several m5 muscarinic receptors are shown. The wild-type receptor responds to carbachol with an EC_{50} of 42 ± 16 nM. Also shown is the response of a constitutively active mutant form of the m5 receptor that was isolated from a library of randomly mutated receptors.[21] The constitutively active receptor has significant functional activity even in the absence of added carbachol. The EC_{50} of the receptor for carbachol is shifted to the left (2.8 ± 1.0 nM), and the constitutive activity can be fully blocked by atropine. The m5 receptors amplified from DNA extracted from a patient with schizophrenia and Tourette's had functional responses that are indistinguishable from those of the wild-type receptor — EC_{50}s of 29.6 ± 12.3 and 30.8 ± 11.8 nM, respectively. We have now developed procedures where the functional analysis can be performed in a 96-well format, allowing a very high-throughput analysis of the functional properties of a wide range of receptor genes.

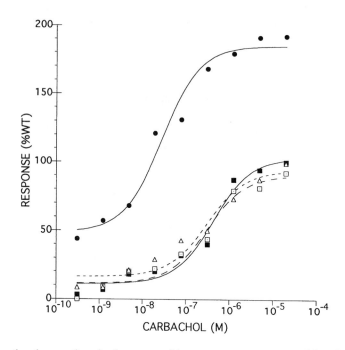

FIGURE 7-1 Functional properties of m5 receptors. Lines are computer-generated fits of the data to a mass-action relationship. The data are normalized to the maximum response observed for wild-type receptors, (■) = 100%. Zero % response for each receptor was evaluated in the presence of 1 μM atropine. We have previously shown that atropine is an inverse agonist for muscarinic receptors. (●) Is the constitutively active receptor isolated from a random library,[21] (□) are receptors amplified from a patient with Tourette's, and (Δ) is amplified from a patient with schizophrenia. Receptors were tested from the total mixture of potential receptors amplified from each patient, thus both alleles were effectively screened for mutant phenotypes.

D. CONCLUSIONS

There are several methods available for evaluating molecular changes in neurotransmitter receptors in patients with neuropsychiatric disease.

- Early biochemical studies were confounded by the lack of the specificity of the pharmacological probes that were used. With the recent development of subtype-selective antibodies and more selective drugs, these studies may show more promise in the future.
- Association studies will ultimately lead to important breakthroughs in molecular diagnostics. Unfortunately, these approaches are dependent on the availability of polymorphisms that are directly related to disease.
- Linkage studies are a powerful way of testing for a major role of a receptor gene, regardless of the location of the mutation. Negative data should be interpreted with caution.

- Sequence analysis is an extremely rigorous and powerful way of identifying mutations. The approach is expensive and has limited utility for finding mutations outside of the coding regions. Time diverted to testing functionally irrelevant mutations is a major consideration.
- Combining genetic testing with phenotypic analysis is now practical. The approach is very cost effective and precise for finding functionally relevant mutations in coding regions. This approach is not useful for finding mutations in noncoding regions that may alter receptor expression. Evidence for mutations of the later type may best be found by measuring receptor subtypes in tissues from patients.

REFERENCES

1. Gilman, A., Rall, T., Nies, A. S., and Taylor, P., Eds., Goodman and Gilman's: *The Pharmacological Basis of Therapeutics*, Pergamon Press, New York, 1991.
2. Brann, M. R., Ed., *Molecular Biology of G-Protein-Coupled Receptors*, Birkahuser, Boston, 1992.
3. Watson, S. and Girdlestone, D., 1995 Receptor and ion channel nomenclature supplement. 6th ed. *Trends Pharmacol. Sci.*, 16, 1995.
4. Brandt, B., Vogt, U., Schlotter, C. M., Jackisch, C., Werkmeister, R., Thomas, M., von Eiff, M., Bosse, U., Assmann, G., and Zanker, K. S., Prognostic relevance of aberrations in the erbB oncogenes from breast, ovarian, oral and lung cancers: double-differential polymerase chain reaction (ddPCR) for clinical diagnosis, *Gene*, 159, 35-42, 1995.
5. Maeda, S., Namba, H., Takamura, N., Tanigawa, K., Takahashi, M., Noguchi, S., Nagataki, S., Kanematsu, T., and Yamashita, S., A single missense mutation in codon 918 of the RET proto-oncogene in sporadic medullary thyroid carcinomas, *Endocrinol. J.*, 42, 245-250, 1995.
6. Rosenthal, A., Seibold, W., Antaramian, A., Gilbert, S., Birnbaumer, M., Bichet, D. G., Arthus, M. F., and Lonergan, M., Mutations in the vasopressin V2 receptor gene in families with nephrogenic diabetes insipidus and functional expression of the Q-2 mutant, *Cell. Mol. Biol.*, 40, 429-436, 1994.
7. Raymound, J. R., Hereditary and acquired defects in signaling through the hormone-receptor-G protein complex, *Am. J. Physiol.*, 266, 163-174, 1994.
8. Sherrington, R., Rogaev, E. I., Liang, Y., Rogaeva, E. A., Levesque, G., Ikeda, M., Chi, H., Lin, C., Li, G., and Holman, K., Cloning of a gene bearing missense mutations in early-onset familial Alzheimer's disease, *Nature*, 375, 754-760, 1995.
9. Blum, K., Sheridan, P. J., Wood, R. C., Braverman, E. R., Chen, T. J., and Comings, D. E., Dopamine D2 receptor gene variants: association and linkage studies in impulsive-addictive-compulsive behaviour, *Pharmacogenetics*, 5, 121-141, 1995.
10. Gejman, P. V., Ram, A., Gelernter, J., Friedman, J. E., Cao, Q., Pickar, D., Blum, K., Noble, E. P., Kranzler, H. R., and O'Malley, S., No structural mutation in the dopamine D2 receptor gene in alcoholism or schizophrenia. Analysis using denaturing gradient gel electrophoresis, *JAMA*, 271, 204-8, 1994.
11. Liu, Q., Sobell, J. L., Heston, L. L., and Sommer, S. S., Screening the dopamine D1 receptor gene in 131 schizophrenics and eight alcoholics: identification of polymorphism but lack of functionally significant sequence changes, *Am. J. Med. Genet.*, 60, 165-171, 1995.
12. Asghari, V., Schoots, O., Van Kats, S., Ohara, K., Jovanovic, V., Guan, H.-C., Bunzow, J. R., Petronis, A., and Van Tol, H., Dopamine D4 receptor repeat: analysis of different native and mutant forms of the human and rat genes, *Mol. Pharmacol.*, 46, 364-373, 1994.
13. Lam, S., Shen, Y., Nguyen, T., Messier, T. L., Brann, M. R., Comings, D., George, S. R., and O'Dowd, B. F., A serotonin receptor gene (5HT1A) variant found in a Tourette's syndrome patient, *Biochem. Biophys. Res. Commun.*, 219, 853-858, 1996.
14. Arinami, T., Itokawa, M., Enguchi, H., Tagaya, H., Yano, S., Shimizu, H., Hamaguchi, H., and Toru, M., Association of dopamine D2 receptor molecular variant with schizophrenia, *Lancet*, 343, 703-4, 1994.
15. Epstein, R. P., Novick, O., Umansky, R., Priel, B., Osher, Y., Blaine, D., Bennett, E. R., Nemanov, L., Katz, M., and Belmaker, R. H., Dopamine D4 receptor (D4DR) exon III polymorphism associated with the human personality trait of novelty seeking, *Nat. Genet.*, 12, 78-80, 1996.
16. Benjamin, J., Li, L., Patterson, C., Greenberg, B. D., Murphy, D. L., and Hamer, D. H., Population and familial association between the D4 dopamine receptor gene and measures of novelty seeking, *Nat. Genet.*, 12, 81-84, 1996.
17. Kalsi, G., Mankoo, B. S., Curtis, D., Brynjolfsson, J., Read, T., Sharma, T., Murphy, P., Patursson, H., and Guring, H. M., Exclusion of linkage of schizophrenia to the gene for the dopamine D2 receptor (DRD2) and chromosome 11q translocation sites, *Psychol. Med.*, 25, 531-537, 1995.
18. Erdmann, J., Shimron-Abarbanell, D., Cichon, S., Albus, M., Maier, W., Lichtermann, D., Minges, J., Reuner, U., Franzek, E., and Ertl, M. A., et al., Systematic screening for mutations in the promoter and the coding region of the 5-HT1A gene, *Am. J. Med. Genet.*, 60, 393-399, 1995.

19. Brauner-Osborne, H. and Brann, M. R., Pharmacology of muscarinic acetylcholine receptor subtypes (m1-m5): high throughput assays in mammalian cells, *Eur. J. Pharmacol.*, 295, 93-102, 1996.
20. Messier, T. L., Dorman, C. M., Brauner-Osborne, H., Eubanks, D., and Brann, M. R., High throughput assays of cloned adrenergic, muscarinic, neurokinin, and neurotrophin receptors in living mammalian cells, *Pharmacol. Toxicol.*, 76, 308-311, 1995.
21. Spalding, T. A., Burstein, E. S., Brauner-Osborne, H., Hill-Eubanks, D., and Brann, M. R., Pharmacology of a constitutively active muscarinic receptor generated by random mutagenesis, *J. Pharmacol. Exp. Ther.*, 275, 1274-1279, 1995.
22. Burstein, E. S., Spalding, T. A., and Brann, M. R., Amino acid side chains that define muscarinic receptor/G-protein coupling, *J. Biol. Chem.*, 271, 2882-2885, 1996.
23. Latronico, A. C., Anasti, J., Arnhold, I. J., Mendonea, B. B., Domenice, S., Albano, M. C., Zachman, K., Wajchenberg, B. L., and Tsigor, C., A novel mutation of the luteinizing hormone receptor gene causing male gonadotropin-independent precocious puberty, *J. Clin. Endocrinol. Metab.*, 80, 2490-2494, 1995.
24. Rim, J. and Oprian, D. D., Constitutive activation of opsin: interaction of mutants with rhodopsin kinase and arrestin, *Biochemistry*, 34, 11938-11945, 1995.

8 Studies on Dopamine Receptors and Role in Drug Addiction

Gordon Y.K. Ng, Susan R. George, and Brian F. O'Dowd

CONTENTS

A. Introduction 95
B. Characterization of Dopamine Receptors 96
    1. Biochemical Classification of Dopamine Receptors 96
    2. Pharmacological Classification of Dopamine Receptors 96
    3. Structural Classification of Dopamine Receptors 96
    4. Brain Distribution of Dopamine Receptors 97
C. Dopamine Receptors and Drug Addiction 99
    1. Drug Abuse Research Using Inbred Mouse Strains 100
    2. Medical Genetics and the Association of Dopamine Receptor Genes with Drug Abuse 103
D. The Study of Dopamine Receptor Regulation *In Vitro* 103
E. Conclusion 106
References 106

A. INTRODUCTION

Dopamine is a neurotransmitter present in nervous systems ranging in complexity from the human to the snail.[1] Dopamine-containing neurons in mammalian brain comprise two major ascending mesencephalic projections with cell bodies located in the substantia nigra and ventral tegmental area of the midbrain.[2] The nigrostriatal pathway has projections terminating primarily in the caudate nucleus of the striatal complex, whereas the mesolimbic pathway has projections terminating in the globus pallidus, olfactory tubercle, nucleus accumbens, septum, amygdala, and cortex. In mammals, dopamine has been shown to play an important role in posture and motor control, as well as in generating emotions such as feelings of well-being, pleasure, and euphoria. Dopamine's diverse biological activities may owe in part to the multiplicity of dopamine receptors, and these receptors belong to the G-protein-coupled receptor family. They have been classified on the basis of biochemical and pharmacological criteria and structure into D1-like receptors which include the D1 and D5 receptors, and D2-like receptors which include the isoforms of the D2 and D3 receptors, and the D4 receptor and its variants. Molecular genetic studies have yet to show a clear linkage between any of the dopamine receptors and disease. In the case of drug addiction, it is well known that drugs of abuse such as cocaine, nicotine, and alcohol activate the dopaminergic system.[3-5]

B. CHARACTERIZATION OF DOPAMINE RECEPTORS

1. Biochemical Classification of Dopamine Receptors

Dopamine receptors were classically categorized into D1 and D2 receptors based on the ability of D1 receptors to stimulate and D2 receptors to inhibit adenylyl cyclase activity, and the generation of cAMP.[6] D1 and D2 receptors were subsequently reported to also differentially modulate the levels of other second messengers such as IP_3[7,8] and Ca^{+2}.[9,10] Activation of D2 receptors may also modulate Na^+/H^+ exchange.[11] Molecular cloning has revealed the existence of multiple other dopamine receptors. To date, dopamine receptors have been classified as D1-like receptors (D1 and D5 receptors) that mediate dopamine activation of G_s-like G-proteins to stimulate adenylyl cyclase activity,[12,13] and D2-like receptors (D2, D3, and D4 receptors) that mediate dopamine activation of $G_{i/o}$-like G-proteins to inhibit adenylyl cyclase.[14,15] This pleiotropy might explain the complexity of dopamine receptor-mediated transmembrane signaling underlying complex behaviors.

2. Pharmacological Classification of Dopamine Receptors

Radioligand binding assays have been a popular tool for identifying receptors since each receptor may be predicted to have an unique pharmacological rank order of affinities for agonists and antagonists. Among the first selective D1-like receptor ligands to be characterized were the benzazepine compounds such as the partial agonist <SKF-38393>[16] and the antagonist SCH-23390.[17] The tritium-labeled SCH-23390 is routinely used to identify D1-like receptors, exhibiting high affinity at picomolar concentrations.[18] Multiple serotonin receptor subtypes can also be detected with [^{3}H]SCH-23390, but with lower affinity at nanomolar concentrations.[19] For D2-like receptors, agonist activity was demonstrated for the ergot derivative bromocriptine[20] and for quinpirole, a quinoline.[21] D2-like receptor antagonist activity is characteristic of neuroleptic drugs. Among the neuroleptics, [^{3}H]spiperone, a butyrophenone, has emerged as the ligand of choice for labeling D2-like receptors due to its high affinity and low nonspecific binding.[22] Spiperone, however, will weakly bind D1-like and serotonin receptors at high concentrations. (-)Sulpiride and eticlopride, substituted benzamides, have also been used as selective D2-like receptor antagonists.[23] The atypical neuroleptic clozapine, a diazepine, has lower affinity for D2 receptors than spiperone, but shows greater affinity at D3 and D4 receptors which may be useful for labeling these subtypes.[24] However, subsequent studies have shown that clozapine demonstrates high affinity at serotonin and muscarinic receptors, limiting the usefulness of this ligand for the labeling of a single receptor subtype in tissues. Thus radioligands show selectivity but not specificity for a receptor subtype, and as a result, past radioligand binding studies failed to identify the heterogeneity in dopamine receptor subtypes in tissues until the recent application of molecular cloning techniques, which identified separate genes for D1, D2, D3, D4, and D5 receptors.

3. Structural Classification of Dopamine Receptors

Molecular cloning of dopamine receptors has revealed the existence of a multiplicity of D1-like and D2-like receptors not revealed previously by traditional pharmacological and biochemical techniques. D1-like receptors include the D1 and D5[12,13] receptors whose genes

TABLE 8-1
Biochemical, Pharmacological, and Structural Classification of Dopamine Receptors

Receptor Type	Receptor Subtype	Second Messenger	Antagonist /Agonist	% aa Identity	Gene Structure	Chromosome
D1-like	D1	↑cAMP, ↑IP$_3$, ↑Ca^{+2}	SCH-23390 /SKF-38393	100	Intronless	5
	D5	↑cAMP	SCH-23390 /SKF-38393	55	Intronless	4
D2-like	D2	↓cAMP, ↓IP$_3$, ↓Ca^{+2}, Na$^+$/H$^+$	Eticlopride /quinpirole	100	Intron	11
	D3	↓cAMP	Clozapine /7-OHDPA	50	Intron	3
	D4	↓cAMP	Clozapine /quinpirole	37	Intron	11

Note: The distinguishing characteristics of dopamine receptor subtypes include the coupling to second messengers, selective antagonist and agonist, percentage amino acid (aa) identity, the gene structure, and chromosome localization.

are intronless in the coding regions. It should also be noted that two transcribed D5 receptor pseudogenes have been identified.[25,26] Although the physiological significance of this finding remains obscure, pseudogenes likely mark the evolution of dopamine receptor subtypes. D2-like receptors include the D2, D3, and D4 receptors whose gene coding sequences are separated by introns. D2 receptors may be further categorized into D2$_{Long}$ (D2$_L$) and D2$_{Short}$ (D2$_S$) isoforms that arise from alternative mRNA splicing of a single D2 receptor gene.[27,28] Similarly, D3 receptor mRNAs encoding distinct functional receptors and truncated versions are thought to arise from alternative splicing from a single D3 receptor pre-mRNA.[29-31] In contrast, D4 receptors and variants arise from a genetic polymorphism of a single D4 receptor gene.[32,33]

Analysis of the amino acid sequences of the dopamine receptors show that they all belong to the family of G-protein-coupled receptors that have seven highly conserved membrane-spanning regions which are linked by intracellular and extracellular loops. Like other cationic amine receptors, dopamine receptors have a conserved aspartic acid residue in transmembrane region III and a serine residue in transmembrane region V, suggested to be involved in dopamine binding. Dopamine receptors differ most in the homology and size of hydrophilic intracellular domains. D1 and D5 dopamine receptors have a shorter intracellular third loop and a longer carboxyl tail when compared to D2, D3, and D4 receptors. In particular, the third intracellular loop and carboxyl tail have been shown for other structurally related receptors to be involved with G-protein interaction and contain consensus serine and threonine residues for phosphorylation by regulatory kinase(s). In the carboxyl tail, there is a putative cysteine residue for palmitoylation which may play a role in functional coupling. We have recently shown that D1 and D2 receptors are phosphorylated and palmitoylated,[34,35] however, it remains to be shown whether all dopamine receptors are posttranslationally modified and thus regulated in a similar manner. An extensive review of dopamine receptor structure has been reported elsewhere.[36] Such structural differences in these important regions may contribute to the functional differences observed for dopamine receptors (see Figures 8-1 and 8-2).

4. BRAIN DISTRIBUTION OF DOPAMINE RECEPTORS

D1 and D2 receptors share overlapping distributions in the brain; receptor proteins and mRNAs are predominantly located in the striatum (caudate-putamen, nucleus accumbens) and olfactory tubercle with lower densities for the D1 receptor in the cortex, hippocampus, limbic, and midbrain regions.[37-39] At the cellular level, these receptor have been localized by radioligand autoradiography to synaptic nerve terminals, but only the D2 receptors has been identified presynaptically, where they may function as autoreceptors regulating the synthesis

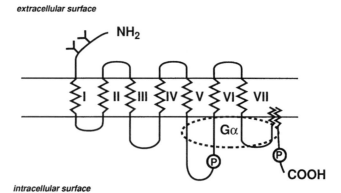

FIGURE 8-1 Schematic model of the proposed membrane topology of a G-protein-coupled receptor. The extracellular amino (NH$_2$), and intracellular carboxy (COOH) termini, putative palmitoylation («), and phosphorylation (P) sites of the receptor, and areas of interaction with G-protein (Gα) are indicated. Glycosylation sites (Y) and the transmembrane-spanning domains are numbered I to VII.

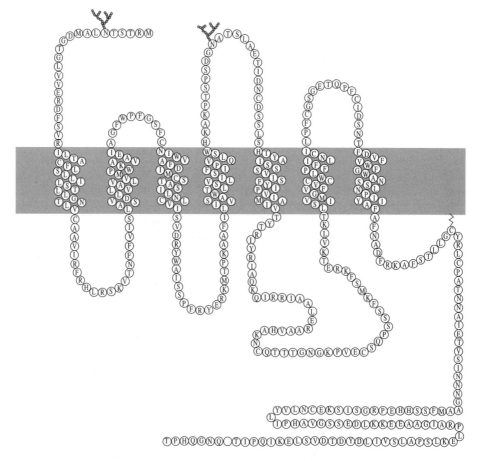

FIGURE 8-2 Proposed topology for the human D1 (A) and D4 (B) dopamine receptors.

and/or release of dopamine.[40] D2 receptor mRNAs for long and short isoforms are colocalized,[28] with the long isoform being present at higher levels, the significance of which remains to be determined. In contrast, less is known about the D3 receptor proteins believed also to have autoreceptor functions since mRNAs for D3 receptors are found in highest density in limbic brain regions and cerebellum.[41,42] Similarly, the distributions of D4 and D5 receptor proteins remain to be determined although their mRNAs have been localized primarily to limbic regions.[13,32,39,43] Thus the differential distribution of dopamine receptor mRNAs and receptor proteins is highly suggestive of the different activities, organization and, possibly, functions of the brain dopamine receptor systems.

C. DOPAMINE RECEPTORS AND DRUG ADDICTION

Brain dopamine, noradrenalin, serotonin, and opioid systems have all been shown to be involved in drug-seeking behavior. In fact, it can be agreed that multiple factors, both biological and environmental, are involved in drug addiction. However, it is attractive to speculate that one pathway is likely more determinant in this complex behavior. A large body of evidence has clearly shown an increased activity of the central dopaminergic system following intake of most abused substances, and it has been postulated that dopamine may be the neural substrate for drug reward.[4] Although this evidence does not support a causative link, it is provocative. Research into the genetic vulnerability of drug abuse provides the strongest evidence supporting a dopamine theory of drug addiction. For this review, we will focus on the dopaminergic system because it is our opinion that it has an intimate relationship with the mechanisms of drug abuse.

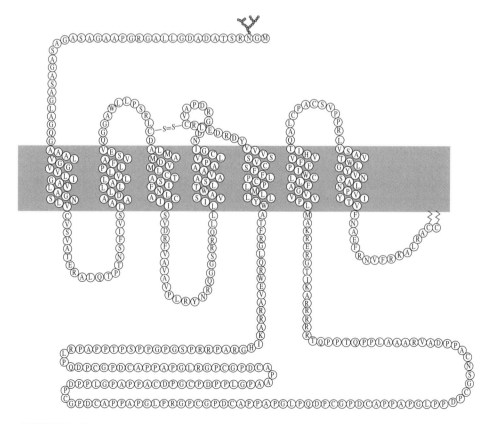

FIGURE 8-2B

1. DRUG ABUSE RESEARCH USING INBRED MOUSE STRAINS

Although the C57BL/6J and DBA/2J inbred mouse strains were originally developed for cancer research, it is well known they exhibit markedly different susceptibilities for abused substances such as cocaine, morphine, and ethanol.[44] The C57 mouse exhibits an innate high-abuse vulnerability for all of these substances, whereas the DBA/2J mouse is almost completely drug avoiding. The neural basis of these strain-related drug predispositions are incompletely known. Genetic studies have precluded a role for serotonin in ethanol-drinking behavior in these mouse strains,[45] and raise the possibility that a candidate neural basis for the vulnerability for abused drugs may be a genetically determined difference in dopaminergic function. Evidence will be presented herein to support this hypothesis.

Studies have shown that although dopamine receptor pharmacology and the sizes and distributions of D1 and D2 receptor mRNAs were similar in the DBA and the C57 mouse strains, higher forebrain D1 and D2 receptor densities and mRNA levels in the C57 mouse were detected.[46-48] These findings were consistent with elevated striatal dopamine-sensitive adenylyl cyclase activity,[50] and enhanced dopamine receptor gene sensitivity following antagonist treatments in the C57 mouse compared to the DBA mouse.[46] Further, D2 receptor binding and receptor mRNA, possibly representing autoreceptors in the midbrain and medulla pons of the C57 mouse, were accordingly lower.[46] Although alternative interpretations of these findings are possible, we interpreted these data to collectively show that striatal dopamine receptor-linked function is upregulated in the C57 compared to the DBA mouse, possibly to compensate for a deficient nigrostriatal/mesolimbic dopaminergic activity. We have proposed that this neurochemical profile may be a possible basis for an increased vulnerability for abused drugs in the C57 mouse[46] (see Figure 8-3).

Interestingly, a deficit in forebrain dopamine has also been reported for the outbred P rat selected for ethanol preference.[51] It should be noted that no similar correlation has been shown in the AA and ANA outbred rat strains.[52] These data are consistent with the notion that a deficiency in the dopaminergic system is linked to an increased risk for at least one form of voluntary high ethanol consumption. Experiments in our laboratory indicate that the nature of the impairment in the C57 mouse appears to be related to levels of synaptic dopamine turnover that is determined by dopamine synthesis, release, and breakdown.[53]

While the notion that the dopaminergic system is the sole determinant in the susceptibility for ethanol abuse can be argued, it is clear that the dopaminergic system is involved in the actions of ethanol. In the C57 mouse, ethanol intake has been reported to increase striatal and mesolimbic dopamine metabolism[53,54] and accordingly decreased striatal D1 and D2 receptor densities, D1 and D2 receptor mRNAs in the olfactory tubercle, and D2 receptor mRNA in the medulla pons.[46] Similar effects of ethanol on the rat brain dopaminergic system have been reported, suggesting possibly a common mechanism of ethanol action. Ethanol intake has been shown to increase dopamine release in the forebrain of outbred rats selected for ethanol preference,[55,56] and ethanol has been shown to increase dopamine release[57] in the nucleus accumbens,[58] enhance the accumulation of DOPA,[59] and increase the formation of dopamine metabolites[58] while downregulating striatal dopamine receptors in rats.[60] In the inbred mouse strains, an ethanol-promoted increase in forebrain dopaminergic activity was accompanied by a selective upregulation of D2 receptors in the midbrain, presumably representing autoreceptors. Consistent with this interpretation, others have shown that ethanol enhances the activity of dopamine neurons in the substantia nigra[61] and the ventral tegmental area[62] in rats. Furthermore, ethanol treatment has been shown to selectively increase the expression of G_i but not G_s in the pons and cerebellum of two other (short and long sleep) inbred ethanol-sensitive mouse strains.[63] Taken together, these data support the conclusion that the neural basis of ethanol drinking and ethanol action involves multiple components of the dopaminergic pathway.

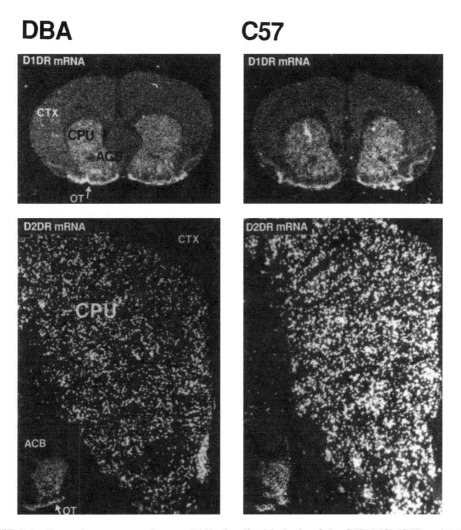

FIGURE 8-3 Dopamine receptor subtype mRNAs localized in brain of the C57BL/6J (C57) and DBA/2J (DBA) inbred mouse strains by receptor-specific DNA probes using *in situ* hybridization and dark-field autoradiography techniques. Top panels show the distribution of D1 dopamine receptor (D1DR) mRNA in coronal sections. Bottom panels show the distribution of D2 dopamine receptor (D2DR) mRNA in sagittal sections. Inset shows the D2DR mRNA signal in the nucleus accumbens. The caudate-putamen (CPU), nucleus accumbens (ACB), and olfactory tubercle (OT) are indicated. The receptor mRNA hybridization signal appears as white. (Modified from Ng, G.Y.K., O'Dowd, B.F., and George, S.R., *Eur. J. Pharmacol.*, 7, 67, 1994.)

We speculate that the actions of ethanol owe to synaptic concentrations of dopamine released upon ethanol intake. It is likely that at low levels of ethanol drinking, primarily the D1 receptor is involved owing to its higher affinity for dopamine than the D2 receptor. At higher doses of ethanol, resulting in higher synaptic dopamine levels, the D2 receptor may be activated and may mediate the more intense rewarding and/or side effects of ethanol. Alternatively, the mechanism of ethanol actions may involve D1-D2 receptor synergism as evidenced in the ability of the D1 receptor antagonist to upregulate D2 receptor mRNA in the C57 mouse.[46] A synergistic interaction between these receptor subtypes at the biochemical[64-66] and behavioral levels[67] has been demonstrated. A role for a D1-D2 link in human brain[68] has also been suggested, but the molecular mechanism of this interaction is presently unknown. Given that each receptor has overlapping and distinct tissue expression,[39] such a D1-D2 receptor link is likely a brain region-specific phenomenon. These region-specific ethanol effects are in keeping with the functional organization of dopamine receptors in the brain and suggest that multiple brain loci and dopamine receptors are involved in modulating ethanol drinking behaviors.

Other evidence consistent with the hypothesis that an endogenous dopaminergic deficit may mediate the susceptibility for high ethanol drinking comes from pharmacological studies. Pretreatments with dopamine receptor agonists that augment endogenous synaptic dopamine levels mediate marked reductions in voluntary high ethanol drinking in a number of animal models. In the C57 inbred mouse, the dopamine receptor agonists bromocriptine, quinpirole, SKF-81297, or (+)-SKF-38393 all reduce the predisposition for high ethanol preference and consumption.[53,69] In these animals, striatal D1 and D2 receptor densities and D1 and D2 receptor mRNAs in the olfactory tubercle are downregulated, accompanied by an upregulation of D2 autoreceptors (activation of feedback mechanisms), confirming a role for the dopaminergic system in these drug effects on ethanol drinking. Behavioral pharmacological studies have also shown that apomorphine, a nonselective dopamine receptor agonist, can reduce the response for ethanol reinforcement in free-feeding genetically unselected Long-Evans rats.[70]

In another study, it was found that dopamine receptor agonists reduced ethanol consumption in the high-alcohol-drinking HAD line of rats[71] and in alcohol-preferring P and unselected Wistar rats.[72] Recently, aberophine, a synthetic nonselective dopamine receptor agonist with no serotonergic activity was shown to reduce ethanol self-administration in ethanol-trained rats.[73] Further, microinjection of sulpiride into the nucleus accumbens has been shown to increase ethanol drinking in alcohol-preferring P rats, in keeping with the dopamine hypothesis of drug addiction. Interestingly, bromocriptine administration has been shown to promote reductions in cocaine administration in rats,[75] suggestive that dopaminergic deficit(s) may confer susceptibility to abuse substances. Since receptor agonists are selective rather than specific for a receptor subtype, the biological activity of these drugs may owe to the relative activity at D1 and D2 receptors.

It should be noted that ethanol-seeking behavior in the inbred mouse has also been linked to a deficiency of endogenous opioids.[76,77] It is well known that opioid and dopamine neural systems are integrally linked and, accordingly, opioid deficits may be reflected in reduced dopaminergic activity (unpublished observations). Any discrepancies with the dopamine data are readily reconciled in the inbred mouse model. However, additional evidence indicates that still other neural systems such as GABA may act as determinants in ethanol-seeking behavior.[78] However, taken together the evidence convinces us to suggest that a genetic basis for the susceptibility for high ethanol abuse may involve an endogenous deficiency in central dopaminergic function which is restored by ethanol drinking in the inbred mouse strains.

However, it is unclear whether the same or other dopamine mechanisms are in place to mediate the continued pattern of ethanol drinking following the development of ethanol tolerance and dependence.[79] In ethanol-sensitive C57 mice, bromocriptine does not reduce ethanol-seeking behavior. This may owe to ethanol-induced changes in the steady state of the dopaminergic system as chronic ethanol consumption has been shown to alter the synthesis, metabolism, and release of dopamine, deplete dopamine stores, and depress dopamine neurons and dopamine receptor sensitivity in the nigro-striatal ventral tegmental-accumbens dopamine pathway.[3,5,46,60,80-83] Collectively, these findings clearly indicate that different mechanisms, although dopaminergic, likely mediate ethanol drinking after sensitization to ethanol, or once ethanol tolerance and dependence are established. This may account for the reduction in voluntary ethanol preference but not in voluntary ethanol consumption in haloperidol-treated ethanol-sensitized C57 mice.[46] Alternatively, a haloperidol-induced blockade at D2 receptors but not D1 receptors may explain the partial effect. Others have postulated from studies in rats that dopamine receptor antagonists such as haloperidol block the reinforcing properties of ethanol.[70,84] The extrapyramidal motor side effects of neuroleptics are well known and could also be a contributing factor. This further does not exclude haloperidol blockade at other neurotransmitter systems such as noradrenalin or serotonin to mediate ethanol consumption under these conditions.[85-87] Recent developments in the manipulation of selective gene expression employing antisense strategies, or gene ablation models such as the genetically engineered receptor knock-out animals, may offer the specificity needed to better elucidate the function of D1 and D2 receptor systems in ethanol-drinking and drug-consuming behaviors.

2. MEDICAL GENETICS AND THE ASSOCIATION OF DOPAMINE RECEPTOR GENES WITH DRUG ABUSE

Growing molecular genetic evidence suggests a role for dopamine receptor gene variants marked by RFLPs in the susceptibility to drug addiction. The *Taq I* A1 RFLP of the D2 receptor gene has been reported by some, but not others, to be associated with severe alcoholism.[88-92] Similarly, cigarette smokers demonstrate significantly higher *Taq I* A1 frequencies than nonsmokers.[93] A higher prevalence of *Taq I* A1 and B1 RFLPs of the D2 receptor gene has been identified in most subjects who abuse several addictive substances (cocaine, amphetamine, opiates) including alcohol.[94-96] Genetic typing of the D4 receptor gene in alcoholic subjects have shown greater prevalence of the D4(3) and D4(6) alleles than has been reported in normals.[97] These molecular genetic findings suggest that dopamine receptor genes are plausible candidate determinants that may confer, at least in part, the vulnerability to some forms of drug abuse. However, the pathophysiological consequences for these genetic findings remains unclear. Significantly reduced D2 receptor density has been reported in alcoholics[90] and smokers[93] compared to controls, and it has been postulated by these authors that a mutation in linkage disequilibrium with *Taq I* A polymorphism was associated with a decrease in the function of the D2 receptor gene, resulting in a decrease in number, without a change in the structure, of the D2 receptor. The hypothesis is intriguing, but it remains to be proven whether reduced D2 receptor density was not a consequence of chronic drug use. Nevertheless, this research provides the basis for the speculation that genes involved in dopaminergic neurotransmission, conferring a brain region-specific dopaminergic deficit, may be a common basis for drug-craving behaviors.

D. THE STUDY OF DOPAMINE RECEPTOR REGULATION *IN VITRO*

Aberrations in dopamine receptor-linked signaling may be a candidate basis of certain drug addictions. Further, differential regulation of dopamine receptor systems in the brain has been reported.[98-100] This is intriguing and may indicate some mechanism for separate drug actions. However progress in understanding dopamine receptor-linked signaling at the molecular level has been hampered owing to the inability to distinguish the biology of individual receptor subtypes in tissues. To better understand this and in turn, the effects of drugs on each receptor subtype, cloned receptors have been expressed in foster cell lines for detailed molecular pharmacological studies (see Figure 8-4).

Indeed, the D1 and D2 receptor-coupled adenylyl cyclase systems have been shown *in vitro* as *in vivo* to be subject to distinct patterns of regulation following activation by dopamine. Dopamine treatment has been shown to induce the desensitization of D1 receptor-stimulated adenylyl cyclase activity[34,101-103] accompanied by a dopamine-promoted increase in the phosphorylation and palmitoylation of the D1 receptor.[34] These biochemical modifications of the receptor are believed to play a role in the functional coupling of the receptor. The desensitization of the D1 receptor-coupled adenylyl cyclase system appears to proceed concomitant to a loss of surface D1 receptors binding.[103] This has been suggested to involve a conformational change or aggregation of surface D1 receptors, rendering receptors incapable

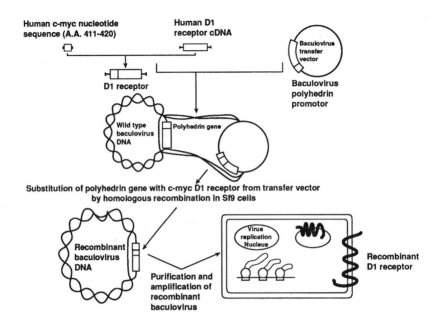

FIGURE 8-4 A schematic showing the expression of an epitope-tagged D1 receptor in the baculovirus/Sf9 cell system. The human D1 receptor cDNA is linked to a c-myc nucleotide sequence and inserted into the baculovirus transfer vector, and through the process of homologous recombination with wild-type baculovirus DNA, which results in a recombinant baculovirus DNA encoding an epitope-tagged D1 receptor.

of ligand binding. This may represent a possible obligatory stage for internalization since prolonged agonist exposure produces receptor internalization and downregulation long after agonist-induced functional uncoupling of the receptor.[101-103] Agonist-induced D1 receptor internalization may be a process leading to resensitization mechanisms, since it has been shown that desensitized β-adrenergic receptors are sequestered and subsequently dephosphorylated, reactivated, and recycled back to the plasma membrane.[104] These studies show that agonist-induced rapid desensitization of the D1 receptor-coupled adenylyl cyclase system, and the slower in onset receptor internalization, constitute distinct mechanisms regulating the biology of the D1 receptor (see Figure 8-5).

In contrast, sustained exposure of the D2 receptor-coupled adenylyl cyclase system to dopamine does not readily result in desensitization.[105-108] Having been demonstrated for D2 receptors expressed in a prolactin-secreting pituitary cell line, 293, CHO, and Sf9 cells,[106-108,113] these findings indicate that this may owe to a property of the D2 receptor protein. We have shown that the D2 receptor forms receptor dimers,[35,109] and is phosphorylated and palmitoylated,[35] supporting the hypothesis that such modifications may contribute to the regulation of the D2 receptor. Interestingly, receptor dimer formation may be a common property of G-protein-coupled receptors since it has been reported to occur for D1 and D2 receptors, D3 and D4 receptors (unpublished observations), M1 and M2 muscarinic,[110] 5-HT$_{1B}$,[111] and metabotropic glutamate receptors.[112] Research, however, has yet to show a physiological role for receptor dimers.

Sustained agonist exposure has also been shown to increase cell surface D2 receptor density.[106,107,113] This may involve the translocation of receptors from intracellular pools and the activation of G-proteins (see Figure 8-6).[114,115] Alternatively, agonist-promoted upregulation of surface D2 receptors in CHO cells has been reported to be linked to increased levels of mRNA. D2 receptor upregulation and behavioral supersensitivity have also been observed in animals treated with indirect dopamine agonists.[98-100]

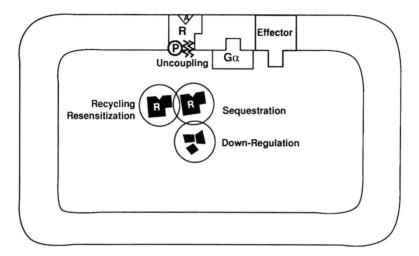

FIGURE 8-5 A schematic illustrating the multiple levels of receptor regulation following agonist activation. The biochemical processes involved in the agonist-induced functional uncoupling (desensitization) of the receptor-coupled G-protein/effector system may include phosphorylation and palmitoylation of the receptor. Sequestration (internalization), downregulation (receptor degradation), and receptor recycling have also been shown to regulate receptor activity. In the figure: receptor (R), agonist (A), phosphorylation (P), palmitoylation (>>), G protein (Gα).

The differences observed in the ability of dopamine to distinctly regulate D1 and D2 receptor-transmembrane signaling suggest possibly important differences in the physiological roles of dopamine mediated through these receptor types. To conceptualize what may be occurring at the molecular level, we speculate the following. Since D1 and D2 receptors are found colocalized in many of the same brain regions[39,116] and perhaps even on the same neurons,[117] conceivably, as discussed earlier, at low concentrations of dopamine the D1 receptor would be preferentially activated owing to its higher affinity for dopamine, whereas for D2 receptor activation higher concentrations of dopamine or longer exposure may be required. This notion is interesting since drugs of abuse are known to increase dopamine release, and the onset of drug action may be mediated by the D1 receptor, whereas the rewarding properties may be mediated by the D2 receptor.

Alternatively, the differential functions of D1 and D2 receptors may serve as a mechanistic basis to explain multiple drug actions at the dopamine neuron where D1 receptors may be involved to a greater extent in cocaine reward, whereas the converse may be true for other abused substances. Still possibly, the proposed temporal and concentration-dependent activation of D1 and then D2 receptors may contribute to mechanisms regulating the start and termination of the dopamine signal. The rapid pattern of D1 receptor desensitization would ensure that D1 receptor-linked cellular signaling events are rapidly inactivated again, whereas D2 receptor activation will result in a longer, more sustained pattern of activation of D2 receptor-linked intracellular events. In keeping with this hypothesis in a model of cocaine self-administration in rats, we have found desensitization of the D1 receptor but not of the D2 receptor.[118] These findings suggest that in tissues containing both D1 and D2 receptors, the repertoire of dopamine-mediated postreceptor events may be modulated in a temporal manner as well, with a changing ratio or shifting balance of D1:D2 receptor effects. In

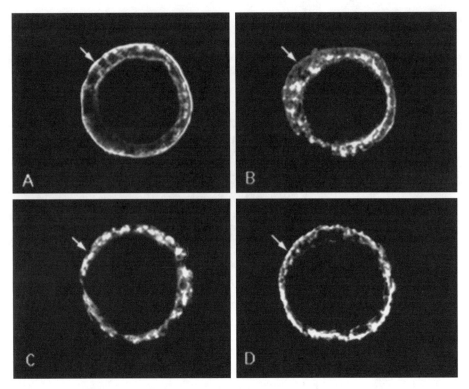

FIGURE 8-6 Agonist-induced differential redistribution of dopamine D1 and D2 receptors expressed in Sf9 cells imaged by confocal laser microscopy. The monoclonal 9E10 antibody recognizing the c-myc epitope-tagged D1 receptor and the polyclonal AL-26 antibody recognizing the D2 receptor were used for immunocytochemical labeling of fixed cells before and after exposure to dopamine. Panel A: vehicle-treated c-myc D1 receptors; panel B: dopamine-treated c-myc D1 receptors (1-h treatment); panel C: vehicle-treated D2 receptors; panel C: dopamine-treated D2 receptors (4-h treatment).

particular, since D2 receptor-mediated mechanisms are important for autoreceptor function in dopamine neurons, the relative insensitivity to desensitization by agonist activation would form an important aspect of the functional role of D2 receptors. It must be recognized that such conceptualizations may be an uncertain exercise, given the complexity of the workings of the brain.

E. CONCLUSION

Although it is clear that multiple factors are involved in drug-seeking behaviors, the dopaminergic system is a plausible candidate basis of drug addictions. While the mechanisms remain as yet unclear, molecular biological techniques have resulted in recent identification of multiple dopamine receptor genes, and progress has been made in the study of the molecular pharmacology of receptors. These advances should further the understanding of dopamine receptor-linked transmembrane signaling and how this may play a role in drug addiction.

REFERENCES

1. Syed, N.I., Roger, I., Ridgway, R.L., Bauce, L.G., Lukowiak, K., and Bulloch, A.G., Identification, characterization and in vitro reconstruction of an interneuronal network of the snail *Melisoma trivolvis*, *J. Exp. Biol.* 174, 19, 1993.

2. Lindvall, O. and Bjorklund, A., Dopamine and norepinephrine-containing neuron systems: their anatomy in the rat brain, in *Chemical Neuroanatomy*, Emson, P.C., Ed., Raven Press, New York, 1983, 229.
3. Bozarth, M.A., Neural basis of psychostimulant and opiate reward. Evidence suggesting the involvement of a common dopaminergic pathway, *Behav. Brain Res.*, 22, 107, 1986.
4. Wise, R.A. and Rompre, P.P., Brain dopamine and reward, *Annu. Rev. Psychol.*, 40, 191, 1989.
5. DiChiara, G. and Imperato, A., Drugs abused by humans preferentially increase synaptic dopamine concentrations in the mesolimbic system of freely moving rats, *Proc. Natl. Acad. Sci. U.S.A.*, 85, 5274, 1988.
6. Kebabian, J.W. and Calne, D.B., Multiple receptors for dopamine, *Nature*, 277, 93, 1979.
7. Berridge, M.J. and Irvine, R.F., Inositol triphosphate, a novel second messenger in cellular signal transduction, *Nature*, 312, 315, 1984.
8. Vallar, L., Muca, C., Magni, M., Albert, P., Bunzow, J., Meldolesi, J., and Civelli, O., Differential coupling of dopaminergic D2 receptors expressed in different cell types. Stimulation of phosphatidylinositol 4,5-bisphosphate hydrolysis in LtK$^-$ fibroblasts, hyperpolarization, and cytosolic-free Ca^{+2} concentration decrease in GH$_4$C$_1$ Cells, *J. Biol. Chem.*, 265, 10320, 1990.
9. Malgaroli, A., Vallar, L., Elahi, F.R., Pozzan, T., Spada, A., and Meldolesi, J.J., Dopamine inhibits cytosolic Ca^{+2} increases in rat lactotroph cells. Evidence of a dual mechanism of action, *J. Biol. Chem.*, 262, 139020, 1987.
10. Mahan, L.C., Burch, R.M., Monsma, F.J., Jr., and Sibley, D.R, Expression of striatal D1 dopamine receptor coupled to inositolphosphate production and Ca^{+2} mobilization in *Xenopus* oocytes, *Proc. Natl. Acad. Sci. U.S.A.*, 87, 2196, 1990.
11. Neve, K.A., Kozlowski, M., and Rosser, M.P., Dopamine D2 receptor stimulation of Na$^+$/H$^+$ exchange assessed by quantification of extracellular acidification, *J. Biol. Chem.*, 267, 25748, 1992.
12. Sunahara, R.K., Niznkik, H.B., Weiner, D.M., Stormann, T.M., Brann, M.R., Kennedy, J.L., Gerlenter, J.E., Rozmahel, R., Yang, Y., Israel, Y., Seeman, P., and O'Dowd, B.F., Human dopamine D1 receptor encoded by an intronless gene on chromosome 5, *Nature*, 347, 80, 1990.
13. Sunahara, R.K., Guan, H.C., O'Dowd, B.F., Seeman, P., Ng, G., George, S.R., Torchia, J., Van Tol, H.H.M., and Niznik, H., Cloning a human dopamine receptor gene (D5) with higher affinity for dopamine than D1, *Nature*, 350, 614, 1991.
14. Chio, C.L., Drong, R.F., Riley, D.T., Gill, G.S., Slightom, J.L., and Huff, R.M., D4 dopamine receptor-mediated signalling events determined in transfected chinese hamster ovary cells, *J. Biol. Chem.*, 269, 11813, 1994.
15. Chio, C.L., Lajiness, M.E., and Huff, R.M., Activation of heterologously expressed D3 dopamine receptors: comparison with D2 dopamine receptors, *Mol. Pharmacol.*, 45, 51, 1994.
16. Setler, P.E., Sarau, H.M., Zirkle, C.L., and Saunders, H.L., The central effects of a novel dopamine agonist, *Eur. J. Pharmacol.*, 50, 419, 1978.
17. Irorio, L.C., Barnett, A., Leitz, F.H., Houser, V.P., and Korduba, C.A., SCH-23390, a potential benzazepine antipsychotic with unique interaction on dopaminergic systems, *J. Pharmacol. Exp. Ther.*, 226, 462, 1983.
18. Billard, W., Ruperto, V., Crosby, G., Iorio, L.C., and Barnett, A., Characterization of the binding of [^{3}H]SCH-23390, a selective D1 receptor antagonist ligand, in rat striatum, *Life Sci.*, 35, 1885, 1984.
19. Hess, E.J., Battaglia, G., Norman, A.B., Iorio, L.C., and Creese, I., Guanine nucleotide regulation of agonist interactions at [^{3}H]SCH-23390 labelled D1 dopamine receptors in rat striatum, *Eur. J. Pharmacol.*, 121, 31, 1986.
20. Fuxe, K., Fredholm, B.B., Ogren, S.O., Agnati, L.F., Hokfelt, T., and Gustafsson, J.A., Pharmacological and biochemical evidence for the dopamine agonistic effect of bromocriptine, *Acta Endocrinol.*, Suppl. 2166, 27, 1978.
21. Seeman, P. and Schaus, J.M., Dopamine receptors labelled by [^{3}H]quinpirole, *Eur. J. Pharmacol.*, 203, 105 1991.
22. Leyson, J.E., Gommeren, W., and Laduron, P.M., Spiperone: a ligand of choice for neuroleptic receptors, kinetics and characteristics of in vitro binding, *Biochem. Pharmacol.*, 27, 307, 1978.
23. Hall, H., Kohler, C., and Gawell, L.L., Some in vitro receptor binding properties of [^{3}H]eticlopride, a novel substituted benzamide, selective for dopamine D2 receptors in the rat brain, *Eur. J. Pharmacol.*, 111, 191, 1985.
24. Baldessarini, R.J. and Frankenburg, F.R., Clozapine: a novel antipsychotic agent, *N. Engl. J. Med.*, 324, 746, 1991.
25. Nguyen, T., Bard, J., Jin, H., Taruscio, D., Ward, D.C., Kennedy, J.L., Weinshank, R., Seeman, P., and O'Dowd, B.F., Dopamine D5 receptor human pseudogenes, *Gene*, 109, 211, 1991.
26. Nguyen, T., Sunahara, R., Marchese, A., Van Tol, H.M., Seeman, P., and O'Dowd, B.F., Transcription of a human dopamine D5 pseudogene, *Biochem. Biophys. Res. Commun.*, 181, 16, 1991.
27. Bunzow, J.R., Van Tol, H.H.M., Grandy, D.K., Albert, P., Salon, J., Christie, M., Machida, C.A., Neve, K.A., and Civelli, O., Cloning and expression of a rat D2 dopamine receptor cDNA, *Nature*, 336, 783, 1988.
28. Dal Toso, R., Sommer, B., Ewert, M., Herb, A., Pritchett, D.B., Bach, A., Shivers, B.D., and Seeburg, P.H., The dopamine D2 receptor: two molecular forms generated by alternative splicing, *EMBO J.*, 8, 4025, 1989.

29. Sokoloff, P., Giros, B., Martres, M.P., Bouthenet, M.L., and Schwartz, J.C., Molecular cloning and characterization of a novel dopamine receptor (D3) as a target for neuroleptics, *Nature*, 347, 146, 1990.
30. Fishburn, C.S., Bellei, D., David, C., Carmon, S., and Fuchs, S., A novel short isoform of the D3 dopamine receptor generated by alternative splicing in the third cytoplasmic loop, *J. Biol. Chem.*, 268, 5872, 1993.
31. Liu, K., Bergson, C., Levenson, R., and Schmauss, C., On the origin of mRNA encoding the truncated dopamine D3-type receptor $D3_{nf}$ and detection of $D3_{nf}$-like immunoreactivity in human brain, *J. Biol. Chem.*, 269, 29220, 1994.
32. Van Tol, H.H.M., Bunzow, J.R., Guan, H.-C., Sunahara, R.K., Seeman, P., Niznik, H.B., and Civelli, O., Cloning of the gene for a human dopamine D4 receptor with high affinity for the antipsychotic clozapine, *Nature*, 350, 610, 1991.
33. Van Tol, H.H.M., Wu, C.M., Guan, H.-C., Ohara, K., Bunzow, J.R., Civelli, O., Kennedy, J., Seeman, P., Niznik, H.B., and Jovanovic, V., Multiple dopamine D4 receptor variants in he human population, *Nature*, 358, 149, 1992.
34. Ng, G.Y.K., Mouillac, B., George, S.R., Caron, M., Dennis, M., Bouvier, M., and O'Dowd, B.F., Desensitization, phosphorylation and palmitoylation of the human dopamine D1 receptor, *Eur. J. Pharmacol.*, 7, 267, 1994.
35. Ng, G.Y.K., Caron, M., Dennis, M., Brann, M.R., O'Dowd, B.F., and George, S.R., Phosphorylation and palmitoylation of the human D2L dopamine receptor in Sf9 cells, *J. Neurochem.*, 63, 1589, 1994.
36. O'Dowd, B.F., Structures of dopamine receptors, *J. Neurochem.*, 60, 804, 1993.
37. Bouthenet, M.-L., Martres, M.-P., Sales, N., and Schwartz, J.-C., A detailed mapping of dopamine D-2 receptors in rat central nervous system by autoradiography with [^{125}I]iodosulpiride, *Neuropharmacology*, 26, 117, 1987.
38. Wamsley, J.K., Gehlert, D.R., Filloux, F.M., and Dawson, T.M., Comparison of the distribution of D1 and D2 dopamine receptors in the rat brain, *J. Comp. Neuroanat.*, 2, 119, 1989.
39. Mansour, A., Meador-Woodruff, J.H., Bunzow, J.R., Civelli, O., Akil, H., and Watson, S.J., Localization of dopamine D2 receptor mRNA and D1 and D2 receptor binding in the rat brain and pituitary: an in situ hybridization-receptor autoradiographic analysis, *J. Neurosci.*, 10, 2587, 1990.
40. Stoof, J., De Boer, T., Sminia, P., and Mulder, A.H., Stimulation of D2 dopamine receptors in rat neostriatum inhibits the release of acetylcholine and dopamine but does not affect the release of gamma-aminobutyric acid, glutamate or serotonin, *Eur. J. Pharmacol.*, 84, 211, 1982.
41. Levesque, D., Diaz, J., Pilon, C., Martres, M.-P., Giros, B., Souil, E., Schott, D., Morgat, J.-L., Schwartz, J.-C., and Sokoloff, P., Identification, characterization, and localization of the dopamine D3 receptor in rat brain using 7-[3H]hydroxy-N-N-di-n- propyl-2-aminotetralin, *Proc. Natl. Acad. Sci. U.S.A.*, 89, 8155, 1992.
42. Bouthenet, M.-L., Souil, E., Martres, M.-P., Sokoloff, P., Giros, B., and Schwartz, J.-C., Localization of dopamine D3 receptor mRNA in the rat brain using in situ hybridization histochemistry: comparison with dopamine D2 receptor mRNA, *Brain Res.*, 564, 203, 1991.
43. Laurier, L.G., O'Dowd, B.F., and George, S.R., Heterogeneous tissue-specific transcription of dopamine receptor subtype messenger RNA in rat brain, *Mol. Brain Res.*, 25, 344, 1994.
44. George, F.R. and Goldberg, S.R., Genetic approaches to the analysis of addiction processes, *TIPS*, 10, 78, 1989.
45. Pickett, R.A. and Collins, A.C., Use of genetic analysis to test the potential role of serotonin in alcohol preference, *Life Sci.*, 17, 1291, 1975.
46. Ng, G.Y.K., O'Dowd, B.F., and George, S.R., Genotypic differences in brain dopamine receptor function in the DBA/2J and C57BL/6J inbred mouse strains, *Eur. J. Pharmacol.*, 269, 349, 1994.
47. Severson, J.A., Randall, P.K., and Finch, C.E., Genotypic influences on striatal dopaminergic regulation in mice, *Brain Res.*, 210, 201, 1981.
48. Boehme, R.E. and Ciaranello, R.D., Genetic control of dopamine and serotonin receptors in brain regions of inbred mice, *Brain Res.*, 266, 51, 1982.
49. Erwin, V.G., Womer, D.E., Campbell, A.D., and Jones, B.C., Pharmacogenetics of cocaine. II. Mesocorticolimbic and striatal dopamine and cocaine receptors in C57BL and DBA mice, *Pharmacogenetics*, 3, 189, 1993.
50. Cotzias, G.C. and Tang, L.C., An adenylyl cyclase of brain reflects propensity for breast cancer in mice, *Science*, 197, 1095, 1977.
51. Murphy, J.M., McBride, W.J., Lumeng, L., and Li, T.K., Regional brain levels of monoamines in alcohol-preferring and nonpreferring lines of rats, *Pharmacol. Biochem. Behav.*, 16, 145, 1981.
52. Sinclair, J.D., Le, A.D., and Kiianmaa, K., The AA and ANA rat lines, selected for differences in voluntary alcohol consumption, *Experientia*, 45, 797, 1989.
53. George, S.R., Fan, T., Ng, G.YK., Jung, S.Y., O'Dowd, B.F., and Naranjo, C.A., Low endogenous dopamine function in brain predisposes to high alcohol preference and consumption: reversal by increasing synaptic dopamine, *J. Pharmacol. Exp. Ther.*, 273, 373, 1995.
54. Barbaccia, M.L., Reggiani, A., Spano, P.F., and Trabucchi, M., Ethanol-induced changes of dopaminergic function in three strains of mice characterized by a different population of opiate receptors, *Psychopharmacology*, 74, 260, 1981.

55. Khatib, S.A., Murphy, J.M., and McBride, W.J., Biochemical evidence for activation of specific monoamine pathways by ethanol, *Alcohol*, 5, 295, 1988.
56. Fadda, F., Mosca, E., Colombo, G., and Gessa, G.L., Effects of spontaneous ingestion of ethanol on brain dopamine metabolism, *Life Sci.* 44, 281, 1989.
57. Kornetsky, C., Bain, G.T., Unterwald, E.M., and Lewis, M.J., Brain stimulation reward: effects of ethanol, *Alcoholism*, 12, 609, 1988.
58. Imperato, A. and DiChiara, G., Preferential stimulation of dopamine release in the nucleus accumbens of freely moving rats by ethanol, *J. Pharmacol. Exp. Ther.*, 239, 219, 1986.
59. Tabakoff, B. and Hoffman, P.L., Development of functional dependence on ethanol in dopaminergic systems, *J. Pharmacol. Exp. Ther.*, 208, 216, 1978.
60. Lucchi, L., Moresco, R.M., Govoni, S., and Trabucchi, M., Effect of chronic ethanol treatment on dopamine receptor subtypes in rat striatum, *Brain Res.*, 449, 347, 1988.
61. Mereu, G., Fadda, F., and Gessa, G.L., Ethanol stimulates the firing rate of nigral dopaminergic neurons in anaesthetized rats, *Brain Res.*, 292, 63, 1984.
62. Gessa, G.L., Muntoni, F., Collu, M., Vargiu, L., and Mereu, G., Low doses of ethanol activate dopaminergic neurons in the ventral tegmental area, *Brain Res.*, 348, 201, 1985.
63. Wand, G.S., Dieh, A.M., Levine, M.A., Wolfgang, D., and Samy, S., Chronic ethanol treatment increases expression of inhibitory G-proteins and reduces adenylyl cyclase activity in the central nervous system of two lines of ethanol-sensitive mice, *J. Biol. Chem.*, 268, 2595, 1993.
64. Walters, J.R., Bergstrom, D.A., Carlson, J.H., Chase, T.N., and Braun, A.R., D1 dopamine receptor activation required for postsynaptic expression of D2 agonist effects, *Science*, 236, 719, 1987.
65. Piomelli, D., Pilon, C., Giros, B., Sokoloff, P., Martres, M.P., and Schwartz, JC., Dopamine activation of the arachidonic acid cascade as a basis for D1/D2 receptor synergism, *Nature*, 353, 164, 1991.
66. LaHoste, J. and Marshall, J.F., Dopamine supersensitivity and D1/D2 synergism are unrelated to changes in striatal receptor density, *Synapse*, 12, 14, 1992.
67. Starr, M.S. and Starr, B.S., Behavioral synergism between the dopamine agonists SKF38393 and LY 171555 in dopamine-depleted mice. Antagonism by sulpiride reveals only stimulant postsynaptic D2 receptors, *Pharmacol. Biochem. Behav.*, 33, 41, 1989.
68. Seeman, P. and Niznik, H.B., Dopamine receptors and transporters in Parkinson's disease and schizophrenia, *FASEB J.*, 4, 2737, 1990.
69. Ng, G.Y.K. and George, S.R., Dopamine receptor agonist reduces ethanol self- administration in the ethanol-preferring C57BL/6J inbred mouse, *Eur. J. Pharmacol.*, 269, 365, 1994.
70. Pfeffer, A.O. and Samson, H.H., Haloperidol and apomorphine effects on ethanol reinforcement in free feeding rats, *Pharmacol. Biochem. Behav.*, 29, 343, 1988.
71. Dyr, W., McBride, W.J., Lumeng, L., Li, T.-K., and Murphy, J.M., Effects of D1 and D2 dopamine receptor agents on ethanol consumption in the High-Alcohol-Drinking (HAD) line of rats, *Alcohol*, 10, 207, 1993.
72. Weiss, F., Mitchiner, M., Bloom, F.E., and Koob, G.F., Free-choice responding for ethanol versus water in alcohol preferring (P) and unselected Wistar rats is differentially modified by naloxone, bromocriptine, and methysergide, *Psychopharmacology*, 101, 178, 1990.
73. Rassnick, S., Pulvirenti, L., and Koob, G.F., SDZ-205,152, a novel dopamine receptor agonist, reduces oral ethanol self-administration in rats, *Alcohol*, 10, 127, 1993.
74. Levy, A.D., Murphy, J.M., McBride, W.J., Lumeng, L., and Li, T.K., Microinjections of sulpiride into the nucleus accumbens increases ethanol drinking in alcohol-preferring (P) rats, *Alcohol Alcoholism*, Suppl. 1, 417, 1991.
75. Hubner, C.B. and Koob, G.F., Bromocriptine produces decreases in cocaine self- administration in the rat, *Neuropsychopharmacology*, 3(2), 101, 1989.
76. Blum, K. and Briggs, A.H., Opioid peptides and genotypic responses to ethanol, *Biogenic Amines*, 5, 527, 1988.
77. George, S.R., Roldan, L., Lui, A., and Naranjo, C.A., Endogenous opioids are involved in the genetically determined high preference for ethanol consumption, *Alcoholism: Clin. Exp. Res.*, 15, 668, 1991.
78. McBride, W.J., Murphy, J.M., Lumeng, L., and Li, T.K., Serotonin, dopamine and GABA involvement in alcohol drinking of selectively bred rats, *Alcohol*, 7, 1990.
79. Tabakoff, B., Dissociation of alcohol tolerance and dependence, *Nature*, 263, 418, 1976.
80. Wise, R.A., Action of drugs of abuse on brain reward systems, *Pharmacol. Biochem. Behav.*, 24, 291, 1980.
81. Karoum, F., Wyatt, R.J., and Majchrowicz, E., Brain concentrations of biogenic amine metabolites in acutely treated and ethanol-dependent rats, *Br. J. Pharmacol.*, 56, 403, 1976.
82. Hoffman, P.L. and Tabakoff, B., Alterations in dopamine receptor sensitivity by chronic ethanol treatment, *Nature*, 268, 551, 1977.
83. Rabin, R.A., Wolfe, B.B., Dibner, M.D., Zahniser, N.R., Melchoir, C., and Molinoff, P.B., Effects of ethanol administration and withdrawal on neurotransmitter receptor systems in C57 mice, *J. Pharmacol. Exp. Ther.*, 213, 491, 1980.
84. Wise, R.A., Neuroleptics and operant behaviour. The anhedonia hypothesis, *Behav. Brain Sci.*, 5, 39, 1982.

85. Myers, R.D. and Melchior, C.L., Alcohol drinking in the rat after destruction of serotonergic and catecholaminergic neurons in the brain, *Res. Commun. Chem. Pathol. Pharmacol.*, 10, 363, 1975.
86. Kiianmaa, K., Andersson, K., and Fuxe, K., On the role of ascending dopamine systems in the control of voluntary ethanol intake and ethanol intoxication, *Pharmacol. Biochem. Behav.*, 10, 603, 1979.
87. Amit, Z. and Brown, Z.W., Actions of drugs of abuse on brain reward systems. A reconsideration with specific attention to alcohol, *Pharmacol. Biochem. Behav.*, 17, 233, 1982.
88. Blum, K., Noble, E.P., Sheridan, P.J., Montgomery, A., Ritchie, T., Jagadeeswaran, P., Nogami, H., Briggs, A.H., and Cohn, J.B., Allelic association of human dopamine D2 receptor gene in alcoholism, *JAMA*, 263, 2055, 1990.
89. Blum, K., Noble, E.P., Sheridan, P.J., Finley, O., Montgomery, A., Ritchie, T., Ozkaragoz, T., Fitch, R.J., Sadlack, F., Sheffield, D., Dahlmann, T., Halbardier, S., and Nogami, H., Association of the A1 allele of the D2 receptor gene with severe alcoholism, *Alcohol*, 8, 409, 1991.
90. Noble, E.P., Blum, K., Ritchie, T., Montgomery, A., and Sheridan, P.J., Allelic association of the D2 dopamine receptor gene with receptor binding characteristics in alcoholism, *Arch. Gen. Psychiatry*, 48, 648, 1991.
91. Bolos, A.M., Dean, M., Lucas-Derse, S., Ramsburg, M., Brown, G.L., and Goldman, D., Population and pedigree studies reveal a lack of association between the dopamine D2 receptor gene and alcoholism, *JAMA*, 264, 3156, 1990.
92. Gelernter, J., O'Malley, S., Risch, N., Kranzler, H.R., Krystal, J., Merikangas, K., Kennedy J.L., and Kidd, K., No association between an allele at the D2 dopamine receptor gene (DRD2) and alcoholism, *JAMA*, 266, 1801, 1991.
93. Noble, E.P., St. Jeor, S.T., Ritchie, T., Syndulko, K., St. Jeor, S.C., Fitch, R.J., Brunner, R.L., and Sparkes, R.S., D2 dopamine receptor gene and cigarette smoking: a reward gene?, *Med. Hypotheses*, 42, 257, 1994.
94. Comings, D.E., Comings, B.G., Muhleman, D., Dietz, G., Shahbahrami, B., Tast, D., Knell, E., Kocsis, P., Baumgarten, R., Kovacs, B.W., and Levy, D.L., The dopamine D2 receptor gene locus as a modifying gene in neuropsychiatric disorders, *JAMA*, 266, 1793, 1991.
95. Smith, S.S., O'Hara, B.F., Persico, A.M., Gorelick, D.A., Newlin, D.N., Vlahov, D., Solomon, L., Pickens, R., and Uhl, G.R., Genetic vulnerability to drug abuse, *Arch. Gen. Psychiatry*, 49, 723, 1992.
96. Uhl, G.R., Persico, A.M., and Smith, S.S., Current excitement with D2 dopamine receptor gene alleles in substance abuse, *Arch. Gen. Psychiatry*, 49, 157, 1992.
97. George, S.R., Cheng, R., Nguyen, T., Israel Y., and O'Dowd, B.F., Polymorphisms of the D4 dopamine receptor alleles in chronic alcoholism, *Biochem. Biophys. Res. Commun.*, 196, 107, 1993.
98. Rouillard, C., Bedard, P.J., Falardeau, P., and Dipaolo, T., Behavioral and biochemical evidence for a different effect of repeated administration of L-DOPA and bromocriptine on denervated versus non-denervated striatal dopamine receptors, *Neuropharmacology*, 26, 1601, 1978.
99. Vassout, A., Bruinink, A., Krauss, J., Waldmeier, P., and Bischoff, S., Regulation of dopamine receptors by bupropion. Comparison with antidepressants and CNS stimulants, *J. Receptor Res.*, 13, 341, 1993.
100. Bischoff, S., Krauss, J., Grunenwald, C., Gunst, F., Heinrich, M., Schaub, M., Stocklin, K., Vassout, A., Waldmeier, P., and Maitre, L., Endogenous dopamine (DA) modulates [^{3}H]spiperone binding in vivo in rat brain, *J. Receptor Res.*, 11, 163, 1991.
101. Barton, A.C. and Sibley, D.R., Agonist-induced desensitization of D1-dopamine receptors linked to adenylyl cyclase activity in cultured NS20Y neuroblastoma cells, *Mol. Pharmacol.*, 38, 531, 1990.
102. Bates, M.D., Olsen, C.L., Becker, B.N., Albers, F. J., Middleton, J.P., Mulheron, J.G., Jin, S.-L., Conti, M., and Raymond, J.R., Elevation of cAMP is required for downregulation, but not agonist-induced desensitization, of endogenous dopamine D1 receptors in opossum kidney cells. Studies in cells that stably express a rat cAMP phosphodiesterase (rPDE3) cDNA, *J. Biol. Chem.*, 268, 14757, 1993.
103. Ng, G.Y.K., Trogadis, J., Stevens, J., Bouvier, M., O'Dowd, B.F., and George, S.R., Agonist-induced desensitization of the dopamine D1 receptor coupled adenylyl cyclase system is temporally and biochemically separate from receptor internalization, *Proc. Natl. Acad. Sci. U.S.A.*, 92, 10157, 1995.
104. Perkins, J.P., Hausdorff, W.P., and Lefkowitz, R.J., Mechanisms of ligand-induced desensitization of the β-adrenergic receptor, in *The β-Adrenergic Receptor*, Perkins, J.P., Ed., Humana Press, New York, 1990, 1.
105. Barton, A.C., Black, L.E., and Sibley, D.R., Agonist-induced desensitization of D2 dopamine receptors in human Y-79 retinoblastoma cells, *Mol. Pharmacol.*, 39, 650, 1991.
106. Ivins, K.J., Luedtke, R.R., Arthmyshyn, R.P., and Molinoff, P.B., Regulation of dopamine D2 receptors in a novel cell line (SUP1), *Mol. Pharmacol.*, 39, 531, 1991.
107. Filtz, T.M., Artymyshyn, R.P., Guan, W., and Molinoff, P.B., Paradoxical regulation of dopamine receptors in transfected 293 cells, *Mol. Pharmacol.*, 44, 371, 1993.
108. Zhang, L.-J., Lachowicz, J.E., and Sibley, D.R., The D2S and D2L dopamine receptor isoforms are differentially regulated in Chinese hamster ovary cells, *Mol. Pharmacol.*, 45, 878, 1994.
109. Ng, G.Y.K., Varghese, G., Bouvier, M., Brann, M., O'Dowd, B.F., and George, S.R., Coexpression of dopamine D1 and D2 receptors in Sf9 cells, *Soc. Neurosci.*, 21, 617, 1995.
110. Parker, E.M., Kameyama, K., Higsijima, T., and Ross, E.M., Reconstitutively active G protein-coupled receptors purified from baculovirus-infected insect cells, *J. Biol. Chem.*, 266, 519, 1991.

111. Ng, G.Y.K., George, S.R., Zastawny, R., Caron, M., Dennis, M., and O'Dowd, B.F., Human serotonin$_{1B}$ receptor expression in Sf9 cells: phosphorylation, palmitoylation, and adenylyl cyclase inhibition, *Biochemistry*, 32, 11727, 1993.
112. Pickering, D.S., Thomsen, C., Suzdak, P.D., Fletcher, E.J., Robitaille, R., Salter, M.W., MacDonald, J.F., Huang, X.-P., and Hampson, D.R., A comparison of two alternatively spliced forms of a metabotropic glutamate receptor coupled to phosphoinositide turnover, *J. Neurochem.*, 61, 85, 1993.
113. Ng, G.Y.K., Trogadis, J., Stevens, J., Bouvier, M., O'Dowd, B.F., and George, S.R., Agonist-induced dopamine receptor subtype-specific redistribution in Sf9 cells, *Eur. J. Neurosci.*, Suppl. 7, 139, 1995.
114. Neer, E.J. and Clapham, D.E., Roles of G protein subunits in transmembrane signalling, *Nature*, 333, 129, 1988.
115. Milligan, G., Agonist regulation of cellular G protein levels and distribution: mechanisms and functional implications, *TIPS*, 14, 413, 1993.
116. Levey, A.I., Hersch, S.M., Rye, D.B., Sunahara, R., Niznik, H., Kitt, C.A., Prize, D.L., Maggio, R., Brann, M.R., and Ciliax, B.J., Localization of D1 and D2 dopamine receptors in brain with subtype-specific antibodies, *Proc. Natl. Acad. Sci. U.S.A.*, 90, 8861, 1993.
117. Surmeier, D.J., Eberwine, J., Wilson, C.J., Cao, Y., Stefani, A., and Kitai, S.T., Dopamine receptor subtypes colocalize in rat striatonigral neurons, *Proc. Natl. Acad. Sci. U.S.A.*, 89, 10178, 1992.
118. Laurier, L.G., Corrigall, W.A., and George, S.R., Dopamine receptor density, sensitivity and mRNA levels are altered following self-administration of cocaine in the rat, *Brain Res.*, 634, 31, 1994.

9 Serotonin Receptors: Role in Psychiatry

Joanne M. Scalzitti and Julie G. Hensler

CONTENTS:

A. Introduction 114
    1. Discovery 114
    2. Classification of Serotonin Receptor Subtypes 115
B. Subtypes of Receptors for Serotonin 118
    1. The Serotonin-1A ($5-HT_{1A}$) Receptor 118
        a. Pharmacology 118
        b. Distribution and Function in Brain 118
        c. Clinical Correlates 119
        d. Molecular Biology and Genetics 120
    2. The Serotonin-1B ($5-HT_{1B}$) Receptor 121
        a. Pharmacology 121
        b. Distribution and Function in Brain 122
    3. The Serotonin-1D ($5-HT_{1D}$) Receptor 122
        a. Pharmacology 122
        b. Distribution and Function in Brain 123
        c. Clinical Correlates 123
        d. Molecular Biology and Genetics 123
    4. The Serotonin-2A ($5-HT_{2A}$) Receptor 124
        a. Pharmacology 124
        b. Distribution and Function in Brain 125
        c. Clinical Correlates 125
        d. Molecular Biology and Genetics 126
    5. The Serotonin-2B ($5-HT_{2B}$) Receptor 127
        a. Pharmacology 127
        b. Distribution and Function in Brain 127
        c. Molecular Biology 127
    6. The Serotonin-2C ($5-HT_{2C}$) Receptor 128
        a. Pharmacology 128
        b. Distribution and Function in Brain 128
        c. Clinical Correlates 128
        d. Molecular Biology and Genetics 129
    7. The Serotonin-3 ($5-HT_3$) Receptor 129
        a. Pharmacology 129
        b. Distribution and Function in Brain 130
        c. Clinical Correlates 130
        d. Molecular Biology 131

8. The Serotonin-4 (5-HT$_4$) Receptor 131
   a. Pharmacology 131
   b. Distribution and Function in Brain 132
   c. Molecular Biology 132
C. Recombinant Serotonin Receptors 133
   1. The 5-HT$_{1E}$ Receptor 133
   2. The 5-HT$_{1F}$ Receptor 133
   3. The 5-HT$_5$ Receptor 133
   4. The 5-HT$_6$ Receptor 134
   5. The 5-HT$_7$ Receptor 134
Acknowledgments 135
References 135

A. INTRODUCTION

Although previously identified in the periphery, the discovery of serotonin (5-hydroxytryptamine, 5-HT) in brain by Twarog and Page,[1] followed by studies using histofluorescence describing its distribution in brain,[2] indicated that this indolalkylamine had an important role in brain function. The hallucinogen (+) lysergic acid diethylamide (LSD) was shown to block responses mediated by serotonin, specifically contractile responses of intestinal smooth muscle, further suggesting an important role of serotonin in brain and behavior.[3] Various theories arose linking abnormalities of central serotonergic function with the etiology of a number of psychiatric disorders. The involvement of serotonin in brain function and behavior, and a potential role for serotonin in a variety of psychiatric disorders, has been further supported by the development of psychotherapeutic drugs which modulate central serotonergic neurotransmission.

1. DISCOVERY

The initial suggestion of subtypes of receptor for serotonin came from the early work of Gaddum and Picarelli.[4] In experiments using guinea pig ileum, they demonstrated that only a portion of the contractile response to serotonin was antagonized by high concentrations of morphine. The remaining response to serotonin was blocked by low concentrations of dibenzyline (phenoxybenzamine). In the presence of maximally effective concentrations of dibenzyline, the contractile response to 5-HT was antagonized by low concentrations of morphine. Gaddum and Picarelli[4] proposed that there were two receptors for serotonin in the ileum, one blocked by morphine (termed the M receptor) and one blocked by dibenzyline (termed the D receptor).

The development of radioligand binding methodology in the 1970s served to further our understanding of subtypes of receptors for serotonin, particularly in brain. Farrow and Van Vunakis[5,6] observed high-affinity, stereospecific binding of ^{3}H-LSD in cortex that was inhibited more potently by serotonin than by other neurotransmitters. The neuroleptic ^{3}H-spiroperidol, but not ^{3}H-haloperidol, was shown by Leysen and associates[7] to bind to receptors for serotonin in frontal cortex. In experiments using several radioligands, ^{3}H-5-HT, ^{3}H-LSD, and ^{3}H-spiroperidol, Peroutka and Snyder[8] demonstrated the presence of subtypes of receptor for 5-HT in frontal cortex. Binding sites labeled with high affinity by ^{3}H-5-HT were designated 5-HT$_1$; binding sites labeled with high affinity by ^{3}H-spiroperidol were designated 5-HT$_2$. Both classes of serotonin receptor were labeled with similar affinity by ^{3}H-LSD.[8] Although pharmacological and physiological criteria have contributed to the definition and classification of receptor subtypes for serotonin, the application of techniques used in molecular biology to the study of serotonin receptors has led to the recent discovery of a number of additional serotonin receptor subtypes and furthered our understanding of the structure and function of serotonin receptors.[9,10]

2. CLASSIFICATION OF SEROTONIN RECEPTOR SUBTYPES

The recent and rapid discovery of additional subtypes of receptor for serotonin has made it necessary to establish an unambiguous system of nomenclature for serotonin receptors. This classification scheme takes into account not only operational criteria (drug-related characteristics), but also information about intracellular signal transduction mechanisms and molecular structure (amino acid sequence of the receptor protein). In addition, the amino acid sequence of several new serotonin receptors has been reported. However, the classification of these receptors remains tentative due to limited knowledge of their operational and transductional characteristics, which have only been described for these recombinant receptors in transfected cell systems. Because the functions mediated by these serotonin receptors in intact tissue are unknown, lower-case appellations are presently used.[9,10]

The multiplicity of serotonin receptor subtypes is the direct result of the evolutionary age of the serotonin system. The primordial G-protein-coupled serotonin receptor appears to have evolved over 750 million years ago. It was at this point that the initial divergence of the primordial serotonin receptor into three receptor families began. The three receptor families, the 5-HT_1 family, the 5-HT_2 family, and the family that includes the 5-HT_4, 5-HT_6, and 5-HT_7 receptors represent the three major classes of serotonin receptor.[11,12] These three serotonin receptor families belong to the G-protein-coupled receptor superfamily. Members of the G-protein-coupled receptor superfamily of proteins all contain seven transmembrane domains, an intracellular carboxy-terminus, and an extracellular amino-terminus in their characteristic structure (Figure 9-1). It is the interaction of the receptor with the G-protein that allows the receptor to modulate the activity of different effector systems (e.g., ion channels, phospholipase C, adenylyl cyclase).

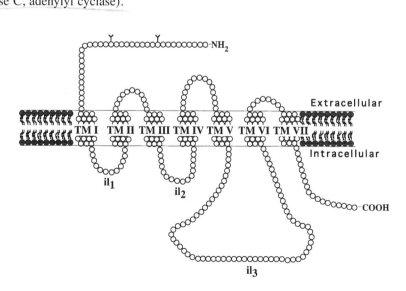

FIGURE 9-1 Generalized schematic diagram of a G-protein-coupled receptor. The amino-terminus of the receptor, containing potential sites for N-linked glycosylation (Y), projects into the extracellular space. The polypeptide chain spans the membrane seven times as an alpha-helix, resulting in transmembrane domains (TM). Structural features also include three intracellular loops (il), and an intracellular carboxy-terminus.

The 5-HT$_1$ receptor family contains receptors that are negatively coupled to adenylyl cyclase and includes the 5-HT$_{1A}$, 5-HT$_{1B}$, 5-HT$_{1D}$, 5-HT$_{1E}$, and 5-HT$_{1F}$ receptors. The 5-HT$_2$ receptor family stimulates phospholipase C and includes the 5-HT$_{2A}$, 5-HT$_{2B}$, and 5-HT$_{2C}$ (formerly the 5-HT$_{1C}$) receptors. The family of serotonin receptors that are coupled to the stimulation of adenylyl cyclase include the 5-HT$_4$, 5-HT$_6$, and 5-HT$_7$ receptors. The 5-HT$_{5A}$ and 5-HT$_{5B}$ receptors may constitute a new family of serotonin receptors since neither are coupled to adenylyl cyclase or phospholipase C; their effector systems are currently unknown.[13]

The 5-HT$_3$ receptor is a homomeric receptor that belongs to the ligand-gated ion channel superfamily (Figure 9-2). Members of this receptor superfamily consist of five subunits, each of which possesses four transmembrane segments and a large, extracellular N-terminal region. Unlike serotonin receptors of the G-protein-coupled receptor superfamily, the 5-HT$_3$ receptor is a serotonin-gated cation channel that causes the rapid depolarization of neurons.

Although each serotonin receptor can be potently activated by serotonin, differences in signal transduction mechanisms, neuroanatomical distribution, and affinities for synthetic chemicals creates opportunities for drug discovery and makes each serotonin receptor subtype a potential therapeutic target (see Table 9.1).

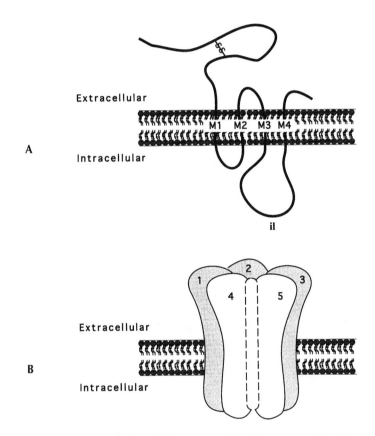

FIGURE 9-2 Structural diagram of the 5-HT$_3$ receptor. (A) Schematic model of the 5-HT$_3$ receptor subunit. Hydrophobicity analysis of the deduced amino acid sequence suggests that the polypeptide chain spans the membrane four times. The long amino-terminus contains a characteristic disulfide bond. Transmembrane domains M3 and M4 are separated by a long intracellular loop. (B) Hypothesized pentameric structure of the 5-HT$_3$ receptor ligand-gated ion channel. The five individual subunits are numbered 1 through 5. The internal ion channel is represented by the dotted lines.

TABLE 9-1
Serotonin Receptors Present in the Central Nervous System

Receptor	Human Locus	mRNA Distribution	Receptor Protein Distribution
5-HT_{1A}	5q11.2-13	Hippocampus, amygdala, septum, raphe nuclei	Hippocampus, amygdala, septum, entorhinal cortex, hypothalamus, raphe nuclei
$5\text{-HT}_{1D\alpha}$	1p34.3-36.3	Striatum, nucleus accumbens, dorsal raphe, hippocampus	Not distinguishable from $5\text{-HT}_{1D\beta}$
$5\text{-HT}_{1D\beta}$	6q13	Striatum, hippocampus, cingulate cortex, subthalamic nuclei, basal ganglia, entorhinal cortex, cerebellum raphe nuclei, spinal cord	Substantia nigra, basal ganglia, superior colliculus
5-HT_{1E}	?	Caudate putamen, parietal cortex, frontoparietal motor cortex, olfactory tubercle	?
5-HT_{1F}	3p11	Dentate gyrus, hippocampus, neocortex, nucleus of the solitary tract, cingulate and piriform cortex, spinal cord	?
5-HT_{2A}	13q14-21	Frontal cortex, hippocampus, olfactory bulb, caudate nucleus, entorhinal and piriform cortex, nucleus accumbens, brainstem nuclei	Claustrum, cerebral cortex, olfactory tubercle, striatum, nucleus accumbens
5-HT_{2B}	?	Cerebral cortex, caudate nucleus, thalamus, hypothalamus, amygdala, cerebellum, substantia nigra, retina	?
5-HT_{2C}	Xq24	Choroid plexus, hippocampus, amygdala, lateral habenula, cingulate and piriform cortex, thalamic nuclei, subthalamic nuclei, hypothalamus, substantia nigra, suprachiasmatic nuclei, locus ceruleus, spinal cord	Choroid plexus, globus pallidus, cerebral cortex, hypothalamus septum, substantia nigra
5-HT_3	?	Hippocampus, amygdala, olfactory bulb, piriform, cingulate and entorhinal cortex, several cranial nerve nuclei, dorsal tegmental area, spinal cord, dorsal root ganglia	Hippocampus, entorhinal cortex, amygdala, nucleus accumbens, solitary tract nerve, trigeminal nerve, motor nucleus of the dorsal vagal nerve, area postrema, spinal cord
5-HT_4	?	Hippocampus, striatum, thalamus, olfactory bulb, brainstem	Hippocampus, caudate nucleus, globus pallidus, olfactory tubercle, substantia nigra
5-HT_{5A}	7q36	Cerebral cortex, hippocampus, septum, amygdala, thalamic nuclei, olfactory bulb cerebellum, medial habenula	?
5-HT_{5B}	2q11-13	Hippocampus, habenula, dorsal raphe	?
5-HT_6	?	Cerebral cortex, hippocampus, striatum, hypothalamus, nucleus accumbens, olfactory tubercles and bulb	?
5-HT_7	10q23.3-24.3	Cerebral cortex, hippocampus, thalamus, amygdala, striatum, cerebellum, substantia nigra, superior colliculus, raphe nuclei	?

B. SUBTYPES OF RECEPTORS FOR SEROTONIN

1. THE SEROTONIN-1A (5-HT$_{1A}$) RECEPTOR

a. Pharmacology

The observation that the binding of ^{3}H-5-HT to 5-HT$_1$ receptor sites in brain homogenates was displaced by the neuroleptic spiperone in a biphasic manner led to the suggestion that what was originally termed the 5-HT$_1$ receptor was a heterogeneous population of receptors. ^{3}H-5-HT binding sites which displayed high affinity for spiperone were called the 5-HT$_{1A}$ receptor; ^{3}H-5-HT binding sites with low affinity for spiperone were called the 5-HT$_{1B}$ receptor.[14] The 5-HT$_{1A}$ receptor is perhaps one of the most well-characterized of the serotonin receptor subtypes, due largely to the availability of a selective agonist with high affinity for the 5-HT$_{1A}$ receptor, 8-OH-DPAT.[15] Although several compounds have been proposed for labeling 5-HT$_{1A}$ sites, the tritiated derivative of 8-OH-DPAT has been the most useful radioligand.[16] The substituted azapirones, buspirone, gepirone, ipsapirone, and tandospirone, exhibit high affinity for the 5-HT$_{1A}$ receptor,[16-19] and have antidepressant- and anxiolytic-like effects in animal models of anxiety and depression.[20] These compounds are generally considered partial agonists at the 5-HT$_{1A}$ receptor,[19-23] although it is important to note that this characteristic is dependent upon the particular assay system used to determine agonist intrinsic activity. In contrast, 8-OH-DPAT is considered a full agonist.

Our understanding of the role of the 5-HT$_{1A}$ receptor in brain has been somewhat limited by the lack of a selective, silent antagonist for this receptor. Potent antagonists of the 5-HT$_{1A}$ receptor such as spiperone, pindolol, or methiothepin have high affinity for other neurotransmitter receptors or other serotonin receptors.[24,25] Compounds such as SDZ-216525, BMY 7378, and NAN-190, which were originally considered selective antagonists at the 5-HT$_{1A}$ receptor, have been found to be weak partial agonists in tests of presynaptic 5-HT$_{1A}$ receptor activation.[25,26] This may be due to a larger receptor reserve for agonists at presynaptic, as compared with postsynaptic, 5-HT$_{1A}$ receptor sites.[27,28] Recently, several silent and selective antagonists for the 5-HT$_{1A}$ receptor have been described. WAY 100135 initially appeared to be an antagonist at both presynaptic and postsynaptic 5-HT$_{1A}$ receptors;[25] however, a number of more recent studies suggest that WAY 100135 may have partial agonist activity at 5-HT$_{1A}$ receptors.[29-31] WAY 100635, a second-generation achiral analogue of WAY 100135, has higher affinity for 5-HT$_{1A}$ receptors than WAY 100135 and appears to be a silent, highly selective, and potent antagonist at the 5-HT$_{1A}$ receptor.[32-35] WAY 100635 has been tritiated and proven to be a useful radioligand to label 5-HT$_{1A}$ receptors.[35,36] Two phenylpiperazine derivatives, p-MPPI and p-MPPF, also appear to be selective 5-HT$_{1A}$ receptor antagonists.[37-39] p-MPPI has the advantage over other selective 5-HT$_{1A}$ receptor antagonists in that it is readily iodinated with ^{125}I or ^{123}I to enable its use as a radioligand both *in vivo* and *in vitro*.[38,40]

b. Distribution and Function in Brain

The 5-HT$_{1A}$ receptor is present in high density in cortical and limbic structures (i.e., hippocampus, entorhinal cortex, septum, amygdala, frontal cortex).[35,41,42] The distribution of the 5-HT$_{1A}$ receptor in brain suggests that this serotonin receptor subtype may have a role in cognitive or integrative functions, as well as in emotional states. Destruction of serotonergic neurons with the neurotoxin 5,7-dihydroxytryptamine (5,7-DHT) does not reduce 5-HT$_{1A}$ receptor number in forebrain areas, indicating that the 5-HT$_{1A}$ receptors in terminal field areas of serotonergic innervation are located postsynaptically.[41,42] 5-HT$_{1A}$ receptors are also present in high density in serotonergic cell body areas in the brainstem, in particular the dorsal and median raphe nuclei. Here they function as somatodendritic autoreceptors which are involved in the negative feedback modulation of serotonergic neuronal activity.[43,44]

Destruction of serotonergic cell bodies dramatically reduces the number of 5-HT$_{1A}$ receptors in this area,[41,42] consistent with their location on serotonergic soma as determined by immunohistochemistry using antibodies to the 5-HT$_{1A}$ receptor.[45]

The 5-HT$_{1A}$ receptor is coupled via G-proteins to two distinct effector systems: (1) the inhibition of adenylyl cyclase activity,[46,47] or (2) the opening of potassium channels which results in neuronal hyperpolarization.[48,49] In terminal field areas of serotonergic innervation such as the hippocampus, the 5-HT$_{1A}$ receptor is coupled to both effector systems. However, in the dorsal raphe nucleus 5-HT$_{1A}$ receptors are coupled only to the opening of potassium channels.[50]

c. **Clinical Correlates**

In animal studies, the sensitivity of behavioral and electrophysiological responses mediated by the 5-HT$_{1A}$ receptor is decreased after chronic administration of 5-HT$_{1A}$ receptor agonists or antidepressant drugs.[21,42,51,52] The desensitization of 5-HT$_{1A}$ receptor-mediated responses does not appear to be accompanied by a decrease in 5-HT$_{1A}$ receptor number or downregulation. Although some investigators have reported changes in 5-HT$_{1A}$ receptor number following such treatments,[53,54] the downregulation of 5-HT$_{1A}$ receptors has not been a consistent observation.[42,55,56] Changes in 5-HT$_{1A}$ receptor second-messenger function have not been consistently reported to follow administration of antidepressants or 5-HT$_{1A}$ receptor agonists.[57-60]

A tremendous amount of attention has been focused on the 5-HT$_{1A}$ receptor since the discovery that buspirone, a clinically effective nonbenzodiazepine anxiolytic, possessed high affinity for the 5-HT$_{1A}$ receptor. The clinical efficacy of other azapirones (also referred to as pyrimidinylpiperazines), ipsapirone, gepirone, and tandospirone, in general anxiety disorder and in major depressive disorder have been demonstrated.[61,62] The possibility that 5-HT$_{1A}$ receptor agonists may have clinical efficacy for both anxiety and depression raises interesting issues concerning the clinical distinction of these disorders and their pharmacological management.[63]

Activation of 5-HT$_{1A}$ receptors can decrease serotonergic neurotransmission through an autoinhibitory mechanisms (somatodendritic autoreceptor) and influence neuronal activity in the limbic system (postsynaptic receptors). Preclinical studies of the mechanism of action of these compounds indicate that activation of both pre- and postsynaptic 5-HT$_{1A}$ receptors contribute to the anxiolytic-like effect of these 5-HT$_{1A}$ receptor agonists in animal models of anxiety.[20,63-65] This is somewhat paradoxical. Activation of presynaptic 5-HT$_{1A}$ receptors would be expected to result in a reduction of serotonergic neurotransmission; activation of postsynaptic 5-HT$_{1A}$ receptors would be equivalent to an increase in serotonergic neurotransmission. As mentioned above, buspirone, gepirone, ipsapirone, and tandospirone are partial agonists at postsynaptic 5-HT$_{1A}$ receptors. However, these compounds act as full agonists at presynaptic 5-HT$_{1A}$ receptors due to larger receptor reserve for agonists at presynaptic, as compared with postsynaptic 5-HT$_{1A}$ receptor sites.[27,28] Because the postsynaptic 5-HT$_{1A}$ receptor has little or no receptor reserve, partial agonists can act as antagonists at these sites. Thus, it has been proposed that anxiolytic-like activity is a result of a reduction in serotonergic neurotransmission mediated by (1) agonist activation of presynaptic receptors, and (2) antagonism at postsynaptic receptors.[64,65]

Antidepressant-like activity of 5-HT$_{1A}$ receptor agonists in a variety of animal models has been observed.[20,64,66] Studies on the mechanism of action for the antidepressant-like effects of these compounds have also addressed whether it is activation of pre- or postsynaptic 5-HT$_{1A}$ receptors mediating this effect. Because antidepressant-like activity of 5-HT$_{1A}$ receptor agonists appears to correlate with their intrinsic activity at postsynaptic 5-HT$_{1A}$ receptors, and because these agents exert their antidepressant-like effects when applied locally into the septum, but not into the raphe nuclei, it has been proposed that the therapeutic antidepressant effect of these compounds is a result of activation of postsynaptic 5-HT$_{1A}$ receptors.[64]

The therapeutic anxiolytic/antidepressant effects of 5-HT$_{1A}$ receptor agonists are typically observed after 2 to 4 weeks of administration in humans.[61,62] This is also the case for antidepressant drugs such as monoamine oxidase inhibitors, tricyclic antidepressants, and selective serotonin reuptake inhibitors (SSRIs), where delayed onset of a significant therapeutic effect is customary but equally problematic.[67] It has been demonstrated that acute administration of antidepressant drugs which preferentially block reuptake of serotonin decrease the firing of serotonergic neurons in the dorsal raphe nucleus,[68-71] a result of increased extracellular serotonin levels within the raphe nuclei[71-74] and increased activation of 5-HT$_{1A}$ somatodendritic autoreceptors. The extracellular concentrations of serotonin in terminal field areas of serotonergic innervation are affected little by acute systemic administration of these drugs.[71-74] Only after repeated administration do inhibitors of serotonin uptake enhance serotonin neurotransmission.[74,75] The increase in serotonergic neurotransmission after chronic treatment with SSRIs is believed to be due to a desensitization of 5-HT$_{1A}$ somatodendritic autoreceptors. Thus, it has been proposed that coadministration of a 5-HT$_{1A}$ receptor antagonist with SSRIs may accelerate the onset of increased serotonergic neurotransmission by blocking 5-HT$_{1A}$ somatodendritic autoreceptors.

A variety of serotonin receptor antagonists, including nonselective agents that are antagonists at 5-HT$_{1A}$ receptors, have been shown not to have antidepressant-like effects in animal models predictive of antidepressant-like drug activity.[66] However, coadministration of 5-HT$_{1A}$ receptor antagonists and SSRIs prevents the acute inhibition of serotonergic neuronal firing and increases serotonin release in terminal field areas of serotonergic innervation.[69,70] Administration of 5-HT$_{1A}$ receptor antagonists to rats treated chronically with the serotonin uptake inhibitor, citalopram, increases the activity of serotonergic neurons in the dorsal raphe nucleus.[69] Thus, 5-HT$_{1A}$ receptor antagonists may reduce the latency in therapeutic effect and perhaps increase the clinical efficacy of SSRIs. This hypothesis has been supported by a recent clinical study indicating that treatment with an SSRI combined with pindolol, a β-adrenergic receptor antagonist with potent 5-HT$_{1A}$ receptor antagonist properties, results in the rapid improvement of patients, including a group considered therapy resistant.[76]

5-HT$_{1A}$ receptor agonists have also been reported to decrease alcohol intake and preference in rodent models of alcoholism.[77] In several animal models of aggression, antiaggressive effects of 5-HT$_{1A}$ receptor agonists have been reported.[64] Because aggression and alcoholism can be considered to be instances of uncontrolled impulsivity, it has been proposed that the efficacy of 5-HT$_{1A}$ receptor agonists in animal models of aggression and alcoholism is related to possible antiimpulsivity effects of these compounds.[64]

5-HT$_{1A}$ receptor agonists may prove useful in the treatment of extrapyramidal symptoms that develop as a result of neuroleptic administration. In animal studies, neuroleptic-induced catalepsy is believed to reflect extrapyramidal side effects experienced with chronic administration of typical antipsychotics. 5-HT$_{1A}$ receptor agonists have been shown to block the catalepsy induced by D2 dopamine receptor antagonists.[78-81] These observations suggest that the combination of 5-HT$_{1A}$ receptor agonists with a typical neuroleptic may result in antipsychotic treatment with reduce liability for extrapyramidal side effects.[82]

d. **Molecular Biology and Genetics**

The human 5-HT$_{1A}$ receptor gene was the first of the serotonin receptors to be successfully cloned.[83] The intronless genomic clone, G21, was found by screening a human genomic library with probes for the β$_2$-adrenoreceptor. The gene has been localized to human chromosome 5q11.2-13 and encodes a 422-amino-acid protein.[83-85] 5-HT$_{1A}$ receptor mRNA has been localized to the hippocampus, raphe nuclei, amygdala, and septum.

The 5' flanking sequence of the gene, the untranslated portion of the gene upstream from the protein synthesis initiation site, has been isolated. This cloned fragment possesses a GC-rich region, but does not contain a TATA box.[86] The 5-HT$_{1A}$ promoter is similar to promoters found in housekeeping genes, such as the cytoskeletal protein actin and cell growth control genes, and contrasts with the promoters of receptors in the 5-HT$_2$ receptor family (i.e., the 5-HT$_{2A}$ or 5-HT$_{2C}$ receptor) (see below).

The rat 5-HT$_{1A}$ receptor was first cloned, expressed, and characterized in 1990 by Albert and associates.[87] The rat 5-HT$_{1A}$ receptor gene and protein display significant sequence homology (89%) and structural homology to the human gene and 5-HT$_{1A}$ receptor. Surprisingly, three mRNA species, 3.9, 3.6, and 3.3 kb, have been detected for the rat 5-HT$_{1A}$ receptor.[89] Early biochemical characterization of the 5-HT$_{1A}$ receptor complex isolated from rat hippocampus demonstrated that the receptor is a glycoprotein with an apparent molecular weight (MW) of 155 kDa. In the presence of GTP, the receptor complex dissociates into two components of 60 kDa (receptor) and 80 kDa (G-protein).[88,89] The pharmacology, biochemistry, and anatomical distribution of the rat and human 5-HT$_{1A}$ receptor are essentially identical.

Mutagenesis studies have shown that an asparagine (amino acid 396) in the seventh transmembrane segment of 5-HT$_{1A}$ receptor protein is important for binding of the β-adrenergic antagonist ^{125}I-cyanopindolol,[90] as well as serotonin and 8-OH-DPAT.[85] Other mutagenesis studies have identified the role of other amino acid residues in the second, third, and fifth transmembrane segments that are important for agonist binding.[91] The 5-HT$_{1A}$ receptor sequence has a leucine zipper (a protein sequence in which every seventh amino acid is a leucine over a repeat length of four to five leucines) present in the third transmembrane region.[92] Leucine zippers are thought to facilitate protein dimerization, and exist in DNA binding proteins and oncogenes as well as in potassium, calcium, and sodium channels.[93]

Although neuropharmacological evidence implicates the 5-HT$_{1A}$ receptor in anxiety and depression, molecular genetic analysis has not established any heritable mutations in the 5-HT$_{1A}$ receptor gene linking it to either disorder.[94] Using single-strand conformational polymorphism (SSCP) analysis, Nakhai and colleagues[95] identified two rare polymorphisms that alter the structure of the N-terminus of the 5-HT$_{1A}$ receptor. The characterization and significance of these polymorphisms has not been established.

2. THE SEROTONIN-1B (5-HT$_{1B}$) RECEPTOR

a. Pharmacology

As discussed above, the neuroleptic spiperone displaces the binding of ^{3}H-5-HT to 5-HT$_1$ receptors in rat brain homogenates in a biphasic manner, leading to the suggestion that what was originally termed the 5-HT$_1$ receptor was a heterogeneous population of receptors. ^{3}H-5-HT binding sites with high affinity for spiperone were called the 5-HT$_{1A}$ receptor; ^{3}H-5-HT binding sites with low affinity for spiperone were called the 5-HT$_{1B}$ receptor.[14] Because of the lack of selective agonists and antagonists for the 5-HT$_{1B}$ receptor, 5-HT$_{1B}$ sites are best labeled with ^{3}H-5-HT, in the presence of 8-OH-DPAT and mianserin to block 5-HT$_{1A}$ and 5-HT$_2$ receptors, respectively, or with ^{125}I-cyanopindolol in the presence of isoproterenol to block β-adrenergic receptors.[96,97] A number of agonists have been shown to display high affinity but limited selectivity for the 5-HT$_{1B}$ receptor including 5-carboxamidotryptamine (5-CT), RU 24969, TFMPP, mCPP, and CGS 12066B. Several drugs such as methiothepin, pindolol, cyanopindolol, and isamaltane, are potent, but not selective, antagonists at the 5-HT$_{1B}$ receptor.[10,24]

b. **Distribution and Function in Brain**

The 5-HT$_{1B}$ receptor is present in high density in the substantia nigra and basal ganglia (i.e., caudate nucleus, globus pallidus) of rat and mouse brain.[98,99] Intrastriatal injection of the neurotoxin kainic acid and the destruction of striatonigral projections results in a significant reduction in the density of 5-HT$_{1B}$ receptors in the substantia nigra, suggesting that 5-HT$_{1B}$ receptors are located on terminals of fibers originating in the striatum.[100] 5-HT$_{1B}$ receptors function as autoreceptors on serotonergic nerve terminals modulating the release of serotonin.[101,102] Despite the lack of autoradiographic evidence for 5-HT$_{1B}$ binding sites in the dorsal raphe nucleus, the modulation of serotonin release from rat dorsal raphe nucleus slices by 5-HT$_{1B}$ receptor agonists and antagonists indicates presence of functional 5-HT$_{1B}$ autoreceptors in the dorsal raphe nucleus.[103] In addition, 5-HT$_{1B}$ receptors are located postsynaptically where they function as heteroreceptors, modulating the release of other neurotransmitters such as acetylcholine and glutamate.[104,105] The 5-HT$_{1B}$ receptor is also a member of the G-protein-coupled receptor superfamily, and in rat substantia nigra has been shown to be linked to the inhibition of adenylyl cyclase.[106]

Recently, transgenic mice have been generated that lack the 5-HT$_{1B}$ receptor.[107] These mice do not exhibit any obvious developmental defects and are fertile. However, the 5-HT$_{1B}$ receptor-deficient mice are more aggressive than wild-type or heterozygous mice of the same strain in the isolation-induced aggression test. The increased aggression and impulsivity in these transgenic mice suggests the possible involvement of the 5-HT$_{1B}$ receptor in aggression and impulse control.[107]

The pharmacological characteristics of ^{3}H-5-HT binding sites in the basal ganglia of other mammalian species such as guinea pig, dog, cow, or human is distinct from that of 5-HT$_{1B}$ sites in rodents such as rats and mice (see below). These ^{3}H-5-HT binding sites are not sensitive to β-adrenergic receptor antagonists and display a pharmacological profile different from that characteristic of the 5-HT$_{1B}$ receptor,[10,24] and were designated 5-HT$_{1D}$ sites. Functional, pharmacological and biochemical data suggest that the 5-HT$_{1B}$ and 5-HT$_{1D}$ receptor are functionally equivalent species homologues — the 5-HT$_{1B}$ receptor in rodents, the 5-HT$_{1D}$ receptor in other mammalian species.

3. THE SEROTONIN-1D (5-HT$_{1D}$) RECEPTOR

a. **Pharmacology**

Another distinct subtype of the 5-HT$_1$ receptor "family" was identified in bovine brain by Heuring and Peroutka in 1987. ^{3}H-5-HT binding in bovine caudate was determined in the presence of 8-OH-DPAT and mesulergine to block 5-HT$_{1A}$ and 5-HT$_{2C}$ receptors, respectively.[108] Under these conditions, the remaining ^{3}H-5-HT binding sites displayed a pharmacological profile different from that previously described for other serotonin receptor subtypes and was designated the 5-HT$_{1D}$ binding site.[108] A number of agonists display high affinity but limited selectivity for 5-HT$_{1D}$ receptor binding sites, including 5-CT, metergoline, and 5-methoxytryptamine. RU 24969, pindolol, and cyanopindolol are approximately two orders of magnitude less potent at 5-HT$_{1D}$ sites that at 5-HT$_{1B}$ sites, and are useful in differentiating pharmacologically the 5-HT$_{1D}$ receptor from the 5-HT$_{1B}$ receptor. Drugs with nanomolar affinity for the 5-HT$_{1A}$ receptor such as 8-OH-DPAT, ipsapirone, and buspirone display micromolar affinity for 5-HT$_{1D}$ sites.[10,24] Sumatriptan, a potent agonist at the 5-HT$_{1D}$ receptor is 100-fold more selective for the 5-HT$_{1D}$ receptor than the 5-HT$_{1A}$ or 5-HT$_{1B}$ receptor and has been used clinically to treat migraines.[109] Because no selective antagonists had been available for the 5-HT$_{1D}$ receptor until very recently, the characterization of the 5-HT$_{1D}$

receptor relied on the use of nonselective antagonists such as methiothepin and metergoline. A potent and selective antagonist, GR 127935, has recently been described.[10]

b. **Distribution and Function in Brain**

The 5-HT$_{1D}$ receptor is present in high density in the substantia nigra and basal ganglia (i.e., caudate nucleus, globus pallidus) in brains of all mammalian species except rats and mice (i.e., cat, guinea pig, cow, pig, human).[100] A marked reduction in the density of 5-HT$_{1D}$ binding sites within the substantia nigra of patients dying with Huntington's disease has led to the proposal that these sites may be located on striatonigral projections, i.e., on terminals of fibers originating in the striatum.[110] 5-HT$_{1D}$ receptors have also been found on bovine and human cerebral arteries where they appear to mediate the contraction of these vessels.[111,112]

Recent studies in higher mammalian species including humans indicate that terminal autoreceptors modulating the release of serotonin appear to be of the 5-HT$_{1D}$ subtype.[113,114] Despite the lack of autoradiographic evidence for 5-HT$_{1D}$ binding sites in the dorsal raphe nucleus, the modulation of serotonin release from guinea pig dorsal raphe nucleus slices by 5-HT$_{1D}$ receptor agonists and antagonists indicates the presence of functional 5-HT$_{1D}$ autoreceptors in dorsal raphe nucleus.[115] In addition, 5-HT$_{1D}$ receptors also appear to be located postsynaptically where they function as heteroreceptors, modulating the release of other neurotransmitters such as acetylcholine.[116] The 5-HT$_{1D}$ receptor has been shown to be coupled to the inhibition of adenylyl cyclase.[117] Thus, several lines of investigation support the functional equivalence between 5-HT$_{1B}$ sites in rodents and 5-HT$_{1D}$ sites in other species.

c. **Clinical Correlates**

The presence of 5-HT$_{1D}$ receptors in high density in the basal ganglia raises the interesting possibility that these receptors may be involved in diseases of the basal ganglia such as Huntington's disease or Parkinson's disease. To date, the major clinical application of 5-HT$_{1D}$ agonists has been in the acute treatment of migraine headaches. Although the exact pathogenesis of migraine is not well understood, migraine appears to involve vasoconstriction followed by vasodilatation, as well as neurogenic inflammation. Activation of 5-HT$_{1D}$ receptors by the 5-HT$_{1D}$ receptor agonist sumatriptan is thought to produce vasoconstriction of intracranial blood vessels, which may be dilated during migraine attacks, and decrease inflammation around sensory nerves.[109,118] The relative absence of side effects such as nausea, vomiting, and peripheral vasoconstriction may be due to sumatriptan's lack of activity at other serotonin receptors.

d. **Molecular Biology and Genetics**

Although biochemical, pharmacological, and functional data indicate that the 5-HT$_{1B}$ receptor found in rats and mice and the 5-HT$_{1D}$ receptor found in other species, including humans, are functionally equivalent species homologues, the story is somewhat complicated by the discovery of two genes encoding the 5-HT$_{1D}$ receptor, 5-HT$_{1D\alpha}$ and 5-HT$_{1D\beta}$. It should be noted that radioligand binding studies do not currently allow the differentiation of 5-HT$_{1D\alpha}$ and 5-HT$_{1D\beta}$ receptors. The 5-HT$_{1D\beta}$ receptor is the counterpart of what has been described in functional and biochemical studies as the 5-HT$_{1D}$ receptor.[10] Because there are no compounds currently available to differentiate between the 5-HT$_{1D\alpha}$ and the 5-HT$_{1D\beta}$ receptor, we refer to them throughout this review as 5-HT$_{1D}$. Furthermore, because of distinct pharmacological profiles of the 5-HT$_{1B}$ receptor found in rat and the 5-HT$_{1D}$ receptor found in other species, we will *not* refer to the 5-HT$_{1D\beta}$ as the human 5-HT$_{1B}$ receptor.

The human gene for the 5-HT$_{1D}$ receptor was identified simultaneously by numerous independent groups through polymerase chain reaction (PCR) amplification using degenerate oligonucleotides corresponding to consensus sequences of certain G-protein transmembrane domains. Although all groups reported an intronless gene with an identical 390-amino-acid sequence, the gene was designated the human 5-HT$_{1B}$,[119-121] 5-HT$_{1D\beta}$,[122,123] 5-HT$_{S12}$,[124] or 5-HT$_{1D}$-like[125] receptor. The gene has been localized to chromosome 6q13.[120] The distribution of 5-HT$_{1B}$ and 5-HT$_{1D\beta}$ receptor mRNA in the brain is similar across species.[120,126] The 5-HT$_{1D\beta}$ receptor mRNA has been localized to the striatum, hippocampus, subthalamic nuclei, entorhinal and cingulate cortex, cerebellum, nucleus accumbens, and raphe nuclei, but not in the substantia nigra.[13]

The amino acid sequence of the human 5-HT$_{1D\beta}$ receptor is 93% homologous to that of the rat 5-HT$_{1B}$ receptor.[126,127] Surprisingly, the amino acid sequence of the human 5-HT$_{1D\beta}$ receptor is only 61% homologous to that of the human 5-HT$_{1D\alpha}$ receptor. In mutagenesis studies, when amino acid threonine 355 of the human 5-HT$_{1D\beta}$ receptor protein is replaced with an asparagine, as found in the rodent 5-HT$_{1B}$ receptor, the pharmacology of the mutated human 5-HT$_{1D\beta}$ receptor is identical to the rodent 5-HT$_{1B}$ receptor.[128-130]

A polymorphism of the human 5-HT$_{1D\beta}$ receptor gene has been identified. Digestion of genomic DNA with *Hinc*II results in a two-allele polymorphism exhibiting a codominant inheritance pattern. This polymorphism was identified in a group of five unrelated Caucasian families of northern European descent. Allele A1 (3.2 kb) had an allele frequency of 0.7 while the frequency of allele A2 (2.5 kb) was 0.3 (n = 40 chromosomes).[131]

The human 5-HT$_{1D\alpha}$ gene has been isolated[132] based on its homology with the canine clone of Libert et al.[133] The predicted 377-amino-acid protein is encoded by an intronless gene localized to chromosome 1p34.3-36.3. The amino acid sequence of the 5-HT$_{1D\alpha}$ receptor protein is 88% homologous to that of the canine receptor and 43% identical to that of the 5-HT$_{1A}$ receptor.[132,134,135] The 5-HT$_{1D\alpha}$ gene has also been identified in rat; it is also intronless and displays 90% homology with the human clone.[136,137] 5-HT$_{1D\alpha}$ receptor mRNA has been found in the striatum, nucleus accumbens, dorsal raphe nucleus, and hippocampus, but not in globus pallidus and substantia nigra.[136,137]

Interestingly, a human 5-HT$_{1D}$ receptor pseudogene has been identified that is most homologous to the human 5-HT$_{1D\alpha}$ receptor gene.[138] Pseudogenes are nonfunctional genes with sequence homology to a known structural gene present elsewhere in the genome. Pseudogenes are thought to arise by duplication and later acquire inactivating mutations, but because the gene is duplicated the organism's survival is not affected. In addition to numerous stop codons, frame shifts, and deletions that appear in the human 5-HT$_{1D}$ receptor pseudogene, a 283-bp Alu repeat (the most common short interspersed element in the human genome) has been inserted into the coding region[138]

4. THE SEROTONIN-2A (5-HT$_{2A}$) RECEPTOR

a. Pharmacology

The 5-HT$_{2A}$ receptor (formerly the 5-HT$_2$) was originally described in brain by Leyson and associates in radioligand binding studies using the neuroleptic ^{3}H-spiperone as "the serotonergic component of neuroleptic receptors".[7] In experiments using several radioligands, ^{3}H-5-HT, ^{3}H-LSD, and ^{3}H-spiroperidol, Peroutka and Snyder[8] demonstrated the presence of subtypes of a receptor for serotonin in frontal cortex. Binding sites labeled with high affinity by ^{3}H-5-HT were designated 5-HT$_1$; binding sites labeled with high affinity by ^{3}H-spiroperidol were designated 5-HT$_2$.[8] There are several radioligands that can be used to label 5-HT$_{2A}$ sites (i.e., ^{125}I-LSD, ^{3}H-spiperone), however, ^{3}H-ketanserin is more selective. To date, there are no selective agonists for the 5-HT$_{2A}$ receptor. Although LSD and the hallucinogenic amphetamine derivatives, DOI, DOB, and DOM, have high affinity for the 5-HT$_{2A}$ receptor, these compounds are partial agonists at the 5-HT$_{2A}$ receptor in many systems and have high affinity for the 5-HT$_1$ receptor. The 5-HT$_1$ receptor agonist 5-CT and the 5-HT$_{1A}$ agonist 8-OH-DPAT have very low affinity for the 5-HT$_{2A}$ receptor.[10] Although the pharmacological profiles of the 5-HT$_{2A}$ and 5-HT$_{2C}$ receptors are very similar, there are several antagonists which can be used to distinguish the 5-HT$_{2A}$ and 5-HT$_{2C}$ receptors: altanserin, risperidone, pirenperone, and ketanserin all have 100- to 300-fold selectivity for the 5-HT$_{2A}$ receptor.[139] The D receptor of Gaddum and Picarelli and the 5-HT$_{2A}$ receptor are pharmacologically indistinguishable.

b. **Distribution and Function in Brain**

The 5-HT$_{2A}$ receptor is enriched in many brain areas including the frontal cortex, the claustrum, an area of brain connected to the visual cortex, nucleus accumbens, olfactory tubercle, and striatum. In neocortex, 5-HT$_{2A}$ receptors are concentrated in layers I and IV of the rat and layers I and V of human cortex.[140,141] In the cortex, 5-HT$_{2A}$ receptors are thought to be located postsynaptically on intrinsic GABAergic or somatostatin-containing neurons, as destruction of projections to the cortex does not reduce 5-HT$_{2A}$ receptor number.[142] A decrease in cortical 5-HT$_{2A}$ receptor density has been reported in senile dementia of the Alzheimer type, paralleling the loss of somatostatin immunoreactivity and the decrease in GABA concentration in this region.[143]

5-HT$_{2A}$ receptors are coupled by means of a G-protein to the stimulation of phospholipase C and the hydrolysis of membrane phosphoinositides (PI). The 5-HT$_{2A}$ receptor-mediated stimulation of PI hydrolysis results in the formation of two second messengers: (1) IP$_3$ which releases calcium from intracellular stores, and (2) diacylglycerol which activates protein kinase C. The stimulation of PI hydrolysis by serotonin in cortex is not dependent on the activity of lipoxygenase or cyclooxygenase pathways.[144] Activation of 5-HT$_{2A}$ receptors also mediates neuronal depolarization, a result of decreased potassium conductance.[44,145] Although 5-HT$_{2A}$ receptor-mediated depolarization is believed to involve the closing of potassium channels, the exact transduction pathway involved (i.e., whether 5-HT$_{2A}$ receptors are coupled to potassium channels via a G-protein) has not been established.

c. **Clinical Correlates**

5-HT$_{2A}$ receptors have been implicated in the actions of hallucinogenic drugs, antipsychotic drugs, and antidepressants. Numerous studies have indicated that 5-HT$_{2A}$ receptors are involved in the actions of hallucinogenic drugs such as LSD. The affinities of hallucinogenic compounds for 5-HT$_2$ receptors correlate significantly with their potencies in humans.[146,147] Furthermore, there is a significant correlation between the affinity of antagonists at the 5-HT$_{2A}$ receptor and blockade of LSD or DOM stimuli in drug discrimination studies with rats. 5-HT$_{2C}$ receptor affinity does *not* correlate significantly with either LSD or DOM stimuli in drug discrimination tests.[148]

Several compounds with high affinity for 5-HT$_{2A}$ receptors are currently marketed or are under investigation for the treatment of schizophrenia. These include ritanserin, risperidone, sertindole, and amperozide.[149] Except for ritanserin, these antagonists are selective among the 5-HT$_2$ receptors for the 5-HT$_{2A}$ receptor subtype — evidence for a role of the 5-HT$_{2A}$ receptor in the treatment of schizophrenia. These compounds appear to be more effective against the negative symptoms of schizophrenia (i.e., withdrawal, lack of motivation, anhedonia) and have a lower propensity for the development of extrapyramidal side effects or tardive dyskinesia than classical or typical neuroleptic drugs.[150,151]

Chronic administration of a wide variety of antidepressant drugs results in the desensitization or downregulation of 5-HT$_{2A}$ receptors in brain.[51,152] Downregulation of 5-HT$_{2A}$ receptor number and function not only occurs following chronic administration of antidepressant drugs, but after administration of 5-HT$_2$ receptor antagonists as well. Chronic and sometimes acute administration of 5-HT$_2$ receptor antagonists results in a paradoxical decrease in the density of cortical 5-HT$_{2A}$ receptors and in 5-HT$_{2A}$ receptor-mediated responses.[51] The mechanism(s) responsible for the anomalous regulation of the 5-HT$_{2A}$ receptor are unclear.

Historically, studies of receptor regulation have focused on effects of chronic over- or underexposure of a receptor to its neurotransmitter. Apparent interactions between 5-HT$_{1A}$ and 5-HT$_{2A}$ receptors may be an important factor in their regulation. In animal studies 5-HT$_{1A}$ receptor-mediated responses are potentiated by coadministration of 5-HT$_{2A}$ receptor antagonists, as well as by coadministration of the 5-HT$_2$ receptor agonist DOI.[153,154] Conversely, 5-HT$_{2A}$ receptor-mediated headshake behavior in rats is reduced by coadministration of 5-HT$_{1A}$ receptor agonists.[153,155,156] In light of the existence of these acute interactions, it is not unexpected that there are regulatory interactions between the two receptor subtypes as well. As described above, antidepressant treatments regulate both 5-HT$_{1A}$ and 5-HT$_{2A}$ receptors. Chronic administration of 5-HT$_{1A}$ receptor agonists not only results in a desensitization of 5-HT$_{1A}$ receptor-mediated responses, but also has been shown to produce a decrease in 5-HT$_{2A}$ receptor-mediated behaviors and a decrease in 5-HT$_{2A}$ receptor number.[56,157] Recently, we have investigated the effect of 5-HT$_{2A}$ receptor downregulation on 5-HT$_{1A}$ receptor sensitivity *in vivo* and have demonstrated that chronic administration of 5-HT$_2$ receptor antagonists does not alter the sensitivity of 5-HT$_{1A}$ receptor-mediated behavioral responses.[158]

d. **Molecular Biology and Genetics**

The 5-HT$_{2A}$ receptor was first cloned in the rat[159,160] by homology with the rat 5-HT$_{2C}$ (formerly the 5-HT$_{1C}$) receptor.[161] The rat 5-HT$_{2A}$ receptor is 49% homologous to the rat 5-HT$_{2C}$ receptor.[159,160] The human 5-HT$_{2A}$ receptor gene, cloned by Saltzman and colleagues in 1991,[162] encodes a 471-amino-acid protein that is 91% homologous to the rat 5-HT$_{2A}$ receptor. In both the rat and human receptor sequences a leucine zipper is present in transmembrane region I.[92] The human gene is located on chromosome 13q14-21.[163,164] Unlike members of the 5-HT$_1$ receptor family, the 5-HT$_{2A}$ receptor gene contains 3 exons separated by 2 introns and spans over 20 kb.[165] The gene exists in at least two allelic forms, but both substitutions are in the wobble position and thus are silent.[92]

5-HT$_{2A}$ receptor mRNA has been detected in the rat in frontal cortex, piriform cortex, entorhinal cortex, hippocampus, caudate nucleus, nucleus accumbens, olfactory bulb, and several brainstem nuclei.[160,166] However, later studies in rat[167] and human[168] brain have not detected mRNA in the nucleus accumbens. There are two discrete 5-HT$_{2A}$ receptor mRNA species in the rat, indicating that the mRNA may be differentially spliced.[160]

Because the 5-HT$_{2A}$ receptor has been implicated in anxiety, depression, migraine, schizophrenia, and in the action of hallucinogens, numerous mutagenesis studies have been performed in an attempt to understand structure-function relationships for this receptor. In the human receptor, a serine residue is present at position 242 (transmembrane segment V), whereas an alanine residue is present at this position in the rat receptor. When a serine residue is substituted for the alanine residue at position 242 in the rat receptor, the mutant receptor displays a pharmacological profile almost identical to that of the human receptor.[169] When the alanine residue is substituted for the serine residue at position 242 in the human receptor, the mutant receptor had an increased affinity for mesulergine, which is characteristic of the rat receptor.[170] Thus, as is seen in the 5-HT$_1$ receptor family, replacement of one amino acid can drastically alter the binding characteristics of the 5-HT$_{2A}$ receptor.

Because of the anomalous regulation of the 5-HT$_{2A}$ receptor, the 5' flanking sequence of the 5-HT$_{2A}$ receptor gene has been cloned and extensively studied in order to better understand the control of 5-HT$_{2A}$ receptor gene expression. The promoter has at least 11 clustered start sites, does not have a TATA motif, but does contain ATA elements before the major and minor transcription initiation sites.[171-173] Upstream of the dominant start site, an initiator consensus sequence, a cyclic AMP response element-like sequence, two GC boxes (SP-1 sites), E-boxes,

Egr1, NF1, PEA3, AP-1, and AP-2 sites, as well as repressor sites have been identified.[171-174] Thus, cell-specific expression of the 5-HT$_{2A}$ receptor gene appears to be regulated by complex interactions between upstream repressor and activator domains.[171-173] This contrasts with the apparently less complex promoter region of the 5-HT$_{1A}$ receptor gene.

5. THE SEROTONIN-2B (5-HT$_{2B}$) RECEPTOR

a. Pharmacology

Although the 5-HT$_{2B}$ receptor is the most recently cloned of the 5-HT$_2$ receptor class, it was among the first of the serotonin receptors to be characterized using pharmacological criteria. The first report of the sensitivity of rat stomach fundus to serotonin was published by Vane in 1959.[175] This receptor, whose activation results in the contraction of fundus smooth muscle, was originally placed in the 5-HT$_1$ receptor class by Bradley et al.[176] because of its sensitivity to serotonin and because responses mediated by this receptor were not blocked by the 5-HT$_{2A}$ receptor antagonist ketanserin or by 5-HT$_3$ receptor antagonists. It has been reclassified as a 5-HT$_2$ receptor because of its similar pharmacological profile to the 5-HT$_{2C}$ receptor and its coupling to the stimulation of PI hydrolysis. Once cloned, this receptor was termed the 5-HT$_{2F}$ (fundus) receptor.[177] It is now referred to as the 5-HT$_{2B}$ receptor.[10] Binding data obtained from the recombinant receptor in transfected cell systems using ^{125}I-DOI or ^{3}H-5-HT are highly correlated with functional data obtained from rat fundus strips.[178,179] The 5-HT$_{2B}$ receptor has very low affinity for the 5-HT$_{1A}$ agonist 8-OH-DPAT and for the 5-HT$_{2A}$ receptor antagonists ketanserin and spiperone.[179] DOI has similar affinity for the 5-HT$_{2A}$, 5-HT$_{2B}$, and 5-HT$_{2C}$ receptors.[10,179]

b. Distribution and Function in Brain

Although compelling evidence for the existence of the 5-HT$_{2B}$ receptor in brain is not yet available, the presence of 5-HT$_{2B}$ receptor mRNA in many regions of human brain implies the expression of this receptor protein in brain (see below). A potential role for the 5-HT$_{2B}$ receptor in brain function and psychiatric disorders previously attributed to the 5-HT$_{2A}$ and 5-HT$_{2C}$ receptors has been suggested.[149] Several compounds with high affinity for 5-HT$_2$ receptors, currently being developed for the treatment of psychiatric disorders, sleep disorders, drug abuse, and migraine prophylaxis, are nonselective, making it difficult to attribute their therapeutic effects to one of the 5-HT$_2$ receptor subtypes. Moreover, mCPP, an agonist probe of serotonergic function widely administered to humans, is a partial agonist at both 5-HT$_{1B}$ and 5-HT$_{2C}$ receptors. In addition, the antagonist SB 200646A, which has somewhat higher affinity (10-fold) for the 5-HT$_{2B}$ than 5-HT$_{2C}$, and almost 100-fold higher affinity for the 5-HT$_{2B}$ than 5-HT$_{2A}$ receptor, has anxiolytic-like properties in the social interaction test.[180] Thus, many of the functional and clinical effects attributed to the 5-HT$_{2A}$ or 5-HT$_{2C}$ receptors may actually involve the 5-HT$_{2B}$ receptor.[149]

The rat and human 5-HT$_{2B}$ receptor expressed in transfected cell systems couples to phospholipase C and the stimulation of PI hydrolysis.[10,177] This, however, has not been demonstrated for the endogenous receptor in rat stomach fundus[181] or in brain.

c. Molecular Biology

The 5-HT$_{2B}$ receptor is the most recently cloned of the 5-HT$_2$ receptor family. The gene has been cloned in the rat,[177,178] mouse,[182,183] and human.[184,185] The rat gene encodes a 479-amino-acid protein; the mouse gene encodes a 504-amino-acid protein (MW = 56,508 Da).[183] The rat 5-HT$_{2B}$ receptor shares 45% amino acid sequence homology with the rat 5-HT$_{2A}$ receptor and 51% with the rat 5-HT$_{2C}$ receptor.[185] The human gene has been reported to encode a 481-amino-acid protein[184] or a 483-amino-acid protein[185] that is 82% homologous to the rat receptor. The human 5-HT$_{2B}$ receptor shares 58% sequence homology with the human 5-HT$_{2A}$ receptor and 51% with the human 5-HT$_{2C}$ receptor.[185] Both the rat and human receptor genes contain the same exon-intron boundaries as the 5-HT$_{2A}$ and 5-HT$_{2C}$ receptors.[177,185]

Using quantitative PCR, the 5-HT$_{2B}$ mRNA transcript (2.3 kb) has been detected in the stomach fundus, intestine, kidney, heart, lung, and dura mater of rat but it is not detected in rat brain.[177,178,183] In humans, 5-HT$_{2B}$ receptor mRNA is found peripherally and has been detected in cerebellum, cerebral cortex, amygdala, substantia nigra, caudate, thalamus, hypothalamus, and retina.[184]

6. THE SEROTONIN-2C (5-HT$_{2C}$) RECEPTOR

a. Pharmacology

5-HT$_{2C}$ receptor (formerly the 5-HT$_{1C}$) was originally identified in choroid plexus[186] and was classified as a member of the 5-HT$_1$ receptor family because of its high affinity for serotonin. However, because of the close similarities between the pharmacological properties of the 5-HT$_{1C}$ and 5-HT$_{2A}$ receptors, the high degree of sequence homology, and their coupling to the same effector system (phospholipase C), the 5-HT$_{1C}$ receptor has been reclassified as a member of the 5-HT$_2$ receptor family.[10] In general, compounds claimed to be 5-HT$_{2A}$ receptor selective show similar affinity for 5-HT$_{2C}$ receptors. To date, there are no selective agonists for the 5-HT$_{2C}$ receptor. LSD and the hallucinogenic amphetamine derivatives DOI, DOB, and DOM, which are agonists at the 5-HT$_{2C}$ receptor, also have high affinity for the 5-HT$_{2A}$ receptor.[10] Although there are several antagonists with high affinity for the 5-HT$_{2C}$ receptor (e.g., mianserin, metergoline, mesulergine, methysergide, ritanserin), none are selective for this serotonin receptor subtype.[10] Because of the lack of selective agonists and antagonists for the 5-HT$_{2C}$ receptor, 5-HT$_{2C}$ sites in brain are best labeled with ^{3}H-mesulergine in the presence of spiperone to block 5-HT$_{2A}$ receptors.

b. Distribution and Function in Brain

5-HT$_{2C}$ receptors are present in high density in the choroid plexus.[186] High-resolution autoradiography has shown that they are enriched on the epithelial cells of the choroid plexus.[187] It has been postulated that activation of 5-HT$_{2C}$ receptors may regulate the composition and volume of the cerebrospinal fluid.[186] 5-HT$_{2C}$ receptors are also found throughout the brain but in much lower densities than in the choroid plexus; in particular in areas of the limbic system (hypothalamus, hippocampus, septum, neocortex) and in those areas associated with motor behavior (substantia nigra, globus pallidus).[98] Interestingly, 5-HT$_{2C}$ receptors appear to be more abundant in the basal ganglia of humans, particularly in the substantia nigra and globus pallidus.[188] However, because of the lack of truly selective 5-HT$_{2C}$ receptor agonist and antagonists, our knowledge about the functional role of the 5-HT$_{2C}$ receptor in brain is severely limited.

c. Clinical Correlates

Because of the lack of selective agonists to differentiate between the 5-HT$_{2A}$ and 5-HT$_{2C}$ receptors, and because the 5-HT$_{2A}$ receptor antagonists risperidone, pirenperone, and ketanserin are only 100- to 300-fold selective for the 5-HT$_{2A}$ receptor,[139] many of the functional and clinical correlates of the 5-HT$_{2A}$ receptor may very well involve or be attributed to the 5-HT$_{2C}$ receptor. mCPP, a metabolite of the antidepressant drug trazadone, has been a useful tool in the generation of clinical hypotheses for the involvement of the 5-HT$_{2C}$ receptor in a number of psychiatric disorders, as well as migraine.[149] mCPP has partial agonist activity at the 5-HT$_{2C}$ receptor and is inactive at the 5-HT$_{2A}$ receptor. When administered to animals, it has anxiogenic-like properties and induces hypophagia and hypolocomotion.[189] The effects of mCPP in animals can be blocked by antagonists with high affinity for the 5-HT$_{2C}$ receptor.[180,190-192] When administered to humans, mCPP increases anxiety and can cause panic attacks, precipitate migraine, disrupt sleep, and induce hypophagia.[149]

In many animal models of anxiolytic-like drug efficacy, antagonists with high affinity for the 5-HT$_{2C}$ receptor, such as mianserin and LY 53857, have anxiolytic-like effects.[193,194] This contrasts with ketanserin, which does not exhibit anxiolytic-like drugs efficacy.[193,194] These observations suggest a role for the 5-HT$_{2C}$ receptor blockade induced anxiolysis.

Several antagonists with high affinity for 5-HT$_{2C}$ receptors are currently marketed, or are under investigation for the management of migraine. The observations that all of these compounds are effective migraine prophylactic agents, and that ketanserin is not, suggest that the 5-HT$_{2C}$ receptor is involved. These include pizotifen, methysergide, mianserin, and cyproheptadine.[149] Cyproheptadine and pizotifen are also appetite stimulants in humans. Mianserin, a clinically effective antidepressant, is equipotent at all of the 5-HT$_2$ receptor subtypes, suggestive of a potential role for the 5-HT$_{2A}$, 5-HT$_{2B}$, and/or 5-HT$_{2C}$ receptor blockade in its antidepressant action. Compounds with high affinity for 5-HT$_{2C}$ receptors are also being developed for the treatment of anxiety.[149]

Very recently, transgenic mice have been generated that lack the 5-HT$_{2C}$ receptor.[195] 5-HT$_{2C}$ receptor-deficient mice are fertile and exhibit no obvious developmental defects. However, the mice are overweight as a result of abnormal control of feeding behavior. 5-HT$_{2C}$ receptor knockout mice are also prone to spontaneous death from seizures and display a lowered threshold for metrazol-induced seizures. These results suggest a role for 5-HT$_{2C}$ receptors in the control of appetite and feeding, and suggest that 5-HT$_{2C}$ receptors may mediate tonic inhibition of neural network excitability.[195]

d. **Molecular Biology and Genetics**

The 5-HT$_{2C}$ (originally the 5-HT$_{1C}$) receptor was first isolated by functional expression of rat choroid plexus RNA in *Xenopus* oocytes.[161,196] The gene encodes a protein of 458 amino acids in humans,[162] 460 amino acids in the rat,[161] and 459 amino acids in the mouse.[197] The human receptor gene is localized to chromosome Xq24.[198] In contrast to other G-protein-coupled receptors, the rat and mouse 5-HT$_{2C}$ receptors appear to possess an eighth hydrophobic domain at the N-terminus.[197] Rat 5-HT$_{2C}$ mRNA has been detected in choroid plexus, lateral hebenula, hippocampus, amygdala, cingulate and piriform cortices, thalamic nuclei, subthalamic nuclei, hypothalamus, substantia nigra, suprachiasmatic nuclei, locus ceruleus, and spinal cord.[161,199-202]

Interestingly, the 5-HT$_{2C}$ receptor gene contains three introns, with the second and third introns corresponding to the introns in the 5-HT$_{2A}$ and 5-HT$_{2B}$ receptor genes.[182] Promoter analysis of the mouse 5-HT$_{2C}$ receptor gene revealed the presence of a TATA box and binding sites for AP-1 and AP-2 transcription factors.[203] The similarities in their gene structure (i.e., presence of introns) and promoter sequences are further support for the classification of the 5-HT$_{2A}$, 5-HT$_{2B}$, and 5-HT$_{2C}$ receptors as members of the same serotonin receptor family.

7. THE SEROTONIN-3 (5-HT$_3$) RECEPTOR

a. **Pharmacology**

The M receptor of Gaddum and Picarelli,[4] originally described in guinea pig ileum and considered to be on nerve ganglia or on nerves within the smooth muscle, is pharmacologically distinct from all of the previously described serotonin receptors. Bradley and associates[176] in their classification scheme using pharmacological criteria and functional responses primarily in peripheral tissues, renamed this receptor the 5-HT$_3$ receptor. The observation that cocaine is an effective and selective antagonist at the 5-HT$_3$ receptor led to the development of potent and selective 5-HT$_3$ receptor antagonists, MDL 72222 (tropanserin) and ICS 205-930 (tropisetron).[204,205] Many other selective and potent 5-HT$_3$ receptor antagonists have been described including granisetron, zacopride and ondansetron.[10] The first selective agonist for the 5-HT$_3$ receptor was 2-methyl-5-HT,[206] however, higher affinity and selectivity make m-chlorophenylbiguanide the current agonist of choice.[207] The 5-HT$_1$ receptor agonist 5-CT, the 5-HT$_{1A}$ receptor agonist 8-OH-DPAT, and the 5-HT$_{1B}$ receptor agonist RU 24969 are inactive at the 5-HT$_3$ receptor.[10]

b. **Distribution and Function in Brain**

In the central nervous system, 5-HT$_3$ receptors are relatively sparse in comparison to other serotonin receptors. The 5-HT$_3$ receptor is present in the highest densities in the entorhinal cortex, amygdala, hippocampus, nucleus accumbens, caudate nucleus, dorsal motor nucleus of the vagus nerve, nucleus tractus solitarius, and area postrema.[207-209] 5-HT$_3$ receptors modulate neuronal firing and the metabolism and release of various neurotransmitters in brain. In striatal slices, 5-HT$_3$ receptor activation stimulates both basal and potassium-evoked dopamine release.[210] 5-HT$_3$ receptor antagonists block the activation of dopamine turnover in the mesolimbic dopaminergic pathway by morphine, nicotine, or ethanol.[211] 5-HT$_3$ receptor activation has also been shown to enhance electrically evoked release of ^{3}H-norepinephrine in slices of hippocampus, hypothalamus, or frontal cortex,[212] and to modulate the release of acetylcholine from the entorhinal cortex and dorsal hippocampus.[213,214] Activation of 5-HT$_3$ receptors enhances the electrically evoked release of ^{3}H-serotonin from slices of frontal cortex, hippocampus, and hypothalamus.[215] However, 5-HT$_3$ receptor antagonists and agonists appear to have no effect on the firing rate of serotonergic neurons in the dorsal raphe.[216] Studies utilizing selective neurotoxins to lesion neurons in the central nervous system indicate that the majority of 5-HT$_3$ receptors are not found on noradrenergic, dopaminergic, or serotonergic nerve terminals.[212,217] Thus, the action of 5-HT$_3$ receptors on the release of monoamines may be through the modulation of glutamatergic and GABAergic neurotransmission.[218]

The 5-HT$_3$ receptor is a ligand-gated ion channel, i.e., the receptor protein subunits form an ion channel and the response is not mediated by a second messenger or through G-proteins (see Figure 9-2). The depolarization mediated by 5-HT$_3$ receptor activation is caused by a transient inward current, specifically the opening of a channel for cations Na$^+$ and K$^+$.[207]

c. **Clinical Correlates**

The neuroanatomical localization of 5-HT$_3$ receptors in cortical and limbic structures of the brain is consistent with behavioral studies in animals which suggest that 5-HT$_3$ receptor antagonists may have potential anxiolytic, antidepressant, and cognitive effects.[63,211] The observation that 5-HT$_3$ receptor activation stimulates the release of dopamine in the mesolimbic system has led to the suggestion that this receptor may have a role in drug abuse or dependence. Although there appears to be some correlation between decreased mesolimbic dopamine levels as a result of 5-HT$_3$ receptor blockade and attenuation of the psychomotor stimulant effects of drugs such as cocaine, amphetamine, and morphine, 5-HT$_3$ receptor antagonists have little effect on discriminative stimulus and reinforcing effects of these drugs of abuse.[219,220]

In the periphery, 5-HT$_3$ receptors are located exclusively on neurons where their activation stimulates transmitter release from parasympathetic, sympathetic, sensory, and enteric neurons. The functional consequences of 5-HT$_3$ receptor activation are diverse and include cardioinhibition, or activation, vasodilatation, pain, and initiation of the vomiting reflex.[10] In animal studies 5-HT$_3$ receptor antagonists appear to have antinociceptive effects at the level of the spinal cord. 5-HT$_3$ receptor antagonists have been reported to have therapeutic utility in the treatment of migraine, as well as relief of visceral discomfort associated with irritable bowel syndrome.[220]

The only clearly established therapeutic use of 5-HT$_3$ receptor antagonists is as an antiemetic. 5-HT$_3$ receptor antagonists are effective antiemetics when nausea is induced by cytotoxic chemotherapy or radiation, or when vomiting follows general anesthesia.[211]

d. Molecular Biology

The cloning and molecular characterization of the 5-HT$_3$ receptor have been greatly facilitated by the endogenous expression of this serotonin receptor subtype in a number of clonal cell lines including the mouse neuroblastoma cell lines N1E-115 and N18, the mouse neuroblastoma x rat glioma hybrid NG108-15, and the mouse neuroblastoma x Chinese hamster embryonic brain cell hybrid NCB-20 hybrid.[207] The 5-HT$_3$ receptor has been isolated by expression cloning in *Xenopus* oocytes which involved injection of size-fractionated mRNA from NCB20 cells into the oocytes. Oocytes were then tested for the presence of serotonin-gated currents characteristic of the 5-HT$_3$ receptor, and the positive mRNA pools were used to construct a cDNA library that was serially diluted until a single positive clone remained.[221] The 2.2-kb mRNA (5HT$_3$R-A) transcript encodes a 487-amino-acid protein (MW = 55,966 Da) showing high sequence homology to other ligand-gated ion channel family members.[221] A 5-HT$_3$ receptor genomic mouse clone is composed of 12 kb, which are organized into 9 exons.[222] The receptor sequence is 27% identical to the α subunit of the nicotinic acetylcholine receptor and is 22% homologous to the β subunit of the GABA$_A$ receptor[221] and other members of the ligand-gated ion channel superfamily.

A 2-kb splice variant of the 5HT$_3$R-A receptor has been cloned from a N1E-115 neuroblastoma cell line.[223] This variant (483 amino acids, MW = 53,178 Da) is the result of alternative splicing at two adjacent splice acceptor sites in intron 8.[224] The amino acid sequence is 98% homologous to 5HT$_3$R-A and has a six-amino-acid deletion in the large cytoplasmic loop between the putative M3 and M4 transmembrane regions (see Figure 9-2A). The long and short isoforms are present in both N1E-115 and NCB-20 cell lines. However, the short isoform is approximately five times more abundant than the long form in both cell lines and neuronal tissue.[224]

5-HT$_3$ receptor mRNA transcripts have been identified by quantitative PCR in mouse cortex, brainstem, midbrain, spinal cord, and heart.[221,225] In the central nervous system, *in situ* hybridization has revealed 5-HT$_3$ receptor mRNA in interneurons of the hippocampal formation, the piriform, cingulate and entorhinal cortices, amygdala, olfactory bulb, trochlear nerve nucleus, dorsal tegmental region, the facial nerve nucleus, the nucleus of the spinal tract of the trigeminal nerve, the dorsal horn of the spinal cord, and the dorsal root ganglia.[226] Peripherally, the mRNA has been localized in rat submucosal and myenteric ganglia of the duodenum, jejunum, and ileum.[227]

8. THE SEROTONIN-4 (5-HT$_4$) RECEPTOR

a. Pharmacology

The 5-HT$_4$ receptor was originally described in cultured mouse collicular neurons as a serotonin receptor coupled to the stimulation of adenylyl cyclase activity, possessing pharmacological characteristics distinct from those of the 5-HT$_1$, 5-HT$_2$, or 5-HT$_3$ receptors.[228] In the absence of a selective, high-affinity radioligand, its identification and characterization have depended primarily on functional analyses. Agonists and antagonists fall into three major classes: 5-HT and related indoles, benzamides (zacopride, renzapride, and cisapride), and benzimidazolones (BIMU-1, BIMU-8).[10] Surprisingly, many of the compounds that are antagonists at the 5-HT$_3$ receptor such as zacopride, renzapride, and metoclopramide, have agonist activity at the 5-HT$_4$ receptor. Cisapride and 5-methoxy-N,N-dimethyltryptamine are potent agonists at the 5-HT$_4$ receptor but are not selective for this serotonin receptor subtype.[10] The lack of selective agonists for the 5-HT$_4$ receptor has hindered the study of 5-HT$_4$ receptor function *in vivo*. Recently, two selective partial agonists have been described, RS 67333 and RS 67506, which may prove to be useful tools *in vivo*.[229]

The 5-HT$_3$ receptor antagonist ICS 205-930 (tropisetron) is also a potent antagonist of the 5-HT$_4$ receptor,[228] but its lack of selectivity for the 5-HT$_4$ receptor has limited its usefulness. Several other antagonists are now available; of these, GR 113808 and SB 204070 are the most potent and selective.[10] However, these compounds have a short half life *in vivo*, which limits their use for elucidation of 5-HT$_4$ receptor function in brain. Several antagonists, RS 39604, and RS 67532 have recently been reported to have both enhanced 5-HT$_4$ receptor selectivity and *in vivo* stability.[229] RS 67532 may enter the central nervous system, making it particularly useful in the study of 5-HT$_4$ receptor function in brain.

b. **Distribution and Function in Brain**

Studies of the 5-HT$_4$ receptor have been hampered by the absence of a high-affinity radioligand. The recent synthesis and development of specific radioligands, ^{3}H-GR 113808 and ^{125}I-SB 207710, has provided the necessary tools for the study and characterization of the 5-HT$_4$ receptor.[230,231] The 5-HT$_4$ receptor is localized with high densities in the striatum, globus pallidus, substantia nigra, and olfactory tubercle of rat and guinea pig brain[230] and has been reported in the hippocampus and caudate of pig brain.[231] The 5-HT$_4$ receptor in colliculi neurons and hippocampus of mouse, rat, and guinea pig is coupled to the stimulation of adenylyl cyclase and to the inhibition of potassium channels. The inhibition of potassium conductance in colliculi neurons and increases in calcium currents in cardiac cells are associated with the activation of cAMP-dependent protein kinase A.[10,232] The 5-HT$_4$ receptor, coupled to the stimulation of adenylyl cyclase, has been identified in human cerebral cortex.[233]

Studies utilizing the technique of *in vivo* microdialysis have demonstrated a role of the 5-HT$_4$ receptor in the modulation of neurotransmitter release. Acetylcholine release in rat frontal cortex, but not dorsal hippocampus or striatum, is increased by activation of 5-HT$_4$ receptors.[234] Activation of 5-HT$_4$ receptors also enhances dopamine release in the striatum.[235,236] The localization of high densities of the 5-HT$_4$ receptor in limbic structures of brain suggests a role of this receptor in memory and learning. Activation of 5-HT$_4$ receptors by zacopride and renzapride appears to increase electroencephalogram (EEG) energy in the rat and the effect may involve, at least in part, release of acetylcholine,[237] a neurotransmitter involved in cognitive function. In patients with Alzheimer's disease there is a marked decrease in the density of 5-HT$_4$ receptors in the hippocampus.[238] Recent findings from behavioral studies using animal models of learning and anxiety suggest a potential role of the 5-HT$_4$ receptor in cognitive function and anxiolysis which warrants further investigation.[229]

c. **Molecular Biology**

The 5-HT$_4$ receptor gene has recently been cloned from rat brain RNA by reverse transcriptase-polymerase chain reaction (RT-PCR).[239] Two different cDNA clones, 5-HT$_{4l}$ (long isoform, 5.5 kb) and 5-HT$_{4s}$ (short isoform, 4.5 kb) were isolated, and are most likely the result of alternative splicing of 5-HT$_4$ receptor mRNA. The long isoform of the 5-HT$_4$ receptor protein contains a phosphorylation site for protein kinase C which is not present in the short isoform. In COS-7 cells transiently transfected with 5-HT$_{4l}$ cDNA, the 5-HT$_4$ receptor agonists BRL-24924, cisapride, and zacopride are equipotent in activating adenylyl cyclase. However, in COS-7 cells transiently transfected with 5-HT$_{4s}$ cDNA, the order of potency to stimulate adenylyl cyclase is as follows: cisapride > BRL-24924 > zacopride.[239] In brain, RT-PCR has detected 5-HT$_{4l}$ cDNA in striatum, thalamus, hippocampus, olfactory bulb, and brainstem. 5-HT$_{4s}$ cDNA was found only in the striatum.[239]

C. RECOMBINANT SEROTONIN RECEPTORS

The classification of recombinant receptors for serotonin remains tentative due to limited knowledge of the operational and transductional characteristics of these receptors. As discussed above, because the functions mediated by these new receptors in intact tissue are unknown, lower-case appellations are presently used.[9,10]

1. THE 5-HT$_{1E}$ RECEPTOR

The 5-HT$_{1E}$ receptor was originally identified in homogenates of human frontal cortex by radioligand binding studies using ^{3}H-5-HT in the presence of 5-CT to block 5-HT$_{1A}$ and 5-HT$_{1D}$ receptor sites.[240] Because of the lack of specific radioligands for the 5-HT$_{1E}$ receptor, the overall distribution of this serotonin receptor in brain is unknown. The function of the 5-ht$_{1E}$ receptor in intact tissue is not known due to the lack of selective agonists or antagonists.[10] In transfected cells, the 5-HT$_{1E}$ receptor is coupled to the inhibition of adenylyl cyclase activity.[241,242]

The intronless 5-HT$_{S31}$ gene was initially cloned by Levy et al.[242] from a human genomic library under low stringency conditions using oligonucleotides derived from the human 5-HT$_{1A}$ and rat 5-HT$_{2C}$ receptors. The 5-HT$_{S31}$ receptor protein consists of 365 amino acids and has the characteristic structure of a member of the G-protein-coupled receptor superfamily (Figure 9-1).[241,243,244] The 5-HT$_{1E}$ receptor mRNA has been found in the caudate putamen, parietal cortex, and olfactory tubercle.[13] The 5-HT$_{1E}$ receptor displays a higher degree of homology with the 5-HT$_{1D}$ receptor (64%) than any other 5-HT$_1$ receptor.[243]

2. THE 5-HT$_{1F}$ RECEPTOR

Using degenerate oligonucleotides derived from the 5-HT$_{1B}$ receptor, Amlaiky et al.[245] isolated the 5-HT$_{1F}$ receptor (called 5-HT$_{1E\beta}$ by the authors) by screening a mouse brain cDNA library at low stringency. The human and rat 5-HT$_{1F}$ receptor was cloned and sequenced in 1993.[246,247] The 366-amino-acid receptor protein is encoded by an intronless gene that displays the greatest homology with the 5-HT$_{1E}$ receptor (61%). The gene is localized to chromosome 3p11.[13] The 5-HT$_{1F}$ receptor mRNA is found in cortex, hippocampus, dentate gyrus, nucleus of the solitary tract, spinal cord, uterus, and mesentery.[246] The rat gene is highly homologous (93%) to the human species homologue and contains one intron in the 5' untranslated region.[247]

Little is known about the distribution and function of the 5-HT$_{1F}$ receptor in brain. In transfected cells the 5-HT$_{1F}$ receptor is coupled to the inhibition of adenylyl cyclase.[245,246] No selective agonists or antagonists for the 5-HT$_{1F}$ receptor are known.[10]

3. THE 5-HT$_5$ RECEPTOR

Both the 5-HT$_{5A}$ and 5-HT$_{5B}$ receptors were cloned by using degenerate oligonucleotides derived from transmembrane domains III and VI of G-protein-coupled serotonin receptors.[248-250] Both genomic clones possess one intron in the middle of the third cytoplasmic loop.[249] The receptor proteins are 77% identical to each other whereas the homology to other serotonin receptors is low.[248]

The human 5-HT$_{5A}$ receptor gene encodes a 357-amino-acid protein, and has been localized to chromosome 7q36. This locus also contains the holoprosencephaly type III mutation which causes abnormal forebrain development.[249] Northern blots reveal three mRNA transcripts in mouse brain and cerebellum (5.8, 5.0, and 4.5 kb)[248] while in rat, two transcripts of 3.8 and 4.5 kb were detected.[250] The 5-HT$_{5A}$ receptor mRNA transcripts have been detected by *in situ* hybridization in the cerebral cortex, hippocampus, granule cells of the cerebellum, medial habenula, amygdala, septum, several thalamic nuclei, and the olfactory bulb of the rat and mouse.[248,250]

The human 5-HT$_{5B}$ receptor gene is localized to chromosome 2q11-13 and encodes a protein of 370 amino acids.[249] In mouse, three mRNA transcripts of 1.5, 1.8, and 3.0 kb have been identified, presumably the result of alternative mRNA splicing.[250] The 5-HT$_{5B}$ mRNA has been detected by *in situ* hybridization in the hippocampus, habenula, and the dorsal raphe nucleus of rat and human.[249,250]

In transfected cells, these recombinant receptors exhibit a pharmacological profile distinct from other serotonin receptors.[10] At the present time functional correlates for these receptors and their transductional properties are unknown.

4. THE 5-HT$_6$ RECEPTOR

The 5-HT$_6$ receptor was cloned by RT-PCR from rat striatal mRNA utilizing degenerate primers derived from transmembrane domains III and VI of G-protein-coupled serotonin receptors.[251] The cDNA clone encodes a protein of 437 amino acids (MW = 46.8 kDa).[251] The protein is approximately 30% homologous to other serotonin receptors. In Northern blots, expression of the 5-HT$_6$ receptor mRNA transcript has been detected in the striatum, with lower expression in the olfactory tubercle, hippocampus, and cerebral cortex.[251] Ruat and colleagues[252] cloned the 5-HT$_6$ receptor by screening a rat genomic library with a probe derived from the rat histamine H$_2$ receptor. This clone predicted a 436-amino-acid protein with a MW of 46,922 Da. Northern analysis revealed two mRNA transcripts of 4.1 and 3.2 kb which are expressed in striatum, cerebral cortex, hippocampus, and hypothalamus. *In situ* hybridization has detected additional expression in the olfactory tubercles, nucleus accumbens, and olfactory bulb. Both groups report the presence of at least one intron. Interestingly, the third cytoplasmic loop of the 5-HT$_6$ receptor is short compared to other serotonin receptors, while the C-terminal tail is long.[251,252]

The 5-HT$_6$ receptor when expressed in transfected cells shows high affinity for ^{125}I-LSD and ^{3}H-5-HT. The pharmacology of this recombinant receptor is unique. Methiothepin is the compound with the highest affinity for this receptor; the receptor has low affinity for the 5-HT$_1$ receptor agonist 5-CT. Interestingly, this receptor has high affinity for various antipsychotic and antidepressant drugs such as clozapine, amitriptyline, clomipramine, mianserin, and ritanserin. The 5-HT$_6$ receptor stimulates adenylyl cyclase when expressed in some, but not all, cell systems.[10]

5. THE 5-HT$_7$ RECEPTOR

The 5-HT$_7$ receptor has been cloned by RT-PCR in rat,[247,253-255] mouse,[256] and human.[247,257] The human gene is localized to chromosome 10q23.3-24.3 and encodes a 445-amino-acid protein (MW = 49 kDa). The rat gene reported by Lovenberg et al.[247] encodes a 435-amino-acid protein; Ruat et al.[254] have isolated a cDNA encoding a 448-amino-acid protein in rat. These two rat receptor clones differ only in their C-terminals and presumably result from alternative mRNA splicing. Both groups report the presence of two introns in the 5-HT$_7$ gene: one after the third cytoplasmic domain or intracellular loop, and the second close to the end of the coding region. All 5-HT$_7$ clones show the highest amino acid sequence homology with the *Drosophila* 5-HT1A receptor (42%) and approximately 35% homology with all other serotonin receptors. The 5-HT$_7$ mRNA transcript is found in brain in the cortex, cerebellum, brainstem, hippocampus, hypothalamus, thalamus, amygdala, striatum, substantia nigra, superior colliculus, dorsal, and median raphe nuclei.[247,253-256]

To date, no selective agonists or antagonists have been described for the 5-HT$_7$ receptor. In transfected cells, the 5-HT$_7$ receptor stimulates adenylyl cyclase. Serotonin, 5-CT, 5-MeOT, RU 24969, and 8-OH-DPAT are potent agonists at the 5-HT$_7$ receptor in transfected cell systems. Methysergide, mesulergine, clozapine, butaclamol, and methiothepin are potent antagonists at this recombinant receptor.[10]

ACKNOWLEDGMENTS

We are grateful to Ms. Patti Lairsey, Ms. Rosie Ortiz, and Mr. Craig Smith for assistance in preparing the manuscript. Supported by USPHS grant MH-52369.

REFERENCES

1. Twarog, B. M. and Page, I. H., Serotonin content of some mammalian tissues and urine and a method for its determination, *Am. J. Physiol.*, 175, 157, 1953.
2. Dahlstrom, A. and Fuxe, K., Evidence for the existence of monoamine-containing neurons in the central nervous system. I. Demonstration of monoamines in the cell bodies of brain stem neurons, *Acta Physiol. Scand.*, Suppl., 232, 1, 1965.
3. Gaddum, J. H. and Hameed, K. A., Drugs which antagonize 5-hydroxytryptamine, *Br. J. Pharmacol.*, 9, 240, 1954.
4. Gaddum, J. H. and Picarelli, Z. P., Two kinds of tryptamine receptor. *Br. J. Pharmacol.*, 12, 323, 1957.
5. Farrow, J. T. and Van Vunakis, H., Binding of *d*-Lysergic acid diethylamide to subcellular fractions from rat brain, *Nature*, 237, 164, 1972.
6. Farrow, J. T. and Van Vunakis, H., Characteristics of D-lysergic acid diethylamide binding to subcellular fractions derived from rat brain, *Biochem. Pharmacol.*, 22, 1103, 1973.
7. Leysen, J. E., Niemegeers, C. J. E., Tollenaere, J. P., and Laduron, P. M., Serotonergic component of neuroleptic receptors, *Nature*, 272, 168, 1978.
8. Peroutka, S. J. and Snyder S. H., Multiple serotonin receptors: differential binding of [^{3}H]5-hydroxtryptamine, [^{3}H]lysergic acid diethylamide and [^{3}H]spiroperidol, *Mol. Pharmacol.*, 16, 687, 1979.
9. Martin, G. R. and Humphrey, P. P. A., Receptors for 5-hydroxytryptamine: current perspectives on classification and nomenclature, *Neuropharmacology*, 33, 261, 1994.
10. Hoyer, D., Clarke, D. E., Fozard, J. R., Hartig, P. R., Martin, G. R., Mylecharane, E. J., Saxena, P. R., and Humphrey, P. P. A., VII. International union of pharmacology classification of receptor 5-hydroxytryptamine, *Pharmacol. Rev.*, 46, 157, 1994.
11. Peroutka, S. J. and Howell, T. A., The molecular evolution of G protein-coupled receptors. Focus on 5-hydroxytryptamine receptors, *Neuropharmacology*, 33, 319, 1994.
12. Saudau, F. and Hen, R., 5-Hydroxytryptamine receptor subtypes in vertebrates and invertebrates, *Neurochem. Int.*, 25, 503, 1994.
13. Lucas, J. J. and Hen, R., New players in the 5-HT receptor field: genes and knockouts, *Trends Pharmacol. Sci.*, 16, 246, 1995.
14. Pedigo, N. W., Yamamura, H. I., and Nelson, D. L., Discrimination of multiple ^{3}H-5-hydroxytryptamine binding sites by the neuroleptic spiperone in rat brain, *J. Neurochem.*, 36, 220, 1981.
15. Middlemiss, D. N. and Fozard J. R., 8-Hydroxy-2-(DI-n-propylamino)-tetralin discriminates between subtypes of the 5-HT$_1$ recognition site, *Eur. J. Pharmacol.*, 90, 151, 1983.
16. Gozlan, H., El Mestikawy, S., Pichat, L., Glowinski, J., and Hamon, M., Identification of presynaptic serotonin autoreceptors by a new ligand: ^{3}H-PAT, *Nature*, 305, 140, 1983.
17. Glaser, T. and Traber, J., Binding of the putative anxiolytic TVXQ 7821 to hippocampal 5-hydroxytryptamine (5-HT) recognition sites, *Naunyn-Schmiedeberg's Arch. Pharmacol.*, 329, 211, 1985.

18. Traber, J. and Glaser, T., The 5-HT$_{1A}$ receptor-related anxiolytics, *Trends Pharmacol. Sci.*, 8, 432, 1987.
19. Hamik, A., Oksenberg, D., Fischette, C., and Peroutka, S. J., Analysis of tandospirone (SM-3997) interactions with neurotransmitter receptor binding sites, *Biol. Psychiatry*, 28, 99, 1990.
20. DeVry, J., Schreiber, R., Glaser, T., and Trabel, J., Behavioural pharmacology of 5-HT$_{1A}$ agonists: animal models of anxiety and depression, in *Serotonin$_{1A}$ Receptors in Depression and Anxiety*, Stahl, S. M. et al., Eds., Raven Press, New York, 1992, p. 55.
21. Blier, P. and de Montigny, C., Modification of 5-HT neuron properties by sustained administration of the 5-HT$_{1A}$ agonist gepirone: electrophysiological studies in the rat brain, *Synapse*, 1, 470, 1987.
22. Bockaert, J., Dumuis, A., Bouhelal, R., Sebben, M., and Cory, R. N., Piperazine derivatives including the putative anxiolytic drugs, buspirone and ipsapirone, are agonists at 5-HT$_{1A}$ receptors negatively coupled with adenylate cyclase in hippocampal neurons, *Arch. Pharmacol.*, 335, 588, 1987.
23. Godbout, R., Chaput, Y., Blier, P., and de Montigny, C., Tandospirone and its metabolite, 1-(2-pyrimidinyl)-piperazine. I. Effects of acute and long-term tandospirone on serotonin neurotransmission, *Neuropharmacology*, 30, 679, 1991.
24. Hoyer, D. and Schoeffter, P., 5-HT receptors: subtypes and second messengers, *J. Receptor Res.*, 11, 197, 1991.
25. Fletcher, A., Cliffe, I. A., and Dourish, C. T., Silent 5-HT$_{1A}$ receptor antagonists: utility as research tools and therapeutic agents, *Trends Pharmacol. Sci.*, 14, 441, 1993.
26. Lanfumey, L., Haj-Dahmane, S., and Hamon, M., Further assessment of the antagonist properties of the novel and selective 5-HT$_{1A}$ receptor ligands (+)- WAY 100135 and SDZ 216-525, *Eur. J. Pharmacol.*, 249, 25, 1993.
27. Meller, E., Goldstein, M., and Bohmaker, K., Receptor reserve for 5-hydroxytryptamine$_{1A}$-mediated inhibition of serotonin synthesis: possible relationship to anxiolytic properties of 5-hydroxytryptamine agonists, *Mol. Pharmacol.*, 37, 231, 1989.
28. Yocca, F. D., Iben, L., and Meller, E., Lack of apparent receptor reserve at postsynaptic 5-hydroxytryptamine$_{1A}$ receptors negatively coupled to adenylyl cyclase activity in rat hippocampal membranes, *Mol. Pharmacol.*, 41, 1066, 1992.
29. Millan, M. J., Canton, H., Gobert, A., Lejeune, F., Rivet, J.-M., Bervoets, K., Brocco, M., Widdowson, P., Mennini, T., Audinot, V., Honoré, P., Renouard, A., Le Marouille-Girardon, S., Verriéle, L., Gressier, H., and Peglion, J.-L., Novel benzodioxopiperazines acting as antagonists at postsynaptic 5-HT$_{1A}$ receptors and as agonists at 5-HT$_{1A}$ autoreceptors: a comparative pharmacological characterization with proposed 5-HT$_{1A}$ antagonists, *J. Pharmacol. Exp. Ther.*, 268, 337, 1993.
30. Fornal, C. A., Metzler, C. W., Veasey, S. C., McCreary, A. C., Dourish, C. T., and Jacobs, B. L., Single-unit recordings from freely-moving animals provide evidence that WAY-100635, but not (S)-WAY-100135, blocks the action of endogenous serotonin at the 5-HT autoreceptor, *Br. J. Pharmacol.*, 112, 92, 1994.
31. Assie, M.-B. and Koek, W., Way 100635 reverses the decrease of 5-HT levels produced by the putative 5-HT1A antagonist, WAY 100135, *Soc. Neurosci.*, 21, 1854, 1995.
32. Fletcher, A., Bill, D. J., Cliffe, I. A., Forster, E. A., Jones, D., and Reilly, Y., A pharmacological profile of WAY-100635, a potent and selective 5-HT$_{1A}$ receptor antagonist, *Br. J. Pharmacol.*, 112, 91, 1994.
33. Gurling, J., Ashworth-Preece, M. A., Dourish, C. T., and Routledge, C., Effects of acute and chronic treatment with the selective 5-HT$_{1A}$ receptor antagonist WAY-100635 on hippocampal 5-HT release *in vivo*, *Br. J. Pharmacol.*, 112, 299, 1994.
34. Forster, E. A., Cliffe, I. A., Bill, D. J., Dover, G. M., Jones, D., Reilly, Y., and Fletcher, A., A pharmacological profile of the selective silent 5-HT1A receptor antagonist, WAY-100635, *Eur. J. Pharmacol.*, 281, 81, 1995.
35. Khawaja, X., Evans, N., Reilly, Y., Ennis, C., and Minchin, M. C. W., Characterisation of the binding of [^{3}H]WAY-100635, a novel 5-hydroxytryptamine$_{1A}$ receptor antagonist, to rat brain, *J. Neurochem.*, 64, 2716, 1995.
36. Khawaja, X., Quantitative autoradiographic characterisation of the binding of [^{3}H]WAY-100635, a selective 5-HT$_{1A}$ receptor antagonist, *Brain Res.*, 673, 217, 1995.
37. Kung, H. F., Kung, M. P., Clarke, W., Maayani, S., and Zhuang, Z. P., A potential 5-HT$_{1A}$ receptor antagonist: p-MPPI, *Life Sci.*, 55, 1459, 1994.
38. Kung, M.-P., Frederick, D., Mu, M., Zhuang, Z.-P., and Kung, H. F., 4-(2'-Methoxy-phenyl)-1-[2'-(n-2'-pyridinyl)-*p*-iodobenzamido]-ethyl-piperazine ([^{125}I]p-MPPI) as a new selective radioligand of serotonin-1A sites in rat brain: *in vitro* binding and autoradiographic studies, *J. Pharmacol. Exp. Ther.*, 272, 429, 1995.
39. Thielen, R. J., Fangon, N. B., and Frazer, A., 4-(2'-Methoxyphenyl)-1-[2'[N-(2'-pyridinyl)-p-iodobenzamido]ethyl]piperazine (p-MPPI) and 4-(2'-methoxyphenyl)-1-[2'-[N-(2'-pyridinyl)-p-fluorobenzamido]ethyl]piperazine (p-MPPF), two new antagonists at pre- and postsynaptic serotonin-1A receptors, *J. Pharmacol. Exp. Ther.*, 277, 661, 1996.
40. Zhuang, Z.-P., Kung, M.-P., and Kung, H. F., Synthesis and evaluation of 4-(2'-methoxyphenyl)-1-[2'[N-(2'-pyridinyl)-p-iodobenzamido]piperazine (p-MPPI): a new iodinated 5-HT$_{1A}$ ligand, *J. Med. Chem.*, 37, 1406, 1994.

41. Vergé, D., Daval, G., Marcinkiewicz, M., Patey, A., El Mestikawy, S., Gozlan, H., and Hamon, M., Quantitative autoradiography of multiple 5-HT_1 receptor subtypes in the brain of control of 5,7-dihydroxytryptamine-treated rats, *J. Neurosci.*, 6, 3474, 1986.
42. Hensler, J. G., Kovachich G. B., and Frazer A., A quantitative autoradiographic study of serotonin$_{1A}$ receptor regulation. Effect of 5,7-dihydroxytryptamine and antidepressant treatments, *Neuropsychopharmacology*, 4, 131, 1991.
43. deMontigny, C., Blier, P., and Chaput, Y., Electrophysiologically-identified serotonin receptors in the rat CNS, *Neuropharmacology*, 23, 1511, 1984.
44. Aghajanian, G. K., Sprouse, J. S., Sheldon, P., and Rasmussen, K., Electrophysiology of the central serotonin system. Receptor subtypes and transducer mechanisms, *Ann. N.Y. Acad. Sci.*, 600, 93, 1990.
45. Sotelo, C., Cholley, B., El Mestikawy, S., Gozlan, H., and Hamon, M., Direct immunohistochemical evidence of the existence of 5-HT_{1A} autoreceptors on serotoninergic neurons in the midbrain raphe nuclei, *Eur. J. Neurosci.*, 2, 1144, 1990.
46. Devivo, M. and Maayani, S., Inhibition of forskolin-stimulated adenylate cyclase activity by 5-HT receptor agonists, *Eur. J. Pharmacol.*, 119, 231, 1985.
47. Markstein, R., Hoyer, D., and Engel, G., 5-HT_{1A}-receptors mediate stimulation of adenylate cyclase in rat hippocampus, *Arch. Pharmacol.*, 333, 335, 1986.
48. Andrade, R., Malenka, R. C., and Nicoll, R. A., A G protein couples serotonin and $GABA_B$ receptors to the same channels in hippocampus, *Science*, 234, 1261, 1986.
49. Clarke, W. P., De Vivo, M., Beck, S. G., Maayani, S., and Goldfarb, J., Serotonin decreases population spike amplitude in hippocampal cells through a pertussis toxin substrate, *Brain Res.*, 410, 357, 1987.
50. Clarke, W. P., Yocca, F. D., and Maayani, S., Lack of 5-HT_{1A}-mediated inhibition of adenylyl cyclase in dorsal raphe of male and female rats, *J. Pharmacol. Exp. Ther.*, in press6.
51. Frazer, A., Offord, S. J., and Lucki, I., Regulation of serotonin receptors and responsiveness, in Brain, in *The Serotonin Receptors*, Sanders-Bush, E., Ed., Humana Press, New York, 1988, p. 319.
52. Goodwin, G. M., DeSouza, R. J., and Green, A. R., Attenuation by electroconvulsive shock and antidepressant drugs of the 5-HT_{1A} receptor-mediated hypothermia and serotonin syndrome produced by 8-OH-DPAT in the rat, *Psychopharmacology*, 91, 506, 1987.
53. Welner, S. A., deMontigny, C., Desroches, J., Desjardins, P., and Suranyi-Cadotte, B. E., Autoradiographic visualization of serotonin (5-HT)$_{1A}$ binding sites in rat brain following antidepressant drug treatment, *Synapse*, 4, 347, 1989.
54. Fanelli, R. J. and McMonagle-Strucko, K., Alteration of 5-HT_{1A} receptor binding sites following chronic treatment with ipsapirone measured by quantitative autoradiography, *Synapse*, 12, 75, 1992.
55. Larsson, L. G., Renyi, L., Ross, S. B., Svensson, B., and Angeby-Moller, K., Different effects on the responses of functional pre-and postsynaptic 5-HT_{1A} receptors by repeated treatment of rats with the 5-HT_{1A} receptor agonist 8-OH-DPAT, *Neuropharmacology*, 29, 85, 1990.
56. Schechter, L. E., Bolanos, F. J., Gozlan, H., Lanfumey, L., Haj-Dahmane, S., Laporte, A.-M., Fattaccini, C.-M., and Hamon, M. J., Alterations of central serotonergic and dopaminergic neurotransmission in rats chronically treated with ipsapirone. Biochemical and electrophysiological studies, *J. Pharmacol. Exp. Ther.*, 255, 1335, 1990.
57. Newman, M. E. and Lerer, B., Chronic electroconvulsive shock and desimipramine reduce the degree of inhibition by 5-HT and carbachol of forskolin-stimulated adenylate cyclase in rat hippocampal membranes, *Eur. J. Pharmacol.*, 148, 257, 1988.
58. Sleight, A. J., Marsden, C. A., Palfreyman, M. G., Mir, A. K., and Lovenberg, W., Chronic MAOA and MAOB inhibition decreases the 5-HT_{1A} receptor-mediated inhibition of forskolin-stimulated adenylate cyclase, *Eur. J. Pharmacol.*, 154, 255, 1988.
59. Varrault, A., Leviel, V., and Bockaert, J., 5-HT_{1A}-sensitive adenylyl cyclase of rodent hippocampal neurons. Effects of antidepressant treatments and chronic stimulation with agonists, *J. Pharmacol. Exp. Ther.*, 257, 433, 1991.
60. Newman, M. E., Shapira, B., and Lerer, B., Regulation of 5-hydroxytryptamine$_{1A}$ receptor function in rat hippocampus by short- and long-term administration of 5-hydroxytryptamine$_{1A}$ agonists and antidepressants, *J. Pharmacol. Exp. Ther.*, 260, 16, 1992.
61. Keppel Hesselink, J. M., Promising anxiolytics? A new class of drugs: the azapirones, in *Serotonin 1A Receptors In Depression and Anxiety*, Stahl, S. M., Gastpar, M., Hesselink, J. M. K., and Traber, J., Eds., Raven Press, New York, 1992, p. 171.
62. Kurtz, N., Efficacy of azapirones in depression, in *Serotonin 1A Receptors In Depression and Anxiety*, Stahl, S. M., Gastpar, M., Hesselink, J. M. K., and Traber, J., Eds., Raven Press, New York, 1992, 163.
63. Barrett, J. E. and Vanover, K. E., 5-HT receptors as targets for the development of novel anxiolytic drugs: models, mechanisms and future directions, *Psychopharmacology*, 112, 1, 1993.
64. Schreiber, R. and De Vry, J., 5-HT_{1A} receptor ligands in animal models of anxiety, impulsivity and depression: multiple mechanisms of action?, *Prog. Neuro-Psychopharmacol. Biol. Psychiatr.*, 17, 87, 1993.

65. Handley, S. L. and McBlane, J. W., 5HT drugs in animal models of anxiety, *Psychopharmacology*, 112, 13, 1993.
66. Lucki, I., Singh, A., and Kreiss, D. S., Antidepressant-like behavioral effects of serotonin receptor agonists, *Neurosci. Biobehav. Rev.*, 18, 85, 1994.
67. Frazer, A., Antidepressant drugs, *Depression*, 2, 1, 1994.
68. Blier, P. and de Montigny, C., Electrophysiological investigations on the effect of repeated zimelidine administration on serotonergic neurotransmission in the rat, *J. Neurosci.*, 3, 1270, 1983.
69. Chaput, Y., de Montigny, C., and Blier, P., Effects of a selective 5-HT reuptake blocker, citalopram, on the sensitivity of 5-HT autoreceptors: electrophysiological studies in the rat brain, *Naunyn-Schmiedeberg's Arch. Pharmacol.*, 33, 342, 1986.
70. Arborelius, L., Nomikos, G. G., Grillner, P., Hertel, P., Backlund Höök, B., Hacksell, U., and Svensson, T. H., 5-HT$_{1A}$ receptor antagonists increase the activity of serotonergic cells in the dorsal raphe nucleus in rats treated acutely or chronically with citalopram, *Naunyn-Schmiedeberg's Arch. Pharmacol.*, 352, 157, 1995.
71. Gartside, S. E., Umbers, V., Hajós, M., and Sharp, T., Interaction between a selective 5-HT$_{1A}$ receptor antagonist and an SSRI *in vivo*: effects on 5-HT cell firing and extracellular 5-HT, *Br. J. Pharmacol.*, 115, 1064, 1995.
72. Adell, A. and Artigas, F., Differential effects of clomipramine given locally or systemically on extracellular 5-hydroxytryptamine in raphé nuclei and frontal cortex, *Naunyn-Schmiedeberg's Arch. Pharmacol.*, 343, 237, 1991.
73. Invernizzi, R., Belli, S., and Samanin, R., Citalopram's ability to increase the extracellular concentrations of serotonin in the dorsal raphé prevents the drug's effect in the frontal cortex, *Brain Res.*, 584, 322, 1992.
74. Bel, N. and Artigas, F., Fluvoxamine preferentially increases extracellular 5-hydroxytryptamine in the raphé nuclei: an *in vivo* microdialysis study, *Eur. J. Pharmacol.*, 229, 101, 1992.
75. Rutter, J. J., Gundlah, C., and Auerbach, S. B., Increase in extracellular serotonin produced by uptake inhibitors is enhanced after chronic treatment with fluoxetine, *Neurosci. Lett.*, 171, 183, 1994.
76. Artigas, F., Perez, V., and Alvarez, E., Pindolol induces a rapid improvement of depressed patients treated with serotonin reuptake inhibitors, *Arch. Gen. Psychiatry*, 51, 248, 1994.
77. DeVry, J., 5-HT$_{1A}$ receptor agonists: recent developments and controversial issues, *Psychopharmacology*, 121, 1, 1995.
78. Wadenberg, M. L. and Ahlenius, S., Antipsychotic-like profile of combined treatment with raclopride and 8-OH-DPAT in the rat: enhancement of antipsychotic-like effects without catalepsy, *J. Neural Transm. Gen. Sect.*, 83, 43, 1991.
79. Neal-Beliveau, B. S., Joyce, J. N., and Lucki, I., Serotonergic involvement in haloperidol-induced catalepsy, *J. Pharmacol. Exp. Ther.*, 265, 207, 1993.
80. Simiand, J., Keane, P. E., Barnouin, M. C., Keane, M., Soubrie, P., and LeFur, G., Neuropsychopharmacological profile in rodents of SR 57746A, a new, potent 5-HT$_{1A}$ receptor agonist, *Fundam. Clin. Pharmacol.*, 7, 413, 1993.
81. Wadenberg, M. L., Cortizo, L., and Ahlenius, S., Evidence for specific interactions between 5-HT$_{1A}$ and dopamine D2 receptor mechanisms in the mediation of extrapyramidal motor functions in the rat, *Pharmacol. Biochem. Behav.*, 47, 509, 1994.
82. Seeger, T. F., Seymour, P. A., Schmidt, A. W., Zorn, S. H., Schulz, D. W., Lebel, L. A., McLean, S., Guanowsky, V., Howard, H. R., and Lowe, J. A., Ziprasidone (CP-88,059): a new antipsychotic with combined dopamine and serotonin receptor antagonist activity, *J. Pharmacol. Exp. Ther.*, 275, 101, 1995.
83. Kobilka, B. K., Frielle, T., Collins, S., Yang-Feng, T., Kobilka, T. S., Francke, U., Lefkowitz, R. J., and Caron, M.G., An intronless gene encoding a potential member of the family of receptors coupled to guanine nucleotide regulatory proteins, *Nature*, 329, 75, 1987.
84. Fargin, A., Raymond, J. R., Lohse, M. J., Kobilka, B. K., Caron, M. G., and Lefkowitz, R. J., The genomic clone G-21 which resembles a β-adrenergic receptor sequence encodes the 5-HT$_{1A}$ receptor, *Nature*, 335, 358, 1988.
85. Chanda, P. K., Minchin, M. C. W., Davis, A. R., Greenberg, L., Reilly, Y., McGregor, W. H., Bhat, R., Lubeck, M. D., Mizutani, S., and Hung, P. P., Identification of residues important for ligand binding to the human 5-hydroxytryptamine$_{1A}$ serotonin receptor, *Mol. Pharmacol.*, 43, 516, 1993.
86. Parks, C. L., Chang, L.-S., and Shenk, T., A polymerase chain reaction mediated by a single primer: cloning of genomic sequences adjacent to a serotonin receptor protein coding region, *Nucleic Acid Res.*, 19, 7155, 1991.
87. Albert, P. R., Zhou, Q.-Y., Van Tol, H. H. M., Bunzow, J. R., and Civelli, O., Cloning, functional expression, and mRNA tissue distribution of the rat 5-hydroxytryptamine$_{1A}$ receptor gene, *J. Biol. Chem.*, 265, 5825, 1990.
88. El Mestikawy, S., Cognard., C., Gozlan, H., and Hamon, M., Pharmacological and biochemical characterization of rat hippocampal 5-hydroxytryptamine$_{1A}$ receptors solubilized by 3-[3-(cholamidopropyl)dimethylammonio]-1-propane sulfonate (CHAPS), *J. Neurochem.*, 51, 1031, 1988.

89. El Mestikawy, S., Taussig, D., Gozlan H., Emerit, M. B., Ponchant, M., and Hamon, M., Chromatographic analyses of the serotonin 5-HT$_{1A}$ receptor solubilized from the rat hippocampus, *J. Neurochem.*, 53, 1555, 1989.
90. Guan, X.-M., Peroutka, S. J., and Kobilka, B. K., Identification of a single amino acid residue responsible for the binding of a class of β-adrenergic receptor antagonists to 5-hydroxytryptamine$_{1A}$ receptors, *Mol. Pharmacol.*, 41, 695, 1992.
91. Peroutka, S. J., Molecular biology of serotonin (5-HT) receptors, *Synapse*, 18, 241, 1994.
92. Hartig, P., Kao, H.-T., Macchi, M., Adham, N., Zgombick, J., Weinshank, R., and Branchek, T., The molecular biology of serotonin receptors, *Neuropsychopharmacology*, 3, 335, 1990
93. McCormack, K., Campanelli, J. T., Ramaswani, M., Mathew, M. K., and Tanouye, M. A., Leucine-zipper motif update, *Nature*, 340, 103, 1989.
94. Xie, D.-W., Deng, Z.-L., Ishigaki, T., Nakamura, Y., Suzuki, Y., Miyasato, K., Ohara, K., and Ohara, K., The gene encoding the 5-HT$_{1A}$ receptor is intact in mood disorders, *Neuropsychopharmacology*, 12, 263, 1995.
95. Nakhai, B., Nielsen, D. A., Linnoila, M., and Goldman, D., Two naturally occurring amino acid substitutions in the human 5-HT$_{1A}$ receptor: glycine 22 to serine 22 and isoleucine 28 to valine 28, *Biochem. Biophys. Res. Commun.*, 210, 530, 1995.
96. Hoyer, D., Engel, G., and Kalkman, H. O., Characterization of the 5-HT$_{1B}$ recognition site in rat brain: binding studies with (–)[^{125}I]iodocyanopindolol, *Eur. J. Pharmacol.*, 118, 1, 1985.
97. Peroutka, S. J., Pharmacological differentiation and characterization of 5-HT$_{1A}$, 5-HT$_{1B}$, and 5-HT$_{1C}$ binding sites in rat frontal cortex, *J. Neurochem.*, 47, 529, 1986.
98. Pazos, A. and Palacios, J. M., Quantitative autoradiographic mapping of serotonin receptors in the rat brain. I. Serotonin-1 receptors, *Brain Res.*, 346, 205, 1985.
99. Pazos, A., Engel, G., and Palacios, J. M., β-Adrenoreceptors blocking agents recognize a subpopulation of serotonin receptors in brain, *Brain Res.*, 343, 403, 1985.
100. Hamon, M., Lanfumey, L., El Mestikawy, S., Boni, C., Miguel, M.-C., Bolaños, F., Schechter, L., and Gozlan, H., The main features of central 5-HT$_1$ receptors, *Neuropsychopharmacology*, 3, 34a, 1990.
101. Engel, G., Göthert, M., Hoyer, D., Schlicker, E., and Hillenbrand, K., Identity of inhibitory presynaptic 5-hydroxytryptamine (5-HT) autoreceptors in the rat brain cortex with 5-HT$_{1B}$ binding sites, *Arch. Pharmacol.*, 332, 1, 1986.
102. Kikvadze, I. and Foster, G. A., Action potential-dependent output of 5-hydroxytryptamine in the anaesthetized rat amygdalopiriform cortex is strongly inhibited by tonic 5-HT$_{1B}$-receptor stimulation, *Brain Res.*, 692, 111, 1995.
103. Davidson, C. and Stamford, J. A., Evidence that 5-hydroxytryptamine release in rat dorsal raphé nucleus is controlled by 5-HT$_{1A}$, 5-HT$_{1B}$ and 5-HT$_{1D}$ autoreceptors, *Br. J. Pharmacol.*, 114, 1107, 1995.
104. Maura, G. and Raiteri, M., Cholinergic terminals in rat hippocampus possess 5-HT$_{1B}$ receptors mediating inhibition of acetylcholine release, *Eur. J. Pharmacol.*, 129, 333, 1986.
105. Raiteri, M., Maura, G., Bonanno, G., and Pittaluga, A., Differential pharmacology and function of two 5-HT$_1$ receptors modulating transmitter release in rat cerebellum, *J. Pharmacol. Exp. Ther.*, 237, 644, 1986.
106. Bouhelal, R., Smounya, L., and Bockaert, J., 5-HT$_{1B}$ receptors are negatively coupled with adenylate cyclase in rat substatia nigra, *Eur. J. Pharmacol.*, 151, 189, 1988.
107. Saudou, F., Amara, D., Dierich, A., LeMeur, M., Ramboz, S., Segu, L., Buhot, M.-C., and Hen, R., Enhanced aggressive behavior in mice lacking 5-HT$_{1B}$ receptor, *Science*, 265, 1875, 1994.
108. Heuring, R. E. and Peroutka, S. J., Characterization of a novel ^{3}H-5-hydroxytryptamine binding site subtype in bovine brain membranes, *J. Neurosci.*, 7, 894, 1987.
109. Deliganis, A. V. and Peroutka, S. J., 5-Hydroxytryptamine$_{1D}$ receptor agonism predicts antimigraine efficacy, *Headache*, 31, 228, 1991.
110. Waeber, C. and Palacios, J. M., Serotonin-1 receptor binding sites in the human basal ganglia are decreased in Huntington's chorea but not in Parkinson's disease: a quantitative in vitro autoradiography study, *Neuroscience*, 32, 337, 1989.
111. Hamel, E. and Bouchard, D., Contractile 5-HT$_1$ receptors in human isolated pial arterioles: correlation with 5-HT$_{1D}$ binding sites, *Br. J. Pharmacol.*, 102, 227, 1991.
112. Hamel, E., Fan, E., Linville, D., Ying, V., Villemure, J. G., and Chia, L. S., Expression of mRNA for the serotonin 5-hydroxytryptamine$_{1D\beta}$ receptor subtype in human and bovine cerebral arteries, *Mol. Pharmacol.*, 44, 242, 1993.
113. Schlicker, E., Fink, K., Göthert, M., Hoyer, D., Molderings, G., Roschke, I., and Schoeffter, P., The pharmacological properties of the presynaptic serotonin autoreceptor in the pig brain cortex conform to the 5-HT$_{1D}$ receptor subtype, *Naunyn-Schmiedeberg's Arch. Pharmacol.*, 340, 45, 1989.
114. Limberger, N., Deicher, R., and Starke, K, Species differences in presynaptic serotonin autoreceptors: mainly 5-HT$_{1B}$ but possibly in addition 5-HT$_{1D}$ in the rat, 5-HT$_{1D}$ in the rabbit and guinea-pig brain cortex, *Naunyn-Schmiedeberg's Arch. Pharmacol.*, 343, 353, 1991.

115. Starkey, S. J. and Skingle, M., 5-HT$_{1D}$ as well as 5-HT$_{1A}$ autoreceptors modulate 5-HT release in the guinea-pig dorsal raphé nucleus, *Neuropharmacology*, 33, 393, 1994.
116. Harel-Dupas, C., Cloez, I., and Fillion, G. The inhibitory effect of trifluoromethylphenylpiperazine on [^{3}H]acetylcholine release in guinea-pig hippocampal synaptosomes is mediated by a 5-hydroxytryptamine$_1$ receptor distinct from 1A, 1B, and 1C subtypes, *J. Neurochem.*, 56, 221, 1991.
117. Hoyer, D., Schoeffter, P., Waeber, C., and Palacios, J. M., Serotonin 5-HT1D receptors, *Ann. N.Y. Acad. Sci.*, 600, 168, 1990.
118. Cady, R. K., Wendt, J. K., Kirchner, J. R., et al., Treatment of acute migraine with subcutaneous sumatriptan, *JAMA*, 265, 2831, 1991.
119. Hamblin, M. W., Metcalf, M. A., McGuffin, R. W., and Karpells, S., Molecular cloning and functional characterization of a human 5-HT$_{1B}$ serotonin receptor: a homologue of the rat 5-HT$_{1B}$ receptor with 5-HT$_{1D}$-like pharmacological specificity, *Biochem. Biophys. Res. Commun.*, 184, 752, 1992.
120. Jin, H., Oskenberg, D., Ashkenazi, A., Peroutka, S. J., Duncan, A. M. V., Rozmahel, R., Yang, Y., Mengod, G., Palacios, J. M., and O'Dowd, B. F., Characterization of the human 5-hydroxytryptamine$_{1B}$ receptor, *J. Biol. Chem.*, 267, 5735, 1992.
121. Mochizuki, D., Yuyama, Y., Tsujita, R., Komaki, H., and Sagai, H., Cloning and expression of the human 5-HT1B-type receptor gene, *Biochem. Biophys. Res. Commun.*, 185, 517, 1992.
122. Demchyshyn, L., Sunahara, R. K., Miller, K., Teitler, M., Hoffman, B. J., Kennedy, J. L., Seeman, P., Van Tol, H. H. M., and Niznik, H.B., A human serotonin 1D receptor variant (5HT1Dβ) encoded by an intronless gene on chromosome 6, *Proc. Natl. Acad. Sci. U.S.A.*, 89, 5522, 1992.
123. Weinshank, R. L., Zgombick, J. M., Macchi, M. J., Branchek, T. A., and Hartig, P. R., Human serotonin 1D receptor is encoded by a subfamily of two distinct genes: 5-HT$_{1D\alpha}$ and 5-HT$_{1D\beta}$, *Proc. Natl. Acad. Sci. U.S.A.*, 89, 3630, 1992.
124. Levy, F. O., Gudermann, T., Perez-Reyes, E., Birnbaumer, M., Kaumann, A. J., and Birnbaumer, L., Molecular cloning of a human serotonin receptor (S12) with a pharmacological profile resembling that of the 5-HT$_{1D}$ subtype, *J. Biol. Chem.*, 267, 7553, 1992.
125. Veldman, S. A. and Bienkowski, M. J., Cloning and pharmacological characterization of a novel human 5-hydroxytryptamine$_{1D}$ receptor subtype, *Mol. Pharmacol.*, 42, 439, 1992.
126. Voigt, M. M., Laurie, D. J., Seeburg, P. H., and Bach, A., Molecular cloning and characterization of a rat brain cDNA encoding a 5-hydroxytryptamine$_{1B}$ receptor, *EMBO J.*, 10, 4017, 1991.
127. Adham, N., Romanienko, P., Hartig, P., Weinshank, R. L., and Branchek, T., The rat 5-hydroxytryptamine$_{1B}$ receptor is the species homologue of the human 5-hydroxytryptamine$_{1Db}$ receptor, *Mol. Pharmacol.*, 41, 1, 1992.
128. Metcalf, M. A., McGuffin, R. W., and Hamblin, M. W., Conversion of the human 5-HT$_{1D\beta}$ serotonin receptor to the rat 5-HT$_{1B}$ ligand-binding phenotype by THR^{355} ASN site directed mutagenesis, *Biochem. Pharmacol.*, 44, 1917, 1992.
129. Oksenberg, D., Marsters, S. A., O'Dowd, B. F., Jin, H., Havlik, S., Peroutka, S. J., and Ashkenazi, A., A single amino-acid difference confers major pharmacological variation between human and rodent 5-HT$_{1B}$ receptors, *Nature*, 360, 161, 1992.
130. Parker, E. M., Grisel, D. A., Iben, L. G., and Shapiro, R. A., A single amino acid difference accounts for the pharmacological distinctions between the rat and human 5-hydroxytryptamine$_{1B}$ receptors, *J. Neurochem.*, 60, 380, 1993.
131. Sidenberg, D. G., Bassett, A. S., Demchyshyn, L., Niznik, H. B., Macciardi, F., Kamble, A. B., Honer, W. G., and Kennedy, J. L., New polymorphism for the human serotonin 1D receptor variant (5-HT$_{1D\beta}$) not linked to schizophrenia in five Canadian pedigrees, *Hum. Hered.*, 43, 315, 1993.
132. Hamblin, M. W. and Metcalf, M. A., Primary structure and functional characterization of a human 5-HT$_{1D}$-type serotonin receptor, *Mol. Pharmacol.*, 40, 143, 1991.
133. Libert, F., Passage, E., Parmentier, M., Simons, M.-J., Vassart, G., and Mattei, M.-G., Chromosomal mapping of A1 and A2 adenosine receptors, VIP receptor and a new subtype of serotonin receptor, *Genomics*, 11, 225, 1991.
134. Maenhaut, C., Van Sande, J., Massart, C., Dinsart, C., Libert, F., Monferini, E., Giraldo, E., Ladinsky, H., Vassart, G., and Dumont, J. E., The orphan receptor cDNA RDC4 encodes a 5HT$_{1D}$ serotonin receptor, *Biochem. Biophys. Res. Commun.*, 180, 1460, 1991.
135. Zgombick, J. M., Weinshank, R. L., Macchi, M., Schechter, L. E., Branchek, T. A., and Hartig, P. R., Expression and pharmacological characterization of a canine 5-hydroxytryptamine$_{1D}$ receptor subtype, *Mol. Pharmacol.*, 40, 1036, 1991.
136. Hamblin, M. W., McGuffin, R. W., Metcalf, M. A., Dorsa, D. M., and Merchant, K. M., Distinct 5-HT$_{1B}$ and 5-HT$_{1D}$ serotonin receptors in rat: structural and pharmacological comparison of the two cloned receptors, *Mol. Cell. Neurosci.*, 3, 578, 1992.
137. Bach, A. W., Unger, L., Sprengel, R., Mengod, G., Palacios, J., Seeburg, P. H., and Voigt, M. M., Structure functional expression and spatial distribution of a cloned cDNA encoding a rat 5-HT$_{1D}$-like receptor, *J. Receptor Res.*, 13, 479, 1993.

138. Nguyen, T., Marchese, A., Kennedy, J. L., Petronis, A., Peroutka, S. J., Wu, P. H., and O'Dowd, B. F., An *alu* sequence interrupts a human 5-hydroxytryptamine$_{1D}$ receptor pseudogene, *Gene*, 124, 295, 1993.
139. Leysen, J. E., Gaps and peculiarities in 5-HT$_2$ receptor studies, *Neuropsychopharmacology*, 3, 361, 1990.
140. Pazos, A., Cortes, R., and Palacios, J. M., Quantitative autoradiographic mapping of serotonin receptors in the rat brain. II. Serotonin-2 receptors, *Brain Res.*, 346, 231, 1985.
141. Pazos, A., Probst, A., and Palacios, J. M., Serotonin receptors in the human brain. IV. Autoradiographic mapping of serotonin-2 receptors, *Neuroscience,* 21, 123, 1987.
142. Leysen, J. E., Van Gompel, P., Verwimp, M., and Niemegeers, C. J. E., Role and localisation of serotonin (S2)-receptor binding sites: effects of neuronal lesions, in *CNS Receptors — From Molecular Pharmacology to Behaviour,* Mandel, P. and de Feuedis, F. V., Eds., Raven Press, New York, 1983, p. 373.
143. Cross, A. J., Slater, P., Perry, E. K., and Perry, R. H., An autoradiographic analysis of serotonin receptors in human temporal cortex changes in Alzheimer-type dementia, *Neurochem. Int.,* 13, 89, 1988.
144. Sanders-Bush, E., Tsutsumi, M., and Burris, K. D., Serotonin receptors and phosphatidylinositol turnover, *Ann. N.Y. Acad. Sci.,* 600, 224, 1990.
145. North, R. A. and Uchimura, N., 5-Hydroxytryptamine acts at 5-HT$_2$ receptors to decrease potassium conductance in rat nucleus accumbens neurones, *J. Physiol.,* 417, 1, 1989.
146. Glennon, R. A., Titeler, M., and McKenney, J. D., Evidence for 5-HT$_2$ involvement in the mechanism of action of hallucinogenic agents, *Life Sci.*, 35, 2505, 1984.
147. Titeler, M., Lyon, R. A., and Glennon, R. A., Radioligand binding evidence implicates the brain 5-HT$_2$ receptor as a site of action for LSD and phenylisopropylamine hallucinogens, *Psychopharmacology*, 94, 213, 1988.
148. Fiorella, D., Rabin, R. A., and Winter, J. C., The role of the 5-HT$_{2A}$ and 5-HT$_{2C}$ receptors in the stimulus effects of hallucinogenic drugs. I. Antagonist correlation analysis, *Psychopharmacology*, 121, 347, 1995.
149. Baxter, G., Kennett, G., Blaney, F., and Blackburn, T., 5-HT$_2$ receptor subtypes: a family re-united?, *Trends Pharmacol. Sci.*, 16, 105, 1995.
150. Meltzer, H. Y. and Nash, J. F., VII. Effects of Antipsychotic drugs on serotonin receptors, *Pharmacol. Rev.*, 43, 587, 1991.
151. Leysen, J. E., Janssen, P. M. F., Schotte, A., Luyten, W. H. M. L., and Megens, A. A. H. P., Interaction of antipsychotic drugs with neurotransmitter receptor sites *in vitro* and *in vivo* in relation to pharmacological and clinical effects: role of 5HT$_2$ receptors, *Psychopharmacology*, 112, S40, 1993.
152. Sanders-Bush, E., Breeding, M., Knoth, K., and Tsutsumi, M., Sertraline-induced desensitization of the serotonin 5HT-2 receptor transmembrane signaling system, *Psychopharmacology*, 99, 64, 1989.
153. Arnt, J. and Hyttel, J., Facilitation of 8-OH-DPAT-induced forepaw treading of rats by the 5-HT$_2$ agonist DOI, *Eur. J. Pharmacol.*, 161, 45, 1989.
154. Backus, L. I., Sharp, T., and Grahame-Smith, D. G., Behavioral evidence for a functional interaction between central 5-HT$_2$ and 5-HT$_{1A}$ receptors, *Br. J. Pharmacol.*, 100, 793, 1990.
155. Darmani, N. A., Martin, B. R., Pandey, U., and Glennon, R. A., Do functional relationships exist between 5-HT$_{1A}$ and 5-HT$_2$ receptors?, *Pharmacol. Biochem. Behav.*, 36, 901, 1990.
156. Yocca, F. D., Wright, R. N., Margraf, R. R., and Eison, A. S., 8-OH-DPAT and busipirone analogs inhibit the ketanserin-sensitive quipazine-induced head shake response in rats, *Pharmacol. Biochem. Behav.*, 3, 251, 1990.
157. Eison, A. S. and Yocca, F. D., Reduction in cortical 5-HT$_2$ receptor sensitivity after continuous gepirone treatment, *Eur. J. Pharmacol.*, 111, 389, 1985.
158. Truett, K. A. and Hensler, J. G., Effect of 5-HT$_{2A}$ receptor downregulation on 5-HT$_{1A}$ receptor sensitivity in vivo, *Soc. Neurosci. Abstr.*, 21, 1126, 1995.
159. Pritchett, D. B., Bach, A. W. J., Wozny, M., Taleb, O., Dal Toso, R., Shih, J. C., and Seeburg, P. H., Structure and functional expression of cloned rat serotonin 5HT-2 receptor, *EMBO J.*, 7, 4135, 1988.
160. Julius, D., Huang, K. N., Livelli, T. J., Axel, R., and Jessell, T. M., The 5HT2 receptor defines a family of structurally distinct but functionally conserved serotonin receptors, *Proc. Natl. Acad. Sci. U.S.A.*, 87, 928, 1990.
161. Julius, D., MacDermott, A. B., Axel, R., and Jessell, T. M., Molecular characterization of a functional cDNA encoding the serotonin 1c receptor, *Science*, 241, 558, 1988.
162. Saltzman, A. G., Morse, B., Whitman, M. M., Ivanshchenko, Y., Jaye, M., and Felder, S., Cloning of the human serotonin 5-HT2 and 5-HT1C receptor subtypes, *Biochem. Biophys. Res. Commun.*, 181, 1469, 1991.
163. Hsieh, C.-L., Bowcock, A. M., Farrer, L. A., Herbert, J. M., Huang, K. N., Cavalli-Storza, L. L., Julius, D., and Francke, U., The serotonin receptor subtype 2 locus HTR2 is on human chromosome 13 near genes for esterase D and retinoblastoma-1 and on mouse chromosome 14, *Somat. Cell Mol. Genet.*, 16, 567, 1990.
164. Sparkes, R. S., Lan, N., Klisak, I., Mohandas, T., Diep, A., Kojis, T., Heinzmann, C., and Shih, J. C., Assignment of a serotonin 5HT-2 receptor gene (HTR2) to human chromosome 13q14-q21 and mouse chromosome 14, *Genomics*, 9, 461, 1991.
165. Chen, K., Yang, W., Grimsby, J., and Shih, J. C., The human 5-HT$_2$ receptor is encoded by a multiple intron-exon gene, *Mol. Brain Res.*, 14, 20, 1992.

166. Mengod, G., Pompeiano, M., Martínez-Mir, M. I., and Palacios, J. M., Localization of the mRNA for the 5-HT$_2$ receptor by in situ hybridization histochemistry. Correlation with the distribution of receptor sites, *Brain Res.*, 524, 139, 1990.
167. Wright, D. E., Seroogy, K. B., Lundgren, K. H., Davis, B. M., and Jennes, L., Comparative localization of serotonin$_{1A, 1C,}$ and $_2$ receptor subtype mRNAs in rat brain, *J. Compar. Neurol.*, 351, 357, 1995.
168. Burnet, P. W. J., Eastwood, S. L., Lacey, K., and Harrison, P. J., The distribution of 5-HT$_{1A}$ and 5-HT$_{2A}$ receptor mRNA in human brain, *Brain Res.*, 676, 157, 1995.
169. Johnson, M. P., Loncharich, R. J., Baez, M., and Nelson, D. L., Species variations in transmembrane region V of the 5-hydroxytryptamine type 2A receptor alter the structure-activity relationship of certain ergolines and tryptamines, *Mol. Pharmacol.*, 45, 277, 1993.
170. Kao, H.-T., Adham, N., Olsen, M. A., Weinshank, R. L., Branchek, T. A., and Hartig, P. R., Site-directed mutagenesis of a single residue changes the binding properties of the serotonin 5-HT$_2$ receptor from a human to a rat pharmacology, *FEBS Lett.*, 307, 324, 1992.
171. Ding, D., Toth, M., Zhou, Y., Parks, C., Hoffman, B. J., and Shenk, T., Glial cell-specific expression of the serotonin 2 receptor gene: selective reactivation of a repressed promoter, *Mol. Brain Res.*, 20, 181, 1993.
172. Du, Y.-L., Wilcox, B. D., and Jeffrey, J. J., Regulation of rat 5-hydroxytryptamine type 2 receptor gene activity: identification of *cis* elements that mediate basal and 5-hydroxytryptamine-dependent gene activation, *Am. Soc. Pharmacol. Exp. Ther.*, 47, 915, 1995.
172a. Du, Y.-L., Wilcox, B. D., Tietler, M., and Jeffrey, J. J., Isolation and characterization of the rat 5-hydroxytryptamine type 2 receptor promoter: constitutive and inducible activity in myometrial smooth muscle cells, *Mol. Pharmacol.,* 45, 1125, 1994.
173. Garlow, S. J., Chin, A. C., Marinovich, A. M., Heller, M. R., and Ciaranello, R. D., Cloning and functional promoter mapping of the rat serotonin-2 receptor gene, *Mol. Cell. Neurosci.*, 5, 291, 1994.
174. Zhu, Q.-S., Chen, K., and Shih, J. C., Characterization of the human 5-HT$_{2A}$ receptor gene promoter, *J. Neurosci.*, 15, 4885, 1995.
175. Vane, J. R., The relative activities of some tryptamine analogues on the isolated rat stomach strip preparation, *Br. J. Pharmacol. Chemother.*, 14, 87, 1959.
176. Bradley, P. B., Engel, G., Feniuk, W., Fozard, J. R., Humphrey, P. P. A., Middlemiss, D. N., Mylecharane, E. J., Richardson, B. P., and Saxena, P. R., Proposal for the classification and nomenclature of functional receptors for 5-hydroxytryptamine, *Neuropharmacology*, 25, 563, 1986.
177. Kursar, J. D., Nelson, D. L., Wainscott, D. B., Cohen, M. L., and Baez, M., Molecular cloning, functional expression and pharmacological characterisation of a novel serotonin receptor (5-hydroxytryptamine$_{2F}$) from rat stomach fundus, *Mol. Pharmacol.,* 42, 549, 1992.
178. Foguet, M., Hoyer, D., Pardo, L. A., Parekh, A., Kluxen, F. W., Kalkman, H. O., Stühmer, W., and Lübbert, H., Cloning and functional characterization of the rat stomach fundus serotonin receptor, *EMBO J.*, 11, 3481, 1992.
179. Wainscott, D. B., Cohen, M. L., Schenck, K. W., Audia, J. E., Niessen, J. S., Baez, M., Kursar, J. D., Lucaites, V. L., and Nelson, D. L., Pharmacological characteristics of the newly cloned rat 5-hydroxytryptamine$_{2F}$ receptor, *Mol. Pharmacol.,* 43, 419, 1993.
180. Kennett, G. A., Wood, M. D., Glen, A., Grewal, S., Forbes, I., Gadre, A., and Blackburn, T. P., *In vivo* properties of SB 200646A, a 5-HT$_{2C/2B}$ receptor antagonist, *Br. J. Pharmacol.*, 111, 797, 1994.
181. Secrest, R. J., Lucaites, V. L., Mendelsohn, L. G., and Cohen, M. L., Protein kinase C translocation in rat stomach fundus: effects of serotonin, carbamylcholine and phorbol dibutyrate, *J. Pharmacol. Exp. Ther.*, 256, 103, 1991.
182. Foguet, M., Nguyen, H., Le, H., and Lubbert, H., Structure of the mouse 5-HT$_{1C}$, 5-HT$_2$ and stomach fundus serotonin receptor genes, *Neuroreport*, 3, 345, 1992.
183. Loric, S., Launay, J.-M., Colas, J.-F., and Maroteaux, L., New mouse 5-HT2-like receptor. Expression in brain, heart and intestine, *FEBS Lett.*, 312, 203, 1992.
184. Kursar, J. D., Nelson, D. L., Wainscott, D. B., and Baez, M., Molecular cloning, functional expression, and mRNA tissue distribution of the human 5-hydroxytryptamine$_{2B}$ receptor, *Mol. Pharmacol.*, 46, 227, 1994.
185. Schmuck, K., Ullmer, C., Engels, P., and Lübbert, H., Cloning and functional characterization of the human 5-HT$_{2B}$ serotonin receptor, *FEBS Lett.*, 342, 85, 1994.
186. Pazos, A., Hoyer, D., and Palacios, J. M., The binding of serotonergic ligands to the porcine choroid plexus: characterisation of a new type of serotonin recognition site, *Eur. J. Pharmacol.,* 106, 539, 1984.
187. Yagaloff, K. A. and Hartig, P. R., [^{125}I]Lysergic acid diethylamide binds to a novel serotonin site on rat choroid plexus epithelial cells, *J. Neurosci.,* 5, 3178, 1985.
188. Pazos, A., Probst, A., and Palacios, J. M., Serotonin receptors in the human brain. III. Autoradiographic mapping of serotonin-1 receptors, *Neuroscience,* 21, 97, 1987.
189. Kennett, G. A., 5-HT$_{1C}$ receptors and their therapeutic relevance, *Curr. Opin. Invest. Drugs*, 2, 317, 1993.
190. Kennett, G. A. and Curzon, G., Potencies of antagonists indicate that 5-HT$_{1C}$ receptors mediate 1-3(chlorophenyl) piperazine-induced hypophagia, *Br. J. Pharmacol.*, 103, 2016, 1991.
191. Lucki, I., 5-HT$_1$ receptors and behavior, *Neurosci. Biobehav. Rev.*, 16, 83, 1992.

192. Kennett, G. A., Whitton, P., Shah, K., and Curzon, G., Anxiogenic-like effects of mCPP and TFMPP in animal models are opposed by 5-HT$_{1C}$ receptor antagonists, *Eur. J. Pharmacol.*, 164, 445, 1989.
193. Kennett, G. A., 5-HT$_{1C}$ receptor antagonists have anxiolytic-like actions in the rat social interaction model, *Psychopharmacology*, 107, 379, 1992.
194. Kennett, G. A., Pittaway, K., and Blackburn, T. P., Evidence that 5-HT$_{2C}$ receptor antagonists are anxiolytic in the rat Geller-Seifter model of anxiety, *Psychopharmacology*, 114, 90, 1994.
195. Tecott, L. H., Sun, L. M., Akana, S. F., Strack, A. M., Lowenstein, D. H., Dallman, M. F., and Julius, D., Eating disorder and epilepsy in mice lacking 5-HT$_{2C}$ serotonin receptors, *Nature*, 374, 542, 1995.
196. Lübbert, H., Hoffman, B. J., Snutch, T. P., Van Dyke, T., Levine, A. J., Hartig, P. R., Lester, H. A., and Davidson, N., cDNA cloning of a serotonin 5-HT$_{1C}$ receptor by electrophysiological assays of mRNA-injected *Xenopus* oocytes, *Neurobiology*, 84, 4332, 1987.
197. Yu, L., Nguyen, H., Le, H., Bloem, L. J., Kozak, C. A., Hoffman, B. J., Snutch, T. P., Lester, H. A., Davidson, N., and Lübbert, H., The mouse 5-HT$_{1C}$ receptor contains eight hydrophobic domains and is X-linked, *Mol. Brain Res.*, 11, 143, 1991.
198. Milatovich, A., Hsieh, C. L., Bonaminio, G., Tecott, L., Julius, D., and Francke, U., Serotonin receptor 1c gene assigned to X chromosome in human (band q24) and mouse (bands D-F4), *Hum. Mol. Genet.*, 1, 681, 1992.
199. Hoffman, B. J. and Mezey, E., Distribution of serotonin 5-HT$_{1C}$ receptor mRNA in adult rat brain, *FEBS Lett.*, 247, 453, 1989.
200. Molineaux, S. M., Jessell, T. M., Axel, R., and Julius, D., 5-HT1c receptor is a prominent serotonin receptor subtype in the central nervous system, *Neurobiology*, 86, 6793, 1989.
201. Mengod, G., Nguyen, H., Le, H., Waeber, C., Lübbert, H., and Palacios, J. M., The distribution and cellular localization of the serotonin 1C receptor mRNA in the rodent brain examined by *in situ* hybridization histochemistry. Comparison with receptor binding distribution, *Neuroscience*, 35, 577, 1990.
202. Roca, A. L., Weaver, D. R., and Reppert, S. M., Serotonin receptor gene expression in the rat suprachiasmatic nuclei, *Brain Res.*, 608, 159, 1993.
203. Bloem, L. J., Chen, Y., Liu, J., Bye, L. S., and Yu, L., Analysis of the promoter sequence and the transcription initiation site of the mouse 5-HT$_{1C}$ serotonin receptor gene, *Mol. Brain Res.*, 17, 194, 1993.
204. Fozard, J. R., MDL 72222, a potent and highly selective antagonists at neuronal 5-hydroxytryptamine receptors, *Naunyn-Schmiedeberg's, Arch. Pharmacol.*, 326, 36, 1984.
205. Richardson, B. P., Engel, G., Donatsch, P., and Stadler, P. A., Identification of serotonin M-receptor subtypes and their specific blockade by a new class of drugs, *Nature*, 316, 126, 1985.
206. Watling, K. J., Radioligand binding studies identify 5-HT$_3$ recognition sites in neuroblastoma cell lines and mammalian CNS, *Trends Pharmacol. Sci.*, 9, 227, 1988.
207. Jackson, M. B. and Yakel, J. L., The 5-HT$_3$ receptor channel, *Annu. Rev. Physiol.*, 57, 447, 1995.
208. Kilpatrick, G. J., Jones, B. J., and Tyers, M. B., Identification and distribution of 5-HT$_3$ receptors in rat brain using radioligand binding, *Nature*, 330, 746, 1987.
209. Abi-Dargham, A., Laruelle, M., Wong, D. T., Robertson, D. W., Weinberger, D. R., and Kleinman, J. E., Pharmacological and regional characterization of [^{3}H]LY-278584 binding sites in human brain, *J. Neurochem.*, 60, 730, 1993.
210. Blandina, P., Goldfarb, J., Craddock-Royal, B., and Green, J. P., Release of endogenous dopamine by stimulation of 5-hydroxytryptamine receptors in rat striatum, *J. Pharmacol. Exp. Ther.*, 251, 803, 1989.
211. Apud, J. A., The 5-HT3 receptor in mammalian brain: a new target for the development of psychotropic drugs?, *Neuropsychopharmacology*, 8, 117, 1993.
212. Mongeau, R., de Montigny, C., and Blier, P., Activation of 5-HT$_3$ receptors enhances the electrically evoked release of [^{3}H]noradrenaline in rat brain limbic structures, *Eur. J. Pharmacol.*, 256, 269, 1994.
213. Barnes, J. M., Barnes, N. M., Costall, B., Naylor, R. J., and Tyers, M. B., 5-HT$_3$ receptors mediate inhibition of acetylcholine release in cortical tissue, *Nature*, 338, 762, 1989.
214. Consolo, S., Bertorelli, R., Russi, G., Zambelli, M., and Ladinsky, H., Serotonergic facilitation of acetylcholine release *in vivo* from rat dorsal hippocampus via serotonin 5-HT$_3$ receptors, *J. Neurochem.*, 62, 2254, 1994.
215. Blier, P. and Bouchard, C., Functional characterization of a 5-HT$_3$ receptor which modulates the release of 5-HT in the guinea-pig brain, *Br. J. Pharmacol.*, 108, 13, 1993.
216. Adrien, J., Tissier, M. H., Lanfumey, L., Haj-Dahmane, S., Jolas, T., Franc, B., and Hamon, H., Central action of 5-HT$_3$ receptor ligands in the regulation of sleep-wakefulness and raphe neuronal activity in the rat, *Neuropharmacology*, 31, 519, 1992.
217. Kidd, E. J., Laporte, A. M., Langlois, X., Fattaccini, C.-M., Doyen, C., Lombard, M. C., Gozlan, H., and Hamon, M., 5-HT$_3$ receptors in the rat central nervous system are mainly located on nerve fibres and terminals, *Brain Res.*, 612, 289, 1993.
218. Zeise, M. L., Batsche, K., and Wang, R. Y., The 5-HT$_3$ receptor agonist 2-methyl-5-HT reduces postsynaptic potentials in rat CA1 pyramidal neurons of the hippocampus in vitro, *Brain Res.*, 651, 337, 1994.
219. Grant, K. A., The role of 5-HT$_3$ receptors in drug dependence, *Drug Alcohol Depend.*, 38, 155, 1995.

220. Greenshaw, A. J., Behavioral pharmacology of 5-HT$_3$ receptor antagonists: a critical update on therapeutic potential, *Trends Pharmacol. Sci.*, 14, 265, 1993.
221. Maricq, A. V., Peterson, A. S., Brake, A. J., Myers, R. M., and Julius, D., Primary structure and functional expression of the 5HT$_3$ receptor, a serotonin-gated ion channel, *Science*, 254, 432, 1991.
222. Uetz, P., Abdelatty, F., Villarroel, A., Rappold, G., Weiss, B., and Koenen, M., Organization of the murine 5-HT$_3$ receptor gene and assignment to human chromosome 11, *FEBS Lett.*, 339, 302, 1994.
223. Hope, A. G., Downie, D. L., Sutherland, L., Lambert, J. J., Peters, J. A., and Burchell, B., Cloning and functional expression of an apparent splice variant of the murine 5-HT$_3$ receptor A subunit. *Eur. J. Pharmacol. Mol. Pharmacol. Sect.*, 245, 187, 1993.
224. Werner, P., Kawashima, E., Reid, J., Hussy, N., Lundstrom, K., Buell, G., Humbert, Y., and Jones, K. A., Organization of the mouse 5-HT$_3$ receptor gene and functional expression of two splice variants, *Brain Res. Mol. Brain Res.*, 26, 233, 1994.
225. Kia, H. K., Miquel, M. C., McKernan, R. M., Laporte, A. M., Lombard, M. C., Bourgoin, S., Hamon, M., and Verge, D., Localization of 5-HT$_3$ receptors in the rat spinal cord: immunohistochemistry and *in situ* hybridization, *Neuroreport*, 6, 257, 1995.
226. Tecott, L. H., Maricq, A. V., and Julius, D., Nervous system distribution of the serotonin 5-HT$_3$ receptor mRNA, *Proc. Natl. Acad. Sci. U.S.A.*, 90, 1430, 1993.
227. Johnson, D. S. and Heinemann, S. F., Detection of 5-HT$_3$R-A, a 5-HT$_3$ receptor subunit, in submucosal and myenteric ganglia of rat small intestine using *in situ* hybridization, *Neurosci. Lett.*, 184, 67, 1995.
228. Dumuis, A., Bouhelal, R., Sebben, M., Cory, R., and Bockaert, J., A nonclassical 5-hydroxytryptamine receptor positively coupled with adenylate cyclase in the central nervous system, *Mol. Pharmacol.*, 34, 880, 1988.
229. Eglen, R. M., Wong, E. H. F., Dumuis, A., and Bockaert, J., Central 5-HT$_4$ receptors, *Trends Pharmacol. Sci.*, 16, 391, 1995.
230. Grossman, C. J., Kilpatrick, G. J., and Bunce, K. T., Development of a radioligand binding assay for 5-HT$_4$ receptors in guinea-pig and rat brain, *Br. J. Pharmacol.*, 109, 618, 1993.
231. Brown, A. M., Young, T. J., Patch, T. L., Cheung, C. W., Kaumann, A., Gaster, L., and King, F. D., [^{125}I]SB 207710, a potent, selective radioligand for 5-HT$_4$ receptors, *Br. J. Pharmacol.*, 110, 10P, 1993.
232. Fagni, L., Dumuis, A., Sebben, M., and Bockaert, J., The 5-HT$_4$ receptor subtype inhibits K+ current in colliculi neurones via activation of a cyclic AMP-dependent protein kinase, *Br. J. Pharmacol.*, 105, 973, 1992.
233. Monferini, E., Gaetani, P., Rodriguez, Y., Baena, R., Giraldo, E., Parenti, M., Zocchetti, A., and Rizzi, C. A., Pharmacological characterisation of the 5-hydroxytryptamine receptor coupled to adenylyl cyclase stimulation in human brain, *Life Sci.*, 52, 61, 1993.
234. Consolo, S., Arnaboldi, S., Giorgi, S., Russi, G., and Ladinsky H., 5-HT$_4$ receptor stimulation facilitates acetylcholine release in rat frontal cortex, *Neuroreport*, 5, 1230, 1994.
235. Benloucif, S., Keegan, M. J., and Galloway, M. P., Serotonin-facilitated dopamine release *in vivo*: pharmacological characterization, *J. Pharmacol. Exp. Ther.*, 265, 373, 1993.
236. Bonhomme, N., De Deurwaerdere, P., Le Moal, M., and Spampinato, U., Evidence for 5-HT$_4$ receptor subtypes involvement in the enhancement of striatal dopamine release induced by serotonin: a microdialysis study in the halothane-anesthetized rat, *Neuropharmacology*, 34, 269, 1995.
237. Boddeke, H. W. G. M. and Kalkman, H. P., Zacopride and BRL 24924 induce an increase in EEG-energy in rats, *Br. J. Pharmacol.*, 101, 281, 1990.
238. Reynolds, G. P., Mason, S. L., Meldrum, A., De Keczer, S., Parnes, H., Eglen, R. M., and Wong, E. H. F., 5-Hydroxytryptamine (5HT)$_4$ receptors in *post mortem* human brain tissue: distribution, pharmacology and effects of neurodegenerative diseases, *Br. J. Pharmacol.*, 114, 993, 1994.
239. Gerald, C., Adham, N., Kao, H.-T., Olsen, M. A., Laz, T. M., Schechter, L. E., Bard, J. A., Vaysse, P. J.-J., Hartig, P. R., Branchek, T. A., and Weinshank, R. L., The 5-HT$_4$ receptor: molecular cloning and pharmacological characterization of two splice variants, *EMBO J.*, 14, 2806, 1995.
240. Leonhardt, S., Herrick-Davis, K., and Teitler, M., Detection of a novel serotonin receptor subtype (5-HT$_{1E}$) in human brain: interaction with a GTP-binding protein, *J. Neurochem.*, 53, 465, 1989.
241. McAllister, G., Charlesworth, A., Snodin, C., Beer, M. S., Noble, A. J., Middlemiss, D. N., Iversen, L. L., and Whiting, P., Molecular cloning of a serotonin receptor from human brain (5HT1E): a fifth 5HT1-like subtype, *Proc. Natl. Acad. Sci. U.S.A.*, 89, 5517, 1992.
242. Levy, F. O., Gudermann, T., Birnbaumer, M., Kaumann, A. J., and Birnbaumer, L., Molecular cloning of a human gene (S31) encoding a novel serotonin receptor mediating inhibition of adenylyl cyclase, *FEBS Lett.*, 296, 201, 1991.
243. Zgombick, J. M., Schechter, L. E., Macchi, M., Hartig, P. R., Branchek, T. A., and Weinshank, R. L., Human gene S31 encodes the pharmacologically defined serotonin 5-hydroxytryptamine$_{1E}$ receptor, *Mol. Pharmacol.*, 42, 180, 1992.
244. Gudermann, T., Levy, F. O., Birnbaumer, M., Birnbaumer, L., and Kaumann, A. J., Human S31 serotonin receptor clone encodes a 5-hydroxytryptamine$_{1E}$-like serotonin receptor, *Mol. Pharmacol.*, 43, 412, 1993.

245. Amlaiky, N., Ramboz, S., Boschert, U., Plassat, J.-L., and Hen, R., Isolation of a mouse "5HT1E-like" serotonin receptor expressed predominantly in hippocampus, *J. Biol. Chem.*, 267, 19761, 1992.
246. Adham, N., Kao, H.-T., Schechter, L. E., Bard, J., Olsen, M., Urquhart, D., Durkin, M., Hartig, P. R., Weinshank, R. L., and Branchek, T. A., Cloning of another human serotonin receptor (5-HT$_{1F}$): a fifth 5-HT$_1$ receptor subtype coupled to the inhibition of adenylate cyclase, *Proc. Natl. Acad. Sci. U.S.A.*, 90, 408, 1993.
247. Lovenberg, T. W., Erlander, M. G., Baron, B. M., Racke, M., Slone, A. L., Siegel, B. W., Craft, C. M., Burns, J. E., Danielson, P. E., and Sutcliffe, J. G., Molecular cloning and functional expression of 5-HT$_{1E}$-like rat and human 5-hydroxytryptamine receptor genes, *Proc. Natl. Acad. Sci. U.S.A.*, 90, 2184, 1993.
248. Plassat, J.-L., Boschert, U., Amlaiky, N., and Hen, R., The mouse 5HT5 receptor reveals a remarkable heterogeneity within the 5HT1D receptor family, *EMBO J.*, 11, 4779, 1992.
249. Matthes, H., Boschert, U., Amlaiky, N., Grailhe, R., Plassat, J.-L., Muscatelli, F., Mattei, M.-G., and Hen, R., Mouse 5-hydroxytryptamine5A and 5-hydroxytryptamine5B receptors define a new family of serotonin receptors. Cloning, functional expression, and chromosomal localization, *Mol. Pharmacol.*, 43, 313, 1993.
250. Erlander, M. G., Lovenberg, T. W., Baron, B. M., De Lecea, L., Danielson, P. E., Racke, M., Slone, A. L., Siegel, B. W., Foye, P. E., Cannon, K., Burns, J. E., and Sutcliffe, J. G., Two members of a distinct subfamily of 5-hydroxytryptamine receptors differentially expressed in rat brain, *Proc. Natl. Acad. Sci. U.S.A.*, 90, 3452, 1993.
251. Monsma, F. J., Jr., Shen, Y., Ward, R. P., Hamblin, M. W., and Sibley, D. R., Cloning and expression of a novel serotonin receptor with high affinity for tricyclic psychotropic drugs, *Mol. Pharmacol.*, 43, 320, 1993.
252. Ruat, M., Traiffort, E., Arrang, J.-M., Tardivel-Lacombe, J., Diaz, J., Leurs, R., and Schwartz, J.-C., A novel rat serotonin (5-HT$_6$) receptor: molecular cloning, localization and stimulation of cAMP accumulation, *Biochem. Biophys. Res. Commun.*, 193, 268, 1993.
253. Meyerhof, W., Obermuller, F., Fehr, S., and Richter, D., A novel rat serotonin receptor: primary structure, pharmacology, and expression pattern in distinct brain regions, *DNA Cell Biol.*, 12, 401, 1993.
254. Ruat, M., Traiffort, E., Leurs, R., Tardivel-Lacombe, J., Diaz, J., Arrang, J.-M., and Schwartz, J.-C., Molecular cloning, characterization, and localization of a high-affinity serotonin receptor (5-HT$_7$) activating cAMP formation, *Proc. Natl. Acad. Sci. U.S.A.*, 90, 8547, 1993.
255. Shen, Y., Monsma, F. J., Jr., Metcalf, M. A., Jose, P. A., Hamblin, M. W., and Sibley, D. R., Molecular cloning and expression of a 5-hydroxytryptamine$_7$ serotonin receptor subtype, *J. Biol. Chem.*, 268, 18200, 1993.
256. Plassat, J.-L., Amlaiky, N., and Hen, R., Molecular cloning of a mammalian serotonin receptor that activates adenylate cyclase, *Mol. Pharmacol.*, 44, 229, 1993.
257. Bard, J. A., Zgombick, J., Adham, N., Vaysse, P., Brancheck, T. A., and Weinshank, R. L., Cloning of a novel human serotonin receptor (5-HT$_7$) positively linked to adenylate cyclase, *J. Biol. Chem.*, 268, 23422, 1993.

10 Brain Adrenergic Receptors

David A. Morilak

CONTENTS

A. Introduction 147
B. Pharmacologic Classification of Adrenergic Receptors...148
C. Molecular Biology of Adrenergic Receptors 149
D. Second Messenger Effector Systems 151
E. Regulation of Receptor Function 152
F. Distribution of Brain Adrenergic Receptor Subtypes...154
G. Cellular and Physiological Functions 155
H. Conclusions...157
References 157

A. INTRODUCTION

The catecholamines norepinephrine (NE) and epinephrine (E) act as neuromodulators and neurotransmitters throughout the brain. These adrenergic monoamines, NE in particular, are among the most thoroughly studied of central transmitters and are broadly implicated in a diverse range of physiological, neuroendocrine, behavioral, cognitive, and pathological processes. Like all other neurotransmitters, the biological effects of NE and E are transduced by integral membrane proteins called receptors; like other neurotransmitter receptors, the adrenergic receptors have been grouped and classified into subtypes based on a number of functional, structural, and biochemical criteria.[1] Traditionally, initial classification of receptor subtypes has been made based on pharmacological profiles of agonist and antagonist potency in bioassays, and later in ligand binding assays. Additional receptor subtype differentiation has been made according to the biochemical effector responses to which they are coupled, i.e., the cellular second messenger systems activated upon agonist binding. Most recently, with the advent of molecular cloning and controlled expression of recombinant receptors in cell lines *in vitro*, a host of new receptor subtypes, many of which had not been recognized previously by pharmacological definition, have been identified and classified based on sequence and structural homology with known receptors, and by pharmacological similarity upon transfection and expression in cell lines.

The adrenergic receptors are all members of the superfamily of G-protein-coupled receptors which includes a large and growing number of proteins that transduce the biological effects of a variety of drugs, hormones, and neurotransmitters. These receptor proteins share a number of structural and functional features.[2,3] Common structural features include an extracellular N-terminal domain containing a number of N-linked glycosylation sites, an intracellular C-terminal tail, and a series of 7 hydrophobic stretches of 25 to 28 amino acids that presumably form alpha-helices spanning the plasma membrane in a barrel-like configuration linked by alternating loops of intracellular and extracellular segments. The transmembrane domains of the G-protein-linked receptors are highly conserved, with a higher degree of amino acid similarity and identity observed between members of related receptor subtypes.[4] It is thought that the pocket formed by the transmembrane segments represents the binding site for the adrenergic neurotransmitters. Specific residues within the transmembrane segments and extracellular loops determine ligand specificity and affinity. Portions of the third intracellular loop and the proximal segment of the C-terminal tail are believed to be involved in the coupling of receptor stimulation to activation of the appropriate G-protein. These regions also contain a number of potential phosphorylation sites that may be important in the mechanisms of receptor regulation. Such regulation, considered in more detail below, is an important process in limiting the degree of activation of intracellular second messenger cascades by the continued presence of agonists.

The evolutionary and physiological significance of having multiple, closely related receptor subtypes that bind the same endogenous ligands and elicit similar effector responses is currently unknown. However, understanding the diversity of adrenergic receptor subtypes and their distinct functions and regulation will help us to further understand the potential involvement of the adrenergic neurotransmitters in the etiology of various physiological and psychopathologies, and may thereby direct the development of more specific and targeted therapeutic tools and strategies.

B. PHARMACOLOGIC CLASSIFICATION OF ADRENERGIC RECEPTORS

Adrenergic receptors were first categorized into two general classes, α and β, based on rank-order agonist potencies of a number of chemical derivatives of epinephrine in a series of bioassays.[5] With later pharmacologic refinements, the α-adrenergic receptors were further differentiated into α_1 and α_2 subtypes.[6] Both α- and β-adrenergic receptors are activated by the endogenous adrenergic transmitters epinephrine and norepinephrine, but are differentiated by the pharmacological profiles of a few key synthetic reagents.[7] α_1-Receptors are activated by the selective agonist phenylephrine and blocked by the antagonist prazosin, while α_2-receptors are activated by the agonists clonidine and UK-14,304 and blocked by the antagonists yohimbine, rauwolscine, and idazoxan.[8,9] The distinct pharmacologic characterization of the two α-adrenoreceptor classes was further substantiated when it was subsequently demonstrated that they couple to different G-proteins and elicit different second messenger responses (see below).

The prototypic β-adrenoreceptor agonist is isoproterenol, with propranolol and related reagents acting as antagonists. The β-adrenoreceptors have also been further subclassified into β_1 and β_2 subtypes.[10] The major characteristic by which the two β-receptors initially were discriminated was the difference in relative potencies of the endogenous catecholamines: NE and E are approximately equipotent at β_1-receptors, while E possesses a higher potency than NE at the β_2 subtype.[10] A number of antagonist drugs have since been developed that are selective for β_1-receptors (e.g., atenolol, betaxolol) or β_2-receptors (e.g., ICI-118,551).[1,11,12] Nonetheless, the β-adrenoreceptor subtypes are much more closely related to each other than are the α_1 and α_2 subtypes, both structurally and functionally. β_1- and β_2-adrenoreceptors show very similar binding profiles for most drugs and couple to similar if not identical effector responses (see below).

Thus, the three-part subdivision of adrenergic receptors into α_1, α_2, and β subtypes remained the major classification scheme until very recently. With further pharmacological refinement, including the development of more selective drugs and more sensitive bioassays, it became obvious that a more extensive and subtle subdivision of these major receptor classes would be necessary (see Table 10-1). For example, different populations of α_1-receptor binding sites were observed, termed α_{1A} and α_{1B}, showing differential affinity for the α_1 antagonist WB-4101 ($\alpha_{1A} > \alpha_{1B}$),[13] and differential sensitivity to irreversible alkylation by chlorethylclonidine ($\alpha_{1A} < \alpha_{1B}$).[14] A distinct subset of β-receptor-like binding sites showed an atypical binding profile, with unusually low affinity for a number of classical β-receptor agonists and antagonists. These sites have since been equated with the β_3-receptor.[15] Additional α_2-receptor subtypes have also been described based on the differential affinity of prazosin ($\alpha_{2A} < \alpha_{2B}$), a drug that is generally considered a selective α_1 antagonist.[8,16] With the advent of molecular cloning, expression and analyses of recombinant receptors has substantiated and typically extended these often subtle subclassifications to include an increasing number of related receptors subtypes (see below). However, despite the addition of several new subtypes within each of these groups, the major categorization of adrenergic receptors into α_1-, α_2- and β-receptor families still remains a valid conceptual and functional classification scheme,[16] which has been further reinforced by subsequent biochemical and molecular approaches.

C. MOLECULAR BIOLOGY OF ADRENERGIC RECEPTORS

The molecular cloning of a growing number of genes and cDNAs encoding G-protein-linked neurotransmitter receptors has led to the development of novel criteria by which to classify receptors based on sequence homology and by ligand binding profiles and second messenger responses elicited after expression of recombinant receptors in cell lines *in vitro*. These advances have in turn produced a dramatic increase in the number of identified (and in some cases unidentified) G-protein-coupled receptor subtypes, and a growing appreciation of the complexity and diversity with which these closely related proteins have evolved. It is clear that the receptor subtypes themselves remain well conserved across species, with a much higher degree of sequence homology observed between homologs of the same receptor subtypes in different species than is observed between different subtypes of the same family within the same species.[4]

However, with the growing number of cloned receptors, there has often been confusion and uncertainty regarding the correspondence between these recombinant receptor subtypes and native receptor subtypes as defined by pharmacologic criteria. The picture has become somewhat clearer recently as more subtypes have been cloned, and as many apparent pharmacologic inconsistencies have been attributed to species differences or to anomalies associated with *in vitro* expression of recombinant receptors in normally nonexpressing cell lines.

The current consensus is that there are three α_1-adrenergic receptor subtypes, termed α_{1A}, α_{1B}, and α_{1D}.[17] The α_{1B} receptor cDNA was the first to be cloned from hamster vas deferens smooth muscle cells,[18] and it was shown to correspond very clearly to the pharmacologically defined α_{1B} subtype, showing a relatively high sensitivity to irreversible alkylation by chlorethylclonidine and low affinity for the antagonist WB-4101.[14] The human α_{1B}-receptor gene is located on chromosome 5.[19] Like the other α_1-receptors cloned to date, the α_{1B}-gene is interrupted by a single intron, allowing for the possibility of splice variants of these receptors in the N-terminal region, though experimental evidence for such isoform variations is lacking. Shortly after cloning the α_{1B}-receptor cDNA, the same group cloned two additional α_1-adrenergic receptors.[20,21] Initially, one of these clones derived from a rat brain library was thought to correspond to the pharmacologically defined α_{1A} receptor.[21] However, additional pharmacologic analyses of that and a nearly identical clone from another group[22] revealed important discrepancies in the affinity profiles for a number of α_{1A}-selective antagonists, and it was concluded that this clone represented a novel subtype distinct from but very closely

TABLE 10-1
Adrenergic Receptor Classification and Characteristics

Adrenergic Receptor Class	Selective Agonist	Selective Antagonist	G-Protein	Major Effector Response	Subtypes	Human Chromosome	Introns	Sites of Prominent CNS Expression
α_1	Phenylephrine	Prazosin	$G\alpha_q$	PI hydrolysis	α_{1A}	8	Yes	Olfactory bulb, neocortex, hippocampus, brainstem motor nuclei, spinal cord
					α_{1B}	5	Yes	Thalamus, neocortex, hypothalamus
					α_{1D}	20	Yes	Neocortex, hippocampus, amygdala, basal forebrain, olfactory bulb
α_2	Clonidine UK 14,304	Yohimbine Rauwolscine	$G\alpha_i$	Inhibition of adenylyl cyclase	α_{2A}	10	No	Cortex, hypothalamus, NTS, locus coeruleus, hippocampus
					α_{2B}	2	No	Thalamus
					α_{2C}	4	No	Cortex, basal forebrain, striatum, hippocampus
β	Isoproterenol	Propranolol	$G\alpha_s$	Activation of adenylyl cyclase	β_1	10	No	Neocortex, amygdala, hippocampus, basal forebrain, thalamus, medulla/pons
					β_2	5	No	Cerebellum, thalamus, medulla/pons
					β_3	8	Yes	None

related to the native α_{1A}. This receptor subtype has now been designated as the α_{1D}, the gene for which is located on human chromosome 20.[23,24] A third α_1 receptor subtype also showed similarity to the pharmacologically defined α_{1A} receptor. This clone, originally derived from a bovine cortical cDNA library[20] was initially thought to represent another novel subtype, termed α_{1C}, based on an apparent lack of detectable expression and several unique pharmacological characteristics. However, subsequent analyses of both the distribution and binding profiles of this receptor, aided by the cloning of the rat and human homologs, have provided strong evidence that this indeed represents the native α_{1A} receptor subtype.[25] The α_{1A} receptor gene has been localized to human chromosome 8.[20]

Understanding the correspondence between the recombinant α_2-receptor subtypes and those defined by pharmacological criteria has also been difficult due to a lack of clearly discriminating pharmacological reagents. Consensus again suggests that there are three α_2-receptor subtypes.[1] The human α_2 receptor cDNA clone that has been designated α_2-C10 because of its localization to human chromosome 10, corresponds to the pharmacologic α_{2A} binding site[26] showing low affinity for prazosin (approximately 1 μM). The remaining α_2 subtypes show a relatively higher affinity for prazosin (approximately 30 to 60 nM), which is nonetheless still quite low compared to that of the α_1-receptors (0.1 to 0.5 nM). The recombinant human α_2-C2 and α_2-C4 cDNA clones, localized to chromosomes 2 and 4, respectively, correspond to the pharmacologically defined α_{2B} and α_{2C} subtypes.[27,28] All of the α_2 receptor genes apparently lack introns. Thus, any diversity in α_2-receptor characteristics can not be attributed to splice variants of the mRNA transcript encoding the receptor proteins.

Correspondence between recombinant and pharmacologically defined β-adrenergic receptor subtypes has been much more straightforward than for either of the α-receptor classes. The characteristics of expressed clones for both the β_1 and β_2 subtypes correlate extremely well with the characteristics of the native receptors.[29,30] Further, a third human genomic clone, when expressed in CHO cells, shows the "atypical" β-receptor binding profile associated with the native β_3 subtype,[15] including a very low affinity for a number of classical β-receptor antagonists. The β_1 and β_2 genes are both intronless and are localized to human chromosomes 10 and 5, respectively.[19,30] The β_3 receptor gene is apparently interrupted by introns, thus allowing for the possibility of expression of C-terminal splice variants.[31,32] The β_3-receptor gene has been localized to human chromosome 8.[33]

D. SECOND MESSENGER EFFECTOR SYSTEMS

Binding of agonist to any of the adrenergic receptors induces dissociation of GDP from the specific trimeric G-protein to which the receptor couples.[34] This is followed by the association of GTP with the α-subunit of the G-protein, and subsequent activation and dissociation of the catalytic α-GTP complex from the βγ-subunits. Depending on the identity of the α-subunit, an enzymatic cascade of events is then triggered, resulting in the production of specific effector molecules affecting such parameters as the phosphorylation state of target proteins, the opening or closing of ion channels, and liberation of intracellular calcium. In addition, the βγ-complex can elicit responses distinct from those initiated by the α-subunit.

Characterization of the second messenger effector systems activated by the adrenergic receptors, and identification of the specific G-proteins to which they couple, has reinforced the classification of these receptors into three major families, α_1, α_2, and β.[1,7] The α_1-adrenoreceptors associate with a pertussis toxin-insensitive G-protein, the α subunit of which is part of the Gq family. Stimulation of α_1 receptors induces activation of phospholipase C (PLC) through the actions of either $G\alpha_q$ or $G\alpha_{11}$,[35,36] resulting in an increase in intracellular calcium concentration. PLC catalyzes the hydrolysis of membrane phosphoinositol (PI), producing several second messenger molecules, including inositol 1,4,5-trisphosphate and diacylglycerol[37] which, in turn, activate protein kinase C (PKC) and increase intracellular calcium levels by activating membrane calcium channels, thus stimulating influx of extracellular calcium by eliciting the liberation of calcium from intracellular stores. A number of other cell-specific effects are also invoked that may involve cross-talk with other second messenger systems, and α_1-adrenoreceptors have been shown to interact with several minor second messenger pathways, including PLA_2 and inhibition of cAMP. It is not clear, however, whether these are direct effects or rather are secondary to activation of the primary PLC cascade.[7,37]

The α_2-adrenoreceptor subtypes are coupled to the inhibition of adenylyl cyclase and a reduction in cAMP formation through activation of $G\alpha_{i2}$.[38] However, α_2 receptors have also been shown to couple more promiscuously to several G-protein subtypes and to elicit a variety of cellular responses, especially when they are expressed at artificially high levels. For instance, recombinant α_2-receptors have been shown to couple to $G\alpha_s$ and stimulate cAMP production when expressed at high levels in CHO cells and exposed to high agonist concentration.[39] In similar circumstances, α_2 receptors have also been shown to couple to PLA_2, PLC, PLD, and other kinase cascades less well associated with G-protein-coupled receptor mechanisms.[1,7] In contrast, the effector system activated by the β-adrenoreceptors is much more straightforward than for either of the two α-adrenoreceptor families. All of the β-adrenoreceptor subtypes stimulate the production of cAMP by adenylyl cyclase via activation of $G\alpha_s$.[34] This in turn results in activation of the cAMP-dependent protein kinase A (PKA), and subsequent phosphorylation of protein substrates of PKA.

E. REGULATION OF RECEPTOR FUNCTION

Regulation of receptor function occurs in several contexts and over a broad range of time frames. The most thoroughly studied aspect of adrenergic receptor regulation has been in the context of agonist-induced alterations in receptor function, in which agonist stimulation of a receptor induces a three-part sequence of regulatory processes (see Figure 10-1).[3,40] The initial and most rapid (seconds to minutes) of these responses is the functional uncoupling of the receptor from the effector system (i.e., the G-protein). The next response, occurring over several minutes, involves a reduction in accessibility of existing receptors to either the ligand or the effector system by physical removal and recycling of the receptors from the plasma membrane into an intracellular vesicular compartment ("sequestration"). In the longer term, the process of downregulation takes place, representing a reduction in the absolute number of receptor molecules after repeated or prolonged exposure to agonist, occurring over hours to days.

The mechanisms of downregulation, involving long-term, steady state changes in receptor number are mediated, as for any cellular protein, by a number of complementary processes.[41-43] These include changes in the rate of transcription of the receptor gene, stability and degradation of mature mRNA and rate of translation of mRNA into protein, or in the stability and rate of degradation of existing functional receptors which is affected at least in part by phosphorylation of the receptor protein. The cellular mechanisms involved in such long-term downregulation are enacted by a number of independent and sometimes interrelated factors and processes, including activation or inhibition of synthetic and degrading enzymes, second messenger systems, protein-protein interactions, protein-RNA interactions, changes in hormonal milieu, and the induction of gene transcription factors. These processes are complex and numerous, and are only beginning to be understood with respect to regulation of adrenergic receptors. Indeed, such processes are more generally involved in the long-term cellular regulation of proteins of many classes.

The shorter-term agonist-induced changes in receptor coupling and sequestration, on the other hand, are not only more rapid, but seem also to involve specific biochemical modification of existing receptor protein, primarily via phosphorylation.[40,44,45] The β_2-adrenergic receptor

Brain Adrenergic Receptors

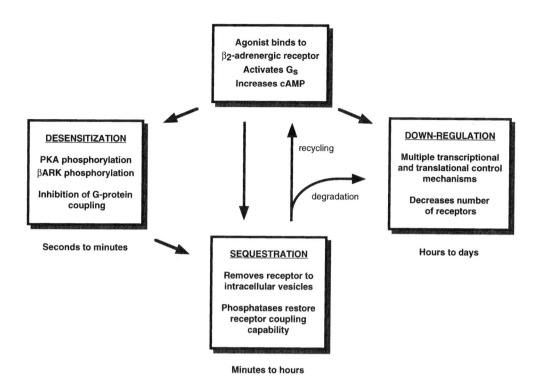

FIGURE 10-1 The three-part process of agonist-induced receptor regulation, for which the β_2-adrenergic receptor has served as a model system. Short-term desensitization depends on phosphorylation of the receptor protein by cAMP-dependent PKA, or by the receptor-specific βARK, inhibiting further G-protein coupling. Sequestration removes ligand-bound receptors from the membrane, then proceeds either to restoration and recycling of receptors back to the membrane, or to lysosomal degradation of the receptor protein, one of the initial steps in the process of long-term downregulation. See text for further details.

has served as a prototype model system for investigating phosphorylation-induced alterations in receptor function elicited by agonist activation of the receptor.[3,40,41] With agonist stimulation of the β_2-adrenergic receptor, cAMP-dependent protein kinase A (PKA) is activated. The β_2-receptor itself contains consensus recognition sequences for phosphorylation by PKA in the region that presumably interacts with the G-protein. With sufficient stimulation of PKA activity, the receptor is phosphorylated at these sites, inhibiting subsequent G-protein coupling and activation. This mechanism of desensitization is particularly effective at low agonist concentration, since phosphorylation of the target β_2-receptor protein by PKA does not depend on occupation of that receptor by agonist. That is to say, agonist-induced activation of PKA will initiate phosphorylation and desensitization of any β-receptor protein accessible to PKA, whether or not that particular receptor molecule itself has been stimulated by the agonist.[3] Both β_1- and β_2-adrenoreceptors appear to be regulated by PKA-mediated phosphorylation. However, β_3-adrenoreceptors lack a PKA recognition sequence, and are thereby resistant to short-term agonist-induced desensitization.[46]

At higher agonist concentrations, a second mechanism of β_2-receptor desensitization is recruited, involving phosphorylation of the receptor protein by another kinase, the β-adrenergic receptor kinase (βARK).[47] This enzyme is highly selective for the agonist-occupied form of the receptor, and because desensitization via βARK activity depends on occupation of the target receptor protein by ligand, it appears to be of more importance at higher agonist concentrations than is necessary for regulation by PKA.[3,45] Moreover, desensitization by βARK requires the participation of an additional cytoplasmic protein or family of proteins called the β-arrestins.[48,49] The βARK-phosphorylated β-adrenergic receptor is recognized and bound by β-arrestin, inhibiting further interaction of the receptor with the G-protein.[48,49] It is not clear whether this mechanism of regulation is unique to the β-adrenoreceptors, or whether other adrenergic receptors may also induce specific receptor-kinases and associated arrestin-like proteins. However, βARK and β-arrestin are closely related to rhodopsin kinase and its associated arrestin, which are involved in the light-induced phosphorylation and regulation of rhodopsin.[48,49] Thus, βARK is but one member of a family of related receptor kinases, and this mode of regulation may be common to many G-protein-coupled receptor systems.

Occupation of the receptor by agonist also induces sequestration of the receptor protein away from the plasma membrane and into a vesicular endosomal compartment, rendering it inaccessible to hydrophilic ligands.[50] The nature of this trafficking process is poorly understood, but at least a portion of the sequestered receptors appear to be recycled back to the membrane. It is also likely that some of the vesicularized receptors are ultimately subjected to degradation as an initial step in the long-term process of downregulation. What regulates and directs sequestered receptors through these alternative pathways is currently unknown. Interestingly, it has recently been suggested that the process of sequestration may serve not only to temporarily remove receptors from the membrane, but may more importantly be a mechanism by which the phosphorylated (and therefore functionally uncoupled) receptors are dephosphorylated and recycled back to the membrane in a restored, active state.[51] Thus, even though sequestration appears to be a desensitizing process, it may actually represent a mechanism for maintaining an adequate level of receptor responsivity in the face of prolonged stimulation.

F. DISTRIBUTION OF BRAIN ADRENERGIC RECEPTOR SUBTYPES

Adrenergic receptors typically have been localized to specific regions of the brain through radioligand binding studies and autoradiography. However, as mentioned above with reference to early pharmacologic approaches to receptor classification employing bioassays and ligand binding, such studies have been limited in the degree to which the distributions of closely related subtypes could be distinguished. More recently, *in situ* hybridization approaches have been adopted for the localization of neuronal cell bodies exhibiting receptor subtype-specific messenger RNA expression.

Autoradiographic descriptions of brain α_1-adrenergic receptor binding sites have been obtained using ^{3}H-prazosin or ^{125}I-HEAT (BE2254) as radioligands.[52,53] Distributions of cell bodies that synthesize the different α_1-receptor subtypes have been obtained by *in situ* hybridization for α_{1B} and α_{1D} mRNA,[54] and more recently for the α_{1A} mRNA (Domyancic and Morilak, unpublished data) in rat brain. In the forebrain, α_1 receptors are found throughout all layers of the neocortex, being particularly dense in the deep subgranular layers. The bulk of cortical labeling appears to be of the α_{1D} subtype, but there is also a more restricted distribution of α_{1B} mRNA in middle layers of cortex, and a moderate and dispersed representation of cells expressing α_{1A} mRNA. α_1-receptors are found in several primary sensory nuclei, including the olfactory bulb and sensory thalamic nuclei such as the medial and lateral geniculate. Binding sites as well as mRNA expression of α_{1B}-receptors are most dense in the thalamus. α_1-Adrenergic receptors are also found in high concentration in several limbic areas, including pyriform cortex, amygdala (central and lateral subnuclei), and in many areas of the basal forebrain. There is a moderate representation in hippocampus, comprised of α_{1A} and α_{1D} subtypes. The density of all three subtypes is high in brainstem cranial nerve motor nuclei, especially the facial and trigeminal motor nuclei. Finally, α_1-receptors are found in moderate levels in several autonomic-related brain regions, including the hypothalamus, central gray, and nucleus tractus solitarius, as well as being dispersed broadly throughout the brainstem reticular formation.

Binding sites for α_2 adrenergic receptors have been localized by autoradiography using the radioligand ^{3}H-*para*-aminoclonidine,[55] and the distributions of α_{2A}-, α_{2B}-, and α_{2C}-receptor subtype mRNA expression have been described by *in situ* hybridization.[56,57] High-affinity α_2 agonist binding is observed on neurons within all noradrenergic cell groups, including NTS, locus coeruleus, and ventrolateral pons and medulla. This distribution, which is correlated almost exclusively with the α_{2A} subtype, most likely reflects the role of α_2-adrenoreceptors in serving an inhibitory negative feedback autoreceptor function on adrenergic terminals and cell bodies (see below). Other brain regions showing high levels of α_2-receptor expression include the superficial layers of neocortex (i.e., supragranular layers I to III), ventral hippocampus, and amygdala. The mRNA for the α_{2C}-receptor subtype is expressed at highest levels in the basal forebrain and striatum. α_2-receptors are also expressed in many brain regions related to autonomic regulation, including the nucleus tractus solitarius, dorsal motor nucleus of the vagus, ventrolateral medulla, and the arcuate and paraventricular nuclei of the hypothalamus. These also appear to be mostly of the α_{2A} subtype.[56, 57] α_2-Receptor expression is evident in areas involved in pain processing and nociception, including the substantia gelatinosa of both the spinal cord and the spinal trigeminal nucleus, and in the brainstem periaqueductal gray. Curiously, except for a limited expression in thalamus, there appears to be almost a complete absence of α_{2B}-receptor mRNA expression in the brain.[56,57]

The distribution of brain β-adrenergic receptors has been described using radioligand binding with ^{3}H-dihydroalprenolol[58] or ^{125}I-pindolol.[59] Subtype-specific distribution has been described by competition with β_1- and β_2-specific antagonists,[59] and also by *in situ* hybridization.[60] Both β_1- and β_2-subtypes are found throughout the brain, though there are regional differences in the relative density of binding sites and mRNA hybridization for the two subtypes. In general, the β_1-receptor predominates in many forebrain regions, including the neocortex (superficial layers I to II, and deep layers IV to Vb), cingulate cortex, caudate-putamen, amygdala, hippocampus, basal forebrain, and many regions of thalamus. The β_2-receptor, on the other hand, predominates in cerebellum. β_2-receptors are also found in thalamic nuclei in a manner that is complementary to that of the β_1-receptors, and both subtypes are found in substantia nigra/VTA, olfactory tubercle, medial preoptic area, superior colliculus, and throughout the medulla and pons. Brain β-adrenergic receptors are found on glial cells as well as neurons.[61-63] It has even been suggested that cortical β-adrenoreceptor expression is predominantly if not exclusively glial.[64] There are no detectable β_3-adrenoreceptors in brain.[65]

G. CELLULAR AND PHYSIOLOGICAL FUNCTIONS

Central noradrenergic neurotransmission has been implicated in a diverse range of behavioral, cognitive, and physiological processes from selective attention,[66] sleep,[67] and vigilance[68,69] to regulation of autonomic function[70] and the behavioral and endocrine response to stress.[71,72] Ambiguity as to the precise physiological "functions" of the central noradrenergic system has been compounded by the fact that noradrenergic innervation of the entire CNS arises from relatively few neurons in the brainstem with extremely divergent and widespread projections.[73-75]

The major postsynaptic adrenergic receptors in the brain are α_1-, β_1- and β_2-adrenoreceptors. The cellular actions of NE in the brain, attributable to activation of these postsynaptic adrenergic receptors, have been described as modulating rather than mediating.[76] In other words, rather than having a direct inhibitory or excitatory influence on target neurons, NE acts to facilitate synaptic activity evoked in its target circuits by other afferents, both excitatory and inhibitory. The result is an increase in the "signal-to-noise" ratio of evoked activity relative to basal neuronal activity in many brain targets.[76-78] Noradrenergic modulation of evoked synaptic responses has been most thoroughly described by Woodward and colleagues in recordings of cerebellar Purkinje cells,[79] where this facilitatory effect was attributed to increases in cAMP elicited by β-adrenergic receptor stimulation.[80] Interestingly, very similar facilitatory effects have also been reported for α_1-adrenoreceptors in many other brain regions[81-84] even though these two receptor families are coupled to different effector systems.

Together with the divergent innervation pattern of noradrenergic fibers arising from relatively few brainstem neurons, the modulatory effects of NE suggest that noradrenergic neurotransmission is neither necessary nor sufficient for the mediation of any given response, but may act instead to alter the operating characteristics of many response systems throughout the brain in a context-specific fashion. It has been hypothesized that NE is primarily involved in a global resetting of overall brain function associated with alterations in behavioral state, as might occur with arousal[69] or stress.[71] Activation of central α_1-receptors induces increases in EEG activity and behavioral indices of arousal, while selective pharmacological blockade of α_1-receptors produces sedation and EEG synchronization.[85,86] Further, the modulatory effects described above for NE on cellular activity have been extended to more complex behavioral and cognitive processes. Stimulation of α_1-adrenoreceptors facilitates simple and complex sensorimotor reflex responses,[84,87-89] and such behavioral facilitation occurs in response to presentation of stressful stimuli associated with increases in noradrenergic neuronal activity.[90,91]

Hypotheses relating noradrenergic function to global state changes associated with stress and arousal are supported by the observation that the level of electrical activity recorded in central noradrenergic neurons in the locus coeruleus *in vivo* are highly correlated with behavioral arousal.[69,71,92] Further, these cells are activated by stressful stimuli,[93-97] and the release of NE is elevated in a number of brain regions during stress.[98-101] Thus, the activation of noradrenergic neurons during arousal or stress, and the subsequent modulatory influence exerted by postsynaptic adrenergic receptors on targets throughout the brain, presumably would serve to facilitate the various behavioral, cognitive, endocrine, or autonomic responses mediated by these targets.[71] This perhaps may explain why noradrenergic neurotransmission has been implicated in such a diverse array of behavioral and physiological functions.

In contrast to the postsynaptic β- and α_1-adrenergic receptors, α_2-adrenoreceptors appear to exert a primarily inhibitory effect on brain cellular activity. Further, the observation that α-adrenoreceptor blockade facilitated the evoked release of ^{3}H-NE from noradrenergic terminals, both peripheral and central, led to the hypothesis that an α-adrenoreceptor located on the noradrenergic nerve terminal itself acted as an inhibitory autoreceptor, reducing the release of transmitter in a homeostatic negative feedback fashion.[102] Differences in the potency of α-antagonist drugs in facilitating evoked release relative to blockade of postsynaptic adrenergic responses led to the suggestion that these effects were mediated by different receptor subtypes. This was in fact the first recognition of α-adrenoreceptor subtypes; postsynaptic α-receptors were termed α_1-, whereas presynaptic effects were attributed to α_2-adrenergic receptors.[102] This differentiation was later supported by the selectivity of agonists such as prazosin and clonidine.[6] While it was believed for some time that α_2 receptors were located exclusively on presynaptic noradrenergic terminals, it has since been shown that they also serve as somatodendritic autoreceptors. These autoreceptors, located on the cell bodies of noradrenergic neurons themselves, inhibit the subsequent activation of these neurons following the release of NE.[103] Given the hypothesized role of the other postsynaptic adrenergic receptor subtypes in maintaining vigilance and arousal (see above), it is likely that this autoreceptor-mediated inhibition of noradrenergic neurotransmission accounts for the sedative properties of α_2 agonists such as clonidine[104] and the behaviorally activating and even anxiety-provoking effects of α_2 antagonist drugs such as yohimbine.[105]

In addition to their role as presynaptic terminal and somatodendritic autoreceptors, α_2-adrenergic receptors are also localized postsynaptically on nonadrenergic targets. Perhaps the most intensively studied postsynaptic effect of central α_2-adrenoreceptor activation is a decrease in blood pressure associated with inhibition of sympathetic outflow.[104,106] This hypotensive effect of α_2-adrenoreceptors has been localized to the medulla oblongata, most likely including both the nucleus tractus solitarius and the ventrolateral medullary pressor region as sites of action.[106] Indeed, the α_2 agonist clonidine was used clinically for some time as an antihypertensive agent, though it has become clear that at least part of the central hypotensive action of clonidine can be attributed to its action at imidazoline receptors located in these same medullary regions.[107]

H. CONCLUSIONS

Since it was first suggested that adrenergic receptor subtypes subserve different biological functions,[5] the question of the physiological significance and evolutionary advantages of having multiple receptor classes and subclasses has been open to considerable debate. Different receptor classes for a given transmitter (e.g., α_1- vs. α_2- vs. β-receptors) may display different binding affinities for the endogenous ligand, may couple to different effector systems, and may be regulated differently by virtue of the presence or absence of key consensus amino acid or nucleotide sequences. This diversity obviously allows for a great degree of flexibility in linking the activity of a neurotransmitter pathway or chemical system with a variety of postsynaptic effects depending on the innervated target site, the concentration of transmitter released, and baseline physiological conditions that can influence relative receptor expression or responsivity. More difficult to understand, however, are those examples where closely related subtypes within a receptor class display similar binding profiles for endogenous ligands and couple to identical effector responses. Such within-family subtypes may be regulated by different factors within a given cell, allowing for the selective maintenance, attenuation, or facilitation of biological responses in different hormonal or metabolic states. Such subtypes may be expressed differentially in cells of different phenotype, allowing the process of receptor regulation to have evolved in conjunction with the evolution of function in distinct neural pathways.

At present, very little is understood about the different physiological roles played by these closely related receptor subtypes, but it is clear that the maintenance of a number of diverse and related subtypes is the rule rather than the exception for all G-protein-linked neurotransmitter systems, and the debate regarding the significance of multiple receptor subtypes is not limited to the adrenergic receptors. As the differential regulation and function of neurotransmitter receptor subtypes becomes clearer, so will our understanding of the impact of receptor dysregulation, dysfunction, or genetic mutation in the wide range of physiological and psychological pathologies in which these systems have been implicated — from hypertension and other cardiovascular diseases, to depression, schizophrenia, and drug addiction. Clearly, an emerging awareness of the potential roles played by specific neurotransmitter receptor subtypes in these disorders will be an essential component of future strategies aimed at developing targeted and effective therapeutic approaches toward their cure and prevention.

REFERENCES

1. Bylund, D. B., Eikenberg, D. C., Hieble, J. P., Langer, S. Z., Lefkowitz, R. J., Minneman, K. P., Molinoff, P. B., Ruffolo, R. R., and Trendelenburg, U., IV. International Union of Pharmacology nomenclature of adrenoceptors, *Pharmacol. Rev.*, 46, 121, 1994.
2. O'Dowd, B., Lefkowitz, R. J., and Caron, M. G., Structure of the adrenergic and related receptors, in *Annual Review of Neuroscience*, Cowan, W. L., et al., Eds., Annual Reviews, Palo Alto, CA, 1989, 67.

3. Caron, M. G. and Lefkowitz, R. J., Catecholamine receptors: structure, function, and regulation, *Recent Prog. Horm. Res.*, 48, 277, 1993.
4. Strosberg, A. D., Structural and functional diversity of β-adrenergic receptors, *Ann. N. Y. Acad. Sci.*, 757, 253, 1995.
5. Ahlquist, R. P., A study of the adrenotropic receptors, *Am. J. Physiol.*, 153, 586, 1948.
6. Starke, K., α-Adrenoceptor subclassification, *Rev. Physiol. Biochem. Pharmacol.*, 88, 199, 1981.
7. Milligan, G., Svoboda, P., and Brown, C. M., Why are there so many adrenoceptor subtypes?, *Biochem. Pharmacol.*, 48, 1059, 1994.
8. Bylund, D. B., Subtypes of α_1- and α_2-adrenergic receptors, *FASEB J.*, 6, 832, 1992.
9. Ruffolo, R. R. and Hieble, J. P., α-Adrenoceptors, *Pharm. Ther.*, 61, 1, 1994.
10. Lands, A. M., Arnold, A., McAuliff, J. P., Luduena, F. P., and Brown, T. G., Differentiation of receptor systems activated by sympathomimetic amines, *Nature*, 214, 597, 1967.
11. O'Donnell, S. R. and Wanstall, J. C., Evidence that ICI 118,551 is a potent, highly beta$_2$-selective adrenoceptor antagonist and can be used to characterize beta-adrenoceptor populations in tissues, *Life Sci.*, 27, 671, 1980.
12. Mauriege, P., De Pergola, G., Berlan, M., and Lafontan, M., Human fat cell beta-adrenergic receptors. Beta-agonist-dependent lipolytic responses and characterization of beta-adrenergic binding sites on human fat cell membranes with highly selective beta$_1$-antagonists, *J. Lipid Res.*, 29, 587, 1988.
13. Morrow, A. L. and Creese, I., Characterization of α_1-adrenergic receptor subtypes in rat brain: a reevaluation of [^{3}H]WB4101 and [^{3}H]prazosin binding, *Mol. Pharmacol.*, 29, 321, 1986.
14. Minneman, K. P., Han, C., and Abel, P. W., Comparison of α_1-adrenergic receptor subtypes distinguished by chlorethylclonidine and WB 4101, *Mol. Pharmacol.*, 33, 509, 1988.
15. Emorine, L. J., Marullo, S., Briend-Sutren, M.-M., Patey, G., Tate, K., Delavier-Klutchko, C., and Strosberg, A. D., Molecular characterization of the human β$_3$-adrenergic receptor, *Science*, 245, 1118, 1989.
16. Bylund, D. B., Subtypes of α_2-adrenoceptors. Pharmacological and molecular biological evidence converge, *Trends Pharmacol. Sci.*, 9, 356, 1988.
17. Hieble, J. P., Bylund, D. B., Clarke, D. E., Eikenburg, D. C., Langer, S. Z., Lefkowitz, R. J., Minneman, K. P., and Ruffolo, R. R., International Union of Pharmacology. X. Recommendation for nomenclature of α_1-adrenoceptors: consensus update, *Pharmacol. Rev.*, 47, 267, 1995.
18. Cotecchia, S., Schwinn, D. A., Randall, R. R., Lefkowitz, R. J., Caron, M. G., and Kobilka, B. K., Molecular cloning and expression of the cDNA for the hamster α_1-adrenergic receptor, *Proc. Natl. Acad. Sci. U.S.A.*, 85, 7159, 1988.
19. Yang-Feng, T. L., Xue, F., Zhong, W., Cotecchia, S., Frielle, T., Caron, M. G., Lefkowitz, R. J., and Francke, U., Chromosomal organization of adrenergic receptor genes, *Proc. Natl. Acad. Sci. U.S.A.*, 87, 1516, 1990.
20. Schwinn, D. A., Lomasney, J. W., Lorenz, W., Szklut, P. J., Fremeau, R. T., Yang-Feng, T. L., Caron, M. G., Lefkowitz, R. J., and Cotecchia, S., Molecular cloning and expression of the cDNA for a novel α_1-adrenergic receptor subtype, *J. Biol. Chem.*, 265, 8183, 1990.
21. Lomasney, J. W., Cotecchia, S., Lorenz, W., Leung, W.-Y., Schwinn, D. A., Yang-Feng, T. L., Brownstein, M., Lefkowitz, R. J., and Caron, M. G., Molecular cloning and expression of the cDNA for the α_{1A}-adrenergic receptor, the gene for which is located on human chromosome 5, *J. Biol. Chem.*, 266, 6365, 1991.
22. Perez, D. M., Piascik, M. T., and Graham, R. M., Solution-phase library screening for the identification of rare clones: Isolation of an α_{1D}-adrenergic receptor cDNA, *Mol. Pharmacol.*, 40, 876, 1991.
23. Loftus, S. K., Shiang, R., Warrington, J. A., Bengtsson, U., McPherson, J. D., and Wasmuth, J. J., Genes encoding adrenergic receptors are not clustered on the long arm of human chromosome 5, *Cytogenet. Cell Genet.*, 667, 69, 1994.
24. Yang-Feng, T. L., Han, H., Lomasney, J. W., and Caron, M. G., Localization of the cDNA for an alpha1-adrenergic receptor subtype (ADRA1D) to chromosome band 20p13, *Cytogenet. Cell Genet.*, 66, 170, 1994.
25. Perez, D. M., Piascik, M. T., Malik, N., Gaivin, R., and Graham, R. M., Cloning, expression, and tissue distribution of the rat homolog of the bovine α_{1C}-adrenergic receptor provide evidence for its classification as the α_{1A} subtype, *Mol. Pharmacol.*, 46, 823, 1994.
26. Kobilka, B. K., Matsui, H., Kobilka, T. S., Yang-Feng, T. L., Francke, U., Caron, M. G., Lefkowitz, R. J., and Regan, J. W., Cloning, sequencing, and expression of the gene coding for the human platelet α_2-adrenergic receptor, *Science*, 238, 650, 1987.
27. Regan, J. W., Kobilka, T. S., Yang-Feng, T. L., Caron, M. G., Lefkowitz, R. J., and Kobilka, B. K., Cloning and expression of a human kidney cDNA for an α_2-adrenergic receptor subtype, *Proc. Natl. Acad. Sci. U.S.A.*, 85, 6301, 1988.
28. Lomasney, J. W., Lorenz, W., Allen, L. F., King, K., Regan, J. W., Yang-Feng, T. L., Caron, M. G., and Lefkowitz, R. J., Expansion of the α_2-adrenergic receptor family. Cloning and characterization of a human α_2-adrenergic receptor subtype, the gene for which is located on chromosome 2, *Proc. Natl. Acad. Sci., U.S.A.*, 87, 5094, 1990.
29. Frielle, T., Collins, S., Daniel, K. W., Caron, M. G., Lefkowitz, R. J., and Kobilka, B. K., Cloning of the cDNA for the human β$_1$-adrenergic receptor, *Proc. Natl. Acad. Sci. U.S.A.*, 84, 7920, 1987.

30. Kobilka, B. K., Dixon, R. A. F., Frielle, T., Dohlman, H. G., Bolanowski, M. A., Sigal, I. S., Yang-Feng, T. L., Francke, U., Caron, M. G., and Lefkowitz, R. J., cDNA for the human β_2-adrenergic receptor: a protein with multiple membrane-spanning domains encoded by a gene whose chromosomal localization is shared with that of the receptor for platelet-derived growth factor, *Proc. Natl. Acad. Sci. U.S.A.*, 84, 46, 1987.
31. van Spronsen, A., Nahmias, C., Krief, S., Briend-Sutren, M.-M., Strosberg, A. D., and Emorine, L. J., The promoter and intron/exon structure of the human and mouse β_3-adrenergic-receptor genes, *Eur. J. Biochem.*, 213, 1117, 1993.
32. Granneman, J. G., Lahners, K. N., and Chaudry, A., Characterization of the human β_3-adrenergic receptor gene, *Mol. Pharmacol.*, 44, 264, 1993.
33. Nahmias, C., Blin, N., Elalouf, J. M., Mattei, M. G., Strosberg, A. D., and Emorine, L. J., Molecular characterization of the mouse beta 3-adrenergic receptor. Relationship with the atypical receptor of adipocytes, *EMBO J.*, 10, 3721, 1991.
34. Gilman, A. G., G proteins: transducers of receptor-generated signals, *Annu. Rev. Biochem.*, 56, 615, 1987.
35. Han, C., Abel, P. W., and Minneman, K. P., α_1-Adrenoceptor subtypes linked to different mechanisms for increasing intracellular Ca^{2+} in smooth muscle, *Nature*, 329, 333, 1987.
36. Wu, D., Katz, A., Lee, C.-H., and Simon, M. I., Activation of phospholipase C by α_1-adrenergic receptors is mediated by the α subunits of Gq family, *J. Biol. Chem.*, 267, 25798, 1992.
37. Minneman, K. P., α_1-Adrenergic receptor subtypes, inositol phosphates, and sources of cell Ca^{2+}, *Pharmacol. Rev.*, 40, 87, 1988.
38. Simonds, W. F., Goldsmith, P. K., Codina, J., Unson, C. G., and Spiegel, A. M., G_{i2} mediates α_2-adrenergic inhibition of adenylyl cyclase in platelet membranes. *In situ* identification with G_α C-terminal antibodies, *Proc. Natl. Acad. Sci. U.S.A.*, 86, 7809, 1989.
39. Eason, M. G., Kurose, H., Holt, B. D., Raymond, J. R., and Liggett, S. B., Simultaneous coupling of α_2-adrenergic receptors to two G-proteins with opposing effects: subtype-selective coupling of α_2C10, α_2C4, and α_2C2 adrenergic receptors to G_i and G_s, *J. Biol. Chem.*, 267, 15795, 1992.
40. Hausdorff, W. P., Caron, M. G., and Lefkowitz, R. J., Turning off the signal: desensitization of β-adrenergic receptor function, *FASEB J.*, 4, 2881, 1990.
41. Collins, S., Lohse, M. J., O'Dowd, B., Caron, M. G., and Lefkowitz, R. J., Structure and regulation of G protein-coupled receptors: the β_2-adrenergic receptor as a model, *Vitam. Horm.*, 46, 1, 1991.
42. Collins, S., Caron, M. G., and Lefkowitz, R. J., Regulation of adrenergic receptor responsiveness through modulation of gene expression, *Annu. Rev. Physiol.*, 53, 497, 1991.
43. Collins, S., Caron, M. G., and Lefkowitz, R. J., From ligand binding to gene expression. New insights into the regulation of G-protein-coupled receptors, *Trends Biochem. Sci.*, 17, 37, 1992.
44. Bouvier, M., Collins, S., O'Dowd, B. F., Campbell, P. T., deBlaski, A., Kobilka, B. K., MacGregor, C., Irons, G. P., Caron, M. G., and Lefkowitz, R. J., Two distinct pathways for c-AMP mediated down-regulation of the beta-2 adrenergic receptor. Phosphorylation of the receptor and regulation of its mRNA level, *J. Biol. Chem.*, 264, 16786, 1989.
45. Hausdorff, W. P., Bouvier, M., O'Dowd, B. F., Irons, G. P., Caron, M. G., and Lefkowitz, R. J., Phosphorylation sites in two domains of the β_2-adrenergic receptor are involved in distinct pathways of receptor desensitization, *J. Biol. Chem.*, 264, 12657, 1989.
46. Nantel, F., Bonin, H., Emorine, L. J., Zilberfarb, V., Strosberg, A. D., Bouvier, M., and Marullo, S., The human β_3-adrenergic receptor is resistant to short term agonist-promoted desensitization, *Mol. Pharmacol.*, 43, 548, 1993.
47. Benovic, J. L., Strasser, R. H., Caron, M. G., and Lefkowitz, R. J., β-Adrenergic receptor kinase: identification of a novel protein kinase that phosphorylates the agonist-occupied form of the receptor, *Proc. Natl. Acad. Sci. U.S.A.*, 83, 2797, 1986.
48. Lohse, M. J., Benovic, J. L., Codina, J., Caron, M. G., and Lefkowitz, R. J., β-Arrestin: a protein that regulates β-adrenergic receptor function, *Science*, 248, 1547, 1990.
49. Lohse, M. J., Andexinger, S., Pitcher, J., Trukawinski, S., Codina, J., Fuare, J.-P., Caron, M. G., and Lefkowitz, R. J., Receptor-specific desensitization with purified proteins. Kinase dependence and receptor specificity of β-arrestin and arrestin in the β_2-adrenergic receptor and rhodopsin systems, *J. Biol. Chem.*, 267, 8558, 1992.
50. von Zastrow, M. and Kobilka, B. K., Ligand-regulated internalization and recycling of human β_2-adrenergic receptors between the plasma membrane and endosomes containing transferrin receptors, *J. Biol. Chem.*, 267, 3530, 1992.
51. Yu, S. S., Lefkowitz, R. J., and Hausdorff, W. P., β-Adrenergic receptor sequestration: a potential mechanism of receptor resensitization, *J. Biol. Chem.*, 268, 337, 1993.
52. Young, W. S. and Kuhar, M. J., Noradrenergic $\alpha 1$ and $\alpha 2$ receptors: light microscopic autoradiographic localization, *Proc. Natl. Acad. Sci. U.S.A.*, 77, 1696, 1980.
53. Jones, L. S., Gauger, L. L., and Davis, J. N., Anatomy of brain alpha$_1$-adrenergic receptors. In vitro autoradiography with [^{125}I]-HEAT, *J. Comp. Neurol.*, 190, 1985.
54. Pieribone, V. A., Nicholas, A. P., Dagerlind, A., and Hokfelt, T., Distribution of α_1 adrenoceptors in rat brain revealed by *in situ* hybridization experiments utilizing subtype-specific probes, *J. Neurosci.*, 14, 4252, 1994.

55. Unnerstall, J. R., Kopajtic, T. A., and Kuhar, M. J., Distribution of α_2 agonist binding sites in the rat and human central nervous system. Analysis of some functional, anatomic correlates of the pharmacologic effects of clonidine and related adrenergic agents, *Brain Res. Rev.*, 7, 69, 1984.
56. Nicholas, A. P., Pieribone, V. A., and Hokfelt, T., Distributions of mRNAs for alpha-2 adrenergic receptor subtypes in rat brain: an in situ hybridization study, *J. Comp. Neurol.*, 328, 575, 1993.
57. Scheinin, M., Lomasney, J. W., Hayden-Hixson, D. M., Schambra, U. B., Caron, M. G., Lefkowitz, R. J., and Fremeau, R. T., Distribution of α_2-adrenergic receptor subtype gene expression in rat brain, *Mol. Brain Res.*, 21, 133, 1994.
58. Palacios, J. M. and Kuhar, M. J., Beta-adrenergic-receptor localization by light microscopic autoradiography, *Science*, 208, 1378, 1980.
59. Rainbow, T. C., Parsons, B., and Wolfe, B. B., Quantitative autoradiography of β_1- and β_2-adrenergic receptors in rat brain, *Proc. Natl. Acad. Sci. U.S.A.*, 81, 1585, 1984.
60. Nicholas, A. P., Pieribone, V. A., and Hokfelt, T., Cellular localization of messenger RNA for beta-1 and beta-2 adrenergic receptors in rat brain: an in situ hybridization study, *Neuroscience*, 56, 1023, 1993.
61. Atkinson, B. N. and Minneman, K. P., Multiple adrenergic receptor subtypes controlling cyclic AMP formation: comparison of brain slices and primary neuronal and glial cultures, *J. Neurochem.*, 56, 587, 1991.
62. Aoki, C. and Pickel, V. M., C-terminal tail of β-adrenergic receptors. Immunocytochemical localization within astrocytes and their relation to catecholaminergic neurons in N. tractus solitarii and area postrema, *Brain Res.*, 571, 35, 1992.
63. Aoki, C., β-Adrenergic receptors. Astrocytic localization in the adult visual cortex and their relation to catecholamine axon terminals as revealed by electron microscopic immunocytochemistry, *J. Neurosci.*, 12, 781, 1992.
64. Stone, E. A. and John, S. M., Further evidence for a glial localization of rat cortical β-adrenoceptors. Studies of in vivo cyclic AMP responses to catecholamines, *Brain Res.*, 549, 78, 1991.
65. Emorine, L. J., Blin, N., and Strosberg, A. D., The human β_3-adrenoceptor: the search for a physiological function, *Trends Pharmacol. Sci.*, 15, 3, 1994.
66. Harley, C., Noradrenergic and locus coeruleus modulation of the perforant path-evoked potential in rat dentate gyrus supports a role for the locus coeruleus in attentional and memorial processes, *Prog. Brain Res.*, 88, 307, 1991.
67. Jones, B. E., The role of noradrenergic locus coeruleus neurons and neighboring cholinergic neurons of the pontomesencephalic tegmentum in sleep-wake states, *Prog. Brain Res.*, 88, 533, 1991.
68. Aston-Jones, G., Behavioral functions of locus coeruleus derived from cellular attributes, *Physiol. Psychol.*, 13, 118, 1985.
69. Aston-Jones, G., Chiang, C., and Alexinsky, T., Discharge of noradrenergic locus coeruleus neurons in behaving rats and monkeys suggests a role in vigilance, *Prog. Brain Res.*, 88, 501, 1991.
70. Guyenet, P. G., Central noradrenergic neurons: the autonomic connection, *Prog. Brain Res.*, 88, 365, 1991.
71. Jacobs, B. L., Abercrombie, E. D., Fornal, C. A., Levine, E. S., Morilak, D. A., and Stafford, I. L., Single-unit and physiological analyses of brain norepinephrine function in behaving animals, *Prog. Brain Res.*, 88, 159, 1991.
72. al-Damluji, S., Adrenergic control of the secretion of anterior pituitary hormones, *Bailliere's Clin. Endocrinol. Metab.*, 7, 355, 1993.
73. Moore, R. Y. and Bloom, F. E., Central catecholamine neuron system: anatomy and physiology of the norepinephrine and epinephrine systems, *Annu. Rev. Neurosci.*, 2, 113, 1979.
74. Nagai, T., Satoh, K., Imamoto, K., and Maeda, T., Divergent projections of catecholamine neurons of the locus coeruleus as revealed by fluorescent retrograde double labeling technique, *Neurosci. Lett.*, 23, 117, 1981.
75. Fallon, J. H. and Loughlin, S. E., Monoamine innervation of the forebrain: collateralization, *Brain Res. Bull.*, 9, 295, 1982.
76. Woodward, D. J., Moises, H. C., Waterhouse, B. D., Yeh, H. H., and Cheun, J. E., Modulatory actions of norepinephrine on neural circuits, *Adv. Exp. Med. Biol.*, 287, 193, 1991.
77. Woodward, D. J., Moises, H. C., Waterhouse, B. D., Hoffer, B. J., and Freedman, R., Modulatory actions of norepinephrine in the central nervous system, *Fed. Proc.*, 38, 2109, 1979.
78. Waterhouse, B. D., Sessler, F. M., Cheng, J. T., Woodward, D. J., Azizi, S. A., and Moises, H. C., New evidence for a gating action of norepinephrine in central neuronal circuits of mammalian brain, *Brain Res. Bull.*, 21, 425, 1988.
79. Woodward, D. J., Moises, H. C., Waterhouse, B. D., Yeh, H. H., and Cheun, J. E., The cerebellar norepinephrine system: inhibition, modulation, and gating, *Prog. Brain Res.*, 88, 331, 1991.
80. Yeh, H. H. and Woodward, D. J., Beta-1 adrenergic receptors mediate noradrenergic facilitation of purkinje cell responses to gamma-aminobutyric acid in cerebellum of rat, *Neuropharmacology*, 22, 629, 1983.
81. Rogawski, M. A. and Aghajanian, G. K., Activation of lateral geniculate neurons by norepinephrine. Mediation by an α-adrenergic receptor, *Brain Res.*, 182, 345, 1980.

82. Waterhouse, B. D., Moises, H. C., and Woodward, D. J., Alpha-receptor mediated facilitation of somatosensory cortical neuronal responses to excitatory synaptic inputs and iontophoretically applied acetylcholine, *Neuropharmacology*, 20, 907, 1981.
83. Waterhouse, B. D., Azizi, S. A., Burne, R. A., and Woodward, D. J., Interactions of norepinephrine and serotonin with visually evoked responses of simple and complex cells in area 17 of rat cortex, *Soc. Neurosci. Abstr.*, 9, 1001, 1983.
84. White, S. R. and Neuman, R. S., Pharmacological antagonism of facilitatory but not inhibitory effects of serotonin and norepinephrine on excitability of spinal motoneurones, *Neuropharmacology*, 22, 489, 1983.
85. Hilakivi, I. and Leppavuori, A., Effects of methoxamine, an alpha-1 adrenoceptor agonist, and prazosin, an alpha-1 antagonist, on the stages of the sleep-waking cycle in the cat, *Acta Physiol. Scand.*, 120, 363, 1984.
86. Hilakivi-Clarke, L. A., Turkka, J., Lister, R. G., and Linnoila, M., Effects of early postnatal handling on brain β-adrenoceptors and behavior in tests related to stress, *Brain Res.*, 542, 286, 1991.
87. Davis, M., Astrachan, D., Kehne, J., Commissaris, R., and Gallager, D., Catecholamine modulation of sensorimotor reactivity measured with acoustic startle, in *Catecholamines: Neuropharmacology and Central Nervous System — Theoretical Aspects*. Alan R. Liss, New York, 1984, 245.
88. Morilak, D. A. and Jacobs, B. L., Noradrenergic modulation of sensorimotor processes in intact rats. The masseteric reflex as a model system, *J. Neurosci.*, 5, 1300, 1985.
89. Fung, S. J., Manzoni, D., Chan, J. Y. H., Pompeiano, O., and Barnes, C. D., Locus coeruleus control of spinal motor output, *Prog. Brain Res.*, 88, 395, 1991.
90. Stafford, I. L. and Jacobs, B. L., Noradrenergic modulation of the masseteric reflex in behaving cats. I. Pharmacological studies, *J. Neurosci.*, 10, 91, 1990.
91. Stafford, I. L. and Jacobs, B. L., Noradrenergic modulation of the masseteric reflex in behaving cats. II. Physiological studies, *J. Neurosci.*, 10, 99, 1990.
92. Jacobs, B. L., Single unit activity of locus coeruleus neurons in behaving animals, *Prog. Neurobiol.*, 27, 183, 1986.
93. Abercrombie, E. D. and Jacobs, B. L., Single-unit response of noradrenergic neurons in the locus coeruleus of freely moving cats. I. Acutely presented stressful and nonstressful stimuli, *J. Neurosci.*, 7, 2837, 1987.
94. Morilak, D. A., Fornal, C. A., and Jacobs, B. L., Effects of physiological manipulations on locus coeruleus neuronal activity in freely moving cats. II. Cardiovascular challenge, *Brain Res.*, 422, 24, 1987.
95. Morilak, D. A., Fornal, C. A., and Jacobs, B. L., Effects of physiological manipulations on locus coeruleus neuronal activity in freely moving cats. III. Glucoregulatory challenge, *Brain Res.*, 422, 32, 1987.
96. Svensson, T. H., Peripheral, autonomic regulation of locus coeruleus noradrenergic neurons in brain. Putative implications for psychiatry and psychopharmacology, *Psychopharmacology*, 92, 1, 1987.
97. Page, M. E., Akaoka, H., Aston-Jones, G., and Valentino, R. J., Bladder distension activates noradrenergic locus coeruleus neurons by an excitatory amino acid mechanism, *Neuroscience*, 51, 555, 1992.
98. Tanaka, M., Kohno, Y., Nakagawa, R., Ida, Y., Takeda, S., Nagasaki, N., and Noda, Y., Regional characteristics of stress-induced increases in brain noradrenaline release in rats, *Pharmacol. Biochem. Behav.*, 19, 543, 1983.
99. Nisenbaum, L. K., Zigmond, M. J., Sved, A. J., and Abercrombie, E. D., Prior exposure to chronic stress results in enhanced synthesis and release of hippocampal norepinephrine in response to a novel stressor, *J. Neurosci.*, 11, 1478, 1991.
100. Pacak, K., Armando, I., Fukuhara, K., Kvetnansky, R., Palkovits, M., Kopin, I. J., and Goldstein, D. S., Noradrenergic activation in the paraventricular nucleus during acute and chronic immobilization stress in rats, *Brain Res.*, 589, 91, 1992.
101. Pacak, K., Palkovits, M., Kvetnansky, R., Fukuhara, K., Kopin, I. J., and Goldstein, D. S., Effects of single or repeated immobilization on release of norepinephrine and its metabolites in the central nucleus of the amygdala in conscious rats, *Neuroendocrinology*, 57, 623, 1993.
102. Langer, S. Z., Presynaptic regulation of catecholamine release, *Biochem. Pharmacol.*, 23, 1793, 1974.
103. Cedarbaum, J. M. and Aghajanian, G. K., Catecholamine receptors on locus coeruleus neurons: pharmacological characterization, *Eur. J. Pharmacol.*, 44, 375, 1977.
104. Timmermans, P. B. M. W. M., Schoop, A. M. C., Kwa, H. Y., and van Zwieten, P. A., Characterization of α-adrenoceptors participating in the central hypotensive and sedative effects of clonidine using yohimbine, rauwolscine and corynanthine, *Eur. J. Pharmacol.*, 70, 7, 1981.
105. Charney, D. S., Woods, S. W., Nagy, L. M., Southwick, S. M., Krystal, J. H., and Heninger, G. R., Noradrenergic function in panic disorder, *J. Clin. Psychiatry*, 51 (Suppl. A), 5, 1990.
106. Gillis, R. A., Gatti, P. J., and Quest, J. A., Mechanism of the antihypertensive effect of alpha$_2$-agonists, *J. Cardiovasc. Pharmacol.*, 7 (Suppl. 8), S38, 1985.
107. Bousquet, P., Feldman, J., Tibirica, E., Bricca, G., Molines, A., Doutenwill, M., and Belcourt, A., New concepts on the central regulation of blood pressure: alpha$_2$-adrenoceptors and "imidazoline receptors", *Am. J. Med.*, 87 (Suppl. 3C), 10, 1989.

11 Emerging Bacterial Models for GABA Receptors and Transporters

Steven C. King

CONTENTS

A. Introduction 164
B. Fundamental Properties and Molecular Physiology 165
    1. Kinetics and Ion Dependence 165
    2. Uncoupling and Channel Activity 167
    3. Chloride, "Leak Current", and Transport Rate 168
    4. Role of Transporter in GABA Release 170
C. Sites for Posttranslational Modification 170
    1. Role of Glycosylation Sites 170
    2. Regulation by Kinase Activity 171
D. Clinical Relevance of GABA 172
E. Cloning of GABA Transporters 173
    1. GAT-1 From Rat 174
    2. The NTT Superfamily 174
    3. The *gab* Permease From *E. coli* 175
        a. Cloning *gabP* for Inducible Overexpression 175
        b. Relationship of *gabP* to Mammalian GAT Genes 175
F. Inherent Advantages of *E. coli* Genetics 175
    1. Natural Selection of Point Mutants 175
    2. Site-Directed Amber Suppression in *E. coli* 177
    3. Topology by Transposition 179
        a. C-Terminal Reporter Enzymes 180
        b. Epitope Insertions 180
        c. *PacI*-Mediated Gene Fusion 181
G. Characterization of GabP 184
    1. Pharmacological Inhibition of GabP 184
        a. Open-Chain Amino Acids 184
        b. Conformationally Constrained Heterocycles 184
        c. Planar Heterocycles 185
        d. Comparison of Active Compounds 185
    2. Substrate Specificity of GabP 186
        a. GabP-Mediated Counterflow of [^{3}H]GABA 186
        b. Counterflow Related to ACHC and Nipecotic Acid 188
        c. Counterflow Related to TACA and CACA 189

    d. Counterflow Related to Guvacine and Isoguvacine 190
H. Epilogue 190
Acknowledgments 192
References 193

A. INTRODUCTION

More than four decades have passed since Roberts and Frankel[1] recognized the importance of GABA (γ-aminobutyric acid) in the brain. The role of GABA as a central neurotransmitter became more fully established when Hayashi[2] demonstrated the inhibitory nature of GABA on cerebral cortical neurons in 1956, and Krnjevic and Phillips[3] presented early evidence that application of GABA to neurons caused a rapid, highly reversible, and general blocking action on cortical neurons. Such observations were certainly consistent with the idea that GABA could be an inhibitory neurotransmitter. Krnjevic and Schwartz[4] later used twin micropipettes to record the effects in a single cell following local application of GABA by microiontophoresis. The techniques were possible to monitor membrane potential and resistance during an IPSP (inhibitory postsynaptic potential) and compare these parameters to the effects induced by application of GABA. Like an evoked IPSP, application of GABA was associated with hyperpolarization and decreased membrane resistance. The IPSP and GABA effects were reversed at similar potentials. Introduction of chloride to the intracellular compartment (by KCl electrode) converted the hyperpolarizing effects of both GABA and the evoked IPSP to depolarizing effects. The authors concluded that GABA was quite likely to be the main cortical inhibitory neurotransmitter.

Thus, cellular recordings now more than 30 years old make it obvious that the biochemical basis for GABAergic synaptic transmission could not be understood from any gross biochemical characterization of brain tissue in the aggregate. Instead, methods were needed to detect — in an individual cell — those particular biochemical attributes that would suit it to subserve a *specific role* at a *specific synapse* or related group of synapses. Working independently, Lam and Steinman[5] and Drujan and Svaetichin[6] recognized that the retina, consisting of only seven well-defined cellular layers, is anatomically simple compared to the brain. The retina would therefore be an excellent model system in which to discern the fundamental neurochemical attributes of individual cells and defined categories of cells whose anatomical relationships to one another were known. Such studies are impossible in the brain where neurons are packed too tightly to be clearly distinguished as individuals.

Drujan and Svaetichin[6] treated retinas from marine teleost fishes with enzymes (hyaluronidase and papain) in order to dissociate individual cells from the tissue. Such isolated retinal cells were viable and retained their distinctive morphological attributes so that photoreceptors; external, medial, and internal cone horizontal cells; rod horizontal cells; interstitial and piriform amiacrine cells; cone-and-rod bipolar cells, and ganglion cells could all be distinguished in the dissociated cell preparation. General metabolic characteristics (e.g., oxygen consumption, membrane potential) as well as specific neurochemical specializations (e.g., acetylcholinesterase activity) could thus be quantitated in individual cells of defined morphology. This capability, together with the relative simplicity of the retina, made it feasible to begin the task of correlating the specific biochemical/neurochemical characteristics of the individual cells with the well-known anatomical organization of these cells within the intact retina.

Similar to biochemical work done with brain tissue,[7] the retina was also found to contain a robust and specific mechanism for uptake of extracellular GABA.[8] Intact retinas were treated with 50 nM [^{3}H]GABA and then assayed to determine the total radioactivity associated with the tissue. It was found that the tissue had actually concentrated (50:1) [^{3}H]GABA from the medium within 30 min. More than 90% of the accumulated GABA was found to be unmetabolized GABA. Additionally, the uptake process was found to be saturable, temperature dependent, sensitive to removal of Na[+] or addition of the Na[+]-pump inhibitor (ouabain), and unaffected by addition of other amino acids. Thus, the retinal GABA uptake system appeared to be specific, saturable, energy dependent, and sodium dependent. Lam and Steinman[5] went on to show not only that GABA was taken up by the retina, but that it was taken up by defined cell types (suggesting biochemical specialization), and that the uptake could be modulated by exposing the retina to light (suggesting biological relevance in the context of information processing). In short, it can be said today that the importance of GABA (γ-aminobutyric acid; 4-aminobutyrate) and GABA transport in the nervous system has become well established through decades of work at the chemical,[1,9,10] biochemical,[8,11-15] physiological,[16,17] histological,[5,18,19] developmental,[20-22] and molecular levels.[23-28] Today, it is widely accepted that GABA is the major inhibitory neurotransmitter of the mammalian central nervous system.

Although the forgoing discussion makes it clear that the nervous system has evolved highly specialized apparatus (transporters, storage vesicles, and receptors) to utilize GABA as a neurotransmitter, it is also the case that GABA occupies a pivotal place in cellular metabolism. The chemical structure of GABA suits it to link the metabolism of nitrogen-containing compounds to that of carbohydrates. In fact, the metabolism of GABA in the mammalian brain (via the glutamate to succinic semialdehyde shunt) is well conserved in evolution and can be found in simple, unicellular prokaryotes such as common *E. coli* K-12 laboratory strains. Interestingly, these bacteria utilize a cluster of genes which encode not only the enzymes of the glutamate succinic semialdehyde shunt, but also a carrier designed to mediate active transport of GABA across the *E. coli* plasma membrane. The *E. coli* GABA transporter (GabP) has an overall structure and function (Figure 11-1) which is quite similar to its counterparts from the mammalian nervous system. Moreover, many ligand recognition properties of GABAergic membrane proteins (transporters and receptors) from the mammalian brain seem to be preserved in the bacterial transporter. Thus, bacteria stand to provide a simple system from which to tease out links between GABA transport, GABAergic ligand recognition, and the fundamental protein structural motifs that subserve these functions throughout the phylogenetic tree.

A. FUNDAMENTAL PROPERTIES AND MOLECULAR PHYSIOLOGY

1. KINETICS AND ION DEPENDENCE

Characteristics of GABA transport studied in various eukaryotic systems (brain slices, synaptosomes, membrane vesicles, artificial liposomes) have been remarkably consistent. Transport is found to be mediated by a saturable, high-affinity (K_m = 2 to 4 μM) carrier [29] that depends for its function upon the presence of Na[+] and Cl[-] in the medium.[30] Where these ion dependencies have been studied in detail, the data are consistent with the notion that both Na[+] and Cl[-] are cotransported with GABA and that concentrative GABA uptake is driven at least in part by the transmembrane Na[+] electrochemical gradient.[31] Both ions are also required at the inner aspect of the membrane during efflux.[32] These fundamental properties are retained following purification and reconstitution into artificial liposomes,[13,33] and also in heterologous systems[25,28,34-36] used to express the cloned GABA transporters.

Stoichiometry of transport — This is an important point which has been difficult to investigate due to the inherent complexity of GABA transport reaction. Assuming that GABA, Na[+], and Cl[-] are cotransported by an electrogenic mechanism, four transmembrane thermodynamic gradients (the electrical gradient as well as the chemical gradients of Na[+], Cl[-], and

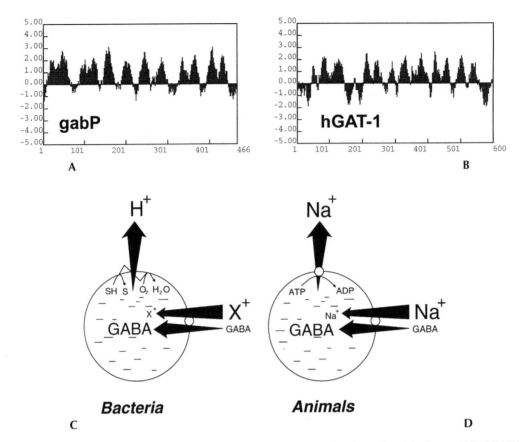

FIGURE 11-1 Structural and functional comparisons between the *gabP* of *E. coli* and the human GAT-1GABA transporters. Panel A and B are Kyte-Doolittle hydropathy plots of the *gabP* and hGAT-1 amino acid sequences. Each protein has 12 regions that exhibit a positive index of hydrophobicity. Absent evidence to the contrary, these domains are presumed to be membrane-spanning α-helices. Panels C and D are schematic representations of the GABA transport energetics in bacterial and animal cells. In these schemes expanding arrows represent a solute heading thermodynamically uphill, whereas the contracting arrows imply the opposite. The diagrams also represent that GABA transport is tightly coupled to a cation electrochemical gradient. As a result, concentrative GABA uptake can be driven either by membrane potential or by a transmembrane cation chemical gradient. In *E. coli*, the coupled cation (here represented as X^+) appears to be a proton. Addition of GABA acidified the lumen of inside-out *E. coli* plasma membrane vesicles (Panel E) whereas addition of Na^+ (activated the Na^+/H^+ antiporter) had the opposite effect. The K^+/H^+ ionophore, nigericin, collapsed the transmembrane pH (Brechtel, C., Boyarsky, C. and King, S.C., unpublished data). The human GAT-1 transporter is electrogenic so that addition of GABA induces an inward Na^+ current in *Xenopus* oocytes injected with hGAT-1cRNA (Panel F). No current is observed with sham-injected oocytes (Brechtel C. and King, S. C., unpublished data).

GABA) will necessarily contribute to the equilibrium state. How does one design an experimental protocol in which these thermodynamic parameters can be simultaneously measured and/or clamped in order to accurately assess the stoichiometric coefficients, m and n, in the equilibrium expression (Equation 11-1) that should obtain for a tightly coupled cotransporter? Due to this technical dilemma, it has been possible only to obtain estimates of the stoichiometry and to place lower limits on the true values. There is an emerging consensus that the stoichiometry may be at least 2 Na^+:1 GABA:1 Cl^-.[31,35-38]

$$\frac{[GABA]_o}{[GABA]_i} = \left(\frac{[Na^+]_i}{[Na^+]_o}\right)^n \left(\frac{[Cl^\pm]_i}{[Cl^\pm]_o}\right)^m \exp\left(\frac{\Delta\psi F}{RT}\right) \qquad (11\text{-}1)$$

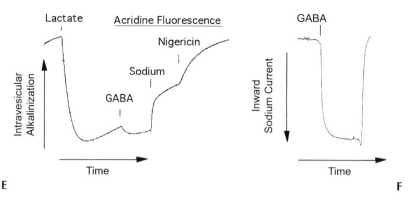

FIGURE 11-1 (Continued)

Biochemical evidence is consistent with the Na$^+$:GABA stoichiometry being greater than unity.[31,37,38] The difficulty with this interpretation is that the sodium electrochemical potential could not be controlled over the time-course of these biochemical experiments utilizing vesicles or synaptosomes. In fact, membrane potential could be known with some precision in these systems only at time zero, before significant GABA uptake occurred. This difficulty has been overcome recently with the availability of cloned GABA transporter molecules. Cloned transporters may be expressed in *Xenopus* oocytes or transfected cells. GABA-induced currents can be measured in these expression systems under voltage-clamp conditions, and stoichiometry of at least 2 Na$^+$:1 GABA:1 Cl$^-$ can be obtained.[35,36] On the other hand, there is a suggestion from voltage-clamped cells (HEK293 stably transfected with GAT-1) that the system is quite complex, and that stoichiometry may in fact be variable.[39]

Even in its simplest form, GABA transport is mechanistically complex. As many as ten steps might have to be considered:

1. The binding of Cl$^-$ to the outwardly oriented carrier
2. The binding of Na$^+$ to the outwardly oriented carrier
3. The binding of a second Na$^+$ to the outwardly oriented carrier
4. The binding of GABA to the outwardly oriented carrier
5. The transmembrane isomerization of the fully loaded carrier (ternary complex) to form the fully loaded carrier with inward orientation
6. The dissociation of GABA from the inwardly oriented carrier
7. The dissociation of Na$^+$ from the inwardly oriented carrier
8. The dissociation of a second Na$^+$ from the inwardly oriented carrier
9. The dissociation of Cl$^-$ from the inwardly oriented carrier
10. Completion of the transport cycle by transmembrane isomerization of the inwardly oriented unloaded carrier to reform the outwardly oriented unloaded carrier

Any of these ten partial reactions that cause uncompensated charge movement within the transmembrane electrical gradient should exhibit voltage-dependent characteristics. For example, the binding affinity of Na$^+$ appears to be voltage dependent,[35] and associated with a measured current transient that can be blocked with SKF-89976A, a high-affinity uptake blocker. Sodium ion may also be able to interact with a channel-like state of the GABA transporter which engages in forbidden uncoupled transport reactions.[39]

2. UNCOUPLING AND CHANNEL ACTIVITY

Although consideration of the various "normal", well-coupled steps in transport is important, it is also important (even critical) to consider "forbidden" steps which — in an ideal world — should not occur in a cotransport cycle.[40] "Forbidden" reactions are essentially short-circuit currents or fluxes, involving net translocation of GABA, Na+, or Cl- across the membrane independently of its cosubstrates. Such uncoupled "slip" reactions compromise the ability of an energy-transducing protein to operate efficiently since energy is expended to create solute gradients that are allowed to "slip" back across the membrane without doing work on another solute molecule. Interestingly, certain neurotransmitter transporters may be poised to catalyze a "slip" reaction despite the apparent inefficiency of this capability. Both the serotonin transporter[41] and the GABA transporter[39] can be induced to pass a current in the absence of the organic substrate. These currents appear to be specific since (1) they are blocked by transport inhibitors, and (2) they are not observed in host cells not expressing the transporter. But, are these uncoupled leakage currents physiologically significant?

Whether seemingly anomalous conductance states actually occur *in vivo* may not be as significant as the fact that they can be observed to occur in the laboratory under prescribed conditions. The observation of uncoupled currents (i.e., in the absence of neurotransmitter) could be telling us a great deal about the catalytic mechanism of these transporters. For example, the uncoupled leakage currents could be construed to mean that the ion-bound transporters are poised in a conformation that has a free-energy level quite near the transition state for transmembrane isomerization (or channel opening).[39,41,42] Consideration of transition states may allow us to gain some insight not only into the basis of leakage currents, but also the relationship of these leakage currents to the kinetics of "normal", well-coupled transport. What, for example, is the role of chloride in the transport of GABA and other neurotransmitters? Could leakage currents and coupling to chloride serve a common purpose in neurotransmitter transporters?

3. Chloride, "Leak Current", and Transport Rate

Why would evolution have established in the nervous system multiple neurotransmitter transporter families[43] that utilize not only sodium, but an additional coupling ion? What, for example, can a (Na$^+$ + Cl$^-$)-coupled GABA carrier do that a hypothetical (Na$^+$)-coupled GABA carrier could not do, and why might this be important in the case of neurotransmitter transporters? One possibility is that neurotransmitter transporters are required to simultaneously: (1) recognize transmitters with very high affinity, (2) pump the transmitters against extremely high thermodynamic gradients, and (3) do the pumping at high rates (finite rates). There are theoretical reasons (discussed below) why doing all of these things at once could be problematic for a "sodium-only" transporter.

The hypothesis is that introduction of chloride into the GABA transport mechanism allows the transporter:

- To operate at higher affinity without sacrificing turnover number
- To pump a larger GABA gradient
- To do so without suffering ruinous levels of current leakage

The theoretical considerations (briefly) are as follows. First, high substrate affinity (by dropping the substrate into an energy well) works against high rates of catalysis.[40,44-49] Neurotransmitter transporters could use chloride as a tool to stabilize the transition state of the transport reaction so that (relative to the sodium-only carrier) a higher substrate affinity (lower K_m) could be had without sacrificing turnover number (again, relative to the sodium-only carrier). How chloride could accomplish this effect on transport kinetics is summarized in Figure 11-2. Second, how can chloride allow the carrier to pump a larger GABA gradient? The proposed answer is that chloride allows the carrier to utilize a second Na$^+$ in its mechanism. Without chloride, a second Na$^+$ would dispose the carrier to development of a ruinous

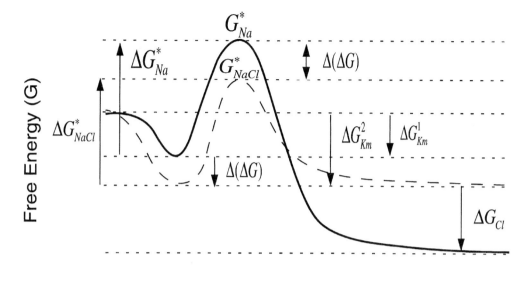

FIGURE 11-2 Role of chloride in the GABA transporter transition state. This scheme is a complicated way to illustrate the hypothesis that nature has chosen to use chloride as a tool to maximize both the affinity for GABA and the transport rate without causing an intolerable leakage current in the absence of GABA. The rationale is that when the membrane potential is held at the chloride equilibrium potential, then the translocation of chloride across the membrane is not accompanied by a free energy change ($\Delta G_{Cl} = 0$). Thus, the only possible contribution of chloride is to the kinetics of transport. The kinetics of a catalyzed reaction is determined by the difference in free energy between the ground state and the transition state. The hypothetical example compares the energy barrier for a transport reaction catalyzed by a sodium-only carrier and the reaction catalyzed by a carrier which incorporates chloride into the mechanism. It will be noted that the affinity of the sodium-only carrier for its substrate (e.g., GABA) is low compared to the affinity of the carrier that uses chloride. However, the turnover numbers for both carriers are equal since chloride is being used to stabilize the transition state by $\Delta(\Delta G)$, an amount exactly equal to the difference in the free energies of GABA binding in the ground state, i.e., $\Delta(\Delta G)$. It is to be emphasized that the chloride effect does not depend on the transmembrane chloride electrochemical gradient (ΔG_{Cl}) which can be either zero or non-zero. This scheme is highly simplistic and does not consider the tendency of the carrier to catalyze uncoupled sodium leakage currents in the absence of GABA (see text discussion of the role of chloride in minimizing such leaks).

"uncoupled" leakage current. But, how mechanistically does a second sodium ion predispose the transporter to current leakage, and how does chloride plug the leak. A sodium current leak would arise — according to transition state theory[40] — if in the absence of neurotransmitters a relatively stable 2Na$^+$:carrier transition state occurred that could isomerize readily (gate) across the membrane. The key consideration is not in the net transfer of NaCl across the membrane, but rather in the canceling (by bound Cl$^-$) of the contribution that bound Na$^+$ makes to transition state stabilization in the absence of neurotransmitter. It is of great interest in this regard that Mager et al.[41] have described evidence consistent with chloride having precisely this effect: namely, that leakage current increases when chloride is removed from a solution already lacking the neurotransmitter (in this case serotonin).

It bears emphasizing that the argument made here is for the effect of chloride to be understood entirely in kinetic terms, i.e., the contribution of chloride to the energy of the transition state has profound kinetic consequences even when the membrane potential is set equal to the chloride equilibrium potential so that no thermodynamic contribution is possible. Obviously, chloride must also contribute to the thermodynamics of the transport reaction when the anion is not at equilibrium. Potassium ions may have kinetic effects in the glutamate family[91] of transporters that is analogous to the above postulated effect of chloride on the GABA/biogenic amine family of transporters.

4. Role of Transporter in GABA Release

Can the synaptic plasma membrane GABA transporter play dual roles, mediating both the release of GABA as well as its reuptake? The general consensus — relying fundamentally on the observation that pharmacologic inhibitors of neurotransmitter reuptake will potentiate postsynaptic actions following evoked release[50] — evolving over more than three decades of experience, is that high-affinity neurotransmitter transport functions to terminate the postsynaptic response. Profound support for this position has come from recent experiments with a "knock-out" mouse, lacking the dopamine transporter.[51] These mice survive, but are behaviorally hyperactive. Paradoxically, they have below-normal dopamine levels, owing to compensatory downregulation of dopamine-β-hydroxylase (rate-limiting step in biosynthesis). These biochemical changes are ostensibly to compensate for loss of the dopamine transporter which would normally sequester the transmitter to the cytoplasm. Thus, at least for the dopaminergic system, the results imply that the transporter (1) plays a pivotal role in regulating transmitter availability, and (2) that mechanisms other than the transporter are available to affect release of the neurotransmitter to cause hyperactivity and compensatory downregulation of the dopamine system.

Indeed, it is conventionally held that the evoked release of a neurotransmitter occurs by exocytosis of synaptic vesicles — and certainly GABA is stored in synaptic vesicles.[52-54] Nevertheless, there is biophysical[17] and biochemical evidence[55,56] that depolarizing conditions cause the release of GABA via reversal of the "reuptake" transporter. Such GABA release is calcium independent (i.e., nonexocytotic) and sensitive to classical GABA transport inhibitors such as nipecotic acid. Whether transporter-mediated release is physiologically significant in all GABAergic cells is not known. Certainly, horizontal cells from fish retina are unique, undergoing long-term cycles of depolarization-repolarization in response to changing light levels.[17] It has been suggested recently that transporter-mediated GABA release, in a kinetic sense, is under dual control of membrane potential plus the internal Na^+ concentration. Efflux would be significant only under circumstances in which significant levels of cytoplasmic Na^+ are accumulated.[39] Although there is no consensus as to the relative importance of vesicular vs. nonvesicular release, interest in this topic certainly continues.[57]

C. SITES FOR POSTTRANSLATIONAL MODIFICATION

1. Role of Glycosylation Sites

Molecular cloning of GAT-1 has revealed the presence of several N-linked glycosylation sites in the second extracellular loop (Figure 11-3). One or more of the putative glycosylation sites may be important in the eukaryotic transporters since neuraminidase treatment (removal of sialic acid) reduces the [^{3}H]GABA transport in synaptosomes as a linear function of the amount of sialic acid liberated.[58] Similarly, GABA transport in cells transfected with GAT-1 cDNA could be abolished with tunicamycin.[34] Neither the use of neuraminidase nor the use

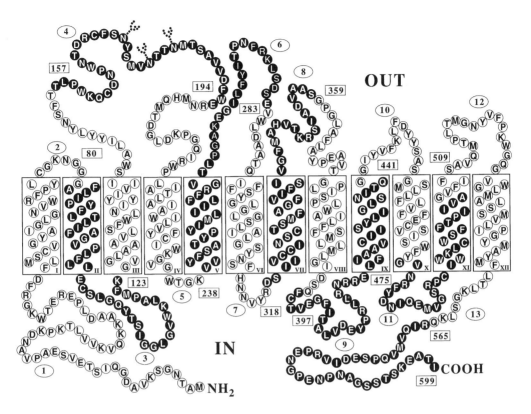

FIGURE 11-3 Topological model and genomic structure of the human GAT-1 GABA transporter. The deduced amino acid sequence is represented by 599 beads. The putative transmembrane domains (presumably α-helices) are enclosed within boxes and labeled I through XII. The black and white segments on the beaded chain represent portions of the polypeptide encoded by distinct exons. The human GAT-1 gene spans about 25 kb and contains 15 introns, 2 of which precede the initiating methionine.[25] Thus, the translation start site is in exon 3 and the putative glycosylation sites in the second extracellular loop are in exon 6. Compared to the original publication,[25] 14 residues from loop 11 have been pushed into helix XI. This manipulation makes helix XI more hydrophobic. The manipulation also increases the already strong tendency for each of the transmembrane domain to be encoded by a separate exon. Note that the 12 α-helical transmembrane domains are almost entirely encoded by 12 separate exons.

of tunicamycin directly addresses whether glycosylation of the GABA transporter per se is important to GABA transport. Due to the nonspecific nature of these agents, it remains a formal possibility that GABA transporter glycosylation could be irrelevant, and that glycosylation of some accessory protein is crucial.

The cloning of the GABA transporters has made it possible to take advantage of the exquisite selectivity of site-directed mutagenesis as a means to "deglycosylate" only the GABA transporter — and at specific sites on the transporter. Information should soon be available detailing not only whether glycosylation is necessary to support a transport activity, but also information bearing on precisely which glycosylation sites are crucial and which are not — at least in host-vector expression systems. To make this determination in neurons remains a difficult problem.

2. REGULATION BY KINASE ACTIVITY

Consensus protein kinase A phosphorylation sites are also present in the rat GAT-1 sequence, but not in loop regions predicted to have cytoplasmic exposure. On the other hand, consensus protein kinase C phosphorylation sites are present in the amino- and carboxy-terminal tails of rat GAT-1.[23] While the presence of these sites could theoretically enable phosphorylation as an important form of direct GABA transporter regulation, current evidence favors the notion that transporter activity is affected only secondarily by the battery of experimental manipulations which have been deployed to activate protein kinase signaling pathways in cells expressing GABA transporter. For example, agents that activate protein kinase C can either decrease GABA transport activity (glial cells),[59] increase transport activity (*Xenopus* oocytes injected with GAT-1 cRNA),[60] or have no effect (primary neurons).[59] Corey et al. suggest that these variable effects are probably cell specific, and related to modulation of the subcellular distribution of the GABA transporter (i.e., the mole fraction of transporter trafficked to the plasma membrane in active form). Similarly, the effects of cAMP (and by implication protein kinase A) appear to be indirect, affecting GAT-1 mRNA levels or subcellular distribution.[61]

D. CLINICAL RELEVANCE OF GABA

The link between GABAergic neurotransmission and suppression of seizure activity is incontrovertible. GABA-mediated inhibition controls neuronal excitability, prevents the generation of burst firing, and contributes to termination of epileptiform burst discharges.[62] Accordingly, several clinically effective antiepileptic drugs are based upon actions that facilitate GABAergic neurotransmission. The benzodiazepines and the barbiturates modify $GABA_A$ chloride channel gating,[63] thereby increasing the effectiveness of inhibitory postsynaptic potentials (IPSPs). The benzodiazepines are known to be active against myoclonic seizures in humans and pentylenetetrazol-induced seizures in animals,[64] at low nanomolar concentrations consistent with activity at the $GABA_a$ receptor. Valproate increases the concentration of GABA in neurons[65] and additionally appears to facilitate the actions of GABA itself.[66] The GABA transaminase inhibitor, γ-vinyl GABA, also elevates the concentration of GABA in nerve terminals[67] and has significant activity in 50% of the patients with seizures who resisted treatment by other drugs.[68]

It is an important observation that many of the above-mentioned anticonvulsants have different mechanisms of action, have different spectrums of activity against the various categories of seizure, and exhibit different side effects. The implication is that new anticonvulsants, working through entirely new mechanisms, may be expected to exhibit important differences in terms of desired activities and undesired side effects. Such novel compounds could become useful not only in treating clinically recalcitrant epilepsies, but also in the design of pharmacological countermeasures for seizures that arise in unusual circumstances involving exposure to proconvulsant compounds (e.g., agricultural or military exposure to cholinesterase inhibitors).

Certainly, the GABA transporters from presynaptic neurons and glia have in recent years become promising targets for a new generation[69] of mechanistically distinct compounds with clinically significant antiepileptic activity.[70-74] These GABA-mimetic anticonvulsants act by inhibiting[28,75-77] the presynaptic and glial GABA transporters. Certain of these new presynaptically acting GABA-mimetics exert anxiolytic and anticonvulsant effects[78] similar to the benzodiazepines which facilitate GABAergic transmission postsynaptically.[63] Thus, it is conceivable that other presynaptically acting compounds might eventually be found to exert both the sedative and the anticonvulsant properties of the barbiturates — which, like the benzodiazepines, modify chloride channel gating to facilitate GABAergic transmission involving the $GABA_A$ receptors.[63]

The current generation of antiepileptics that target GABA transporters are all hydrophobic (cross the blood-brain barrier for oral activity) derivatives of classical uptake inhibitors

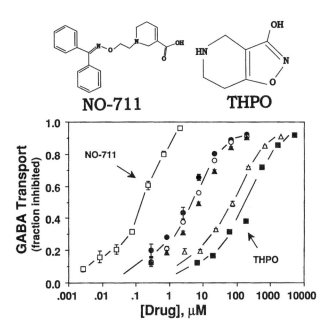

FIGURE 11-4 Pharmacological profile of rat GAT-1 expressed in mouse fibroblasts. The rank order of potency (NO-711, □ << GABA, ● < guvacine, ○ < nipecotic acid, ▲ << cis-4-hydroxy-nipecotic acid, △ < THPO, ■) is consistent with assignment of GAT-1 as the "neuronal" subtype in the classical pharmacological scheme[86,151] in which transporters were recognized as being either of neural or glial origin. With benefit of knowledge from molecular cloning and the ability to express the clones separately, it has become possible to pharmacologically distinguish four GABA transporter subtypes.[26] Classical GABA uptake inhibitors such as guvacine and nipecotic acid have limited utility as anticonvulsants because they are highly charged and cannot cross the blood-brain barrier. THPO crosses the blood-brain barrier, but is not a very potent GABA transport inhibitor. The attachment of hydrophobic groups to classical GABA transport inhibitors tend to produce compounds (like NO-711) that are of far greater potency and are hydrophobic enough to cross the blood-brain barrier. Compounds in this family are showing clinical promise. (King, S. C., unpublished data.)

(Figure 11-4). The classical inhibitors do not exhibit much specificity for GABA transporter subtypes, and this could present a significant limitation in terms of generating new drug actions and broader spectrums of activity against the different epilepsies.[79] The recent cloning of the GABA transporter in multiple subtypes and from multiple species has refocused attention on the potential to obtain a broader spectrum of pharmacological actions by generating greater specificity for the subtypes. Certainly, the availability of cloned GABA transporters affords for the first time an opportunity not only to manipulate the structure of drugs, but also to manipulate the structure of the proteins to which the drugs bind.

E. CLONING OF GABA TRANSPORTERS

The feasibility of using molecular biology as a tool to study neurotransmitter transport systems became apparent as early as the mid-1980s, when it was shown that *Xenopus* oocytes could be used as an expression system [80,81] for the mammalian GABA transporter. The injection of rat brain mRNA into the oocyte led to expression of a Na^+-dependent [^{3}H]GABA transport activity that was more than tenfold higher than controls not receiving the mRNA. The GABA transport activity was sensitive to a number of classical GABA transport inhibitors. The results implied that oocytes could adequately translate the message, carry out any required posttranslational modifications, and productively insert the GABA transporter polypeptide into the cytoplasmic membrane. Likewise, nonneuronal mammalian cells were also shown to be capable of expressing neurotransmitter transporters.[82] Thus, by the late 1980s expression systems capable of verifying the identity of candidate neurotransmitter transporter clones had been described. Indeed, the *Xenopus* oocyte was to eventually play an important role in characterizing GAT-1 GABA transporter from rat brain.[23]

1. GAT-1 FROM RAT

Rat GAT-1 was the first neurotransmitter transporter cDNA to be cloned,[23] and this was possible because this transporter was also the first to be purified. Sequencing of peptides derived from the purified GABA transporter enabled synthesis of degenerate oligonucleotide probes that were used to identify candidate cDNA clones by hybridization screening of a λ phage library. The probe hybridized to a clone harboring a cDNA more than 4 kb in length which, in turn, hybridized to a 4.2-kb transcript from brain. Sequencing of the cDNA revealed an open reading frame sufficient to specify a protein of 599 amino acids. The predicted open reading frame encoded a number of the peptides that had been obtained by CNBr-cleavage of the purified GABA transporter. Additionally, hydropathy analysis indicated that the protein could be modeled as a membrane protein containing 12 transmembrane α-helical domains with the amino- and carboxy-termini protruding into the cytoplasmic compartment. Consistent with reports that GABA transport activity might be subject to inhibition by protein kinase C,[59] several consensus protein kinase C phosphorylation sites were predicted to lie in regions modeled to be cytoplasmic. Likewise, hydropathy analysis predicted the existence of a large extracellular loop containing three consensus glycosylation sites, a prediction consistent with the fact that the GABA transporter is a glycoprotein [13] subject to inactivation by neuraminidase.[58]

RNA synthesized *in vitro* from the GAT-1 clone provided two additional lines of evidence that the newly identified clone encoded a rat brain GABA transporter. First, *in vitro* translation produced a 67-kDa protein which cross-reacted with antibodies[83] prepared against the GABA transporter. Additionally, injection of this RNA into *Xenopus* oocytes greatly enhanced the uptake of [^{3}H]GABA. It was shown that:

1. K_m for GABA transport was around 7 μM, a value in reasonable agreement with that of the high-affinity transporter from brain[29]
2. Transport depended upon the presence of both Na$^+$ and Cl$^-$, a well-established ionic requirement[30]
3. The GAT-1 transporter was sensitive to ACHC (an inhibitor of neuronal GABA uptake)[84] and relatively insensitive to THPO and β-alanine, two compounds thought to be most effective against the variety of GABA transporter expressed in glial cells[85,86]

2. THE NTT SUPERFAMILY

The cloning of GAT-1 from rat was followed by the identification of a neurotransmitter transporter (NTT) superfamily[87,88] consisting not only of the highly homologous biogenic amine transporters (Figure 11-5A and B), but also of multiple GABA transporter subtypes (Figure 11-5C)[26] and several other categories of transporter — some of them orphan transporters for which the substrate remains unknown.[27,89] The recently cloned — and mechanistically distinct[90] — glutamate/aspartate transporters are part of an evolutionarily distinct family (Figure 11-5D) of excitatory amino acid carriers.[91-93]

3. THE gab PERMEASE FROM E. COLI

a. Cloning gabP for Inducible Overexpression

The *gabP* was first mobilized from the *E. coli* chromosome by Metzer and Halpern,[94] who showed that a BamHI-HindIII restriction fragment (1.9 kb) could complement a GabP-negative mutant. The underlying basis for this complementation was substantiated by Niegemann et al.[95] who cloned and sequenced the gene, confirming the presence of *gabP* on a BamHI-HindIII fragment. These workers also cloned the *gabP* for expression under control of an IPTG-inducible promoter, but found the gene (and even smaller fragments derived from it) to be highly toxic (killed the cells when IPTG was added) in the *E. coli* host strain, W3110. For reasons that are not presently clear, the *E. coli* host strain SK35 (Figure 11-6) has been more successful in supporting IPTG-inducible overexpression of *gabP* from a plasmid. Derivatives of SK35 have therefore become the strains of choice for functional characterization of GabP expressed from a plasmid.[96-98]

b. Relationship of gabP to Mammalian GAT Genes

Polypeptide sequence alignments (Figure 11-7) suggest that if the *E. coli* GABA transporter gene (*gabP*) and the eukaryotic GABA transporters are indeed derived from a common ancestral gene, then their point of divergence in evolution must have been quite distant compared to the points at which individual members of the neurotransmitter transporter family began to diverge from one another.[89] Because of this considerable phylogenetic divergence of the primary sequences, it came as a surprise that the bacterial GABA transporter shares many ligand binding characteristics in common with GABAergic transporters from the mammalian brain. These functional similarities made the *E. coli* GabP an attractive model system capable of enhancing our ability to gain insight into essential features of the interaction between GABAergic neurochemicals and the proteins to which they bind. The *E. coli* GabP offers many technical advantages over cloned neuronal proteins that require eukaryotic expression systems. The remainder of this chapter is concerned with describing the advantages and fundamental properties of the *E. coli* GABA transporter (GabP) in relation to some of its neuronal counterparts such as the human GAT-1 transporter.

F. INHERENT ADVANTAGES OF E. COLI GENETICS

The use of an *E. coli* model system to study fundamental aspects of GABAergic ligand recognition has the potential to afford investigators a considerable technical advantage over analogous mammalian proteins that are expressed from eukaryotic cells. The cloning of *gabP*, the construction of appropriate host strains, and the successful expression of the gene under *lac* control makes it possible to realize major advantages in the areas of (1) point mutagenesis, (2) topological analysis, and (3) large-scale growth and overexpression.

1. NATURAL SELECTION OF POINT MUTANTS

An important advantage of *E. coli* is the potential to obtain GABA transporter mutants by natural selection — since in this bacterium GABA is a nutrient that can be made necessary for growth and survival. Whereas the mammalian brain uses GABA primarily as a neurotransmitter, *E. coli* will consume GABA as the sole source of carbon and nitrogen when grown in a minimal medium containing simple salts that provide only a source of phosphate (required

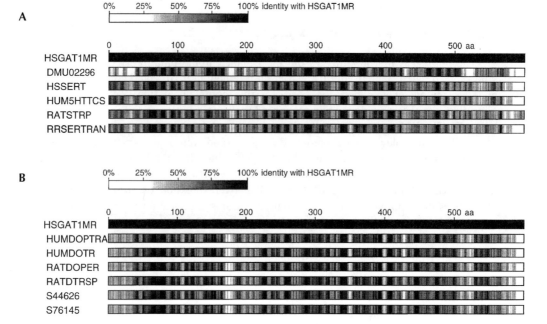

FIGURE 11-5 Gray-scale representation of homologies among transporters that recognize serotonin, dopamine, GABA, and glutamate/aspartate. The amino acid sequence alignments were calculated with Clustal V[152] and then converted to a gray-scale representation using the Sequence Similarity Presenter.[153] It is apparent that the mammalian and *Torpedo* GABA transporters are highly homologous to one another as well as to the biogenic amine transporters. These eukaryotic GABA transporters are evolutionarily far away from the family of glutamate/aspartate transporters and the GABA transporter from *E. coli*. Access to the primary sequence data was obtained with the following Genbank accession numbers: HSGAT1MR, X54673; DMU02296, U02296; HSSERT, X70697; HUM5HTTCS, L05568; RATSTRP, M79450; RRSERTRAN, X63253; HUMDOPTRA, M95167; HUMDOTR, L24178; RATDOPER, M80570; RATDTRSP, M80233; S44626, S44626; S76145, S76145; HSGAT1MR, X54673; MUSGAGATR, M97632; MUSGATIII, L04663; MUSGATIV, L04662; MUSGABAX, M92378; S42358, S42358; RATGABAQ; M95738; RAT2GAT, M95762; RAT3GAT, M95763; TCTGAT1, X77139; gabP, X65104; RABGLUTRAN, L12411; RNGLT, X67857 S49853; RNGLUAS, X63744 S49018; S59158, S59158.

for DNA synthesis) and sulfate (required for protein synthesis). Under such conditions, growth depends upon the GABA transporter to deliver carbon and nitrogen to the cell at rates sufficient to meet the total anabolic demand. Neurochemicals capable of blocking GABA transport would be expected to deprive the cells of carbon and nitrogen. Under these conditions, cells expressing a normal GabP could not grow, and the medium would thus be "selective" for structurally altered transporter mutants capable of rejecting the inhibitor without losing the ability to transport GABA.

Such ligand recognition mutants are potentially useful probes for residues at or near the active site. Mutants isolated by this technique cannot be an irrelevant null phenotype since the culture condition imposed by the investigator selects for a structural alteration that discriminates the substrate (GABA) from the inhibitor (a related but unique structure). The technique is powerful because literally billions and billions of bacteria are subjected to selection, and only a very few are able to survive — having acquired appropriate mutations at positions that could not have been predicted in advance. It is likely that results from this kind of experimentation will soon be contributing to our understanding of GABAergic ligand binding since a number of structurally diverse GABAergic neurochemicals (see Section G) have been found to interact with the GabP active site.

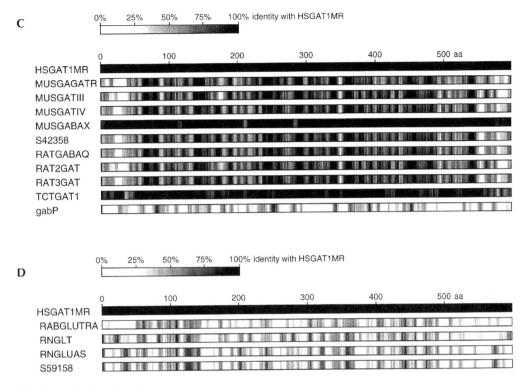

FIGURE 11-5 (Continued)

2. SITE-DIRECTED AMBER SUPPRESSION IN *E. COLI*

In contrast to the situation with the *E. coli* GabP, neuronal GABA transport proteins must be altered by site-directed mutagenesis and one must count on successfully "guessing" both the relevant locations and the relevant amino acid substitutions prior to engineering the protein. If survival of cells used in eukaryotic expression systems could be made to depend upon transport of GABA, this might enable approaches based on mutant selection. However, no such strategies have yet been developed, and thus site-directed mutagenesis remains a powerful and important technique for structure-function studies in these systems.

Such analysis of structure-function relationships can be accelerated tenfold by combining site-directed mutagenesis with another powerful genetic technique called amber suppression.[99] Moreover, the strategy is applicable in *E. coli*, but not in eukaryotic expression systems. What is site-directed amber suppression? Amber is the name given to the stop codon TAG. A battery of 12 *E. coli* strains have been developed that fail to reliably recognize TAG as a stop codon. Instead, these suppressor strains will frequently insert a particular amino acid when the ribosome encounters the amber codon. Thus, by engineering the TAG triplet at specified locations within *gabP*, it is possible to use the 12 suppressor strains[99-103] to rapidly screen the functional effect of 12 different amino acid substitutions at a defined position. In the case of *gab* permease, it was necessary in advance of applying amber suppressor technology to remove the naturally occurring amber stop codon found in the unmodified gene, replacing it with the ochre stop codon, TAA (Figure 11-7B).

With the modified *gabP* in hand, the author's laboratory has taken an interest in helix 8 of *gab* permease. This helix contains the only membrane-imbedded histidine (His-284) in the protein (see Figure 11-7B). Helix 8 also contains a nearby acidic (Asp-290) residue and

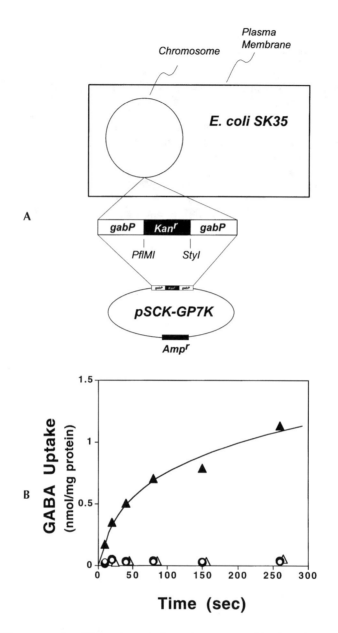

FIGURE 11-6 Chromosomal *gabP* knock-out by homologous recombination. The diagram illustrates schematically the relationship between the engineered plasmid and the chromosomal *gabP* defect created in *E. coli* strain SK35, which is used as a host cell for expression of the plasmid-borne, IPTG-inducible *gabP*. To create SK35, a portion of *gabP* was deleted from the *E. coli* DW1 by homologous recombination between the chromosome and pSCK-GP7K, a derivative of pSCK-GP7 in which the StyI-PflMI restriction fragment within *gabP* was replaced by a gene encoding kanamycin resistance. The pSCK-GP7K plasmid was initially placed in DW1 by transformation to create strain SK25. Subsequently SK25 was cured of pSCK-GP7K by serial passage in LB medium containing kanamycin but not ampicillin. A kanamycin-resistant (but ampicillin-sensitive) strain called SK35 was isolated and found to be highly defective for [³H]GABA uptake — compare wild type (▲) with SK35 (●) in Panel B. (From *J. Biol. Chem.*, 270, 19893, 1995. With permission.)

a nearby basic residue (Lys-286). The only other polar residue in GabP helix 8 is Asn-281. The codons at all four of these positions were changed to TAG to enable analysis by amber suppression. Although the analysis is not complete at the time of this writing, it can be said that the His and Asn residues are nonessential. On the other hand, both charged residues

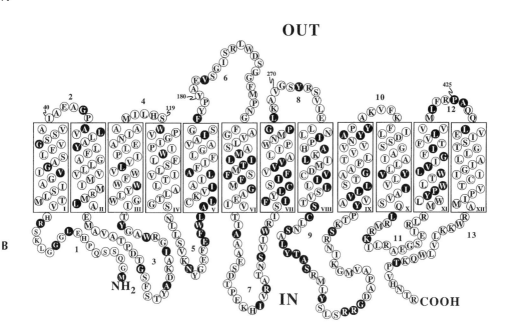

FIGURE 11-7 Clustering of homology between the *E. coli gabP* and other GABA transporters. (*Panel A*) Clustal V[152] sequence alignments showing the regions from 377–423 of human GAT-1 (top-most sequence) and from 282–326 of the *E. coli gabP* (bottom-most sequence). Shaded residues denote regions of agreement between gabP and other GABA transporters. (*Panel B*) Topological model of the *E. coli* GabP. Regions of homology to other GABA transporters (colored gray in Panel A) are denoted as black beads. Clearly, the greatest preservation of primary structure is in the central portion of the molecule between putative α-helices IV to IX, and there is more homology in the internal loops than in the external loops.

appear to be more important. Interestingly, the juxtapositioning of the high pK_a and the low pK_a residues on the same helix[104-106] together with an intermediate pK_a residue is reminiscent of the situation on helix 10 of the *E. coli lac* permease [107,108] and other transporters.[109,110] The *lac* permease helix 10 has been implicated in various aspects of the transport process including substrate recognition and energy coupling. It remains to be seen where the seat of energy coupling resides in any of the GABA transporters found in nature.

3. TOPOLOGY BY TRANSPOSITION

What is known about the topography of mammalian GABA transporters has been discerned mainly from computer-based hydropathy analysis of cloned cDNA sequences and from biochemical work aimed at defining the location of terminal domains, antibody epitopes, and sites sensitive to proteolysis and deglycosylation.[83,111-113] What is known thus far about the topology of bacterial GABA transporters has been gleaned entirely from hydropathy analysis — initial proposals for the GabP topology are shown here (e.g., Figure 11-7B). Fortunately, there are reliable and comprehensive approaches that can be exploited to rapidly confirm putative topological features ascribed to *E. coli* membrane proteins. The methods are based on the transposon-mediated random insertion of a topological reporter gene into a target gene, i.e., a gene such as *gabP* which encodes the *E. coli* GABA transporter.[114-120]

a. C-Terminal Reporter Enzymes

The *lacZ* and *phoA* are two topological reporter genes which encode the enzymes β-galactosidase (LacZ) and alkaline phosphatase (PhoA). From the perspective of topological analysis, these two enzymes are complimentary (Figure 11-9A).[120] Both enzymes have been engineered to lack topogenic sequences of their own. As a result they are destined after synthesis to remain in the cytoplasm unless peptide sequences to which they are fused contain topogenic signals for export across the membrane. When a reporter enzyme is fused in-frame to a membrane protein, the reporter enzyme will, with few exceptions,[118] be topologically localized based on information contained in portions of the membrane protein that are N-terminal to the fusion junction. Analysis of a large number of random gene fusions will allow assignment of each hydrophilic loop in the membrane protein to either the extracellular or the intracellular space, based upon colorimetrically measured activity of the reporter enzyme.

b. Epitope Insertions

A potential pitfall of the analysis described in the previous section and in Figure 11-9A is that the reporter enzyme literally replaces that portion of the target membrane protein which lies C-terminal to the fusion junction. If by chance these missing sequences would have been topologically influential, then obviously that influence is lost and the reporter enzyme could be topologically misdirected. The latest generation of transposons help to overcome this potential problem by allowing the fusion construct to be converted in vitro to a different construct which preserves all of the target protein sequence, i.e., all topogenic sequences are potentially preserved.

The *in vitro* manipulation needed to effect this conversion is simple, requiring only a restriction enzyme digestion to remove the reporter gene and ligation of the construct to create a new gene fusion that results in a 31-amino-acid epitope insertion at the original fusion junction, and with preservation of the target protein sequence C-terminal to the fusion junction. The 31-amino-acid insert contains a proteolysis site and antibody epitope to facilitate topological analysis using what are essentially traditional strategies.[114]

Apart from allowing one to answer the above-mentioned criticism of reporter enzyme technology, the new transposons also allow the creation of insert-resistance maps. That is to say, it is possible to score a collection of random 31 amino acid insertion according to whether they abolish or preserve functional activity of the transport protein. Although this kind of work is as yet unpublished, this author believes that epitope insertion experiments have implications regarding the structural organization of the membrane protein. Apparently, domains resistant to inactivation by epitope insertion are likely to be devoid of important secondary and tertiary structural elements. For example, transmembrane helices must be around 20 residues in length. Therefore, fusions within helices are highly detrimental to transport function. In contrast, many fusions within hydrophilic loop domains are well tolerated.

c. PacI-Mediated Gene Fusion

The random nature of transposon-mediated gene fusion assures a statistical probability of failing to obtain a transposon "hit" within a region where a gene fusion would be particularly informative. For that reason some site-directed fusion approach is desirable. In order to simultaneously enable a novel site-directed *PacI*-mediated gene fusion strategy and a means to perform site-directed amber suppression (Section VI.2), a *gabP* was synthesized by polymerase chain reaction to create the modified 5′ and 3′ regions shown in Figure 11-8. The *PacI* site is uniquely suited to carry out a novel cassette-based attachment of epitope tags and reporter enzymes to the C-terminus of a cloned gene that utilizes TAA as its stop codon.

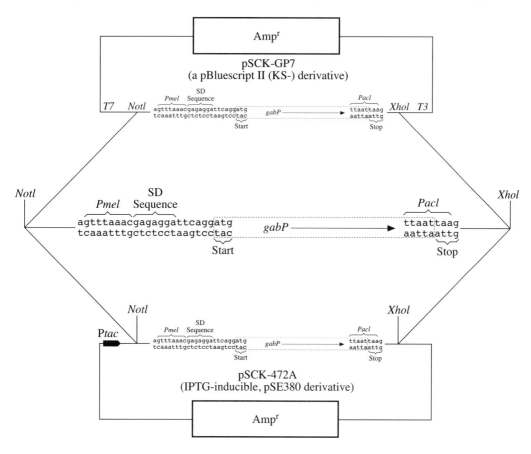

FIGURE 11-8 Cloning of an end-modified derivative of the *E. coli gabP* gene. The polymerase chain reaction (PCR) was used to simultaneously mutate the *gabP* stop codon from the wild-type amber (TAG) to ochre (TAA), introduce a *PmeI* site 5′ of *gabP*, and introduce a *PacI* site that overlaps the TAA stop codon. The resulting PCR product was blunt-end cloned in the *EcoRV* site of pBluescript II (KS-). The removal of amber from *gabP* was crucial to implementation of site-directed amber suppression (Section F.2). The placement of *PacI* over the TAA ochre stop codon was crucial for implementation of a novel cassette-based gene fusion strategy used to tag the C-terminus of GabP with reporter enzymes and peptide epitopes (Section VI.C.2). Placement of the *gabP* under *lac* control was crucial to avoid toxicity related to overexpression. For reasons which are not truly understood, overexpression of GabP from a plasmid appears to be toxic in some *E. coli* strains.[95] One possibility is that the full complement of *lac* genes present in wild type *E. coli* contribute to toxicity since IPTG would necessarily induce the *lac* genes along with *gabP*. Regardless, it is certainly the case that *gabP* induction is not overtly toxic in the *E. coli* strain SK35 (Figure 11-6) which is simultaneously *gabP*-negative and devoid of IPTG-inducible *lac* genes.

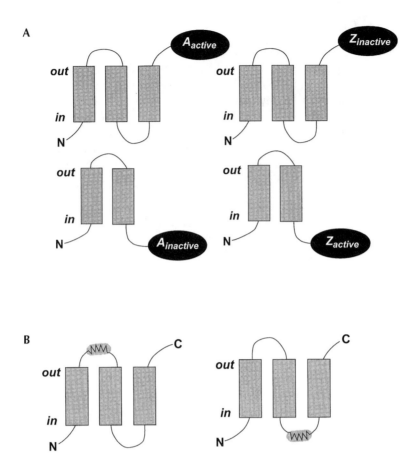

FIGURE 11-9 Tools for topological analysis of membrane proteins in *E. coli*. The method of analysis is based on the use of topological reporter enzymes which are differentially active when localized to one side of the membrane or the other. In other words, enzyme activity becomes the measured index of topological localization. *Panel A*. The β-galactosidase (LacZ) is a cytoplasmic enzyme that is inactive ($Z_{inactive}$) when exported to the outside. In contrast, alkaline phosphatase (PhoA) is a periplasmic enzyme that is inactive ($A_{inactive}$) when retained on the inside. Thus, the two topological reporter enzymes are complementary,[120] and results from one can be used to validate results from the other. Neither reporter enzyme has signals to specify its own export, and therefore retention in the cytoplasm vs. export to the outside is determined by topogenic sequences contributed by the membrane protein to which the enzymes are fused. In practice, these enzymes are fused at random sites within the membrane protein, and the investigator tests each in-frame fusion construct for enzyme activity. The gene fusions are constructed *in vivo* by infecting cells with λ phage particles that carry the reporter genes on a transposon. Transpositions into the plasmid-borne membrane protein gene are easily isolated by powerful selection techniques based on antibiotic resistances which are also carried by the transposon.[115,116,118-120] An ampicillin-resistant plasmid receiving a transposon "hit" would become resistant to both ampicillin and a second antibiotic such as chloramphenicol. This combination of resistances is readily selectable, and the exact position of the gene fusion can be easily determined by DNA sequencing. *Panel B*. A new generation of transposon is now available which allows selection for random, in-frame insertions (31 residue epitope that can be recognized by an antibody and a protease). Clearly, the application of transposon-mediated gene fusion is highly efficient compared to methods used to analyze topology of eukaryotic membrane proteins.[154-156] These techniques will be important for structural analysis of GABA transport proteins that prove to be expressible in *E. coli*. On the other hand, the random nature of the method could make it difficult to obtain fusions at exactly the desired location. Such information voids can be filled with in vitro gene fusion techniques (see Figure 11-10).

Sequences encoding epitope tags and reporter genes were then cloned 3′ of the *gabP* in order to generate fusion constructs by the methods indicated schematically in Figure 11-10. Several C-terminal tagged derivatives of *gabP* have been prepared using cassettes encoding (1) Green Fluorescent Protein, (2) hexahistidine or -(His)$_6$, (3) β-galactosidase, and (4) alkaline phosphatase. The latter enzymes have great utility in the topological analysis of membrane proteins expressed in *E. coli*. Since *PacI*-mediated fusion involves engineering of the construct *in vitro*, the method provides an excellent complement to the transposon-mediated fusion of reporter enzymes to gabP. The latter method, being random in nature, will leave information gaps in the topological analysis that can be filled in with *PacI*-mediated fusion.

The cassette nature of *PacI*-mediated gene fusion affords considerable analytical flexibility. For example, using *PacI*-mediated gene fusion to attach the ETSQVAPA monoclonal antibody epitope[121] to the GabP C-terminus, it was possible to demonstrate a single Western blot band having a mobility equivalent to 38 kDa on SDS-PAGE, i.e., far more mobile than might be expected based on the calculated GabP molecular mass of 51.1 kDa.[95] On the other

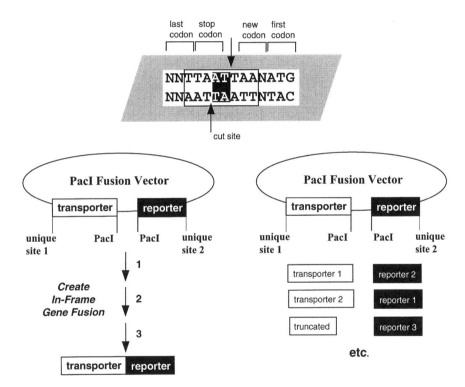

FIGURE 11-10 The *PacI*-mediated gene fusion method. The method requires that *PacI* sites be engineered (e.g., by PCR) appropriately into the end of the target gene and into the beginning of the reporter gene. *Top*. The procedure deletes two base pairs (black box), destroys the stop codon, and causes in-frame fusion between upstream and downstream genes with two "extra" amino acids in between. *Bottom Left*. The three required steps include (1) *PacI* restriction; (2) T4 polymerase "blunt-off"; (3) intramolecular blunt-end ligation. *Bottom Right*. The "cassette" nature of the system allows different combinations of reporter genes and transporters (point mutants, deletions, etc.) to be readily "swapped" in site-directed fashion. This inherent flexibility makes the method broadly applicable to many different projects. For example, several monoclonal antibody epitopes, a hexahistidine tag, green fluorescent protein, and topological reporter enzymes like alkaline phosphatase have been linked to the C-terminus for different applications such as topological analysis, protein purification, and immunological tagging.

hand, it is characteristic of transport proteins to exhibit anomalously high mobility on SDS-PAGE owing to high charge-to-mass ratios attributable to excess detergent binding to hydrophobic domains.[122]

G. CHARACTERIZATION OF GabP

1. Pharmacological Inhibition of GabP

There is relatively little information on the structural selectivity of the ligand recognition domain(s) in GabP. GABA is recognized with relatively high affinity (12 μM),[123] and of the 20 common α-amino acids, only threonine[123] and aspartate[95] are weakly competitive ($K_i > 1$ mM) with GABA. The above studies were fundamentally important, establishing that GabP is specific and able to reject common amino acids from the cellular environment. On the other hand, information on ligand structures rejected by the transporter are not informative as to the molecular recognition features built into the *E. coli* GabP, i.e., is the prokaryotic ligand binding site a good model for familiar GABAergic sites from the central nervous system? Information on this question is now surfacing in the literature.

The SK45 and SK55 strains have been used in the studies described below to show that *E. coli* GAB transporter recognizes a variety of GABAergic neurochemicals.

Despite the fact that *gabP* has diverged extensively from the eukaryotic GAT transporters (Figure 11-5C), ligands developed through study of the latter tend to be well recognized by the former, suggesting considerable three-dimensional conservation of the ligand binding domain. Two *E. coli* strains have been utilized to study the molecular recognition properties of *gab* permease (GabP).[96-98] These strains, SK55 and SK45, were derived from the gabP-negative SK35 strain (Figure 11-6) by transformation with either pSCK-472A (Figure 11-8) or the expression vector lacking an insert. Strain SK55 exhibits high-level, IPTG-inducible gabP expression, whereas SK45 does not transport [^{3}H]GABA at all.[98]

a. Open-Chain Amino Acids

The ability of a series of open-chain aminocarboxylic acids to interact with the cloned *E. coli* GABA transporter (GabP) has been investigated using the recently constructed SK55 strain.[98] Four of the analogs studied were found to be inhibitors of *gabP*-dependent [^{3}H]GABA transport. The series of active compounds ranged from three to six carbons in length. Within this series the C4 and C5 compounds were the most potent transport inhibitors (IC$_{50}$ = 10 μM). The C4 compound is, of course, the native substrate, GABA. The C5 compound is 5-aminovaleric acid, a GABA$_B$ receptor antagonist.[124,125] Overall, the data suggested that molecular size and/or separation between the amino and carboxyl groups might play important roles in molecular recognition since it was found that the C3 and C6 compounds were 10- to 20-fold less potent. Moreover, the data provide a basis for the suggestion that the *E. coli* GabP may have a binding site architecture that can mimic certain properties of GABA receptors as well as transporters from brain. Other data point to the same conclusion.

b. Conformationally Constrained Heterocycles

GABA is a flexible molecule in which all bonds may rotate freely. GABA may thus assume many independent structures that might be recognized by transporters and receptors. In contrast, cyclic GABA analogs may adopt far fewer structures since many bond rotations are necessarily restricted. Two conformationally restricted GABA analogs active in the central nervous system are also potent inhibitors of GabP. Nipecotic acid (3-piperidine carboxylic acid), ACHC (cis-3-aminocyclohexyl carboxylic acid), and guvacine (1,2,3,6-tetrahydro-3-pyridinecarboxylic acid) are able to inhibit uptake of [^{3}H]GABA by 50% at 50 µM.[97,98] This is about fourfold lower affinity than GABA itself — K_m = 12 µM.[95,123]

Nipecotic acid[126,127] and guvacine[128,129] are classical GABA transport inhibitors. ACHC is likewise a known inhibitor of GABA transport[130] and is thought to have specificity for transporters expressed by neurons.[23,34,131] Thus, without conserving the bulk of the primary amino acid sequence (Figure 11-5C), nature seems to have conserved in GabP the classical features of the GABA ligand binding domain found in GABA transporters from the nervous system — at least by pharmacological criteria. One interpretation of this pharmacological similarity is that the three dimensional architecture of the binding sites have been conserved.

c. **Planar Heterocycles**

The carbon skeleton of GABA, as well as the above-mentioned cyclic and open-chain transport inhibitors, exhibits geometry which is essentially tetrahedral at each carbon center. The puckered appearance shared in common by all of these molecules does not appear to be a necessary feature for recognition by GabP.[98] Highly planar heterocyclic compounds are also able to inhibit GABA transport. The planar molecule, 3-hydroxy-5-aminomethylisoxazole (muscimol) was recognized by GabP with an apparent affinity (IC_{50} = 10 µM) similar to the affinity of GABA itself. Two other planar compounds, THIP (4,5,6,7-tetrahydroisoxazolo[5,4-c]pyridin-3-ol) and THPO (4,5,6,7-tetrahydroisoxazololo[4,5-c]pyridin-3-ol) were less potent, having IC_{50} values of 200 and 2000 µM, respectively. Muscimol and THIP are well known as $GABA_A$ receptor agonists,[132-136] whereas THPO is known classically as a glial-selective GABA transport inhibitor.[28,128,136-139] The ability of muscimol, THIP, and THPO to inhibit [^{3}H]GABA transport indicates that neither extreme flexibility nor the carboxyl function are required for a molecule to be recognized by GabP.

d. **Comparison of Active Compounds**

Three categories of [^{3}H]GABA transport inhibitor seem to be active against the *E. coli* GabP: (1) the open-chain analogs of GABA, (2) the conformationally constrained, nonplanar heterocycles, and (3) the highly planar, unsaturated heterocycles. What have the actions of these well-known neurochemicals taught us about the molecular architecture of the GabP ligand recognition domain?

Carbon chain length — The zwitterionic GABA molecule displays as its most prominent feature a pair of charged groups separated by three carbon atoms. It seems reasonable *a priori* to assume that these groups interact with GabP (as well as with transporters and receptors from brain). Because GABA has a defined carbon chain length (four carbons), there is necessarily an upper limit on the possible separation of the carboxyl and amino functions when bound by the protein. If the GABA transporter was optimized to bind a maximally extended GABA conformer, then a shorter-chain analog might not be a very effective inhibitor. On the other hand, the inherent flexibility of long-chain analogs might allow them to adjust the charge separation distance appropriately — provided there was room within the binding site to accommodate ligand folding. The data indicate minimally that the GabP ligand binding domain has to incorporate features which (1) provide space to accommodate GABA as well as the larger five-carbon analog, 5-aminovaleric acid, and (2) provide a mechanism to exclude smaller short-chain analogs such as 3-aminopropanoate and the naturally occurring α-amino acids.

Constrained nonplanar heterocycles — An issue which could not be addressed using flexible, open-chain molecules is whether four- and five-carbon amino acid structures are likely to be recognized in their extended form, or in a more folded form in which the amino and carboxyl groups are closer to one another in space. Work with conformationally constrained heterocyclic molecules has provided some insight into this question. Cyclic GABA analogs such as nipecotic acid and ACHC have structures that (1) contain considerable bulk compared with open-chain structures, and (2) lack the rotational freedom inherent in open–chain molecules. These features place severe constraints on the ability of the amino and carboxyl groups to move in space relative to one another. Despite these differences from open-chain GABA analogs, both ACHC and nipecotic acid are recognized with apparent affinities (50 μM) that are comparable to the K_m for GABA (12 μM). It may be worth noting (Figure 11-11) that both GABA and 6-aminocaproate (6-ACA) are "incorporated" within the ACHC structure. Thus, the cyclic ACHC has the potential to be recognized as an analog of either (or both) of these open-chain structures. Similarly, the potent (IC_{50} = 10 μM) inhibitor 5-aminovaleric acid (5-AVA) and the less potent (IC_{50} = 100 μM) 3-aminopropanoic acid (3-APA) are "incorporated" into the structure of nipecotic acid (Figure 11-11). Thus, the cyclic nipecotic acid molecule has the potential to be recognized as an analog of either (or both) of the "incorporated" linear molecules. Judging from the potency of the individual open-chain transport inhibitors, it appears to be the case that the 6-ACA moiety within ACHC detracts from the intrinsic favorability of the interaction contributed by the GABA moiety. By similar reasoning, it would appear that the 3-APA and 5-AVA moieties contribute about equally to the potency of nipecotic acid.[96]

Constrained planar heterocycles — Work with nonplanar heterocycles — such as nipecotic acid and ACHC established that the GabP ligand recognition domain was somehow spacious enough or flexible enough to accommodate extra bulk beyond the minimal carbon chain required to mimic GABA (e.g., compare the ACHC structure to that of GABA and 6-ACA (Figure 11-11). Further probing of the active-site space with highly unsaturated GABAergic ligands (muscimol, THIP, and THPO) revealed that GabP has no absolute requirement for either a carboxyl group or for the corrugated structure of the saturated carbon backbone in GABA.

In summary, inhibition studies have provided cogent evidence that the ligand recognition domain of GabP should not be viewed as highly selective and incapable of recognizing substrates other than GABA. Instead, there appears to be sufficient flexibility and/or space within the GabP ligand binding domain to accommodate molecules smaller than GABA, bulkier than GABA, and molecules which are highly planar and lack a carboxyl group. On the other hand, an issue not addressed by inhibition studies is whether any of the recognized analogs are in fact transported substrates of GabP.

2. Substrate Specificity of GabP

Molecules that inhibit [^{3}H]GABA transport could do so via either competitive or allosteric mechanisms. Although some of the transport inhibitors bear close structural resemblance to GABA, others are sufficiently different (Figure 11-11) to warrant speculation that their actions are allosteric. However, if it could be shown that a given inhibitor was translocated by GabP, then this would certainly rank competition as the leading mechanism for inhibition of [^{3}H]GABA transport. Unfortunately, since many of the inhibitors described in the previous sections are unavailable in radiolabeled form it was impossible to demonstrate GabP-mediated uptake directly. Instead, methods had to be developed that permitted analysis of GabP-mediated transport without reliance upon direct uptake of unavailable radiolabeled compounds.[96,97]

a. GabP-Mediated Counterflow of [^{3}H]GABA

The applicability of the counterflow assay[140] to the nonradiolabeled substrate problem was validated with initial studies which relied on the few [^{3}H]GABA transport inhibitors that

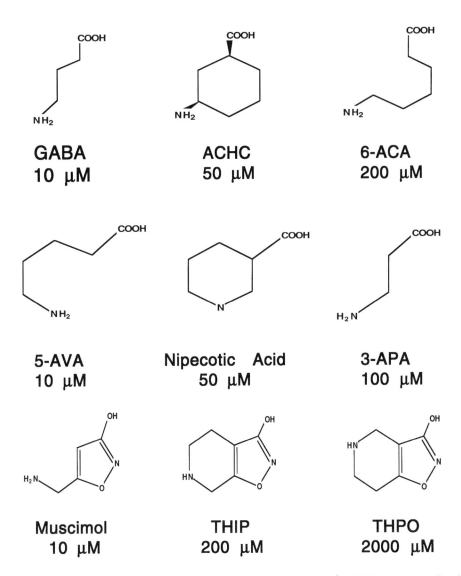

FIGURE 11-11 Structures and potency for inhibition of GabP-mediated [^{3}H]GABA transport. It will be noted that the structures of 6-ACA and GABA can be seen "imbedded" within the ACHC ring. Similarly, the structures of 3-APA and 5-AVA can be seen "imbedded" in the nipecotic acid ring. The IC$_{50}$ (defined as the drug concentration producing 50% inhibition) for inhibition of [^{3}H]GABA (10 μM) transport is given below the name of each compound.

were themselves available in tritiated form. For example, [^{3}H]muscimol is commercially available, and its uptake by GabP (Figure 11-12B) correlates with the ability of unlabeled muscimol to drive transient accumulation of [^{3}H]GABA in the counterflow assay (Figure 11-12C). Likewise, [^{3}H]nipecotic acid provided similar validation of the counterflow technique.[96] It is of interest to note that both muscimol and nipecotic acid appear to GABA transporter substrates in neurons and glia.[38,141-146]

Since no other [^{3}H]GABA uptake inhibitors could be obtained in labeled form, the counterflow assay provided not only the means to test GabP-mediated translocation of a many nonradiolabeled compounds (summarized in Figure 11-13), it also pointed to an interesting correlation between ligand structure and the ability to function as a well-recognized GabP substrate.[96,97] GABA is a highly flexible model that can adopt many conformations. Results

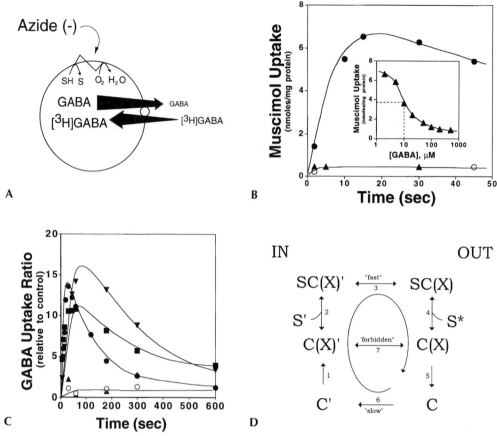

FIGURE 11-12 Basis and validation of the GabP counterflow assay. Counterflow provides a means by which to test whether a nonradiolabeled molecule can be translocated by GabP. *Panel A*. The scheme depicts a concentrated suspension of bacteria that are poisoned with azide and allowed to equilibrate with a high concentration of the test substrate (here GABA itself). Azide is included to prevent active substrate accumulation. A sample of the azide-poisoned cells is greatly diluted into medium containing a low concentration of [^{3}H]GABA. Thus, the test compound (unlabeled) undergoes net GabP-mediated transport out of the cell (down its concentration gradient). The [^{3}H]GABA becomes transiently concentrated within the cell, owing to a rapid exchange reaction catalyzed by GabP. *Panel B*. Direct demonstration that [^{3}H]muscimol is a GabP substrate. Uptake of [^{3}H]muscimol occurs in the IPTG-induced cell (●), but not in the control cell which was not induced with IPTG (○). The inset demonstrates that [^{3}H]muscimol uptake is inhibited by GABA. Similar results were obtained with [^{3}H]nipecotic acid and [^{3}H]GABA. *Panel C*. Demonstration that nonradiolabeled GABA (●) nipecotic acid (■), or muscimol (▼) undergo transient accumulation of [^{3}H]GABA. Transient accumulation does not occur if the test compound is omitted (○), or if the cell is GabP-negative (▲). *Panel D*. Mechanisms of efflux and exchange reactions used in counterflow. GabP catalyzes net efflux (IN → OUT) which requires cycling through all six steps of the transport cycle in numerical order. Exchange requires only steps 2 to 4 and is thus rapid compared to efflux. Rapid isotope equilibration by exchange, followed by slow net efflux, leads to the characteristic rise and fall of the counterflow time-course seen in Panel C. (Panels B and C from *J. Biol. Chem.*, 271, 783, 1996. With permission.)

to date suggest a general conclusion that the GabP prefers to transport substrates that mimic a nonextended conformation of GABA.

    b.   **Counterflow Related to ACHC and Nipecotic Acid**

Nipecotic acid and ACHC are cyclic compounds that inhibit [^{3}H]GABA transport (Section VII.1) and which incorporate into their rings the structure of certain open-chain GABA

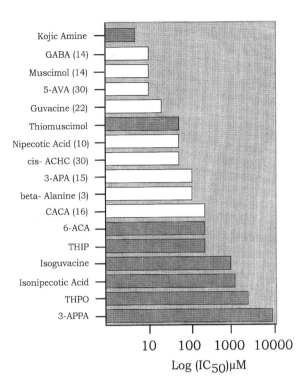

FIGURE 11-13 Summary of inhibitory potency and counterflow of GABAergic ligands tested for activity on *E. coli* GabP. The inhibitory potency against [^{3}H] GABA transport is indicated by the length of the bars. The compounds that were transported substrates (i.e., supported counterflow) are represented with white bars. The peak GABA uptake ratio (e.g., the peaks in Figure 11-12C) in counterflow is given in parentheses next to the compound name. Compounds which acted only as inhibitors of [^{3}H]GABA uptake are represented with dark bars.

analogs (Figure 11-11). Among these open-chain GABA uptake inhibitors, only 6-ACA is not transported by GabP in the counterflow assay. These and other data led to a general conclusion that compounds having poor affinity for the transporter cannot participate in the counterflow reaction (Figure 11-13). However, the converse of this statement was proved false by the actions of Kojic amine and thiomuscimol.

Kojic amine and thiomuscimol function as GABA receptor agonists in the mammalian nervous system.[132,147] These two compounds stood as obvious exceptions among the various compounds that were not GabP counterflow substrates (Figure 11-13, dark bars). The majority of nonsubstrates were also poor inhibitors (low affinity) of GabP-mediated [^{3}H]GABA transport. In contrast, Kojic amine and thiomuscimol were outstanding inhibitors (high affinity), but nevertheless behaved as nonsubstrates in the counterflow assay. What factors might be held to account for the "anomalous" behavior of these particular drugs? The actions of several simpler compounds on GabP provided additional insight that allows the actions of Kojic amine and thiomuscimol to be rationalized in the context of broader structure-activity relationships.

c. **Counterflow Related to TACA and CACA**

TACA (*trans*-4-aminocrotonic acid) mimics the extended conformation of GABA while CACA (*cis*-4-aminocrotonic acid) — active at GABA$_C$ receptors in the nervous system — mimics the nonextended conformation of GABA (Figure 11-14). TACA is neither an inhibitor of [^{3}H]GABA transport (no GabP transport inhibition at 10 m*M*) nor is it a counterflow substrate. In contrast, CACA is both an inhibitor of [^{3}H]GABA transport as well as an excellent counterflow substrate.[96] How can the differences between TACA and CACA be related to observations made with other inhibitors and counterflow substrates? First, it will be recalled (Section VII.A.4) that nipecotic acid and 3-APA appear to mimic nonextended conformations of GABA and both of these compounds are active against GabP (Figure 11-13). Second, model building confirms that the amine and carboxy functions in 5-AVA, 3-APA, nipecotic acid, and CACA can be made nearly isosteric. Perhaps not coincidentally, all of these compounds are well recognized by GabP. In contrast, TACA is not recognized by GabP, and model building shows that the amino and carboxyl groups of TACA cannot be made isosteric with those of the well-recognized compounds. Finally, it is perhaps not coincidental that the "inactive" TACA structure is apparent in both Kojic amine as well as thiomuscimol.

d. Counterflow Related to Guvacine and Isoguvacine

The "inactive" TACA structure is also seen to be "imbedded" within the structure of isoguvacine acid (Figure 11-14). Isoguvacine is quite a poor inhibitor of GabP-mediated [^{3}H]GABA transport (IC$_{50}$ = 1000 μM). Moreover, isoguvacine and its saturated analog, isonipecotic acid, are not counterflow substrates.[97] In contrast, guvacine, like its saturated analog nipecotic acid, is a good inhibitor (IC$_{50}$ = 50 μM) and an effective counterflow substrate. Thus, GabP prefers the analogs which resemble a nonextended conformation of GABA (modeled by CACA, nipecotic acid, and guvacine). Structures such as isoguvacine and isonipecotic acid which mimic an extended conformation of GABA, modeled by TACA, are poorly recognized by GabP.[97]

Thus, although it remains unclear precisely why Kojic amine and thiomusimol are good inhibitors but poor substrates, it is conceivable that this is related somehow to the TACA moiety imbedded within their structures (Figure 11-14). It is perhaps relevant to point out that the ligand binding site in active transporters must undergo a series of "processing" events that lead to changes in the chemical potential of the bound ligand during the course of a catalytic cycle.[148-150] Such processing is probably associated with changes in the conformation of the protein and its bound ligand. Relatively inflexible molecules such as Kojic amine and thiomuscimol may probe these structural alterations since the ligands have the capacity to bind GabP without displaying any capacity to undergo the additional processing events that would lead eventually to translocation across the membrane.

In summary, inhibition and counterflow studies have led to identification of nine GABAergic ligands that are able to interact productively with the *E. coli* GabP, and traverse the membrane under counterflow conditions (Figure 11-11; white bars). However, two novel GABAergic ligands (both receptor agonists) were found which could only cause transport inhibition and not counterflow. These planar molecules cannot be fully processed through the transport channel (Figure 11-15), whereas analogs that supported counterflow must traverse the lipid bilayer and are therefore imagined to interact with the core of the transport channel.

H. EPILOGUE

After more than 40 years of experience with GABA in the nervous system, it is now becoming clear that bacterial membrane proteins may be able to provide simple model systems that mimic the ligand binding properties of multiple GABAergic proteins from the nervous system. During the course of its catalytic cycle, the *Escherichia coli* GabP active site clearly displays conformations which can be probed pharmacologically with ligands having specificity for receptors and transporters from the central nervous system. The structural simplicity

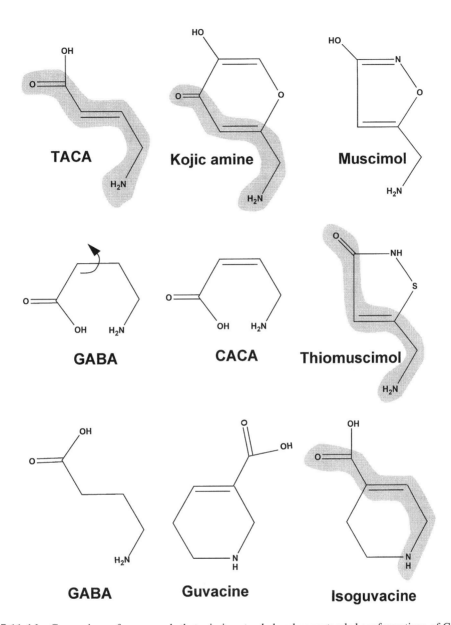

FIGURE 11-14 Comparison of compounds that mimic extended and nonextended conformations of GABA. TACA (*trans*-4-aminocrotonic acid) mimics the extended conformation of GABA. Neither TACA nor any of the compounds containing "imbedded" TACA moieties (shaded) function as substrates that can be translocated by GabP. In contrast, CACA (*cis*-4-aminocrotonic acid) mimics the nonextended conformation of GABA, and is a transported GabP substrate. Guvacine likewise mimics a nonextended GABA conformation, and is a transported substrate. Interestingly, Kojic amine and thiomuscimol — though not transported — are excellent inhibitors of [^{3}H]GABA transport (see Figure 11-13), a set of observations consistent with the idea that the "imbedded" TACA moiety may interfere with translocation across the membrane more than with binding per se. (From *J. Biol. Chem.*, 271, 783, 1996. With permission.)

of GABA may have left nature with few ways to construct GABAergic sites, thus enabling the relatively simple bacterial systems to provide reasonable approximations of GABA binding in the nervous system. Although the cloning of GabP is recent, and much fundamental characterization remains to be done, the inherent advantages of *E. coli* genetics — particularly in areas of mutant selection, topological analysis, and overexpression — bode well for

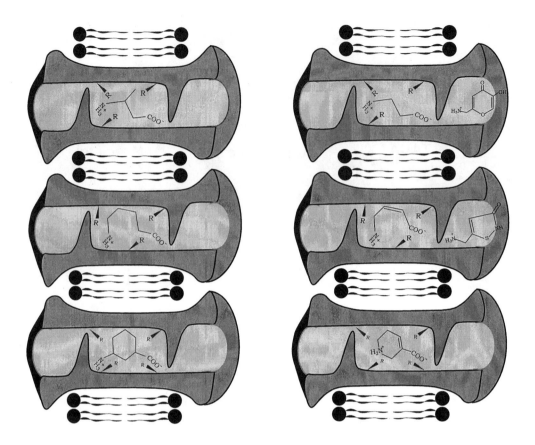

FIGURE 11-15 A model for *GabP* structure and function. GABA is a known substrate [95,123] which necessarily enters the permease core to occupy an "occluded" or "gated" space. Some mechanism of bidirectional gating is necessary for counterflow to occur. Accordingly, several molecules which behave as counterflow substrates are depicted in the core of the *GabP* transport channel, e.g., 3-aminobutyric acid (BABA), 5-AVA, ACHC, 3-APA, CACA, and guvacine. On the other hand, inhibitors that fail to support counterflow, e.g., Kojic amine and thiomuscimol, are depicted at the mouth of the transport channel for illustrative purposes and without any implication as to whether or not the permease can bind these molecules simultaneously with an occluded substrate. The possibility that any of these compounds might exert an inhibitory effect via some allosteric site distinct from the transport channel cannot be formally excluded. The amino acid side chains (R-groups) which line the transport channel or the channel mouth are not known. However, the ionic nature of the counterflow substrates strongly suggests that polar residues will be important. Regardless of detailed mechanism, this newly identified array of *GabP* ligands stands to enable new biochemical and/or genetic studies that could not previously have been contemplated.

continued rapid progress with the prokaryotic GABAergic system. Future synergy with the more fully characterized eukaryotic GABAergic models is anticipated.

ACKNOWLEDGMENTS

The author thanks Casey Brechtel who provided many of the original drawings. This work was supported in part by a U.S. Army Young Investigator Award DAAH04-94-G-0014 and by an award from the John Sealy Memorial Endowment Fund for Biomedical Research.

REFERENCES

1. Roberts, E. and Frankel, S., γ-Aminobutyric acid in brain, *Fed. Proc.*, 9, 219, 1950.
2. Hayashi, T. *Chemical Physiology of Excitation in Muscle and Nerve*, Nakayama-Shoten Ltd., Tokyo, 1956, Vol. 166.
3. Krnjevic, K. and Phillips, J. W., Iotophoretic studies of neurones in the mammalian cerebral cortex, *J. Physiol. (London)*, 165, 274, 1963.
4. Krnjevic, K. and Schwartz, S., The action of γ-aminobutyric acid on cortical neurones, *Exp. Brain Res.*, 3, 1967.
5. Lam, D. M. K. and Steinman, L., The uptake of [γ-^{3}H] aminobutyric acid in the goldfish retina, *Proc. Natl. Acad. Sci. U.S.A.*, 68, 2777, 1971.
6. Drujan, B. D. and Svaetichin, G., Characterization of different classes of isolated retinal cells, *Vision Res.*, 12, 1777, 1972.
7. Iversen, L. L. and Neal, M. J., The uptake of [^{3}H]GABA by slices of rat cerebral cortex, *J. Neurochem.*, 15, 1141, 1968.
8. Goodchild, M. and Neal, M. J., Uptake of ^{3}H-γ-aminobutyric acid (GABA) by rat retina, *J. Physiol. (London)*, 210, 1970.
9. Roberts, E., γ-Aminobutyric acid, *Neurochemistry*, K. A. Elliot, I. H. Page, and J. H. Quastel, Eds., Charles C Thomas, Springfield, IL, 1962, 636.
10. Roberts, E., Wein, J., and Simonsen, D. G., γ-Aminobutyric acid (γABA), vitamin B$_6$, and neuronal function — a speculative synthesis, *Vitam. Horm.*, 22, 503, 1964.
11. Kanner, B. I., Active transport of gamma-aminobutyric acid by membrane vesicles isolated from rat brain, *Biochemistry*, 17, 1207, 1978.
12. Kanner, B. I., Solubilisation and reconstitution of the gamma-aminobutyric acid transporter from rat brain, *FEBS Lett.*, 89, 47, 1978.
13. Radian, R., Bendahan, A., and Kanner, B. I., Purification and Identification of the functional sodium- and chloride-coupled γ-aminobutyric acid transport glycoprotein from rat brain, *J. Biol. Chem.*, 261, 15437, 1986.
14. Wood, J. D., Watson, W. J., and Ducker, A. J., The effect of hypoxia on brain γ-aminobutyric acid levels, *J. Neurochem.*, 15, 603, 1968.
15. Wood, J. D. and Watson, W. J., Gamma-aminobutyric acid levels in the brain of rats exposed to oxygen at high pressures, *Can. J. Biochem. Physiol.*, 41, 1907, 1963.
16. Roberts, E. and Kuriyama, K., Biochemical-physiological correlations in studies of the γ-aminobutyric acid system, *Brain Res.*, 8, 1, 1968.
17. Schwartz, E. A., Depolarization without calcium can release γ-aminobutyric acid from a retinal neuron, *Science*, 238, 350, 1987.
18. Brandon, C., Lam, D. M. K., Su, Y. Y. T., and Wu, J.-Y., Immunocytochemical localization of GABA neurons in the rabbit and frog retina, *Brain Res. Bull.*, 5 (Suppl. 2), 21, 1980.
19. Marc, R. E., Stell, W. K., Bok, D., and Lam, D. M. K., GABA-ergic pathways in the goldfish retina, *J. Comp. Neurol.*, 182, 221, 1978.
20. Lam, D. M. K., Fung, S.-C., and Kong, Y.-C., Postnatal development of GABA-ergic neurons in the rabbit retina, *J. Comp. Neurol.*, 193, 89, 1980.
21. Fung, S.-C., Kong, Y.-C., and Lam, D. M. K., Prenatal development of GABAergic, glycinergic, and dopaminergic neurons in the rabbit retina, *J. Neurosci.*, 2, 1623, 1982.
22. Hollyfield, J. G., Rayborn, M. E., Sarthy, P. V., and Lam, D. M. K., Retinal development: time and order of appearance of specific neuronal properties, *Neurochemistry*, 1, 93, 1980.
23. Guastella, J., Nelson, N., Nelson, H., Czyzyk, L., Keynan, S., Miedel, M. C., Davidson, N., Lester, H. A., and Kanner, B. I., Cloning and expression of a rat brain GABA transporter, *Science*, 249, 1303, 1990.
24. Borden, L. A., Smith, K. E., Hartig, P. R., Branchek, T. A., and Weinshank, R. L., Molecular heterogeneity of the gamma-aminobutyric acid (GABA) transport system. Cloning of two novel high affinity GABA transporters from rat brain, *J. Biol. Chem.*, 267, 21098, 1992.
25. Lam, D. M. K., Fei, J., Zhang, X.-Y., Tam, A. C. W., Zhu, L.-H., Huang, F., King, S. C., and Guo, L.-H., Molecular cloning and structure of the human (GABATHG) GABA transporter gene, *Mol. Brain Res.*, 19, 227, 1993.
26. Liu, Q. R., Lopez, C. B., Mandiyan, S., Nelson, H., and Nelson, N., Molecular characterization of four pharmacologically distinct γ-aminobutyric acid transporters in mouse brain, *J. Biol. Chem.*, 268, 2106, 1993.
27. Liu, Q.-R., Mandiyan, S., Nelson, H., and Nelson, N., A family of genes encoding neurotransmitter transporters, *Proc. Natl. Acad. Sci. U.S.A.*, 89, 6639, 1992.
28. King, S. C. and Lam, D. M.-K., Characterization of a cloned human high-affinity GABA transporter expressed in mouse fibroblasts: glucocorticoid-inducibility, coupling stoichiometry, and inhibition by NO-711, *Adv. Sci. Res.*, 1, 5, 1994.

29. Mabjeesh, N. J. and Kanner, B. I., Low-affinity gamma-aminobutyric acid transport in rat brain, *Biochemistry*, 28, 7694, 1989.
30. Kanner, B. I., Active transport of γ-aminobutyric acid by membrane vesicles isolated from rat brain, *Biochemistry*, 17, 1207, 1978.
31. Radian, R. and Kanner, B. I., Stoichiometry of sodium- and chloride-coupled gamma-aminobutyric acid transport by synaptic plasma membrane vesicles isolated from rat brain, *Biochemistry*, 22, 1236, 1983.
32. Kanner, B. I. and Kifer, L., Efflux of γ-aminobutyric acid by synaptic plasma membrane vesicles isolated from rat brain, *Biochemistry*, 20, 3354, 1981.
33. Keynan, S. and Kanner, B. I., γ-Aminobutyric acid transport in reconstituted preparation from rat brain: coupled sodium and chloride fluxes, *Biochemistry*, 27, 12, 1988.
34. Keynan, S., Suh, Y-J., Kanner, B. I., and Rudnick, G., Expression of a cloned γ-aminobutyric acid transporter in mammalian cells, *Biochemistry*, 31, 1974, 1992.
35. Mager, S., Naeve, J., Quick, M., Labarca, C., Davidson, N., and Lester, H. A., Steady states, charge movements, and rates for a cloned GABA transporter expressed in Xenopus oocytes, *Neuron*, 10, 177, 1993.
36. Kavanaugh, M. P., Arriza, J. L., North, R. A., and Amara, S. G., Electrogenic uptake of γ-aminobutyric acid by a cloned transporter expressed in *Xenopus* oocytes, *J. Biol. Chem.*, 267, 22007, 1992.
37. Pastuszko, A., Wilson, D. F., and Erecinska, M., Energetics of γ-aminobutyrate transport in rat brain synaptosomes, *J. Biol. Chem.*, 257, 7514, 1982.
38. Larsson, O. M., Drejer, J., Hertz, L., and Schousboe, A., Ion dependency of uptake and release of GABA and (RS)-nipecotic acid studied in cultured mouse brain cortex neurons, *J. Neurosci. Res.*, 9, 291, 1983.
39. Cammack, J. N., Rakhilin, S. V., and Schwartz, E. A., A GABA transporter operates asymmetrically and with variable stoichiometry, *Neuron*, 13, 949, 1994.
40. King, S. C. and Wilson, T. H., Toward understanding the structural basis of "forbidden" transport pathways in the *Escherichia coli* lactose carrier: mutations probing the energy barriers to uncoupled transport, *Mol. Microbiol.*, 4, 1433, 1990.
41. Mager, S., Min, C., Henry, D. J., Chavkin, C., Hoffman, B. J., Davidson, N., and Lester, H., Conducting states of a mammalian serotonin transporter, *Neuron*, 12, 845, 1994.
42. Fairman, W. A., Vandenberg, R. J., Arriza, J. L., Kavanaugh, M. P., and Amara, S. G., An excitatory amino-acid transporter with properties of a ligand-gated chloride channel, *Nature*, 375, 599, 1995.
43. Kanner, B. I., Structure and function of sodium-coupled neurotransmitter transporters, *Renal Physiol. Biochem.*, 17, 208, 1994.
44. Eyring, H., The activated complex in chemical reactions, *J. Chem. Phys.*, 3, 107, 1935.
45. Fersht, A. R., Leatherbarrow, R. J., and Wells, N. C., Binding energy and catalysis: a lesson from protein engineering of the tyrosyl-tRNA synthetase, *Trends Biochem. Sci.*, 11, 321, 1986.
46. Jencks, W. P., Binding energy, specificity, and enzymic catalysis: the Cirece effect, *Adv. Enzymol.*, 43, 219, 1975.
47. Jencks, W. P., The utilization of binding energy in coupled vectorial processes, *Adv. Enzymol.*, 51, 75, 1980.
48. Krupka, R. M., Role of substrate binding forces in exchange-only transport systems. I. Transition-state theory, *J. Membr. Biol.*, 109, 151, 1989.
49. Krupka, R. M., Role of substrate binding forces in exchange-only transport systems. II. Implications for the mechanism of the anion exchanger of red cells, *J. Membr. Biol.*, 109, 159, 1989.
50. Glowinski, J. and Axelrod, J., Inhibition of uptake of tritiated-noradrenaline in the intact rat brain by imipramine and structurally related compounds, *Nature*, 204, 1318, 1964.
51. Giros, B., Jaber, M., Jones, S. R., Wightman, R. M., and Caron, M. G., Hyperlocomotion and indifference to cocaine and amphetamine in mice lacking the dopamine transporter, *Nature*, 379, 606, 1996.
52. Hell, J. W., Edelmann, L., Hartinger, J., and Jahn, R., Functional reconstitution of the γ-aminobutyric acid transporter from synaptic vesicles using artificial ion gradients, *Biochemistry*, 30, 11795, 1991.
53. Hell, J. W., Maycox, P. R., and Jahn, R., Energy dependence and functional reconstitution of the γ-aminobutyric acid carrier from synaptic vesicles, *J. Biol. Chem.*, 265, 2111, 1990.
54. Thomas, R. A., Hell, J. W., During, M. J., Walch, S. C., Jahn, R., and De, C. P., A γ-aminobutyric acid transporter driven by a proton pump is present in synaptic-like microvesicles of pancreatic beta cells, *Proc. Natl. Acad. Sci. U.S.A.*, 90, 5317, 1993.
55. Schwartz, E. A., Calcium-independent release of GABA from isolated horizontal cells of the toad retina, *J. Physiol. (London)*, 323, 211, 1982.
56. Turner, T. J. and Goldin, S. M., Multiple components of synaptosomal [^{3}H]-γ-aminobutyric acid release resolved by a rapid superfusion system, *Biochemistry*, 28, 586, 1989.
57. Attwell, D., Barbour, B., and Szatkowski, M., Nonvesicular release of neurotransmitter, *Neuron*, 11, 401, 1993.
58. Zaleska, M. M. and Erecinska, M., Role of sialic acid in synaptosomal transport of amino acid transmitters, *Proc. Natl. Acad. Sci. U.S.A.*, 84, 1709, 1987.
59. Gomeza, J., Casado, M., Gimenez, C., and Aragon, C., Inhibition of high-affinity gamma-aminobutyric acid uptake in primary astrocyte cultures by phorbol esters and phospholipase C, *Biochem J.*, 275, 435, 1991.

60. Corey, J. L., Davidson, N., Lester, H. A., Brecha, N., and Quick, M. W., Protein kinase C modulates the activity of a cloned γ-aminobutyric acid transporter expressed in *Xenopus* oocytes via regulated subcellular redistribution of the transporter, *J. Biol. Chem.*, 269, 14759, 1994.
61. Gomeza, J., Gimenez, C., and Zafra, F., Cellular distribution and regulation by cAMP of the GABA transporter (GAT-1) mRNA, *Mol. Brain Res.*, 21, 150, 1994.
62. Dichter, M. A., Cellular mechanism of epilepsy and potential new treatment strategies, *Epilepsia*, 30, S3, 1989.
63. Twyman, R., Rogers, C., and MacDonald, R., Differential regulation of $GABA_A$ receptor channels by diazepam and phenobarbital, *Ann. Neurol.*, 25, 213, 1989.
64. Meldrum, B. S. and Chapman, A. G., Benzodiazepine receptors and their relationship to the treatment of epilepsy, *Epilepsia*, 27, S3, 1986.
65. Godin, Y., Heiner, L., Mark, J., and Mandel, P., Effects of di-n-propylacetate and anticonvulsive compounds on GABA metabolism, *J. Neurochem.*, 16, 869, 1969.
66. MacDonald, R. L. and Bergey, G. K., Valproic acid augments GABA-mediated postsynaptic inhibition in cultured mammalian neurons, *Brain Res.*, 170, 558, 1979.
67. Gale, K., GABA in epilepsy: the pharmacological basis, *Epiliepsia*, 30 (Suppl. 3), S1, 1989.
68. Treiman, D. M., γ-Vinyl GABA: current role in the management of drug-resistant epilepsy, *Epilepsia*, 30, S31, 1989.
69. Krogsgaard, L. P., Johnston, G. A., Lodge, D., and Curtis, D. R., A new class of GABA agonist, *Nature*, 268, 53, 1977.
70. Richens, A., Chadwick, D., Duncan, J., Dam, M., Morrow, J., Gram, L., Mengel, H., Shu, V., Pierce, M., Rask, C., and Hightower, B., Safety and efficacy of Tiagabine HCl as adjunctive treatment for complex partial seizures, *Epilepsia*, 33 (Suppl. 3), 119, 1992.
71. Rowan, A. J., Ahmann, P., Wannamaker, B., Schacter, S., Rask, C., and Uthman, B., Safety and efficacy of three dose levels of Tiagabine HCL versus placebo as adjuctive treatment for complex partial seizures, *Epilepsia*, 34 (Suppl. 2), 157, 1993.
72. Mengel, H. B., Houston, A., and Back, D. J., Tiagabine: evaluation of the risk of interaction with oral contraceptive pills in female volunteers, *Epilepsia*, 34 (Suppl. 2), 157, 1993.
73. Sedman, A. J., Gilment, G. P., Sayed, A. J., and Posvar, E. L., Initial human safety and tolerance study of a GABA uptake inhibitor, CI-966, *Drug Dev. Res.*, 21, 235, 1990.
74. Snel, S., Mukherjee, S., Richens, A., and Mengel, H. B., Pharmacokinetics of Tiagabine in the elderly, *Epilepsis*, 34 (Suppl. 2), 157, 1993.
75. White, S. H., Hunt, J., Wolf, H. H., Swinyard, E. A., Falch, E., Krogsgaard-Larsen, P., and Schousboe, A., Anticonvulsant activity of the γ-aminobutyric acid uptake inhibitor N-4,4-diphenyl-3-butenyl-4,5,6,7-tetrahydroisoxazolo[4,5-c]pyridin-3-ol, *Eur. J. Pharmacol.*, 236, 147, 1993.
76. Borden, L. A., Dhar, T. G. M., Smith, K. E., Weinshank, R. L., Branchek, T. A., and Gluchowski, C., Tiagabine, SK&F 89976-A, CI-966, and NNC-711 are selective for the cloned GABA transporter GAT-1, *Eur. J. Pharmacol.*, 269, 219, 1994.
77. Andersen, K. E., Braestrup, C., Gronwald, F. C., Jorgensen, A. S., Nielsen, E. B., Sonnewald, U., Sorensen, P. O., Suzdak, P. D., and Knutsen, L. J., The synthesis of novel GABA uptake inhibitors. I. Elucidation of the structure-activity studies leading to the choice of (R)-1-ç4,4-bis(3-methyl-2-thienyl)-3-butenyl:-3-piperidinecarboxylic acid (tiagabine) as an anticonvulsant drug candidate, *J. Med. Chem.*, 36, 1716, 1993.
78. Kovacs, I., Maksay, G., and Simonyi, M., Inhibition of high-affinity synaptosomal uptake of γ-aminobutyric acid by a bicyclo-heptane derivative, *Drug Res.*, 39, 295, 1989.
79. International League Against Epilepsy, Proposal for revised classification of epilepsies and epileptic syndromes, *Epilepsia*, 30, 389, 1989.
80. Sarthy, V., γ-Aminobutyric acid (GABA) uptake by *Xenopus* oocytes injected with rat brain mRNA, *Mol. Brain Res.*, 1, 97, 1986.
81. Blakely, R. D., Robinson, M. B., and Amara, S. G., Expression of neurotransmitter transport from rat brain mRNA in *Xenopus laevis* oocytes, *Proc. Natl. Acad. Sci. U.S.A.*, 85, 9846, 1988.
82. Chang, A. S.-C., Frnka, J. V., Chen, D., and Lam, D. M.-K., Characterization of a genetically reconstituted high-affinity system for serotonin transport, *Proc. Natl. Acad. Sci. U.S.A.*, 86, 9611, 1989.
83. Kanner, B. I., Keynan, S., and Radian, R., Structural and functional studies on the sodium- and chloride-coupled γ-aminobutyric acid transporter: deglycosylation and limited proteolysis, *Biochemistry*, 28, 3722, 1989.
84. Mabjeesh, N. J., Frese, M., Rauen, T., Jeserich, G., and Kanner, B. I., Neuronal and glial gamma-aminobutyric acid + transporters are distinct proteins, *FEBS Lett.*, 299, 99, 1992.
85. Schousboe, A., Thorbek, P., Hertz, L., and Krogsgaard, L. P., Effects of GABA analogues of restricted conformation on GABA transport in astrocytes and brain cortex slices and on GABA receptor binding, *J. Neurochem.*, 33, 181, 1979.
86. Krogsgaard-Larsen, P., Falch, E., Larsson, O. M., and Schousboe, A., GABA uptake inhibitors: relevance to antiepileptic drug research, *Epilepsy Res.*, 1, 77, 1987.

87. Uhl, G. R., Neurotransmitter transporters (plus): a promising new gene family, *Trends Neurosci.*, 15, 265, 1992.
88. Clark, J. A. and Amara, S. G., Amino acid neurotransmitter transporters: structure, function, and molecular diversity, *Bioessays*, 15, 323, 1993.
89. Nelson, N. and Lill, H., Porters and neurotransmitter transporters, *J. Exp. Biol.*, 196, 213, 1994.
90. Bouvier, M., Szatkowski, M., Amato, A., and Attwell, D., The glial cell glutamate uptake carrier countertransports pH-changing anions, *Nature*, 360, 471, 1992.
91. Kanner, B. I., Glutamate transporters from brain: a novel neurotransmitter transporter family, *FEBS Lett.*, 325, 95, 1993.
92. Storck, T., Schulte, S., Hofmann, K., and Stoffel, W., Structure, expression, and functional analysis of a Na^+-dependent glutamate/aspartate transporter from rat brain, *Proc. Natl. Acad. Sci. U.S.A.*, 89, 10955, 1992.
93. Kanai, Y. and Hediger, M. A., Primary structure and functional characterization of a high-affinity glutamate transporter, *Nature*, 360, 467, 1992.
94. Metzer, E. and Halpern, Y. S., In vivo cloning and characterization of the gabCTDP gene cluster of *Escherichia coli* K-12, *J. Bacteriol.*, 172, 3250, 1990.
95. Niegemann, E., Schulz, A., and Bartsch, K., Molecular organization of the *Escherichia coli* gab cluster: nucleotide sequence of the structural genes *gabD* and *gabP* and expression of the GABA permease gene, *Arch. Microbiol.*, 160, 454, 1993.
96. Brechtel, C. E., Hu, L., and King, S. C., Substrate specificity of the *Escherichia coli* 4-aminobutyrate carrier encoded by *gabP*. Uptake and counterflow of structurally diverse molecules, *J. Biol. Chem.*, 271, 783, 1996.
97. King, S. C., Fleming, S. R., and Brechtel, C., Pyridine carboxylic acids as inhibitors and substrates of the *Escherichia coli gab* permease encoded by *gabP*, *J. Bacteriol.*, 177, 5381, 1995.
98. King, S. C., Fleming, S. R., and Brechtel, C., Ligand recognition properties of the *Escherichia coli* 4-aminobutyrate transporter encoded by *gabP*: specificity of *Gab* permease for heterocyclic inhibitors, *J. Biol. Chem.*, 270, 19893, 1995.
99. Huang, A. M., Lee, J. I., King, S. C., and Wilson, T. H., Amino acid substitution in the lactose carrier protein with the use of amber suppressors, *J. Bacteriol*, 174, 5436, 1992.
100. Normanly, J., Masson, J. M., Kleina, L. G., Abelson, J., and Miller, J. H., Construction of two *Escherichia coli* amber suppressor genes: $tRNA^{Phe}_{CUA}$ and $tRNA^{Cys}_{CUA}$, *Proc. Natl. Acad. Sci. U.S.A.*, 83, 6548, 1986.
101. Kleina, L. G., Masson, J. M., Normanly, J., Abelson, J., and Miller, J. H., Construction of *Escherichia coli* amber suppressor tRNA genes. II. Synthesis of additional tRNA genes and improvement of suppressor efficiency, *J. Mol. Biol.*, 213, 705, 1990.
102. Kleina, L. G. and Miller, J. H., Genetic studies of the *lac* repressor. XIII. Extensive amino acid replacements generated by the use of natural and synthetic nonsense suppressors, *J. Mol. Biol.*, 212, 295, 1990.
103. Normanly, J., Kleina, L. G., Masson, J. M., Abelson, J., and Miller, J. H., Construction of *Escherichia coli* amber suppressor tRNA genes. III. Determination of tRNA specificity, *J. Mol. Biol.*, 213, 719, 1990.
104. Franco, P. J. and Brooker, R. J., Evidence that the asparagine 322 mutant of the lactose permease transports protons and lactose with a normal stoichiometry and accumulates lactose against a concentration gradient, *J. Biol. Chem.*, 266, 6693, 1991.
105. Franco, P. J. and Brooker, R. J., Functional roles of Glu-269 and Glu-325 within the lactose permease of *Escherichia coli*, *J. Biol. Chem.*, 269, 7379, 1994.
106. Carrasco, N., Antes, L. M., Poonian, M. S., and Kaback, H. R., *Lac* permease of *Escherichia coli*: histidine-322 and glutamic acid-325 may be components of a charge-relay system, *Biochemistry*, 25, 4486, 1986.
107. King, S. C. and Wilson, T. H., Galactoside-dependent proton transport by mutants of the *Escherichia coli* lactose carrier. Replacement of histidine 322 by tyrosine or phenylalanine, *J. Biol. Chem.*, 264, 7390, 1989.
108. King, S. C. and Wilson, T. H., Sensitivity of efflux-driven carrier turnover to external pH in mutants of the *Escherichia coli* lactose carrier that have tyrosine or phenylalanine substituted for histidine-322. A comparison of lactose and melibiose, *J. Biol. Chem.*, 265, 3153, 1990.
109. Poolman, B., Modderman, R., and Reizer, J., Lactose transport system of *Streptococcus thermophilus*. The role of histidine residues, *J. Biol. Chem.*, 267, 9150, 1992.
110. Poolman, B., Knol, J., and Lolkema, J. S., Kinetic analysis of lactose and proton coupling in Glu379 mutants of the lactose transport protein of *Streptococcus thermophilus*, *J. Biol. Chem.*, 270, 12995, 1995.
111. Mabjeesh, N. J. and Kanner, B. I., Neither amino nor carboxyl termini are required for function of the sodium- and chloride-coupled gamma-aminobutyric acid transporter from rat brain, *J. Biol. Chem.*, 267, 2563, 1992.
112. Bendahan, A. and Kanner, B. I., Identification of domains of a cloned rat brain GABA transporter which are not required for its functional expression, *FEBS Lett.*, 318, 41, 1993.
113. Mabjeesh, N. J. and Kanner, B. I., The substrates of a sodium- and chloride-coupled γ-aminobutyric acid transporter protect multiple sites throughout the protein against proteolytic cleavage, *Biochemistry*, 32, 8540, 1993.
114. Jennings, M. L., Topography of membrane proteins, *Annu. Rev. Biochem.*, 58, 999, 1989.

115. Calamia, J. and Manoil, C., *lac* Permease of *Escherichia coli*: topology and sequence elements promoting membrane insertion, *Proc. Natl. Acad. Sci. U.S.A.*, 87, 4937, 1990.
116. Boyd, D., Manoil, C., and Beckwith, J., Determinants of membrane protein topology, *Proc. Natl. Acad. Sci. U.S.A.*, 84, 8525, 1987.
117. Botfield, M. C., Naguchi, K., Tsuchia, T., and Wilson, T. H., Membrane topology of the melibiose carrier of *Escherichia coli*, *J. Biol. Chem.*, 267, 1818, 1992.
118. Calamia, J. and Manoil, C., Membrane protein spanning segment as export signals, *J. Mol. Biol.*, 224, 539, 1992.
119. Seligman, L. and Manoil, C., An amphipathic sequence determinant of membrane protein topology, *J. Biol. Chem.*, 269, 19888, 1994.
120. Manoil, C., Analysis of protein localization by use of gene fusions with complementary properties, *J. Bacteriol.*, 172, 1035, 1990.
121. Suzuki, H., Prado, G. N., Wilkinson, N., and Navarro, J., The N terminus of interleukin-8 (IL-8) receptor confers high affinity binding to human IL-8, *J. Biol. Chem.*, 269, 18263, 1994.
122. Newman, M. J., Foster, D. L., Wilson, T. H., and Kaback, H. R., Purification and reconstitution of functional lactose carrier from *Escherichia coli*, *J. Biol. Chem.*, 256, 11804, 1981.
123. Kahane, S., Levitz, R., and Halpern, Y. S., Specificity and regulation of γ-aminobutyrate transport in *Escherichia coli*, *J. Bacteriol.*, 135, 295, 1978.
124. Schwarz, M., Klockgether, T., Wullner, U., Turski, L., and Sontag, K. H., δ-Aminovaleric acid antagonizes the pharmacological actions of baclofen in the central nervous system, *Exp. Brain Res.*, 70, 618, 1988.
125. Early, S. L., Michaelis, E. K., and Mertes, M. P., Pharmacological specificity of synaptosomal and synaptic membrane gamma-aminobutyric acid (GABA) transport processes, *Biochem. Pharmacol.*, 30, 1105, 1981.
126. Johnston, G. A., Krogsgaard, L. P., Stephanson, A. L., and Twitchin, B., Inhibition of the uptake of GABA and related amino acids in rat brain slices by the optical isomers of nipecotic acid, *J. Neurochem.*, 26, 1029, 1976.
127. Krogsgaard, L. P. and Johnston, G. A., Inhibition of GABA uptake in rat brain slices by nipecotic acid, various isoxazoles and related compounds, *J. Neurochem.*, 25, 797, 1975.
128. Krogsgaard, L. P., Inhibitors of the GABA uptake systems, *Mol. Cell. Biochem.*, 31, 105, 1980.
129. Johnston, G. A., Krogsgaard, L. P., and Stephanson, A., Betel nut constituents as inhibitors of gamma-aminobutyric acid uptake, *Nature*, 258, 627, 1975.
130. Hitzemann, R. J. and Loh, H. H., Effects of some conformationally restricted GABA analogues on GABA membrane binding and nerve ending transport, *Brain Res.*, 144, 63, 1978.
131. Kanner, B. I. and Bendahan, A., Two pharmacologically distinct sodium- and chloride-coupled high-affinity gamma-aminobutyric acid transporters are present in plasma membrane vesicles and reconstituted preparations from rat brain, *Proc. Natl. Acad. Sci. U.S.A.*, 87, 2550, 1990.
132. Krogsgaard-Larsen, P., Hjeds, H., Curtis, D. R., Lodge, D., and Johnston, G. A., Dihydromuscimol, thiomuscimol and related heterocyclic compounds as GABA analogues, *J. Neurochem.*, 32, 1717, 1979.
133. Krogsgaard-Larsen, P. and Johnston, G. A., Structure-activity studies on the inhibition of GABA binding to rat brain membranes by muscimol and related compounds, *J. Neurochem.*, 30, 1377, 1978.
134. Arnt, J. and Krogsgaard-Larsen, P., GABA agonists and potential antagonists related to muscimol, *Brain Res.*, 177, 395, 1979.
135. Arnt, J., Scheel-Kruger, J., Magelund, G., and Krogsgaard-Larsen, P., Muscimol and related GABA receptor agonists: the potency of GABAergic drugs in vivo determined after intranigral injection, *J. Pharm. Pharmacol.*, 31, 306, 1979.
136. Krogsgaard, L. P., γ-Aminobutyric acid agonists, antagonists, and uptake inhibitors. Design and therapeutic aspects, *J. Med. Chem.*, 24, 1377, 1981.
137. Gonsalves, S. F., Twitchell, B., Harbaugh, R. E., Krogsgaard, L. P., and Schousboe, A., Anticonvulsant activity of intracerebroventricularly administered glial GABA uptake inhibitors and other GABAmimetics in chemical seizure models, *Epilepsy Res.*, 4, 34, 1989.
138. Seiler, N., Sarhan, S., Krogsgaard-Larsen, P., Hjeds, H., and Schousboe, A., Amplification by glycine of the anticonvulsant effect of THPO, a GABA uptake inhibitor, *Gen. Pharm.*, 16, 509, 1985.
139. Schousboe, A., Larsson, O. M., Wood, J. D., and Krogsgaard, L. P., Transport and metabolism of gamma-aminobutyric acid in neurons and glia: implications for epilepsy, *Epilepsia*, 24, 531, 1983.
140. Rosenberg, T. and Wilbrandt, W., Uphill transport induced by counterflow, *J. Gen. Physiol.*, 41, 289, 1957.
141. Yazulla, S. and Brecha, N., Binding and uptake of the GABA analogue, 3H-muscimol, in the retinas of goldfish and chicken, *Invest. Ophthalmol. Vis. Sci.*, 19, 1415, 1980.
142. Johnston, G. A., Kennedy, S. M., and Lodge, D., Muscimol uptake, release and binding in rat brain slices, *J. Neurochem.*, 31, 1519, 1978.
143. Lodge, D., Curtis, D. R., and Johnston, G. A., Does uptake limit the actions of GABA agonists in vivo? Experiments with muscimol, isoguvacine and THIP in cat spinal cord, *J. Neurochem.*, 31, 1525, 1978.
144. Johnston, G. A., Stephanson, A. L., and Twitchin, B., Uptake and release of nipecotic acid by rat brain slices, *J. Neurochem.*, 26, 83, 1976.

145. Larsson, O. M., Krogsgaard, L. P., and Schousboe, A., High-affinity uptake of (RS)-nipecotic acid in astrocytes cultured from mouse brain. Comparison with GABA transport, *J. Neurochem.*, 34, 970, 1980.
146. Larsson, O. M. and Schousboe, A., Comparison between (RS)-nipecotic acid and GABA transport in cultured astrocytes: coupling with two sodium ions, *Neurochem. Res.*, 6, 257, 1981.
147. Kendall, D. A., Browner, M., and Enna, S. J., Comparison of the antinociceptive effect of γ-aminobutyric acid (GABA) agonists: evidence for a cholinergic involvement, *J. Pharmacol. Exp. Ther.*, 220, 482, 1982.
148. Tanford, C., Chemical potential of bound ligand, an important parameter for free energy transduction, *Proc. Natl. Acad. Sci. U.S.A.*, 78, 270, 1981.
149. Tanford, C., Mechanism of free energy coupling in active transport, *Annu. Rev. Biochem.*, 52, 379, 1983.
150. Tanford, C., Mechanism for the interconversion of chemical and osmotic free energy, *Structure and Function of Sarcoplasmic Reticulum,* S. Fleischer and Y. Tonomura, Ed., Academic Press, Orlando, FL, 1985, 259.
151. Krogsgaard-Larsen, P., GABA synaptic mechanisms: stereochemical and conformational requirements, *Med. Res. Rev.*, 8, 27, 1988.
152. Higgins, D. G., Bleasby, A. J., and Fuchs, R., CLUSTAL V: improved software for multiple sequence alignment, *Comput. App. Biosci.*, 8, 189, 1992.
153. Frohlich, K. U., Sequence similarity presenter: a tool for the graphic display of similarities of long sequences for use in presentations, *Comput. Appl. Biosci.*, 10, 179, 1994.
154. Zhang, J. T. and Ling, V., Membrane orientation of transmembrane segments 11 and 12 of MDR and non-MDR-associated P-glycoproteins, *Biochim. Biophys. Acta*, 1994.
155. Zhang, J.-T., Duthie, M., and Ling, V., Membrane topology of the N-terminal half of the hamster P-glycoprotein molecule, *J. Biol. Chem.*, 268, 15101, 1993.
156. Zhang, J. T. and Ling, V., Study of membrane orientation and glycosylated extracellular loops of mouse P-glycoprotein by in vitro translation, *J. Biol. Chem.*, 266, 18224, 1991.

Section IV

Psychiatric Genetics

12 Identification of Phenotypes for Molecular Genetic Studies of Common Childhood Psychopathology

James J. Hudziak

CONTENTS

A. Introduction 201
B. Protocols and Methods for Identification 202
    1. "Genetic Nosologic Approach" to Identifying Phenotypes for Molecular
       Genetic Studies of Childhood Psychopathology 202
       a. Developmental Sensitivity 203
       b. Gender Sensitivity 204
       c. Multi-Informant Sensitivity 205
    2. Importance of Accuracy Once a Phenotype is Selected 206
    3. DSM and Empirically Based Taxonomy as Phenotypic Indicators 206
    4. Establishing Spectrum Phenotypes Combining DSM and Empirically
       Based Approaches 208
    5. Genetics of ADHD, CD, AP, and AGG 209
    6. Genetics of the AP Syndrome of the Empirically Based Taxonomy 210
    7. Genetics of ODD and CD 210
    8. Genetics of the AGG Syndrome of the Empirically Based Taxonomy 211
    9. Genetic Studies Combining the DSM and Quantitative Approaches 211
C. New Molecular and Statistical Genetic Methods and Techniques 211
D. Summary 213
Acknowledgment 213
References 213

A. INTRODUCTION

There is tremendous heuristic value to identifying disease genes (or trait genes) for the common child psychopathologic conditions. Such conditions as Attention-Deficit-Hyperactivity Disorder, Oppositional Defiant Disorder, Conduct Disorder, and the childhood manifestations of pathologic anxiety and depression are extremely prevalent and cause extraordinary suffering. The long-term benefits of identifying genes that predict vulnerability, and subsequently developing early identification and intervention strategies, are obvious. Because of these benefits, there has been heightened interest in identifying genes that are involved in the etiology of the common child psychiatric disorders. Twin, family, and adoptee studies have demonstrated strong genetic components to childhood disorders. Molecular genetic studies have yet to identify the specific genes, but some promising results have been forthcoming. Prior to addressing these results, a major obstacle facing child psychiatric molecular geneticists — the lack of a valid phenotypic taxonomy for genetic studies — will be discussed.

The molecular geneticist needs two tools to discover genes for psychiatric disorders. The first, an advanced molecular genetic technology, has been achieved. The second, a valid, specific, and accurate taxonomy to identify phenotypes for molecular genetic studies, has not.

This discussion will review the need for a more discriminating taxonomic framework for childhood psychiatric conditions. This most basic need, a taxonomy that will enable us to distinguish between children who do vs. those who do not have the disorders of interest, must be met in order to identify disease genes.

After discussing the need for a more discriminating taxonomy, findings from twin, family, adoption, and molecular genetic studies of Attention Deficit Hyperactivity Disorder (ADHD) and Oppositional Defiant Disorder/Conduct Disorder (ODD/CD) identified via the DSM taxonomic approach will be reviewed.[1] What is known about the corresponding taxonomic constructs Attention Problems (AP), and Aggressive Behavior (AGG) will be identified via the empirically based taxonomy.[2] The lessons we can learn from findings on Attention-Deficit-Hyperactivity Disorder (ADHD)/Conduct Disorder (CD) and Attention Problems (AP)/Aggressivity (AGG) can be applied to the study of other common childhood disorders.

The empirically based taxonomy comprises factor analytically derived syndromes that describe children's problems in quantitative terms. The DSM-IV taxonomy describes children's problems in categorical terms. They represent the two main paradigms used to conceptualize child psychopathology. The two systems will be compared and contrasted regarding their utility in genetic studies. Recent results from association studies as well as from new molecular genetic techniques such as Quantitative Trait Locus approaches (QTL) will be presented.

B. PROTOCOLS AND METHODS FOR IDENTIFICATION

1. "Genetic Nosologic Approach" to Identifying Phenotypes for Molecular Genetic Studies of Childhood Psychopathology

There are two requirements for identifying disease genes in medicine. They are advanced molecular genetic techniques with enough sophistication to identify disease genes (whether they be structurally different or functionally mutated) and a taxonomy that is specific enough to validly discriminate between genetic and nongenetic manifestations of the trait (disease) to be studied.

Molecular genetic techniques have advanced to a point where subjects can be genotyped, cloned, and in many cases, function quantified. However nosologic approaches to identifying subjects for molecular genetic studies of childhood psychopathology have not achieved the same degree of sophistication. Simply stated, an investigation cannot identify a relation between a gene and a trait if the trait is not accurately phenotyped.

Investigators face several obstacles when attempting to elucidate the mode of genetic transmission of any disorder. As Rice et al.[3,4] state "the first step in a family-genetic study is to define an appropriate phenotype (the observed trait to be studied), and to delineate the covariates that mediate the risk of illness." They suggest that etiologic heterogeneity, resulting from multiple pathways leading to an indistinguishable clinical disorder, is one of the main obstacles to the genetic classification and understanding of the transmission of common illnesses. ADHD and CD may be examples of such common illnesses in which several genetic and nongenetic disorders share a common clinical phenotype. Other common childhood conditions, such as ODD and pathological manifestations of anxiety and affect, are also examples of common disorders that may suffer from misclassification due to etiologic heterogeneity. As Tsuang[5] points out, "Statistical procedures and molecular genetic techniques have attained a fine degree of resolution. Their ability to find disease genes has revolutionized medicine and raised hopes for breakthroughs in psychiatry. However, such breakthroughs may require an equally discriminating psychiatric nosology — a nomenclature that can more validly discriminate genetic and non-genetic subtypes of illness (and genetic subtypes as well)." We are urged to develop strategies for constructing a "genetic nosology", one that seeks to classify patients into categories corresponding to distinct genetic entities rather than relying on diagnostic constructs created for other purposes.[6] Because all psychiatric genetic studies begin with diagnoses, the success of such studies depends on the diagnostic criteria used to identify probands.

Tsuang et al.[6] offer six specific guidelines for creating phenotypic identification strategies for genetic studies of psychopathology. The phenotypic indicator should have the following qualities:

1. **Specificity:** the indicator is more strongly associated with the diseases of interest than with other psychiatric conditions.
2. **State independence:** the indicator is stable over time and not an epiphenomenon of the illness or its treatment.
3. **Heritability:** the indicator shows familial transmission.
4. **Familial association:** the indicator is more prevalent among relatives of ill probands than among relatives of appropriate control subjects.
5. **Cosegregation:** the indicator is more prevalent among the ill relatives of the ill probands than among well relatives of the ill probands.
6. **Biological plausibility:** the indicator bears some conceptual relationship to the disease.[6]

These guidelines have been applied to a number of psychiatric conditions in order to elucidate the genetic etiology of complex psychiatric disorders. They have provided a base for defining "phenotypes" that may reflect genetic liability in nonaffected relatives of affected probands. Many such studies have been done in schizophrenia and have led to some promising results.[7]

The **Tsuang guidelines** apply to phenotyping strategies in all areas of psychopathology. There are, however, some additional requirements for phenotypic indicators that are especially important in the study of child psychopathology. These are developmental sensitivity, gender sensitivity, and multi-informant sensitivity. Each is discussed below. The phenotypic indicator should have the following qualities:

a. **Developmental Sensitivity**

One problem that affects the identification of phenotypes in childhood psychopathology is that phenotypes may change over time. In adult psychiatric genetics, the same criteria are usually applied to adults of all ages. This is based on the assumption that the phenotype of interest, for example schizophrenia, is stable in the adult age group. Even for adult psychopathology, this assumption may not always be valid.

In child psychopathology, however, it is clear that the assumption of a single stable phenotype is absurd. It is clear from work by Piaget, Erickson, Freud, Achenbach[8-11] and others that children progress through different periods of development. Although these authors framed development differently (cognitive periods, psychosocial conflicts, psychosexual phases), all recognized the need for models of normal development. As Achenbach states, we need to "consider the developmental tasks, problems, and competence is marking successive developmental periods."[12] Simply stated, child psychiatric geneticists need to understand normal development in order to design developmentally appropriate phenotypic measures of psychopathology. For example, it may not be appropriate to apply the DSM-IV criteria for ADHD (e.g., often fails to give close attention to details or makes careless mistakes in school work, work, or activities; often does not seem to listen when spoken to directly, etc.) to a 2-year-old.

Irrespective of the developmental framework, there is general agreement that children should be considered as progressing through a number of developmental stages. These stages can be divided into the age groups 0–2, 2–5, 6–11, and 12–20. Each of these age groups has a corresponding developmental period, in Piaget, Freud, and Erickson's view.[13] Psychiatric genetic epidemiologists must develop phenotyping strategies that can relate psychopathology in a 2-year-old to that in a 6-year-old, 11-year-old, and 17-year-old. Longitudinal studies to accomplish this are currently under way — for example, by Biederman[14] in his longitudinal follow-up of ADHD children, by Loeber[15] and his work with conduct disorder children, and by Achenbach et al.[16-18] in research on a national sample. These will all significantly inform the field about the relations of symptoms and behaviors in children and adolescents of different developmental stages.

The work of these investigators clearly demonstrates that children's psychopathology changes with development. If we apply identical criteria for psychopathology to all developmental levels, we may misidentify phenotypes. For example, it is known that scores for Aggressive Behavior (AGG) as measured by the empirically based taxonomy tend to decline over normal development.[19] The score for Delinquent Behavior (DB), however, tends to increase over normal development.[19] Failure to control for these differences could lead to incorrect measurements of the changing phenotypes. Such measurement errors may produce excessive false positives or false negatives, depending on the developmental level of the subjects.

b. **Gender Sensitivity**

The phenotypic indicator should allow for potential gender differences in the manifestations of genotypes in girls vs. boys. In the childhood period, psychiatric disorders are more often diagnosed in boys than girls.[20,21] Boys are more often diagnosed as ADHD and ODD. Girls are more often diagnosed as having anxiety disorders.[22,23] Prepubertal boys and girls have similar rates of diagnosed depression,[24] but during adolescence girls are thought to show a marked increase in the rate of depression over that of boys.[25-29] Finally, it is well known that at least twice as many adult women as men receive psychiatric diagnoses.[30-34]

These "gender differences" are well known, but not clearly explained. Many explanations have been proposed for why males are at a greater risk for psychopathology than females in childhood only to have this trend reverse in adulthood. Undoubtedly, diagnostic bias and referral practices, biological vulnerability, and social expectations and pressures all play a role in these differences. In any case, gender differences in the development of psychopathology need to be recognized. This is doubly true in genetic studies. It is possible that a single genotype may be differently expressed in males and females. Without a gender-sensitive taxonomy, gene-disorder relations may be missed due to invalid assumptions of homogeneity of phenotypes between genders.

Research is now addressing gender differences in childhood psychopathology. Faraone and colleagues have demonstrated greater familiality of ADHD in girls vs. boys.[35] They are currently engaged in a 5-year longitudinal study which will undoubtedly increase what is known about the differences between boys vs. girls with ADHD. Twin studies with enough power to compare boys and girls at different developmental levels will also yield important evidence.

Perhaps the most elegant demonstration of the need for a gender-specific and -sensitive taxonomy comes from the longitudinal follow-up studies of Achenbach et al.[16-18] These reports demonstrate the different developmental pathways that girls and boys follow.

> For syndromes having the clearest DSM counterparts, cross-informant predictive paths revealed similar traitlike patterns for aggressive behavior in both sexes; delinquent behavior was less traitlike, with greater sex differences in predictive paths; the attention problems syndrome was developmentally stable, but, surprisingly, it was associated with more diverse difficulties among girls than boys. Conversely, Anxious/depressed was associated with more diverse difficulties among boys than girls.[16]

The results indicate that quantification of problems via empirically based syndromes can detect important sex, age, and developmental variations that may be masked by uniform diagnostic cutoff points for both sexes. This may be especially true for diagnostic cutpoints derived mainly from clinical cases of one sex, such as depression for girls vs. attention and conduct disorders for boys.[16]

The results of these studies indicate there are gender differences in the developmental expression of psychopathology. They support phenotypic indicators that are gender sensitive and specific. Failure to use such a taxonomic approach may lead to both false positives and false negatives when selecting subjects for psychiatric genetic studies.

c. **Multi-Informant Sensitivity**

A potential strength of genetic studies of childhood psychopathology is the ability to use multiple informants to create the phenotype of interest. When describing the phenotype of interest in studies of adult psychopathology, the traditional methods include an interview of the patient and chart reviews. This approach is also the typical model used in adult clinical psychiatry.

In child psychiatry, however, both clinical practice and the study of psychopathology can be enhanced through the use of multiple informants. This approach has been highly successful and is based on the assumption that the primary parent is the best informant regarding her/his child.[36,37] Faraone et al.[38] concluded in a recent study that maternal reports of their children's psychopathology provided an accurate and reliable means of assessment.

There are other useful informants in the study of childhood behaviors. Teachers are tremendously useful in describing their students' behaviors. In fact, agreement between parental reports and teacher reports are often high for DSM-derived diagnoses.[39]

Traditionally, obtaining DSM diagnostic interviews has been viewed as too time consuming, too expensive, and of questionable diagnostic reliability. The multi-informant approach to obtaining interview-based diagnoses has, therefore, been considered impractical by many. However, reports from parents, teachers, and children are routinely and inexpensively collected when using the empirically based taxonomy. There are abundant data on the agreement between the different informants and these may provide valuable "phenotypic information" that ultimately can be used in phenotypic identification.[40]

To demonstrate the utility of the multi-informant approach in selecting subjects for genetic studies, compare and contrast the two following vignettes. When both parents, two teachers, and the youth all report high scores on Attention Problems (AP) on the Child Behavior Checklist (CBCL), Teacher Report Form (TRF), and Youth Self-Report Form (YSR), the geneticist will feel confident that the phenotype "AP" is manifested in all areas of the child's life, i.e., the child has attention problems at home, at school, and recognizes it himself.

This vignette, which provides strong evidence for the presence of the phenotype, needs to be contrasted with the following one. Reports obtained from paternal CBCL, TRFs, and the YSR reveal that the child did not score high on AP in any of these informant reports. The mother's CBCL revealed high scores on AP, as well as on all seven other syndromes. If only maternal reports of AP were obtained in both vignettes, the researcher would conclude, perhaps erroneously, that the children from both vignettes manifest a similar phenotype. By employing the multi-informant approach, the researcher can see there is general agreement with the maternal report in the first scenario and disagreement with the maternal report in the second scenario.

Given the high negative impact of including a false positive in a genetic study, the child from vignette two would likely be excluded from the study until the disagreement between the multiple informants could be clarified.

Work is currently under way relating DSM-IV diagnoses of ADHD, CD, ODD, childhood anxiety, and depression to the multi-informant approaches of the empirically based taxonomy.[41]

In psychiatric genetic studies of childhood pathology, the multi-informant approach is necessary to reduce false positives and thereby increase the likelihood of identifying genes that contribute to the development of child psychiatric disorders. Relying on a single informant may lead to inclusion and exclusion errors. Because of the high cost of false positives, it is important to select probands on the basis of multi-informant agreement on the presence of phenotypes.

2. IMPORTANCE OF ACCURACY ONCE A PHENOTYPE IS SELECTED

Once a phenotypic indicator has met the rigor of the Tsuang criteria, plus filled the additional requirements of being developmentally gender and multi-informant sensitive, it must be exposed to more analyses. As Faraone et al. point out,[7] the choice of a phenotype is vulnerable to other confounds that have surfaced when a diagnosis is used as a phenotypic identifier. These are in the area of ***diagnostic accuracy***. They argue that before a phenotypic indicator is employed in genetic studies, it is necessary to test the specificity, sensitivity, and positive and negative predictive powers of the indicator, and that diagnostic accuracy analyses must take into account the costs and benefits of false positives and false negatives. For genetic studies, false positives are subjects classified as having the genotype when they do not; false negatives are those classified as not having the genotype when they do. For linkage studies, false negatives ***do not*** diminish statistical power. In contrast, false positives ***reduce the power*** to detect linkage and for that matter association. Phenotypic indicators with high specificity are necessary to minimize false positive rates.

For a phenotypic indicator to be useful in the study of childhood psychopathology, it should meet the Tsuang guidelines (specificity, state independence, heritability, familial association, cosegregation, and biological plausibility), be developmentally sensitive, be gender sensitive, make effective use of data from multiple informants, and be highly specific (low rates of false positives).

3. DSM AND EMPIRICALLY BASED TAXONOMY AS PHENOTYPIC INDICATORS

When designing a genetic study of common childhood psychopathology, the researcher has a choice of two diagnostic systems. The first, the DSM-IV, has a large body of clinical studies and treatment research to support its utility. The second, the empirically based taxonomy, is quantitative, has a large body of clinical studies and treatment research, plus longitudinal and statistical data to support its utility. What follows is a discussion of the strengths and weaknesses of each as they relate to genetic studies. Thereafter, ways of combining the two systems to capitalize on their respective strengths for genetic research on child psychopathology will be proposed.

Misclassification is a common problem in psychiatric genetic studies. It is an especially important problem in the study of child psychopathology. For example, the current diagnostic system used to classify childhood psychopathology is the DSM-IV.[1] The conceptual framework of the DSM assumes that disorders are discrete categorical entities. Because of this, the DSM is often referred to as the "categorical approach", which requires that an individual must accrue a minimum number of symptoms lasting a specific period of time and causing impairment in order to be placed in a diagnostic "category". A weakness in the DSM approach that particularly affects genetic studies is the lack of quantitative differentiation within disorders. To illustrate, by the rules of the DSM-IV a child with 12 symptoms of ADHD is considered to be categorically the same as a child with 18 symptoms of ADHD. Furthermore, a child with ten symptoms (five of nine symptoms of inattention and five of nine symptoms of hyperactivity-impulsivity) of ADHD (missing the required cutpoint by one symptom in each category), does not qualify for the category at all. However, a child with six symptoms of inattention and four symptoms of hyperactivity-impulsivity does qualify for a diagnosis of ADHD. Both children have ten symptoms of ADHD, but one child is diagnosed as having ADHD whereas the other is not. To meet the DSM-IV diagnostic requirements for Conduct Disorder, a child must have three symptoms that have lasted at least six months. Whether male or female, whether 7 or 17, the child would be categorized the same as a child having all 15 DSM-IV Conduct Disorder criteria. Perhaps even more illustrative of the weakness of the categorical approach is that the child with 3 symptoms is considered to be categorically distinct from the child with 2 symptoms, but categorically similar to the child with 15 symptoms of conduct disorder.

The DSM approach limits analysis of family, twin, and adoptee studies to the categorization of subjects as either having or not having a particular disorder and ignores the possible variations in the degree to which subjects may manifest the disorder.

The categorical approach of the DSM may be useful for everyday clinical services, but it may not be adequate for identifying phenotypes for psychiatric genetic studies. Quantitative approaches have been suggested for developing a more discriminating nosology for genetic studies.

The strength of a quantitative approach can be summarized by saying that although some patients clearly have a disorder and others clearly do not, many more fall between these extremes of diagnostic clarity. A quantitative approach to cases should facilitate the genetic analysis to the degree that a case corresponds to the true illness status of that subject.[6] Geneticists who study child psychiatric conditions have a potential advantage over those studying adult conditions, in that a quantitative paradigm for the assessment of child psychopathologic conditions already exists.

The empirically based taxonomy is an approach that emphasizes the quantitative assessment.[40] The quantitative paradigm for child psychopathology, embodied in empirically based taxonomy,[40] stems from the psychometric tradition and seeks to quantify characteristics that distinguish among children.[11] It quantifies the degree to which children manifest a particular syndrome. To assess children in terms of empirically derived syndromes, the Child Behavior Checklist (CBCL) is filled out by parents, the Teacher's Report Form (TRF) is filled out by teachers, and the Youth Self Report (YSR) is filled out by adolescents. These are well-standardized procedures for assessing behavioral/emotional problems. Factor analytic methodology has been used to derive the following eight syndromes from these instruments: Anxious/Depressed, Withdrawn, Somatic Complaints, Social Problems, Thought Problems, Delinquent Behavior, Aggressive Behavior, and Attention Problems.[42-44] The assessment instruments are widely used for clinical and research purposes. They have been translated into 50 languages and have been used in over 1700 studies.[45]

Strengths of the empirically based taxonomy include its ability to quantify the degree to which a child has particular problems, the fact that the syndromes are normed according to age and gender, the ability to compare data from multiple informants, and a large data base supporting the integrity of the taxonomy.[11,40] The syndromes have been shown to predict behavior problems in 3- and 6-year assessments of a national epidemiological sample.[16-18,46,47] In recent developmental studies, the taxonomy has been used to demonstrate the differential developmental expression of childhood psychopathology.[48] Additionally, similar studies have been done in children from other cultures, allowing for international comparisons.[49]

There are advantages to conceptualizing psychiatric disorders on a continuum via the quantitative approach rather than conceptualizing them as discrete disorders by the all-or-none categorical approach. Nowhere are these advantages more evident than in genetic studies of psychiatric illness.

An obvious weakness of the empirically based taxonomy is that the proponents of the categorical DSM approach have difficulty translating the results of quantitative studies into terms that relate to the DSM. This is changing. The two systems are not incompatible. Bringing them together to advance the study of common childhood psychopathology may advance the understanding of child psychiatric genetics.

4. Establishing Spectrum Phenotypes Combining DSM and Empirically Based Approaches

Much work has already been done to establish relations between the DSM and the quantitative approach. Numerous studies have shown strong associations between the empirically derived syndromes and specific DSM diagnoses. Principal contributors to these relations include Biederman et al., Bird et al., Edelbrock and Costello, Jensen et al., Steingard et al., and Weinstein et al.[50-56] Several studies established associations between attention deficit disorder and the Attention Problems syndrome.[47,50,53,57] Chen et al. recently reported that the CBCL Attention Problems scale provides an excellent diagnostic accuracy for predicting ADHD.[58] In this study, children with an AP T-score of 70 all met criteria for ADHD. In addition, Hudziak et al. found that 100% of children with AP greater than 67 met DSM-III-R criteria for ADHD, that 100% of children with an Aggressive Behavior score above 67 met criteria for Oppositional Defiant Disorder, and that 100% of children who were deviant on both AP and AGG met criteria for both ADHD and ODD.[62] Along with the Chen et al. study,[58] these findings establish a taxonomic approach whereby children can be described with high specificity to minimize the false positives that undermine molecular genetic investigations. In a separate study, Hudziak et al. demonstrated that these relations also hold true for the DSM-IV categories.[59]

Hudziak et al. found that when starting with DSM-ADHD only 70% of children were deviant on AP. Similarly, when starting with ODD only 70% of children were deviant on AGG. The presence of false positives, when beginning with DSM diagnoses to predict CBCL syndromes, may indicate the heterogeneity of the DSM diagnoses. Although it can be argued that AP does not identify all children who meet DSM criteria for ADHD, it can be said that they identify a large percentage of these children, with few false positives. This may be an example of how a genetic nosology differs from a clinical one. A clinician would be unwilling to use a taxonomy with low sensitivity if it missed 30% of ADHD cases. A geneticist, however, would gladly use a technique that was less sensitive, but highly specific, as in this example where AP T-scores ≥ 67 almost always predict a diagnosis of ADHD.[59]

Similar associations have been reported between Conduct Disorder and Delinquent Behavior [55,56,60] and between Conduct Disorder and Oppositional Defiant Disorder and Aggressive Behavior reflecting both the delinquent and aggressive components of the DSM-III-R conduct disorder criteria.[53,55,56,60] A recent study by Bernstein et al. reports relations between childhood anxiety disorders and the Anxious Depressed syndrome of the CBCL.[61] Similarly, Compas et al. have reported relations between DSM-III-R depressive symptoms and the Anxious Depressed syndrome.[62] Other studies have indicated that the quantitative approach is a good indicator of DSM comorbidity.[55] A growing body of research is thus constructing crosswalks between the DSM and empirically based approaches to assessing child psychopathology.[54]

While crosswalks are being constructed, it is nevertheless important to emphasize that the needs of a genetic taxonomy are different from those of a clinical taxonomy. Work such as that of Hudziak et al.[59] and Chen et al.[58] suggest that by combining the quantitative approach with the DSM approach, we may be able to identify subsets of DSM probands free of the confounds of false positives and therefore ideal for psychiatric genetic investigations. This approach to identifying probands for genetic studies may help to solve a major dilemma facing psychiatric genetics today. The question as to whether it is better to conceptualize psychiatric disorders as categorically discrete or as quantitative extremes can't be answered by using the DSM or the quantitative approach alone. By combining the two approaches, we can determine whether children meet the DSM criteria, as well as determining to what degree they manifest a particular quantitative syndrome. As Rutter points out, "the issue cannot be the choice between a dimension or a category... rather, the query is whether, in addition, is there a category of disorder that is quantitatively distinct from the extreme of the dimensions."[63]

5. GENETICS OF ADHD, CD, AP, AND AGG

Because of confounds such as penetrance, variable expressivity, false positives and negatives, and phenotypic heterogeneity, the current categorical approach alone to identifying subjects for genetic studies is less than ideal. Employing an approach that considers not only if an individual has a disorder (DSM), but also quantifies the degree to which the individual has a disorder (empirically based taxonomy), can make child psychiatric genetic studies more informative.

What follows is a discussion of what is known about the twin, family, and adoptee studies of ADHD and CD from the DSM approach and the AP and AGG syndromes from the empirically based approach.

As classified by the DSM-III-R (APA, 1987), ADHD is a syndrome of unknown etiology with a prevalence between 2 and 6%.[51,64] It involves impaired attention skills, impulsivity, and hyperactivity. Family, twin, and adoption studies suggest that genetic factors are important in at least some cases of ADHD.[65-77] However, no agreement has been reached on the mode of transmission of ADHD. Perhaps the confound that has attracted the most attention from psychiatric geneticists is that ADHD is often comorbid with other problems.[54,78-83] A comprehensive review of the literature reported the following rates of comorbidity: ADHD and oppositional defiant disorder 35%,[84,85] ADHD and conduct disorder 50%,[84,85] ADHD and mood disorders 15 to 75%,[66,69,85] ADHD and anxiety disorders 25%,[68-85] and ADHD and learning disabilities 10 to 92%.[86,87] ADHD also has been followed by psychiatric disorders in adulthood[88] and is associated with psychiatric disorders in family members.[35,66-69,71,85,89-91] Because of these high rates of overlap with other disorders, some authors have questioned the validity of the diagnosis of ADHD.[36,92] Others report that ADHD is distinct from other disorders, including CD.[93] Because of the high rates of comorbidity and the diagnostic heterogeneity associated with the DSM-III-R diagnosis of ADHD, numerous studies have sought to clarify the familiality of ADHD by subtyping according to comorbidity.

The results of these studies, although preliminary, have been instructive. A series of family-genetic analyses suggests that the presence of conduct disorder signals a discrete subtype of ADHD and that major depression is a variable expression of the disorder, but anxiety and learning disorders appear to be transmitted independently of ADHD.[68,69,84,94] Similar studies have subtyped ADHD by gender of proband and whether or not a proband's parent has ADHD.[35,94] All the studies have subtyped the heterogeneous diagnostic category of ADHD in order to establish more specific phenotypes for family studies. This subtyping is one of the ways to cope with the challenge that phenotypic heterogeneity presents to psychiatric genetics.

One approach to solving the dilemma of heterogeneity is to define phenotypes more specifically by comorbidity subtyping, as is done in the studies above. However, this approach is confounded by the possibility that each comorbid category used to subtype ADHD (i.e., Mood, Anxiety, ODD, CD, and Learning Disabilities) is also etiologically heterogeneous. An alternative approach has been to move away from psychiatric criteria and towards noncategorical laboratory measures. Measures of attention and overactivity have been proposed as potential ways to differentiate subtypes of ADHD.[95-97] Biological markers obtained by PET studies,[98,99] dopamine receptor studies,[100] dopamine transporter gene studies,[101] and relating ADHD to genetic-medical disorders such as a generalized resistance to thyroid hormone offer promising strategies for identifying more homogenous phenotypes.[102]

6. GENETICS OF THE AP SYNDROME OF THE EMPIRICALLY BASED TAXONOMY

Three twin studies have yielded high heritability estimates for AP of the empirically based taxonomy.[103-105] In a study of MZ twins vs. DZ twins, using standardized regression coefficients, Edelbrock et al. found a significant genetic contribution of .64 to Attention Problems and no significant contribution of shared environment.[104] In a study of 3-year-old twins, van den Oord et al. reported a heritability coefficient of .50 for the Dutch Overactive syndrome that correlates with the AP syndrome in the Achenbach version of the CBCL.[105] In a study of problem behaviors in 10- to 15-year-old biologically related and unrelated international adoptees, van den Oord et al. reported that genetic influences accounted for 47% of the variance in the AP with little influence of shared environment.[106,107]

In summary, the AP syndrome has been shown to have high construct validity, to be highly heritable in twin and adoptee studies, to be highly related to the categorical DSM diagnosis of ADHD, to be stable over time, and to predict future serious emotional and behavioral problems.

7. GENETICS OF ODD AND CD

As Hinshaw et al. point out,[108] two studies have reported that 96% of boys who met DSM criteria for CD also met DSM criteria for ODD.[109,110] In addition, in the DSM-III-R field trials, 84% of clinic-referred children who met criteria for ODD also met criteria for CD.[111] The high rates of diagnostic overlap suggest that these two categories do not represent separate entities. This is likely due to the fact that all eight of the DSM-IV criteria behaviors for ODD are of the kind that typically correlate with the aggressive CD behaviors.

Conduct problems are the most common reason for referral of children for psychiatric treatment.[112] It is estimated that the prevalence of CD is much higher for boys (9%) than for girls (2%). It is one of the most studied of the behavioral disorders and has been associated with a great deal of impairment and suffering in children,[93] and also with the development of adult antisocial personality disorder, criminality, and substance abuse.[113] Evidence from twin studies and adoption studies supports a genetic contribution to antisocial behavior, but the precise genetic contribution is unclear.[114,115] The role of genetic influences is even less clear for childhood antisocial behavior. This may be due to phenotypic heterogeneity, which is particularly germane to the discussion of CD because prior work indicates different heritabilities, the aggressive vs. delinquent components of CD.[2] It is clear that the DSM criteria for Conduct Disorder include individuals who differ according to whether they manifest primarily aggressive vs. delinquent phenotypes.

8. GENETICS OF THE AGG SYNDROME OF THE EMPIRICALLY BASED TAXONOMY

The existence of two types of conduct disorder is borne out by empirical studies.[2,116] These two distinct behavioral syndromes are often designated as socialized delinquent or delinquent behavior vs. undersocialized aggressive or aggressive behavior.[40] The delinquent behavior scale (DB) has counterparts in the DSM-III-R and DSM-IV criteria: has stolen, has run away, lies, fire-setting, and truancy from school. The aggressive behavior syndrome (AGG) also has counterparts in the DSM-III-R and DSM-IV criteria: has deliberately destroyed other's property, cruelty to animals, initiates physical fights, and physically cruel to people. These two syndromes have been shown to have distinctly different heritability patterns.[104] This may explain some of the difficulties in the genetic study of conduct disorder as a unitary category. Child psychiatric geneticists may choose to study the aggressive behavior syndrome (AGG) of the quantitative taxonomy in relation to the DSM diagnoses of conduct disorder.

There is evidence for stronger biological influences on the AGG syndrome than on the DB syndrome. Correlations of $-.50$ and $-.72$ between the AGG syndrome and measures of imipramine binding have been reported in two studies.[117,118] In a study of six neurotransmitters, metabolites, and enzymes in relation to the eight syndromes of the CBCL, a Pearson correlation of $-.81$ was obtained between dopamine beta hydroxylase activity and AGG.[119] Finally, and most germane to this discussion, the evidence includes higher heritability coefficients for AGG than DB.[104,105,120] In their twin study, Edelbrock et al. reported a genetic effect of .60 for AGG, but a nonsignificant effect of .35 for DB. Conversely, they reported a shared environmental effect of .15 for AGG and .37 for DB.[104] Ghodsian-Carpey and Baker report a heritability of .94 on the CBCL AGG scale in their study of MZ and DZ twins.[120] Van den Oord reported a genetic contribution of .69 for boys and .57 for girls for aggressive behavior in a study of 3-year-old MZ and DZ twins.[105]

In summary, the AGG syndrome is one of the most universally identified and statistically robust syndromes of childhood behavior. It has been shown to be highly heritable in twin and adoptee studies, to be biologically influenced, developmentally stable, and to be highly related to the DSM diagnoses of ODD/CD.

9. GENETIC STUDIES COMBINING THE DSM AND QUANTITATIVE APPROACHES

Considerable evidence supports relations between CBCL AP and DSM ADHD, as well as between CBCL AGG and DSM ODD/CD. To date, no convincing molecular genetic evidence relating a disease gene to these disorders has been forthcoming. Perhaps the discoveries will follow the creation of a combined taxonomic approach that emphasizes the best features of both. This combined approach will be developmentally, gender, and multi-informant sensitive, as well as generalizable in DSM terms. Such a taxonomy may then be useful in sophisticated molecular genetic investigations.

C. NEW MOLECULAR AND STATISTICAL GENETIC METHODS AND TECHNIQUES

Once appropriate identification of a phenotype for study has been achieved, the child psychiatric geneticist has a host of exciting and powerful techniques available to implement.

Traditional methods such as family, twin, and adoptee studies will be enhanced by the refinement of phenotypes via the quantitative taxonomy.

To date, twin studies of the empirically based taxonomy have yielded some of the highest heritability scores in behavioral genetics. Twin studies using this taxonomy are underway at many centers and will also advance our understanding of the heritability of these syndromes. The familiality of the empirically based taxonomy is currently being studied as well.[59] Combined with what is known from the twin and adoptee studies, these studies may yield data on the familiality of these phenotypes and thus be useful in molecular strategies.

This proposed refinement of phenotype identification will make it possible to perform the types of association studies and quantitative trait loci studies that have recently led to replicated discoveries linking genes to behavior. Both of these findings were achieved using phenotypes identified in quantitative rather than categorical terms.

The first was the work of Cardon et al.[121] They were able to establish a Quantitative Trait Locus (QTL) on chromosome 6 p21.3. They used interval mapping from two independent samples of sib pairs, one of which had a reading disability defined by quantitatively extreme deficits. Their method is based on the sib-pair approach of Haseman and Elston,[122] but extended to accommodate interval mapping.[123]

The method involves squaring the difference between phenotypic scores of a pair of sibs (sib 1 – sib 2^2 = Y) and then regressing value Y onto an estimate of the proportion of alleles that the sib-pair share identity by descent at the marker locus.[122] This technique, combined with a recent report by Risch et al. on choosing sib pairs who are extremely discordant on the phenotypic measure of interest, dramatically increases the power for finding quantitative trait locus that link genes and behaviors.[124]

Ebstein et al. and Benjamin et al. both reported a positive association between the dopamine D_4 receptor (D_4DR) exon III polymorphism and measures of Novelty Seeking.[125,126] Novelty Seeking is one of three quantitative domains from the Tridimensional Personality Questionnaire developed by Cloninger et al.[127] Measured as a quantitative domain, Novelty Seeking consists of items characterized as impulsive, exploratory, fickle, excitable, quick-tempered, and extravagant.

The significance of the D_4R and the Chr 6 findings are that they are the only replicated findings linking genes and behavior (with the exception of some forms of Alzheimer's disease).[128] That both were achieved through the use of a quantitative approach to phenotype identification is a testament to the utility of quantitative taxonomies for common psychiatric conditions.

Another potentially powerful technique that can benefit from improved phenotype identification is relating constitutively active mutant receptors to phenotypes. In Chapter 7 of this book, the authors discuss the impact of relating constitutively active receptors of the m5 receptor to psychopathology.[129] As powerful as this technique is, it will not lead to a conclusive link between genes and behavior without proper phenotypic identification. Work relating abnormal receptor function, rather than architecture, is under way and will be important to the understanding of receptor psychopathology relationships. Because these mutations are undoubtedly present from birth, the techniques will be empirically useful in developmentally sensitive studies of childhood psychopathology.

Finally, and perhaps most importantly, are the techniques of developmental twin studies. Using behavioral genetic techniques, psychiatric geneticists will be able to increase the fund of knowledge about the genetic and environmental effects on twins of different ages. Twins can be empirically sampled at different ages and quantitative genetic analysis of the developmental pathways of monozygotic, dizygotic, same sex, and opposite sex twins will lead to an increased understanding of the genetics of the phenotypic indicators. Such studies may reveal differential genetic influences at different ages, will yield gender specific and developmentally specific genetic and environmental data that can then be tested by powerful association, segregation, QTL, and linkage techniques.

D. SUMMARY

The genetic study of common childhood psychopathology is at an important point in its development. With the refining of phenotypic identification techniques, it may be possible to develop a discriminating nosology that can classify patients into discrete genetic and nongenetic types for use in twin, family, adoptee, and ultimately molecular genetic studies.

This refinement will emphasize the minimization of false positives. The taxonomy will need to be developmentally sensitive, gender sensitive, sensitive to multiple informants, quantifiable, easy and inexpensive to apply, and generalizable in DSM terms. This "genetic nosology" will aid studies of childhood psychopathology in the search for disease genes for common childhood psychiatric disorders.

ACKNOWLEDGMENT

This work was supported by NIMH Grant MHO 1265-01A1.

REFERENCES

1. American Psychiatric Association, *Diagnostic and Statistical Manual of Mental Disorders*, 4th ed., (DSM-IV), American Psychiatric Association, Washington, D.C., 1994.
2. Achenbach, T. M., Taxonomy and comorbidity of conduct problems. Evidence from empirically based approaches, *Dev. Psychopathol.*, 5, 51, 1993.
3. Rice, J. P., Reich, T., Cloninger, C. R., and Wette, R., An approximation to the multivariate normal integral: its application to multifactorial qualitative traits, *Biometrics*, 35, 451, 1979.
4. Rice, J., Reich, T., Andreasen, N. C., Endicott, J., Van Eerdewegh, P., Fishman, R., Hirschfield, R. M. A., and Klerman, G. L., The familial transmission of bipolar illness, *Arch. Gen. Psychiatry*, 44, 441, 1987.
5. Tsuang, M. T., Genetics, epidemiology, and the search for causes of schizophrenia, *Am. J. Psychiatry*, 151, 3, 1994.
6. Tsuang, M. T., Faraone, S. V., and Lyons, M. J., Identification of the phenotype in psychiatric genetics, *Eur. Arch. Psychiatr. Clin. Neurosci.*, 243, 131, 1993.
7. Faraone, S. V., Kremen, W. S., Lyons, M. J., Peeple, J. R., Seidman, L. J., and Tsuang, M. T., Diagnostic accuracy and linkage analysis: how useful are schizophrenia spectrum phenotypes?, *Am. J. Psychiatry*, 152, 1286, 1995.
8. Piaget, J., Development and learning, in *Piaget Rediscovered*, Ripple, R. E. and Rockcastle, V. N., Eds., Cornell University Press, Ithaca, NY, 1964.
9. Erickson, E. H., *Childhood and Society*, 2nd ed., Norton, New York, 1963.
10. Freud, A., Child analysis as the study of mental growth, in *The Course of Life: Psychoanalytic Contributions Toward Understanding Personality Development*, Greenspan, S. E. and Pollack, G. H., Eds., (Vol 1.): Infancy and early childhood, NIMH Mental Health Study Center, Adelphi, MD, 1980.
11. Achenbach, T. M., The classification of children's psychiatric symptoms: a factor-analytic study, *Psychol. Monogr.*, 80(No. 615), 1966.
12. Achenbach, T. M., *Developmental Psychopathology*, 2nd ed., Perspectives on development, John Wiley & Sons, New York, 1982, chap. 3.
13. Achenbach, T. M. and Edelbrock, C. S., Behavioral problems and competencies reported by parents of normal and disturbed children aged 4 through 16, *Monogr. Soc. Res. Child Dev.*, 46: Serial No. 188, 1981.
14. Biederman, J., Continuity of attention deficit hyperactivity (ADHD) into adulthood, *J. Am. Acad. Child Adolescent Psychiatry*, 1995.
15. Loeber, R., The stability of antisocial and delinquent child behavior: a review, *Child Dev.*, 53, 1431, 1982.
16. Achenbach, T. A., Howell, C. A., McConaughy, S. H., and Stanger, C., Six year predictors of problems in a national sample of children and youth. I. Cross-informant syndromes, *J. Am. Acad. Child Adolescent Psychiatry*, 34(3), 336, 1995.
17. Achenbach, T. A., Howell, C. A., McConaughy, S. H., and Stanger, C., Six year predictors of problems in a national sample of children and youth. II. Signs of disturbance, *J. Am. Acad. Child Adolescent Psychiatry*, 34(4), 488, 1995.

18. Achenbach, T. A., Howell, C. A., McConaughy, S. H., and Stanger, C., Six year predictors of problems in a national sample of children and youth. III. Transitions to young adult syndromes, *J. Am. Acad. Child Adolescent Psychiatry,* 34(5), 658, 1995.
19. Stanger, C., Achenbach, T. M., and Verhulst, F. C., Accelerated longitudinal comparisons of aggressive versus delinquent syndromes, *Dev. Psychopathol.*, submitted.
20. Offord, D., Boyle, M., Szatmari, P., Rae-Grant, N., Links, P., Cadman, D., Byles, J., Crawford, J., Munroe Blum, H., Byrne, L., Thomas, H., and Woodward, C., Ontario Child Health Study: Six month prevalence of disorder and rates of service utilization, *Arch. Gen. Psychiatry,* 44, 832, 1987.
21. Rutter, M., Tizard, J., and Whitmore, K., *Education, Health, and Behavior,* Longman, London, 1970.
22. Last, C. G., Anxiety disorders, in *Handbook of Child Psychopathology,* Ollendick, T. H. and Hersen, M., Eds., Plenum Press, New York, 1989.
23. Miller, S. M., Boyer, B. A., and Rodoletz, M., Anxiety in children. Nature and development, in *Handbook of Developmental Psychopathology,* Lewis, M. and Miller, S. M., Eds., Plenum Press, New York, 1990.
24. Fleming, J. E., Offord, D. R., and Boyle, M. H., Prevalence of childhood and adolescent depression in the community, *Br. J. Psychiatry,* 155, 647, 1989.
25. Brooks-Gunn, J. and Petersen, A., Studying the emergency of depression and depressive symptoms during adolescence, *J. Youth Adolescence,* 20, 115, 1991.
26. Cairns, R. B., Cairns, D. R., Neckerman, H. J., Ferguson, L. L., and Gariepy, J., Growth and aggression. I. Childhood to early adolescence, *Dev. Psychol.,* 25, 320, 1989.
27. Forehand, R., Neighbors, B., and Wierson, M., The transition to adolescence: the role of gender and stress in problem behavior and competence, *J. Child Psychiatry Psychol.,* 32, 929, 1991.
28. Links, P. S., Boyle, M. H., and Offord, D. R., The prevalence of emotional disorder in children, *J. Nerv. Ment. Dis.,* 177, 85, 1989.
29. Rutter, M., The development of psychopathology in depression. Issues and perspectives, in *Depression in Young People,* Rutter, M., Izard, C., and Read, P. B., Eds., Guilford Press, New York, 1986.
30. Brisco, M., Sex differences in psychological well-being, *Psychol. Med.,* Monogr. Suppl. I, 1982.
31. Gove, W. R., Mental illness and psychiatric treatment among women, *Psychol. Women Q.,* 4, 345, 1980.
32. McGrath, E., Keita, G. P., Strickland, B. R., and Russo, N. F., *Women and Depression: Risk Factors and Treatment Issues,* American Psychological Association, Washington, D.C., 1990.
33. Nolen-Hocksema, S., Sex differences in unipolar depression: evidence and theory, *Psychol. Bull.,* 101, 259, 1987.
34. Weissman, M. M. and Klerman, G. L., Sex differences in the epidemiology of depression, *Arch. Gen. Psychiatry,* 34, 98, 1985.
35. Faraone, S. V., Biederman, J., Keenan, K., and Tsuang, M. T., A family-genetic study of girls with DSM-III attention deficit disorder, *Am. J. Psychiatry,* 148, 112, 1991.
36. Rutter, M., Syndromes attributed to "minimal brain dysfunction" in childhood, *Am. J. Psychiatry,* 139, 21, 1982.
37. Rutter, M., DSM-III-R: A postscript, in *Assessment and Diagnosis in Child Psychopathology,* Rutter, M., Tuma, A. H., and Lann, I. S., Eds., Guilford Press, New York, 1988, 453.
38. Faraone, S. V., Biederman, J., and Milberger, S., How reliable are maternal reports of their children's psychopathology? One-year recall of psychiatric diagnoses of ADHD children, *J. Am. Acad. Child Adolescent Psychiatry,* 34, 8, 1001, 1995.
39. Biederman, J., Keenan, K., and Faraone, S. V., Parent based diagnosis of attention deficit disorder predicts a diagnosis based on teacher report, *J. Am. Acad. Child Adolescent Psychiatry,* 29, 69, 1990.
40. Achenbach, T. M., Empirically Based Taxonomy. How to use syndromes and profile types derived from the CBCL/4-18, TRF, and YSR, University of Vermont Department of Psychiatry, Burlington, 1993.
41. Hudziak, J. J, Family Studies Combining the DSM and Quantitative Approaches, National Institute of Mental Health, funded grant, August, 1995.
42. Achenbach, T. M., Integrative guide for the 1991 CBCL/4-18, YSR, and TRF profiles, University of Vermont Department of Psychiatry, Burlington, 1991.
43. Achenbach, T. M., Manual for the Youth Self-Report and 1991 Profile. University of Vermont Department of Psychiatry, Burlington, 1991.
44. Achenbach, T. M., Manual for the Teacher's Report Form and 1991 Profile, University of Vermont Department of Psychiatry, Burlington, 1991.
45. Brown, J. S. and Achenbach, T. M., Bibliography of Published Studies Using the Child Behavior Checklist and Related Materials, University of Vermont Department of Psychiatry, Burlington, 1996.
46. McConaughy, S. H., Stanger, C., and Achenbach, T. M., Three-year course of behavioral/emotional problems in a national sample of 4- to 16-year-olds. I. Agreement among informants, *J. Am. Acad. Child Adolescent Psychiatry,* 31, 932, 1992.
47. Stanger, C., Achenbach, T. M., and Verhulst, F. C., Accelerating longitudinal research on child psychopathology: a practical example, *Psychol. Assessment,* 6, 102, 1994.
48. Verhulst, F. C. and van der Ende, J., Six-year developmental course of internalizing and externalizing problem behaviors, *J. Am. Acad. Child Adolescent Psychiatry,* 31, 924, 1992.
49. Verhulst, F. C. and Achenbach, T. M., Empirically based assessment and taxonomy of psychopathology: cross-cultural applications, *Eur. Child Adolescent Psychiatry,* 4, 61, 1995.

50. Biederman, J., Faraone, S. V., Doyle, A., Lehman Krifcher, B., Kraus, I., Perrin, J., and Tsuang, M. T., Convergence of the Child Behavior Checklist with Structured Interview-based Psychiatric Diagnoses of ADHD children with and without comorbidity, *J. Child Psychol. Psychiat.,* 34(7), 1241, 1993.
51. Bird, H. R., Canino, G., Rubio-Stipec, M., Gould, M. S., Ribera, J., Sesman, M., Woodbury, M., Huertas-Goldman, S., Pagan, A., Sanchez-Lacay, A., and Moscoso, J., Estimates of the prevalence of childhood maladjustment in a community survey in Puerto Rico, *Arch. Gen. Psychiatry,* 45, 1120, 1988.
52. Bird, H. R., Gould, M. S., Rubio-Stipec, M., Staghezza, B. M., and Canino, G., Screening for childhood psychopathology in the community using the Child Behavior Checklist, *J. Am. Acad. Child Adolescent Psychiatry,* 30, 116, 1991.
53. Edelbrock, C. and Costello A. J., Convergence between statistically derived behavior problem syndromes and child psychiatric diagnoses, *J. Abnorm. Child Psychol.,* 16, 219, 1988.
54. Jensen, P. S., Shervette, R. E., III, Xenakis, S. N., and Richters, J., Anxiety and depressive disorders in attention deficit disorder with hyperactivity. New findings, *Am. J. Psychiatry,* 150, 1203, 1993.
55. Steingard, R., Biederman, J., Doyle, A., and Sprich-Buckminster, S., Psychiatric comorbidity in attention deficit disorder: impact on the interpretation of child behavior checklist results, *J. Am. Acad. Adolescent Psychiatry,* 31, 449, 1992.
56. Weinstein, S. R., Noam, G. G., Grimes, K., Stone, K., and Schwab-Stone, M., Convergence of DSM-III diagnoses and self-reported symptoms in child and adolescent inpatients, *J. Am. Acad. Child Adolescent Psychiatry,* 29, 627, 1990.
57. Edelbrock, C., Costello, A. J., and Kessler, M. D., Empirical corroboration of attention deficit disorder, *J. Am. Acad. Child Psychiatry,* 23, 285, 1984.
58. Chen, W. J., Faraone, S. V., Biederman, J., and Tsuang, M. T., Diagnostic accuracy of the Child Behavior Checklist scales for Attention-Deficit Hyperactivity Disorder: a receiver-operating characteristic analysis, *J. Consult. Clin. Psychol.,* 62, 1017, 1994.
59. Hudziak, J., J., Stanger, C. S., Shih, F., and Blais-Hall, D., The use of the Empirically Based Taxonomy to predict the DSM III-R and IV diagnoses of ADHD, ODD, and CD, manuscript in preparation.
60. Kazdin, A. E. and Heidish, I. E., Convergence of clinically derived diagnoses and parent checklists among inpatient children, *J. Abnorm. Child Psychol.,* 12(3), 421, 1984.
61. Bernstein, G. A., Massie, E. D., Crosby, R. D., and Borchardt, C. M., Somatic symptoms in adolescents with comorbid anxiety and depressive disorders, submitted.
62. Compas, B. E., Ey, S., and Grant, K. E., Taxonomy, assessment, and diagnosis of depression during adolescence, *Psychol. Bull.,* 114, 323, 1993.
63. Zoccolillo, M., Pickles, A., Quinton, D., and Rutter, M., The outcome of childhood conduct disorder: implications for defining adult personality disorder and conduct disorder, *Psychol. Med.,* 22, 971, 1992.
64. Costello, E. J., Costello, A. J., Edelbrock, C., Burns, B. J., Dulcan, M. K., Brent, D., and Janszewki, S., Psychiatric disorders in pediatric primary care, *Arch. Gen. Psychiatry,* 45, 1107, 1988.
65. Biederman, J., Munir, K., Knee, D., Habelow, W., Armentano, M., Autor, S., Hoge, S. K., and Waternaux, C., A family study of patients with attention deficit disorder and normal controls, *J. Psychiatr. Res.,* 20, 263. 1986.
66. Biederman, J., Munir, K., Knee, D., Armentano, M., Autor, S., Waternaux, C., and Tsuang, M., High rate of affective disorders in probands with attention deficit disorder and in their relatives. A controlled family study, *Am. J. Psychiatry,* 144, 330, 1987.
67. Biederman, J., Faraone, S. V., Keenan, K., Knee, D., and Tsuang, M. T., Family-genetic and psychosocial risk factors in DSM-III attention deficit disorder, *J. Am. Acad. Child Adolescent Psychiatry,* 29, 526, 1990.
68. Biederman, J., Faraone, S. V., Keenan, K., Steingard, R., and Tsuang, M. T., Familial association between attention deficit disorder and anxiety disorders, *Am. J. Psychiatry,* 148, 251, 1991.
69. Biederman, J., Faraone, S. V., Keenan, K., and Tsuang, M. T., Evidence of familial association between attention deficit disorder and major affective disorders, *Arch. Gen. Psychiatry,* 48, 633, 1991b.
70. Biederman, J., Faraone, S. V., Keenan, K., Benjamin, J., Krifcher, B., Moore, C., Sprich, S., Ugaglia, K., Jellinek, M. S., Steingard, R., Spencer, T., Norman, D., Kolodny, R., Klaus, I., Perrin, J., Keller, M. B., and Tsuang, M. T., Further evidence for family-genetic risk factors in Attention Deficit Hyperactivity Disorder (ADHD). Patterns of comorbidity in probands and relatives in psychiatrically and pediatrically referred samples, *Arch. Gen. Psychiatry,* 49, 728, 1992.
71. Cantwell, D. P., Genetics of hyperactivity, *J. Child Psychol. Psychiatry,* 16, 261, 1975.
72. Faraone, S. V., Biederman, J., Chen, W. J., Krifcher, B., Keenan, K., Moore, C., Sprich, S., and Tsuang, M., Segregation analysis of attention deficit hyperactivity disorder: evidence for single gene transmission, *Psychiatr. Genet.,* 2, 257, 1992.
73. Goodman, R. and Stevenson, J., A twin study of hyperactivity. I. An examination of hyperactivity scores and categories derived from Rutter teacher and parent questionnaires, *J. Child Psychol. Psychiatry,* 30, 671, 1989a.
74. Goodman, R. and Stevenson, J., A twin study of hyperactivity. II. The aetiological role of genes, family relationships and perinatal adversity, *J. Child Psychol. Psychiatry,* 30, 691, 1989b.
75. Pauls, D. L., Shaywitz, S. E., Kramer, P. L., Shaywitz, B. A., and Cohen, D. J., Demonstration of vertical transmission of attention deficit disorder, *Ann. Neurol.,* 14, 363, 1983.

76. Pauls, D. L., Genetic factors in the expression of attention-deficit hyperactivity disorder, *J. Child Adolescent Psychopharmacol.*, 1, 353, 1991.
77. Welner, Z., Welner, A., Stewart, M., Palkes, H., and Wish, E., A controlled study of siblings of hyperactive children, *J. Nerv. Ment. Dis.*, 165, 110, 1977.
78. Biederman, J., Faraone, S. V., and Lapey, K., Comorbidity of diagnosis in attention deficit hyperactivity disorder, *Child Adolescent Psychiatr. Clin. N. Am.*, 1(2), 335, 1992.
79. Caron, C. and Rutter, M., Comorbidity in child psychopathology. Concepts, issues and research strategies, *J. Child Psychol. Psychiatry*, 32, 1063, 1991.
80. Hudziak, J. J. and Todd, R. D., Familial subtyping of ADHD, *Curr. Op. Psychiatry*, 6, 489, 1993.
81. Pliszka, S. R., Comorbidity of attention-deficit hyperactivity disorder and overanxious disorder, *J. Am. Acad. Child Adolescent Psychiatry*, 31, 197, 1992.
82. Semrud-Clikeman, M., Biederman, J., Sprich-Buckminster, S., Krifcher Lehman, B., Faraone, S.V., and Norman, D., Comorbidity between ADHD and learning disability: a review and report in a clinically referred sample, *J. Am. Acad. Child Adolescent Psychiatry*, 31, 439, 1992.
83. Stewart, M. A., Cummings, C., Singer, S., and deBlois, C. S., The overlap between hyperactive and unsocialized aggressive children, *J. Child Psychol. Psychiatry*, 22, 35, 1981.
84. Faraone, S. V., Biederman, J., Keenan, K., and Tsuang, M. T., Separation of DSM-III attention deficit disorder and conduct disorder: evidence from a family-genetic study of American child psychiatric patients, *Psychol. Med.*, 21, 109, 1991.
85. Biederman, J., Newcorn, J., and Sprich, S., Comorbidity of attention deficit hyperactivity disorder with conduct, depressive, anxiety, and other disorders, *Am. J. Psychiatry*, 148, 564, 1991c.
86. August, G. J. and Holmes, C. S., Behavior and academic achievement in hyperactivity subgroups and learning-disabled boys, *Am. J. Dis. Child.*, 138, 1025, 1984.
87. Silver, L. B., The relationship between learning disabilities, hyperactivity, and behavioral problems, *J. Am. Acad. Child Adolescent Psychiatry*, 20, 385, 1981.
88. Mannuzza, S., Klein, R. G., Bessler, A., Malloy, P., and Lapadula, M., Adult outcome of hyperactive boys: educational achievement, occupational rank, and psychiatric status, *Arch. Gen. Psychiatry*, 50, 565, 1993.
89. Morrison, J., Adult psychiatric disorders in parents of hyperactive children, *Am. J. Psychiatry*, 137, 825, 1980.
90. Schachar, R. and Wachsmuth, R., Hyperactivity and parental psychopathology, *J. Child Psychol. Psychiatry*, 31, 381, 1990.
91. Stewart, M. A., deBlois, C. S., and Cummings, C., Psychiatric disorder in the parents of hyperactive boys and those with conduct disorder, *J. Child Psychol. Psychiatry*, 21, 283, 1980.
92. Prior, M. and Sanson, A., Attention deficit disorder with hyperactivity, *J. Child Psychol. Psychiatry*, 27, 307, 1986.
93. Szatmari, P., Boyle, M., and Offord, D. R., ADHD and conduct disorder: Degree of diagnostic overlap and differences among correlates, *J. Am. Acad. Child Adolescent Psychiatry*, 28, 865, 1989.
94. Faraone, S. V., Biederman, J., Chen, W. J., and Tsuang, M. T., Genetic heterogeneity in attention deficit hyperactivity disorder: gender, psychiatric comorbidity and parental illness, *J. Abnorm. Psychol.*, 104(2), 334, 1995.
95. Barkley, R. A., The ecological validity of laboratory and analogue assessment methods of ADHD symptoms, *J. Abnorm. Child Psychol.*, 19, 149, 1991.
96. Halperin, J. M., Matier, K., Bedi, G., Sharma, V., and Newcorn, J., Specificity of inattention, impulsivity, and hyperactivity to the diagnosis of attention-deficit hyperactivity disorder, *J. Am. Acad. Child Adolescent Psychiatry*, 31, 191, 1992.
97. Melnyk, L. M. and Das, J. P., Measurement of attention deficit: correspondence between rating scales and tests of sustained and selective attention, *Am. J. Ment. Retard.*, 96, 599, 1992.
98. Zametkin, A. J., Nordahl, T. E., Gross, M., King, C., Semple, W. E., Rumsey, J., Hamburger, S., and Cohen, R. M., Cerebral glucose metabolism in adults with hyperactivity of childhood onset, *N. Engl. J. Med.*, 323, 1361, 1990.
99. Zametkin, A. J., Liebenauer, L. L., Fitzgerald, G. A., King, A. C., Minkunas, D. V., Herscovitch, P., Yamada, E. M., and Cohen, R. M., Brain metabolism in teenagers with attention deficit hyperactivity disorder, *Arch. Gen. Psychiatry*, 50, 333, 1993.
100. Comings, D. E., Comings, B. G., Muhleman, M. S., Dietz, G., Shabahrami, M. S., Tast, D., Knell, E., Kocsis, P., Baumgarten, R., Kovacs, B. W., Levy, D. L., Smith, M., Borison, R. L., Evans, D., Klein, D. N., MacMurray, J., Tosk, J. M., Sverd, J., Gysin, R., and Flanagan, S. D., The dopamine D2 receptor locus as a modifying gene in neuropsychiatric disorders, *JAMA*, 266, 1793, 1991.
101. Cook, E. H., Jr., Stein, M. A., Krasowski, M. D., Cox, N. J., Olkon, D. M., Kieffer, J. E., and Leventhal, B. L., Association of attention-deficit disorder and the dopamine transporter gene, *Am. J. Hum. Genet.*, 56, 993, 1995.
102. Hauser, P., Zametkin, A. J., Martinex, P., Benedetto, V., Matochik, J. A., Mixson, J., and Weintraub, B. D., Attention deficit-hyperactivity disorder in people with generalized resistance to thyroid hormone, *N. Engl. J. Med.*, 328, 997, 1993.

103. Schmitz, S., Fulker, D. W., and Mrazek, D. A., Problem behavior in early and middle childhood: an i... behavior genetic analysis, *J. Child Psychol. Psychiatry*, 36, 1443, 1995.
104. Edelbrock, C., Rende, R., Plomin, R., and Thompson, L. A., A twin study of competence and problem behavior in childhood and early adolescence, *J. Child Psychol. Psychiatry Allied Discip.*, 36(5), 775, 1995.
105. Van Den Oord, E. J. C. G., Verhulst, F. C., and Boomsma, D. I., A genetic study of maternal and paternal ratings of problem behaviors in three-year-old twins, *J. Abnorm. Psychol.*, in press.
106. Van Den Oord, E., Boomsma, D. I., and Verhulst, F. C., A study of problem behaviors in 10- to 15-year-old biologically related and unrelated international adoptees. A Genetic Study of Problem Behaviors in Children, Erasmus University, Rotterdam, 1993, 33.
107. Van Den Oord, E. J. C. G., Boomsma, D. I., and Verhulst, F. C., A study of problem behaviors in 10- to 15-year-old biologically related and unrelated international adoptees, *Behav. Genet.*, 24(3), 193, 1994..
108. Hinshaw, S. P., Lahey, B. B., and Hart, E. L., Issues of taxonomy and comorbidity in the development of conduct disorder, *Dev. Psychopathol.*, 5, 31, 1993.
109. Biederman, J., Munir, K., and Knee, D., Conduct and oppositional disorder in clinical referred children with attention deficit disorder. A controlled family study, *J. Am. Acad. Child Adolescent Psychiatry*, 26, 724, 1987.
110. Walker, J. L., Lahey, B. B., Russo, M. F., Christ, M. A. G., McBurnett, K., Loeber, R., Stouthamer-Loeber, M., and Green, S. M., Anxiety, inhibition, and conduct disorder in children. I. Relations to social impairment, *J. Am. Acad. Child Adolescent Psychiatry*, 30, 187, 1991.
111. Spitzer, R. L., Davies, M., and Barkley, R. A., The DSM-III-R field trial of disruptive behavior disorders, *J. Am. Acad. Child Adolescent Psychiatry*, 29, 690, 1990.
112. Robins, L. N., Conduct Disorder, *J. Child Psychol. Psychiatry,* 32(1), 193, 1991.
113. Hesselbrock, M. N., Childhood behavior problems and adult antisocial personality disorder in alcoholism, in *Psychopathology and Addictive Disorders*, Meyer, R. E. Ed., Guilford Press, New York, 1986, 79.
114. Centerwall, B. S. and Robinette, D. C., Twin concordance for dishonorable discharge from the military: with a review of the genetics of antisocial behavior, *Compr. Psychiatry*, 30, 442, 1989.
115. Cadoret, R. J., Psychopathology in adopted-away offspring of biologic parents with antisocial behavior, *Arch. Gen. Psychiatry,* 35, 176, 1978.
116. Quay, H. C., Classification, in *Psychopathological Disorders of Childhood,* 3rd ed., Quay, H. C. and Werry J. S., Eds., John Wiley & Sons, New York, 1986, 1.
117. Birmaher, B., Stanley, M., Greenhill, L., Twomey, J., Gavrilescu, A., and Rabinovich, H., Platelet imipramine binding in children and adolescents with impulsive behavior, *J. Am. Acad. Child Adolescent Psychiatry*, 29, 914, 1990.
118. Stoff, D. M., Pollack, L., Vitiello, B., Behar, D., and Bridger, W. H., Reduction of ^{3}H-imipramine binding sites on platelets of conduct disordered children, *Neuropsychopharmacology,* 1, 55, 1987.
119. Gabel, S., Stadler, J., Bjorn, J., Shindledecker, R., and Bowden, C., Dopamine-beta-hydroxylase in behaviorally disturbed youth. Relationship between teacher and parent ratings, *Biol. Psychiatry,* 34, 434, 1993.
120. Ghodsian-Carpey, J. and Baker, L. A., Genetic and environmental influences on aggression in 4- to 7-year-old twins, *Aggressive Behav.*, 13, 173, 1987.
121. Cardon, L. R., Smith, S. D., Fulker, D. W., Kimberling, W. J., Pennington, B. F., and DeFries, J. C., Quantitative Trait Locus for reading disability on chromosome 6, *Science,* 266, 276, 1994.
122. Haseman, J. K. and Elston, R. C., *Behav. Genet.*, 2, 3, 1972.
123. Lander, E. S. and Botstein, D., *Genetics,* 121, 185, 1989.
124. Risch, N. and Zhang, H., Extreme discordant sib pairs for mapping quantitative trait loci in humans, *Science,* 268, 1584, 1995.
125. Ebstein, R. P., Novick, O., Umansky, R., Priel, B., Osher, Y., Blain, D., Bennett, E. R., Nemanov, L., Katz, M., and Belmaker, R. H., Dopamine D4 receptor (D4DR) exon III polymorphism associated with the human personality trait of Novelty Seeking, *Nat. Genet.*, 12, 78, 1996.
126. Benjamin, J., Li, L., Patterson, C., Greenberg, B. D., Murphy, D. L., and Hamer, D. H., Population and familial association between the D4 dopamine receptor gene and measures of Novelty Seeking, *Nat. Genet.*, 12, 81, 1996.
127. Cloninger, C. R., Przybeck, T. R., Svrakic, D. M., and Wetzel, R. D., *The Temperament Guide and Character Inventory (TCI): A Guide to Its Development and Use*, Cloninger, C. R., Ed., Center for Psychobiology of Personality, St. Louis, MO, 1994.
128. Sherrington, R. V., Rogaev, E. I., Liang, Y., Rogaeva, E. A., Levesque, G., Ikeda, M., Chi, H., Lin, C., Li, G., and Holman, K., Cloning of a gene bearing missense mutations in early onset familial Alzheimer's disease, *Nature*, 375, 754, 1995.
129. Spalding, T. A., Burstein, E. S., Brauner-Osborne, H., Hill-Eubanks, D., and Brann, M. R., Pharmacology of constitutively active muscarinic receptor generated by random mutagenesis, *J. Pharmacol. Exp. Ther.*, 275, 1274, 1995.

13 Molecular Genetics of Alzheimer's Disease

Gerard D. Schellenberg

CONTENTS

Abstract 219
A. Introduction 220
B. Amyloid Precursor Protein Gene And Chromosome 21 221
    1. Mutations in the APP Gene 221
    2. APP Mutations and AD Pathogenesis 224
C. Early-Onset FAD Loci 225
    1. The Chromosome 14 FAD Gene 225
    2. The Volga German Kindreds and the Chromosome 1 FAD Gene 226
    3. Late-Onset AD and the Chromosome 14 Locus 226
D. Chromosome 19 AD Locus 227
    1. Interaction of APOE ε4 and Other AD Loci 228
    2. APOE and Potential Pathogenic Mechanisms 229
        a. ApoE as a Molecular Chaperon for the Aβ Protein 229
        b. ApoE Interactions with Tau 229
        c. ApoE as a Neuronal Injury-Response Protein 229
E. Conclusions 230
References 230

ABSTRACT: Defective genes play an important role in some, if not all cases of Alzheimer's disease (AD). Epidemiologic case control studies, family pedigree analysis, and recent twin studies clearly implicate inherited gene defects in development of the disease. In addition to defective genes, trisomy 21 also results in the neuropathology of AD and an increased risk of early dementia. The genetics of AD have been partially resolved. In some rare kindreds, AD is inherited by an autosomal dominant mechanism. In some of these rare families, mutations in the amyloid precursor protein (APP) gene on chromosome 21 are responsible for AD. APP mutations appear to account for approximately 5% of early-onset familial AD (FAD). Linkage analysis was used with a genomic scanning strategy to localize an early-onset FAD locus to chromosome 14q24.3. The gene for the chromosome 14 locus, called Presenilin 1, has been recently cloned and encodes a membrane protein of unknown function. The chromosome 14 FAD locus is found in ethnically diverse populations including European Caucasians, Hispanics from Mexico, and in at least one Japanese family. However, the chromosome 14 locus is not responsible for FAD in the Volga German FAD families, a group of ethnically related kindreds with family age of onset means ranging from 50 to 79 years. Also, the chromosome 14 locus does not appear to be responsible for late-onset FAD. The locus responsible for AD in the Volga Germans was recently mapped by linkage analysis to chromosome 1q31-42. The gene, called Presenilin 2, was cloned and is partially homologous to Presenilin 1. A locus on chromosome

19 appears to be a risk factor for AD in at least some late-onset FAD kindreds. Early genetic association studies implicated the apolipoprotein (apo) gene cluster on chromosome 19 which includes the apoCII, apoE, and apoCI genes. More recent genetic studies indicate that the ε4 allele of the APOE gene is a possible risk factor for late-onset disease. However, some FAD individual kindreds do not carry the ε4 allele. Thus, additional late-onset FAD genes remain to be identified. The chromosome 19 locus is not responsible for AD in the Volga Germans or the other early-onset kindreds. In summary, AD is genetically heterogeneous; while four loci involved in AD have been identified, additional contributing genes remain to be found.

A. INTRODUCTION

The role of inheritance in the occurrence of Alzheimer's disease (AD) has come under intense scrutiny during the past decade. Epidemiologic survey studies have repeatedly shown that families of AD probands have more cases of AD in close relatives compared to families of control subjects.[1] Twin studies not only show that many identical twins are often concordant for AD,[2] but also that the families of concordant twins have more AD cases then the families of discordant twins.[3] These types of studies are highly suggestive that the inheritance of defective genes is important in the pathogenesis of AD. However, population-based studies cannot clearly determine the mode of inheritance responsible for the observed family clustering.

Some have suggested that all AD is autosomal dominant.[4] This hypothesis is difficult to prove by family studies since the onset of AD typically occurs late in life, and relatives of probands often die of other causes before they reach the age of appropriate risk (age censoring). Thus, even if all AD was the result of autosomal dominant inheritance for randomly ascertained AD probands, the number of secondary cases in close relatives would be underestimated due to age censoring. In addition, single cases in a family could be the result of somatic cell mutations and thus would have a genetic basis but would not be heritable, and close relatives would not be expected to have an increased risk for AD. An equally plausible model is that only some cases of AD are genetic while most are caused by other factors such as infectious agents or environmental factors. Moreover, mechanisms related to the intrinsic biological aging may be important in AD. The fact that AD is a common disease in the elderly further complicates the study of AD genetics. Thus, family clustering could occur by chance, or multiple cases in a kindred could be due to a mixture of genetic and nongenetic causes.

The most convincing evidence that defective genes are involved in AD comes from the many early-onset families described in the literature[5-9] in which the disease appears to be inherited as an autosomal dominant trait. In some of the most dramatic kindreds, onset ages range from the early 30s to the mid-40s.[7] In these rare kindreds, approximately 50% of those at risk develop the disease. Since AD onset in the third and fourth decade of life is rare, clustering of AD in multiple generations in these families is clearly the result of inheritance and not chance.

The goal of molecular genetic studies of AD is to identify the genes responsible for the disease. In the past several years, remarkable progress has been made in identifying mutations responsible for familial AD (FAD) in at least some early-onset families.[10-15] Also, a new locus for early-onset FAD has been identified on chromosome 14.[16] This success in the study of early-onset cases has come, in part, because the mode of inheritance is unambiguous in these rare but important families. Late-onset families are also being studied with dramatically encouraging results.[17-20] Though study of these kindreds is more difficult due to the problems outlined above, late-onset AD is tremendously important since it is the most common form of AD. The recent identification of the ε4 allele of the apolipoprotein (apo) E gene as a risk factor for late-onset AD in unrelated subjects is a significant advance in unraveling the genetics of this disease.[21-25]

B. AMYLOID PRECURSOR PROTEIN GENE AND CHROMOSOME 21

The neuropathologic feature most characteristic of AD is amyloid deposits in the brain, either associated with blood vessels or found as extracellular deposits termed plaques. A principal component of AD amyloid is the Aβ protein or β-amyloid protein (Figure 13-1), a 39- to 43-amino-acid protein whose partial sequence was first determined by Glenner and Wong in 1984.[26] This same peptide is also a component of the amyloid deposits found in Down's syndrome.[27,28] The Aβ protein amino acid sequence was used to obtain a cDNA clone[29] derived from the messenger RNA transcript of the amyloid precursor protein (APP) gene. This gene encodes 3 major isoforms, 695, 751, and 770 amino acids in length,[30-32] which are normal cellular proteins found in the central nervous system as well as in many other peripheral tissues. These isoforms contain the Aβ sequence in the carboxyl terminal portion of the protein (Figure 13-2). Other minor alternative splice variants have also been observed (Figure 13-2); not all of these splice variants contain the Aβ protein sequence.[33] The APP gene maps to chromosome 21q21.1 (Figure 13-3);[34] this location has profound though poorly understood implications for the high prevalence of AD in Down's syndrome subjects (who have an extra copy of chromosome 21).

```
Asp-Ala-Glu-Phe-Arg-His-Asp-Ser-Gly-Try-Glu-Val-His-His-Gln-
 1   2   3   4   5   6   7   8   9  10  11  12  13  14  15

Lys-Leu-Val-Phe-Phe-Ala-Glu-Asp-Val-Gly-Ser-Asn-Lys-Gly-Ala-
16  17  18  19  20  21  22  23  24  25  26  27  28  29  30
        ↑
    Exon junction

Ile-Ile-Gly-Leu-Met-Val-Gly-Gly-Val-Val-Ile-Ala-Thr
31 32  33  34  35  36  37  38  39  40  41  42  43
```

FIGURE 13-1 Amino acid sequence of the Aβ protein. The sequence shown is for the 43-amino-acid form of the Aβ protein. Amino acid 1 is encoded by codon 672 of the APP gene (numbering based on the 770-amino-acid splice variant of the APP gene). The junction between exons 16 and 17 is indicated with an arrow. The solid line indicates the transmembrane domain portion of the protein.

1. Mutations in the APP Gene

Cloning of the APP gene permitted it to be tested as a candidate gene for FAD. Initial work appeared to eliminate the gene as being responsible for FAD.[7,35-37] DNA polymorphisms

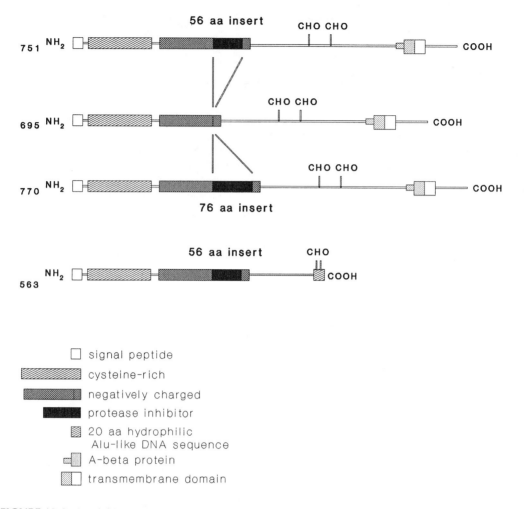

FIGURE 13-2 Amyloid precursor protein isoforms. Four splice variants of the APP gene are shown including the major forms (695, 751, and 770 amino acids)[30-33] and an isoform lacking the Aβ protein sequence (563 amino acids).[33] Other isoforms lacking the Aβ protein sequence[89] and an Aβ protein-containing form lacking exon 15[90] have also been identified. The protease inhibitor domain, consisting of 56 amino acids, is encoded by exon 7. The alternatively spliced exon 8, found in the 770-amino-acid form, encodes for 20 amino acids. The Aβ protein sequence is encoded for by exons 16 and 17; these 2 exons also encode for sequences flanking the amino- and carboxyl-terminal ends of the Aβ protein sequence.

associated with the gene did not show the same pattern of inheritance as the disease in a number of large early-onset families, and obligate recombinants between APP polymorphisms and FAD were observed.[7,35-37] When genetic linkage results for either early or late-onset families[36, 37] were pooled, the results were quite negative for the two groups, indicating that in most FAD families mutations in the APP gene were not responsible for AD. Because of these negative findings, the APP gene was ignored as an FAD candidate gene until Levy et al.[38] in 1990 identified a mutation in this gene responsible for hereditary cerebral hemorrhage with amyloidosis of the Dutch type (HCHWA-D). In this disease, amyloid deposits containing the Aβ-protein are found associated with cerebral blood vessels.[39] Polymorphisms in the APP gene cosegregate with the disease[40] and a point mutation in codon 693 (codon numbering based on the 770-amino-acid form of APP) was found in all affected subjects studied.[38] The mutation results in a glutamate to glutamine change at amino acid position 22 of the

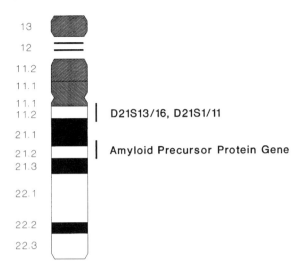

FIGURE 13-3 Chromosome 21 and the APP gene. Localization of the APP gene and genetic markers D21S13/16 and D21S1/11 are shown by the solid lines.

Aβ-protein (Figure 13-4). Identification of the HCHWA-D mutation demonstrated that a mutation in the APP gene could cause Aβ-protein deposition.

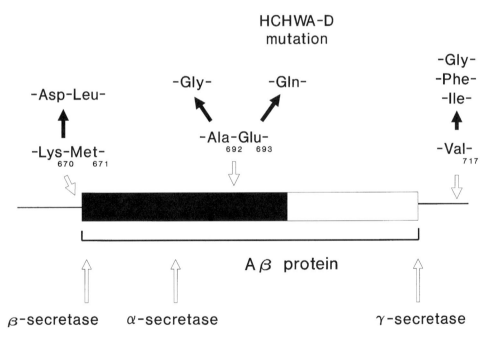

FIGURE 13-4 APP gene mutations. The thick bar represents the Aβ protein, with the hatched portion being the transmembrane domain segment. The lys-met → asp-leu AD double mutation is at codon 670/671. The ala → gly AD/cerebral hemorrhage disease mutation is at codon 692 (amino acid 21 of the Aβ protein). The glu → gln hereditary cerebral hemorrhage disease with amyloidosis mutation is at codon 693 (amino acid 22 of the Aβ protein). The val → ile, val → phe, and val → gly mutations are at codon 717, corresponding to the third amino acid past the last amino acid of the long Aβ protein sequence shown in Figure 13-1.

In 1991, Goate et al.[10] reevaluated the APP gene in an AD family in which APP polymorphisms cosegregated with the disease. Direct sequencing of APP exon 17 (which encodes part of the Aβ peptide sequence) revealed a mutation at codon 717 resulting in a valine to isoleucine change. Codon 717 is 3 amino acids past the C-terminal to the end of the Aβ-protein sequence (43-amino-acid form shown in Figure 13-4). The APP$_{717}$ mutation has subsequently been found to cosegregate with AD in a number of other early-onset kindreds[10-12,41] and has not been found in a large number of "sporadic" AD subjects or normal controls.[10,42-46] The absence of this mutation in normal controls indicates that it is not a rare polymorphism. The fact that this mutation is only observed in early-onset FAD kindreds clearly demonstrates that it is sufficient to initiate the pathogenesis of FAD. Two other APP mutations have been subsequently identified at the same codon in single early-onset FAD kindreds (Figure 13-4); one results in a valine to glycine substitution[15] and the other a valine to phenylalanine change.[14] Mullan and co-workers[12] subsequently identified a double FAD mutation which changes a lysine to asparagine and a methionine to leucine at codons 670 and 671, respectively (Figure 13-4). These changes occur at the two amino acids immediately before the N-terminal amino acid of the Aβ-protein. Another APP mutation has recently been described[13] in a Dutch family with both presenile dementia and cerebral hemorrhage disease in cognitively normal subjects. The mutation in this family is at codon 692 and results in the substitution of a glycine in place of the normal alanine at this position (Figure 13-4). The relationship of this disorder to FAD remains to be determined as only biopsy material has been studied to date.

Once specific APP mutations were defined, families in which AD was caused by APP mutations could be identified and compared for common clinical and neuropathologic similarities. Perhaps the most unifying clinical feature of these families is the age of onset. Onset ages range from approximately 42 to 61 years. Thus APP mutations cause a relatively early-onset form of AD. Clinical and neuropathologic examinations of APP-mutation AD cases have yet to reveal a clear subset of features unique to this type of AD.

2. APP Mutations and AD Pathogenesis

The location of the FAD mutations within the APP gene may provide information concerning the steps leading to Aβ-protein production and eventual amyloid deposition. The APP protein is normally processed by at least two proteolytic events. In one mechanism, termed the α-secretase pathway, a proteolytic cleavage event occurs at approximately amino acids 16 to 17 of the Aβ-protein sequence (Figure 13-1), and a soluble, truncated form of APP is secreted.[47,48] Because this cleavage event cuts the Aβ-protein sequence, no Aβ-protein is produced by this process. The presenile dementia/cerebral hemorrhage mutation described by Hendriks et al.[13] and the HCHWA-D mutations occur within the Aβ sequence at amino acids 21 and 22, respectively, close to the α-secretase site. Conceivably, these mutations could retard the destruction of the Aβ-protein and thus acceleration of its production, leading to AD.

A second normal APP processing mechanism leads to production and release of Aβ-protein from normal cells and tissues.[49,50] The Aβ-protein is released *in vitro* from culture cells derived from both normal and AD subjects and is also found in the cerebral spinal fluid from patients and controls. Thus, the Aβ-protein appears to be produced in the absence of disease and is probably a normal breakdown product generated when APP is degraded or processed. The possibility also exists that the Aβ-protein, subsequent to release from cells, has a specific, as-yet undefined function *in vivo*.

Proteolytic cuts must occur for the Aβ-peptide to be excised from APP. The FAD mutations immediately before the N-terminal amino acid and immediately after the C-terminal amino acids of the Aβ-protein could accelerate the proteolytic steps which excise the Aβ-protein from its precursor, APP. Recent *in vitro* work, using cell lines transfected with constructs expressing either the wild-type APP cDNA sequence or the APP cDNA sequence containing the 670 to 671 double mutation, indicates that there is a 5- to 8-fold increase in production of the Aβ-protein when the mutation is present.[51,52] Thus, the 670 to 671 mutation may cause AD by increasing the amount of Aβ-protein produced. Similarly, *in vitro* experiments have demonstrated that the APP_{717} isoleucine to valine mutation results in a 1.5- to 1.9-fold increase in secretion of a form of the Aβ protein which is 43 amino acids long.[53] Again, this work suggests that APP mutations may cause AD by increased production of the Aβ-protein.

C. EARLY-ONSET FAD LOCI

1. THE CHROMOSOME 14 FAD GENE

The work described above demonstrates that defects in the APP gene can cause AD in at least some early-onset families. However, linkage data[7,35-37,42,43] and mutation analysis[10,42-46,54] also clearly demonstrates that numerous early-onset kindreds do not have mutations in the APP gene. Despite an exhaustive search for APP mutation families, presently less than 15 have been identified worldwide. Perhaps, at most, 5% of early-onset kindreds have APP mutations. No APP mutations have been identified in late-onset families, although the possibility that a rare late-onset APP mutation exists cannot be excluded. Thus, the combined data indicate that FAD is genetically heterogeneous and that genes at locations other than the APP locus are capable of causing FAD.

Additional FAD genes were sought by genetic linkage analysis using markers located on all chromosomes. Highly polymorphic markers were tested for cosegregation with AD in early-onset families which did not have APP mutations. Using this genomic scanning approach, a locus responsible for early-onset FAD was identified on chromosome 14 at band q24 (Figure 13-5).[16] This gene is responsible for an autosomal dominant form of the disease in families where the mean ages of onset are very early, ranging between 35 and 52 years of age.[16,55-58] From the combined data, the chromosome 14 locus appears to be responsible for FAD in most early-onset kindreds in which FAD is not caused by APP mutations.

Positional cloning methods have recently been used to identify the chromosome 14 FAD gene. This approach to gene discovery uses the genetically determined location of the target gene to identify candidate genes. The region surrounding marker D14S77 was cloned using yeast artificial chromosomes. Genes contained within these clones were identified by a variety of methods and subsequently screened for mutations. By this approach, a gene initially called S182 was cloned and mutations found which co-segregated with FAD in a number of large early-onset kindreds.[59] This gene is also called Presenilin 1. The cDNA clone corresponding to S182 encodes a predicted protein of 467 amino acids. From the amino acid sequence, the protein is a membrane protein with 7 or more transmembrane domains. The normal function of the S182 gene product is unknown, and its role in AD pathogenesis remains to be determined. The amino acid sequence shows striking homology to a *Caenorhabditis elegans* gene product called *sel-12* which performs an undefined function in the Notch signaling pathway during development.[60]

2. THE VOLGA GERMAN KINDREDS AND THE CHROMOSOME 1 FAD GENE

APP mutations and the chromosome 14 locus do not explain all of early-onset FAD. In this respect, the Volga German families are a particularly interesting group of FAD pedigrees. These families are Germans who emigrated to the Volga River region of Russia and subsequently to the U.S.[5,6] FAD in these families is most probably the result of a common genetic

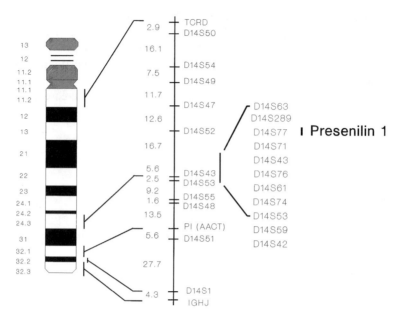

FIGURE 13-5 Genetic map of chromosome 14 and the AD locus. Chromosome 14 is shown along with the genetic map of polymorphic markers.[16] Genetic distances are given in centiMorgans. The approximate location of the chromosome 14 AD locus is given by a solid bar on the right.

founder and thus is likely to be genetically homogeneous. Age of onset in the group being used for genetic analysis occurs between 52 and 79 years for different families; the mean age of onset for the group is approximately 57 years. The chromosome 14 locus is not responsible for FAD in this group of families.[16]

The location of the gene responsible for FAD in the Volga Germans was identified by linkage analysis. As in the initial localization of the chromosome 14 FAD gene, markers on all chromosomes were tested for co-segregation with AD in the families. Markers on chromosome 1q31-42 showed evidence of linkage to AD with a maximum LOD score of 6.29 at a recombination fraction of 0.10.[61] The chromosome 1 FAD gene was identified as a homologue to the chromosome 14 FAD gene. When the sequence of S182 was used to search the GenBank database of DNA and protein sequences, an expressed sequence tagged (EST) sequence was identified which encoded a portion of a predicted protein which is closely homologous to S182. (ESTs are cDNA clones which have been randomly selected for DNA sequence analysis and the resulting sequences deposited in GenBank). When this EST was mapped, it was found to be located at chromosome 1q31-42,[62] the same location as the markers linked to AD in the Volga German kindreds. The gene was cloned using the EST sequence and was found to encode a 448 amino acid protein which was 67% identical to S182 and was initially designated STM2. This gene is also known as Presenilin 2. A missense mutation was identified in STM2 which changed an asparagine to an isoleucine. This mutation co-segregated with AD in the Volga German kindreds. STM2, like S182, is predicted to encode a transmembrane protein, whose function is presently unknown.

3. LATE-ONSET AD AND THE CHROMOSOME 14 LOCUS

Late-onset families have also been examined using chromosome 14 markers.[57,62,63] By linkage and affected pedigree member analysis, no evidence for involvement of the chromosome 14 locus in late-onset disease could be detected.[63] However, these methods are limited in that if a locus responsible for disease is acting in only a subset of the kindreds (genetic heterogeneity), the presence of that locus might not be detected. The ability to detect a locus in the presence of heterogeneity depends on the sample size and the fraction of families in which the locus is responsible for the disease. Thus, the chromosome 14 AD locus could be responsible for a subset of late-onset AD. Once the chromosome 14 locus is identified, late-onset cases can be screened for mutations in this gene.

D. CHROMOSOME 19 AD LOCUS

A combined genetic analysis and candidate gene analysis has identified the apoE gene on chromosome 19 as a late-onset AD gene. This gene is part of an apolipoprotein gene cluster at band q13.2 (Figure 13-6). The gene cluster contains three functional apolipoprotein genes, APOCII, APOCI, and APOE, and an APOCI pseudogene (Figure 13-6). Pericak-Vance and co-workers,[17] using both conventional linkage analysis, sib-pair and affected pedigree member methods, implicated this region of chromosome 19 in late-onset FAD. Subsequently, Strittmatter and co-workers[18] found that the Aβ protein could bind to the gene product of the apoE gene. Thus APOE became a prime candidate gene which could be tested by using known polymorphisms. In the coding region of APOE there are two sites which are commonly variable or polymorphic (Figure 13-7). At amino acid 112, either a cysteine or an arginine is found. At amino acid 158, again, either a cysteine or an arginine is present. Three APOE haplotypes are commonly observed; ε2 which is cys_{112}-cys_{158}, ε3 which is cys_{112}-arg_{158}, and ε4 which is arg_{112}-arg_{158}. These are found in Caucasian populations at allele frequencies of 0.08, 0.77, and 0.15, respectively. This polymorphic system was used to evaluate the role of the APOE in AD by genetic association analysis. In late-onset familial AD pedigrees, the ε4 allele frequency was 0.52 in cases, which was significantly different from the control frequency of 0.15.[18,19] This work has been confirmed in other collections of late-onset FAD families[20,21] and provides strong evidence for a genetic association between the ε4 allele and late-onset FAD.

The association between the ε4 allele and AD has also been observed in groups of unrelated AD patients[21-25] in clinic-based,[21,22,24,25] community-based,[23,64] and population-based[23,64] studies. In Caucasians, ε4 allele frequencies for AD cases range from 0.24[65] to 0.40.[21] In a Japanese population, the ε4 frequency was 0.276 in AD cases and 0.093 in controls.[22] Thus, ε4 allele is clearly a risk factor for late-onset AD, the most common form of the disease. Finally, an allelic association between the ε4 allele and early-onset AD has also been reported in a population-based study.[66] The early-onset cases in this study were not from APP mutation or chromosome 14 families. While the ε4 allele is a strong risk factor for AD, the ε4 allele is not necessary for AD to occur. For example, in the study of Poirier et al.[24] approximately 40% of all AD cases did not have a copy of the ε4 allele.

The ε4 allele appears to confer increased risk of AD by lowering the age-of-onset of the disease. Corder et al.,[19] studying a group of families selected to have multiple cases of AD (familial AD), found that ε4/ε4 homozygotes had an age of onset of 68.4 years, compared to 75.5 years for subjects having only a single allele and 84.3 years for subjects having no ε4 alleles. In this study of AD families, the risk of an ε4/ε4 homozygote subject developing AD approached 90% by age 80 years. However, since the subjects were from families heavily loaded with AD cases, caution must be used in extrapolating conclusions from this study to AD in the general population. While the ε4 allele increases risk of developing AD, the ε2 allele appears to have a protective effect.[67-70] AD cases from both familial AD series and from unrelated probands have a lower ε2 frequency compared to controls. In studies of unrelated AD probands, the ε2 frequency ranged from 0.005[69] to 0.025[67] compared to control frequencies of 0.06[68] to

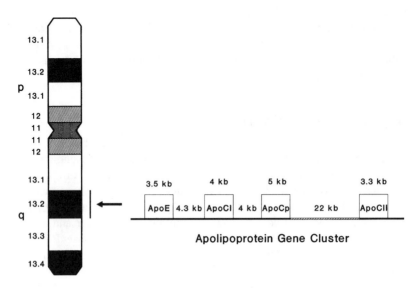

FIGURE 13-6 Chromosome 19 and the apolipoprotein gene cluster. The organization of the chromosome 19 apolipoprotein gene cluster is given, with approximate sizes of the genes and the distances between the genes given in kilobases (kb).

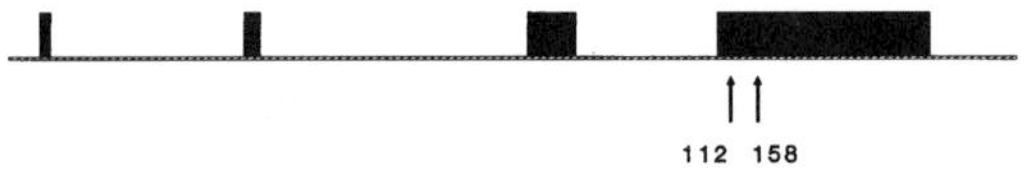

Allele	Amino Acid 112	Amino Acid 158	Frequencies
e2	Cys	Cys	0.080
e3	Cys	Arg	0.769
e4	Arg	Arg	0.150

FIGURE 13-7 ApoE gene structure and polymorphic sites. Amino acids 112 and 158 are variable in the general population. Typical Caucasian frequencies for the ε2, ε3, and ε4 alleles are given in the table.

0.105[67]. The same protective effect was also observed in AD cases from familial AD kindreds, with affected subjects having an ε2 frequency of 0.013.

The combined genetic evidence suggests that the apoE gene is involved in AD. However, another possible interpretation of the genetic findings is that the APOE polymorphic sites are in linkage disequilibrium with a gene or DNA segment which is the true risk factor. Several lines of evidence suggests that the chromosome 19 risk factor is the APOE gene itself. **First**, the genetic association between the ε4 allele and AD has been observed in a large number of different populations, including different racial groups (Japanese and Caucasian). **Second**, since ε4 appears to be a risk factor and ε2 a protective factor, a second locus in linkage disequilibrium would also have to have both protective and risk-incurring alleles. **Third**, since the ε2, ε3, and ε4 isoforms are in the coding region of APOE and are known to have different biologic properties with respect to LDL receptor binding, the hypothesis that these alleles have different properties in AD pathogenesis is more believable.

Molecular Genetics of Alzheimer's Disease

1. Interaction of APOE ε4 and Other AD Loci

The potential interaction of APOE genotypes with expression of AD caused by the chromosome 14 locus and APP mutations has been explored. No allelic association between the ε4 allele and AD was observed in families in which the chromosome 14 locus is responsible for AD.[21,71,72] Also, unlike late-onset AD, the APOE genotype does not appear to influence the onset age in these kindreds.[71,73] In contrast, the APOE genotype may influence onset ages in families with some of the APP mutations. In families with the $APP_{670/671}$ mutation, 3 individuals with an ε2/ε3 genotype had the latest ages of onset (57 to 60 years), 3 subjects with an ε3/ε3 genotype had an intermediate onset age (51 to 54 years) and 1 subject who was ε3/ε4 had the youngest onset age (44 years).[74] Similarly, in a family with a APP_{717} Val → Ileu mutation, 3 affected subjects with an ε3/ε4 genotype had an onset mean of 47.6 ± 3.0 years; a single mutation carrier with an APOE genotype of ε2/ε3 remained unaffected at an age 2 standard deviations greater than the mean for the family.[75] Thus, in APP families, APOE alleles may influence age of onset. Unfortunately, due to the scarcity of APP-mutation families, the subjects available for study are very limited. In contrast, APOE genotypes do not appear to affect age of onset in subjects with APP mutations in either codons 692 (mixed dementia and cerebral hemorrhage) or codon 693 (hereditary cerebral hemorrhage with amyloidosis, Dutch type).[76] APOE genotypes also do not appear to affect the age of onset in the Volga German families.[73]

2. APOE and Potential Pathogenic Mechanisms

The ε4 allele of the APOE gene is clearly a risk factor for AD while the ε2 allele may be protective. However, the role of the gene product, the apoE protein, in AD remains to be determined. The following are several possible mechanisms for the role of apoE in AD pathogenesis.

a. ApoE as a Molecular Chaperon for the Aβ Protein

Several studies have demonstrated by immunohistochemical methods that amyloid plaques in AD brain contain apoE.[18,25,77-79] This finding led Wisniewski and Frangione[79] to propose that apoE is a molecular chaperon protein for various amyloidogenic proteins. Strittmatter et al.[18,80] and others[80] have reported that *in vitro*, apoE binds the Aβ-protein and that the ε4 isoform of apoE binds to the Aβ-protein with a higher avidity than ε3 or ε4. Also, plaque density appears to be higher in brains for ε4/ε4 homozygous individuals.[25,77] Thus an apoE-Aβ complex could be an intermediate in plaque formation. As an extension of this molecular chaperon hypothesis, Hyman and co-workers[25] proposed that apoE binds Aβ and removes it from the extracellular space via uptake mediated by either the LDL or the LDL receptor-related protein (LRP) receptor. Allele-specific interactions with these receptors or subsequent steps in degradation could explain the genetic association of AD with the ε4 allele.

b. ApoE Interactions With Tau

Strittmatter and co-workers[82] reported that apoE ε3 binds to tau while the ε4 isoform does not. Tau is a microtubule-associated protein which, when abnormally phosphorylated, forms the paired helical filaments which make neurofibrillary tangles (NFTs), one of the hallmarks of AD neuropathology. ApoE ε3, through its binding to tau, could prevent formation of NFTs.

c. **ApoE as a Neuronal Injury-Response Protein**

ApoE is a nerve injury-response protein. Its production is elevated several hundredfold in sciatic nerve following injury.[83-85] ApoE levels are also elevated, to a lesser extent, in a model of nerve injury in the central nervous system.[86] The presumed role of apoE in nerve repair, at least in the peripheral nervous system, is as a lipid transport protein which mobilizes cholesterol after injury.[83-85] During regeneration, apoE lipid particles are taken up via the LDL receptor at the growth cone of extending neurites.[87] The levels of apolipoproteins apoD, apoA-I, and apoA-IV are also elevated in the sciatic nerve injury model.[85] The ability of apoE complexed to lipids to promote neuronal growth may be isoform dependent. In experiments with cultures of dorsal root ganglion neurons in the presence of β-VLDL, the ε3 allele promoted neurite outgrowth while the ε4 allele inhibited both outgrowth and branching.[88] Thus the observed genetic association of ε4 with AD could be the result of reduced neurite extension and repair in response to some type of injury.

E. CONCLUSIONS

The work summarized above demonstrates that FAD is genetically heterogeneous. Early-onset FAD can be caused either by mutations in the APP gene, or by mutations in either Presenilin 1 or Presenilin 2. There may be at least one other locus which remains to be identified which can cause early-onset disease. Late-onset FAD may be caused in part by a locus on chromosome 19. As with early-onset disease, there are probably additional late-onset loci at chromosomal locations which remain to be identified.

Identification of all the FAD genes is important for several reasons. First, completely resolving the genetics of FAD by identifying all the genes and mutations involved will permit addressing the question of whether all AD is genetic and determining the role environmental factors play in the disease. Second, identification of genes which when defective cause FAD will yield critical information concerning the pathway(s) which change the normal brain into an AD brain with resulting neuropathology and dementia. Understanding the normal function of the Presenilin gene products as well as understanding how mutations in these genes cause AD should greatly expand our understanding of AD pathogenesis.

REFERENCES

1. Van Duijn, C. M., Clayton, D., Chandra, V., Fratiglioni, L., Graves, A. B., Heyman, A., Jorm, A. F., Kokmen, E., Kondo, K., Mortimer, J. A., Rocca, W. A., Shalat, S. L., Soininen, H., and Hofman, A. F., Familial aggregation of Alzheimer's disease and related disorders — a collaborative re-analysis of case-control studies, *Int. J. Epidemiol.*, 20, S13-S20, 1991.
2. Nee, L. E., Eldridge, R., Sunderland, T., Thomas, C. B., Katz, D., Thompson, K. E., Weingartner, H., Weiss, H., Julian, C., and Cohen, R., Dementia of the Alzheimer type. Clinical and family studies of 22 twin pairs, *Neurology*, 37, 359-363, 1987.
3. Rapoport, S. I., Pettigrew, K. D., and Schapiro, M. B., Discordance and concordance of dementia of the Alzheimer type (DAT) in monozygotic twins indicate heritable and sporadic forms of Alzheimer's disease, *Neurology*, 41, 1549-1553, 1991.
4. Mohs, R. C., Breitner, J. C. S., Silverman, J. M., and Davis, K. L., Alzheimer's disease. Morbid risk among first-degree relatives approximates 50% by age 90, *Arch. Gen. Psychiatry*, 44, 405-408, 1987.
5. Bird, T. D., Lampe, T. H., Nemens, E. J., Miner, G. W., Sumi, S. M., and Schellenberg, G. D., Familial Alzheimer's disease in American descendants of the Volga Germans: probable genetic founder effect, *Ann. Neurol.*, 23, 25-31, 1988.
6. Bird, T. D., Sumi, S. M., Nemens, E. J., Nochlin, D., Schellenberg, G. D., Lampe, T. H., Sadovnick, A., Chui, H., Miner, G. W., and Tinklenberg, J., Phenotypic heterogeneity in familial Alzheimer's disease: a study of 24 kindreds, *Ann. Neurol.*, 25, 12-25, 1989.

7. Van Broeckhoven, C., Genthe, A. M., Vandenberghe, A., Horsthemke, B., Backhovens, H., Raeymaekers, P., Van Hul, W., Wehnert, A., Gheuens, J., Cras, P., Bruyland, M., Martin, J. J., Salbaum, M., Multhaup, G., Masters, C. L., Beyreuther, K., Gurling, H. M. D., Mullan, M. J., Holland, A., Barton, A., Irving, N., Williamson, R., Richards, S. J., and Hardy, J. A., Failure of familial Alzheimer's disease to segregate with the A-4 amyloid gene in several European families, *Nature*, 329, 153-155, 1987.
8. St. George-Hyslop, P. H., Tanzi, R. E., Polinsky, R. J., Haines, J. L., Nee, L., Watkins, P. C., Myers, R. H., Feldman, R. G., Pollen, D., Drachman, D., Growdon, J., Bruni, A., Foncin, J.-F., Salmon, D., Frommelt, P., Amaducci, L., Sorbi, S., Piacentini, S., Stewart, G. D., Hobbs, W. J., Conneally, P. M., and Gusella, J. F., The genetic defect causing familial Alzheimer's disease maps on chromosome 21, *Science*, 235, 885-890, 1987.
9. Heston, L. L., Orr, H. T., Rich, S. S., and White, J. A., Linkage of an Alzheimer disease susceptibility locus to markers on human chromosome 21, *Am. J. Med. Genet.*, 40, 449-453, 1991.
10. Goate, A. M., Chartier-Harlin, C. M., Mullan, M., Brown, J., Crawford, F., Fidani, L., Giuffra, L., Haynes, A., Irving, N., James, L., Mant, R., Newton, P., Rooke, K., Roques, P., Talbot, C., Pericak-Vance, M. A., Roses, A. D., Williamson, R., Rossor, M., Owen, M., and Hardy, J., Segregation of a missense mutation in the amyloid precursor protein gene with familial Alzheimer's disease, *Nature*, 349, 704-706, 1991.
11. Naruse, S., Igarashi, S., Aoki, K., Kaneko, K., Iihara, K., Miyatake, T., Kobayashi, H., Inuzuka, T., Shimizu, T., Kojima, T., and Tsuji, S., Mis-sense mutation Val-Ile in exon 17 of amyloid precursor protein gene in Japanese familial Alzheimer's disease, *Lancet*, 337, 978-979, 1991.
12. Mullan, M., Crawford, F., Axelman, K., Houlden, H., Lulius, L., Winblad, B., and Lannfelt, L., A pathogenic mutation for probable Alzheimer's disease in the APP gene at the N-terminus of β-amyloid, *Nat. Genet.*, 1, 345-347, 1992.
13. Hendriks, L., Van Duijn, C. M., Cras, P., Cruts, M., Van Hul, W., Van Harskamp, F., Warren, A., McInnis, M. G., Antonarakis, S. E., Martin, J.-J., Hofman, A., and Van Broeckhoven, C., Presenile dementia and cerebral haemorrhage linked to a mutation at codon 692 of the ß-amyloid precursor gene, *Nat. Genet.*, 1, 218-221, 1992.
14. Murrell, J., Farlow, M., Ghetti, B., and Benson, M. D., A mutation in the amyloid precursor protein associated with hereditary Alzheimer's disease, *Science*, 254, 97-99, 1991.
15. Chartier-Harlin, M.-C., Crawford, F., Houlden, H., Warren, A., Hughes, D., Fidani, L., Goate, A., Rossor, M., Roques, P., Hardy, J., and Mullan, M., Early-onset Alzheimer's disease caused by mutation at codon 717 of the β-amyloid precursor protein gene, *Nature*, 353, 844-846, 1991.
16. Schellenberg, G. D., Bird, T. D., Wijsman, E. M., Orr, H. T., Anderson, L., Nemens, E., White, J. A., Bonnycastle, L., Weber, J. L., Alonso, M. E., Potter, H., Heston, L. L., and Martin, G. M., Genetic linkage evidence for a familial Alzheimer disease locus on chromosome 14, *Science*, 258, 668-671, 1992.
17. Pericak-Vance, M. A., Bebout, J. L., Gaskell, P. C., Yamaoka, L. A., Hung, W.-Y., Alberts, M. J., Walker, A. P., Bartlett, R. J., Haynes, C. S., Welsh, K. A., Earl, N. L., Heyman, A., Clark, C. M., and Roses, A. D., Linkage studies in familial Alzheimer's disease: evidence for chromosome 19 linkage, *Am. J. Hum. Genet.*, 48, 1034-1050, 1991.
18. Strittmatter, W. J., Saunders, A. M., Schmechel, D., Pericak-Vance, M., Enghild, J., Salvesen, G. S., and Roses, A. D., Apolipoprotein-E: high-avidity binding to beta-amyloid and increased frequency of type 4 allele in late-onset familial Alzheimer disease, *Proc. Natl. Acad. Sci. U.S.A.*, 90, 1977-1981, 1993.
19. Corder, E. H., Saunders, A. M., Strittmatter, W. J., Schmechel, D. E., Gaskell, P. C., Small, G. W., Roses, A. D., Haines, J. L., and Pericak-Vance, M. A., Gene dose of apolipoprotein-E type-4 allele and the risk of Alzheimer's disease in late onset families, *Science*, 261, 921-923, 1993.
20. Payami, H., Kaye, J., Heston, L. L., Bird, T. D., and Schellenberg, G. D., Apolipoprotein-E genotype and Alzheimer's disease, *Lancet*, 342, 738, 1993.
21. Saunders, A. M., Strittmatter, W. J., Schmechel, D., St. George-Hyslop, P. H., Pericak-Vance, M. A., Joo, S. H., Rosi, B. L., Gusella, J. F., Crapper-MacLachlan, D. R., Alberts, M. J., Hulette, C., Crain, B., Goldgaber, D., and Roses, A. D., Association of apolipoprotein-E Allele epsilon-4 with late-onset familial and sporadic Alzheimer's disease, *Neurology*, 43, 1467-1472, 1993.
22. Noguchi, S., Murakami, K., and Yamada, N., Apolipoprotein-E genotype and Alzheimer's disease, *Lancet*, 342, 737, 1993.
23. Mayeux, R., Stern, Y., Ottman, R., Tatemichi, T. K., Tang, M.-X., Maestre, G., Nagai, C., Tycko, B., and Ginsberg, H., The apolipoprotein e4 allele in patients with Alzheimer's disease, *Ann. Neurol.*, 34, 752-754, 1993.
24. Poirier, J., Davignon, J., Bouthillier, D., Kogan, S., Bertrand, P., and Gauthier, S., Apolipoprotein-E polymorphism and Alzheimer's disease, *Lancet*, 342, 697-699, 1993.
25. Rebeck, G. W., Reiter, J. S., Strickland, D. K., and Hyman, B. T., Apolipoprotein E in sporadic Alzheimer's disease: allelic variation and receptor interactions, *Neuron*, 11, 575-580, 1993.
26. Glenner, G. G. and Wong, C. W., Alzheimer's disease. Initial report of the purification and characterization of a novel cerebrovascular amyloid protein, *Biochem. Biophys. Res. Commun.*, 120, 885-890, 1984.
27. Wong, C. W., Quaranta, V., and Glenner, G. G., Neuritic plaques and cerebrovascular amyloid in Alzheimer disease are antigenically related, *Proc. Natl. Acad. Sci. U.S.A.*, 82, 8729-8732, 1985.
28. Masters, C. L., Simms, G., Weinman, N. A., Multhaup, G., McDonald, B. L., and Beyreuther, K., Amyloid plaque core protein in Alzheimer disease and Down syndrome, *Proc. Natl. Acad. Sci. U.S.A.*, 82, 4245-4249, 1985.

29. Kang, J., Lemaire, H.-G., Unterbeck, A., Salbaum, J. M., Masters, C. L., Grzeschik, K.-H., Multhaup, G., Beyreuther, K., and Muller-Hill, B., The precursor of Alzheimer's disease amyloid A4 protein resembles a cell-surface receptor, *Nature*, 325, 733-736, 1987.
30. Tanzi, R. E., McClatchey, A. I., Lamperti, E. D., Villa-Komaroff, L., Gusella, J. F., and Neve, R. L., Protease inhibitor domain encoded by an amyloid protein precursor mRNA associated with Alzheimer's disease, *Nature*, 331, 528-530, 1988.
31. Kitaguchi, N., Takahashi, Y., Tokushima, Y., Shiojiri, S., and Ito, H., Novel precursor of Alzheimer's disease amyloid protein shows protease inhibitory activity, *Nature*, 331, 530-532, 1988.
32. Ponte, P., Gonzalez-DeWhitt, P., Schilling, J., Miller, J., Hsu, D., Greenberg, B., Davies, K., Wallace, W., Lieberburg, I., Fuller, F., and Cordell, B., A new A4 amyloid mRNA contains a domain homologous to serine protease inhibitors, *Nature*, 331, 525-527, 1988.
33. De Sauvage, F. and Octave, J.-N., A novel mRNA of the A4 amyloid precursor gene coding for a possible secreted protein, *Science*, 245, 651-653, 1989.
34. Tanzi, R. E., Gusella, J. F., Watkins, P. C., Bruns, G. A. P., St. George-Hyslop, P., Van Keuren, M. L., Patterson, D., Pagan, S., Kurnit, D. M., and Neve, R. L., Amyloid beta protein gene: cDNA, mRNA distribution, and genetic linkage near the Alzheimer locus, *Science*, 235, 880-884, 1987.
35. Tanzi, R. E., St. George-Hyslop, P. H., Haines, J. L., Polinsky, R. J., Nee, L., Foncins, J.-F., Neve, R. L., McClatchey, A. I., Conneally, P. M., and Gusella, J. F., The genetic defect in familial Alzheimer's disease is not tightly linked to the amyloid beta-protein gene, *Nature*, 329, 156-157, 1987.
36. Pericak-Vance, M. A., Yamaoka, L. H., Haynes, C. S., Speer, M. C., Haines, J. L., Gaskell, P. C., Hung, W.-Y., Clark, C. M., Heyman, A. L., Trofatter, J. A., Eisenmenger, J. P., Gilbert, J. R., Lee, J. E., Alberts, M. J., Dawson, D. V., Bartlett, R. J., Earl, N. L., Siddique, T., Vance, J. M., Conneally, P. M., and Roses, A. D., Genetic linkage studies in Alzheimer's disease families, *Exp. Neurol.*, 102, 271-279, 1988.
37. Schellenberg, G. D., Bird, T. D., Wijsman, E. M., Moore, D. K., Boehnke, M., Bryant, E. M., Lampe, T. H., Sumi, S. M., Deeb, S. M., Beyreuther, K., and Martin, G. M., Absence of linkage of chromosome 21 q21 markers to familial Alzheimer's disease in autopsy-documented pedigrees, *Science*, 241, 1507-1510, 1988.
38. Levy, E., Carman, M. D., Fernandez-Madrid, I. J., Power, M. D., Lieberburg, I., van Duinen, S. G., Bots, G. Th. A. M., Luyendijk, W., and Frangione, B., Mutation of the Alzheimer's disease amyloid gene in hereditary cerebral hemorrhage, Dutch type, *Science*, 248, 1124-1126, 1990.
39. Van Duinen, S. G., Castano, E. M., Prelli, F., Bots, T. A. B. G., Luyendijk, W., and Frangione, B., Hereditary cerebral hemorrhage with amyloidosis in patients of Dutch origin is related to Alzheimer's disease, *Proc. Natl. Acad. Sci. U.S.A.*, 84, 5991-5994, 1987.
40. Van Broeckhoven, C., Haan, J., Bakker, E., Hardy, J. A., Van Hul, W., Wehnert, A., Vegter-Van der Vlis, M., and Roos, R. A. C., Amyloid β protein precursor gene and hereditary cerebral hemorrhage with amyloidosis (Dutch), *Science*, 248, 1120-1122, 1991.
41. Karlinsky, H., Vaula, G., Haines, J. L., Ridgley, J., Bergeron, C., Mortilla, M., Tupler, R. G., Percy, M. E., Robitaille, Y., Noldy, N. E., Yip, T. C. K., Tanzi, R. E., Gusella, J. F., Becker, R., Berg, J. M., Crapper-McLachlan, D. R., and St. George-Hyslop, P. H., Molecular and prospective phenotypic characterization of a pedigree with familial Alzheimer's disease and a missense mutation in codon 717 of the beta-amyloid precursor protein gene, *Neurology*, 42, 1445-1453, 1992.
42. Schellenberg, G. D., Anderson, L.-J., O'dahl, S., Wijsman, E. M., Sadovnick, A. D., Ball, M. J., Larson, E. B., Kukull, W. A., Martin, G. M., Roses, A. D., and Bird, T. D., APP_{717}, APP_{693}, and PRIP gene mutations are rare in Alzheimer disease, *Am. J. Hum. Genet.*, 49, 511-517, 1991.
43. Kamino, K., Orr, H. T., Payami, H., Wijsman, E. M., Alonso, M. E., Pulst, S. M., Anderson, L., O'Dahl, S., Nemens, E., White, J. A., Sadovnick, A. D., Ball, M. J., Kaye, J., Warren, A., McInnis, M., Antonarakis, S. E., Korenberg, J. R., Sharma, V., Kukull, W., Larson, E., Heston, L. L., Martin, G. M., Bird, T. D., and Schellenberg, G. D., Linkage and mutational analysis of familial Alzheimer disease kindreds for the APP gene region, *Am. J. Hum. Genet.*, 51, 998-1014, 1992.
44. Chartier-Harlin, M.-C., Crawford, F., Hamandi, K., Mullan, M., Goate, A., Hardy, J., Backhovens, H., Martin, J.-J., and Van Broeckhoven, C., Screening for the β-amyloid precursor mutation (APP717: Val → Ile) in extended pedigrees with early-onset Alzheimer's disease, *Neurosci. Lett.*, 129, 134-135, 1991.
45. Tanzi R. E., Vaula, G., Romano, D. M., Mortilla, M., Huang, T. L., Tupler, R. G., Wasco, W., Hyman, B. T., Haines, J. L., Jenkins, B. J., Kalaitsidaki, M., Warren, A. C., McInnis, M. C., Antonarakis, S. E., Karlinsky, H., Percy, M. E., Connor, L., Growden, J., Crapper-Mclachlan, D. R., Gusella, J. F., and St. George-Hyslop, P. H., Assessment of amyloid β-protein precursor gene mutations in a large set of familial and sporadic Alzheimer disease cases, *Am. J. Hum. Genet.*, 51, 273-282, 1992.
46. Van Duijn, C. M., Hendriks, L., Cruts, M., Hardy, J. A., Hofman, A., and Van Broeckhoven, C., Amyloid precursor protein gene mutation in early-onset Alzheimer's disease, *Lancet*, 337, 978, 1991.
47. Sisodia, S. S., Koo, E. H., Beyreuther, K., Unterbeck, A., and Price, D. L., Evidence that the β-amyloid protein in Alzheimer's disease is not derived by normal processing, *Science*, 248, 492-495, 1990.
48. Esch, F. S., Keim, P. S., Beattie, E. C., Blacher, R. W., Culwell, A. R., Oltersdorf, T., McClure, D., and Ward, P. J., Cleavage of amyloid β peptide during constitutive processing of its precursor, *Science*, 248, 1122-1124, 1990.

49. Shoji, M., Golde, T. E., Ghiso, J., Cheung, T. T., Estus, S., Shaffer, L. M., Cai, X.-D., McKay, D. M., Tintner, R., Frangione, B., and Younkin, S. G., Production of the Alzheimer amyloid beta protein by normal proteolytic processing, *Science*, 258, 126-129, 1992.
50. Seubert, P., Vigo-Pelfrey, C., Esch, F., Lee, M., Dovey, H., Davis, D., Sinha, S., Schlossmacher, M., Whaley, J., Swindlehurst, C., McCormack, R., Wolfert, R., Selkoe, D., Lieberburg, I., and Schenk, D., Isolation and quantification of soluble Alzheimer's beta peptide in biological fluids, *Nature*, 359, 325-327, 1992.
51. Cai, X.-D., Golde, T. E., and Younkin, S. G., Release of excess amyloid beta protein from a mutant amyloid beta protein precursor, *Science*, 259, 514-516, 1993.
52. Citron, M., Oltersdorf, T., Haass, C., McConlogue, L., Hung, A. Y., Seubert, P., Vigo-Pelfrey, C., Lieberburg, I., and Selkoe, D. J., Mutation of the beta-amyloid precursor protein in familial Alzheimer's disease increases beta-protein production, *Nature*, 360, 672-674, 1992.
53. Suzuki, N., Cheung, T. T., Cai, X.-D., Odaka, A., Otvos, L., Eckman, C., Golde, T. E., and Younkin, S. G., An increased percentage of long amyloid beta protein secreted by familial amyloid beta protein precursor (beta APP(717)) mutants, *Science*, 264, 1336-1340, 1994.
54. Schellenberg, G. D., Pericak-Vance, M. A., Wijsman, E. M., Moore, D. K., Gaskell, P. C., Yamaoka, L. A., Bebout, J. L., Anderson, L., Welsh, K. A., Clark, C. M., Martin, G. M., Roses, A. D., and Bird, T. D., Linkage analysis of familial Alzheimer's disease using chromosome 21 markers, *Am. J. Hum. Genet.*, 48, 563-583, 1991.
55. Mullan, M., Houlden, H., Windelspecht, M., Fidani, L., Lombardi, C., Diaz, P., Rossor, M., Crook, R., Hardy, J., Duff, K., and Crawford, F., A locus for familial early-onset Alzheimer's disease on the long arm of chromosome 14, proximal to the α1-antichymotrypsin gene, *Nat. Genet.*, 2, 340-342, 1992.
56. Van Broeckhoven, C., Backhovens, H., Cruts, M., De Winter, G., Bruyland, M., Cras, P., and Martin, J-J., Mapping of a gene predisposing to early-onset Alzheimer's disease to chromosome 14q24.3, *Nat. Genet.*, 2, 335-339, 1992.
57. St. George-Hyslop, P., Haines, J., Rogaev, E., Mortilla, M., Vaula, G., Pericak-Vance, M., Foncin, J.-F., Montesi, M., Bruni, A., Sorbi, S., Rainero, I., Pinessi, L., Pollen, D., Polinsky, R., Nee, L., Kennedy, J., Macciardi, F., Rogeava, E., Liang, Y., Alexandrova, N., Lukiw, W., Schlumpf, K., Tanzi, R., Tsuda, T., Farrer, L., Cantu, J.-M., Duara, R., Amaducci, L., Bergamini, L., Gusella, J., Roses, A. D., and Crapper McLachlan, D., Genetic evidence for a novel familial Alzheimer's disease locus on chromosome 14, *Nat. Genet.*, 2, 330-334, 1992.
58. Nechiporuk, A., Fain, P., Kort, E., Nee, L. E., Frommelt, E., Polinsky, R. J., Korenberg, J. R., and Pulst, S. M., Linkage of familial Alzheimer disease to chromosome-14 in 2 large early-onset pedigrees. Effects of marker allele frequencies on LOD scores, *Am. J. Med. Genet.*, 48, 63-66, 1993.
59. Sherrington, R. E., Rogaev, I., Liang, Y., Rogaeva, E. A., Levesque, G., Ikeda, M., Chi, H., Lin, C., Li, G., Holman, K., Tsuda, T., Mar, L., Foncin, J. F., Bruni, A. C., Montesi, M. P., Sorbi, S., Rainero, I., Pinessi, L., Nee, L., Chumakov, I., Pollen, D., Brookes, A., Sanseau, P., Polinsky, R. J., Wasco, W., Dasilva, H. A. R., Haines, J. L., Pericak-Vance, M. A., Tanzi, R. E., Roses, A. D., Fraser, P. E., Rommens, J. M., and St George-Hyslop, P. H., Cloning of a gene bearing missense mutations in early-onset familial Alzheimer's disease, *Nature*, 375, 754-760, 1995.
60. Levitan, D. and Greenwald, I., Facilitation of lin-12-mediated signalling by sel-12, a Caenorhabditis elegans S182 Alzheimer's disease gene, *Nature*, 377, 351-354, 1995.
61. Levy-Lahad, E., Wijsman, E. M., Nemens, E., Anderson, L., Goddard, K. A. B., Weber, J. L., Bird, T. D., and Schellenberg, G. D., A familial Alzheimer's disease locus on chromosome 1, *Science*, 269, 970-973, 1995.
62. Levy-Lahad, E., Wasco, W., Poorkaj, P., Romano, D. M., Oshima, J., Pettingell, W. H., Yu, C. E., Jondro, P. D., Schmidt, S. D., Wang, K., Crowley, A. C., Fu, Y. H., Guenette, S. Y., Galas, D., Nemens, E., Wijsman, E. M., Bird, T. D., Schellenberg, G. D., and Tanzi, R. E., Candidate gene for the chromosome 1 familial Alzheimer's disease locus, *Science*, 269, 973-977, 1995.
63. Schellenberg, G. D., Payami, H., Wijsman, E. M., Orr, H. T., Goddard, K. A. B., Anderson, L., Nemens, E., White, J. A., Alonso, M. E., Ball, M. J., Kaye, J., Morris, J. C., Chui, H., Sadovnick, A. D., Heston, L. L., Martin, G. M., and Bird, T. D., Chromosome-14 and late-onset familial Alzheimer disease (FAD), *Am. J. Hum. Genet.*, 53, 619-628, 1993.
64. Tsai, M.-S., Tangalos, E. G., Petersen, R. C., Smith, G. E., Schaid, D. J., Kokmen, E., Ivnik, R. J., and Thibodeau, S. N., Apolipoprotein E: risk factor for Alzheimer disease, *Am. J. Hum. Genet.*, 54, 643-649, 1994.
65. Brousseau, T., Legrain, S., Berr, C., Gourlet, V., Vidal, O., and Amouyel, P., Confirmation of the epsilon 4 allele of the apolipoprotein E gene as a risk factor for late-onset Alzheimer's disease, *Neurology*, 44, 342-344, 1994.
66. van Duijn, C. M., de Kniff, P., Cruts, M., Wehnert, A., Havekes, L. M., Hofman, A., and Van Broeckhoven, C., Apolipoprotein e4 allele in a population-based study of early-onset Alzheimer's disease, *Nat. Genet.*, 7, 74-78, 1994.
67. Corder, E. H., Saunders, A. M., Risch, N. J., Strittmatter, W. J., Schmechel, D. E., Gaskell, P. C., Rimmler, J. B., Locke, P. A., Conneally, P. M., Schmader, K. E., Small, G. W., Roses, A. D., Haines, J. L., and Pericak-Vance, M. A., Protective effect of apolipoprotein E type 2 allele for late onset Alzheimer disease, *Nat. Genet.*, 7, 180-184, 1994.
68. Smith, A. D., Johnston, C., Sim, E., Nagy, Z., Jobst, K. A., Hindley, N., and King, S. C., Protective effect of apoE epsilon 2 in Alzheimer's disease, *Lancet*, 344, 473-474, 1994.

69. Talbot, C., Lendon, C., Craddock, N., Shears, S., Morris, J. C., and Goate, A., Protection against Alzheimer's disease with apoE epsilon 2, *Lancet*, 343, 1432-1433, 1994.
70. Benjamin, R., Leake, A., McArthur, F. K., Ince, P. G., Candy, J. M., Edwardson, J. A., Morris, C. M., and Bjertness, E., Protective effect of apoE epsilon 2 in Alzheimer's disease. *Lancet*, 344, 473, 1994.
71. Van Broeckhoven, C., Backhovens, H., Cruts, M., Martin, J. J., Crook, R., Houlden, H., and Hardy, J., APOE genotype does not modulate age of onset in families with chromosome-14 encoded Alzheimer's disease, *Neurosci. Lett.*, 169, 179-180, 1994.
72. Yu, C., Payami, H., Olson, J. M., Boehnke, M., Wijsman, E. M., Orr, H. T., Kukull, W. A., Goddard, K. A. B., Nemens, E., White, J. E., Alonso, M. E., Taylor, T. D., Ball, M. J., Kaye, J., Morris, J., Chui, H., Sadovnick, A. D., Martin, G. M., Larson, E. B., Heston, L. L., Bird, T. D., and Schellenberg, G. D., The apolipoprotein E/CI/CII gene cluster and late-onset Alzheimer's disease, *Am. J. Hum. Genet.*, 54, 631-642, 1994.
73. Levy-Lehad, E., Lahad, A., Wijsman, E. M., Bird, T. D., and Schellenberg, G. D., Apolipoprotein E genotypes and age-of-onset in early-onset familial Alzheimer's disease, *Am. Neurol.*, 38, 678, 1995.
74. Hardy, J., Houlden, H., Collinge, J., Kennedy, A., Newman, S., Rossor, M., Lannfelt, L., Lilius, L., Winblad, B., Crook, R., and Duff, K., Apolipoprotein-E genotype and Alzheimer's disease, *Lancet*, 342, 737-738, 1993.
75. St. George-Hyslop, P., Crapper McLachlan, D., Tuda, T., Rogaev, E., Karlinsky, H., Lippa, C. F., and Pollen, D., Alzheimer's disease and possible gene interaction, *Science*, 263, 537, 1994.
76. Haan, J., Van Broeckhoven, C., van Duijn, C. M., Voorhoeve, E., Van Harskamp, F., van Swieten, J. C., Maats-Chieman, M. L. C., Roos, R. A. C., and Bakker, E., The apolipoprotein E epsilon 4 allele does not influence the clinical expression of the amyloid precursor protein gene codon 693 or 692 mutations, *Ann. Neurol.*, 36, 434-437, 1994.
77. Schmechel, D. E., Saunders, A. M., Strittmatter, W. J., Crain, B. J., Hulette, C. M., Joo, S. H., Pericak-Vance, M. A., Goldgaber, D., and Roses, A. D., Increased amyloid beta-peptide deposition in cerebral cortex as a consequence of apolipoprotein-E genotype in late-onset Alzheimer disease, *Proc. Natl. Acad. Sci. U.S.A.*, 90, 9649-9653, 1993.
78. Namba, Y., Tomonaga, M., Kawasaki, H., Otomo, E., and Ikeda, K., Apolipoprotein E immunoreactivity in cerebral amyloid deposits and neurofibrillary tangles in Alzheimer's disease and kuru plaque amyloid in Creutzfeldt-Jakob disease, *Brain Res.*, 541, 163-166, 1991.
79. Wisniewski, T. and Frangione, B., Apolipoprotein E: a pathological chaperon protein, *Neurosci. Lett.*, 135, 235-238, 1992.
80. Strittmatter, W. J., Weisgraber, K. H., Huang, D. Y., Dong, L. M., Salvesen, G. S., Pericak-Vance, M., Schmechel, D., Saunders, A. M., Goldgaber, D., and Roses, A. D., Binding of human apolipoprotein-E to synthetic amyloid beta peptide. Isoform-specific effects and implications for late-onset Alzheimer disease, *Proc. Natl. Acad. Sci. U.S.A.*, 90, 8098-8102, 1993.
81. Wisniewski, T., Golabek, A., Matsubara, E., Ghiso, J., and Frangione, B., Apolipoprotein E: binding to soluble Alzheimer's beta-amyloid, *Biochem. Biophys. Res. Commun.*, 192, 359-365, 1993.
82. Strittmatter, W. J., Weisgraber, K. H., Goedert, M., Saunders, A. M., Huang, D., Corder, E. H., Dong, L., Jakes, R., Alberts, M. J., Gilbert, J. R., Han, S. H., Hulette, C., Einstein, G., Schmechel, D. E., Pericakvance, M. A., and Roses, A. D., Hypothesis: microtubule instability and paired helical filament formation in the Alzheimer disease brain are related to apolipoprotein E genotype, *Exp. Neurol.*, 125, 163-171, 1994.
83. Mahley R. W., Apolipoprotein E: cholesterol transport protein with expanding role in cell biology, *Science*, 240, 622-630, 1988.
84. Boyles J. K., Zoellner, C. D., Anderson, L. J., Kosik, L. M., Pitas, R. E., Weisgraber, K. H., Hui, D. Y., Mahley, R. W., Gebicke-Haerter, P. J., Ignatius, M. J., and Shooter, E. M., A role for apolipoprotein E, apolipoprotein A-1, and low density lipoprotein receptors in cholesterol transport during regeneration and remyelination of the rat sciatic nerve, *J. Clin. Invest.*, 83, 1015-1031, 1989.
85. Boyles, J. K., Notterpek, L. M., and Anderson, L. J., Accumulation of apolipoproteins in the regenerating and remyelinating mammalian peripheral nerve, *J. Biol. Chem.*, 265, 17805-17815, 1990.
86. Poirier, J., Hess, M., May, P. C., and Finch, C. E., Astrocytic apolipoprotein E mRNA and GFAP mRNA in hippocampus after entorhinal cortex lesioning, *Mol. Brain. Res.*, 11, 97-106, 1991.
87. Ignatius, M. J., Shooter, E. M., Pitas, R. E., and Mahley, R. W., Lipoprotein uptake by neuronal growth cones in vitro, *Science*, 236, 959-962, 1987.
88. Nathan, B. P., Bellosta, S., Sanan, D. A., Weisgraber, K. H., Mahley, R. W., and Pitas, R. E., Differential effects of apolipoproteins E3 and E4 on neuronal growth *in vitro*, *Science*, 264, 850-852, 1994.
89. Jacobson, J. S., Muenkel, H. A., Blume, A. J., and Vitek, M. P., A novel species-specific RNA related to alternatively spliced amyloid precursor protein mRNAs, *Neurobiol. Aging*, 12, 575-583, 1991.
90. Konig, G., Monning, U., Czech, C., Prior, R., Banati, R., Schreiter-Gasser, U., Bauer, J., Masters, C. L., and Beyreuther, K., Identification and differential expression of a novel alternative splice isoform of the beta A4 amyloid protein precursor protein (APP) mRNA in leukocytes and brain microglial cells, *J. Biol. Chem.*, 267, 10804-10809, 1992.

14 Polygenic Inheritance In Psychiatric Disorders

David E. Comings

CONTENTS

Abstract 236
A. Introduction 236
B. Inappropriate Use of the Autosomal Dominant Model 236
C. Association Studies 238
    1. Linkage Disequilibrium 238
    2. Some Pedigree Examples 238
D. Genetic Loading — Clinical Genetics 240
E. Genetic Loading — Molecular Genetics 241
    1. General Considerations 241
    2. The *Controls Without, Cases Without, Cases With* Strategy 242
F. Regression Analysis 242
G. Spectrum Disorders 243
H. Examples of Association Studies 243
    1. The DRD2 Gene 243
    2. The DRD2 Gene and Drug Addiction 244
    3. The DRD2 Gene and Impulse Disorders 244
    4. Posttraumatic Stress Disorder 245
    5. Pathological Gambling 247
    6. Smoking 247
    7. Obesity and Height 247
    8. The DRD3 Gene 247
    9. Tryptophan 2,3-Dioxygenase 248
    10. HRAS 249
    11. Dopamine β Hydroxylase 249
    12. Dopamine Transporter 249
    13. Additive Effect of Three Dopamine Genes 249
    14. MAO 250
    15. Serotonin 1A Receptor Gene 250
I. Polygenic Spectrum Model of Behavioral Disorders 250
J. Lessons for Psychiatric Genetics 251
    1. Polygenic Convergence 251
    2. The Ethnicity Issue 251
    3. The "Real" Mutations 253
    4. Mutations in Nonexonic DNA 254
    5. Role of Repeat Polymorphism in Polygenes 254

6. Implications for Linkage Studies 254
  7. Implications for Association Studies 255
K. The Future 255
References 256

ABSTRACT: While family, twin, and adoption studies have clearly demonstrated a role of genes in many human behavioral disorders, there has been little success in the identification of which genes are involved. It is proposed that the reason for this is that the wrong genetic models and thus the wrong techniques are being used. The most popular model is that of a rare, disease-specific, autosomal dominant gene with reduced penetrance and the assumption that the mutations are in exons. The most popular technique, based on this model, is linkage analysis using large families. The author proposes that the correct model, capable of giving identical-appearing pedigrees, is that of polygenic inheritance. The best technique for identifying the genes involved in such a model is the use of association studies with large numbers of severely affected probands compared to unrelated controls. It is also suggested that the *polygenes* (mutant genes involved in polygenic inheritance) are not disease specific but are involved in a spectrum of disorders and are fundamentally different from those involved in single-gene disorders in that they have a much milder effect on gene function and tend to involve non-exon sequences. As such, the carrier rate in the population can be high. Their deleterious effect comes when individuals inherit a greater than threshold number of polygenes. By binding transcription factors, dinucleotide, trinucleotide, and other repeat polymorphisms may affect gene function and thus may be one cause of polygenes. One of the distinctive characteristics of polygenic inheritance is that the genes are contributed by both parents, and in psychiatric disorders the relatives on both the maternal and paternal sides often demonstrate an increased frequency of a spectrum of behavioral disorders.

A. INTRODUCTION

Family, twin, and adoption studies have clearly indicated the important role of genes in a wide range of human behavioral disorders. However, in this era of molecular biology, where the genes for almost every important nonpsychiatric genetic disorder have been discovered, to simply say that genes play a role in behavior is inadequate. The longer we fail to identify *which* genes play a role, the more skeptics will come to believe that no genes play a role in determining how we behave.[1]

The field of psychiatric genetics was, or at least appeared to be, healthier several years ago than it is now. Studies had purported to show that linkage analysis had identified the location of the major genes in manic-depressive disorder,[2,3] schizophrenia,[4] and dyslexia,[5] and similar linkage studies were being vigorously pursued to identify the genes for Tourette's syndrome, panic disorder, obsessive-compulsive disorder, alcoholism, autism, and other major behavioral disorders. However, the claims for linkage to manic-depressive disorder [6,7] and schizophrenia [8] have been withdrawn and that for dyslexia has been considerably modified.[9,10] Over 550 markers have been tested in the search for linkage in Tourette syndrome,[11] and 250 markers have been examined in the Amish manic-depressive pedigrees,[12] with no positive results. The linkage between male homosexuality (a nonpsychiatric disorder) and Xq28 [13] is one apparent exception.

B. INAPPROPRIATE USE OF THE AUTOSOMAL DOMINANT MODEL

Why has the search for genes in psychiatric disorders been so barren while the search for the genes causing other genetic disorders has been so successful? The answer is simple and straightforward. We have been seduced by the very success that the study of other genetic disorders has enjoyed. We may be like the man in the night searching only under the street light for his lost billfold because, "That is where the light is." The model that has worked so well for Huntington's disease, myotonic dystrophy, neurofibromatosis, and many other disorders is that of a single, rare gene always present in those with the disorder and, once corrected for age of onset and penetrance, always absent from those without the disorder. With only minor modifications, this model has been reapplied to the field of psychiatric genetics using the same assumptions. However, in this process a few not so minor details were overlooked or ignored. These are

1. Unlike most of the clear-cut autosomal dominant diseases, in psychiatric disorders many of the family members who are obligatory carriers of the gene are unaffected. This is usually considered to be due to reduced penetrance. However, alternate explanations for reduced penetrance (see below) are rarely cited.
2. Unlike the clear-cut autosomal dominant disorders, for psychiatric disorders it is often necessary to screen many families to find pedigrees suitable for linkage studies that "fit" the autosomal dominant model. If it were not for deceased members, the majority of Huntington's disease families would be informative for linkage studies. By contrast, in Tourette's syndrome, for example, where the autosomal dominant model has been used extensively [14-16] and when tics alone, or tics and obsessive-compulsive behaviors are used as the phenotype, a distinct minority of families are suitable for linkage studies.
3. In clear-cut autosomal dominant disorders, the pedigrees almost always show that the gene is coming from either the mother's or the father's side (unilineal), but not both (bilineal). This is not true of the behavioral disorders. In Tourette's syndrome we reported that in 8.2% of 170 TS families chronic tics were bilineal and in 45.3% of the families some behavioral disorder was present on both sides.[17] Two of the large published TS pedigrees used for linkage studies show bilineal families[18,19] and one is from a religious isolate, a situation well known for enhancement of recessive genes. Others have also reported the presence of bilaterality in many TS cases.[20-22] In the Amish pedigree, where the gene linkage for manic depressive disorder came and went, there is at least one bilineal family. In a genetic study of manic-depressive disorder (Type I) Simpson et al.[23] examined 1800 probands. Most had to be rejected for linkage studies because of bilinearity, small size, or lack of availability for further study. Of 34 useful families, 12 were unilineal, 10 probably unilineal, and 12 probably bilineal. This study showed that the majority of the available bipolar I families were bilineal. This is an important issue because bilineal families imply recessive or polygenic inheritance, while unilineal families imply dominant inheritance. Even though the predominant model for manic depressive disorder is that of autosomal dominant inheritance, these results suggest most receive a genetic contribution from both parents.
4. In clear-cut autosomal dominant disorders the phenotype has well-defined borders with little or no comorbidity for a wide range of other disorders. By contrast, as much as we might like to believe the behavioral disorders we are studying also have sharply defined borders, they do not. Of individuals with manic-depression, depression, panic attacks, phobias, attention deficit hyperactivity disorder (ADHD), obsessive-compulsive disorder, drug dependence, dyslexia, and Tourette syndrome, up to 80% have other comorbid behavioral disorders. While many researchers claim to have eliminated this problem by selecting for study only those with the "pure" disorder, this may be an act of double deception. First, those with no comorbid conditions are often the least severely affected. If the genes that are being sought actually cause a spectrum of disorders, this selective process eliminates individuals with the greatest degree of expression and with the greatest degree of genetic loading for the genes in question. Secondly, by selecting out those with comorbid conditions, we may get in the habit of believing in the artifact we have produced, i.e., that these disorders are due to disease-specific genes. In a similar vein, it is tempting to perform segregation analysis on the pedigrees collected for linkage studies, and in the process forget they have been artificially selected to show a specific mode of inheritance. If families are collected because they show an autosomal dominant mode of inheritance, it is a foregone conclusion that segregation analysis will show an autosomal dominant mode of inheritance.

5. In clear-cut autosomal dominant disorders, the unaffected members of the family don't show an increased prevalence of psychiatric disorders such as alcoholism, depression, anxiety disorder, and others. By contrast, in our experience, for psychiatric disorders, regardless of the diagnosis in the proband, the relatives show a significantly increased frequency of a spectrum of affective, addictive, mood, and anxiety disorders. Others report the same results.[24-27] This suggests that most psychiatric disorders share many genes in common.[28-30]

The lesson of the above is that several other genetic models can produce pedigrees that look exactly like those resulting from an autosomal dominant trait with reduced penetrance. If autosomal dominant inheritance is assumed, when in fact a different genetic model is involved, linkage studies may be so severely impacted as to be totally useless. The failure to find genes in psychiatric disorders may not be because they are not there, but simply because the wrong models and thus the wrong techniques are being used to search for them. What is needed is a light with batteries that allow us to look more carefully into the darkness.

C. ASSOCIATION STUDIES

We will argue here that: (1) association studies provide the light that is needed; (2) the disease-specific autosomal dominant reduced penetrance model is seriously flawed; (3) classical LOD score linkage studies may never work for some psychiatric disorders; and (4) polygenic inheritance of a spectrum of behaviors is the correct model for most or all psychiatric disorders.

Association studies involve determining if one of the alleles of a genetic polymorphism, located within or close to a specific gene, are present more often in a specific behavioral disorder than in racially matched controls. Even a short time ago, this technique was unlikely to succeed because so few polymorphism were available and few neurobiologically important genes had been cloned. Now the situation is almost totally reversed. So many genes that may have an important role in behavior have been cloned that it is difficult for one laboratory to test them all.

Unlike linkage studies, association studies are independent of the mechanism of inheritance. If the chosen candidate gene is involved and if enough polymorphisms are examined, association studies will give a positive result regardless of nature of the disorder, the frequency of the gene, or the mechanism of inheritance.

1. Linkage Disequilibrium

Association studies are possible because of a phenomenon called *linkage disequilibrium*. This refers to a context wherein two mutations are relatively so close together that crossovers rarely occur between them. Most genes, including introns and exons, are 10 to 40 kb in length and these distances are well within those involved in linkage disequilibrium. The power of the technique comes from the fact that it is relatively easy, once a gene is cloned, to find a genetic polymorphism within or close to the gene. Even if that polymorphism is only in partial linkage disequilibrium with another as-yet undiscovered mutation that affects the function of the gene, if variations in function of that gene play a role in the expression of the behavior disorder being tested and if enough probands and enough suitably matched controls are tested, there will be a difference in the frequency of one of the alleles in the affected individuals compared to controls.

2. Some Pedigree Examples

Figure 14-1 illustrates some of the principles discussed. Figure 14-1A illustrates a small pedigree typical of that seen in some TS and other behavior disorder families. The proband,

II-1 (arrow), has severe TS, his brother, II-2, has chronic motor tics. Neither parent has tics, but the father's brother has severe TS. The identical pedigree is shown in B, C, and D.

Figure 14-1B is an *autosomal dominant incomplete penetrance model*. Here the pattern of inheritance is assumed to be due to a rare, autosomal dominant gene with incomplete penetrance. Thus, a single mutant *Gts* gene, X is segregating in the family and coming from the father's side, since the father's brother has TS. However, the father has no symptoms and thus appears to represent incomplete penetrance of the *Gts* gene. The same gene is fully expressed in the proband but more mildly expressed in the brother with chronic motor tics. If the model is correct, as it has been assumed to be in TS linkage studies, this type of pedigree is well suited for such linkage studies. In larger families, with many mildly affected relatives, this approach essentially represents a comparison of mildly affected individuals, such as II-2, who are assumed to carry the gene, against unaffected individuals, such as II-3, who are assumed to not carry the gene.

Figure 14-1C is a *semidominant-semirecessive inheritance model*. The identical inheritance pattern can be produced by a common, semidominant-semirecessive gene.[17] Here the severely affected proband (arrow), and the severely affected uncle, are assumed to carry a double dose of the gene (recessive-like), while the mildly affected brother carries a single dose (dominant-like), and the "nonpenetrant" father carries a single dose (recessive-like). If one assumes that this pedigree was produced by an autosomal dominant trait with reduced penetrance as in A, when in fact it was produced by a semidominant-semirecessive trait, linkage studies would fail since so many individuals would be mislabeled. For example, the mother and the unaffected brother would be assumed to not carry the *Gts* gene, when in fact they were heterozygous for it, and the proband and the uncle would be assumed to be heterozygous for the *Gts* gene when in fact they were homozygous for it. Only the unaffected father would be correctly identified as a heterozygous *Gts* gene carrier. Linkage studies using a recessive model would also fail since the brother with chronic motor tics would be assumed to be homozygous when in fact he was heterozygous for the *Gts* gene. For this form of inheritance, the most reliable technique to identify the affected genes would be the use of association studies comparing many probands such as II-1 to many unrelated controls.

Figure 14-1D is a *polygenic model*. The identical pedigree can also be produced by polygenic inheritance. A polygenic threshold model might require, as an example, the presence of five mutant genes (capital letters) with homozygosity for two of them (I-1), or six mutant genes in the heterozygous state (II-2). Different affected members do not necessarily have the same set of mutant genes. Here the proband (II-1) is affected because he carries the B and D genes in a heterozygous state and the AA, CC, and FF genes in a homozygous state. The uncle (I-1) is affected because he carries the B, E, and F genes in a heterozygous state and the AA and DD genes in a homozygous state. The brother (II-2) with chronic motor tics is mildly affected because although he has A, B, C, D, E, and F in the heterozygous state, none are homozygous. All the remaining members of the pedigree are unaffected because they do not meet threshold conditions. Controls unrelated to these families have significantly fewer of the relevant mutant genes.

Simple inspection of this pedigree shows why linkage studies would fail to identify any of the mutant genes involved. For example, if the A gene was being studied and the autosomal dominant model assumed, the proband, uncle, and brother would be assumed to carry the mutant A gene, which they do. However, since the three unaffected members also carry the mutant A gene, linkage would be negative, when in fact the mutant A gene was involved. The reader can progressively move through all the other B to F genes and see that linkage studies, using either the autosomal dominant or recessive model, would consistently fail.

While association studies would show a tendency for affected members to have a higher frequency of the mutant gene than unaffected members, the high genetic loading of these unaffected members makes this very inefficient. The most reliable and sensitive technique would be association studies comparing the frequency of each mutant gene in severely affected

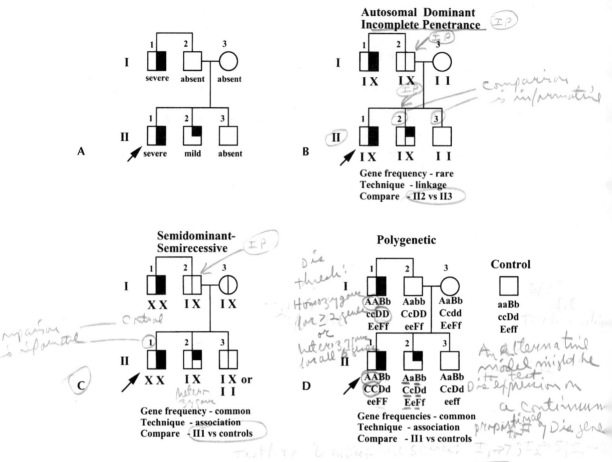

FIGURE 14-1 Comparison of three modes of inheritance autosomal dominant models with reduced penetrance, semidominant-semirecessive, and polygenic. See text for implications for linkage and association studies.

probands to unrelated controls carefully screened to exclude TS and many of its associated behaviors.

The purpose of this exercise is to show that the same type of pedigree could be produced by three different mechanisms of inheritance. If the investigator is assuming the first to be in operation, when in fact either of the latter two are occurring, it is very unlikely that linkage studies would find the genes involved. This appears to be exactly what is occurring in the field of psychiatric genetics, i.e., massive numbers of negative linkage studies are purporting to have ruled out genes that may in fact be playing an important role in the diseases being studied.

E. GENETIC LOADING — CLINICAL GENETICS

The classical methods of examining the role of genetic factors in human behavior have been family, twin, and adoption studies. As powerful as these are they provide no information on the specific genes involved. We have explored a new technique that we call genetic loading. The concept is to start with a disorder that is already accepted as being genetic, then examine the frequency of various behaviors in groups of individuals with varying degrees of genetic loading for that disorder. Tourette syndrome is an ideal disorder for this purpose because it is an established genetic disorder with a high frequency of a wide range of behavioral disorders in probands.[31,32] The groups with progressively less loading for the Tourette syndrome (*Gts*) genes were TS probands, relatives with TS, relatives without TS, and controls. A significant progressive decrease in the frequency of a given behavior across these groups would be a strong indication that the behavior was due, at least in part, to the *Gts* genes. However, one could still complain that the results were due to a high frequency of the behavior in probands, due to proband ascertainment bias,[33] or that there was inappropriate matching of the controls for socioeconomic and other factors. Both of these potential problems can be eliminated by comparing the frequency of the behaviors in relatives with TS vs. relatives without TS. This technique has shown that the *Gts* genes play a role in a wide number of disorders including drug [34] and alcohol abuse,[35] ADHD,[36] sexual disorders,[37] functional GI symptoms,[38] depression,[39] conduct and oppositional defiant disorder,[40] and others.[32] Figure 14-2 illustrates the results using this technique in studies of the genetic factors in conduct and oppositional defiant disorder.[40] These combined results indicate TS is a neuropsychiatric spectrum disorder.

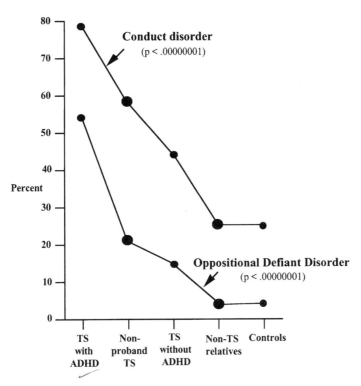

FIGURE 14-2 The results of the genetic loading technique for symptoms of conduct disorder (CD) and oppositional defiant disorder (ODD) in groups with a progressive decrease in genetic loading for *Gts* genes from TS probands with ADHD, to relative with TS (nonproband TS), probands without ADHD, relatives without TS, and controls. The larger dots show the more critical comparison of nonproband TS vs. relative without TS. The differences in the frequency of CD and ODD in these two groups were also highly significant.

F. GENETIC LOADING — MOLECULAR GENETICS

1. General Considerations

After we had observed how well the genetic loading technique worked for testing the role of genetic factors in a range of behaviors, we realized that the same approach could provide increased power to molecular genetic association studies. Many association studies have simply compared the frequency of specific alleles in controls vs. patients with various disorders. If there was no difference it was assumed the gene played no role in the disorder in question. However, such an approach can lack power in three different ways.

1. Inadequate number of cases studied. Our experience with over 10 different genes suggests that at least 200 to 300 subjects need to be included in the study (controls + patients) to provide the necessary power to exclude an association. Many studies have claimed exclusion using less than 100 subjects.
2. Failure to examine controls. Since many of the behaviors being studied are very common, it is critical to examine the controls with the same structured instruments used to examine the patients, and to exclude not only controls with the behavior in question, but controls with related comorbid aspects of the spectrum disorder. For example, in studying TS, because of the high comorbidity with ADHD, OCD, and substance abuse, it is important to screen controls to exclude not only tics, but each of these comorbid conditions.
3. Examination of a range of behaviors. Some genes may be preferentially associated with specific behaviors that are present in some patients but not in all. For example, in the study of TS, one might hypothesize that a certain gene associated with serotonin metabolism might be more strongly associated with depression and attempted suicide than with tics. Of 300 TS probands, only 40 may have had depression with one or more suicide attempts. The serotonin gene polymorphism could show no association with tics, but a very strong association with the subset who had attempted suicide.

2. THE *CONTROLS WITHOUT, CASES WITHOUT, CASES WITH* STRATEGY

To accommodate these considerations in our studies of TS, each subject, whether a TS proband, relative, or control, is screened using an extensive questionnaire based on the Diagnostic Interview Schedule and the DSM diagnostic criteria. This allows the examination of different specific sets of behaviors including drug and alcohol abuse, learning disorders, tics, ADHD, depression, stuttering, phobias, obsessive-compulsive, and other behaviors. Over 300 subjects are then placed into 3 groups: controls without alcohol or drug abuse or the behavior in question *(controls without)*, TS probands or relatives without the behavior in question *(cases without)*, and TS probands or relatives with the behavior in question *(cases with)*. The *a priori* hypothesis is that if an allele of a candidate gene is associated with a given behavior, the frequency of that allele should progressively increase with increased genetic loading across these three groups — *controls without*, *cases without*, and *cases with*. The statistic used is a chi square for a linear trend. We have observed numerous instances where there was little or no significant difference in the frequency of specific alleles between controls and TS probands as a whole, but highly significant differences for a number of specific behaviors using the *controls without, cases without, cases with* strategy.

F. REGRESSION ANALYSIS

Another important and powerful type of statistic is the use of regression analysis. Each of the behaviors discussed above can be examined in the form of a continuous variable depending upon the number of a specific set of symptoms present. For example, in examining ADHD, the number of positive responses to 22 different questions relating to the DSM criteria for ADHD are added for each individual to provide an ADHD score. In addition, individuals carrying the allele in question are scored as 1, while those not carrying this allele are scored as 0. A regression analysis then provides a correlation coefficient r, and r^2 represents the percent of the variance of the ADHD score that can be attributed to that allele. This provides an important and sobering figure. In virtually all of the cases we have examined, despite significant results by *the controls without, cases without, cases with* strategy, a specific allele rarely accounts for more that 3 to 4% of the variance of specific behavior score. This is consistent with the polygenic, multifactorial nature of these behaviors and explains why standard linkage analyses fail to identify the role of specific genes. Once the variance for a behavior falls below 30%, linkage studies lose the power required to identify an effect.

G. SPECTRUM DISORDERS

One of the central concepts of the single-gene, autosomal dominant hypothesis is that the mutant gene is disease specific. Thus, a mutation of the *huntingtin* gene causes Huntington's disease and has no effect on muscular dystrophy or neurofibromatosis. When the autosomal dominant model is transported to psychiatric genetics, this same concept tends to travel with it. However, most psychiatric disorders show a great deal of comorbidity for other psychiatric disorders. This is particularly well illustrated in Tourette syndrome. While the presence of chronic motor and vocal tics is required for the diagnosis, there is a high degree of comorbidity with ADHD, obsessive compulsive disorder, phobias, panic attacks, depression, mood swings, sleep disorders, substance abuse, and inappropriate sexual behaviors.[28,41,42] It has been suggested that these are artifacts of proband ascertainment bias.[43] This explanation can be ruled in or out by examining the frequency of the comorbid behavioral disorders in relatives of TS probands who themselves have TS or chronic tics (nonproband TS) and comparing them to relatives without tics. When this is done, we observed a highly significant increase in the frequency of the same wide range of comorbid behaviors in the relatives with TS as in the probands.[34,35,37-39] These observations suggest that the genes responsible for producing a sufficient neurochemical imbalance to cause tics can also cause a wide range of other behavioral disorders. This concept of a range of behavioral disorders having a common pathophysiology has been proposed by others as well.[27,44-49] This concept is supported by studies of specific genes, such as the DRD2 gene, genetic variants of which show an association with a range of impulsive, compulsive, and addictive behaviors (see below).

Many view TS as a "rare" disorder of limited interest to the general psychiatric community. By contrast, we believe that psychiatric disorders are caused by the chance convergence of a sufficient number of polygenes affecting neurotransmitter function to result in a significant disruption of dopamine/serotonin/norepinephrine balance producing a spectrum of symptoms. When the imbalance is sufficiently severe, one of the symptoms can be physical, i.e., tics. Thus, instead of being some largely irrelevant "rare" disorder, TS and the presence of tics should be viewed as observable physical evidence that identifies those individuals who have accumulated a threshold number of neuropsychiatric polygenes. As such, they present an unusually ideal group for studies of psychiatric genetics.

H. EXAMPLES OF ASSOCIATION STUDIES

The following are some examples where association studies have strongly implicated a number of genes that had been "excluded" by classical linkage studies.

1. THE DRD2 GENE

While many association studies in psychiatric disorders have been done with blood groups and other genes,[50,51] the first important marker using molecular biological techniques was reported by Blum et al.[52] They found a significant association between the *Taq I* A1 variant of the dopamine D2 receptor (DRD2) gene and severe alcoholism. Others have both confirmed and refuted this relationship and this story is presented elsewhere in this volume. Some who have questioned the relationship of the D2A1 variant to alcoholism have generalized this to conclude that variants of the DRD2 gene do not play a role in any behavioral disorder.[53] We will present some of our results with other behavioral disorders to illustrate that not only is this an important gene in psychiatric genetics, but that the results provide important clues as to why linkage studies have so far largely failed.

2. THE DRD2 GENE AND DRUG ADDICTION

One of the major reasons that a role of the DRD2 gene in addictive behaviors seems so "right" is that the major reward pathway in humans utilizes dopaminergic neurons.[54-56] A genetic defect in this pathway provides an attractive model for the role of genes in increasing the risk for vulnerability to addictive behaviors. However, DiChiara and Imperato [57] showed that dextroamphetamine and cocaine caused a far greater perturbation of the dopamine metabolism of this pathway than alcohol. This led us to suspect that if the D2A1 variant was involved in addictive behaviors it might be easier to detect, with less controversy, in drug addiction or polysubstance abuse than in alcoholism per se. To test this we examined 200 non-Hispanic Caucasians on a V.A. Addiction Treatment Ward and divided them into four categories: alcohol abuse only, alcohol dependence only, drug and alcohol abuse/dependence, and drug abuse/dependence only.[58] This showed that there was a significant increase in the prevalence of the D2A1 allele (42.3%) only in those with both drug and alcohol abuse/dependence, i.e., polysubstance abuse. Further intragroup comparisons showed that of the more severely affected drug and alcohol addicts, i.e., those spending more than $25/week on two or more substances, 56.9% carried the D2A1 allele vs. 28.2% of those abusing a single drug (p <.0005). By logistic regression analysis there was a significant association with age of onset of drug abuse (p <.0001). There was also a significant increase in the D2A1 allele in those who had been jailed for a violent crime (53.1%) vs. those jailed only for DUI's (28.8%). If they had both been jailed for a violent crime and had a childhood history of having been expelled from school for fighting, the prevalence of the D2A1 allele increased to 69.2%.

In contrast to the story with alcoholism, to date there has been no conflict between the four published studies of drug abuse/dependence, with all four agreeing there is a significant association between severe problems with drug addiction and the prevalence of the D2A1 or D2B1 allele.[58-61]

3. THE DRD2 GENE AND IMPULSE DISORDERS

The DSM-IIIR lists attention deficit hyperactivity disorder (ADHD), TS, and conduct disorder as impulse disorders. The role of the childhood impulse disorders as risk factors in the development of alcoholism and drug addiction in adults has been repeatedly stressed.[34,35] Because of this, and our long-term interest in TS and ADHD, and the frequent hypotheses that defects in dopamine metabolism are involved in both,[28] we were interested in determining if there was an association with the D2A1 allele in these disorders. These studies showed a significant increase in the prevalence of the D2A1 allele in TS, ADHD, conduct disorder, autism, and posttraumatic stress disorder.[29] Associations were not found with depression, schizophrenia, or Parkinson's disease. The more severe cases of TS had the highest prevalence of the D2A1 marker. Devor subsequently reported a similar increase in prevalence of the D2A1 allele in his TS patients and a correlation with comorbid ADHD or obsessive-compulsive disorder.[62] We verified a correlation with global severity in a larger series.[63] This is shown in Table 14-1. The prevalence of the D2A1 allele increased from 35.0% in mild, to 40.4%

in moderate, to 55.5% in severe TS. Gelernter et al.[64] also found a correlation between the presence of the D2A1 allele and severity. However, because of the relatively small number of subjects studied, this was not significant. When compared with our studies, the results were virtually identical (Table 14-1). The prevalence of the D2A1 allele in mild cases in the two studies was 35.0 and 31.8%, and in moderate to severe cases it was 44.5 and 45.8%. In both studies the differences in the moderate to severe and total groups were significant at p <.002. In a subsequent report, Gelernter et al.[65] found that 51.6% of 64 TS or chronic motor tics subjects carried the D2A1 allele. This was significantly higher than the 25.9% carrier rate in 714 nonalcoholic, nondrug-abusing, non-Hispanic Caucasians ($p = .00001$).[66]

TABLE 14-1
Correlation Between the Prevalence of the D2A1 Variant and Severity in Tourette Syndrome

Tourette Syndrome	N	% A1	X^2	p
Controls[a]	444	26.1	—	—
Comings et al.[63]				
Grade 1 (mild)	20	35.0	.77	NS
Grade 2 (moderate)	146	40.4	9.73	.002
Grade 3 (severe)	54	55.5	20.12	.00001
Total	220	43.6	20.75	.00001
Grade 2 + grade 3	200	44.5	21.45	<.00001
Gelernter et al.[64]				
Mild (CMT)	22	31.8	.34	NS
Moderate or severe	59	45.8	9.86	.0017
Total	81	41.9	8.43	.0037
Total				
Mild	42	33.8	1.01	NS
Moderate or severe	259	44.8	25.76	<.00001
Total	301	43.2	23.62	<.00001

[a] Total known non-Hispanic Caucasian controls known to be nonalcoholic or psychiatrically normal.[29,52,58,69,72,73,118-122]

These findings are of particular importance since studies by linkage analysis had claimed that the DRD2 locus *did not* play a role in TS.[67,68] In addition both of the above studies by Gelernter et al.[64,65] were interpreted as indicating the DRD2 locus *was not* involved in TS. However, when large enough numbers of probands were studied to provide sufficient power to the analysis, when the results were stratified by global severity, and when the TS subjects were compared to a sufficiently large number of racially matched controls to eliminate random fluctuations seen in small numbers, a significant association emerged. A similar phenomenon happened with alcoholism. Although there has been a great deal of controversy about the role of the DRD2 gene in alcoholism, and although several studies have shown no linkage to alcoholism,[69,70] when a large number of cases are examined using association studies with stratification by severity, an effect of the DRD2 gene in some forms of alcoholism emerges.[55,71-73]

4. POSTTRAUMATIC STRESS DISORDER

Since individuals with the impulsive and addictive disorders often show poor response to stress, we wondered if a more detailed examination of posttraumatic stress disorder might be revealing. The veterans on a V.A. Addiction Ward were screened for those who had experienced severe combat stress. The diagnosis of PTSD was made, using the DSM-IIIR criteria, by a psychiatrist experienced in the diagnosis and treatment of this disorder. Since severity of PTSD has been shown to vary as a function of intensity of combat experience,[74] and since the diagnosis of PTSD requires exposure to a severe "stressor that would evoke significant symptoms of distress in almost anyone",[75,76] we required that individuals be "battle-hardened" veterans. These criteria were that: (1) the subjects must have been attached to Marine or Army front line units that engaged in combat patrols leading to "firefights"; (2) they must have been directly exposed to hostile fire; (3) they must have killed enemy troops and directly observed the enemy they killed; and (4) they must have seen members of their own unit killed by the enemy. To eliminate the confounding variable of racial variation in the frequency of the D2A1 allele, the study was limited to non-Hispanic Caucasians. The age range was 33 to 66 years, with a mean of 43.6 years.[77]

The investigation was done in two parts, beginning with an initial exploratory study of 32 subjects. Here, of those who developed PTSD 58.3% carried the D2A1 allele vs. 12.5% of those who did not, $p = .041$. To examine the possibility that this difference was due to chance, a replication study was undertaken using an additional 24 subjects. Here 61.5% of those who developed PTSD carried the D2A1 allele while 0% of those without PTSD carried the D2A1 allele, $p = .002$. For the total group of 56 subjects, of those with PTSD 59.5% carried the D2A1 allele while of those without PTSD, 5.3% carried the allele, $p = .0001$.[77] The increased prevalence of the D2A1 allele in those with PTSD suggests this was a risk factor for this disorder. The very low prevalence for those without PTSD suggests that given a severe stressor, the absence of the D2A1 allele provides some protection against the development of PTSD.

The DSM-IIIR criteria for the diagnosis of PTSD depends on the presence of a clustering of symptoms in three areas (1) a reexperiencing of the traumatic events (flashbacks), (2) avoidance of situations that are associated with the traumatic event, and (3) increased hyperreactivity and vigilance. Summing the variables for each of these sets of symptoms produced a score for each. All were significantly higher in the D2A1 allele carriers. The most significant difference was for hyperreactivity and hypervigilance score.[77] This is not surprising since these symptoms most closely resemble those of childhood and adult ADHD.

There was also a significantly increased prevalence of the D2A1 allele in subjects with moderate to very high battle eagerness.[77] This suggests that the reported association between aggression and PTSD [78] may also be linked to D2A1 carrier status. Moreover, the data indicate that heightened aggression predated exposure to combat in that D2A1A2 subjects reported having been expelled from school for fighting significantly more often than D2A2A2 subjects (75.0% vs. 25.0 $\chi^2 = 8.22$, $p = 0.004$).

A potential defect of this study is that all PTSD subjects were recruited from a substance abuse treatment center. However, the studies of Kulka et al.[74] showed that 95% of high stress war zone exposure Vietnam PTSD subjects had a lifetime history of substance abuse. Thus, identification of a group of nonsubstance abusing Vietnam PTSD subjects, large enough to study, would be difficult. However, to examine this possibility we compared the prevalence of the D2A1 allele of 37.4% among a total series of 235 White substance abusers, to the prevalence of the D2A1 allele of 59.5% in the subjects with PTSD ($\chi^2 = 6.4$, $p = .01$).[77] A second factor that tends to negate the role of substance abuse per se in explaining the PTSD results is the remarkably lower prevalence of the D2A1 allele in hardened combat veterans that *did not* meet PTSD criteria, despite the fact that they all had problems with substance abuse.

These findings are consistent with reports of heightened stress response in dopaminergically lesioned animals,[79] and suggest that genetically determined structural receptor differences in the dopaminergic system may have an important role in response to stress among humans. These differences may also contribute to a gene-environment interaction in which the A1 carriers have a more florrid and persistent set of symptoms than A2 carriers.

5. PATHOLOGICAL GAMBLING

Pathological gamblers share many of the features discussed above for impulsive, compulsive, and addictive behaviors, and many have comorbid substance abuse problems and a history of childhood ADHD. In an ongoing study in conjunction with Drs. Rosenthal, Lesieur, and Rugle, we have examined 171 pathological gamblers.[80] Of these, 50.9% carried the D2A1 allele. Compared to the 714 controls $p < 10^{-8}$. Of the upper 50% in severity, 63.8% carried the D2A1 allele, compared to 40% in the lower half. Of those who had no comorbid substance abuse, 44.1% carried the D2A1 allele compared to 60.5% of those who had comorbid substance abuse. Logistic regression analysis indicated that the pathological gambling score showed a stronger correlation with the presence of the D2A1 allele than 13 other behavioral variables.[80]

6. SMOKING

In conjunction with Dr. Linda Ferry and Dr. Susan Bradshaw of the Jerry L. Pettis V.A. Hospital Department of Public Health, we have examined individuals attending a smoking cessation clinic.[81] To be admitted, subjects must have failed in several prior attempts to stop smoking and not have problems with other substance abuse/dependence. Of the smokers, 48.7% carried the D2A1 allele, $p < 10^{-8}$. There was a significant, inverse relationship between the age of onset of smoking and the prevalence of the D2A1 allele, $p = .02$. Of those who started smoking between 5 and 12 years of age 62.2% carried the D2A1 allele, while of those who started smoking after age 18 only 35.1% carried the D2A1 allele. Of those who had been able to stop smoking for only a week or less, 57.4% carried the D2A1 allele, compared to 33.3% for those able to abstain for more than 6 months, $p = .02$. Of those who smoked 2 or more packs per day, 52.5% carried the D2A1 allele compared to 37.5% for those smoking 1 1/2 packs per day or less, $p = .07$.

7. OBESITY AND HEIGHT

Since the D2A1 *Taq I* allele is located a significant distance 3′ to the DRD2 gene, several investigators have searched for polymorphisms within the gene itself. Sarkar et al.[82] identified two polymorphisms within the DRD2 gene by direct sequencing of individuals with schizophrenia. These were close enough together to allow haplotyping by double allele-specific PCR.[83] In Caucasians this resulted in haplotypes 1, 2, and 4 producing 6 genotypes. Since dopamine plays a major role in the regulation of hunger we wondered if some alleles at the DRD2 locus might correlate with weight. This proved to be the case. We gave our subjects a Z score based on their standard deviation from an epidemiologically based average weight by race and sex. There was a highly significant association between the presence of the 4 haplotype and increased Z score (obesity).[84] A similar association was found between the 4 haplotype and height.[84] This is consistent with the fact that growth hormone is regulated by dopaminergic pathways in the hypothalamus.[85]

8. THE DRD3 LOCUS

Since it was increasingly apparent that there were physiologically important alleleomorphic variants at the DRD2 locus, we wondered if there might not be similar behaviorally relevant variants at the other dopamine receptor genes. We examined a polymorphism in the DRD3 gene that results in the substitution of glycine for serine in the first intron producing a *Msc* I polymorphism. After examining 139 probands with marked TS symptoms we noted a significant decrease in 12 heterozygosity and an increase in 11, and to a lesser extent, 22 homozygosity. The report by Crocq et al.[86] of a similar decrease in heterozygosity in 141 probands with schizophrenia compared to controls, in both France and England, prompted us to publish these findings.[87] In a group of TS probands on whom results of both the D2A1 and D3 alleles were available, the relative risk for TS was 2.33 for individuals carrying the D2A1 allele and 2.45 for individuals showing homozygosity for the D3 alleles. These results were additive in that subjects positive for both showed a 2.84 relative risk while subjects positive for either showed a relative risk of 3.11. This suggested an additive oligogenic model in which the DRD2 and DRD3 loci were two of the genes involved.

Subsequent to this, Brett et al.[88] reported a failure to find any evidence for linkage between this polymorphism and TS in a single large family. They also found no decrease in heterozygosity in 19 mildly affected individuals all from this one family. We pointed out that this was not really an adequate test to either confirm or deny an affect of the D3 locus on the expression of TS since not all TS subjects were homozygous for the D3 alleles and the effect was too mild to be detected by linkage analysis.[63] In addition it was necessary to examine different probands, not individuals all from a single family, many of whom were very mildly affected. Hebebrand et al.[89] concurred that linkage was not the appropriate test and that only probands should be used. They reported results on 66 TS probands, not stratified by severity, and found no decrease in heterozygosity in 12. This prompted us to examine the additional cases we had tested since the original report. Of a total of 350 probands, 38.3% were 12 heterozygotes compared to 358 controls of whom 49.7% were 12 heterozygotes ($p = .002$). Of this, a subset had completed a structured interview that allowed us to stratify the cases by severity. There was a progressive decrease in the prevalence of 12 heterozygotes from controls, to mild, moderate, and severe cases ($p = .005$, Cochran-Armitage linear rank test). We also examined the D3 locus in our pathological gamblers and again there was a significant decrease in 12 heterozygosity 33.3%, $p = .0046$.

While more studies clearly need to be done, the significant progressive decrease in 12 heterozygosity when stratified by severity, is reminiscent of the situation with the D2A1 allele, where linkage studies were negative and some association studies using small numbers were reported as negative, yet when large numbers of probands were examined and the cases stratified by severity, the results show a significant effect. By the time this manuscript was submitted we had examined the effect of variants at all five dopamine receptor genes on the phenotypic expression of TS. The DRD3 gene has the least effect on the phenotype.

Further evidence that the *Msc* I D3 variants have a physiological role comes from studies of the treatment of anovulatory women with clomiphene.[90] There was a significant progressive increase in the frequency of the 22 allele as the required dose of clomiphene needed to induce ovulation increased from 50 mg (29% 22) to 250 mg (83% 22) ($p = .002$). Crocq et al.[91] found significantly lower stimulation of ACTH and cortisol by apomorphine in 22 homozygotes compared to 11 homozygotes or 12 heterozygotes.

9. TRYPTOPHAN 2,3-DIOXYGENASE

Tryptophan 2,3-dioxygenase is a candidate gene for psychiatric disorders because it plays a central role in the regulation of serotonin levels in the blood and CNS. We have cloned the gene and sequenced over 9000 bp of exon, intron, and 5' DNA sequences.[92] Two polymorphisms identified in intron 6 have shown a correlation with various behavioral disorders including Tourette syndrome.[93]

10. HRAS

HRAS, a GTPase, is a molecular switch for signal transduction pathways. While there is little direct evidence that it is involved in neurotransmitter function, Hérault and colleagues[94] reported a significant association between autism and the higher kilobase fragments of the *HRAS* gene *Bam* H1 polymorphism ($p = .008$). We have sought to verify this finding in autism and determine if there was any association with Tourette syndrome (TS), attention deficit hyperactivity disorder (ADHD), or schizophrenia using the more sensitive *Msp* I polymorphism. This provided some support for the findings of Hérault et al. in that there was a just significant increase in the prevalence of carriers of ≥2.1 kb alleles in autistics compared to controls.[95] We also observed that ≥2.1/≥2.1 kb homozygotes scored significantly higher on scales for obsessive-compulsive behaviors and phobias than ≥2.1 heterozygotes and homozygous normal subjects. This suggested a recessive effect for these behaviors. Thelu et al.[96] reported a significant elevation of the ≥2.1 kb alleles of *HRAS* in subjects with alcoholic cirrhosis vs. controls.

While these results require further confirmation, they suggest that genetic defects in HRAS, and possibly other components of the G-protein secondary messenger system, may play a role in some psychiatric disorders.

11. DOPAMINE β HYDROXYLASE

Because it converts dopamine to norepinephrine, DβH is an important enzyme in the metabolism of both of these important neurotransmitters. Low plasma DβH levels have been correlated with conduct disorder and ADHD in children.[97-99] In animals, the inhibition of DβH activity results in impaired memory and learning. We have observed a significant association between the presence of the 1 allele of the DβH *Taq I* B allele[100] and ADHD, oppositional defiant disorder, and learning problems in our clinic patients. In typical additive polygenic fashion, the association with ADD is significantly greater in subjects that carry both the DRD2 *Taq* A1 allele and the DβH *Taq* B1 allele.[101]

12. DOPAMINE TRANSPORTER

The dopamine transporter is responsible for the reuptake of dopamine into the presynaptic neuron. The cloning of the dopamine transporter gene (DAT1) and the demonstration of the presence of a 40-bp repeat [102] allowed the examination of the potential role of this gene in TS subjects. We observed a significant association in TS patients with ADHD, conduct disorder, and other behaviors with the 10/10 genotype.[101] In an independent study, Cook et al.[103] also found an association between the 10 allele and ADHD, and Malison et al.[104] observed a significant increase in dopamine transporter activity in the striatum of TS patients by SPECT studies. Finally, Tiihonen et al.,[105] also using SPECT scanning, observed a decrease in dopamine transporter activity in the brains of nonviolent alcoholics and an increase in violent alcoholics.

13. ADDITIVE EFFECT OF THREE DOPAMINE GENES

We examined the additive effect of the D2A1 allele of the DRD2 gene, the B1 allele of the DβH gene, and the 10/10 genotype of the *DAT1* gene in over 300 TS patients, their relatives, and controls. This showed a highly significant additive effect of these three genes, especially in ADHD, conduct disorder, alcohol abuse, manic, schizoid, and obsessive-compulsive symptoms — the core features of Tourette syndrome. Despite the significant results, regression analysis indicated each of the genes contributed only 2 to 5% of the variance, and combined to less than 8%.

14. MAO

In a single pedigree, Brunner et al.[106,107] reported an association between a mutation of the X-linked monamine oxidase A locus (MAO-A) and a form of mild mental retardation associated with aggressive and hypersexual behaviors. This stimulated us to examine the potential role of variants at the MAO-A and MAO-B genes in TS. This showed that variants at dinucleotide repeat polymorphisms were significantly associated with phenotypic features of TS, but again they accounted for only 1 to 4% of the variance. Of especial interest, there was a significant tendency for the longer base pair alleles of the *MAOA* VNTR and *MAOB* polymorphisms and the shorter base pair alleles of the *MAOA* CA-1 polymorphism to be associated with higher scores for ADHD, stuttering, mania, depression, conduct, and learning problems. These results were consistent with the possibility that the repeat polymorphisms themselves might have an effect on the function of the MAO genes.

15. SEROTONIN 1A RECEPTOR GENE

In a study of a complex dinucleotide repeat at the 5-HTR1A gene,[108] we again observed an association between the size of the repeat and various phenotypic characteristics of TS.[109] There was a significant association between the shorter and longer alleles and scores for ADHD, conduct and oppositional defiant behavior, tics, sexual, and other behaviors. As with the three dopamine genes, there was an additive effect between the D2A1 allele of the DRD2 gene and the shorter and longer alleles of the *HTR1A* gene. Again, despite significant associations, the HTR1A variants accounted for less than 5% of the variance of each of these behaviors.

I. POLYGENIC SPECTRUM MODEL OF BEHAVIORAL DISORDERS

Based on the above results, we suggest the following model for psychiatric disorders:

1. Most psychiatric disorders are polygenic in origin. This mode of inheritance can totally mimic autosomal dominant inheritance with reduced penetrance.
2. The number of genes involved ranges from five to dozens.
3. As a result of this number, each gene has a relatively small effect on the total variance. This means the effect of a given gene may be very subtle, requiring the comparison of the frequency of a given genetic variant in a large number of most severely affected probands vs. unrelated controls, or a within a group comparison of the frequency in mild, moderately, and severely affected probands.
4. A phenotypic effect results when a given individual accumulates a threshold number of mutant genes in the heterozygous or homozygous state.
5. Polygenes are common, with a carrier frequency in the population ranging from 5 to 65%.
6. The genes involved play a role in the regulation of specific neurotransmitters, especially dopamine, serotonin, and norepinephrine.
7. A wide range of different types of neurotransmitter regulation genes will be involved including receptors, transporters, modulators, and metabolic and catabolic enzymes.
8. The mutations involved are distinct from those seen in classical single-gene recessive disorders which result in total or almost total inactivation of the gene. Instead, the mutations involved in

polygenic disorders result in a modest 5 to 30% decrease (or increase) in function. This accounts for why they can be so common in the population. When the number of these mutations is below threshold, they have no deleterious effect and may even have a beneficial effect.

9. The mutations will tend not to be in exons, or if they do occur in exons the functional effects will be minimal. Such mutations will be more subtle and more difficult to identify. They will be in introns, 5' regulatory sequences, 3' untranslated sequences, or 5' and 3' intergenic DNA.
10. The set genes involved may be different for different individuals with the same disorder, and the same for individuals with different disorders, both within a given family and in the population.
11. The repeat polymorphisms themselves may play a role in regulating the activity of the genes they are closest to.

J. LESSONS FOR PSYCHIATRIC GENETICS

Some of the differences between single gene and polygenic inheritance are summarized in Table 14-2. Several aspects of this polygenic model deserve further discussion.

TABLE 14-2
Single Gene Vs. Polygenic Inheritance

	Single Gene	Polygenic
Number of genes	1 (or few)	5 to dozens
Carrier frequency	Rare (<1%)	Common (5 to 65%)
Frequency of the disorder	Rare (<1%)	Common
Type of mutation	Exons, splice junctions	Introns, 5' and 3' sequences, repeat polymorphisms, benign if in exons
Effect on gene function	Major	Minor
Identification by	Linkage analysis	Association analysis
Type of subjects studied	Large pedigrees	Probands vs. controls
Variance explained per gene	95–100%	1–10%
Specificity	Disease specific	Spectrum of disorders

1. POLYGENIC CONVERGENCE

Because the variants are common, individuals can easily by chance pick up one or more of the hypo- or hyperfunctional variants. The term "polygenic convergence" describes the concept well, i.e., if enough genes converge in one person, it can cause a disruption of the normal dopamine-serotonin-norepinephrine balance in the brain. When a certain threshold is passed, that balance is sufficiently disturbed to result in an increased susceptibility to a wide spectrum of impulsive, compulsive, addictive, affective, and anxious behaviors, which can occur in almost any combination. This model is diagrammed in Figure 14-3.

This model accounts for the marked increase in a spectrum of disorders in the relatives of probands with Tourette syndrome, ADHD, substance abuse, and other disorders. These families suggest the presence of a "polygenic cloud" surrounding such probands, with the increased density of polygenes in relatives accounting for the high frequency in them of a wide range of behavioral disorders.

2. THE ETHNICITY ISSUE

Those who feel that there is no association between the D2A1 allele and average alcoholism tend to globalize and suggest that since there are striking racial differences in the

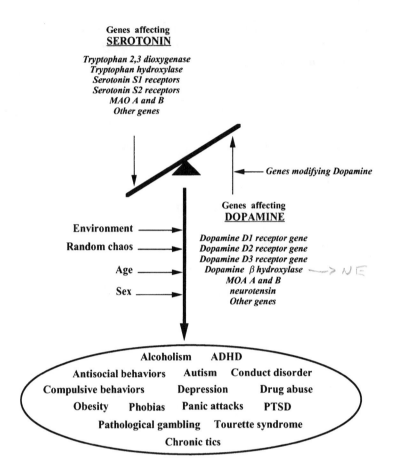

FIGURE 14-3 Hypothesis of the polygenic inheritance of a spectrum of psychiatric disorders. When an individual inherits a threshold number of mutant genes the resulting imbalance in dopamine and serotonin (and norepinephrine) results in increased risk for a spectrum of impulsive, compulsive, addictive, anxious, and affective disorders.

prevalence of the D2A1 allele that all the positive association results are simply due to within-race ethnic differences.[70] However, three observations mitigate against this conclusion.

First, the results to date with the D3 polymorphism[63,86,88,89] and the Hras polymorphism[95] have shown similar gene frequencies in different ethnic groups of controls. It appears that with this polymorphism marked ethnic differences are not present.

Second, within-group comparisons from a single center tend to minimize the effect of ethnic variability and in our hands such comparisons have consistently shown a correlation between the prevalence of the D2A1 allele and impulsive behaviors, the presence or absence of PTSD, and severity in subjects with TS, ADHD, conduct disorder, drug abuse, and pathological gambling.

Third, we have not observed any significant differences in ethnic background to explain these intragroup differences. In our experience most North American non-Hispanic Caucasians are a mixture of 3 to 12 different ethnic groups.

One of the methods used to eliminate false associations due to racial or ethnic stratification is the haplotype relative risk technique of Falk and Rubinstein.[110] This involves the collection of parent-affected child sets. The frequency of the parental alleles present in the affected children is compared to that of a "control" consisting of the other two alleles accumulated across families. However, a disadvantage of this technique is that for a two-allele system, and polygenic inheritance, very large number of samples must be collected. This is illustrated in Table 14-3. If a total of 100 informative sets are required for significance in a single gene disorder, and only 1 in 3 families provides informative data when the frequency of the mutant allele is in the .1 range, and if the gene involved accounts for less than 5% of the variance, many thousands of samples would be required. An additional problem is that in studies of disorders in adults, it is often very difficult to obtain blood of both parents.

TABLE 14-3
Number of Parent-Child Sets Required for a Falk-Rubinstein Study of a Two-Allele System

Father	Mother	A Frequency	B Fraction Informative	A × B
11	11	0.0001	0.0	0.0000
11	12	0.0018	1.0	0.0018
11	22	0.0081	0.0	0.0000
12	11	0.0018	1.0	0.0018
12	12	0.0324	1.0	0.0324
12	22	0.1458	1.0	0.1458
22	11	0.0081	0.0	0.0000
22	12	0.1458	1.0	0.1458
22	22	0.6561	0.0	0.0000
Total		1.0000		0.3276

Note: The genotype for the father and mother are shown in the first two columns. Column A gives frequency of the given mating type of a two-allele system where the frequency of allele 1 = .1 and allele 2 = 0.9. Column B gives the informativeness of each mating type. As can be seen, many matings are uninformative, e.g., 11 × 11, 22 × 22, and 11 × 22. Column A × B gives the relative informativeness of that mating type. The sum (.33) indicates that only one third of the parent-child sets examined for these allele frequencies would be informative. If 50 informative parent-child sets (150 samples) are required for significance in a single gene disorder, and one third of the cases are informative, and the gene in question accounts for less than 5% of the variance, then 150 × 3 × 20 or 9000 samples would be required for the study.

3. THE "REAL" MUTATIONS

One might optimistically hope that most of the markers used to date are only in partial linkage disequilibrium with the "real" mutations affecting the function of the gene. There has been an extensive search in exons for the 'real' mutation in the DRD2 gene,[82,111] without success. However, some "real" mutations may have already been found. The Ala → Ser polymorphism for the DRD3 locus, and the variable repeats in or around the DRD4,[112] dopamine transporter,[113] and Hras[114] genes could very well be the physiologically important mutations. The failure to find a functionally significant mutation in the coding sequence of the DRD2, or other genes, does not mean functional alleomorphic variants do not exist.[53] Studies of the globin and other genes have shown that physiologically important mutations in enhancer regions can be up to several hundred kilobases from the gene they affect.[115] In the HRAS gene, some alleles of a VNTR 3' to the gene preferentially bind to the rel/NF-κB family of transcriptional regulatory factors altering the function of the gene.[116] We anticipate that (1) finding the "real" mutations as opposed to the polymorphisms in linkage disequilibrium with the "real" mutations, will result in only a modest improvement in the strength of statistical correlations, and (2) that finding the "real" mutations may be difficult since they may be hidden in introns or in the 5' or 3' DNA sequences, causing subtle and poorly understood effects on transcription and three-dimensional structure of the gene.

4. MUTATIONS IN NONEXONIC DNA

We often hear the statement that we differ from the chimpanzee by less than 1% of our genes. If, in fact, we have all the genes the chimpanzee does, and vice versa, why do we look so different? Or less mysteriously, why do we look different than our next door neighbor? Clearly, multiple genes are involved. The answer may well lie in the effect of variations in the sequence of intronic and intergenic DNA on the function of the genes. These accumulated differences and the resultant quantitative differences in the relative expression of different genes may be the essence of polygenic inheritance. Some have used the failure to find the "real" mutation in the DRD2 locus as evidence that meaningful, functional, genetic variants of this gene do not exist.[65] In fact, the absence of exon mutations, in the face of extensive evidence that genetic variants of the DRD2 gene play a role in impulsive, compulsive, addictive behaviors, may be giving us some fundamental information about how polygenes work, i.e., they involve variations in nonexonic DNA. Some support for this comes from the observation that some stretches of intergenic DNA show a greater degree of sequence conservation between species than expected.[117]

5. ROLE OF REPEAT POLYMORPHISM IN POLYGENES

Our studies of the repeat polymorphism at the DAT1, HT1A, MAO-A, and MAO-B genes suggest the intriguing possibility that the frequent and widespread dinucleotide, trinucleotide, and other repeat polymorphisms may have the very important effect of causing every gene to occur in multiple functional variants depending on the number and variability of these polymorphisms. This could supply the substrate upon which gene selection, evolution, and polygenic variability can act.

To take this to the extreme, one could easily argue that *every* gene will have one or more common alleles that result in modest functional variations in its expression, and that as a result, *every* gene contributes to the polygenic inheritance of one or more traits affecting appearance, health, or behavior. Whether this contribution is trivial or significant depends upon the relative importance of the function of that gene for the trait in question, the frequency of the allele, and the degree to which the function of the gene is altered. The difference between single gene and polygenic inheritance in relation to the effect of a given gene on explaining the variance in a disorder is illustrated in Figure 14-4.

6. IMPLICATIONS FOR LINKAGE STUDIES

The fact that linkage studies indicated that none of the dopamine receptor genes were important in TS but association studies suggest they are, the fact that virtually 100% of the genome has been screened for linkage in TS and come up empty, and the failure to find reproducible linkage with any of the other psychiatric disorders, suggest that many genes important in these disorders have been inappropriately excluded, i.e., false negative results. If multiple, common genes are involved, if the genes are not disease specific but rather produce susceptibility to a spectrum of interrelated disorders, if some genes are not even present in

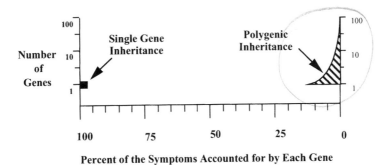

FIGURE 14-4 Comparison of the percent of the variance of the phenotype explained by a specific gene for the single gene vs. the polygenic hypothesis.

the majority of cases, and if the same genes can be present in unaffected relatives, then classic linkage studies may never work, and even sib pair analyses will be compromised.

7. IMPLICATIONS FOR ASSOCIATION STUDIES

Since linkage studies do not have the power to detect these subtle effects, and since the power of sib pair analyses will also be diminished, association studies may be the only remaining viable technique for identifying the relevant genes in psychiatric disorders. Even here the road will be rocky with many conflicting claims. It will be critical to study the most severely affected probands, to study large numbers, to stratify the cases by severity, and to do the studies in racially homogenous populations. The identification of the relatively small effect of each polygene will lead to many conflicting results before the truth is finally sorted out. Definitive results may require the comparison of a number of different studies by meta-analysis. Rather than leading to discouragement, the presence of these expected conflicting results should be considered as part of the process.

K. THE FUTURE

If the above considerations are correct, then the whole strategy of psychiatric genetics needs to be changed. Instead of searching for a few large families with multiple, often mildly affected members, sample acquisition should focus on a large number of the most severely affected probands from many different families. One of the major advantages of using only probands is that it avoids agonizing over whether a more mildly affected relative is, or is not, affected, since all such individuals would be excluded. Structured interviews should be directed toward all DSM-IIIR or DSM-IV diagnoses, not just those of interest to the investigator, and standardized scales should be developed to allow a stratification by severity that is reproducible for different centers. Stratification by sex, age, and maternal vs. paternal origin of the mutation (imprinting) may also be important.

One might argue that if multiple genes are involved this would fatally diminish the potential for using genetic tests for diagnostic purposes, or for understanding the fundamental cause of the disorders in question. We don't see this as problem. With increasing automation, and multiplexing of PCR studies, it will be almost as easy to test for 20 loci as for 1. In addition, if many of the genes fall into specific families, such as the family of dopamine receptor genes, the implications may not be greatly different than if there was only one dopamine receptor gene involved. We do predict that the results will tend to be unidirectional. That is, starting with an affected proband, positive results on a number of tests may allow one to conclude that a given set of genes was playing a role in the disorder and may provide insights into the best form of treatment. On the other hand, if a number of positive genes are identified from a population screen, or on prenatal diagnosis, it may be possible to provide only general statements about an increased risk for a spectrum of disorders, and little definite information about the development of a single specific disorder.

In summary, we believe that once the right tools and methods are brought to bear, the field of psychiatric genetics will begin to blossom. On the basis of the sheer numbers of individuals involved, and the help to humankind, the fruit it produces may quickly outpace that seen with the more clear-cut, but much rarer, single-gene genetic disorders.

REFERENCES

1. Horgan, J., Eugenics revisted, *Sci. Am.*, 268, 122-131, 1993.
2. Egeland, J. A., Gerhard, D. S., Pauls, D. L., Sussex, J. N., Kidd, K. K., Allen, C. R., Hostetter, A. M., and Houseman, D. E., Bipolar affective disorders linked to DNA markers on chromosome 11, *Nature*, 325, 783-787, 1987.
3. Baron, M., Risch, N., Hamburger, R., Mandel, B., Kushner, S., Newman, M., Drumer, D., and Belmaker, R. H., Genetic linkage between X-chromosome markers and bipolar affective illness, *Nature*, 326, 289-292, 1987.
4. Sherrington, R., Brynjolfsson, J., Petursson, H., Potter, M., Dudleston, K., Barraclough, B., Wasmuth, J., Dobbs, M., and Gurling, H., Localization of a susceptibility locus for schizophrenia on chromosome 5, *Nature*, 336, 164-167, 1988.
5. Smith, S. D., Kimberling, W. J., Pennington, B. F., and Lubs, H. A., Specific reading disability: Identification of an inherited form through linkage analysis, *Science*, 219, 1345-1347, 1983.
6. Kelsoe, J. R., Ginns, E. I., Egeland, J. A., Gerhard, D. S., Goldstein, A. M., Bale, S. J., Pauls, D. L., Long, R. T., Kidd, K. K., Conte, G., Houseman, D. E., and Paul, S. M., Reevaluation of the linkage relationship between chromosome 11p loci and the gene for bipolar affective disorder in the old order Amish, *Nature*, 342, 238-243, 1989.
7. Baron, M., Freimer, N. F., Risch, N., Lerer, B., Alexander, J. R., Straub, R. E., Asokan, S., Das, K., Peterson, A., Amos, J., Endicott, J., Ott, J., and Gilliam, T. C., Diminished support for linkage between manic depressive illness and X-chromosome markers in three Israeli pedigrees, *Nat. Genet.*, 3, 49-55, 1993.
8. Gurling, H. M. D., New microsatellite polymorphisms fail to confirm chromosome 5 linkage in Icelandic and British schizophrenia families, Am. Psychopathol. Assoc. Meetings, 1992.
9. Pato, C. N., Macciardl, F., Pato, M. T., Verga, M., and Kennedy, J. L., Review of the putative association of DRD2 and alcoholism: a meta-analysis, *Neuropsychiatr. Genet.*, 53, 1994.
10. Rabin, M., Wen, X. L., Hepburn, M., Lubs, H. A., Feldman, E., and Duara, R., Suggestive linkage of developmental dyslexia to chromosome 1p34-p36, *Lancet*, 342, 178, 1993.
11. Heutink, P., Sandkuyl, L. A., van de Wetering, B. J. M., Oostra, B. A., Weber, J., Wilkie, P., Devor, E. J., Pakstis, A. J., Pauls, D., and Kidd, K. K., Linkage and Tourette syndrome, *Lancet*, 337, 122-123, 1991.
12. Ginns, E. I., Egeland, J. A., Allen, C. R., Pauls, D. L., Falls, K., Keith, T. P., and Paul, S. M., Update on the search for DNA markers linked to manic-depressive illness in the old order Amish, *J Psychiatr Res*, 26, 305-308, 1992.
13. Hamer, D. H., Hu, S., Magnuson, V. L., Hu, N., and Pattatucci, A. M. L., A linkage between DNA markers on the X chromosome and male sexual orientation, *Science*, 261, 321-327, 1993.
14. Pakstis, A. J., Heutink, P., Pauls, D. L., Kurlan, R., van de Wetering, B. J. M., Leckman, J. F., Sandkuyl, L. A., Kidd, J. R., Breedveld, G. J., Castiglione, C. M., Weber, J., Sparkes, R. S., Cohen, D. J., Kidd, K. K., and Oostra, B. A., Progress in the search for genetic linkage with Tourette syndrome — an exclusion map covering more than 50% of the autosomal genome, *Am. J. Hum. Genet.*, 48, 281-294, 1991.
15. Pauls, D. L., Pakstis, A. J., Kurlan, R., Kidd, K. K., Leckman, J. F., Cohen, D. J., Kidd, J. R., Como, P., and Sparkes, R., Segregation and linkage analyses of Tourette syndrome and related disorders, *J. Am. Acad. Child Psychiatry*, 29, 195-203, 1990.
16. Heutink, P., van de Wetering, B. J. M., Breedveld, G. J., and Oostra, B. A., Genetic study on Tourette syndrome in the Netherlands, *Adv. Neurol.*, 58, 167-172, 1992.
17. Comings, D. E., Comings, B. G., and Knell, E., Hypothesis: homozygosity in Tourette syndrome, *Am. J. Med. Genet.*, 34, 413-421, 1989.
18. Kurlan, R., Behr, J., Medved, L., Shoulson, I., Pauls, D., Kidd, J. R., and Kidd, K. K., Familial Tourette's syndrome: report of a large pedigree and potential for linkage analysis, *Neurology*, 36, 772-776, 1986.
19. Robertson, M. M. and Gourdie, A., Familial Tourette's syndrome in a large British pedigree associated with psychopathology, severity, and potential for linkage analysis, *Br. J. Psychiatry*, 156, 515-521, 1990.

20. Kurlan, R., Eapen, V., Stern, J., McDermott, M. P., and Robertson, M. M., Bilineal transmission in Tourette's syndrome families, *Neurology*, 44, 2336-2342, 1994.
21. Walup, J. T., LaBuda, M. C., Hurko, O., Riddle, M. A., and Singer, H. S., A family study and segregation analysis of Tourette's syndrome: Evidence for a mixed model of inheritance, *Am. J. Hum. Genet.*, 57, A174, 1995.
22. Hasstedt, S. J., Leppert, M., Filloux, F., van de Wetering, B., and McMahon, W. M., Intermediate inheritance of Tourette syndrome, assuming assortative mating, *Am. J. Hum. Genet.*, 37, 682-689, 1995.
23. Simpson, S. G., Folstein, S. E., Meyers, D. A., and DePaulo, J. R., Assessment of lineality in bipolar I linkage studies, *Am. J. Psychiatry*, 21294, 31100-21292, 1992.
24. Lane, C., Palmour, R. M., Bradewejn, J., and Boulenger, J.-P., Genetic, epidemiological, and comorbidity factors in panic disorder, Am. Psychiatric Assoc. 146th Annu. Meet., A107, 1993.
25. Biederman, J., Faraone, S. V., Keenan, K., and Tsuang, M. T., Evidence of familial association between attention deficit disorder and major affective disorders, *Arch. Gen. Psychiatry*, 48, 633-642, 1991.
26. Biederman, J., Newcorn, J., and Sprich, S., Comorbidity of attention deficit hyperactivity disorder with conduct, depressive, anxiety, and other disorders, *Am. J. Psychiatry*, 148, 564-577, 1991.
27. Maser, J. D. and Cloninger, C. R., *Comorbidity of Mood and Anxiety Disorders*, American Psychiatric Press, Washington D.C., 1990, 1-869.
28. Comings, D. E., *Tourette Syndrome and Human Behavior*, Hope Press, Duarte, CA, 1990, 1-828.
29. Comings, D. E., Comings, B. G., Muhleman, D., Dietz, G., Shahbahrami, B., Tast, D., Knell, E., Kocsis, P., Baumgarten, R., Kovacs, B. W., Levy, D. L., Smith, M., Kane, J. M., Lieberman, J. A., Klein, D. N., MacMurray, J., Tosk, J., Sverd, J., Gysin, R., and Flanagan, S., The dopamine D2 receptor locus as a modifying gene in neuropsychiatric disorders, *J. Am. Med. Assoc.*, 266, 1793-1800, 1991.
30. Comings, D. E., *Search for the Tourette Syndrome and Human Behavior Genes*, Hope Press, Duarte, CA, 1996.
31. Comings D. E. and Comings, B. G., Comorbid behavioral disorders, in *Tourette Syndrome and Related Disorders*, Kurlan R., Ed., Marcel-Decker, New York, 1993, 111-147.
32. Comings, D. E., Tourette syndrome: a hereditary neuropsychiatric spectrum disorder, *Ann. Clin. Psychiatry*, 6, 235-247, 1995.
33. Pauls, D. L., Cohen, D. J., Kidd, K. K., and Leckman, J. F., Tourette syndrome and neuropsychiatric disorders. Is there a genetic relationship?, *Am. J. Hum. Genet.*, 42, 206-209, 1988.
34. Comings, D. E., Genetic factors in substance abuse based on studies of Tourette syndrome and ADHD probands and relatives. I. Drug abuse, *Drug Alcohol Depend.*, 35, 1-16, 1994.
35. Comings, D. E., Genetic factors in substance abuse based on studies of Tourette syndrome and ADHD probands and relatives. II. Alcohol abuse, *Drug Alcohol Depend.*, 35, 17-24, 1994.
36. Knell, E. and Comings, D. E., Tourette syndrome and attention deficit hyperactivity disorder. Evidence for a genetic relationship, *J. Clin. Psychiatry*, 54, 331-337, 1993.
37. Comings, D. E., The role of genetic factors in human sexual behavior based on studies of Tourette syndrome and ADHD probands and their relatives, *Am. J. Med. Genet. (Neuropsychol. Genet.)*, 54, 227-241, 1994.
38. Comings, D. E., *Search for Tourette Syndrome and Human Behavior Genes*, Hope Press, Duarte, CA, 1996.
39. Comings, D. E., Genetic factors in depression based on studies of Tourette syndrome and Attention Deficit Hyperactivity Disorder probands and relatives, *Am. J. Med. Genet. (Neuropsychol. Genet.)*, 60, 111-121, 1995.
40. Comings, D. E., The role of genetic factors in conduct disorder based on studies of Tourette syndrome and ADHD probands and their relatives, *J. Dev. Behav. Pediatr.*, 16, 142-157, 1995.
41. Comings, D. E., A controlled study of Tourette syndrome. VII. Summary: a common genetic disorder causing disinhibition of the limbic system, *Am. J. Hum. Genet.*, 41, 839-866, 1987.
42. Sverd, J., Clinical presentation of the Tourette's syndrome diathesis, *J. Multihandicapped Person*, 2, 311-326, 1989.
43. Pauls, D. L., Leckman, J. F., Raymond, C. L., Hurst, C. R., and Stevenson, J. M., A family study of Tourette's syndrome. Evidence against the hypothesis of association with a wide range of psychiatric phenotypes, *Am. J. Hum. Genet.*, 43, A64, 1988.
44. van Praag, H. M., Kahn, R. S., Asnis, G. M., Wetzer, S., Brown, S. L., Bleich, A., and Korn, M. L., Denosologication of biological psychiatry or the specificity of 5-HT disturbances in psychiatric disorders, *J. Affective Disord.*, 13, 1-8, 1987.
45. Hudson, J. I. and Pope, H. G., Jr., Affective spectrum disorder. Does antidepressant response indentify a family of disorders with a common pathophysiology?, *Am. J. Psychiatry*, 147, 552-564, 1990.
46. Winokur, G., Cadoret, R., Dorzab, J., and Baker, M., Depressive disease. A genetic study, *Arch. Gen. Psychiatry*, 24, 135-144, 1971.
47. Biederman, J., Keenan, K., and Faraone, S., Attention Deficit Hyperactivity Disorder — family-genetic risk factors and comorbidity, *Pediatr. Adolescent Med.*, 1, 70-94, 1991.
48. Last, C. G., Hersen, M., Kazdin, A., Orvaschel, H., and Perrin, S., Anxiety disorders in children and their families, *Arch. Gen. Psychiatry*, 48, 928-934, 1991.

49. Andrews, G., Stewart, G., Morris-Yates, A., Holt, P., and Henderson, S., Evidence for a general neurotic syndrome, *Br. J. Psychiatry*, 157, 6-12, 1990.
50. Wilson, A. F., Elston, R. C., Mallott, D. B., Tran, L. D., and Winokur, G., The current status of genetic linkage studies of alcoholism and unipolar depression, *Psychiatr. Genet.*, 2, 107-124, 1992.
51. Tsuang, M. T. and Faraone, S. V., *The Genetics of Mood Disorders*, Johns Hopkins University Press, Baltimore, 1990, 1-220.
52. Blum, K., Noble, E. P., Sheridan, P. J., Montgomery, A., Ritchie, T., Jadadeeswaran, P., Nogami, H., Briggs, A. H., and Cohn, J. B., Allelic association of human dopamine D2 receptor gene in alcoholism, *J. Am. Med. Assoc.*, 263, 2055-2059, 1990.
53. Gelernter, J., Goldman, D., and Risch, N., The A1 allele at the D2 dopamine receptor gene and alcoholism, *JAMA*, 269, 1673-1677, 1993.
54. Routtenberg, A., The reward system of the brain, *Sci. Am.*, 239, 154-165, 1978.
55. Blum, K., Wood, R. C., Sheridan, P. J., Chen, T., and Comings, D. E., Dopamine D2 receptor gene variants. Association and linkage studies in impulsive, addictive and compulsive disorders, *Pharmacogenetics*, 5, 121-141, 1995.
56. Blum, K., Sheridan, P. J., Gull, J. G., and Comings, D. E., Reward Deficiency Syndrome, *Am. Sci.*, 84, 132-145, 1996.
57. DiChiara, G. and Imperato, A., Drugs abused by humans preferentially increase synaptic dopamine concentrations in the mesolimbic system of freely moving rats, *Proc. Natl. Acad. Sci. U.S.A.*, 85, 5274-5278, 1988.
58. Comings, D. E., Muhleman, D., Ahn, C., Gysin, R., and Flanagan, S. D., The dopamine D2 receptor gene. A genetic risk factor in substance abuse, *Drug Alcohol Depend.*, 34, 175-180, 1994.
59. Smith, S. S., O'Hara, B. F., Persico, A. M., Gorelick, D. A., Newlin, D. B., Vlahov, D., Solomon, L., Pickins, R., and Uhl, G. R., Genetic vulnerability to drug abuse. The D2 dopamine receptor *Taq I* B1 restriction fragment length polymorphism appears more frequently in polysubstance abusers, *Arch. Gen. Psychiatry*, 49, 723-727, 1992.
60. Noble, E. P., Blum, K., Khalsa, M. E., Ritchie, T., Montgomery, A., Wood, R. C., Fitch, R. J., Ozkaragoz, T., Sheridan, P. J., Anglin, M. D., Paredes, A., Treiman, L. J., and Sparkes, R. S., Allelic association of the D2 dopamine receptor gene with cocaine dependence, *Drug Alcohol Depend.*, 33, 271-285, 1993.
61. O'Hara, B. F., Smith, S. S., Bird, G., Persico, A. M., Suarez, B., Cutting, G. R., and Uhl, G. R., Dopamine D2 receptor RFLPs, haplotypes and their association with substance use in Black and Caucasian Research Volunteers, *Hum. Genet.*, 43, 209-218, 1993.
62. Devor, E. J., D2-dopamine receptor and neuropsychiatric illness [letter], *JAMA*, 267, 651, 1992.
63. Comings, D. E., Muhleman, D., Dietz, G., Dino, M., Legro, R., and Gade, R., Tourette's syndrome and homozygosity for the dopamine D3 receptor gene — reply, *Lancet*, 341, 1483-1484, 1993.
64. Gelernter, J., Pauls, D., Leckman, J., and Kurlan, R., Evidence that D2 dopamine receptor alleles do not influence severity of Tourette's syndrome, *Am. J. Psychiatry* (Abstr.), 127-128, 1992.
65. Gelernter, J., Pauls, D. L., Leckman, J., Kidd, K. K., and Kurlan, R., D2 dopamine receptor alleles do not influence severity of Tourette's syndrome, *Arch. Neurol.*, 51, 397-400, 1994.
66. Comings, D. E., Dopamine D2 receptor and Tourette syndrome, *Arch. Neurol.*, 52, 441-442, 1995.
67. Devor, E. J., Grandy, D. K., Civelli, O., Litt, M., Burgess, A. K., Isenberg, K. E., van de Wetering, B. J. M., and Oostra, B., Genetic linkage is excluded for the D2-dopamine receptor lambda-Hd2G1 and flanking loci on chromosome 11Q22-Q23 in Tourette syndrome, *Hum. Hered.*, 40, 105-108, 1990.
68. Gelernter, J., Pakstis, A. J., Pauls, D. L., Kurlan, R., Gancher, S. T., Civelli, O., Grandy, D., and Kidd, K. K., Gilles-de-La-Tourette syndrome is not linked to D2-dopamine receptor, *Arch. Gen. Psychiatry*, 47, 1073-1077, 1990.
69. Parsian, A., Todd, R. D., Devor, E. J., O'Malley, K. L., Suarez, B. K., Reich, T., and Cloninger, C. R., Alcoholism and alleles of the human dopamine D2 receptor locus. Studies of association and linkage, *Arch. Gen. Psychiatry*, 48, 655-663, 1991.
70. Gelernter, J., O'Malley, S. O., Risch, N., Kranzler, H. R., Krystal, J., Merikangas, K., Kennedy, J. L., and Kidd, K. K., No association between an allele at the D2 dopamine receptor gene (DRD2) and alcoholism, *J. Am. Med. Assoc.*, 266, 1801-1807, 1991.
71. Arinami, T., Itokawa, M., Komiyama, T., Mistushio, H., Morei, H., Mifune, H., Hamaguchi, H., and Toru, M., Association between severity of alcoholism and the A1 allele of the dopamine D2 receptor gene *Taq I* A RFLP in Japanese, *Biol. Psychiatry*, 33, 108-114, 1993.
72. Blum, K., Noble, E. P., Sheridan, P. J., Finley, O., Montgomery, A. R., Ritchie, T., Ozkaragoz, T., Fitch, R. J., Sadlack, F., Sheffield, D., Dahlmann, T., Halbardier, S., and Nogami, H., Association of the A1 allele of the D2 dopamine receptor gene with severe alcoholism, *Alcohol*, 8, 409-416, 1991.
73. Noble, E. P., The D2 dopamine receptor gene. A review of association studies in alcoholism, *Behav. Genet.*, 23, 119-129, 1993.
74. Kulka, R. A., Schlenger, W. E., Fairbank, J. A., et al., *The National Vietnam Veterans Readjustment Study*, Brunner/Mazel, New York, 1990, 1-1000.

75. APA, *Diagnostic and Statistical Manual of Mental Disorders*, 3rd ed., American Psychiatric Association, Washington, D.C., 1987.
76. APA, *Diagnostic and Statistical Manual of Mental Disorders*, 3rd ed., American Psychiatric Association, Washington, D.C., 1980.
77. Comings, D. E., Muhleman, D., and Gysin, R., The dopamine D2 receptor (DRD2) gene in posttraumatic stress disorder. A study and replication, *Biol. Psychiatry*, (in press), 1996.
78. Tennant, C., Streimer, J. H., and Temperly, H., Memories of Vietnam: post-traumatic stress disorders in Australian veterans, *Aust. N. Z. J. Psychiatry*, 24, 29-36, 1990.
79. LeMoal, M. and Simon, H., Mesocorticolimbic dopaminergic network: functional and regulatory roles, *Physiol. Rev.*, 71, 155-234, 1991.
80. Comings, D. E., Rosenthal, R. J., Lesieur, H. R., Rugle, L., Muhleman, D., Chiu, C., Dietz, G., and Gade, R., A study of the dopamine D2 receptor gene in pathological gambling, *Pharmacogenetics*, (in press), 1996.
81. Comings, D. E., Ferry, L., Bradshaw-Robinson, S., Burchette, R., Chiu, C., and Muhleman, D., The dopamine D2 receptor (DRD2) gene: a genetic risk factor in smoking, *Pharmacogenetics*, 6, 73-79, 1996.
82. Sarkar, G., Kapelner, S., Grandy, D. K., Marchionni, M., Civelli, O., Sobell, J., Heston, L., and Sommer, S. S., Direct sequencing of the dopamine D2 receptor (DRD2) in schizophrenics reveals three polymorphisms but no structural change in the receptor, *Genomics*, 11, 8-14, 1991.
83. Sarkar, G. and Sommer, S. S., Haplotyping by double PCR amplification of specific alleles, *BioTechniques*, 10, 436-440, 1991.
84. Comings, D. E., Flanagan, S. D., Dietz, G., Muhleman, D., Knell, E., and Gysin, R., The dopamine D2 receptor (DRD2) as a major gene in obesity and height, *Biochem. Med. Metab. Biol.*, 50, 176-185, 1993.
85. Lorenzi, M., Karam, J. H., McIlroy, M. B., and Forsham, P. H., Increased growth hormone response to dopamine infusion in insulin-dependent diabetic subjects, *J. Clin. Invest.*, 65, 146, 1980.
86. Crocq, M.-A., Mant, R., Asherson, P., Williams, J., Hode, Y., Mayerova, A., Collier, D., Lannfelt, L., Sokoloff, P., Schwartz, J.-C., Gil, M., Macher, J.-P., Mcguffin, P., and Owen, M. J., Association between schizophrenia and homozygosity at the dopamine D3 receptor gene, *J. Med. Genet.*, 29, 858-860, 1992.
87. Comings, D. E., Muhleman, D., Dietz, G., Dino, M., Legro, R., and Gade, R., Association between Tourette's syndrome and homozygosity at the dopamine-D3 receptor gene, *Lancet*, 341, 906, 1993.
88. Brett, P., Robertson, M., Gurling, H., and Curtis, D., Failure to find linkage and increased homozygosity for the dopamine D3 receptor gene in Tourette syndrome, *Lancet*, 341, 1225, 1993.
89. Hebebrand, J., Nöthen, M. M., Lehmkuhl, G., Poustka, F., Schmidt, M., Propping, P., and Remschmidt, H., Tourette's syndrome and homozygosity for the dopamine D3 receptor gene, *Lancet*, 341, 1483, 1993.
90. Legro, R. S., Muhleman, D., Comings, D. E., Lobo, R. A., and Kovacs, B. W., A dopamine D3 receptor genotype is associated with hyperandrogenic chronic anovulation and resistance to ovulation induction with clomiphene citrate, *Fertil. Steri.*, (submitted), 1994.
91. Crocq, M-A., Duval, F., Mayerova, A., Sokoloff, P., Natt, E., Lannfelt, L., Mokrani, M.-C., Bailey, P., Schwartz, J.-C., and Macher, J.-P., Dopamine D3 receptor polymorphism and response to apomorphine challenge, *Neuropsychopharmacology*, 10 (Suppl. 2), 16S, 1994.
92. Shih, C., Padhy, L. C., Murray, M., and Weinberg, R. A., Transforming genes of carcinomas and neuroblastomas introduced into mouse fibroblasts, *Nature*, 290, 261-264, 1981.
93. Comings, D. E., Muhleman, D., Gade, R., Chiu, C., Wu, H., Dietz, G., Winn-Dean, E., Ferry, L., Rosenthal, R. J., Lesieur, H. R., Rugle, L., Sverd, J., Johnson, P., and MacMurray, J. P., Exon and intron mutations in the human tryptophan 2,3-dioxygenase gene and their potential association with Tourette syndrome, substance abuse and other psychiatric disorders, *Pharmacogenetics*, (in press), 1996.
94. Hérault, J., Perrot, A., Barthélémy, C., Büchler, M., Cherpi, C., Leboyer, M., Sauvage, D., Lelord, G., Mallet, J., and Müh, J.-P., Possible association of C-Harvey-Ras-1 (HRAS-1) marker with autism, *Psychiatry Res.*, 46, 261-267, 1993.
95. Comings, D. E., Wi, J., Chiu, C., Muhleman, D., and Sverd, J., HRAS polymorphism in autism, Tourette syndrome, ADHD and schizophrenia. A possible recessive effect in obsessive-compulsive behavior and phobias, *Psychiatry Res.*, (submitted), 1996.
96. Thelu, M.-A., Zarski, J.-P., Froissart, B., Rachail, M., and Seigneurin, J.-M., c-Ha-ras polymorphism in patients with hepatocellular carcinoma, *Gastroenterol. Clin. Biol.*, 17, 903-907, 1993.
97. Rogeness, G. A., Hernandez, J. M., Macedo, C. A., Amrung, S. A., and Hoppe, S. K., Near-zero plasma dopamine-b-hydroxylase and conduct disorder in emotionally disturbed boys, *J. Am. Acad. Child Psychiatry*, 25, 521-527, 1986.
98. Rogeness, G. A., Maas, J. W., Javors, M. A., Macedo, C. A., Fischer, C., and Harris, W. R., Attention deficit disorder symptoms and urine catecholamines, *Psychiatry Res.*, 27, 241-251, 1989.
99. Rogeness, G. A., Hernandez, J. M., Macedo, C. A., and Mitchell, E. L., Biochemical differences in children with conduct disorder socialized and undersocialized, *Am. J. Psychiatry*, 139, 307-311, 1982.
100. d'Amato, T., Leboyer, M., Malafosse, A., Samolyk, D., Lamouroux, A., Junien, C., and Mallet, J., Two *Taq I* dimorphic sites at the human b-hydroxylase locus, *Nucleic Acids Res.*, 17, 5871, 1989.

101. Comings, D. E., Wu, H., Chiu, C., Ring, R. H., Dietz, G., and Muhleman, D., Polygenic inheritance of Tourette syndrome, stuttering, ADHD, conduct and oppositional defiant disorder. The Additive and Subtractive Effect of the three dopaminergic genes — DRD2, DBH and DAT1, *Am. J. Med. Genet. (Neuropsychol. Genet.)*, 67, 284-288, 1996.
102. Vandenbergh, D. J., Persico, A. M., Hawkins, A. L., Griffin, C. A., Li, X., Jabs, E. W., and Uhl, G. R., Human dopamine transporter gene (DAT1) maps to chromosome 5p15.3 and displays a VNTR, *Genomics*, 14, 1866-1868, 1992.
103. Cook, E. H., Vandenbergh, D. J., Stein, M., Cox, N. J., Yan, S., Krasowski, M. D., Uhl, G. R., and Leventhal, B. L., Molecular genetic analysis of the dopamine transporter in attention deficit/hyperactivity disorder (ADHD), *Am. J. Hum. Genet.*, 57, A189, 1995.
104. Malison, R. T., McDougle, C. J., van Dyck, C. H., Scahill, L., Baldwin, R. M., Seibyl, J. P., Price, L. H., Leckman, J. F., and Innis, R. B., [^{123}I]b-CIT SPECT imaging of striatal dopamine transporter binding in Tourette's disorder, *Am. J. Psychiatry*, 152, 1359-1361, 1995.
105. Tiihonen, J., Kuikka, J., Bergström, K., Hakola, P., Karhu, J., Ryynänen, O-P., and Föhr, J., Altered striatal dopamine re-uptake site densities in habitually violent and nonviolent alcoholics, *Nat. Med.*, 1, 654-657, 1995.
106. Brunner, H. G., Nelen, M. R., van Zandvoort, P., Abeling, N. G. G. M., van Gennip, A. H., Wolters, E. C., Kuiper, M. A., Ropers, H. H., and van Oost, B. A., X-linked borderline mental retardation with prominent behavioral disturbance. Phenotype, genotic localization and evidence for disturbed monoamine metabolism, *Am. J. Hum Genet.*, 52, 1032-1039, 1993.
107. Brunner, H. G., Helen, M., Breakfiled, X. O., Ropers, H. H., and van Oost, B. A., Abnormal behavior linked to a point mutation in the structural gene for monamine oxidase A, *Psychiatr. Genet.*, 3, 122, 1993.
108. Bolos, A. M., Goldman, D., and Dean, M., Dinucleotide repeat and *alu* repeat polymorphisms at the 5-HT$_{1A}$ (HTR1A) receptor gene, *Psychiatr. Genet.*, 3, 235-240, 1993.
109. Comings, D. E., Gade, R., Muhleman, D., and MacMurray, J., Role of the 5-HT1A serotonin receptor gene in Tourette syndrome, conduct and oppositional defiant disorder: implications for polygenic inheritance, (submitted), 1996.
110. Falk, C. T. and Rubinstein, P., Haplotype relative risks: an easy reliable way to construct a proper control sample for risk calculations, *Ann. Hum. Genet.*, 51, 227-233, 1987.
111. Gejman, P. V., Ram, A., Gelernter, J., Friedman, E., Cao, Q., Pickar, D., Blum, K., Noble, E. P., Kranzler, H. R., O'Malley, S., Hamer, D. H., Whitsitt, F., Rao, P., DeLisi, L. E., Virkkunen, M., Linnoila, M., Goldman, D., and Gershon, E. S., No structural mutation in the dopamine D2 receptor gene in alcoholism or schizophrenia, *J. Am. Med. Assoc.*, 271, 204-208, 1994.
112. Petronis, A., Vantol, H. H. M., Livak, K. J., Sidenberg, D. G., Macciardi, F. M., and Kennedy, J. L., Genetic analysis of variable repeat sequence in DRD4 gene exon, *Am. J. Hum. Genet.*, 51, A198, 1992.
113. Vandenbergh, D. J., Persico, A. M., and Uhl, G. R., A human dopamine transporter cDNA predicts reduced glycosylation, displays a novel repetitive element and provides racially-dimorphic *Taq I* RFLPs, *Mol. Brain Res.*, 15, 161-166, 1992.
114. Krontiris, T. G., Devlin, B., Karp, D. D., Robert, N. J., and Risch, N., An association between the risk of cancer and mutations in the Hras1 minisatellite locus, *N. Engl. J. Med.*, 329, 517-523, 1993.
115. Cooper, D. N., Regulatory mutations and human genetic disease, *Ann. Med.*, 24, 427-437, 1992.
116. Trepicchio, W. L. and Krontiris, T. G., Members of the rel/NF-kB family of transcriptional regulatory factors bind the HRAS1 minisatellite DNA sequence, *Nucleic Acids Res.*, 21, 977-985, 1922.
117. Koop, B. F., Rowen, L., Wang, K., Kuo, C. L., Seto, D., Lenstra, J. R., Howard, S., Shan, W., Deshpande, P., and Hood, L., The human T-cell receptor TCRAC/TSRDC (C alpha/C delta) region. Organization, sequence, and evolution of 97.6 kb of DNA, *Genomics*, 19, 478-493, 1994.
118. Grandy, D. K., Litt, M., Allen, L., Bunzow, J. R., Marchionni, M., Makam, H., Reed, L., Magenis, R. E., and Civelli, O., The human dopamine D2 receptor gene is located on chromosome 11 at q22-q23 and identifies a *Taq I* RFLP, *Am. J. Hum. Genet.*, 45, 778-785, 1989.
119. Cloninger, C. R., D2 dopamine receptor gene is associated but not linked with alcoholism, *J. Am. Med. Assoc.*, 266, 1833-1834, 1991.
120. Schwab, S., Soyka, M., Niederecker, N., Ackenheil, M., Scherer, J., and Widenauer, D. B., Allelic association of human dopamine D2-receptor DNA polymorphism ruled out in 45 alcoholics, *Am. J. Hum. Genet.*, 49 (Suppl.), 203A, 1991.
121. Uhl, G. R., Persico, A. M., and Smith, S. S., Current excitement with D2 dopamine receptor gene alleles in substance abuse, *Arch. Gen. Psychiatry*, 49, 157-160, 1992.
122. Amadeo, S., Abbar, M., Fourcade, M. L., Waksman, G., Leroux, M. G., Madec, A., Selin, M., Champiat, J.-C., Brethome, A., Lcclaire, Y., Castelnau, D., Venisse, J.-L., and Mallet, J., D2 dopamine receptor gene and alcoholism, *J. Psychiatr. Res.*, 27, 173-179, 1993.

15 Molecular Linkage Studies of Manic-Depressive Illness

Wade Berrettini

CONTENTS

A. Introduction 261
B. Syndrome Descriptions and Epidemiology 261
C. Molecular Linkage Studies of Bipolar Disorder 264
D. Summary 267
Acknowledgments 267
References 267

A. INTRODUCTION

This chapter presents a brief phenomenological description of bipolar (BP) disorder and its epidemiology, followed by a review of molecular genetic linkage studies of this common and severe mood disorder. The review of BP linkage studies is limited mainly to recent reports employing molecular methods of genotyping, as opposed to those earlier papers which relied on classical genetic markers (such as blood groups, color vision, and enzyme activity).

B. SYNDROME DESCRIPTIONS AND EPIDEMIOLOGY

Mood disorders (also known as affective disorders) are common behavioral syndromes which exist in at least two major forms, bipolar (BP) and unipolar (UP). Unipolar disorder describes individuals who have single or recurrent episodes of depression, a syndrome (lasting at least several weeks) of persistent and pervasive sadness, with decreased energy, suicidal ideation, decreased libido, anhedonia (inability to experience pleasure), decreased cognitive ability, sleep dysfunction (insomnia or hypersomnia), and appetite disturbance (with or without weight change). Bipolar (BP) I disorder is characterized by episodes of mania and depression. Mania is a syndrome of elevated mood or euphoria (often with affective lability) lasting at least several weeks, associated with increased activity, decreased need for sleep, grandiosity, excessive energy, increased libido, racing thoughts, and impulsive, reckless behavior, with accompanying impairment or incapacitation in the patient's major life role. Hypomania is a milder form of mania without impairment or incapacitation. Individuals with hypomanic episodes and depressive episodes (but no frank mania) are diagnosed as BP II.

UP illness affects females twice as often as males, but BP illness affects both sexes equally (Weissman and Myers, 1978). The reported prevalence of these affective disorders depends on diagnostic criteria, but most authorities agree that BP illness affects 1% of the general population, while UP illness troubles at least 10% of the population (Weissman and Myers, 1978; Weissman, 1987). Suicide is the outcome in 10% of cases and is the sole cause of increased mortality (Klerman, 1987).

These disorders have a median age of onset in the 20s (BP) or 30s (UP), although onset in adolescence is becoming common. There is excellent evidence that the age at onset is decreasing among younger generations for both UP (Klerman and Weissman, 1989; Wickramaratne et al., 1989; Klerman et al., 1985; Joyce et al., 1990) and BP (Gershon et al., 1987) disorders. This relatively sharp decrease in age at onset for individuals born after World War II is known as a cohort effect, and no single genetic factor can explain this large and rapid effect. It is likely that both cultural and genetic factors may contribute critical components to the cohort effect.

One possible partial explanation for the decrease in age at onset among recent generations is *anticipation,* in which a familial disorder occurs at earlier ages with greater severity in younger generations. Anticipation occurs in several neurodegenerative diseases, including Huntington's disease, Fragile X, myotonic dystrophy, spinocerebellar ataxias, and others. The molecular explanation for anticipation in these disorders involves unstable trinucleotide repeats, which expand in subsequent generations, giving rise to increasing levels of gene disruption and thus to earlier age at onset and increasingly severe phenotype in younger generations (for review see Caskey et al., 1992; Ross et al., 1993).

Evidence for anticipation has been reported in several family studies of BP illness (McInnis et al., 1993; Lipp et al., 1995; Gershon et al., 1987), but there are difficult problems of ascertainment bias (Penrose, 1948; Hodge and Wickramaratne, 1995). For example, people with earlier age at onset may have reduced capacity to marry and reproduce, so parents with such early-onset disorders may be unusual. Further, people who have BP disorder in their families may come to the attention of medical personnel earlier and more often than others, such that less severe mood disorder episodes are detected medically, and an earlier age at onset is defined. Such individuals (by virtue of their familiarity with mood disorder symptoms) may be more likely to report minor mood disturbance in terms of "diagnosable syndromes". The evidence for anticipation in BP disorder comes from extensive studies of multiplex BP families for linkage studies. Such studies would tend to select for earlier age at onset cases, because linkage studies give preference to densely affected kindreds. Among broader cultural factors possibly underlying the cohort effect, if stigma concerning mood disorders is less among younger affected persons (compared to older individuals), then younger cohorts might describe their experiences more easily in terms of a diagnosable mood disorder, since denial (due to stigma) is less prevalent among the younger cohorts. These potential confounding factors make detection of anticipation in BP disorder difficult.

The possibility that anticipation exists in BP disorder has resulted in genomic searches for unstable, expanding trinucleotide repeat sequences (O'Donovan et al., 1995; Lindblad et al., 1995), although no definitive evidence for causative expansions has been found. The hypothesis that unstable trinucleotide repeats represent BP susceptibility factors deserves continued study.

Initial anecdotal observations of familial aggregation for BP and UP disorders were followed by systematic twin, family, and adoption studies (conducted over the last 50 years), which indicated the importance of genetic predisposition (for review see Gershon et al., 1987; Nurnberger et al., 1986). The firmest evidence for BP genetic susceptibility derives from the twin studies, in which monozygotic twins show concordance, on average, 65% of the time and dizygotic twins 14% of the time (see Table 15-1). On the basis of these twin studies it has been estimated that the heritability (the proportion of total phenotypic variance contributed by additive genetic effects) of BP illness is ~70%. This degree of heritability is consistent with a single major locus.

More recent twin studies (Kendler et al., 1992 and 1993; McGuffin et al., 1996) conducted with modern diagnostic criteria, validated semistructured interviews, and blinded assessments,

TABLE 15-1
Concordance Rates for Affective Illness in Monozygotic and Dizygotic Twins[a]

	Monozygotic twins		Dizygotic twins	
Study	Concordant pairs/ total pairs	Concordance (%)	Concordant pairs/ total pairs	Concordance (%)
Luxemberger, 1930	3/4	75.0	0/13	0.0
Rosanoff et al., 1935	16/23	69.6	11/67	16.4
Slater, 1953	4/7	57.1	4/17	23.5
Kallman, 1954	25/27	92.6	13/55	23.6
Harvald and Hauge, 1965	10/15	66.7	2/40	5.0
Allen et al., 1974	5/15	33.3	0/34	0.0
Bertelsen et al., 1977	32/55	58.3	9/52	17.3
Totals	95/146	65.0	39/278	14.0

* Data not corrected for age. Diagnosis includes both bipolar and unipolar illness.

confirm these earlier reports although not without exception (Andrews et al., 1990). The results from the twin studies are consistent with a complex inheritance of these disorders. The reduced MZ concordance clearly suggests decreased penetrance of inherited susceptibility or the presence of phenocopies (nongenetic cases) among the MZ twins.

Among MZ twin pairs concordant for mood disorder, when one twin has a BP diagnosis, 20% of the ill co-twins have UP (Bertelsen et al., 1977; Allen et al., 1974). This observation supports the hypothesis that BP and UP syndromes share some common genetic susceptibility factors. This observation has considerable clinical relevance. For example, it is a common practice for maintenance treatment of recurrent UP illness to include that antidepressant drug which has helped the patient recover from episodes of UP. When a UP patient has a first-degree BP relative, preventative (maintenance) treatment with lithium is often helpful (Souza and Goodwin, 1991).

Multiple controlled family studies of BP illness have been conducted over the past 25 years. From family studies, a reasonable definition of the BP spectrum of disorders would include BPI, BPII with major depression (hypomania and recurrent UP illness in the same person), schizoaffective disorders, and recurrent unipolar depression. This definition is reasonable because these disorders aggregate among the first-degree relatives of individuals with BP disorders (Gershon et al., 1982; Weissman et al., 1984; Baron et al., 1983; Winokur et al., 1982 and 1995; Heltzer and Winokur, 1974; James and Chapman, 1975; Johnson and Leeman, 1977; Angst et al., 1980; Maier et al., 1993). The bipolar spectrum disorders (which include BP, UP, and schizoaffective syndromes, at least) aggregate in families of BP individuals, but it is unclear if this spectrum represents pleiotropic expression of a single genetic susceptibility. Family studies may support the hypothesis that these disorders represent different disease entities but share some etiological factors.

If a single major locus is present in BP disease, it would operate under complex inheritance, which may be characterized by the following:

1. Reduced penetrance in which some individuals with the disease genotype fail to express the illness, possibly because of polygenic effects or environmental events or variable age of onset (in which the illness may not appear until rather late in life)
2. Phenocopies, in which individuals without a disease genotype (at the locus analyzed) manifest the syndrome from nongenetic causes
3. Genetic heterogeneity in which mutations at varying genetic loci can independently cause clinically indistinguishable disease forms

The mode of inheritance of BP genetic susceptibility has been studied through segregation analysis. One way to detect a single major locus for a non-Mendelian disease (when a major locus is not evident from segregation ratios) is to apply complex models of segregation analysis to clinical diagnostic data in families. These models use maximum likelihood procedures to test for single locus and polygenic components of inheritance and can take into account factors such as decreased penetrance, age of onset variation, and phenocopies. Several BP disorder models including single major locus, oligogenic (two or more susceptibility loci in which the effects of the individual loci are detectable), and polygenic (large number of loci each with small effect) have been proposed. The segregation analyses have not demonstrated consistently a single locus inheritance of BP disorder, even when families are subdivided based on clinical criteria (reviewed in Nurnberger et al., 1986), although some analyses do favor a single major locus model (with or without a multifactorial background) compared to models of no major locus (O'Rourke et al., 1983; Spence et al., 1993). However, Rice et al. (1987), who tested the major locus hypothesis in BP disease (taking into account cohort effects) as a nested hypothesis under the general mixed model, found no conclusive evidence for Mendelian transmission.

Some possible explanations for failure to detect a single locus are that no single locus inheritance exists, or the segregation analyses were not powerful enough to detect single locus inheritance under the true conditions of inheritance (e.g., the sample sizes were not adequate because the data are too heterogeneous, and the correct clinical or biological subdivisions which would generate homogeneous data have not yet been found). The fact that evidence for a single locus has not been found in segregation analysis of BP disease should not deter the search for susceptibility genes (genes of partial effect) through linkage (Greenberg, 1993).

C. MOLECULAR LINKAGE STUDIES OF BIPOLAR DISORDER

A linkage study of Old Order Amish pedigrees described evidence (LOD score >4.0) for a BP locus on 11p15 (Egeland et al., 1987), but this evidence has been weakened by failure to confirm the finding in numerous other pedigrees (Detera-Wadleigh et al., 1994; Mitchell et al., 1991; Hodgkinson et al., 1987; Mendlewicz et al., 1991; Debruyn et al., 1994; Gill et al., 1988; Coon et al., 1993) and by evaluation of newly ascertained individuals in the original pedigree (Kelsoe et al., 1989). However, this hypothesis (that a BP susceptibility gene exists on the tip of the short arm of chromosome 11) remains viable and interesting. The LOD score in the original Old Order Amish pedigree 110 is ~2.0, and similar weakly positive LOD scores are reported for this region by other investigators (Gurling et al., 1995; Smyth et al., 1996). Furthermore, several reports have described evidence for association of tyrosine hydroxylase (located in 11p15) with BP disorder (Leboyer et al., 1990; Kennedy et al., 1993; Meloni et al., 1993; Verga et al., 1993), although other groups have not confirmed this observation (Nothen et al., 1990; Korner et al., 1990 and 1994; Gill et al., 1991; Inayama et al., 1993). The existence of an 11p15 locus of small effect on risk for BP illness remains a tenable hypothesis.

The color vision region of Xq28 also has been reported linked to BP illness (Winokur et al., 1969; Mendlewicz and Fleiss, 1974; Baron, 1977; Baron et al., 1987; Mendlewicz et al., 1979 and 1980; Del Zompo et al., 1984) in studies employing clinically assessed color blindness and G6PD deficiency. Again, however, several independent investigators, employing molecular methods, have not confirmed this linkage (Berrettini et al., 1990; Del Zompo et al., 1991; Smyth et al., 1995). New onsets of illness and molecular genotyping reported by Del Zompo et al. (1991) diminished evidence for Xq28 linkage which was described in their earlier report (Del Zompo et al., 1984). Most of the original evidence for linkage to color blindness and G6PD deficiency in the most prominent positive report (Baron et al., 1987) was **not** confirmed in those same pedigrees by molecular methods employing relevant Xq28 DNA markers (Baron et al., 1993). This failure to replicate with molecular techniques in the same pedigrees is remarkable because the original report (Baron et al., 1987) described a maximal LOD score of 9, a level of statistical significance which has not been associated previously with failure to replicate (in the same pedigrees) among linkage studies of any disorder. The original report (Baron et al., 1987) and the molecular failure to replicate (Baron et al., 1993) were contrasted by investigators responsible for the molecular work (Straub and Gilliam, 1993). They concluded that the original report was flawed methodologically by poor quality control of G6PD assays, failure to maintain a blind between diagnosis and genotype assignment, and genotypic determinations of two individuals based solely on family history (Straub and Gilliam, 1993). This experience leads to the conclusion that independent confirmation of reported linkages by molecular methods in novel kindreds must remain the gold standard, regardless of the level of statistical significance achieved by a single report. There is no published molecular linkage study indicating an Xq28 BP susceptibility locus.

Because these initial BP linkage reports of major loci for Xq28 and 11p15 have not been confirmed, several investigators have screened the genome with sufficient numbers of affected individuals that a major BP locus would have been detected in these studies (Pakstis et al., 1991; Coon et al., 1993; Berrettini et al., 1991 and 1996; Detera-Wadleigh et al., 1992, 1994 and 1996; Gejman et al., 1993). If a major locus is defined as one which increases risk by a factor of 10 or more in a majority of kindreds, then such a locus may not exist. However, it is clear that there are several confirmed reports of loci of smaller effect, which can be termed susceptibility loci (Greenberg, 1993). These loci are neither necessary nor sufficient for disease, but increase risk for the disorder, typically in a non-Mendelian manner. It is to be expected that such susceptibility loci will yield positive LOD scores in dominant and recessive models, depending on the presence of environmental and other genetic risk factors. In this situation, nonparametric statistics may yield greater evidence of linkage than LOD score models (see Berrettini et al., 1994, for example; Goldin and Weeks, 1993).

From simulation research in which a disorder is assumed to be caused by one of six loci (Suarez et al., 1994), it is clear that universal agreement by investigators attempting confirmations of reported BP susceptibility loci will not occur. Failure to detect a previously described susceptibility locus (when power is adequate) may be due to sampling variability, ethnic differences, and random errors in diagnostic and genotypic procedures. If two or more independent investigators (using exacting clinical methods and blinded molecular genotypes) find evidence for linkage to the same region in separate series of pedigrees, then it is reasonable to assume validity. Statistical guidelines for accepting these reported (and confirmed) linkages as valid have been suggested (Lander and Kruglyak, 1995). In general, these guidelines require significance levels of $\sim 10^{-5}$ for the initial report and $\sim 10^{-2}$ for at least one confirmatory study. These levels of significance have been suggested because they may be observed less than 5 times randomly in 100 scans of the human genome.

One of the most critical issues in confirmation of reported linkages is power. Attempts at confirmation of a reported susceptibility locus should state what power has been achieved to detect the locus initially described. For example, if a locus increases risk for BP illness in 25% of BP kindreds, it may be necessary to study ~100 affected sibling pairs in order to have adequate (80%) power to detect such a locus (Goldin and Gershon, 1988). Unfortunately, few studies address this key issue. If 100 affected sibling pairs are required to achieve adequate (80%) power to detect a previously described locus, then a publication with less than 75 sibling pairs does not address the central issue of confirmation. However, such power-limited publications may have an important role in meta-analyses, in that they identify invaluable sources of additional data.

Several confirmed BP linkage studies (according to the guidelines of Lander and Kruglyak [1995]) have been reported recently. Berrettini et al. (1994) reported evidence from ~130 affected sibling pairs ($p < \sim 10^{-4}$) for a BP susceptibility locus near the centromere of chromosome 18. They estimated that this linkage was valid for 25% of the 22 pedigrees studied. More recently Berrettini et al. (1996) have observed increased evidence for linkage to this region with updated diagnoses, such that a multipoint affected pedigree member statistic (Weeks and Lange, 1992) and a multipoint affected sibling pair statistic (Goldgar, 1990; Goldgar et al., 1993) yielded p values of $\sim 10^{-5}$. A linkage study of 30 BP pedigrees (~110 affected sibling pairs) confirmed this finding using several of the same DNA markers and similar analyses (Stine et al., 1995). For example, an affected sibling pair statistic p value was .0007 at D18S37. Stine et al. (1995) noted that most of their positive 18p statistics derived from families in which fathers transmitted illness (termed *paternal* pedigrees). Interestingly, the statistical support for a pericentromeric chromosome 18 BP susceptibility locus, reported by Berrettini et al. (1994), derives from kindreds in which there is evidence for *paternal* transmission of illness (Gershon et al. 1993). These results confirm those of Stine et al. (1995) for a parent-of-origin effect. Thus, evidence for a BP susceptibility gene near the centromere of chromosome 18 seems to meet suggested guidelines for validity (Lander and Krugylak, 1995), and appears to be present more often in kindreds where the father transmits illness. It is recommended strongly that attempts to confirm this pericentromeric chromosome 18 BP susceptibility locus should separate BP kindreds by the sex of the transmitting parent when analyzing results. While the genetic interpretation of this result is unclear, it may be consistent with a paternally imprinted BP susceptibility gene.

Several nonconfirmatory reports exist (Coon et al. 1996; Mynett-Johnson et al.,1996; Pauls et al., 1995), but these do not have adequate statistical power to detect the locus described by Berrettini et al. (1994). However, nonconfirmatory studies with adequate power have been described (Kelsoe et al., 1995; Smyth et al., 1996). According to the simulations of Suarez et al. (1994), some nonconfirmatory reports (with adequate power) are expected when a complex disease with oligogenic additive inheritance is studied.

Stine et al. (1995) also reported evidence for a second chromosome 18 susceptibility locus near D18S41 (18q21). The LOD score assuming heterogeneity was 3.51, while affected sibling pair statistics yielded a p value of .00002 at D18S41). These 18q21 DNA markers are approximately 40 cM more telomeric than the region identified by Berrettini et al. (1994), and these statistics suggest that a second chromosome 18 BP susceptibility locus may exist, although independent confirmation is required. No evidence for this 18q21 locus exists in our BP series (Berrettini et al., 1996).

A third genomic region of interest in the genetics of BP illness is Xq26, near the hypoxanthine phosphoribosyl transferase (HPRT) locus. Mendlewicz et al. (1987), who used a Factor IX RFLP polymorphism in 10 BP kindreds, reported a maximal LOD score of 3.1 at 10% recombination. Several confirmatory reports have appeared in the past few years (Craddock and Owen, 1992; Lucotte et al., 1992; Gill et al., 1992; Pekkarinen et al., 1995). The reports of Craddock and Owen (1992) and Gill et al. (1992) involve a single extended pedigree segregating both Factor IX deficiency and BP disorder. Lucotte et al. (1992) studied a single large French BP pedigree with a Factor IX polymorphism, reporting a LOD score in the range of 3 to 4. In the most recent of these, Pekkarinen et al. (1995) presented microsatellite data for a BP susceptibility gene near the HPRT locus in a large Finnish pedigree. They concluded that linkage of BP illness to the region was present in their pedigree, as the LOD score was ~3.5 at DXS994, assuming dominant inheritance. Given that the genetic distance between HPRT and Factor IX is ~10 cM (NIH/CEPH Collaborative Mapping Group, 1992), these reports are reasonably consistent. While evidence for Xq26 linkage is accumulating, such kindreds must represent a small minority of heritable BP illness, as most pedigrees have evidence of male-to-male transmission (Hebebrand, 1992), making an Xq26 susceptibility locus unlikely for those families. The evidence for linkage near HPRT cannot be seen as support for the reports of linkage to the color vision and G6PD loci, as HPRT is not linked (theta = 1/2) to these Xq28 loci (NIH/CEPH Collaborative Mapping Group, 1992).

McMahon et al. (1995) have documented an excess of maternal transmission in BP multiplex kindreds, meaning that mothers appear to transmit illness more often than expected by chance. We have observed a similar result in our 22 kindreds (Gershon et al., 1996). These results are consistent with mitochondrial inheritance and/or a maternally imprinted BP susceptibility gene.

Straub et al. (1994) described evidence for linkage of BP illness to a region of chromosome 21q21, near the phosphofructokinase locus. A single large pedigree with a LOD score of ~3.5 (for dominant inheritance) was reported from a series of ~57 BP kindreds. Affected sibling pair methods also yielded evidence for linkage. A confirmatory report has been described (Gurling et al., 1995), in which evidence for a two-locus BP disease model included an 11p15 and 21q21 marker data. Straub et al. (1994) estimated that this locus might be present in ~15 to 20% of BP kindreds. In our own BP pedigree series (Detera-Wadleigh et al., 1996), confirmation of linkage to this region was obtained using multipoint affected sibling pair methods ($p = .0002$), although no positive results were seen by affected pedigree member or LOD score methods. Thus, our data (Detera-Wadleigh et al., 1996) provide a confirmation of the report of Straub et al. (1994). Interestingly, the kindreds which yield positive statistical support for the chromosome 21q21 locus are those with *maternal* transmission of illness, suggesting that the pericentromeric chromosome 18 locus and the chromosome 21q21 locus may be risk factors for independent (nonoverlapping) sets of BP kindreds which can be differentiated clinically by the sex of the transmitting parent.

A fifth genomic region of interest lies on chromosome 4p, near the type 2c adrenergic receptor gene. Blackwood et al. (1996) describe a single large Scottish kindred in which linkage to D4S394 (LOD = 4.1) was found. Weak evidence in support of this locus was found in ~10 smaller BP Scottish kindreds by these authors. The validity of this locus must await confirmation.

D. SUMMARY

In summary, these linkage studies indicate that BP susceptibility loci may exist near 18p11.11, 18q21, 21q21, 4p24 and Xq26. These linkage studies are most compatible with the hypothesis that at least several different genes can increase risk for BP disorders, indicating that the inherited susceptibility to these illnesses is genetically heterogeneous. Within the next 5 years, susceptibility genes for these serious disorders will be identified. It is likely that the protein products of these genes will become targets for rational drug development by the pharmaceutical industry. If these efforts are successful, a new age of specific, genetically defined pharmacotherapy for these serious disorders will be developed.

ACKNOWLEDGMENTS

This chapter was prepared with the support of NIMH grant # MH48181. John Nurnberger is thanked for providing the table of twin concordances in bipolar illness.

REFERENCES

Allen, M.G., Cohen, S., Pollin, W., and Greenspan, S.I. (1974). Affective illness in veteran twins: a diagnostic review, *Am. J. Psychiat.*, 131:1234-1239.

Andrews, G., Stewart. G., Allen, R., and Henderson, A.S. (1990). The genetics of six neurotic disorders: a twin study, *J. Affective Disord.*, 19:23-29.

Angst, J., Frey, R., Lohmeyer, R., and Zerben-Rubin, E. (1980). Bipolar manic depressive psychoses: results of a genetic investigation, *Hum. Genet.*, 55:237-254.

Baron, M. (1977). Linkage between an X-chromosome marker (deutancolour blindness) and bipolar affective illness, *Arch. Gen. Psychiatry*, 34:721-725.

Baron, M., Freimer, N.F., Risch, N., Lerer, B., Alexander, J.R., Straub, R.E., Asokan, S., Das, K., Peterson, A., Amos, J., Endicott, J., Ott, J., and Gilliam, T.C. (1993). Diminished support for linkage between manic depressive illness and X-chromosome markers in three Israeli pedigrees. *Nat. Genet.*, 3:49-55.

Baron, M., Freimer, N.F., Risch, N., et al. (1993). Diminished support for linkage between manic-depressive illness and X-chromosome markers in three Israeli pedigrees, *Nat. Genet.*, 3:49-55.

Baron, M., Risch, N., Hamburger, R., Mandel, B., Kushner, S., Newman, M., Drumer, D., and Belmaker, R.H. (1987). Genetic linkage between X-chromosome markers and bipolar affective illness. *Nature*, 326:289-292.

Baron, M., Gruen, R., Anis, L., and Kane, J., (1983). Schizoaffective illness, schizophrenia and affective disorders: morbidity risk and genetic transmission, *Acta Psychiatr. Scand.*, 65:253-262.

Berrettini, W.H., Ferraro, T.N., Goldin, L.R., Weeks, D., Detera-Wadleigh, S., Nurnberger, J.I., Jr., and Gershon, E.S. (1994). Pericentromeric chromosome 18 DNA markers and manic-depressive illness: evidence for a susceptibility gene. *Proc. Natl. Acad. Sci. U.S.A.*, 91:5918-5921.

Berrettini, W.H., Ferraro, T.N., Goldin, L.R., Detera-Wadleigh, S.D., Choi, H., Muniec, D., Hsieh, W.-T., Hoehe, M., Guroff, J.J., Kazuba, D., Nurnberger, J.I., Jr., and Gershon, E.S. (1996). Linkage studies of bipolar illness, *Arch. Gen. Psychiatry* (in press).

Berrettini, W.H., Detera-Wadleigh, S.D., Goldin, L.R., Martinez, M., Hsieh, W.-T., Hoehe, M., Choi, H., Muniec, D., Ferraro, T.N., Guroff, J.J., Kazuba, D., Harris, N., Kron, E., Nurnberger, J.I., Jr., Alexander, R., and Gershon, E.S. (1991). Genomic screening for genes predisposing to bipolar disease, *Psychiatr. Genet.*, 2:191-208.

Berrettini, W.H., Goldin, L.R., Gelernter, J., Gejman, P.V., Gershon, E.S., and Detera-Wadleigh, S. (1990). X-chromosome markers and manic-depressive illness. Rejection of linkage to Xq28 in nine bipolar pedigrees. *Arch. Gen. Psychiatry*, 47:366-373.

Bertelsen, A., Harvald, B., and Hauge, M. (1977). A Danish twin study of manic-depressive disorders. *Br. J. Psychiatry*, 130:330-351.

Blackwood, D.H.R., He, L., Morris, S.W., McLean, A., Whitton, C., Thomson, M.L., Walker, M.T., Woodburn, K.J., Sharp, C.M., Wright, A.F., St. Clair, D.M., Porteous, D.J., and Muir, W.J. (1996). A locus for bipolar affective disorder on chromosome 4p, *Nat. Genet.*, 12, 427-430.

Caskey, C.T., Pizzuti, A., Fu, Y.-H., Fenwick, R.G., and Nelson, D.L. (1992). Triplet repeat mutations in human disease, *Science*, 256:784-789.

Coon, H., Hoff, M., Holik, J., Hadley, D., Fang, N., Reimherr, F., Wender, P., and Byerley, W. (1996). Analysis of chromosome 18 DNA markers in multiplex pedigrees with manic depression, *Biol. Psychiatry*, 39:689-696.

Coon, H., Jensen, S., Hoff, M., Holik, J., Plaetke, R., Reimberr, F., Wender, P., Leppert, P., and Byerley, W. (1993). A genome-wide search for genes predisposing to manic-depression, assuming autosomal dominant inheritance, *Am. J. Hum. Genet.*, 53:1234-1249.

Craddock, N. and Owen, M. (1992). Christmas disease and major affective disorder, *Br. J. Psychiatry*, 160:715.

De bruyn, A., Raeymaekers, P., Mendelbaum, K., Sandkuijl, L.A., Raes, G., Delvenne, V., Hirsch, D., Staner, L., Mendlewicz, J., and VanBroeckhoven, C. (1994). Linkage analysis of bipolar illness with X-chromosome DNA markers: a susceptibility gene in Xq27-q28 cannot be excluded, *Am. J. Med. Genet.*, 54:411-419.

De bruyn, A., Mendlebaum, K., Sandkuijl, L.A., Delvenne, V., Hirsch, D., Staner, L., Mendlewicz, J., and Van Broeckhoven, C. (1994). Nonlinkage of bipolar illness to tyrosine hydroxylase, tyrosinase and D2 and D4 dopamine receptor genes on chromosome 11, *Am. J. Psychiatry*, 151:102-106.

Del Zompo, M., Bocchetta A., Goldin, L.R., and Corsini, G.U. (1984). Linkage between X-chromosome markers and manic-depressive illness: two Sardinian pedigrees. *Acta Psychiatr. Scand.*, 70:282-287.

Del Zompo, M., Bocchetta, A., Ruiu, S., Goldin, L.R., and Berrettini, W.H. (1991). Association and linkage studies of affective disorders. In: *Biological Psychiatry*, Vol 2, Racagni, G., Brunello, N., and Fukuda, T., Eds., Elsevier, New York, 446-448.

Detera-Wadleigh, S.D., Badner, J.A., Goldin, L.R., Berrettini, W.H., Sanders, A.R., Rollins, D.Y., Turner, G., Moses, T., Haerian, H., Muniec, D., Nurnberger, J.I., Jr., and Gershon, E.S. (1996). Analysis of linkage to bipolar illness on chromosome 21q, *Nat. Genet.*, in press.

Detera-Wadleigh, S.D., Berrettini, W.H., Goldin, L.R., Boorman, D., Anderson, S., and Gershon, E.S. (1987). Close linkage of c-Harvey-ras-1 and the insulin gene to affective disorder is ruled out in three North American pedigrees, *Nature*, 325:806-808.

Detera-Wadleigh. S.D., Berrettini, W.H., Goldin, L.R., Martinez, M., Hsieh, W.-T., Hoehe, M., Coffman, D., Rollins, D.Y., Muniec, D., Choi, H., Wiesch, D., Guroff, J., and Gershon, E.S. (1992). A systematic search for a bipolar predisposing locus on chromosome 5, *Neuropsychopharmacology*, 6:219-229.

Detera-Wadleigh, S.D., Hsieh, W.-T., Berrettini, W.H., Goldin, L.R., Rollins, D.Y., Muniec, D., Grewal, R., Guroff, J.J., Turner, G., Hoffman, D., Barrick, J., Mills, K., Murray, J., Donohue, S.J., Klein, D.C., Sanders, J., Nurnberger, J.I., Jr., and Gershon, E.S. (1994). Genetic linkage mapping for a susceptibility locus to bipolar illness: Chromosomes 2, 3, 4, 7, 9, 10p, 11p, 22, and Xpter, *Neuropsychiatr. Genet.*, 54:206-218.

Egeland, J.A., Gerhard, D.S., Pauls, D.L., Sussex, J.N., Kidd, K.K., Allen, C.R., Hostetter, A.M., and Housman, D.E. (1987). Bipolar affective disorder linked to DNA markers on chromosome 11, *Nature*, 325:783-787.

Gejman, P.V., Martinez, M., Qiuhe, C., Friedman, E., Berrettini, W.H., Goldin, L.R., Koroulakis, P., Ames, C., Lerman, M.A. and Gershon, E.S. (1993). Linkage analysis of 57 microsatellite loci to bipolar disorder, *Neuropsychopharmacology*, 9:31-49.

Gershon, E.S., Badner, J.A., Ferraro, T.N., Detera-Wadleigh, S., and Berrettini, W.H. (1996). Maternal inheritance and chromosome 18 allele sharing in unilineal bipolar illness pedigrees, *Neuropsychiatr. Genet.*, 67, 1-8.

Gershon, E.S., Hamovit, J., Guroff, J.J., Dibble, E., Leckman, J.F., Sceery, W., Targum, S.D., Nurnberger, J.I., Jr., Goldin, L.R., and Bunney, W.E. Jr., (1982). A family study of schizoaffective, bipolar I, bipolar II, unipolar, and normal control probands, *Arch. Gen. Psychiatry*, 39:1157-1167.

Gershon, E.S., Hamovit, J.H., Guroff, J.G., and Nurnberger, J.I., Jr. (1987). Birth-cohort changes in manic and depressive disorders in relatives of bipolar and schizoaffective patients, *Arch. Gen. Psychiatry*, 44:314, 1987.

Gershon, E.S., Targum, S.D., Matthysse, S., and Bunney, W.E., Jr. (1979). Color blindness not closely linked to bipolar illness, *Arch. Gen. Psychiatry*, 36:1423-1431.

Gill, M., McKeon, P., and Humphries, P. (1988). Linkage analysis of manic-depression in an Irish family using H-ras 1 and INS DNA markers, *J. Med. Genet.*, 25:634-635.

Gill, M., Castle, D., and Duggan, C. (1992). Cosegregation of Christmas disease and major affective disorder in a pedigree. *Br. J. Psychiatry*, 160:112-114.

Gill, M., Castle, D., Hunt, N., Clements, A., Sham, P., and Murray, R.M. (1991). Tyrosine hydroxylase polymorphisms and bipolar affective disorder, *J. Psychiatr. Res.*, 25:179-184

Goldgar, D.E. (1990). Multipoint analysis of human quantitative variation, *Am. J. Hum. Genet.*, 45:957-967.

Goldgar, D.E., Lewis, C.U., and Gholami, K. (1993). Analysis of discrete phenotypes using a multipoint identity by descent method: application to Alzheimer's disease, *Genet. Epidemiol.*, 10:383-388.

Goldin, L.R. and Gershon, E.S. (1988). Power of the affected sib-pair method for heterogeneous disorders, *Genet. Epidemiol.*, 5:35-42.

Goldin, L.R. and Weeks, D.E. (1993). Two-locus models of disease: comparison of likelihood and non-parametric linkage methods, *Am. J. Hum. Genet.*, 53:908-915.

Greenberg, D.A. (1993). Linkage analysis of "necessary" disease loci versus "susceptibility" loci, *Am. J. Hum. Genet.*, 32:135-143.

Gurling, H., Smyth, C., Kalsi, G., et al. (1995). Linkage findings in bipolar disorder, *Nat. Genet.*, 10:8-9.

Harvald, B. and Hauge, M. (1975). In: Genetics and the Epidemiology of Chronic Diseases, J.V. Neal, M.W. Shaw and W.J. Shull, Eds., PHS Pub. No. 1163, U.S. Department of Health, Education and Welfare, Washington, D.C., 61-76.

Hebebrand, J. (1992). A critical appraisal of X-linked bipolar illness, *Br. J. Psychiatry*, 160:7-11.

Helzer, J.E. and Winokur, G. (1974). A family interview study of male manic-depressives, *Arch. Gen. Psychiatry*, 31, 73-77.

Hodge, S.E. and Wickramaratne, P. (1995). Statistical pitfalls in detecting age-at-onset anticipation: the role of correlation in studying anticipation and detecting ascertainment bias, *Psychiatr. Genet.*, 5:43-47.

Hodgkinson, S., Sherrington, R., Gurling, H.M.D., Marchbanks, R.M., Reeders, S.S.T., Mallet, J., McInnis, M., Petursson, H., and Brynjolfsson, J. (1987). Molecular genetic evidence for heterogeneity in manic depression, *Nature*, 325:805-808.

Inayama, Y., Yoneda, H., Sakai, T., Ishida, T., Kobayashi, S., Nonomura, Y., Kono, Y., Koh, J., and Asaba, H. (1993). Lack of association between bipolar affective disorder and tyrosine hydroxylase DNA marker, *Am. J. Med. Genet.*, 48:87-89.

James, N.M. and Chapman, C.J. (1975). A genetic study of bipolar affective disorder, *Br. J. Psychiatry*, 126:449-456.

Johnson, G.F.S. and Leeman, M.M. (1977). Analysis of familial factors in bipolar affective illness, *Arch. Gen. Psychiatry*, 34:1074-1083.

Joyce, P.R., Oakley-Brown, M.A., Wells, J.E., Bushnell, J.A., and Hornblow, A.R. (1990). Birth cohort trends in major depression: increasing rates and earlier onset in New Zealand, *J. Affective Disord.*, 18:83-89.

Kallman, F. (1954). In: *Depression*, Hoch, P.H. and Zubin, J., Eds., Grune & Stratton, New York, pp. 1-24.

Kelsoe, J.R., Sadovick, A.D., Kristbjarnarson, H., Bergesch, P., Mroczkowski-Parker, Z., Flodman, P., Rapaport, M.H., Mirow, A.L., Egeland, J.A., Spence, M.A., and Remick, R.A. (1995). Genetic linkage studies of bipolar disorder and chromosome 18 markers in North American, Icelandic and Amish pedigrees, *Psychiatr. Genet.*, 5: S17.

Kelsoe, J.R., Ginns, E.I., Egeland, J.A., Gerhard, D.S., Goldstein, A.M., Bale, S.J., Pauls, D.L., Long, R.T., Kidd, K.K., Conte, G., Housman, D.E., and Paul, S.M. (1989). Re-evaluation of the linkage relationship between chromosome 11p loci and the gene for bipolar affective disorder in the Old Order Amish, *Nature*, 342:238-243.

Kendler, K.S., Neale, M.C., Kessler, R.C., Heath, A.C., and Eaves, L.J. (1992). A population based twin study of major depression in women: the impact of varying definitions of illness. *Arch. Gen. Psychiatry*, 49:257-266.

Kendler, K.S., Petersen, N., Johnson, L., Neale, M.C., and Mathe, A.A. (1993). A pilot Swedish twin study of affective illness, including hospital and population ascertained subsamples, *Arch. Gen. Psychiatry*, 50:699-706.

Kennedy, J.L., Sidenberg, D.G., Macciardi, F.M., and Joffe, R.T. (1993). Genetic association study of tyrosine hydroxylase and D4 receptor variants in bipolar I patients, *Psychiatr. Genet.*, 3:120.

Klerman, G.L. (1987). Clinical epidemiology of suicide, *J. Clin. Psychiatry*, 48 (Suppl.), 33-38.

Klerman, G.L., Lavori, P.W., Rice, J., Reich, T., Endicott, J., Andreasen, N.C., Keller, M.B., and Hirschfield, R.M.A. (1985). Birth-cohort trends in rates of major depressive disorder among relatives of patients with affective disorder, *Arch. Gen. Psychiatry*, 42:689, 1985.

Klerman, G.L. and Weissman. M.M. (1989). Increasing rates of depression, *JAMA*, 261:2229-2235.

Korner, J., Fritze, J., and Propping, P. (1990). RFLP allels at the tyrosine hydroxylase locus: no association found to affective disorders, *Psychiatry Res.*, 32:275-280.

Korner, J., Rietschel, M., Hunt, N., Castle, D., Gill, M., Nothen, M., Craddock, N., Daniels, J., Owen, M., Frimmers, R., Fritze, J., Moller, H.-J., and Propping, P. (1994). Association and haplotype analysis at the tyrosine hydroxylase locus in a combined German-British sample of manic-depressive patients and controls, *Psychiatr. Genet.*, 4:167-175.

Lander, E.S. and Kruglyak, L. (1995). Genetic dissection of complex traits: guidelines for interpreting and reporting linkage results, *Nat. Genet.*, 11:241-247.

Leboyer, M., Malafosse, A., Boularand, S., Campion, D., Gheysen, F., Samolyk, D., Henriksson, B., Denise, E., des Lauriers, A., Lepine, J.-P., Zarifian, E., Clerget-Darpoux, F., and Mallet, J. (1989). Tyrosine hydroxylase polymorphisms associated with manic-depressive illness, *Lancet*, 335:1219.

LeBoyer, M., Malafosse, A., Boularand, S., Campion, D., Gheysen, F., Samolyk, D., Henriksson, B., Denies, E., des Lauriers, A., Lepine, J-P., Zarifian, E., Clerget-Darpoux, F., and Mallet, J. (1990). Tyrosine hydroxylase polymorphisms associated with manic-depressive illness, *Lancet*, 335:1219.

Lindblad, K., Nylander, P.-O., De Bruyn, A., et al. (1995). Detection of expanded CAG repeats in bipolar affective disorder using the repeat expansion detection (RED) method, *Neurobiol. Dis.*, 2:55-62.

Lipp, O., Souery, D., Mahieu, B., De Bruyn, A., Van Broeckhoven, C., and Mendlewicz, J. (1995). Anticipation in affective disorders, *Psychiatr. Genet.*, 5:S8.

Lucotte, G., Landoulsi, A., Berriche, S., David, F., and Babron, M.C. (1992). Manic-depressive illness is linked to factor IX in a French pedigree, *Ann. Genet.*, 35:93-95.

Luxenberger, H. (1930). Psychiatrisch-neurologische Zwillings pathologie, *Zentralbl. Gesamte Neurol. Psychiatrie*, 14:56-57, 145-180.

Maier, W., Lichtermann, D., Minges, J., Hallmayer, J., Heun, R., Benkert, O., and Levinson, D.F. (1993). Continuity and discontinuity of affective disorders and schizophrenia. Results of a controlled family study, *Arch. Gen. Psychiatry*, 50:871-83.

McGuffin, P. and Katz, R. (1989). The genetics of depression and manic-depressive disorder, *Br. J. Psychiatry*, 155:294-304.

McGuffin, P., Katz, R., Watkins, S., and Rutherford, J. (1996). A hospital-based twin register of the heritability of DSM-IV unipolar depression, *Arch. Gen. Psychiatry*, 53:129-136.

McInnis, M.G., McMahon, F.J., Chase, G.A., Simpson, S.G., and Ross, C.A. (1993). Anticipation in bipolar affective disorder, *Am. J. Hum, Genet.*, 53:385-390.

McMahon, F. J., Stine, O.C., Chase, G.A., Meyers, D.A., Simpson, S.G., and DePaulo, R.J. (1995). Patterns of maternal transmission in bipolar affective disorder, *Am. J. Hum. Genet.*, 56:1277-1286.

Meloni, R., Leboyer, M., Campion, D., Savoye, C., Poirier, M.-F., Samolyk, D., Malafosse, A., and Mallet, J. (1993). Association of manic depressive illness with the TH locus using a microsatellite marker localized in the tyrosine hydroxylase gene, *Psychiatr. Genet.*, 3:121.

Mendlewicz, J., Fleiss, J.L., and Fieve, R.R. (1972). Evidence for X-linkage in the transmission of manic-depressive illness, *JAMA*, 222:1624-1627.

Mendlewicz, J. and Fleiss, J.L. (1974). Linkage studies with X-chromosome markers in bipolar (manic-depressive) and unipolar (depressive) illnesses, *Biol. Psychiatry*, 9:261-294.

Mendlewicz, J., Linkowski, P., Guroff, J.J., and Van Praag, H.M. (1979). Colour blindness linkage to bipolar manic depressive illness: new evidence, *Arch. Gen. Psychiatry*, 36:1442-1447.

Mendlewicz, J., Linkowski, P., and Wilmotte, J. (1980). Linkage between glucose-6-phosphate-dehydrogenase deficiency and manic-depressive psychosis, *Br. J. Psychiatry*, 137:337-342.

Mendlewicz, J., Leboyer, M., De Bruyn, A., Malafosse, A., Sevy, S., Hirsch, D., Van Broeckhoven, C., and Mallet, J. (1991). Absence of linkage between chromosome 11p15 markers and manic-depressive illness in a Belgian pedigree, *Am. J. Psychiatry*, 148:1683-1687.

Mendlewicz, J., Simon, P., Sevy, S., Charon, F., Brocas, H., Legros, S., and Vassart, G. (1987). Polymorphic DNA marker on X-chromosome and manic depression, *Lancet*, 1:1230-1232.

Mitchell, P., Waters, B., Morrison, N., Shine, J., Donald, J., and Eissman, J. (1991). Close linkage of bipolar disorder to chromosome 11 markers is excluded in two large Australian pedigrees, *J. Affective Disord.*, 20:23-32.

Mynett-Johnson, L.A., Murphy, V.E., Manley, P., Shields, D.C., Humphries, P., and McKeon, P. (1996). Lack of evidence for a major locus for bipolar disorder is the pericentromeric region of chromosome 18 in Irish pedigrees, *Biol. Psychiatry*, in press.

NIH/CEPH Collaborative Mapping Group (1992). A comprehensive genetic linkage map of the human genome, *Science*, 258:67-86.

Nothen, M., Korner, J., Lanczik, M., Fritze, J., and Propping, P. (1990). Tyrosine hydroxylase polymorphisms and manic-depressive illness, *Lancet*, 336:575.

Nurnberger, J.I., Jr., Goldin, L.R., and Gershon, E.S. (1986). Genetics of psychiatric disorders, in *The Medical Basis of Psychiatry*, Winokur, G. and Clayton, P., Eds., W.B. Saunders, New York, 486-521.

O'Donovan. M.C., Guy, C., Craddock, N., Murphy, K.C., Cardno, A.G., Jones, L.A., Owen, M.J., and McGuffin, P. (1995). Expanded CAG repeats in schizophrenia and bipolar disorder, *Nat. Genet.*, 10:380.

O'Rourke, D.H., McGuffin, P., and Reich, T. (1983). Genetic analysis of manic-depressive illness, *Am. J. Phys. Anthropol.*, 62:51-59.

Pauls, D.L., Ott, J., Paul, S.M., Allen, C.R., Fann, C.S.J., Carulli, J.P., Falls, K.M., Bouthillier, C.A., Gravius, T.C., Keith, T.P., Egeland, J.A., and Ginns, E.I. (1995). Linkage analyses of chromosome 18 markers do not identify a major susceptibility locus for bipolar affective disorder in the Old Order Amish, *Am. J. Hum. Genet.* 57:636-643.

Pekkarinen, P., Terwilliger, J., Bredbacka, P.-E., Lunnqvist, J., and Peltonen, L. (1995). Evidence of a predisposing locus to bipolar disorder on Xq24-q27.1 in an extended Finnish pedigree, *Genome Res.*, 5:105-115.29.

Penrose, L.S. (1948). The problem of anticipation in pedigrees of dystrophia myotonica, *Ann. Eugen.*, 14:125.

Reich, T., Clayton, P.J., and Winokur, G. (1969). Family history studies. V. The genetics of mania, *Am. J. Psychiatry*, 125:1358-1368.

Rice, J.P., Reich, T., Andreasen, N.C., Endicott, J., Van Eerdewegh, M., Fisherman, R., Hirschfield, R.M.A., and Klerman, G.L. (1987). The familial transmission of bipolar illness, *Arch. Gen. Psychiatry*, 44:441-447.

Rosanoff, A.J., Handy, L., and Plesset, I.R. (1935). The etiology of manic-depressive syndromes with special reference to their occurrence in twins, *Am. J. Psychiatry*, 91:725-762.

Ross, C.A., McInnis, M.G., Margolis, R.L., and Li, S.-H. (1993). Genes with triplet repeats: candidate mediators of neuropsychiatric disorders, *Trends Neurosci.*, 16:254-260.

Slater, E. (1953). Psychotic and Neurotic Illness in Twins, Medical Research Council Special Report Series No. 278, Her Majesty's Stationery Office, London.

Smyth, C., Kalsi, G., Brynjolfsson, J., Sherrington, R.S., O'Neill, J., Curtis, D., Rifkin, L., Murphy, P., Petursson, H., and Gurling, H.M.D. (1996). Linkage analysis of manic depression (bipolar affective disorder) in Icelandic and British kindreds using markers on the short arm of chromosome 18, Arch. Gen. Psychiatry, in press.

Smyth, C., Kalsi, G., Brynjolfsson, J., O'Neill, J., Curtis, D., Rifkin, L., Moloney, E., Murphy, P., Sherrington, R., Petursson, H., and Gurling, H.M.D. (1996). Further tests for linkage at the tyrosine hydroxylase gene locus on chromosome 11p15 in a new sample of British multiplex manic depression (bipolar and unipolar affective disorder) families, Am. J. Med. Genet., in press.

Smyth, C., Kalsi, G., Brynjolfsson, J., Pertursson, H., Curtis, D., Rifkin, L., Murphy, P., Monoley, E., O'Neill, J., and Gurling, H.M.D. (1995). A test of the Xq27-q28 linkage in bipolar and unipolar families selected for absent male-to-male transmission, *Psychiatric Genet.*, 5:S84 (Suppl.).

Spence, M.A., Ameli, H., Sadovnick, A.D., Remick, F.A., Bailey-Wilson, J.A., Flodman, P., and Yee, I.M.L. (1993). A single major locus is the best explanation for bipolar family data: results of complex segregation analysis, *Am. J. Hum. Genet.*, 53:862-866.

Stine, O.C., Xu, J., Koskela, R., McMahon, F.J., Gschwend, M., Friddle, C., Clark, C.D., McInnis, M.G., Simpson, S.G., Breschel, T.S., Vishio, E., Riskin, K., Feilotter, H., Chen, E., Folstein, S., Meyers, D.A., Botstein, D., Marr, T.G., and DePaul, J.R. (1995). Evidence for linkage of bipolar disorder to chromosome 18 with a parent-of-origin effect, *Am. J. Hum. Genet.*, 57:1384.

Straub, R.E., Lehner, T., Luo, Y., Loth, J.E., Shao, W., Sharpe, L., Alexander, J.R., Das, K., Simon, R., Fieve, R.R., Lerer, B., Endicott, J., Ott, J., Gilliam, C.T., and Baron, M. (1994). A possible vulnerability locus for bipolar affective disorder on chromosome 21q22.3, *Nat. Genet.*, 8:291-296.

Straub, R. and Gilliam, C. (1993). Genetic linkage studies of bipolar affective disorder, *Genome Anal.*, 6:77-99.

Suarez, B., Harpe, C.L., and Van Eerdewegh, P. (1994). Problems of replicating linkage claims in psychiatry, in, *Genetic Approaches to Mental Disorders*, Gershon, E.S. and Cloninger, C.R., Eds., American Psychiatric Press, Washington, D.C., pages 23-46.

Souza, F.G.M. and Goodwin, F.K. (1991). Lithium treatment and prophylaxis in unipolar depression: a meta-analysis. *Br. J. Psychiatry*, 158:666-675.

Verga, M., Marino, C., Petronis, A., Cavallini, M.C., Cauli, G., Smeralki, E., Kennedy, J.L., and Macciardi, F. (1993). Association of tyrosine hydoxylase gene and neuropsychiatric disorders, *Psychiatr. Genet.*, 3:168.

Weeks, D.E. and Lange. K. (1992). A multilocus extension of the affected-pedigree-member method of linkage analysis, *Am. J. Hum. Genet.*, 50:859-868.

Weissman, M. (1987). Advances in psychiatric epidemiology: rates and risks for major depression, *Am. J. Public Health*, 77:445-451.

Weissman, M.M., Gershon, E.S., Kidd, K.K., Prusoff, B.A., Leckman, J.F., Dibble, E., Hamovit, J., Thompson, W.D., Pauls, D.L., and Guroff, J.J. (1984). Psychiatric disorders in the relatives of probands with affective disorder, *Arch. Gen. Psychiatry*, 41:13-21.

Weissman, M.M. and Myers, J.K. (1978). Affective disorders in a U.S. urban community: the use of research diagnostic criteria in an epidemiological survey, *Arch. Gen. Psychiatry*, 35:1304-1311.

Wickramaratne, P.J., Weissman, M.M., Leaf, P.J., and Holford, T.R. (1989). Age, period and cohort effects on the risk of major depression: results from five United States communities, *J. Clin. Epidemiol.*, 42:333.

Winokur, G., Clayton, P.J, and Reich T. (1969). *Manic-Depressive Illness*, C.V. Mosby, St. Louis, 112-125.

Winokur, G., Coryell, W., Keller, M., Endicott, J., and Leon, A. (1995). A family study of manic-depressive (bipolar I) disease. Is it a distinct illness separable from primary unipolar depression?, *Arch. Gen. Psychiatry*, 52(5):367-73.

Winokur, G., Tsuang, M.T., and Crowe, R.R. (1982). The Iowa 500. Affective disorder in relatives of manic and depressed patients, *Am. J. Psychiatry*, 139: 209-212.

16 The Genetics of Personality

Thomas J. Bouchard, Jr.

CONTENTS

A. Introduction 273
B. Psychometric Approaches to Personality 274
    1. The Big Three 274
    2. The Big Five 275
    3. The PDI Nine 275
C. Quantitative Genetic Methods 275
    1. Path Analysis 275
    2. Assumptions 279
D. Behavior Genetic Analyses of Kinship Data 280
    1. MZT and DZT Correlations and Kinship Modeling for the Big Three 281
    2. Loehlin's Modeling of Extended Kinship Data for the Big Five 281
    3. Nichols Meta-Analysis Organized According to the Big Five and the Big Nine 283
    4. Personality Disorders 284
    5. SATSA and MISTRA Big Five Analyses 285
    6. Social Attitudes 287
    7. Love Styles 288
    8. Crime 288
E. Role of Intelligence 289
F. Mapping Genes for Intelligence and Personality 289
G. Behavior Genetics vs. Socialization Theory 290
Acknowledgments 290
References 290

A. INTRODUCTION

The systematic study of personality as a facet of human individual differences began, as did the study of most other individual differences, with Francis Galton.[1] Galton recommended that the dictionary be consulted for terms that characterized the most important individual differences. This "lexical" approach has been widely used over the years.[2-6] Nevertheless, psychologists are not agreed upon a common conceptual scheme for the study of personality. There is simply no widely agreed upon theory of personality to provide a basis for a common research program. One widely used introductory textbook solution to this problem is simply to discuss each of the different approaches to the topic as relatively discrete entities: psychoanalytic theories (Freud and Jung), humanistic theories (Rogers and Maslow), behaviorism (Watson and Skinner), social cognitive theories (Mischel and Bandura), and trait and factor approaches (Eysenck, Cattell, and Costa and McCrae). Behavior geneticists pick and choose among the various points of view, but because of their need for a specifiable phenotype, large sample sizes, and ease of measurement, they focus almost exclusively on the work of investigators who have operationalized their ideas in the form of personality questionnaires or inventories. This does not, however, narrow the field a great deal as a very wide array of personality measures are available. There are, for example, almost as many scales for the Minnesota Multiphasic Personality Inventory (MMPI-1) as there are items.[7] With the arrival of the MMPI-2,[8] the problem will become worse. Some of the instruments embody constructs that flow from a theory — thus the Myers-Briggs Type Indicator (MBTI) is based on Jung's theory of types.[9] Other inventories, like the California Psychological Inventory (CPI), reflect a very pragmatic view of the problem, measuring what ordinary people in most societies tend to think are important dimensions. Gough[10] calls these "folk concepts." The MMPI has sometimes been said to measure the "folk concepts of psychiatry," dimensions of psychopathology that lack solid theoretical foundations.

In recent years there has been a developing consensus among trait and factor psychologists that a five-factor model of personality should perhaps replace the previously dominant three-factor model of Eysenck. There are also those who believe it would be well worthwhile pursuing additional factors. We now turn to a brief review of these perspectives.

B. PSYCHOMETRIC APPROACHES TO PERSONALITY

1. The Big Three

Eysenck[11,12] has for many years propounded a "paradigm" for personality research based on three higher-order dimensions: Extraversion, Neuroticism, and Psychoticism. Extraversion subsumes the following lower order or primary traits: sociable, lively, active, assertive, sensation-seeking, carefree, dominant, surgent, and venturesome. Neuroticism subsumes: anxious, pressed, guilt feelings, low self-esteem, tense, irrational, shy, moody, emotional. Psychoticism subsumes: aggressive, cold, egocentric, impersonal, impulsive, antisocial, unempathic, creative, and tough-minded. This latter dimension also subsumes all the various mental disorders,[13] creativity,[14] and criminality.[15] The big three are based on a blend of psychometrics, experimental studies, and biological theorizing.[12] This scheme has a large number of adherents and there is an extensive literature built around these dimensions.[16,17] The big three are generally measured with the Eysenck Personality Questionnaire (EPQ)[11] (see also Reference 17, Appendices B and C).*

A similar scheme more heavily based on empiricism and psychometrics, but certainly not theory free, is represented in the work of Tellegen[18-20] in his Multidimensional Personality Questionnaire (MPQ), an instrument used in the Minnesota Study of Twins Reared Apart (MISTRA).[21,22] The MPQ measures 11 primary scales and also yields 3 higher-order dimensions, Positive Emotionality (PEM), Negative Emotionality (NEM), and Constraint, 2 of which overlap with the Eysenck scheme (PEM = Extraversion; NEM = Neuroticism). The Eysenck Psychoticism dimension is more poorly mapped against Constraint.

---

* Primary traits are generally measured directly by scoring items on a questionnaire that measure that specific trait (e.g., items that measure how sociable one is). Higher-order factors reflect the correlation among the primary traits and can be derived by subjecting the primary traits to factor analysis and using factor scores as measures. Some tests use regression equations, weighting the scales differentially to obtain scores on higher-order factors. It is also possible to simply sum into a score items from a variety of primary scales to derive higher-order factor scores. This latter procedure is used with the EPI. The item content of scales used to measure higher-order factors, or super factors, is much more heterogeneous than the item content of primary scales.

2. THE BIG FIVE

In recent years considerable consensus has developed around a five-factor scheme that has come to be called the "Big Five".[23–26] The Big Five have their origins in work by Fiske,[27] Tupes and Christal,[28] and Norman.[29]

Costa and McCrae have produced an inventory to measure these dimension, the NEO-PI-R,[30] but it has been used only once, in brief form, in behavior genetics studies.[31] The NEO-PI-R had its origins in Eysenck's Big Three.[32] One version of the Big Five is shown in Table 16-1.

The bipolar markers are from Goldberg,[33] the NEO-PI-R facet scales are from Costa and McCrae,[30] the CPI equations are from Gough (personal communication), and the MPQ markers are from Tellegen (personal communication).

3. THE PDI NINE

Because of our interest in the practical use of psychological tools[34, 35] we are partial to a taxonomic scheme developed by the Personnel Decisions Incorporated (PDI) group — the Big Nine.[36] We have mapped the Big Nine against the Big Five, the Big Three, and some specific scales of the MPQ in Figure 16-1.

Our preference for this scheme is both practical and theoretical. On the practical side there is evidence that more than five factors are necessary for effective prediction,[36] but see Ones[37] and Barrick and Mount for discussion.[38] On the theoretical side the arguments are Darwinian; we take the theoretical stance of evolutionary psychology[39-42] that the human mind is made up of numerous highly specialized information processing mechanisms that have evolved to solve adaptive problems. While there is not sufficient space here to expand on this theoretical perspective, it is important to recognize that the evolutionary psychology perspective is distinctly different from that of sociobiology. Many if not most human sociobiologists argue that human beings evolved largely as fitness maximizers, whereas evolutionary psychologists argue that we are adaptation executors.[43] In the first case the relation of a trait or psychological mechanism to fitness, or a modern surrogate of fitness,[44] is important. In the latter case the adaptive nature of a psychological mechanism is judged by evidence of its design.[45,46] Buss et al.[47] provides an informative evolutionary analysis of human jealousy as an example of this line of reasoning. It seems reasonable that the fundamental dimensions of personality will reflect biological adaptations. This brings us to our last point: although a typological approach to personality has long been out of favor in the study of personality, variation in this domain may, in part, reflect the existence of types.[48] We will not deal with this issue in this chapter.

Basically, the argument regarding the proper number of super factors is the classic one between lumpers and splitters, and there are persuasive arguments for and against each position. Until the issue is resolved in some more satisfactory manner, it is probably worth looking at the genetic evidence from a variety of points of view and that is what we will do in this chapter. We will, however, focus largely on measures of the higher-order factors or their surrogates at the primary level in order to avoid getting bogged down in the plethora of primary measures.

C. QUANTITATIVE GENETIC METHODS

1. PATH ANALYSIS

The vast majority of evidence regarding genetic influence on personality comes from twin and adoption studies. Prior to reviewing that evidence we will briefly review the logic

TABLE 16-1
The Big Five Factors (Boldface), Alternate Names for the Factors (Italics), Sample Bipolar Scales, the Six NEO-PI-R Facet Scales, the California Psychological Inventory Equations for Predicting the Big Five, and the Best Marker of the Factor for the Multidimensional Personality Questionnaires (Boldface Italics)

Extraversion
Surgency, Introversion-Extraversion (–), Dominance

introverted .... extroverted
unenergetic .... energetic
timid .... bold

Warmth, Gregariousness, Assertiveness, Activity, Excitement seeking, Positive emotions
Extraversion = Dominance + Self acceptance – Self control

Social closeness

Neuroticism
Adjustment, Anxiety, Emotional Stability (–), Stress Reactivity (–),

angry .... calm
nervous .... at ease
emotional .... unemotional

Anxiety, Angry hostility, Depression Self-consciousness, Impulsiveness, Vulnerability
Adjustment = Well being + Work orientation – Anxiety

Stress reaction

Conscientiousness
Conformity, Dependability, Authoritarianism (–)

disorganized .... organized
irresponsible .... responsible
careless .... thorough

Competence, Order, Dutifulness, Achievement striving, Self-discipline, Deliberation
Conscientiousness = Responsibility + Achievement via Conformance – Flexibility

Control

Agreeableness
Likability, Friendliness, Pleasant

cold .... warm
selfish .... unselfish
distrustful .... trustful

Trust, Straightforwardness, Altruism, Compliance, Modesty, Tendermindedness
Agreeableness = Socialization + Tolerance – Narcissism

Aggression

Openness
Culture, Intellect, Sophistication

intelligent .... unintelligent
reflective .... unreflective
creative .... uncreative

Fantasy, Aesthetics, Feelings, Actions, Ideas, Values
Culture = Empathy + Achievement via Independence + Creativity

Absorption

The Genetics of Personality

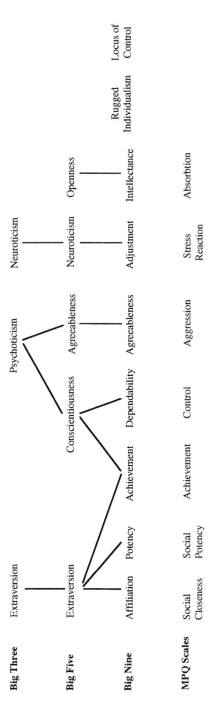

FIGURE 16-1 Relationship between the dimensions of the Big Three, Big Five, Big Nine, and Multidimensional Personality Questionnaire scales.

of these methods. The simplest approach to this issue is via path analysis.[49-52] Figure 16-2. shows the path diagrams for monozygotic twins reared together (MZT), dizygotic twins reared together (DZT), and monozygotic twins and dizygotic twins reared apart (MZA and DZA).

The items in circles indicate latent variables: G = heredity, C = common (shared) family

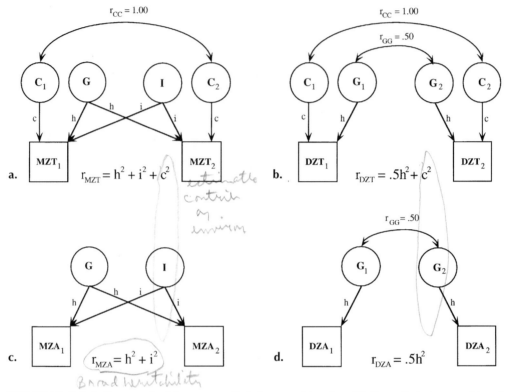

FIGURE 16-2 Path diagrams for (a) monozygotic twins reared together, (b) dizygotic twins reared together, (c) monozygotic twins reared apart, and (d) dizygotic twins reared apart.

environmental influence, and I = nonadditive genetic variance. Items in boxes indicate measurable phenotypes (MZT_1 = personality test score of the first monozygotic twin reared together). The influence of each latent trait is characterized by a path with a directed arrow indicating causation. The standardized path coefficients h, c, and i, on the various paths (arrows) indicate additive genetic, common environmental, and nonadditive genetic influences. We use standardized path coefficients because we are representing correlations not covariances. It is our job to estimate the magnitude of these effects from appropriate data sets. The rules of path analysis allow us to write the following equation for MZT twins (upper left-hand side of the figure):

$$r_{MZT} = h^2 + i^2 + c^2$$

In other words, the causes of similarity between MZT twins is due to genetic factors ($h^2 + i^2$) and common environmental factors (c^2). Following the rules of path analysis,[49, 50] the magnitude of these factors is estimated by multiplying the coefficients of each path connecting the phenotypic measures. For MZT twins the path connecting G is 1.00 because MZ twins share exactly the same heredity. Thus the additive genetic effect is estimated to be (h * 1 * h) or h^2. The other paths are estimated in the same way. Unique environmental influence plus error of measurement can be estimated by subtracting r_{MZT} from 1.00 because we are using standardized coefficients, and by definition the intraclass correlation cannot be higher than 1.00. We have left out the path for the unique environmental contribution (plus error) to variance for each individual because it does not contribute to similarity and it clutters the diagram.

The influence due to i for DZ twins is very modest and consequently has also been left out. Dizygotic twins have half their segregating genes in common by descent and this is shown by the correlation of .50 between the Gs. Thus the equation for dizygotic twins reared together (DZT) is (upper right-hand side of Figure 16-2):

$$r_{DZT} = (.5h^2) + c^2$$

Under the assumptions of additive effect of genes only (no nonadditive gene action), and equal trait relevant environmental influence for the two types of twins, the differences between them is $.5h^2$ or one half the genetic influence. If we multiply this value by 2 we can estimate the full genetic influence. This estimate of heritability is called the Falconer heritability.[53] Under the same assumptions $c^2 = (2r_{DZT} - r_{MZT})$.

The diagram in the lower right-hand corner of the figure is for monozygotic twins reared apart (MZA). By definition if the twins are reared apart there is no common family environmental influence. The equation is

$$r_{MZA} = h^2 + i^2$$

The correlation between MZA twins estimates what is called the *broad heritability*, the combined effects of both the additive and nonadditive genetic variance ($h^2 + i^2$). The *narrow heritability* is simply h^2. The difference between MZA and MZT twins estimates the magnitude of c^2 as does the difference between DZA and DZT twins. If $i^2 = 0$, the correlation for DZA twins estimates one half the heritability.

Notice that r_{MZA} directly estimates the broad heritability. It does not have to be squared.

2. **ASSUMPTIONS**

Our model makes a number of assumptions. It assumes (from genetic theory) that MZ twins share all of their genes and that DZ twins share half of their segregating genes (have half their genes in common by descent). Under a simple genetic model we would assume additivity, that is that the influence of genes in DZ twins is half of that in MZ twins. Empirically, if the DZ correlation is half the MZ correlation this would suggest that the additivity assumption is probably correct. There is sufficient evidence for nonadditive gene action in the personality domain that we must include it in our model.

We also assume that common family environmental influences are the same for both types of twins (the equal environment assumption). This assumption has been constantly challenged by critics of the twin method. It has, however, been studied by twin researchers for many years and the general consensus is that it is well supported. Bouchard[54] has recently reviewed much of this literature. This conclusion is often met with incredulity by non-twin researchers and deserves some comment. First, it is important to recognize that just because a pair of twins are treated alike on some dimension, that treatment does not necessarily cause similarity on an individual difference dimension. A very common argument is that because identical twins are much alike in terms of physical attractiveness this factor influences how people treat them, and it is presumed that they thereby become alike in personality.[55] A careful analysis of the assumptions one needs to make regarding continuing influence (over the course of the years during which personality expresses itself) shows that this argument is implausible.[56,57] It also does not explain the great diversity of personality in individuals of equal physical attractiveness. More importantly, however, causal influence of the variable must be demonstrated, not assumed. Physical attractiveness is virtually uncorrelated with any of the personality traits we will discuss in this chapter[58] so it could not be a cause of similarity in the personality of twins. Other studies demonstrate this point directly.[59,60] Bouchard has discussed this issue in detail both with regard to IQ[56] and with regard to personality.[54] Even if a so-called "environmental variable" is correlated with the personality trait of interest, the mediating effect may not be environmental. Consider the case of socioeconomic status (SES) and IQ. The SES of parents is often thought to be causally related to IQ. The theory is that parents of high SES provide their children with a richer environment, better schools, etc. than parents of children of low SES. The alternative hypothesis is that SES is not causal in the sense described above. Rather, the correlation between the SES of parents and their children's IQs reflects genetic factors to some extent. We need two facts to choose between these competing hypotheses. First, we need the correlation between SES and children's IQ in biological families. This correlation is known from a major meta-analysis of the literature[61] to be about .33 and it is made up of two components, one genetic and one environmental. This correlation alone tells us that parental SES is a very weak determinant of children's IQ as we must square the correlation to estimate variance accounted for (.09% of the variance). Second, we need the same correlation gathered from families of adopted children. This correlation is made up only of an environmental component (unless there is placement of high IQ children in high SES homes and low IQ children in low SES homes). This correlation is very low. Bouchard et al.[22] in their MZA study find it to be .174 (accounts for .03% of the variance in children's IQ). Scarr and Weinberg[62] have shown much the same thing. Bouchard[56] has reviewed this literature with respect to IQ. Plomin[63] has reviewed the growing literature dealing with genetic influence on what are often presumed to be environmental variables. Loehlin and Nichols[64] long ago demonstrated that similarity of treatment during childhood, as reported by the mothers of twins, was unrelated to twin similarity on personality traits (CPI scales). We also note here that while prenatal effects may mediate some similarity in some instances (e.g., fetal alcohol syndrome), the vast majority of prenatal influences are difference-producing and lead to underestimation of genetic influences when one wishes to generalize to the singleton population.[65, 66] More recent discussions of the twinning process are less explicit regarding the direction of bias, but the bulk of the influences are difference-producing, not similarity-producing.[67-70]

These methods also assume no assortative mating. For traits influenced by additive genetic factors positive assortative mating increases the similarity between first-degree relatives (e.g., DZ twins). Surprising as it may seem, assortative mating for personality traits is very modest and often near zero.[17,71-74]

D. BEHAVIOR GENETIC ANALYSES OF KINSHIP DATA

Early twin studies of the genetics of personality appeared to report very inconsistent results. This was largely because of the modest sample sizes utilized. Meta-analyses of such studies are still quite useful[75] and they have been summarized in two different reviews, one by Nichols,[76] organized somewhat arbitrarily, and another by Eaves et al.[17] organized around the Big Three. We begin with the Big Three.

1. MZT AND DZT CORRELATIONS AND KINSHIP MODELING FOR THE BIG THREE

The results of the meta-analysis of the Big Three carried out by Eaves et al.[17] is shown in Table 16-2. Eaves et al.[17] also carried out a detailed analysis of two large twin studies and one study of extended kinships (34 kinships, although some samples were very small). Their results are shown in Table 16-3.

TABLE 16-2
Weighted Mean Correlations for Monozygotic and Dizygotic Twins Organized According to the Big Three Personality Factors

Trait	Number of Studies	r_{MZT}	r_{DZT}	c^2	Falconer h^2
Extraversion	36	.53	.24	–.05	.58
Neuroticism	22	.44	.22	.00	.44
Psychoticism	15	.46	.23	.00	.46
Mean	24.3	.48	.23	–.02	.49

Note: Compiled and calculated from data in Eaves, L. J., Eysenck, H. J., Martin, H. J., and Martin, N. G., *Genes, Culture and Personality: An Empirical Approach,* Academic Press, New York, 1989.

The most striking finding in all these studies is a very simple one. The common family environmental influence is near zero in the vast majority of instances. A consequence of this finding is that the MZT intraclass correlations by themselves are excellent estimates of the broad heritability. The finding holds for vocational interests as well.[77] We will return to this point below.

There is clearly disagreement among studies regarding the role of nonadditive genetic variance for the various factors. For example, the Australian twin study finds significant nonadditive genetic influence for Extroversion but the London twin study and the extended kinship analysis do not. The authors also correctly point out that when fitting the extended kinships their model ignores the possibility of an especially high environmental correlation for MZ twins. We will return to these two points below.

For the record, there is now a modest literature on the heritability of personality questionnaire items.[17,78-80] Heath et al.[81] argue that their multivariate genetic analysis of the EPQ Psychoticism scale refutes the theoretical claims made for the scale by Eysenck. Loehlin,[51] however, has doubt about the reliability of item analyses as does this author.

2. LOEHLIN'S MODELING OF EXTENDED KINSHIP DATA FOR THE BIG FIVE

Loehlin[51] has carried out what will probably remain for many years the definitive analysis of the various kinship correlations for the Big Five. He gathered data points that are scattered across the world literature. One unavoidable difficulty with the Loehlin analysis is that he had to choose from a heterogeneous array of measures. In addition, some of the studies used very brief scales, raising questions about both the reliability and validity of the measures. As a result it is likely that his analysis incorrectly estimates some of the parameters in the models that he fits to the data. His analysis involves MZT, MZA, DZT, DZA, adoption, and family studies. The bottom line on this analysis is that Loehlin was not able to distinguish (statistically) between a model that required the inclusion of special twin and sibling environmental

TABLE 16-3
MZT and DZT Intraclass Correlations, Model Fitting Estimates of Narrow Heritabilities, Common Environmental Influence, Nonadditive Genetic Influence, and Broad Heritabilities for the Big Three From Two Twin Studies and One Extended Kinship Study

Trait	r_{MZT}	r_{DZT}	h^2	c^2	i^2	Broad heritability
London Twin Study						
Extraversion	.50	.18	.49	−.14	.00	.49
Neuroticism	.44	.07	.40	−.30	.00	.40
Psychoticism	.44	.30	.47	.16	.00	.47
Mean	.46		.45	−.09	.00	.45
Australian Twin Study — Females						
Extraversion	.53	.19	.21	−.15	.32	.53
Neuroticism	.52	.26	.51	.00	.00	.51
Psychoticism	.36	.23	.37	.10	.00	.37
Mean	.47		.36	−.02	.11	.47
Australian Twin Study—Males						
Extraversion	.50	.13	.21	−.24	.32	.53
Neuroticism	.46	.18	.46	−.10	.00	.46
Psychoticism	.44	.25	.45	.06	.00	.45
Mean	.47		.37	−.09	.11	.48
Extended Kinship Study						
Extraversion	n/r	n/r	.48	.00	.04	.52
Neuroticism	n/r	n/r	.12	.00	.30	.42
Psychoticism	n/r	n/r	.28	.00	.21	.49
Mean			.29	.00	.18	.48

Note: Compiled and calculated from data in Eaves, L. J., Eysenck, H. J., Martin, H. J., and Martin, N. G., *Genes, Culture and Personality: An Empirical Approach,* Academic Press, New York, 1989.

terms and a model that simply required nonadditive genetic influence. The results of fitting both models are presented in Table 16-4.

Consistent with all previous findings, the amount of common family environmental influence is modest for all five traits under either model. Our own preference is for the model that does not include special environments for twins and sibs. This preference is based on, among other things, the extensive body of evidence in support of the equal environment assumption, cited earlier, and a more recent analysis of the Minnesota Study of Twins Reared Apart (MISTRA) data to be discussed below. We also believe that this model is preferred by Loehlin.[82] Under this latter model the broad heritabilities are very consistent with those discussed earlier. The single exception is Agreeableness, which has a broad heritability of .35 (low) and a c^2 of .11 (larger than zero). We will see replications of this lower heritability for Agreeableness again below. The Loehlin analysis suggests the presence of nonadditive genetic variance for all of the Big Five except Openness.

This is perhaps the place to discuss the topic of nonadditive genetic variance. Only 10 years ago the existence of significant nonadditive genetic variance in the domain of personality was considered unlikely. This was in large part because studies simply did not have the power to detect such effects,[83] not because the topic was considered unimportant.[84] Lykken[85] early on proposed that this phenomenon was more widespread than was believed and coined the term "emergenesis" to emphasize that nonadditive variance included higher-order interactions.

TABLE 16-4
Estimates from Fitting Simple Models to Correlation Data From Multiple Kinships for Big Five Data

	Alternate Models						
	Additive Genes, Special MZ and Sibling Environments			Additive Genes, Nonadditive Genetic Variance and Equal Environments			
Dimension	h^2	c^2_{MZ}	c^2_S	h^2	i^2	c^2	Broad Heritability
Extroversion	.36	.15	.00	.32	.17	.02	.49
Neuroticism	.31	.17	.05	.27	.14	.07	.41
Conscientiousness	.28	.17	.04	.22	.16	.07	.38
Agreeableness	.28	.19	.09	.24	.11	.11	.35
Openness	.46	.05	.05	.43	.02	.06	.45
Mean	.34	.15	.05	.30	.12	.07	.42

From Loehlin, J. C., *Genes and Environment in Personality Development*, Sage Publications, Newbury Park, CA, 1992. With permission.

The arrival of twin studies with much larger sample sizes and studies of extended kinships has confirmed his speculations. The concept is now widely discussed in the personality literature.[86-88] Our own view, however, is more tentative as we believe the power of our models greatly exceeds the quality of our data.

3. NICHOLS META-ANALYSIS ORGANIZED ACCORDING TO THE BIG FIVE AND THE BIG NINE

The Nichols[76] meta-analysis organized according to the Big Five is given in Table 16-5.

TABLE 16-5
MZT and DZT Correlations, Estimated Common Environmental Influence, and Falconer Heritabilities for the Big Five From the Nichols Meta-Analysis

Trait	Number of Studies	r_{MZT}	r_{DZT}	c^2	Falconer Heritability
Extraversion	30	.52	.25	–.02	.54
Neuroticism	23	.51	.22	–.07	.58
Conscientiousness	12	.44	.24	.04	.40
Agreeableness	6	.49	.23	–.03	.52
Openness	7	.43	.17	–.09	.52
Mean	26.2	.48	.22	–.03	.51

Note: Complied and calculated from data in R. C. Nichols, *Homo*, 29, 158-173.

Nancy Breland (personal communication) kindly provided us with the raw data used for the Nichols[76] meta-analysis. This made it possible for us to use data not presented in Nichols' paper and organize those data according to the Big Nine scheme. The results are shown in Table 16-6.

TABLE 16-6
MZT, DZT Correlations, Estimated Common Environmental Influence, and Falconer Heritability for Eight of the Big Nine From the Nichols Meta-Analysis (No Data on Locus of Control)

Trait	Number of Studies and (N) for MZ Sample	r_{MZT}	r_{DZT}	c^2	Falconer Heritability
Affiliation (Extraversion)	30 (3202)	.52	.25	−.02	.54
Potency (Dominance)	13 (1902)	.53	.31	.09	.44
Achievement	5 (853)	.45	.18	−.09	.54
Dependability (Impulsiveness)	6 (1239)	.51	.31	.11	.40
Agreeableness (Aggressiveness)	2 (175)	.27	.13	−.01	.28
Adjustment	23 (2627)	.53	.33	.13	.40
Intellectance	2 (1351)	.52	.34	.16	.36
Rugged Individualism (Masculinity)	7 (1116)	.40	.21	.02	.38
Mean		.47	.26	.05	.42

Note: From raw data provided by Nancy Breland (personal communication).

There was no information regarding heritability of the trait Locus of control. The only behavior genetic data available on this trait are from the Swedish Adoption Twin Study of Aging (SATSA). A 12-item test yielded 3 measures, Luck, Responsibility, and Life Direction. Luck (a three-item scale) showed no genetic influence. Additive genetic variance accounted for 34 and 31% of the variance in the other two measures.[89]

4. PERSONALITY DISORDERS

The field of psychiatric genetics is very large and it would not be possible to review it here, rather we refer the reader to other sources.[90-93] The genetics of personality disorder has recently been reviewed by Nigg and Goldsmith,[94] Dahl,[95] and Torgerson.[96] A recent study directly related to the focus of this chapter and illustrative of findings in the domain was recently carried out by Livesley et al.[97] using the Dimensional Assessment of Personality Pathology—Basic Questionnaire (DAPP—BQ). The reason for presenting these findings is that this is an instrument in which the item content was derived from the disorders described in DSM-III rather than the typical domain from which personality items are ordinarily sampled. As with many other such instruments (e.g., psychiatric checklists and the MMPI), the factor structure does not correspond to the traditional psychiatric categories. Detailed information about this instrument can be found in Livesley et al.[98] and the relation of its dimensions to the Big Five can be found in References 99 and 100. The results of their twin study is presented in Table 16-7; we have indicated, in parentheses, which of the Big Five factors each dimension loads.

The remarkable similarity of these results to those already discussed in this chapter needs no elaboration. We will make a number of comparisons using these data below. One additional feature of the table deserves notice, namely, the very large c^2 for Conduct Problems. A personal communication with a member of this research team at a recent Behavior Genetics Association Meeting reveals that this finding has not held up as the sample of twins has increased in size.

TABLE 16-7
MZ and DZ Twin Intraclass Correlations, Model Fitting Estimates of Additive Genetic Variance, Nonadditive Genetic Variance, Common Environmental Variance, and Broad Heritability for the Dimensional Assessment of Personality Pathology — Basic Questionnaire Organized According to Their Big Five Loadings

Dimension	r_{MZ}	r_{DZ}	h^2	i^2	c^2	Broad Heritability
Stimulus Seeking (E)	.59	.33	.50	.00	.09	.50
Affective Lability (N)	.48	.13	.01	.48	.00	.49
Anxiousness (N)	.54	.33	.49	.00	.06	.49
Identity Problems (N)	.60	.26	.40	.19	.00	.59
Insecure Attachment (N)	.52	.38	.35	.00	.13	.35
Narcissism (N)	.64	.12	.00	.64	.00	.64
Social Avoidance (N)	.57	.26	.47	.10	.00	.57
Submissiveness (N)	.54	.41	.25	.00	.28	.25
Mean (N)	.56	.25	.28	.20	.07	.48
Compulsivity (C)	.42	.23	.39	.00	.03	.39
Oppositionality (C)	.58	.27	.52	.03	.00	.55
Mean (C)	.50	.25	.46	.02	.02	.47
Conduct Problems (A)	.52	.52	.00	.00	.53	.00
Rejection (A)	.49	.22	.45	.00	.00	.45
Self Harm (A)	.30	.10	.14	.15	.00	.29
Suspiciousness (A)	.49	.25	.48	.00	.00	.48
Mean (A)	.45	.27	.27	.04	.13	.31
Intimacy Problems (O)	.40	−.01	.00	.38	.00	.38
Restricted Expression (O)	.51	.25	.47	.00	.00	.47
Mean (O)	.46	.12	.24	.19	.00	.43
Callousness	.63	.29	.56	.00	.00	.56
Cognitive Distortion	.82	.39	.41	.00	.14	.41
Mean (Misc.)	.73	.34	.49	.00	.07	.49

Note: Compiled and calculated from various tables in Livesley, E. J., Jang, K. L., Jackson, D. N., and Vernon, P. A., *Am. J. Psychiatry*, 150, 1826-1831, 1993.

5. SATSA AND MISTRA BIG FIVE ANALYSES

The SATSA data from the four types of twins used by Loehlin are shown in Table 16-8. and compared with the MISTRA data used by Loehlin[51] for the same four groups. More recent MISTRA results with larger sample sizes have been published in graphic form[101] and we present those results in quantitative form in Table 16-9.

We note here that Loehlin has averaged a number of measures to stabilize the SATSA data. If only individual Big Five measures (based on brief scales) were used the data would be quite heterogeneous.[86] The purpose of this table is to provide a direct comparison of the two studies. SATSA regularly reports lower heritabilities than MISTRA and the reason is

TABLE 16-8
MZT, MZA, DZT, and DZA Sample Sizes and Intraclass Correlations for the Big Five Factors for SATSA and MISTRA Used by Loehlin for His Extended Kinship Analysis

	SATSA				MISTRA			
Trait	r_{MZA} 82-95	r_{MZT} 132-151	r_{DZA} 171-218	r_{DZT} 167-204	r_{MZA} 44	r_{MZT} 217	r_{DZA} 27	r_{DZT} 114
Extraversion	.30	.54	.04	.06	.34	.63	−.07	.18
Neuroticism	.25	.41	.28	.24	.61	.54	.29	.41
Conscientiousness	.25	.39	.16	.13	.57	.58	.04	.25
Agreeableness	.18	.37	.09	.32	.46	.43	.06	.14
Openness	.43	.51	.23	.14	.61	.49	.21	.41
Mean	.28	.44	.16	.18	.52	.53	.11	.28

From Loehlin, J. C., *Genes and Environment in Personality Development*. Sage Publications, Newbury Park, CA, 1992. With permission.

clear in the table. The SATSA MZA twins tend to be less similar than the MISTRA twins on all traits. The SATSA MZT, DZA, and DZT groups are quite similar to the MISTRA groups across all the dimensions. The SATSA MZT correlations are slightly lower than the comparable MISTRA correlations, but this is easily explained by the much briefer and consequently less reliable scales they have used. Attenuation due to unreliability has a much more modest effect on small correlations, consequently the DZ correlations are less influenced than the MZ correlations. Pedersen et al.[102] argue that the difference between the SATSA and MISTRA MZA twin correlations lies in the method of ascertainment — SATSA twins having been located from a registry and MISTRA twins having been volunteers where "Pairs may have come to the investigator's (and each other's) attention because of their similarity." This explanation would require that a bias of comparable magnitude influenced all the other data sets as their heritability estimates are closer to those of MISTRA than SATSA. In addition the SATSA results for IQ[103] are identical to MISTRA.[22] In this instance both studies use a highly reliable phenotype. We are at a loss to explain why the SATSA MZA correlations differ so much from the other personality heritability estimates.

The correlations for the four groups and the results of model-fitting more recent data from the MISTRA study are given in Table 16-9. As explained by Bouchard,[101] where these results were first presented the most parsimonious fit was a simple additive genetic model for all five traits with an estimated genetic influence of 46%. We allowed a nonadditive genetic parameter to facilitate a direct comparison with the Loehlin analysis.

The similarity between the updated MISTRA findings and the Loehlin analysis (see Table 16-4) is remarkable. The average broad heritabilities (sum of h^2 and i^2) are almost identical: .41 and .42, respectively. The average c^2 is identical: .07. Both analyses show that Agreeableness has the lowest heritability: .30 and .35, and both studies give it a c^2 value of about .10. For the other four traits the variation from trait to trait for the broad heritability and for c^2 is not very great.

The MISTRA results are also remarkably similar to both the MZT correlations and heritabilities estimated in all of the studies cited above, excluding SATSA, and we present the comparison in Table 16-10.

We have put the smallest correlation in each column in boldface type. As pointed out previously Agreeableness yields, in most instances, the smallest correlation. It should be clear from this table that MZT correlations and the MISTRA MZA correlations appear to be excellent estimators of the heritability of the Big Five. This is true whether the heritability is estimated from twin studies or from extended kinships.

TABLE 16-9
Intraclass Correlation for MZA, DZA, MZT, and DZT Twins and Model Fitting Results for the Big Five

	Twin Type				Proportion of Variance		
Trait	MZA (59)	DZA (47)	MZT (522)	DZT (408)	h^2	i^2	c^2
Extraversion	.41	−.03	.54	.19	.09	.29	.15
Neuroticism	.49	.44	.48	.19	.41	.09	.00
Conscientiousness	.54	.07	.54	.29	.29	.13	.13
Agreeableness	.24	.09	.39	.11	.05	.25	.09
Openness	.57	.27	.43	.14	.29	.15	.00
Mean	.45	.17	.48	.18	.23	.18	.07

Note: The MISTRA measures are the same Multidimensional Personality Questionnaire factors and scales used by Loehlin.[51] Positive Emotionality = Extraversion, Negative Emotionality = Neuroticism, Constraint = Conscientiousness, Adsorbtion = Openness, Aggression = Agreeableness. The MZA and DZA data are based on the most recent MISTRA sample. The MZT and DZT data are from the large Minnesota Twin Registry sample.[21]

TABLE 16-10
MZT, MZA Correlations and Heritabilities from Twin Studies and Extended Kinships for the Big Five

	SATSA		MISTRA		Heritability		
Trait	r_{MZA}	r_{MZT}	r_{MZA}	r_{MZT}	All r_{MZT}[a]	Twin Studies[a]	Loehlin
Extraversion	.30	.54	.41	.54	.44	.52	.49
Neuroticism	.25	.41	.49	.48	**.42**	.48	.41
Conscientiousness	.25	.39	.54	.54	.47	.44	.38
Agreeableness	**.18**	**.37**	**.24**	**.39**	.47	**.42**	**.35**
Openness	.43	.51	.57	.43	.45	.48	.45
Mean	.28	.44	.45	.48	.45	.46	.42

[a] Data from previous tables, excludes SATSA and MISTRA.

Note: Lowest value in each column is in boldface

In conclusion it appears that the broad heritability of four of the Big Five is around .45 and for Agreeableness it may be slightly less.

6. SOCIAL ATTITUDES

Depending on how one defines the field, social attitudes may fall outside the domain of personality. There are very few studies in this domain and we report here on two studies. The first[104] deals with three measures: the Wilson Patterson Conservativism scale,[105] and Radicalism and Toughmindedness from the Eysenck Public Opinion inventory.[17] The heritabilities of the three measures were about .60. Two findings were striking in this study. The first is the high heritabilities for social attitudes. The second is that the DZ correlations (not reported here) were far higher than half the MZ correlations. We have not had to deal with assortative mating for personality traits in this chapter because of the low level of assortative mating. The social attitude domain differs in this regard as there are high levels of assortative mating for these measures. Model fitting by Martin et al.[104] shows that this higher level of correlation between first-degree relatives can be accounted for by assortative mating.

Like many other psychologists, the MISTRA research team found the Martin et al. study difficult to believe, although the results had been foreshadowed a number of years earlier.[106-108] We turned to the MISTRA data base and found that we had a number of measures of religious interests, attitudes, and values from our twins reared apart. Like Martin et al., we also found that these traits are heavily influenced by genetic factors.[109] The heritabilities for various measures of religiousness were between .40 and .50.

These findings are in a slightly different domain than the Martin et al. social attitude data, but being based on twins reared apart and multiple measures they do provide converging evidence for the heritability of attitude-like traits. It is important to emphasize that this is not a study of religious affiliation, a characteristic which has been shown to be largely environmental.[110]

7. LOVE STYLES

This topic is included because the authors gathered both MPQ data and love style data on the same sample.[111] The MPQ correlations were very similar to those gathered by MISTRA,[112] thereby extending the generalizability of the findings of heritability for personality further. The genetic contribution to measured love styles was, however, virtually nil and the common (shared) family environmental component was sizable. This study showed that the twin method has differential sensitivity and directly refutes the commonly made argument that the twin method always finds a heritability between .40 and .60 and little or no common family environmental influence.

8. CRIME

Behavior genetic research on crime is extremely controversial and the relevant literature is very large. Unfortunately, we can only treat it briefly here. The most striking differences in crime rates are between males and females.[113] The difference between men and women are genetic, but not heritable in the usual sense. Men engage in consistently more criminal behavior than women regardless of their race, age, or cultural origins.[114-116] The most recent analysis of a large adoption study of sex differences in property crimes[117] concluded that "heritability estimates were $h^2 = .30$ in families of sons and $h^2 = .27$ in families of daughters. It is clear, then, that heritability of liability towards property-criminal behavior is not greater in females than males, in spite of the significant sex difference in average genetic predisposition."

This latter point is of interest. The authors demonstrate with appropriate comparisons that females who engage in criminal behavior have a greater genetic predisposition than males who engage in crime. They also show that "social class in the adoptive parents of convicted sons and daughters were comparable, further indicating that average shared-family environmental liabilities do not differ between the sexes."[117]

In spite of many claims to the contrary, IQ is very significantly related to delinquency,[118,119] and this influence is strong even with SES removed.[120-123] In addition, personality is also related to crime.[124] Consequently, at least part of the genetic component associated with criminal behavior is probably mediated by these variables. A recent CIBA symposium[125] provides a detailed overview of this domain as well as new twin studies,[126,127] reviews of older work,[128,129] and the application of modern molecular techniques.[130-132]

A point of considerable importance to keep in mind when evaluating studies in this domain is the distinction between juvenile delinquents who engage in delinquent behavior only during the adolescent period and those who begin early and engage in antisocial behavior well into adulthood.[133]

E. ROLE OF INTELLIGENCE

Intelligence as measured by IQ tests is often thought to be a facet of mental abilities and a domain entirely distinct from personality. This is not necessarily the case. Practicing clinicians, in general, do not agree with this view and invariably assess their clients' IQ as part of their personality. The reason is that the trait of intelligence permeates everything we do and is a good predictor of important social behavior.[123,134] We have not reviewed the evidence here but intelligence, in modern Western industrialized societies, has the highest degree of genetic influence of any human psychological character.[135] The consensus estimate is a broad heritability of about .65 to .75. The evidence is now quite strong and flows from multiple designs: twins reared apart,[22,56,103] ordinary twins,[136-138] siblings reared apart,[139] and full adoption studies.[140] Just as striking is the finding that adult unrelated individuals reared together and who grow up in the same home yield a correlation of zero.[141,142] There is still, however, much to learn about the determinants of IQ.[143]

F. MAPPING GENES FOR INTELLIGENCE AND PERSONALITY

The search for genes underlying psychiatric disorders has been a frustrating exercise and few replicable findings exist. Very suggestive findings continue to be reported.[144-147] A similar change appears to be taking place in the study of continuous traits in the normal range. Preliminary findings regarding the mapping of genes for IQ have been reported by Plomin's group,[148-150] but this group has been very careful to not claim to have discovered any genes underlying variance in IQ. The genetics of mental retardation continues to be an active arena of investigation,[151] and new genetic causes continue to be found.[152] The allelic association between Apo-E4 and late-onset Alzheimer's disease,[153] and the demonstration that linguistic ability in early life is related to Alzheimer's disease in late life,[154] suggests additional approaches to the study of genetic influences on cognitive ability.

An association between the D4DR gene on the short arm of chromosome 11 and the personality trait of Novelty Seeking has been reported,[155] as has a constructive replication (different laboratory, different measure of Novelty Seeking).[156] Since Novelty Seeking in Cloninger's theoretical scheme corresponds to Impulsive Sensation Seeking in Zukerman's, Constraint in Tellegen's, Conscientiousness in the Big Five, and Psychoticism in the Eysenck,[157] we can expect to see additional attempts to replicate using a variety of related measures. Perhaps this trait is linked to high levels of self-reported combat experience which has been reported to be heritable (35 to 47%) in the Vietnam era twin registry.[158] Cloninger et al.[159] provide an optimistic discussion of the Novelty-Seeking findings, arguing that the study of genes contributing to normal variation in personality and temperament may be a more productive enterprise than the study of genes relating to complex disorders such as schizophrenia. If indeed this is the case, it will constitute a dramatic paradigm shift in behavior genetics.

G. BEHAVIOR GENETICS VS. SOCIALIZATION THEORY

Socialization theories of personality formation are based on the assumption that personality traits are shaped entirely by proximal environmental influences. In contrast with reviews of the influence of genetic factors such as this one, which contain formal models and quantitative predictions regarding the similarity between relatives under various rearing conditions, socialization theories are still largely formulated at the verbal level and do not make any predictions that are superior to behavior genetic theories. Maccoby,[160] in a recent issue of one of the most important empirical journals in the field — *Developmental Psychology* — has provided a historical overview of the role of parents in the socialization of children. Darling and Steinberg[161] even more recently attempted, in the most important review journal in psychology — *Psychological Bulletin* — to present "an integrative model" of parenting styles. The most notable aspect of these two works is that neither of them presented a single quantitative estimate of anything! Scarr,[162] on the other hand, demonstrates on a quantitative level that behavior genetic theories can explain everything explained by socialization theory, in a more scientific manner, and much more besides. A recent paper by Bronfenbrenner and Ceci,[163] two distinguished developmentalists, acknowledges this fact.

An interesting aside on this point is recent work demonstrating that child-rearing behavior is itself heritable. There are two lines of evidence on this point — reports from the point of view of the recipient and reports from the point of view of the rearing parent. Studies of ordinary twins' perceptions of child-rearing have clearly shown that they contain a genetic component and cannot be used simply as an environmental measure of social influence.[164,165] This work has also been replicated using two different samples of twins reared apart.[166,167] Regarding the second approach, Pérusse and colleagues,[168] using a large adult twin sample and reports by them regarding their child-rearing behavior as measured with the Parental Bonding Instrument, have shown that genetic influence ranges from 19 to 39% on the dimensions measured. The findings from this study are more complex than these simple heritabilities convey and have important implications for socialization theory and theories of gene-culture transmission.

The view that children drive the behavior of their parents also has a great deal of support[169,170] and has generated considerable controversy.[170,171]

The findings reported in this chapter clearly challenge most contemporary theories that consider child-rearing practices and proximal environmental events the primary sources of variance in human personality. More recent studies on the heritability of child-rearing practices and the role of children in driving their parents' behavior demonstrate that the term nature vs. nurture should perhaps be replaced by the term nurture via nature.

ACKNOWLEDGMENTS

The Minnesota Study of Twins Reared Apart has been supported by grants from the Pioneer Fund, the David H. Koch Charitable Trust, the Seaver Institute, the Spencer Foundation, the National Science Foundation (BNS-7926654), the University of Minnesota Graduate School, and the Harcourt Brace Jovanovich Publishing Co. I would like to thank my colleague, Margaret Keyes, for numerous suggestions that greatly improved the manuscript. The errors that remain are mine.

REFERENCES

1. Galton, F., Measurement of Character, *Fortnightly Rev.*, 36 (1884): 179-185.

2. Norman, W. T., 2800 Personality Trait Descriptors: Normative Operating Characteristics for a University Population, Department of Psychology, University of Michigan, Ann Arbor, MI, 1967.
3. Allport, G. W. and H. S. Odbert, Trait names: A psycho-lexical study, *Psychol. Monogr.*, 47.(1, Whole No. 211) (1936).
4. Cattell, R. B., The description of personality: basic traits resolved into clusters, *J. Abnorm. Social Psychol.*, 38 (1943): 476-506.
5. Goldberg, L. R., From Ace to Zombie: some explorations in the language of personality, *Advances in personality assessment*, Vol. 1, C. D. Speilberger and J. N. Butcher, Eds., Erlbaum, Hillsdale, NJ, 1982, 203-234.
6. Waller, N. G., Evaluating the structure of personality, *Personality and Psychopathology*, C. R. Cloninger, Ed., American Psychiatric Press, Washington, D.C., in press.
7. Dahlstrom, W. G., G. S. Welsh, and L. E. Dahlstrom, *An MMPI Handbook, Vol. II: Research Applications*, University of Minnesota Press, Minneapolis, 1975.
8. Grahm, J. R., *MMPI-2: Assessing Personality and Psychopathology*, Oxford, NY, 1990.
9. Myers, I. B. and M. H. McCaulley, *Manual: A Guide to the Development and Use of the Myers-Briggs Type Indicator*, Consulting Psychologists Press, Palo Alto, CA, 1985.
10. Gough, H. G., *California Psychological Inventory Administrator's Guide*, Consulting Psychologists Press, Palo Alto, CA, 1987.
11. Eysenck, H. J. and S. B. G. Eysenck, *Manual of the Eysenck Personality Questionnaire*, Digits, San Diego, 1975.
12. Eysenck, H. J., Four ways five factors are not basic, *Pers. Ind. Diff.*, 13 (1992): 667-673.
13. Eysenck, H. J. and S. B. G. Eysenck, *Psychoticism as a Dimension of Personality*, Hodder & Stroughton, London, 1976.
14. Eysenck, H., Creativity and personality: suggestions for a theory, *Psychol. Inquiry*, 4 (1993): 147-178.
15. Eysenck, H. J. and G. H. Gudjonsson, *The Causes and Cures of Criminality*, Plenum Press, New York, 1989.
16. Modgil, S. and C. Modgil, *Hans Eysenck: Consensus and Controversy*, Falmer Press, London, 1986.
17. Eaves, L. J., H. J. Eysenck, and N. G. Martin, *Genes, Culture and Personality: An Empirical Approach*, Academic Press, New York, 1989.
18. Tellegen, A., Folk concepts and psychological concepts of personality and personality disorder, *Psychol. Inquiry*, 4(1993): 122-130.
19. Tellegen, A. and N. G. Waller, Exploring personality through test construction. Development of the Multidimensional Personality Questionnaire, *Personality Measures: Development and Evaluation*, S. R. Briggs and J. M. Cheek, Eds., JAI Press, Greenwich, CN, in press.
20. Church, A. T., Relating the Tellegen and Five-Factor models of personality structure, *J. Pers. Soc. Psychol.*, 67 (1994): 898-909.
21. Lykken, D. T. et al., The Minnesota twin family registry: some initial findings, *Acta Genet. Med. Gemellol.*, 39 (1990): 35-70.
22. Bouchard, T. J., Jr. et al., Sources of human psychological differences. The Minnesota study of twins reared apart, *Science*, 250 (1990): 223-228.
23. Goldberg, L. R., The structure of phenotypic personality traits, *Am. Psychol.*, 48 (1993): 26-34.
24. Goldberg, L. R., What the hell took so long? Donald Fiske and the big-five factor structure, *Advances in Personality Research, Methods, and Theory: A Festschrift Honoring Donald W. Fiske*, P. E. Shrout and S. T. Fiske, Eds., Erlbaum, New York, 1992.
25. Costa, P. T. and R. R. McCrae, Four ways five factors are basic, *Pers. Individual Differences*, 13 (1992): 653-665.
26. Jackson, D. N. et al., A five-factor versus six-factor model of personality structure, *Pers. Individual Differences*, 20 (1996): 33-45.
27. Fiske, D. W., Consistency of the factorial structures of personality ratings from different sources, *J. Abnorm. Soc. Psychol.*, 44 (1949): 329-344.
28. Tupes, E. C. and R. E. Christal, *Recurrent Personality Factors Based on Trait Ratings*, Tech. Rep. ASD-TR-61-97, Lackland Air Force Base: U.S. Air Force, 1961.
29. Norman, W. T., Toward an adequate taxonomy of personality attributes. Replicated factor structure in peer nomination personality ratings, *J. Abnorm. Soc. Psychol.*, 66 (1963): 574-583.
30. Costa, P. T. and R. R. McCrae, *NEO PI-R, Professional Manual*, Psychological Assessment Resources, Inc., Odessa, FL, 1992.
31. Bergeman, C. S. et al., Genetic and environmental effects on openness to experience, agreeableness, and conscientiousness. An adoption/twin study, *J. Pers.*, 61 (1993): 159-179.
32. Costa, P. T., Jr. and R. R. McCrae, Major contributions to the psychology of personality, *Hans Eysenck: Consensus and Controversy*, S. Modgil and C. Modgil, Eds., Falmer Press, London, 1986.
33. Goldberg, L. R., The development of markers for the Big–Five factor structure, *Psychol. Assess.*, 4 (1992).
34. Arvey, R. D. and T. J. Bouchard, Jr., Genetics, twins, and organizational behavior, *Research in Organizational Behavior*, B. A. Staw and L. L. Cummings, Eds., JAI Press, New York, 1994, chap. 16.

35. Bouchard, T. J., Jr. et al., Genetic influences on Job Satisfaction: a reply to Cropanzano and James, *J. Appl. Psychol.*, 77 (1992): 89-93.
36. Hough, L., The "Big Five" personality variables — construct confusion: description versus prediction, *Hum. Perform.*, 5 (1992): 139-155.
37. Ones, D. S., Bandwidth-fidelity dilemma in personality measurement for personnel selection, *J. Organ. Behav.*, in press.
38. Barrick, M. R. and M. K. Mount, The Big Five personality dimensions and job performance: a meta-analysis, *Pers. Psychol.*, 44 (1991): 1-26.
39. Barkow, J. H., L. Cosmides, and J. Tooby, *The Adapted Mind*, Oxford University Press, Oxford, 1992.
40. DeKay, W. T. and D. M. Buss, Human nature, individual differences, and the importance of context. Perspectives from evolutionary psychology, *Curr. Direct. Psychol. Sci.*, Vol. 1 (1992): 184-189.
41. Bouchard, T. J., Jr. et al., Genes, drives, environment and experience: EPD theory revised, *Psychometrics and Social Issues Concerning Intellectual Talent*, C. P. Benbow and D. Lubinski, Eds., John Hopkins University Press, Baltimore, 1996.
42. Bouchard, T. J., Jr., Longitudinal studies of personality and intelligence: a behavior genetic and evolutionary psychology perspective, *International Handbook of Personality and Intelligence*, D. H. Saklofske and M. Zeidner, Eds., Plenum Press, New York, 1995, 81-106.
43. Tooby, J. and L. Cosmides, The psychological foundations of culture, *The Adapted Mind: Evolutionary Psychology and the Generation of Culture*, J. H. Barkow, L. Cosmides, and J. Tooby, Eds., Oxford University Press, Oxford, 1992, 19-36.
44. Perusse, D., Cultural and reproductive success in industrial societies: testing the relationship at the proximate and ultimate levels, *Behav. Brain Sci.*, 16 (1993): 267-322.
45. Williams, G. C., *Adaptation and Natural Selection: A Critique of Some Current Evolutionary Thought*, Princeton University Press, Princeton, NJ, 1966.
46. Williams, G. C., *Natural Selection: Domains, Levels, and Challenges*, Oxford University Press, New York, 1992.
47. Buss, D. M. et al., Sex differences in jealousy: evolution, physiology, and psychology, *Psychol. Sci.*, 3.4 (1992): 251-255.
48. Gangestad, S. and M. Snyder, To carve nature at its joints: on the existence of discrete classes in personality, *Psychol. Rev.*, 92 (1985): 317-349.
49. Li, C. C., *Path Analysis: A Primer*, Boxwood Press, Pacific Grove, CA, 1975.
50. Loehlin, J. C., *Latent Variable Models: An Introduction to Factor Analysis, Path, and Structural Analysis*, 2nd ed., Erlbaum, Hillsdale, NJ, 1992.
51. Loehlin, J. C., *Genes and Environment in Personality Development*, Sage Publications, Newbury Park, CA, 1992.
52. Neale, M. C. and L. R. Cardon, Eds., *Methodology for Genetic Studies of Twins and Families*, Kluwer Academic, Dordrecht, 1992.
53. Falconer, D. S., *Introduction to Quantitative Genetics*, 3rd ed. Longman, New York, 1990.
54. Bouchard, T. J., Jr., Genetic and environmental influences on adult personality: evaluating the evidence, *Basic Issues in Personality*, I. Deary and J. Hettema, Eds., Kluwer Academic, Dordrecht, 1993.
55. Ford, B. D., Emergenesis: an alternative and a confound, *Am. Psychol.*, 48 (1993): 1294.
56. Bouchard, T. J., Jr., IQ similarity in twins reared apart: findings and response to critics, *Intelligence: Heredity and Environment*, R. J. Sternberg and E. L. Grigorenko, Eds., Cambridge University Press, New York, in press.
57. Hettema, J. M., M. C. Neale, and K. S. Kendler, Physical similarity and the equal-environment assumption in twin studies of psychiatric disorders, *Behav. Genet.*, 25 (1995): 327-336.
58. Feingold, A., Good-looking people are not what we think, *Psychol. Bull.*, 111.2 (1992): 304-341.
59. Rowe, D. C. and M. Clapp, Physical attractiveness and the personality resemblance of identical twins, *Behav. Gene.*, 17 (1977): 191-201.
60. Matheny, A. P., R. S. Wilson, and A. B. Dolan, Relations between twin's similarity of appearance and behavioral similarity: testing an assumption, *Behav. Genet.*, 6 (1976): 343-351.
61. White, R. K., The relation between socioeconomic status and academic achievement, *Psychol. Bull.*, 91 (1982): 461-481.
62. Scarr, S. and R. A. Weinberg, The influence of family background on intellectual attainment, *Am. Sociol. Rev.*, 43 (1978): 674-692.
63. Plomin, R., *Genetics and Experience: The Interplay Between Nature and Nurture*, Sage, Thousand Oaks, CA, 1994.
64. Loehlin, J. C. and R. C Nichols, *Heredity, Environment, & Personality: A study of 850 Sets of Twins*, University of Texas Press, Austin, 1976.
65. Price, B., Primary biases in twin studies, a review of prenatal and natal difference-producing factors in monozygotic pairs, *Am. J. Hum. Genet.*, 2 (1950): 293-352.
66. Price, B., Bibliography on prenatal and natal influences in twins, *AGMG*, 27 (1978): 97-113.
67. Hall, J. G., Twins and twinning, *Am. J. Med. Genet.*, 61 (1996): 202-204.

68. Machin, G. A., Some causes of genotypic and phenotypic discordance in monozygotic twin pairs, *Am. J. Med. Genet.*, 61 (1996): 216-228.
69. Bryan, E. M., Prenatal and perinatal influences on twin children: implications for behavioral studies, *Twins as a Tool of Behavior Genetics*, T. J. Bouchard, Jr. and P. Propping, Eds., John Wiley & Sons, Chichester, 1993, 217-225.
70. Phillips, D. I. W., Twin studies in medical research: can they tell us whether diseases are genetically determined?, *Lancet*, 341 (1993): 1008-1009.
71. Caspi, A. and E. S. Herbener, Continuity and change: assortative marriage and the consistency of personality in adulthood, *JPSP*, 58 (1990): 250-258.
72. Caspi, A. and E. S. Herbener, Marital assortment and phenotypic convergence: longitudinal evidence, *Soc. Biol.*, 40 (1993): 48-60.
73. Nagoshi, C. T., R. C. Johnson, and K. A. M. Honbo, Assortative mating for cognitive abilities, personality, and attitudes, *Pers. Individual Differences*, 13.8 (1992): 883-891.
74. Lykken, D. T. and A. Tellegen, Is human mating adventitious or the result of lawful choice? A twin study of mate selection, *J. Pers. Soc. Psychol.*, 65 (1993): 56-68.
75. Schmidt, F. L., What do data really mean? Research findings, meta-analysis, and cumulative knowledge in psychology, *Am. Psychol.*, 47 (1992): 1173-1181.
76. Nichols, R. C., Twin studies of ability, personality and interests, *Homo*, 29 (1978): 158-173.
77. Lykken, D. T., et al., Heritability of interests: A twin study, *J. Appl. Psychol.*, 78 (1993): 649-661.
78. Neale, M. C., J. P. Rushton, and D. W. Fulker, Heritability of item responses on the Eysenck Personality Questionnaire, *Pers. Individual Differences*, 7 (1986): 771-779.
79. Loehlin, J. C., Are California Psychological Inventory items differentially heritable?, *Behav. Genet.*, 16 (1986): 599-603.
80. Heath, A. C. et al., The genetic structure of personality. II. Genetic item analysis of the EPQ, *Pers. Individual Differences*, 10 (1989): 615-624.
81. Heath, A. C., L. J. Eaves, and N. G. Martin, The genetic structure of personality. III. Multivariate genetic item analysis of the EPQ scales, *Pers. Individual Differences*, 10 (1989): 877-888.
82. Loehlin, J. C., What has behavioral genetics told us about the nature of personality?, *Twins as a Tool of Behavioral Genetics*, T. J. Bouchard, Jr. and P. Propping, Eds., John Wiley & Sons, Chichester, 1993.
83. Hewitt, J. K., Normal components of personality variation, *J. Pers. Soc. Psychol.*, 47 (1984): 671-675.
84. Eaves, L. J. et al., A progressive approach to non-additivity and genotype-environment covariance in the analysis of human differences, *Br. J. Math. Stat. Psychol.*, 30 (1977): 1-42.
85. Lykken, D. T., Research with twins: the concept of emergenesis, *Psychophysiology*, 19 (1982): 361-373.
86. Plomin, R., H. M. Chipuer, and J. C. Loehlin, Behavioral genetics and personality, *Handbook of Personality: Theory and Research.*, L. A. Pervin, Ed., Guilford Press, New York, 1990.
87. Lykken, D. T. et al., Emergenesis: genetic traits that may not run in families, *Am. Psychol.*, 47 (1992): 1565-1577.
88. Li, C. C., A genetical model for emergenesis, *Am. J. Hum. Genet.*, 41 (1987): 517-523.
89. Pedersen, N. L. et al., Individual differences in Locus of Control during the second half of the life span for identical and fraternal twins reared apart and reared together, *J. Gerontol. Psychol. Sci.*, 44 (1989): 100-105.
90. Rutter, M., E. Simonoff, and J. Silberg, How informative are twin studies of child psychopathology, *Twins as a Tool of Behavioral Genetics*, T. J. Bouchard, Jr. and P. Propping, Eds., John Wiley & Sons, Chichester, 1993.
91. Kendler, K. S. et al., Childhood parental loss and adult psychopathology: a twin study perspective, *Arch. Gen. Psychiatry*, 49 (1992): 109-116.
92. Kendler, K. S. et al., The prediction of major depression in women: towards an integrated etiologic model, *Am. J. Psychiatry*, 150 (1993): 1139-1148.
93. Kessler, R. C. et al., Social support, depressed mood, and adjustment to stress: a genetic epidemiological investigation, *J. Pers. Soc. Psychol.*, 62 (1992): 257-272.
94. Nigg, J. T. and H. H. Goldsmith, Genetics of personality disorders: perspectives from personality and psychopathology research, *Psychol. Bull.*, 115 (1994): 346-380.
95. Dahl, A., The personality disorders: a critical review of family, twin and adoption studies, *J. Pers. Disord.*, 7(Suppl.) (1993): 86-99.
96. Torgerson, S., Genetics in borderline conditions, *Acta Psychiatr. Scand.*, 89(Suppl. 379) (1994): 19-25.
97. Livesley, W. J. et al., Genetic and environmental contributions to dimensions of personality disorders, *Am. J. Psychiatry*, 150 (1993): 1826-1831.
98. Livesley, W. J., D. Jackson, and M. Schroeder, A study of the factorial structure of personality pathology, *J. Pers. Disord.*, 3 (1989): 292-306.
99. Schroeder, M. L., J. A. Wormworth, and W. J. Livesley, Dimensions of personality, *Psychol. Assess.*, 4 (1992): 47-53.

100. Schroeder, M. L., J. A. Wormworth, and W. J. Livesley, Dimensions of personality disorder and the five factor model of personality, *Personality Disorders and the Five-Factor Model of Personality*, P. T. Costa and T. A. Widiger, Eds., American Psychological Association, Washington, D.C., 1994, 117-127.
101. Bouchard, T. J., Jr., Genes, environment and personality, *Science*, 264 (1994): 1700-1701.
102. Pedersen, N. L. et al., Neuroticism, extraversion, and related traits in adult twins reared apart and together, *J. Pers. Soc. Psychol.*, 55 (1988): 950-957.
103. Pedersen, N. L. et al., A quantitative genetic analysis of cognitive abilities during the second half of the life span, *Psychol. Sci.*, 3 (1992): 346-353.
104. Martin, N. G. et al., Transmission of social attitudes, *Proc. Natl. Acad. Sci. U.S.A.*, 83 (1986): 4364-4368.
105. Wilson, G. D. and J. R. Patterson, A new measure of conservativism, *Br. J. Soc. Clin. Psychol.*, 7 (1968): 264-269.
106. Eaves, L. J. and H. J. Eysenck, Genetics and the development of social attitudes, *Nature*, 249 (1974): 288-289.
107. Scarr, S. and R. Weinberg, Attitudes, interests and IQ, *Hum. Nat.*, 2.(4) (1978).
108. Scarr, S. and R. Weinberg, The transmission of authoritarianism in families: genetic resemblance in social-political attitudes, *Race, Social Class, and Individual Differences*, S. Scarr, Ed., Erlbaum, Hillsdale, NJ, 1981.
109. Waller, N. G. et al., Genetic and environmental influences on religious interests, attitudes, and values: a study of twins reared apart and together, *Psychol. Sci.*, 1.2 (1990): 1-5.
110. Eaves, L. J., N. G. Martin, and A. C. Heath, Religious affiliation in twins and their parents: testing a model of cultural inheritance, *Behav. Genet.*, 20 (1990): 1-22.
111. Waller, N. G. and P. R. Shaver, The importance of nongenetic influences on romantic love styles: a twin family study, *Psychol. Sci.*, 5 (1994): 268-274.
112. Tellegen, A. et al., Personality similarity in twins reared apart and together, *J. Pers. Soc. Psychol.*, 54 (1988): 1031–1039.
113. Gottfredson, M. R. and T. Hirschi, *A General Theory of Crime*, Stanford University Press, Stanford, CA, 1990.
114. Cloninger, C. R., T. Reich, and S. B. Guze, The multifactorial model of disease transmission. II. Sex differences in the familial transmission of sociopathy (antisocial personality), *Br. J. Psychiatry*, 127 (1975): 11-22.
115. Cloninger, R. C. and I. I. Gottesman, Genetic and environmental factors in antisocial behavior disorders, *The Causes of Crime: New Biological Approaches*, S. A. Mednick, T. E. Moffitt, and S. A. Stack, Eds., Cambridge University Press, New York, 1987.
116. Hindelang, M. J., T. Hirschi, and J. Weiss, Correlates of delinquency. The illusion of discrepancy between self-report and official measure, *Am. Sociol. Rev.*, 44 (1979): 995-1014.
117. Baker, L. A. et al., Sex differences in property crime in a Danish adoption cohort, *Behav. Genet.*, 19 (1989): 355-370.
118. Hirschi, T. and M. J. Hindelang, Intelligence and delinquency: a revisionist review, *Am. Sociol. Rev.*, 42 (1977): 571-587.
119. Gordon, R. A., Research on IQ, race, and delinquency: taboo or not taboo, *Taboos in Criminology*, E. Sagarin, Ed., Sage, Beverly Hills, CA, 1980.
120. Gordon, R. A., SES versus IQ in the Race-IQ Delinquency model, *Int. J. Sociol. Soc. Policy*, 7 (1987): 30-96.
121. Moffitt, T. E. et al., Socioeconomic status, IQ, and delinquency, *J. Abnorm. Psychol.*, 90 (1981): 152-156.
122. Lynam, D., T. E. Moffitt, and M. Stouthamer-Loeber, Explaining the relation between IQ and delinquency: class, race, test motivation, school failure, or self-control?, *J. Abnorm. Psychol.*, 102 (1993): 187-196.
123. Herrnstein, R. J. and C. Murray, *The Bell Curve: Intelligence and Class Structure in American Life*, Free Press, New York, 1994.
124. Krueger, R. et al., Personality traits are linked to crime. Evidence from a birth cohort, *J. Abnorm. Psychol.*, 103 (1994): 328-338.
125. Bock, G. R. and J. A. Goode, Eds., *Genetics of Criminal and Antisocial Behaviour*, John Wiley & Sons, Chichester, 1996, 283.
126. Lyons, M. J., A twin study of self-reported criminal behavior, *Genetics of Criminal and Antisocial Behavior*, G. R. Bock and J. A. Goode, Eds., John Wiley & Sons, Chichester, 1996, 61-70.
127. Silberg, J. et al., Heterogeneity among juvenile antisocial behaviours: findings from the Virginia Twin Study of Adolescent Behavioral Development, *Genetics of Criminal and Antisocial Behaviour*, G. R. Bock and J. A. Goode, Eds., John Wiley & Sons, Chichester, 1996, 76-86.
128. Bohman, M., Predisposition to criminality: Swedish adoption studies in retrospect, *Genetics of Criminal and Antisocial Behaviour*, G. R. Bock and J. A. Goode, Eds., John Wiley & Sons, Chichester, 1996, 99-114.
129. Brennan, P. A., S. A. Mednick, and B. Jacobsen, Assessing the role of genetics in crime using adoption cohorts, *Genetics of Criminal and Antisocial Behaviour*, G. R. Bock and J. A. Goode, Eds., John Wiley & Sons, Chichester, 1996, 115-128.
130. Virkkumen, M., D. Goldman, and M. Linnoila, Serotonin in alcoholic violent offenders, *Genetics of Criminal and Antisocial Behaviour*, G. R. Bock and J. A. Goode, Eds., John Wiley & Sons, Chichester, 1996, 168-177.

131. Goldman, D., J. Lappalainen, and N. Ozaki, Direct analysis of candidate genes in impulsive behaviors, *Genetics of Criminal and Antisocial Behaviour*, G. R. Bock and J. A. Goode, Eds., John Wiley & Sons, Chichester, 1996.
132. Brunner, H. G., MAOA deficiency and abnormal behaviour: perspectives on an association, *Genetics of Criminal and Antisocial Behaviour*, G. R. Bock and J. A. Goode, Eds., John Wiley & Sons, Chichester, 1996, 155-167.
133. Moffitt, T. E., Adolescence-limited and life-course-persistent antisocial behavior: a developmental taxonomy, *Psychol. Rev.*, 100 (1993): 674-701.
134. Barrett, G. V. and R. L. Depinet, A reconsideration of testing for competence rather than for intelligence, *Am. Psychol.*, 46.10 (1991): 1012-1024.
135. Bouchard, T. J., Jr., The genetic architecture of human intelligence, *Biological Approaches to the Study of Human Intelligence*, P. A. Vernon, Ed., Ablex, NJ, 1993, 33-93.
136. McGue, M. et al., Behavior genetics of cognitive ability: a life-span perspective, *Nature, Nurture and Psychology*, R. Plomin and G. E. McClearn, Eds., American Psychological Association, Washington, D.C., 1993, 59-76.
137. Tambs, K., J. M. Sundet, and P. Magnus, Heritability analysis of the WAIS subtests: a study of twins, *Intelligence*, 8 (1984): 283-293.
138. Sundet, J. M. et al., No differential heritability of intelligence test scores across ability levels in Norway, *Behav. Genet.*, 24 (1994): 337-340.
139. Teasdale, T. W. and D. R. Owen, Heritability and family environment in intelligence and educational level — a sibling study, *Nature*, 309 (1984): 620-622.
140. Loehlin, J. C., J. M. Horn, and L. Willerman, Heredity, environment and IQ in the Texas adoption study, *Heredity, Environment and Intelligence*, R. J. Sternberg and E. L. Grigorenko, Eds., Cambridge University Press, New York,
141. Bouchard, T. J., Jr. et al., IQ and heredity, *Science*, 252 (1991): 191-192.
142. Scarr, S. and R. A. Weinberg, Educational and occupational achievements of brothers and sisters in adoptive and biological related families, *Behav. Genet.*, 24 (1994): 301-325.
143. Neisser, U. et al., Intelligence: knowns and unknowns, *Am.x Psychol.*, 51(1996): 77-101.
144. Blum, K. et al., Reward deficiency syndrome, *Am. Sci.*, 84 (1996): 132-145.
145. Straub, R. E. et al., A potential vulnerability locus for schizophrenia on chromosome 6p24-22: evidence for genetic heterogeneity, *Nat. Genet.*, 11 (1995): 287-293.
146. Moises, H. W. et al., An international two-stage genome-wide search for schizophrenia susceptibility genes, *Nat. Genet.*, 11 (1995): 321-324.
147. Schwab, S. G. et al., Evaluation of a susceptibility gene for schizophrenia on chromosome 6p by multipoint affected sib-pair linkage analysis, *Nat. Genet.*, 11 (1995): 325-327.
148. Skuder, P. et al., A polymorphism in mitochondrial DNA associated with IQ?, *Intelligence*, 21 (1995): 1-12.
149. Plomin, R. et al., Allelic associations between 100 DNA markers and high versus low IQ, *Intelligence*, 21 (1995): 31-48.
150. Plomin, R., Molecular genetics and psychology, *Curr. Direct. Psychol. Sci.*, 4 (1995): 114-117.
151. Thapar, A. et al., The genetics of mental retardation, *Br. J. Psychiatry*, 164 (1994): 747-758.
152. Flint, J. et al., The detection of subtelomeric chromosomal rearrangements in idiopathic mental retardation, *Nat. Genet.*, 9 (1995): 132-140.
153. Corder, B. et al., Gene dose of apolipoprotein E type 4 allele and the risk of Alzheimer's disease in late onset families, *Science*, 261 (1993): 921-923.
154. Snowdon, D. A. et al., Linguistic ability in early life and cognitive function and Alzheimer's disease in late life. Findings for the Nun Study, *J. Am. Med. Assoc.*, 275 (1996): 528-532.
155. Ebstein, R. P. et al., Dopamine D4 receptor (D4DR) exon III polymorphism associated with the human personality trait of novelty seeking, *Nat. Genet.*, 12 (1996): 78-80.
156. Benjamin, J. et al., Population and familial association between the D4 dopamine receptor gene and measures of novelty seeking, *Nat. Genet.*, 12 (1996): 81-84.
157. Zuckerman, M., Good and bad humors: biochemical bases of personality and its disorders, *Psychol. Sci.*, 6 (1995): 325-332.
158. Lyons, M. J. et al., Do genes influence exposure to trauma? A twin study of combat, *Am. J. Med. Genet., Neuropsychiatr. Genet.*, 48 (1993): 22-27.
159. Cloninger, C. R., R. Adolfsson, and N. M. Svrakic, Mapping genes for human personality, *Nat. Gene.*, 12 (1996): 3-4.
160. Maccoby, E. E., The role of parents in the socialization of children: a historical overview, *Dev. Psychol.*, 28 (1992): 1006-1017.
161. Darling, N. and L. Steinberg, Parenting style as context: an integrative model, *Psychol. Bull.*, 113 (1993): 487-496.

162. Scarr, S., Behavior genetic and socialization theories of intelligence: truce and reconciliation, *Intelligence: Heredity and Environment*, R. J. Sternberg and E. L. Grigorenko, Eds., Cambridge University Press, New York, in press.
163. Bronfenbrenner, U. and S. J. Ceci, Nature-nurture reconceptualized in developmental perspective: a bioecological model, *Psychol. Rev.*, 101 (1994): 568-586.
164. Rowe, D., *The Limits of Family Influence: Genes, Experience, and Behavior*, Guilford Press, New York, 1994.
165. Rowe, D. C., Environmental and genetic influences on dimensions of perceived parenting: a twin study, *Dev. Psychol.*, 17 (1981): 203-208.
166. Plomin, R. et al., Genetic influence on childhood family environment perceived retrospectively from the last half of the life span, *Dev. Psychol.*, 24 (1988): 738-745.
167. Hur, Yoon-Mi and T. J. Bouchard, Jr. Genetic influence on perceptions of childhood family environment: a reared apart twins study, *Child Dev.*, Vol. 66 (1995): 330-335.
168. Perusse, D. et al., Human parental behavior: evidence for genetic influence and potential implications for gene-culture transmission, *Behav. Genet.*, 24 (1994): 327-336.
169. Lytton, H., Child and parent effects in boys' conduct disorder: a reinterpretation, *Dev. Psychol.*, 26.5 (1990): 683-697.
170. Lytton, H., Child effects — still unwelcome? Response to Dodge and Wahler, *Dev. Psychol.*, 26.5 (1990): 705-709.
171. Dodge, K. A., Nature versus nurture in childhood conduct disorder: it is time to ask a different question, *Dev. Psychol.*, 26.5 (1990): 698-701.

17 Family-Based Association of Attention-Deficit/Hyperactivity Disorder and the Dopamine Transporter

Edwin H. Cook, Jr., Mark A. Stein, and Bennett L. Leventhal

CONTENTS

A. Introduction 297
B. Diagnostic Criteria for ADHD 298
    1. Diagnosis and Classification of ADHD 298
    2. Evidence for Genetic Factors in ADHD 298
    3. Evidence for Nongenetic Etiological Factors in ADHD 298
    4. Comorbidity of ADHD and Other Psychiatric Disorders: Implications for Genetic Analysis 299
    5. Dopamine Transporter as a Candidate Gene for ADHD 299
C. Genetic Association Studies 300
    1. Relationship of ADHD to Previously Identified Genetic Disorders 301
D. Methods 301
    1. Family-Based Association Between ADHD and the Dopamine Transporter Gene 301
    2. Demographic and Descriptive Characteristics of Subjects 302
    3. Genotyping 303
E. Results 304
    1. Results: Family-Based Association 305
    2. Discussion 306
Acknowledgments 307
References 307

A. INTRODUCTION

Attention-deficit/hyperactivity disorder (ADHD) is a heterogeneous syndrome characterized by symptoms of inattention and hyperactivity-impulsivity which are inconsistent with developmental level (Taylor, 1986; Barkley, 1990). It is a relatively common but disabling mental disorder, affecting 3 to 9% of school-aged children (Szatmari et al.,1989; American Psychiatric Association, 1994). ADHD often persists into adulthood and is a risk factor for development of antisocial and drug abuse disorders (Mannuzza et al., 1993; Weiss and Hechtman, 1993).

B. DIAGNOSTIC CRITERIA FOR ADHD

1. Diagnosis and Classification of ADHD

The diagnostic criteria for ADHD/ADD have varied over the years in response to changing conceptualizations of the underlying mechanisms and primary symptoms of the disorder. In the case of ADHD, the diagnosis is based upon the presence of clinically significant behavioral and cognitive symptoms. The diagnosis of ADHD in child probands is reliable after decades of refinement (Schwab-Stone et al., 1993; Lahey et al.,1994). Given the broad spectrum of symptoms incorporated into current diagnostic schemes and the somewhat variable clinical course, it seems likely that ADHD has multiple etiologies.

Issues regarding the fundamental nature of the disorder remain unresolved (Schachar, 1991). One factor impeding progress is the lack of an adequate classification system for reliably identifying etiologically homogeneous subgroups of children with ADHD. As pointed out by Wender, an essential feature of most complex psychiatric disorders is that they are not defined at the etiological or pathophysiological level and that we do not have any specific method of testing for the presence or absence of a disorder (Wender, 1995). As with mental retardation, fever, and diabetes, we hypothesize that our understanding of the genetics of this disorder will be enhanced by refining our diagnostic classifications and, in particular, utilizing additional sources of information such as cognitive functioning, medical history, and dysmorphology to exclude individuals whose ADHD symptoms are due to nongenetic or confounding conditions. Hence, one goal in ascertainment of probands with ADHD for genetic studies is to reduce the frequency of phenocopies.

2. Evidence for Genetic Factors in ADHD

Evidence for the importance of genetic factors in ADHD comes from a variety of family/genetic studies of ADHD. Relatives of probands with ADHD compared to relatives of adoptive parents, normals, or psychiatric controls have revealed an increased prevalence of ADHD in the biological relatives of probands with ADHD (Cantwell, 1972; Cunningham et al., 1975; Alberts-Corush et al., 1986; Biederman et al., 1990, 1992). Similarly, twin studies are consistent with moderate to high heritability of attentional dysfunction (Willerman, 1973; Stevenson, 1992). In addition to an association with clinical characteristics of ADHD, early studies of hyperactive children reported an increased rate of alcoholism, sociopathy, or "hysteria" in first-degree relatives (Morrison and Stewart, 1971; Cantwell, 1972). More recent studies, using more contemporary diagnostic criteria, found similar results for relatives of children with ADHD (Roizen et al., 1996). In addition, studies by Biederman and colleagues also suggest an increase in major depression in the first-degree relatives of individuals with ADHD (Biederman et al., 1987, 1991).

A segregation analysis of ADHD was consistent with autosomal dominant transmission with reduced penetrance of the hypothesized gene (Faraone et al., 1992). In another study, this group examined the cosegregation of ADHD and learning disabilities and found that these two disorders were transmitted independently (Faraone et al., 1993). Altogether, this set of studies implies that ADHD is highly heritable and tends to co-occur with other forms of psychopathology, particularly conduct disorder/antisocial personality disorder, alcoholism, and somatization disorder.

3. Evidence for Nongenetic Etiological Factors in ADHD

A variety of confounding environmental, biological, and social conditions can contribute to ADHD symptoms and diagnosis (Cantwell and Baker, 1987). These include both acquired conditions (e.g., acquired brain damage, postencephalitic encephalopathy, and lead toxicity or exposure to other environmental toxins including alcohol), known genetic syndromes that are associated with learning problems, attention difficulties or hyperactivity, and social factors including access to health care, classroom size, and teacher experience and parenting factors. In a study of genetic aspects of idiopathic ADHD, these factors are viewed as contributors to phenocopy rate whose inclusion would decrease statistical power. Thus, a powerful and valid study of ADHD genetics requires a comprehensive and rigorous approach to the assessment process for probands in order to eliminate potentially confounding factors.

4. COMORBIDITY OF ADHD AND OTHER PSYCHIATRIC DISORDERS: IMPLICATIONS FOR GENETIC ANALYSIS

In clinic-referred samples, ADHD often co-occurs with other psychiatric disorders, especially disruptive behavior disorders, mood disorders, anxiety disorders, and learning disorders (Szatmari et al., 1989; Biederman et al., 1991; Cantwell and Baker, 1991; Hinshaw, 1992; Keller et al., 1992; Pliszka, 1992; Semrud-Clikeman et al., 1992; Wozniak et al., 1995). Relatives of probands with ADHD are at increased risk for several disorders other than ADHD (Cantwell, 1972; Biederman et al., 1991, 1992). As a result, one is left with the question of whether to consider relatives to be affected with an ADHD-related phenotype if they display oppositional-defiant disorder, conduct disorder, substance abuse disorder, or mood disorder without ADHD. In addition, the diagnosis of ADHD in adults is complex and often dependent on retrospective recall of ADHD symptoms of uncertain reliability (Stein et al., 1995). Although follow-up of adults who were diagnosed during childhood and who now have children with ADHD may provide a method of addressing these difficulties (Pauls, 1991; Biederman et al., 1994; Hechtman, 1994), classification issues and bilineal inheritance in some families (Morrison and Stewart, 1974) may continue to limit the applicability of traditional linkage analysis to ADHD. This is not unique to ADHD, but is a general problem for complex disorders including other psychiatric disorders with child and adolescent onset (Lombroso et al., 1994).

Family-based association studies have the advantage of not requiring inclusion of relatives with an uncertain phenotype. One question that is raised by comorbidity is whether ADHD alone differs from ADHD with a comorbid disruptive behavior disorder such as oppositional-defiant disorder or conduct disorder. Data from a family study of relatives of probands with ADD with and without comorbid oppositional-defiant and conduct disorders revealed an increased risk of both ADD and any antisocial disorder in relatives of child probands with ADD with a comorbid conduct disorder, compared to relatives of child probands with ADD alone. These data raise two distinct hypotheses: (1) ADHD with and without comorbid conduct problems are two etiologically distinct disorders; or (2) ADHD with and without comorbid conduct problems lie along a continuum of increasing severity due to increasing familial factors (Faraone et al., 1991). Further research is needed to assess whether poor outcome is a specific consequence of ADHD/ADD, oppositional-defiant disorder, conduct disorder, anxiety disorder, mood disorder, or just a general response to having more severe problems.

5. DOPAMINE TRANSPORTER AS A CANDIDATE GENE FOR ADHD

The most commonly used and well-studied treatment approach for ADHD is pharmacotherapy. At least in the short term, this is effective for most children with ADHD (reviewed in Greenhill, 1992). In numerous double-blind trials, pharmacological agents which inhibit the dopamine transporter (including methylphenidate, dextroamphetamine, pemoline, and bupropion) have been shown to be effective in the treatment of the attentional dysfunction, hyperactivity, and impulsivity of ADHD (Casat et al., 1987; Zametkin and Rapoport, 1987; Greenhill, 1992; Hechtman, 1994; Rapport et al., 1994). These observations made the dopamine transporter (*DAT1* = *SLC643*) a primary candidate gene in ADHD.

The focus on *DAT1* is further sharpened by animal studies. For example, overexpression of mutant rat dopamine transporter (with increased DA transport) has been shown to lead to rapid habituation to a novel environment, increased sensitivity to locomotor effects of the psychostimulant cocaine, and stronger place preference conditioning to cocaine (Miner et al., 1994). Although there has been a negative association study between DAT1 and polysubstance abuse, there are likely to be many factors besides a susceptibility gene involved in development of a substance abuse disorder (Persico et al., 1993).

C. GENETIC ASSOCIATION STUDIES

Genetic association studies have a checkered history, and not all investigators are convinced that the modest likelihood of success is worth the problems and pitfalls inherent in their design and interpretation (e.g., see Kidd, 1993). However, under some circumstances linkage disequilibrium studies offer the best opportunity to detect the effects of loci contributing susceptibility to a complex disorder (Cox and Bell, 1989; Lander and Schork, 1994). It was an association study which demonstrated the association between late-onset Alzheimer's disease and the ε4 allele at the Apo E locus (Strittmatter et al., 1993), using patients without affected relatives. While subsequent association studies have confirmed this finding and the association is widely accepted (Corder et al., 1993), the evidence by LOD score linkage analysis of Apo E to Alzheimer's susceptibility remains equivocal.

The history of the use of association studies in identifying susceptibility loci for insulin-dependent diabetes mellitus (IDDM) may be particularly instructive with respect to complex psychiatric disorders. The first identification of a susceptibility locus for IDDM came through association studies on HLA, and subsequent refinements suggest that variation in the HLA region contributes susceptibility to IDDM (Svejgaard and Ryder, 1989; Davies et al., 1994). In the early 1980s, an association was reported between variation near the insulin gene and IDDM based on a study of unrelated patients with IDDM and controls (Bell et al., 1984). For some time the finding remained controversial. Initially there was no evidence for linkage between IDDM and the insulin gene region (Cox et al., 1988). However, a large dataset was assembled through a massive collaborative effort which included multiplex families from throughout North America and Europe. With these data, it was possible to confirm linkage disequilibrium using the haplotype relative risk method (HRR), and establish that the association was due to linkage by using the transmission/disequilibrium test (TdT), even though conventional parametric linkage analyses and the distribution of insulin gene-sharing in affected sib pairs provided no evidence for linkage (Spielman et al., 1989, 1993). While linkage between IDDM and the insulin gene region was recently confirmed by affected sib pair linkage analysis (Davies et al., 1994), association studies of unrelated patients were essential for identifying the role of the insulin locus in IDDM susceptibility.

In an earlier association study, an increase in the frequency of the *Taq I* A1 allele at the dopamine D2 receptor (*DRD2*) locus was found in groups of subjects with ADHD, alcoholism, Tourette's disorder, or autistic disorder compared to control subjects (Comings et al., 1991). However, the increased frequency of the *DRD2* A1 allele in ADHD compared to controls (and alcoholism, Tourette's disorder, and autistic disorder) may be due to population stratification of marker alleles, even after control for race (Ott, 1991). Another association study in ADHD, with 20 p 11-12 markers, in a region syntenic with a mouse coloboma mutant which is responsive to amphetamine treatment, was negative (Hess et al., 1995).

To avoid false positive findings due to case and control samples having a different population frequency of risk alleles independent of association with a disease susceptibility gene, family-based association analyses were developed (Rubinstein et al., 1981; Ott, 1989). These analyses control for the possibility of an association being an artifact of population stratification. For example, if the individuals with the disorder differ in ethnicity from the individuals without the disorder, and the frequency of allele markers are differentially related to ethnicity, a false positive result may be obtained in a case-control association study. This may help explain a number of failures to replicate associations in psychiatric and other disorders (Kidd, 1993). The advantage of family-based association studies is that each parental allele is considered equally likely to be transmitted to the child with ADHD or not to be transmitted to the child with ADHD. The likelihood of transmission or nontransmission of a given allele is not influenced by the frequency of the allele in the sample. This method has been revised to obtain more power as in the haplotype-based HRR approach (Terwilliger and Ott, 1992). In this test, the frequency of transmission of a marker allele from each parent to an affected child is compared to the frequency of nontransmission of the marker allele and frequency of transmission and nontransmission of other alleles at the same locus. It has been further refined as the TdT, which measures linkage (recombination frequency <0.5) in the presence of linkage disequilibrium (Spielman et al., 1993). Although some consider the HHRR more conservative than the TdT (Thomson, 1995), another group has demonstrated that under certain conditions of population admixture the TdT is preferable to the HHRR (Ewens and Spielman, 1995).

1. Relationship of ADHD to Previously Identified Genetic Disorders

Previously, ADHD was found to be more common in subjects with generalized resistance to thyroid hormone (GRTH) compared to controls (Hauser et al., 1993). However, the prevalence of GRTH in ADHD has been found to be extremely rare (Weiss et al., 1993; Spencer et al., 1995) and subsequent studies have not supported genetic linkage of ADHD and GRTH (Weiss et al., 1994; Stein et al., 1995).

D. METHODS

1. Family-Based Association Between ADHD and the Dopamine Transporter Gene

As discussed above, the dopamine transporter is a primary candidate gene for ADHD. To avoid the potential effects of population stratification, we used the Haplotype-Based Haplotype Relative Risk (HHRR) method to test for association between a VNTR polymorphism at the dopamine transporter locus and DSM-III-R (*Diagnostic and Statistical Manual of Mental Disorders*, 3rd edition, revised) diagnosed ADHD (N = 49) and Undifferentiated Attention-Deficit Disorder (UADD) (N = 8) in trios composed of father, mother, and affected offspring.

Consecutive patients seen in the Hyperactivity, Attention, and Learning Problems (HALP) Clinic at the University of Chicago between April 20, 1993 and October 18, 1994 were screened for possible inclusion in the study. The inclusion criteria were (1) child or adolescent with a DSM-III-R diagnosis of ADHD or UADD, made in a consensus diagnostic conference in which a child psychologist(s), child psychiatrist, and a developmental pediatrician presented findings from each of their evaluations; (2) availability of one or more biological parents; and (3) consent to blood collection and participation in this study from parent(s) and child. (UADD is a DSM-III-R diagnosis in which children have attentional problems and distractibility, but do not have a sufficient number of symptoms of hyperactivity and impulsivity to meet DSM-III-R criteria for diagnosis of ADHD.)

A total of 56 families participated in the study. Due to the unavailability of some of the parents 24 families consisted of the mother and affected child; 4 families consisted of father and affected child; 27 families consisted of mother, father, and affected child; and 1 family consisted of mother, father, and 2 affected children. There were 47 families in which the proband had ADHD, 1 family in which 2 probands had ADHD, and 8 families in which the proband had UADD. The mean age of the probands was 9.4 years (range 4 to 17 years); 47 probands were Caucasian, 9 were African-American, and 1 was Hispanic. Mean family socioeconomic status (Hollingshead and Redlich, 1958) was 2.5 (range 1 to 5). There were 38 children (67%) with a comorbid DSM-III-R diagnosis. Overall intelligence was average, but there were weaknesses in arithmetic and coding which were expected because performance on these subtests is highly influenced by inattention. As summarized in Table 17-2, scores on behavior rating scales were consistent with the clinical diagnoses of ADHD and UADD.

2. DEMOGRAPHIC AND DESCRIPTIVE CHARACTERISTICS OF SUBJECTS

Intelligence was assessed with the Wechsler Preschool and Primary Scale of Intelligence (WPPSI) for the 4 to 5-year-old children, and the Wechsler Intelligence Scale for Children-Third Edition (WISC-III) for the subjects aged 6 to 16 (Wechsler, 1967; Wechsler, 1991). Several parent and teacher rating scales were obtained prior to the evaluation. Parents completed the Achenbach Child Behavior Checklist (CBCL) (Achenbach and Edelbrock, 1983) and Conners Parent Rating Scale, Revised (Goyette et al., 1978). The CBCL and ADD-H: Comprehensive Teacher's Rating Scale (ACTeRS) (Ullmann et al., 1984) were completed by teachers. Comorbid diagnoses and behavior ratings are reported below in Tables 17-1 and 17-2.

TABLE 17-1
Comorbid Diagnoses

Comorbid Diagnosis	Patients	%
Axis I		
Oppositional Defiant Disorder	19	33.3
Conduct Disorder	6	10.2
Anxiety Disorder	5	8.8
Elimination Disorder	4	7.0
Major Depressive Disorder	1	1.8
Dysthymia	1	1.8
Axis II		
Developmental Coordination and/or Expressive Writing Disorder	16	31.4
Developmental Expressive and/or Receptive Language Disorder	8	14.0
Articulation Disorder	4	7.0
Developmental Reading Disorder	6	10.2
Developmental Arithmetic Disorder	3	5.3

TABLE 17-2
Psychometric Testing and Behavior Rating Scales

Tests	N	Mean ± S.D.
WISC-III[a] or WPPSI[b]		
Full-scale	52	97.6 ± 13.0
Verbal	51	98.6 ± 12.2
Performance	51	99.9 ± 14.3
Arithmetic	39	8.5 ± 2.7
Coding	36	8.3 ± 2.9
ACTeRs[c] (T Score)	39	
Attention		38.3 ± 9.0
Hyperactivity		36.7 ± 13.4
Social Skills		39.9 ± 9.4
Oppositional		41.6 ± 9.8
Conners PSQ — Revised[d] (T Score)	44	
Conduct Problems		69.5 ± 20.4
Learning Problem		84.0 ± 18.2
Psychosomatic		62.8 ± 23.9
Impulsive-Hyperactive		69.2 ± 13.2
Anxiety		56.7 ± 13.7
Hyperactivity Index		75.6 ± 16.0
CBCL[e] (T Scores)		
Internalizing (parent rating)	48	62.9 ± 12.4
Externalizing (parent rating)	48	66.2 ± 11.4
Internalizing (teacher rating)	38	58.0 ± 9.0
Externalizing (teacher rating)	38	63.1 ± 10.1

[a] WISC-III — Wechsler Intelligence Scale for Children, Third Edition (Wechsler, 1991).
[b] WPPSI — Wechsler Preschool and Primary Scale of Intelligence (Wechsler, 1967).
[c] ACTeRs — ADD-H: Comprehensive Teacher's Rating Scale (Ullmann et al., 1984).
[d] Conners PSQ — Conners Parent Rating Scale, Revised (Goyette et al., 1978).
[e] CBCL — Child Behavior Checklist (Achenbach and Edelbrock, 1983).

DSM-III-R diagnoses of ADHD or UADD, or either disorder comorbid with conduct disorder or oppositional-defiant disorder, were based upon parental history using a semistructured interview by a clinical psychologist or child and adolescent psychiatrist. Diagnosis was then confirmed by a multidisciplinary team consensus following a 6-h evaluation of each child, which included a physical and neurologic examination performed by a developmental pediatrician and a review of all instruments by at least three experienced clinicians.

3. GENOTYPING

DNA was extracted from whole blood or normal saline mouth rinse by a Tris-EDTA (TE) extraction method. The following primers were used to amplify the region flanking the DAT1 40 bp VNTR: T3-5Long (5'-TGTGGTGTAGGGAACGGCCTGAG-3') and T7-3aLong (5'-CTTCCTGGAGGTCACGGCTCAAGG-3') (Vandenbergh et al., 1992). The primers were synthesized on an Applied Biosystems 380B DNA Synthesizer at the Cancer Research Center at the University of Chicago. Hot-start PCR was carried out in a 75-μl volume containing 400 ng of genomic template, 0.5 μM of each primer, 200 μM of each dNTP (dATP, dCTP, dGTP, dTTP), 1 × PCR buffer, 2 μM MgCl$_2$, and 2 units *Taq* polymerase (Amplitaq, Perkin-Elmer Cetus, Norwalk, CT). PCR gems (Perkin-Elmer Cetus) were added to each sample and a hot-start of 80°C for 5 min, 25°C for 2 min, was performed. The template and *Taq* polymerase were added following the hot-start step. Samples were processed in a Perkin-Elmer Cetus DNA Thermal Cycler through 40 cycles of 30 s at 95°C, 30 s at 68°C, and 1.5 min at 72°C. Samples were stored in a –80°C freezer following the completion of PCR. A 2-μl aliquot of 6-fold-concentrated (Savant vacuum drier) PCR product was mixed with 2 μl of loading buffer (0.05% bromophenol blue, 0.05% xylene cyanol FF) and 1 μl of the resulting products were then separated at 15°C on 20% homogeneous acrylamide PhastGels with native buffer strips (0.88 M L-alanine, 0.25 M Tris, pH 8.8) on PhastSystem (Pharmacia Biotech, Inc., Piscataway, NJ). Each gel contained a 100-bp ladder (Gibco BRL/Life Technologies). The PhastGel was pre-run at 400 V, 10 mA, 2.5 W for 100 Vh. Samples were applied during a step of 400 V, 10 mA, 2.5 W for 2 Vh. Samples were then separated at 400 V, 10 mA, 2.5 W for 100 Vh. Gels were then silver-stained using 75 ml of the following reagents: 20% trichloroacetic acid fixing solution for 5 min. at 20°C; 5% glutardialdehyde sensitization solution for 6 min at 40°C; triple-distilled water for 2 min (twice) at 40°C; 0.4% silver nitrate for 6 min at 30°C; triple-distilled water for 2.5 min and then 0.5 min at 20°C; triple-distilled water for 0.5 min at 30°C; 2.5% sodium carbonate/0.1% formaldehyde developing solution for 0.5 min and then 2.5 min at 30°C; 5% glacial acetic acid for 5 min at 30°C; 12% glycerol/5% acetic acid preserver solution for 3 min at 30°C. For confirmation, 78 of 141 samples were also run on 4% agarose (Perkin-Elmer Cetus) to verify the results observed on the PhastSystem media. Resulting bands on each gel were blindly and independently scored by two individuals. Discrepancies were resolved by repetition of PCR using purified template. DNA extracted from whole blood was spun at 3000 × g for 12 min in a Microcon-100 microconcentrator (Amicon, Inc., Beverly, MA). The vials were then placed upside down and spun at 1000 × g for 3 min. This template was then used in PCR and gel analysis as described above, with the exception that a combined annealing/extension step of 72°C for 2.5 min was substituted for the separate annealing and extension steps of 68°C and 72°C during PCR.

E. RESULTS

Genotypes for each combination of parent(s) and children are listed below. (This table is corrected from the previously published version [Cook et al., 1995]).

TABLE 17-3
Genotypes of Combinations of Parent-Child Trios and Pairs

Observed Parental Genotypes	Child's Genotype				
	200/480	440/440	440/480	480/480	480/520
Trios (2 Parents and Child)					
200/480 × 480/480				1	
440/480 × 440/440			1	4	
440/480 × 480/480			5	10	
480/480 × 480/480				8	
Parent-Child Pairs					
200/480			1		
440/440			3		
440/480	1	3	2	2	
480/480			4	11	1

1. RESULTS: FAMILY-BASED ASSOCIATION

To avoid confounding effects of population stratification, we tested the independence of transmission of each parental allele. Whatever the frequency of alleles in the parents of a group of affected offspring is, the chance that each allele will be transmitted to an offspring is 50%, given the null hypothesis of no linkage disequilibrium ($\delta = 0$; no association). If the transmission of an allele deviates from this ratio, based on a standard chi-square test, evidence for association exists (Rubinstein et al., 1981). A haplotype-based approach was used because of its greater power to detect a genetic association (Ott, 1989). Allele frequency in the total parental sample at the *DAT1* VNTR was 1.2% 200 bp (3 copies of VNTR), 22.6% 440 bp (9 copies), and 76.2% 480 bp alleles (10 copies).

Since the total number of all parental alleles other than the 440- and 480-bp alleles was less than 5, the 200-bp allele was combined with the less common 440-bp allele for HHRR analysis. HHRR analysis revealed significant association between ADHD/UADD and the 480-bp DAT1 allele (χ^2 7.51, 1 df, $p = .006$) (See Table 17-4). (Two mother-child pairs consisted of heterozygotes; therefore the transmission status these two mothers' alleles could not be determined for HHRR). Similar results were found if only the 47 Caucasian probands were included (χ^2 4.55, 1 df, $p = .033$). Furthermore, if only the 49 ADHD probands are considered, the results are essentially unchanged (HHRR χ^2 7.29, 1 df, $p = .007$). In the most conservative analysis (Curtis and Sham, 1995), if only the 23 Caucasian ADHD probands are considered in which both parents are typed, the results are similar (HHRR χ^2 10.3, 1 df, $p = .001$).

TABLE 17-4
Haplotype-Based Haplotype Relative Risk (HHRR) (All Families)

	9 Copy or Other Copy VNTR	10 Copy VNTR	Total
Transmitted	12	72	84
Not transmitted	27	57	84
	39	129	168

Note: HHRR χ^2 7.51, 1 df, $p = .006$.

An alternative, but not independent, method of analysis of the data, preferred by some, is the transmission/disequilibrium test. Table 17-5 demonstrates that this result also provides significant evidence of preferential transmission of the ten-copy allele.

TABLE 17-5
Transmission/Disequilibrium Test (TdT) (All Families)

		Not Transmitted		
		9 Or Other Copy	10 Copy	Total
Transmitted	9 Or other copy	3	9	12
Transmitted	10 Copy	24	48	72
Total		27	57	84

Note: TdT $\chi^2 = (9 - 24)^2/(24 + 9) = 6.82$, 1 df, $p = .009$

2. DISCUSSION

Therefore, we provide preliminary evidence of linkage disequilibrium, which Thomson (1995) points out requires linkage, between ADHD and the dopamine transporter. Caveats about the current study include that 24 fathers and 4 mothers are missing from our pedigrees and the sample size is relatively small. As in any study, replication from an independent sample is necessary. Although it is possible that the association is due to a yet to be discovered ADHD susceptibility gene in linkage disequilibrium with DAT1, this is unlikely. Recently, DAT1 was mapped near D5S678 on the distal end of the small arm of chromosome 5 (Gelernter et al., 1995).

It is likely that other loci are involved in genetic susceptibility to ADHD. The family-based association strategy was chosen because of our assumption that ADHD was a complex disorder involving several genetic loci and interactions between genetic factors and environmental factors.

Previous studies using the same marker did not find a significant association between the 3' UTR VNTR at the dopamine transporter locus and schizophrenia (Byerley et al., 1993; Persico et al., 1995), polysubstance abuse (Persico et al., 1993), or cocaine dependence (Gelernter et al., 1994). The latter study found an increased frequency of the nine-copy allele in cocaine-dependent subjects with paranoid experiences compared to those without paranoid experiences. ADHD is a risk factor for development of substance abuse. Since approximately four fifths of children and adolescents with ADHD do not develop substance use disorders (Mannuzza et al., 1993), any relationship between the dopamine transporter, ADHD, and substance abuse is likely to be complex. Confirmation of family-based association would lead to identification of specific mutation(s) at the dopamine transporter locus, which would assist in the dissection of the complex interaction between ADHD and substance abuse, including the interaction of genetic and environmental factors.

Diagnosis of ADHD for a genetic association study and for other studies differs in emphasis. If one is studying the full spectrum of ADHD cases for epidemiological purposes or for a representative treatment study, it is important to have a representative sample of heterogeneous cases of ADHD. However, for replication of this family-based association finding, it is important to use an ascertainment scheme which is similar to the original study, in which a more homogeneous sample is attained. Figure 17-1 demonstrates the multiple levels of exclusion necessary to obtain a relatively homogeneous subset of subjects with ADHD. Since the original finding included a substantial number of subjects with ADHD and oppositional-defiant disorder or conduct disorder, it is very possible that the association between the dopamine transporter and ADHD is with oppositional-defiant disorder, conduct disorder, or one of the other disorders which frequently occur with ADHD (bottom half of IV in the pyramid).

If our family-based association is replicated, an intensive search for mutations or molecular characterization of the VNTR will be pursued using single-stranded conformational polymorphism analysis and direct sequencing. If identified, susceptibility mutations of a human dopamine transporter cDNA carrying the mutation would be possible. Transfection of this mutated cDNA into heterologous cell lines would address any functional changes in the transporter with respect to binding and uptake of dopamine and other drugs that act at the transporter such as methylphenidate, bupropion, D-amphetamine, and cocaine (Martel et al., 1994). Moreover, gene targeting could be used to produce a mouse model, which would be useful in the study of developmental neurobiological anomalies associated with a putative mutation in the dopamine transporter (Giros, 1996).

Identification of a susceptibility allele would assist analysis of family-genetic data collected by other investigators, particularly in understanding the relationship of genetic susceptibility of ADHD to psychiatric disorders found at increased frequency in relatives of probands with ADHD. Moreover, identification of a mutation conferring some degree of genetic

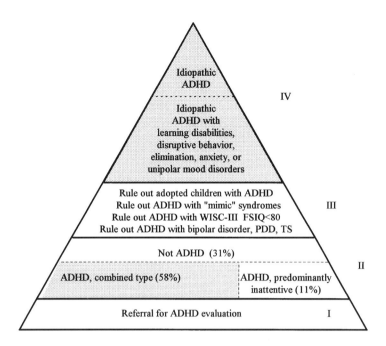

FIGURE 17-1 Diagnosis of ADHD for inclusion in family-based association studies. In this ascertainment system, one codes the highest level of diagnosis met by an individual case. Presumably, the highest level (IV) will consist of individuals whose ADHD is relatively more homogeneous, as children with alternative etiologies and diagnostic uncertainties are excluded. Patients will be excluded after a multidisciplinary team consensus review of all clinical materials if ADHD symptoms are viewed as the result of a medical "mimic" syndrome, including acquired neurologic disorders (head trauma, fetal alcohol effect, lead toxicity) and known genetic syndromes, such as Marfan's syndrome, William's syndrome, Turner syndrome, Fragile X syndrome, resistance to thyroid hormone or other thyroid abnormality, tuberous sclerosis, or neurofibromatosis.

susceptibility to development of ADHD would assist in development of effective prevention and intervention strategies for ADHD (Hechtman, 1993; Ialongo et al., 1993).

ACKNOWLEDGMENTS

We thank N. Cox, M. Krasowski, D. Olkon, S. Yan, T. Blondis, N. Roizen, B. Berget, P. Gardner, and G. Greene for their contribution to this work. This work was funded, in part, by grants from the Smart Family Foundation (M.S.), Shaw Foundation (B.L.), and Harris Foundation (B.L.). We thank the families who have participated in this research and Children and Adults with Attention Deficit Disorders (C.H.A.D.D.).

REFERENCES

Achenbach, T. M. and C. S. Edelbrock (1983). *Manual for the Child Behavior Checklist and Revised Child Behavior Profile*, Thomas A. Achenbach, Burlington, VT.
Alberts-Corush, J., P. Firestone, and J. Goodman (1986). Attention and impulsivity characteristics of the biological and hyperactive parents of hyperactive and normal control children, *Am. J. Orthopsychiatry*, 56(3): 413-423.
APA (1994). *Diagnostic and Statistical Manual of Mental Disorders*, 4th ed., American Psychiatric Association, Washington, D.C.
Barkley, R. (1990). Associated problems, subtyping, and etiologies, *Attention-Deficit Hyperactivity Disorder*, Guilford Press, New York.

Bell, G., S. Horita, and J. Karam (1984). A polymorphic locus near the human insulin gene is associated with insulin-dependent diabetes mellitus, *Diabetes*, 33: 176-183.

Biederman, J., S. Faraone, K. Keenan, J. Benjamin, B. Krifcher, C. Moore, S. Sprich-Buckminster, K. Ugaglia, M. Jellinek, R. Steingard, T. Spencer, D. Norman, R. Kolodny, I. Kraus, J. Perrin, M. Keller, and M. Tsuang (1992). Further evidence for family-genetic risk factors in attention deficit hyperactivity disorder. Patterns of comorbidity in probands and relatives in psychiatrically and pediatrically referred samples, *Arch. Gen. Psychiatry*, 49: 728-738.

Biederman, J., S. Faraone, K. Keenan, D. Knee, and M. Tsuang (1990). Family-genetic and psychosocial risk factors in DSM-III attention deficit disorder, *J. Am. Acad. Child Adolescent Psychiatry*, 29(4): 526-533.

Biederman, J., S. Faraone, E. Mick, T. Spencer, and T. Wilens (1994). Risk for ADHD in children of parents with childhood onset ADHD, *Sci. Proc. Am. Acad. Child Adolescent Psychiatry*, 10: 65.

Biederman, J., S. V. Faraone, K. Keenan, R. Steingard and M. T. Tsuang (1991). Familial association between attention deficit disorder and anxiety disorders, *Am. J. Psychiatry*, 148(2): 251-256.

Biederman, J., S. V. Faraone, K. Keenan, and M. T. Tsuang (1991). Evidence of familial association between attention deficit disorder and major affective disorders, *Arch. Gen. Psychiatry*, 48(7): 633-42.

Biederman, J., K. Munir, D. Knee, M. Armentano, S. Autor, C. Waternaux, and M. Tsuang (1987). High rate of affective disorders in probands with attention deficit disorder and in their relatives: controlled family study, *Am. J. Psychiatry*, 144(3): 330-333.

Biederman, J., J. Newcorn, and S. Sprich (1991). Comorbidity of attention deficit hyperactivity disorder with conduct, depressive, anxiety, and other disorders, *Am. J. Psychiatry*, 148(5): 564-577.

Byerley, W., H. Coon, M. Hoff, J. Holik, M. Waldo, R. Freedman, M. Caron, and B. Giros (1993). Human dopamine transporter gene not linked to schizophrenia in multigenerational pedigrees, *Hum. Hered.*, 43: 319-322.

Cantwell, D. (1972). Psychiatric illness in the families of hyperactive children, *Arch. Gen. Psychiatry*, 27: 414-417.

Cantwell, D. and L. Baker (1991). *Psychiatric and Developmental Disorders in Children with Communication Disorder*, American Psychiatric Association, Washington, D.C.

Cantwell, D. P. and L. Baker (1987). Differential diagnosis of hyperactivity, *Dev. Behav. Pediatrics*, 8: 159-165.

Casat, C., D. Pleasants, and J. Fleet (1987). A double-blind trial of bupropion in children with attention deficit disorder, *Psychopharmacol. Bull.*, 23(1): 120-122.

Comings, D. E., B. G. Comings, D. Muhleman, G. Dietz, B. Shahbahrami, D. Tast, E. Knell, P. Kocsis, R. Baumgarten, B. W. Kovacs, et al. (1991). The dopamine D2 receptor locus as a modifying gene in neuropsychiatric disorders, *JAMA*, 266(13): 1793-800.

Cook, E., M. Stein, M. Krasowski, N. Cox, D. Olkon, J. Kieffer, and B. Leventhal (1995). Association of attention deficit disorder and the dopamine transporter gene, *Am. J. Hum. Genet.*, 56(4): 993-998.

Corder, E. H., A. M. Saunders, W. J. Strittmatter, D. E. Schmechel, P. C. Gaskell, G. W. Small, A. D. Roses, J. L. Haines, and M. A. Pericak-Vance (1993). Gene dose of apolipoprotein E type 4 allele and the risk of Alzheimer's disease in late onset families, *Science*, 261(5123): 921-3.

Cox, N., L. Baker, and R. Spielman (1988). Insulin gene sharing in sib pairs with insulin-dependent diabetes mellitus. No evidence for linkage, *Am. J. Hum. Genet.*, 42: 167-172.

Cox, N. and G. Bell (1989). Disease associations: chance, artifact, or susceptibility genes?, *Diabetes*, 38: 947-950.

Cunningham, L., R. Cadoret, R. Loftus, and J. E. Edwards (1975). Studies of adoptees from psychiatrically disturbed biological parents, *Br. J. Psychiatry*, 126: 534-539.

Curtis, D. and P. Sham (1995). A note on the application of the transmission disequilibrium test when a parent is missing, *Am. J. Hum. Genet.*, 56(3): 811-812.

Davies, J., Y. Kawaguchi, S. Bennett, J. Copeman, H. Cordell, L. Pritchard, P. Reed, S. Gough, S. Jenkins, S. Palmer, K. Balfour, B. Rowe, M. Farrall, A. Barnett, S. Bain, and J. Todd (1994). A genome-wide search for human type 1 diabetes susceptibility genes, *Nature*, 371: 130-136.

Ewens, W. and R. Spielman (1995). The transmission/disequilibrium test: history, subdivision, and admixture, *Am. J. Hum. Genet.*, 57: 455-464.

Faraone, S., J. Biederman, W. Chen, B. Krifcher, K. Keenan, C. Moore, S. Sprich, and M. Tsuang (1992). Segregation analysis of attention deficit hyperactivity disorder, *Psychiatr. Genet.*, 2: 257-275.

Faraone, S. V., J. Biederman, K. Keenan, and M. T. Tsuang (1991). Separation of DSM-III attention deficit disorder and conduct disorder: evidence from a family-genetic study of American child psychiatric patients, *Psychol. Med.*, 21(1): 109-21.

Faraone, S. V., J. Biederman, B. K. Lehman, K. Keenan, D. Norman, L. J. Seidman, R. Kolodny, I. Kraus, J. Perrin, and W. J. Chen (1993). Evidence for the independent familial transmission of attention deficit hyperactivity disorder and learning disabilities: results from a family genetic study, *Am. J. Psychiatry*, 150(6): 891-5.

Gelernter, J., H. Kranzler, S. Satel, and P. Rao (1994). Genetic association between dopamine transporter protein alleles and cocaine-induced paranoia, *Neuropsychopharmacology*, 11(3): 195-200.

Gelernter, J., D. Vandenbergh, S. D. Kruger, D. L. Pauls, R. Kurlan, A. J. Pakstis, K. K. Kidd, and G. Uhl (1995). The dopamine transporter protein gene (SLC6A3). Primary linkage mapping and linkage studies in Tourette syndrome, *Genomics*, 30(3): 459-463.

Giros, B., M. Jaber, S. R. Jones, R. M. Wightman, M. B. Caron (1996). Hyperlocomotion and indifference to cocaine and amphetamine in mice lacking the dopamine transporter, *Nature (London)*, 379: 606-612.

Goyette, C. H., C. K. Conners, and R. F. Ulrich (1978). Normative data on revised Conners parent and teacher rating scales, *J. Abnorm. Child Psychol.*, 6(2): 221-236.

Greenhill, L. L. (1992). Pharmacologic treatment of attention deficit hyperactivity disorder, *Psychiatr. Clinics N. Am.*, 15(1): 1-27.

Hauser, P., A. J. Zametkin, P. Martinez, B. Vitiello, J. A. Matochik, A. J. Mixson, and B. D. Weintraub (1993). Attention deficit-hyperactivity disorder in people with generalized resistance to thyroid hormone, *N. Engl. J. Med.*, 328(14): 997-1001.

Hechtman, L. (1993). Aims and methodological problems in multimodal treatment studies, *Can. J. Psychiatry*, 38(6): 458-464.

Hechtman, L. (1994). Genetic and neurobiological aspects of attention deficit hyperactive disorder: a review. *J. Psychiatr. Neurosci.*, 19(3): 193-201.

Hess, E. J., P. K. Rogan, M. Domoto, D. E. Tinker, R. L. Ladda, and J. C. Ramer (1995). Absence of linkage of apparently single gene mediated ADHD with the human syntenic region of the mouse mutant coloboma, *Am. J. Med. Genet.*, 60(6): 573-579.

Hinshaw, S. (1992). Academic underachievement, attention deficits, and aggression: comorbidity and implications for intervention, *J. Consult. Clin. Psychol.*, 60: 893-903.

Hollingshead, A. and R. C. Redlich (1958). *Social Class and Mental Illness*, John Wiley & Sons, New York.

Ialongo, N., W. Horn, J. Pascoe, G. Greenberg, T. Packard, M. Lopez, A. Wagner, and L. Puttler (1993). The effects of a multimodal intervention with attention-deficit hyperactivity disorder. A 9-month follow-up, *J. Am. Acad. Child Adolescent Psychiatry*, 32(1): 182-189.

Keller, M. B., P. W. Lavori, W. R. Beardslee, J. Wunder, C. E. Schwartz, J. Roth, and J. Biederman (1992). The disruptive behavioral disorder in children and adolescents: comorbidity and clinical course, *J. Am. Acad. Child Adolescent Psychiatry*, 31(2): 204-209.

Kidd, K. (1993). Associations of disease with genetic markers. Déja vu all over again, *Am. J. Med. Genet. (Neuropsychiatr. Genet.)*, 48: 71-73.

Lahey, B., B. Applegate, K. McBurnett, J. Biederman, L. Greenhill, G. Hynd, R. Barkley, J. Newcorn, P. Jensen, J. Richters, B. Garfinkel, L. Kerdyk, P. Frick, T. Ollendick, D. Perez, E. Hart, I. Waldman, and D. Shaffer (1994). DSM-IV field trials for attention deficit/hyperactivity in children and adolescents, *Am. J. Psychiatry*, 151: 1673-1685.

Lander, E. and N. Schork (1994). Genetic dissection of complex traits, *Science*, 265: 2037-2048.

Lombroso, P., D. Pauls, and J. Leckman (1994). Genetic mechanisms in childhood psychiatric disorders, *J. Am. Acad. Child Adolescent Psychiatry*, 33(7): 921-938.

Mannuzza, S., R. Klein, A. Bessler, P. Malloy, and M. LaPadula (1993). Adult outcome of hyperactive boys: educational achievement, occupational rank, and psychiatric status, *Arch. Gen. Psychiatry*, 50(7): 565-576.

Martel, J., D. Vandenbergh, N. Takahashi, and G. Uhl (1994). Expression studies of human dopamine transporter cDNAs with 3'-untranslated region variants, *Soc. Neurosci. Abstr.*, 20: 921.

Miner, L., D. Donovan, J.-B. Wang, J. Mulle, M. Perry, and G. Uhl (1994). Behavioral characterization of mutant dopamine transporter and μ opiate receptor overexpression in catecholaminergic neurons of transgenic mice, *Soc. Neurosci. Abstr.*, 20: 921.

Morrison, J. and M. Stewart (1971). A family study of the hyperactive child syndrome, *Biol. Psychiatry*, 3: 189-195.

Morrison, J. and M. Stewart (1974). Bilateral inheritance as evidence for polygenicity in the hyperactive child syndrome, *J. Nerv. Ment. Dis.*, 158(3): 226-228.

Ott, J. (1989). Statistical properties of the haplotype relative risk, *Genet. Epidemiol.*, 6: 127-130.

Ott, J. (1991). *Analysis of Human Genetic Linkage,* rev. ed., Johns Hopkins University Press, Baltimore, MD.

Pauls, D. (1991). Genetic factors in the expression of attention-deficit hyperactivity disorder, *J. Child Adolescent Psychopharmacol.*, 1(5): 353-360.

Persico, A., Z. Wang, D. Black, N. Andreasen, G. Uhl, and R. Crowe (1995). Exclusion of close linkage of the dopamine transporter gene with schizophrenia spectrum disorders, *Am. J. Psychiatry*, 152(1): 134-136.

Persico, A. M., D. J. Vandenbergh, S. S. Smith, and G. R. Uhl (1993). Dopamine transporter gene polymorphisms are not associated with polysubstance abuse, *Biol. Psychiatry*, 34(4): 265-267.

Pliszka, S. R. (1992). Comorbidity of attention-deficit hyperactivity disorder and overanxious disorder, *J. Am. Acad. Child Adolescent Psychiatry*, 31(2): 197-203.

Rapport, M., C. Denney, G. DuPaul, and M. Gardner (1994). Attention deficit disorder and methylphenidate. Normalization rates, clinical effectiveness, and response prediction in 76 children, *J. Am. Acad. Child Adolescent Psychiatry*, 33(6): 882-893.

Roizen, N., T. Blondis, M. Irwin, A. Rubinoff, J. Kieffer, and M. Stein (1996). Incidence of psychiatric and developmental disorders in families of children with ADHD. Utility of parent report, *Arch. Pediatr. Adolescent Med.*, 150(2): 203-208.

Rubinstein, P., M. Walker, C. Carpenter, C. Carrier, J. Krassner, C. Falk, and F. Ginsburg (1981). Genetics of HLA disease associations. The use of the Haplotype Relative Risk (HRR) and the 'Haplo-Delta' (Dh) estimates in juvenile diabetes from three racial groups, *Hum. Immunol.*, 3: 384.

Schachar, R. (1991). Childhood hyperactivity, *J. Child Psychol. Psychiatry Allied Discip.*, 32(1): 155-191.

Schwab-Stone, M., P. Fisher, J. Piacentini, D. Shaffer, M. Davies, and M. Briggs (1993). The Diagnostic Interview Schedule for Children-Revised Version (DISC-R). II. Test-retest reliability, *J. Am. Acad. Child Adolescent Psychiatry*, 32(3): 651-7.

Semrud-Clikeman, M., J. Biederman, S. Sprich-Buckminster, B. K. Lehman, S. Faraone, and D. Norman (1992). Comorbidity between ADDH and learning disability. A review and report in a clinically referred sample, *J. Am. Acad. Child Adolescent Psychiatry*, 31(3): 439-448.

Spencer, T., J. Biederman, T. Wilens, J. Guite, and M. Harding (1995). ADHD and thyroid abnormalities: a research note, *J. Child Psychol. Psychiatry*, 36: 879-885.

Spielman, R., M. Baur, and F. Clerget-Darpoux (1989). Genetic analysis of IDDM. Summary of GAW5 IDDM results, *Genet. Epidemiol.*, 6: 43-58.

Spielman, R. S., R. E. McGinnis, and W. J. Ewens (1993). Transmission test for linkage disequilibrium: the insulin gene region and insulin-dependent diabetes mellitus (IDDM), *Am. J. Hum. Genet.*, 52(3): 506-16.

Stein, M., R. Sandoval, E. Szumowski, N. Roizen, M. Reinecke, T. Blondis, and Z. Klein (1995). Psychometric properties of the Wender Utah Rating Scale (WURS): Reliability and factor structure for men and women, *Psychopharmacol. Bull.*, 31(2): 425-433.

Stein, M., R. Weiss, and S. Refetoff (1995). Neurocognitive characteristics of individuals with resistance to thyroid hormone. Comparison with individuals with attention deficit hyperactivity disorder, *J. Dev. Behav. Pediatr.*, 16: 406-411.

Stevenson, J. (1992). Evidence for a genetic etiology in hyperactivity in children, *Behav. Genet.*, 22(3): 337-44.

Strittmatter, W., A. Saunders, D. Schmechel, et al. (1993). Apolipoprotein E: high affinity binding to β amyloid and increased frequency of type 4 allele in familial Alzheimer's, *Proc. Natl. Acad. Sci.*, 90: 1977-1981.

Svejgaard, A. and L. Ryder (1989). HLA and insulin-dependent diabetes: an overview, *Genet. Epidemiol.*, 6: 1-14.

Szatmari, P., D. Offord, and M. Boyle (1989). Ontario child health study: prevalence of attention deficit disorder with hyperactivity, *J. Child Psychol. Psychiatry*, 30: 205-217.

Taylor, E. (1986). The causes of hyperactive behavior, *The Overactive Child*, Oxford University Press, London.

Terwilliger, J. and J. Ott (1992). A haplotype based 'Haplotype Relative Risk' approach to detecting allelic associations, *Hum. Hered.*, 42: 337-346.

Thomson, G. (1995). Mapping disease genes: family-based association studies, *Am. J. Hum. Genet.*, 57(8): 487-498.

Ullmann, R. K., E. K. Sleator, and R. L. Sprague (1984). A new rating scale for diagnosis and monitoring of ADD children, *Psychopharmacol. Bull.*, 20: 160-164.

Vanderbergh, D. O., A. M. Persico, A. L. Hawkins, C. A. Griffin, X. Li, E. W. Jabs, G. R. Uhl (1992). Human dopamine transporter gene (DAT1) maps to chromosome 5p15.3 and displays a VNTR, *Genomics*, 14(4): 1104-1106.

Wechsler, D. (1967). *Manual for the Wechsler Preschool and Primary Scale of Intelligence*, The Psychological Corporation, San Antonio, TX.

Wechsler, D. (1991). *Wechsler Intelligence Scale for Children*, 3rd ed., The Psychological Corporation, Chicago.

Weiss, G. and L. Hechtman (1993). *Hyperactive Children Grown Up*, 2nd ed., Guilford Press, New York.

Weiss, R., M. Stein, S. Duck, B. Chyna, W. Phillips, T. O'Brien, L. Gutermuth, and S. Refetoff (1994). Low intelligence but not attention deficit hyperactivity disorder is associated with resistance to thyroid hormone caused by mutation R316H in the thyroid hormone receptor β gene, *J. Clin. Endocrinol. Metab.*, 78(6): 1525-1528.

Weiss, R., M. Stein, B. Trommer, and S. Refetoff (1993). Attention-deficit hyperactivity disorder and thyroid function, *J. Pediatr.*, 123: 539-545.

Wender, P. (1995). *Attention-Deficit Hyperactivity Disorder in Adults*, Oxford University Press, New York.

Willerman, L. (1973). Activity level and hyperactivity in twins, *Child Dev.*, 44(2): 288-293.

Wozniak, J., J. Biederman, K. Kiely, J. Ablon, S. Faraone, E. Mundy, and D. Mennin (1995). Mania-like symptoms suggestive of childhood-onset bipolar disorder in clinically referred children, *J. Am. Acad. Child Adolescent Psychiatry*, 34(7): 867-876.

Zametkin, A. J. and J. L. Rapoport (1987). Neurobiology of attention deficit disorder with hyperactivity: where have we come in 50 years?, *J. Am. Acad. Child Adolescent Psychiatry*, 26(5): 676-86.

18 Reward Deficiency Syndrome: Neurobiological and Genetic Aspects

Kenneth Blum, John G. Cull, Eric R. Braverman, Thomas J. H. Chen, and David E. Comings

CONTENTS

A. Introduction 311
B. The Biology of Reward 312
C. Alcoholism and Genes 314
D. Drug Addiction and Smoking 317
E. Compulsive Bingeing and Gambling 318
F. Attention Deficit Disorder 320
G. The Dopamine D2 Receptor 321
H. Treatment 323
References 325

A. INTRODUCTION

In 1990 one of us published with his colleagues a paper suggesting that a specific genetic anomaly was linked to alcoholism.[1] Unfortunately, it was often erroneously reported that they had found the "alcoholism gene", implying that there is a one-to-one relationship between a gene and a specific behavior. Such misinterpretations are common — readers may recall accounts of an "obesity gene", or a "personality gene". Needless to say, there is no such thing as a specific gene for alcoholism, obesity, or a particular type of personality. However, it would be naive to assert the opposite: that these aspects of human behavior are not associated with any particular genes. Rather, the issue at hand is to understand how certain genes and behavioral traits are connected.

In the past 5 years the authors have pursued the association between certain genes and various behavioral disorders (see Chapter 19). In molecular genetics, an association refers to a statistically significant incidence of a genetic variant (an allele) among genetically unrelated individuals with a particular disease or condition, compared to a control population. In the course of our work, we discovered that the genetic anomaly previously found to be associated with alcoholism is also found with increased frequency among people with other addictive, compulsive, or impulsive disorders. The list is long and remarkable — it comprises alcoholism, substance abuse, smoking, compulsive overeating and obesity, attention deficit disorder, Tourette's disorder, and pathological gambling.

We believe these disorders are linked by a common biological substrate, a "hard-wired" system in the brain (consisting of cells and signaling molecules) that provides pleasure in the process of rewarding certain behavior. Consider how people respond positively to safety, warmth, and a full stomach. If these needs are threatened or are not being met, we experience discomfort and anxiety. An inborn chemical imbalance that alters the intercellular signaling in the brain's reward process could supplant an individual's feeling of well-being with anxiety, anger, or a craving for a substance that can alleviate the negative emotions. This chemical imbalance manifests itself as one or more behavioral disorders for which one of us (Blum) has coined the term "reward deficiency syndrome".[2]

This syndrome involves a form of sensory deprivation of the brain's pleasure mechanisms. It can be manifested in relatively mild or severe forms that follow as a consequence of an individual's biochemical inability to derive reward from ordinary, everyday activities. We believe we have discovered at least one genetic aberration leading to an alteration in the reward pathways of the brain. It is a variant form of the gene for the dopamine D2 receptor, called the A1 allele. This is the same genetic variant that we previously found to be associated with alcoholism. In this review we shall look at evidence suggesting the A1 allele is also associated with a spectrum of impulsive, compulsive, and addictive behaviors. The concept of a reward deficiency syndrome unites these disorders and may explain how simple genetic anomalies give rise to complex aberrant behavior.

B. THE BIOLOGY OF REWARD

The pleasure and reward system in the brain was discovered by accident in 1954.[3] The American psychologist, James Olds, was studying the rat brain's alerting process, when he mistakenly placed the electrodes in a part of the limbic system, a group of structures deep within the brain that are generally believed to play a role in emotions. When the brain was wired so the animal could stimulate this area by pressing a lever, Olds found the rats would press the lever almost nonstop, as many as 5000 times an hour. The animals would stimulate themselves to the exclusion of everything else except sleep. They would even endure tremendous pain and hardship for an opportunity to press the lever. Olds had clearly found an area in the limbic system that provided a powerful reward for these animals.

Research on human subjects revealed that the electrical stimulation of some areas of the brain (the medial hypothalamus) produced a feeling of quasi-orgasmic sexual arousal.[4,5] If certain other areas of the brain were stimulated, an individual experienced a type of light-headedness that banished negative thoughts. These discoveries demonstrated that pleasure is a distinct neurological function that is linked to a complex reward and reinforcement system.[6]

During the past several decades research on the biological basis of chemical dependency has been able to establish some of the brain regions and neurotransmitters involved in reward. In particular, it appears the dependence on alcohol, opiates, and cocaine relies on a common set of biochemical mechanisms.[7,8] A neuronal circuit deep in the brain involving the limbic system and two regions called the nucleus accumbens and the globus pallidus appears to be critical in the expression of reward for people taking these drugs.[9] Although each substance of abuse appears to act on different parts of this circuit, the end result is the same: dopamine is released in the nucleus accumbens and the hippocampus.[10] Dopamine appears to be the primary neurotransmitter of reward at these reinforcement sites.

Although the system of neurotransmitters involved in the biology of reward is complex, at least three other neurotransmitters are known to be involved at several sites in the brain: serotonin in the hypothalamus, the enkephalins (opioid peptides) in the ventral tegmental area and the nucleus accumbens, and the inhibitory neurotransmitter GABA in the ventral tegmental area and the nucleus accumbens.[11,12] Interestingly, the glucose receptor is an important link between the serotonergic system and the opioid peptides in the hypothalamus. An alternative reward pathway involves the release of norepinephrine in the hippocampus from neuronal fibers that originate in the locus coeruleus.

In a normal person, these neurotransmitters work together in a cascade of excitation or inhibition — between complex stimuli and complex responses — leading to a feeling of well-being, the ultimate reward.[7,8,11] In the cascade theory of reward, a disruption of these intercellular interactions results in anxiety, anger, and other "bad feelings", or in a craving for a substance that alleviates these negative emotions. Alcohol, for example, is known to activate the norepinephrine system in the limbic circuitry through an intercellular cascade that includes serotonin, opioid peptides, and dopamine. Alcohol may also act directly through the production of neuroamines that interact with opioid receptors or with dopaminergic systems.[13] In the cascade theory of reward, genetic anomalies, prolonged stress, or long-term abuse of alcohol can lead to a self-sustaining pattern of abnormal cravings in both animals and human beings (Figure 18-1).

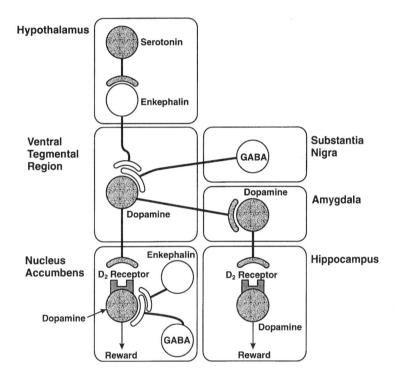

FIGURE 18-1 Reward cascade in the limbic system consists of excitatory and inhibitory connections between neurons that are modulated by neurotransmitters. The activation of the dopamine D2 receptor by dopamine on the cell membranes of neurons in the nucleus accumbens and the hippocampus is hypothesized by the authors to be the "final common pathway" of the reward cascade. If the activity of the dopamine D2 receptor is deficient, the activity of neurons in the nucleus accumbens and the hippocampus is decreased and the individual experiences unpleasant emotions or cravings for substances that can provide temporary relief by releasing dopamine. Alcohol, cocaine, and nicotine are known to promote the release of dopamine in the brain. A simplified version of the cascade is presented here. Disorders of the cells and molecules in the "upstream" part of the cascade may also disrupt the normal activity of the reward system. The cascade begins with the excitatory activity of serotonin-releasing neurons in the hypothalamus. This causes the release of the opioid peptide met-enkephalin in the ventral tegmental area, which inhibits the activity of neurons that release the inhibitory neurotransmitter gamma-aminobutyric acid (GABA). The disinhibition of dopamine-containing neurons in the ventral tegmental area allows them to release dopamine in the nucleus accumbens and in certain parts of the hippocampus, permitting the completion of the cascade.

Support for the cascade theory can be derived from a series of experiments on strains of rats that prefer alcohol to water. Compared to normal rats, the alcohol-preferring rats have fewer serotonin neurons in the hypothalamus, higher levels of enkephalin in the hypothalamus (because less is released), more GABA neurons in the nucleus accumbens (which inhibit the release of dopamine), a reduced supply of dopamine in the nucleus accumbens, and a lower density of dopamine D2 receptors in certain areas of the limbic system.[14-17] Moreover, an association between low contents of dopamine and serotonin in the nucleus accumbens and high alcohol preference was recently reported by Mcbride et al. 1996.[18]

These studies suggest a four-part cascade in which there is a reduction in the amount of dopamine released in a key reward area in the alcohol-preferring rats. The administration of substances that increase the supply of serotonin at the synapse or that directly stimulate dopamine D2 receptors reduce craving for alcohol.[19] For example, D2 receptor agonists reduce the intake of alcohol among rats that prefer alcohol, whereas D2 dopamine-receptor antagonist increase the drinking of alcohol in these inbred animals.[20]

Support for the cascade theory of alcoholism in humans is found in a series of clinical trials. When amino acid precursors of certain neurotransmitters (serotonin and dopamine) and a drug that promotes enkephalin activity were given to alcoholic subjects, the individuals experienced fewer cravings for alcohol, a reduced incidence of stress, an increased likelihood of recovery, and a reduction in relapse rates.[21-23] Furthermore, the notion that dopamine is the "final common pathway" for drugs such as cocaine, morphine, and alcohol is supported by recent studies by Ortiz and associates.[24] These authors demonstrated that the chronic use of cocaine, morphine, or alcohol results in several biochemical adaptations in the limbic dopamine system. They suggest that these adaptations may result in changes in the structural and functional properties of the dopaminergic system.

We believe the biological substrates of reward that underlie the addiction to alcohol and other drugs are also the basis for impulsive, compulsive, and addictive disorders comprising the reward deficiency syndrome.[25]

C. ALCOHOLISM AND GENES

An alteration in any of the genes involved in the expression of the molecules in the reward cascade might predispose an individual to alcoholism. Indeed, the evidence for a genetic basis to alcoholism has accumulated steadily over the past five decades. The earliest report comes from studies of laboratory mice. It was found that, given a choice, certain mice preferred alcohol to water. Gerald McClearn and Rodgers[27] took this a step farther by producing an inbred mouse (the C57 strain) that had a marked preference for alcohol. The alcohol-preferring C57 strain bred true through successive generations — it was the first clear indication that alcoholism has a genetic basis.[27]

The first evidence that alcoholism has a genetic basis in human beings came in 1972 when scientists at the Washington University School of Medicine in St. Louis found that adopted children whose biological parents were alcoholics were more likely to have a drinking problem than those born to nonalcoholic parents.[28] In 1973 Goodwin and colleagues studied 5483 men in Denmark who had been adopted in early childhood. They found that the sons born to alcoholic fathers were three times more likely to become alcoholic than the sons of nonalcoholic fathers.[29]

In the late 1980s research on the inheritance of alcoholism suggested there might be important genetic differences between alcoholics and nonalcoholics.[30,31] Our group suspected that the activity of the chemical signaling molecules in the reward pathways of the brain might be involved. Over the course of two years we compared eight genetic markers associated with various neurotransmitters (including serotonin, endogenous opioids, GABA, transferrin, acetylcholine, alcohol dehydrogenase, and aldehyde dehydrogenase). In each instance we failed to find a direct association between the genetic markers and alcoholism.

The opportunity to investigate a ninth genetic marker arose after Civelli et al.[33] cloned and sequenced the gene for one form of the dopamine D2 receptor (Figure 18-2). The D2 receptor is one of at least five physiologically distinct dopamine receptors (D1, D2a, D2b, D3, D4, and D5) found on the synaptic membranes of neurons in the brain.[32] Previous studies had established that D2 receptors are expressed in neurons within the cerebral cortex and the limbic system, including the nucleus accumbens, the amygdala, and the hippocampus. Because these are the same areas of the brain (with the exception of the cortex) that are believed to be involved in the reward cascade, Civelli's work provided the opportunity to investigate an important molecular candidate for genetic aberrations among alcoholics.

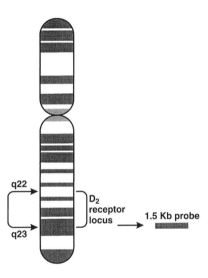

FIGURE 18-2 Human chromosome 11 carries the gene that codes for the dopamine D2 receptor, 1 of 5 known dopamine receptors. The gene is located on the long (*q*) arm of the chromosome and was cloned and sequenced in 1990, providing investigators with the opportunity to test for genetic variations in the population. A 1.5-kb sequence taken from one end of the gene is used as a probe to search for variants.

The technique we use to distinguish between the D2 receptor genes of alcoholics and those of nonalcoholics relies on the detection of restriction-fragment-length polymorphisms (RFLPs). This approach involves the use of DNA-cutting enzymes (restriction endonucleases) that cleave the DNA molecule at specific nucleotide sequences. If there are genetic differences between two individuals such that a restriction enzyme cuts their DNA along different points in (or near) a gene, the resulting fragments of their genes will be of different lengths. These differing fragments, or polymorphisms, are recognized by the use of a radioactively labeled DNA probe — in this case a short sequence of the D2 receptor gene — that binds to a complementary DNA sequence on the fragments. Radiolabeled fragments of different lengths signify a difference in the cleavage sequence recognized by the restriction enzyme[33] (Figure 18-3).

The restriction enzyme (*Taq I*) cuts the nucleotide sequence at a site just outside the coding region for the D2 receptor gene. This produces the *Taq I* A polymorphisms. To date there are four *Taq I* A alleles known, the A1, A2, A3, and A4 alleles. The A3 and A4 alleles are rare, whereas the A2 allele is found in nearly 75% of the general population and the A1 allele in about 25% of the population.

In 1990 we used the *Taq I* enzyme to search for *Taq I* A polymorphisms in the DNA extracted from the brains of deceased alcoholics and a control population of nonalcoholics.

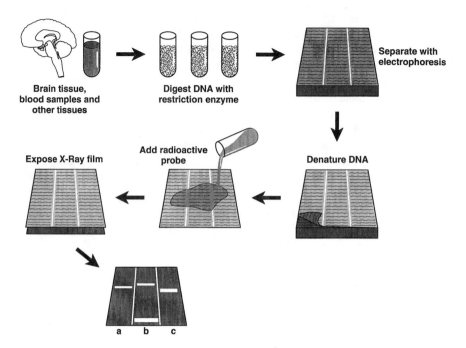

FIGURE 18-3 Method of recognizing genetic variations in the dopamine D2 receptor gene relies on the detection of restriction-fragment-length polymorphisms (RFLPs). DNA is extracted from brain tissue or blood samples and then cut into many fragments with a restriction enzyme (such as *Taq I*), which cleaves the genetic material at a specific nucleotide sequence. The fragments are separated from each other in a gelatin solution by an electric current that carries the fragments different distances according to their lengths and electric charge. The double-stranded DNA molecules are denatured into single strands before being blotted onto a membrane to allow further processing. A radioactive probe (a 1.5-kb sequence of the dopamine D2 receptor gene extracted from a known source) binds to complementary sequences of the single-stranded fragments. The fragments of the single-stranded DNA that bind the probe are visualized when X-ray film is exposed to the membrane. Here the genetic variations are revealed in the X-ray film, producing a DNA fingerprint for each of three individuals (a, b, c).

The results were striking: in our sample of 35 alcoholics we found that 69% had the A1 allele and 31% had the A2 allele. In 35 nonalcoholics we found that 20% had the A1 allele and 80% had the A2 allele.

Since our 1990 study, some laboratories have failed to find a connection between the A1 allele and alcoholism. However, a review of their work shows that their samples were not limited to severe forms of alcoholism, which we believe to be an important distinguishing criterion. In our original study, over 70% of the alcoholics had cirrhosis of the liver, a disease suggestive of severe and chronic alcoholism. Moreover, the negative studies failed to adequately assess controls to eliminate alcoholism, drug abuse and other related "reward behaviors". In this regard, Neiswanger and co-workers recently found a strong association of the A1 allele and alcoholism and suggested that early failures were due to poor assessment of a true phenotype in the controls.[34] To date, 14 independent laboratories have supported the finding that the A1 allele is a causative factor in severe forms of alcoholism, though perhaps not in milder forms.[35] These findings do not prove that the A1 allele of the dopamine D2 receptor gene is the only cause of severe alcoholism, but they are a powerful indication that the A1 allele is involved with alcoholism.

Further evidence for the role of biology in alcoholism comes from efforts to find electrophysiological markers that might indicate a predisposition to the addictive disorder. One such marker is the latency and the magnitude of the Positive 300-ms (P300) wave, an indicator of the general electrical activity of the brain that is evoked by a specific stimulus such as a tone. It turns out abnormalities in the electrical activity of the brain are evident in the young sons of alcoholic fathers. Their P300 waves are markedly reduced in amplitude compared to the P300 waves of the sons of nonalcoholic fathers. These results raised the question as to whether this deficit had been transferred from father to son and whether this deficit would predispose the son to substance abuse in the future.[36]

Experiments carried out since then have answered both questions. The alcoholic fathers had the same P300-wave deficit seen in their sons, and the sons showed increased drug-seeking behaviors (including alcohol and nicotine) compared to the sons of nonalcoholic fathers. Moreover, the sons of alcoholic fathers had an atypical neurocognitive profile.[37] It now appears that children with P300 abnormalities are more likely to abuse drugs and tobacco in later years.[38]

Remarkably, Noble and colleagues found an association between the A1 allele and a prolonged latency of the P300 wave in children of alcoholics.[39] Others extended this work and observed a similar correlation between the A1 allele and a prolonged P300 latency in a neuropsychiatric population. Subjects who were homozygous for the A1 allele showed significantly prolonged P300 latency compared to A1/A2 and A2/A2 carriers.[40]

D. DRUG ADDICTION AND SMOKING

Cocaine can bring intense, but temporary, pleasure to the user. The aftermath is addiction and severe psychological and physiological harm. Various psychosocial theories have been advanced to account for the abuse of cocaine and other illicit drugs. In contrast to alcoholism, where growing empirical evidence is implicating hereditary factors, relatively little has been known about the genetics of human cocaine dependence. However, recent studies have suggested that hereditary factors are involved in the use and abuse of cocaine and other illicit drugs.

Studies of adopted children, for example, show that a biological background of alcohol problems in the parents predicts an increased tendency toward illicit drug abuse in the children.[41] Similarly, family studies of cocaine addicts show a high percentage of first- or second-degree relatives who have been diagnosed as alcoholics.[42,43]

Behavioral anomalies such as conduct disorder (in which children violate social norms and the rights of others) and antisocial personality (the adult equivalent of conduct disorder) are often found to be associated with alcohol and drug problems. Several investigators have noted that sociopathic behavior in children predicts a tendency toward antisocial personality behavior, alcohol abuse, and drug problems later in life. An analysis of 40 studies showed a strong positive correlation between alcoholism and drug abuse, between alcoholism and antisocial personality, and between drug abuse and antisocial personality.[44]

Although there is little known about the genetics of cocaine dependence, extensive scientific data are available on the effects of cocaine on brain chemistry. The current view is that the system that uses dopamine in the brain plays an important role in the pleasurable effects of cocaine. In animals, for example, the principal location where cocaine takes effect is the dopamine D2 receptor gene on chromosome 11.[10] Recently, Caine and Koob found evidence suggesting that the dopamine D3 receptor gene is a primary site of cocaine effects.[45] The exact effect of cocaine on gene expression is unknown. However, we do know that D2 receptors are decreased by chronic cocaine administration, and this may induce severe craving for cocaine and possibly cocaine dreams.

A recent study by Noble et al.[46] found that about 52% of cocaine addicts have the A1 allele of the dopamine D2 receptor gene, compared to only 21% of nonaddicts. The prevalence of the A1 allele increases significantly with three risk factors: parental alcoholism and drug abuse, the potency of the cocaine used by the addict (intranasal vs. "crack" cocaine), and early-childhood deviant behavior such as conduct disorder. In fact, if the cocaine addict has three of these risk factors, the prevalence of the A1 allele rises to 87%. These findings suggest that childhood behavioral disorders may signal a genetic predisposition to drug or alcohol addiction.

A recent survey by the National Institute of Drug Abuse of five independent studies showed that the A1 allele is also associated with polysubstance dependence.[47] The A1 allele is also associated with an increase in the amount of money spent for drugs by polysubstance-dependent people.[48]

Although not viewed in the same light as the use of cocaine and other illicit drugs, cigarette smoking is another form of chemical addiction. Most attempts to stop smoking are associated with withdrawal symptoms typical of the other chemical addictions. Although environmental factors may be important determinants of cigarette use, there is strong evidence that the acquisition of the smoking habit and its persistence are strongly influenced by hereditary factors.

Of particular significance are studies of identical twins, which show that when one twin smokes, the other tends to smoke. This is not the case in nonidentical twins. In one twin study, Swan and associates[49] examined a national sample of male twins who were veterans of World War II. A unique aspect of this study was that the twins were surveyed twice, once in 1967–1968 and again 16 years later. This allowed an examination of genetic factors in all aspects of smoking — initiation, maintenance, and quitting. In general, whatever happened to one identical twin happened to the other — including the long-term pattern of not smoking, smoking, and then quitting smoking. The absence of these similarities in a control population of nonidentical twins suggests a strong biogenetic component in smoking behavior.[49]

Animal studies have suggested that the dopaminergic pathways of the brain may be involved. For example, the administration of nicotine to rodents disturbs dopamine metabolism in the reward centers of the brain to a greater extent than does the administration of alcohol.

With this in mind, Comings and colleagues[50] investigated the incidence of the A1 allele in a population of Caucasian smokers. These smokers did not abuse alcohol or other drugs, but had made at least one unsuccessful attempt to stop smoking. It turned out that 48% of the smokers carried the A1 allele. The higher the prevalence of the A1 allele, the earlier had been the age of onset of smoking, the greater the amount of smoking, and the greater the difficulty experienced in attempting to stop smoking.[50] In another sample of Caucasian smokers and nonsmokers, Noble and colleagues found that the prevalence of the A1 allele was highest in current smokers, lower in those who had stopped smoking, and lowest in those who had never smoked.[51]

E. COMPULSIVE BINGEING AND GAMBLING

Obesity is a disease that comes in many forms. Once thought to be primarily environmental, it is now considered to have both genetic and environmental components. In a Swedish adoption study, for example, the weight of the adult adoptees was strongly related to the body-mass index of the biological parents and to the body-mass index of the adoptive parents. The links to both genetic and environmental factors were dramatic. Other studies of adoptees and twins suggest that heredity is an important contributor to the development of obesity, whereas childhood environment has little or no influence. Moreover, the distribution of fat around the body has also been found to have heritable elements. The inheritance of subcutaneous fat distribution is genetically separable from body fat stored in other compartments (among the viscera in the abdomen, for example). It has been suggested there is evidence for both single and multiple gene anomalies.[52]

Given the complex array of metabolic systems that contribute to overeating and obesity, it is not surprising that a number of neurochemical defects have been implicated. Indeed, at least three such genes have been found: one associated with cholesterol production, one with fat transport, and one related to insulin production.[51] The ob gene and its product the leptin protein have also been implicated in regulating long-term eating behavior.[53] Most recently, another protein, glucagon-like peptide 1 (GLP-1), has been found to be involved in the regulation of short-term eating behavior.[54] The relationship between leptin and GLP-1 is not known. The ob gene may be involved in the animal's selection of fat, but perhaps not in the ingestion of carbohydrates, which appears to be regulated by the dopaminergic system. It may be that the ob gene is functionally linked to the opioid peptodergic systems involved in reward.

Whatever the relationship between these systems, the complexity of compulsive eating disorders suggests that more than one defective gene is involved. Indeed, the relationship between compulsive overeating and drug and alcohol addiction is well documented.[55,56] Neurochemical studies show that pleasure-seeking behavior is a common denominator of addiction to alcohol, drugs, and carbohydrates.[57] Alcohol, drugs, and carbohydrates all cause the release of dopamine in the primary reward area of the brain, the nucleus accumbens. Although the precise localization and specificity of the pleasure-inducing properties of alcohol, drugs, and food are still debated, there is general agreement that they work through the dopaminergic pathways of the brain. Other studies suggest the involvement of at least three other neurotransmitters, serotonin, GABA, and the opioid peptides.

Variants of the dopamine D2 receptor gene appear to be risk factors in obesity. The A1 allele was present in 45% of obese subjects as compared to 19% of nonobese subjects.[58] Furthermore, the A1 allele was not associated with a number of other metabolic and cardiovascular risks, including elevated levels of cholesterol and high blood pressure. In contrast, when the subject's profile included factors such as parental obesity and a later onset of obesity and carbohydrate preference, the prevalence of the A1 allele rose to 85%. More recently, another study found a significant association between genetic variants of the D2 receptor and obese subjects.[59]

There is also an increased prevalence of the A1 allele in obese subjects who have severe alcohol and drug dependence.[60] When obesity, alcoholism, and drug addiction were found in a patient, the incidence of the A1 allele rose to 82%. In contrast, the allele had an incidence of zero % in nonobese patients who were also not substance abusers and did not have a family history of substance abuse. The presence of the dopamine D2 receptor gene variants increases the risk of obesity and related behaviors.

Pathological gambling — in which an individual becomes obsessed with the act of risking money or possessions for greater "payoffs" — occurs at a rate of less than 2% in the general population. Although it is the most socially acceptable of the behavioral addictions, pathological gambling has many affinities to alcohol and drug abuse. Clinicians have remarked on the similarity between the aroused euphoric state of the gambler and the "high" of the cocaine addict or substance abuser. Pathological gamblers express a distinct craving for the "feel" of gambling; they develop tolerance in that they need to take greater risks and make larger bets to reach a desired level of excitement, and they experience withdrawal-like symptoms (anxiety and irritability) when no "action" is available.[61] Indeed, there is a typical course of progression through four stages of the compulsive-gambling syndrome: winning, losing, desperation, and hopelessness — a series not uncommon to other addictive behaviors.

Might the dopamine pathways in the brain be involved with pathological gambling? A recent study of Caucasian pathological gamblers found that 50.9% carried the A1 allele of the dopamine D2 receptor.[62] The more severe the gambling problem, the more likely it was that the individual was a carrier of the A1 allele. Finally, in a population of males with drug problems who were also pathological gamblers the incidence of the A1 allele rose to 76%.

F. ATTENTION DEFICIT DISORDER

This disorder is most commonly found among school-age boys, who are at least four times more likely to express the symptoms than are young girls. These children have difficulty applying themselves to tasks that require a sustained mental effort, they can be easily distracted, they may have difficulty remaining seated without fidgeting, and they may impulsively blurt out answers in the classroom or fail to wait their turn. Although normal children occasionally display these symptoms, attention deficit disorder is diagnosed when the behavior's persistence and severity impedes the child's social development and education.

Early speculation about the causes of attention deficit disorder focused on potential sources of stress within the child's family, including marital discord, poor parenting, psychiatric illness, alcoholism, or drug abuse. It has become progressively clear, however, that stress within the family cannot explain the incidence of the disorder. There is now little doubt that the disorder has a genetic basis.

Evidence in support of this notion comes from patterns of inheritance in the families of children with the disorder and from studies of identical twins. For example, consider instances in which full siblings and half-siblings (who have only half of the genetic identity of full siblings) are both raised in the same family environment. If the behavioral symptoms of attention deficit disorder were "learned" in the family, then the incidence of the disorder should be the same for full siblings as it is for half-siblings. In fact, half-siblings of children with attention deficit disorder have a significantly lower frequency of the disorder than full siblings.[63] In another study, investigators found that if one identical twin had attention deficit disorder, there was a 100% probability that the other also had the disorder. In contrast, the incidence of concordance among nonidentical twins was only 17%. This result has been supported by two other independent studies of identical twins.[64] Finally, Comings and coworkers found that the A1 allele of the dopamine D2 receptor gene was present in 49% of the children with attention deficit disorder compared to only 27% of the controls.[65]

Some other recent work has linked attention deficit disorder with another impulsive disorder: Tourette's disorder. More than 100 years ago the French neurologist Giles de la Tourette described a condition that was characterized by compulsive swearing, multiple muscle tics, and loud noises. He found the disorder usually appeared in children between 7 to 10 years old, with boys more likely to be affected than girls. Tourette suggested that the condition might be inherited.

In the early 1980s Comings and colleagues studied 246 families in which at least one member of the family had Tourette's disorder. The study indicated that virtually all cases of Tourette's disorder are genetic.[65] Subsequent studies also found there was a high incidence of impulsive, compulsive, addictive, mood, and anxiety disorders on both sides of the affected individual's family.[66,67] The A1 allele was implicated in a recent report showing that nearly 45% of the people diagnosed with Tourette's disorder carried the aberrant gene.[65] Moreover, the A1 allele had the highest incidence among people who had the severest manifestations of the disorder.

As mentioned earlier, Tourette's disorder appears to be tightly coupled to attention deficit disorder. In studies of the two disorders, it was found that 50 to 80% of the people with Tourette's disorder also had attention deficit disorder. Furthermore, an increased number of relatives of individuals with Tourette's disorder also had attention deficit/hyperactivity disorder.[68] It now appears that Tourette's disorder is a complex illness that may include attention deficit disorder, conduct disorder, as well as obsessive, compulsive, and addictive disorders and other related disorders. The close coupling between these disorders has led others to propose that Tourette's disorder is a severe form of attention deficit disorder.[69,70]

The high frequency of the A1 allele among people with Tourette's disorder and attention deficit disorder raises the question of whether other genes affecting dopaminergic function might also be involved in these disorders. Two others that have been considered are the gene for the enzyme dopamine B-hydroxylase which converts dopamine to norepinephrine, and the gene for the dopamine transporter which takes dopamine back into the presynaptic terminal after it is released into the synapse. In both cases, variant forms of these genes are associated with Tourette's disorder.[71] The anomalous dopamine B-hydroxylase gene (the "DBH *Taq* B_1" allele) was further associated with learning disabilities, conduct disorder, and substance abuse, whereas the variant of the dopamine transporter (the "10 repeat" allele) was associated with alcohol abuse, depression, and obsessive-compulsive disorder. This observation was supported by other work showing that the 10 repeat allele for the dopamine transporter gene was associated with attention deficit/hyperactivity disorder.[72] Moreover, elevated levels of the dopamine transporter molecule have been found in the brains of patients with Tourette's disorder.[73]

If these dopamine-related molecules are indeed associated with various behavioral disorders, it might be expected that having more than one variant would increase the severity or the likelihood of having a disorder. Indeed, this is the case: the severity of attention deficit disorder, conduct disorder, substance abuse, and mood disorders progressively increased from individuals carrying none of the genes to those who carried all three genes.[71]

Given the widespread prevalence of attention deficit disorder among children, and its frequent association with alcoholism, drug dependence, and other behavioral disorders, it may be that childhood attention deficit disorder is a predisposing cause to various disorders among adults. For example, there is a significant correlation between attention deficit/hyperactivity disorder and adult drug abuse.[74]

G. THE DOPAMINE D2 RECEPTOR

The A1 allele carries a behavioral risk factor that shows up not only in substance addiction and attention deficit disorder, but also antisocial behavior, conduct disorder, and violent or aggressive behavior. In a recent study the A1 allele was present in 60% of a sample population of young adolescents between 12 and 18 years old who were diagnosed as "pathologically violent" subjects. A variant of the dopamine transporter gene (VENT 10 repeat) was present in 100% of the adolescents. Of these, 70% had the so-called 10/10 form whereas 30% carried the 10/9 allelic form. Another study found that 59% of Vietnam veterans with posttraumatic stress disorder also carried the A1 allele, compared to only 5% of veterans who were exposed to similar stress but did not develop the disorder.[75]

Why would carriers of the A1 allele be predisposed to the spectrum of disorders associated with the reward deficiency syndrome? Individuals having the A1 allele have approximately 30% fewer D2 receptors than those with the A2 allele.[76] Since the D2 receptor gene controls the production of these receptors, the finding suggests that the A1 allele is responsible for the reduction in receptors. In some way that we do not yet understand, carrying the A1 allele reduces the expression of the D2 gene compared to carrying the A2 allele. Perhaps a regulatory site for the D2 receptor gene is affected in A1 carriers (Figure 18-4).

Fewer numbers of dopamine D2 receptors in the brains of A1 allele carriers may translate into lower levels of dopaminergic activity in those parts of the brain involved in reward. A1 carriers may not be sufficiently rewarded by stimuli that A2 carriers find satisfying. This may translate into the persistent cravings or stimulus-seeking behavior of A1 carriers. Moreover, because dopamine is known to reduce stress, individuals who carry the A1 allele may have difficulty coping with the normal pressures of life. In response to stress or cravings, A1 carriers may turn to other substances or activities that release additional quantities of dopamine in an attempt to gain temporary relief. Alcohol, cocaine, marijuana, nicotine, and carbohydrates (like chocolate) all cause the release of dopamine in the brain and bring about a

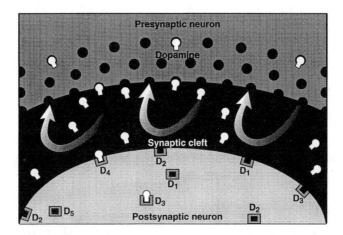

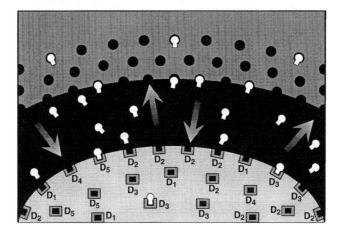

FIGURE 18-4 Individuals who carry the A1 allele (top) of the dopamine D2 receptor gene have a lower density of dopamine D2 receptors compared to individuals who carry the A2 allele (bottom). The authors propose that a decreased number of dopamine D2 receptors in the reward pathways of the brain results in anger, anxiety and a craving for substances such as cocaine, alcohol, or nicotine, that increase the release of the neurotransmitter dopamine in the brain.

temporary relief of craving. These substances can be used singly, in combination, or to some extent interchangeably.

Although we believe that the gene for the D2 receptor plays a critical role in reward deficiency syndrome, other genes (such as the dopamine transporter gene) are undoubtedly involved in the different manifestations of the syndrome. Scientists from Israel and the National Institute of Mental Health recently showed that a genetic variation of the dopamine D4 receptor gene is associated with people who are novelty (or sensation) seekers.[77,78] Both studies set out to test the hypothesis advanced by Robert Cloninger of Washington University (see Chapter 21) that novelty-seeking behavior is modulated by the way brain cells process dopamine. Epstein and colleagues[77] found that novelty seekers — who tended to be compulsive, exploratory, fickle, excitable, quick-tempered, and extravagant — were much more likely to have a longer version of the receptor gene than individuals who were not novelty seekers. Subjects with the shorter version of the gene scored lower on a test of novelty seeking and tended to be reflective, rigid, loyal, stoic, slow-tempered, and frugal. Benjamin and colleagues[78] found similar results in their sample of 315 American subjects.

The work from the laboratories of Benjamin and Epstein provide support of the earlier work of George and associates[79] who found a strong association between variants of the D4 gene and alcoholism and nicotine dependence.[79] The D2 receptor gene and the D4 receptor gene have fairly similar nucleotide sequences and may have similar physiological functions. In this respect, it is intriguing that Compton et al.[80] found an association between the A1 allele and individuals who were classified as "low sensation seekers" and who were characterized by agitation, impulsivity, excitability, and neurotic. All of these studies further support a connection between the reward deficiency syndrome and the dopaminergic system.

H. TREATMENT

In the U.S. alone there are 18 million alcoholics, 28 million children of alcoholics, 6 million cocaine addicts, 14.9 million people who abuse other substances, 25 million people addicted to nicotine, 54 million people who are at least 20% overweight, 3.5 million school-age children with attention deficit disorder or Tourette's disorder, and about 448,000 compulsive gamblers. We believe that recognizing the role of dopamine and the D2 receptor in the manifestation of these addictions and disorders is the first step toward rational treatment for a devastating problem in our society.

There is reason to believe that a pharmacological approach could help people with reward deficiency syndrome. It is tempting to speculate that the pharmacological sensitivity of alcoholics to dopaminergic agonists (bromocryptine, bupropion, and n-propyl-nor-apomorphine) may be partly determined by the individual's D2 genotype. We predict that A1 carriers should be pharmacologically more responsive to D2 agonists, especially in the treatment of alcoholics or stimulant-dependent people. At least one study has already shown that the direct microinjection of the D2 agonist n-propyl-nor-apomorphine into the rat nucleus accumbens significantly suppresses the animal's symptoms after the withdrawal of opiates.[81]

A recent double-blind study demonstrates the utility of this approach in human subjects.[92] The D2 agonist bromocryptine or a placebo was administered to alcoholics who were carriers of the A1 allele (A1/A1 and A1/A2 genotypes) or who only carried the A2 allele (A2/A2). The greatest improvement in the reduction of craving and anxiety was found among the A1 carriers who were treated with bromocryptine. The attrition rate was highest among the A1 carriers who were treated with the placebo (Figure 18-5).

These findings provide an important rationale for DNA testing to detect genetic variants for the D2 receptor or other dopamine-related genetic variants in the tertiary treatment of alcoholism. Unlike certain other complex disorders such as Alzheimer's disease, the early identification and treatment of alcohol and drug abuse can occasionally alter the devastating course of these addictions. Consider the successes of self-help programs such as Alcoholics Anonymous and Narcotics Anonymous, psychopharmacological adjunctive therapy, neuroregulation or brain-wave training, and electrophysiological stimulation. Identifying individuals with the A1 allele offers the possibility of helping individuals before alcoholism or substance abuse affect their lives. We foresee the possibility for better treatment, new forms of prevention, and the removal of the social stigma attached not only to alcoholism but also to related "reward-seeking" behaviors comprising the reward deficiency syndrome[82] based on a dysfunction in the "reward cascade" involving neurotransmitter interactions compromised by both genes and the environment[84] (Figure 18-6).

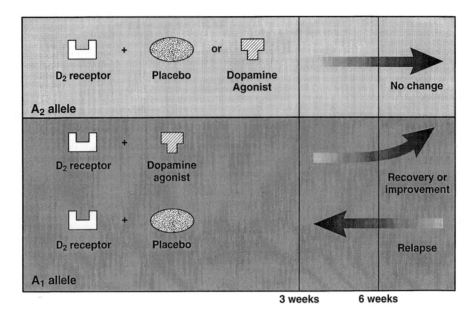

FIGURE 18-5 Effectiveness of dopamine agonists in the treatment of certain forms of alcoholism may depend on the individual's genotype for the dopamine D2 receptor gene. The authors propose that alcoholics who carry the A1 allele are more likely to respond positively to treatment with a dopamine agonist (such as bromocryptine). However, if such individuals are treated with a placebo they are more likely to relapse into alcoholism. Alcoholics with the A2/A2 genotype do not respond to dopamine agonists (or to a placebo) because their alcoholism is not associated with the dopamine D2 receptor. The authors suggest that the use of dopamine agonists to treat alcoholics with the A1 allele initiates a feedback system that produces more dopamine receptors after a period of about 6 weeks.

Reward Deficiency Syndrome			
Addictive behavior	Impulsive behavior	Compulsive behavior	Personality Disorder
Severe alcoholism	Attention-deficit disorder, hyperactivity	Aberrant sexual behavior	Conduct disorder
Polysubstance abuse	Tourette syndrome		Antisocial personality
Smoking		Pathological gambling	
Obesity	Autism		Aggressive behavior

FIGURE 18-6 The reward deficiency syndrome comprises a spectrum of impulsive, compulsive, addictive, and personality disorders that are based on a common genetic deficiency in the dopamine D2 receptor, according to the authors. The type of disorder manifested by any particular individual is determined by other genetic and environmental factors, which are not yet fully understood. A predictive model based on Bayes' Theorem suggests that an individual who carries the A1 allele for the dopamine D2 receptor has a 74% chance of developing one of the disorders of the reward deficiency syndrome.[2]

REFERENCES

1. Blum, K., Noble, E. P., Sheridan, P. J., Montgomery, A., Ritchie, T., Jagadeeswaran, P., Nogami, H., Briggs, A. H., and Cohn, J. B., Allelic association of human dopamine D2 receptor gene in alcoholism, *JAMA*, 263(15), 2055, 1990.
2. Blum, K., Cull, J. G., Braverman, E. R., and Comings, D. E., Reward deficiency syndrome, *Am. Sci.*, 84, 132, 1996.
3. Olds, J. and Milner, P., Positive reinforcement produced by electrical stimulation of septal area and other regions of rat brain, *J. Comp. Physiol. Psychol.*, 419, 4766, 1954.
4. Olds, J., The limbic system and behavioral reinforcement, *Prog. Brain Res.*, 27, 44, 1967.
5. Olds, M. E. and Olds, J., Effects of lesions in medial forebrain bundle on self-stimulation behavior, *Am. J. Physiol.*, 217(5), 1253, 1969.
6. Hall, R. D., Bloom, F. E., and Olds, J., Neuronal and neurochemical substrates of reinforcement, *Neurosci. Res. Prog. Bull.*, 15(2), 131, 1977.
7. Cloninger, C. R., Genetic and environmental factors in the development of alcoholism, *J. Psychiatr. Treat. Eval.*, 5, 487, 1983.
8. Blum, K. and Kozlowski, G. P., Ethanol and neuromodulator interactions: a cascade model of reward, *Prog. Alcohol Res.*, 2, 131, 1990.
9. Wise, R. A. and Bozarth, M. A., Brain reward circuitry: four circuit elements "wired" in apparent series, *Brain Res. Bull.*, 297, 265, 1984.
10. Koob, G. F. and Bloom, F. E., Cellular and molecular mechanisms of drug dependence., *Science*, 242, 715, 1988.
11. Stein, L. and Belluzzi, J. D., Second messenger, natural rewards, and drugs of abuse, *Clin. Neuropharmacol.*, 9(4), 205, 1986.
12. Blum, K., A commentary on neurotransmitter restoration as a common mode of treatment for alcohol, cocaine and opiate abuse, *Integr. Psychiatry*, 6, 199, 1989.
13. Alvaksinen, M. N., Saano, V., Juvonene, H., Huhtikangas, A., and Gunther, J., Binding of beta-carbolines and tetrahydroisoquinolines by opiate receptors of the d-type, *Acta Pharmacol. Toxicol.*, 55, 380, 1984.
14. Russell, V. A., Lanin, M. C. L., and Taljaard, J. F., Effect of ethanol on ^{3}H-dopamine release in rat nucleus accumbens and striatal slices, *Neurochem. Res.*, 13, 487, 1988.
15. McBride, W. J., Guan, X. M., Chernet, E., Lumeng, L., and Li, T.-K., Regional differences in the densities of serotonin 1A receptors between P and NP rats, *Alcoholism: Clin. Exp. Res.*, 14, 316, 1990.
16. McBride, W. J., Chernet, E., Dyr, W., Lumeng, L., and Li, T.-K., Densities of dopamine D2 receptors are reduced in CNS regions of alcohol-preferring P rats, *Alcohol*, 10, 387, 1993.
17. Zhou, F. C., Bledsoe, S., Lumeng, L., and Li, T.-K., Serotonergic immuno-stained terminal fibers are decreased in selected brain areas of alcohol-preferring P rats, *Alcoholism: Clin. Exp. Res.*, 14, 355, 1990.
18. McBride, W. J., Bodart, B., Lumeng, L., and Li, T.-L., Association between low contents of dopamine and serotonin in the nucleus accumbens and high alcohol preference, *Alcoholism: Clin. Exp. Res.*, 19, 1420, 1996.
19. Campbell, A. D. and McBride, W. J., Serotonin-3 receptor and ethanol-stimulated dopamine release in the nucleus accumbens, *Pharmacol. Biochem. Behav.*, 51, 835, 1995.
20. Dyr, W., McBride, W. J., Lumeng, T. K., and Murphy, J. M., Effects of D1 and D2 dopamine receptor agents on ethanol consumption in the high-alcohol-drinking (HAD) line of rats, *Alcohol*, 10, 207, 1993.
21. Brown, R. J., Blum, K., and Trachtenberg, M. C., Neurodynamics of relapse prevention: a neuronutrient approach to outpatient DUI offenders, *J. Psychoactive Drugs*, 22, 173, 1990.
22. Blum, K. and Trachtenberg, M. C., Neurogenic deficits caused by alcoholism: restoration by SAAVE™, *J. Psychoactive Drugs*, 20, 297, 1988.
23. Blum, K., Trachtenberg, M. C., Elliott, C. E., Dingler, M. L., Sexton, R. L., Samuels, A. I., and Cataldie, L., Enkephalinase inhibition and precursor amino acid loading improves inpatient treatment of alcohol and polydrug abusers: double-blind placebo-controlled study of the nutritional adjunct SAAVE™, *Alcohol*, 5, 481, 1989.
24. Ortiz, J., Fitzgerald, L. W., Charlton, M., Lane, S., Trevisan, L., Guitart, X., Shoemaker, W., Duman, R. S., and Nestler, E. J., Biochemical actions of chronic ethanol exposure in the mesolimbic copamine system, *Synapse*, 21, 289, 1995.
25. Blum, K. and Holder, J., *The Disease "Precept" of the Reward Deficiency Syndrome*, Gardner Press, Palm Springs, FL, 1996.
26. Blum, K., *Alcohol and the Addictive Brain*, Free Press, New York, 1991.
27. McClearn, G. E. and Rodgers, D. A., Differences in alcohol preferences among inbred strains of mice, *Q. J. Stud. Alcohol*, 20, 691, 1959.
28. Schuckit, M. A., Goodwin, D. W., and Winokur, G., A study of alcoholism in half-siblings, *Am. J Psychiatr.*, 128, 1132, 1972.
29. Goodwin, D. W., Schulsinger, F., Hermansen, L., Guze, S. B., and Winokur, G., Alcohol problems in adoptees raised apart from alcoholic biological parents, *Arch. Gen. Psychiatry*, 28, 238, 1973.

30. Cloninger, C. R., Bohman, M., and Sigvardsson, S., Inheritance of alcohol abuse: cross-fostering analysis of adopted men, *Arch. Gen. Psychiatry*, 38, 861, 1981.
31. Goodwin, D. S., Alcoholism and heredity, *Arch. Gen. Psychiatry*, 36, 57, 1979.
32. Sibley, D. and Monsma, F. J., Molecular biology of dopamine receptors, *Trends Pharmacol. Sci.*, 13, 61, 1992.
33. Grandy, D. K., Lih, M., Allen, L., Bunzow, J. R., Marchionni, M., Makam, H., Reed, L., Magenis, R. E., and Civelli, D., The human dopamine D2 receptor gene is located on chromosome 11 at q22-q23 and identified as *Taq I* RFLP, *Am. J. Hum. Genet.*, 45, 778, 1989.
34. Neiswanger, K., Kaplan, B. B., and Hill, S. Y., What can the DRD2/alcoholism story teach us about association studies in psychiatric genetics, *Am. J. Med. Genet. (Neuropsychiatr. Genet.)*, 60, 272, 1995.
35. Blum, K. and Noble, E. P., The sobering D2 story, *Science*, 265, 1346, 1994.
36. Begleiter, H., Porjesz, B., Bihari, B., and Kissin, B., Event-related brain potentials in boys at risk for alcoholism, *Science*, 225, 1493, 1984.
37. Whipple, S. C., Parker, E. S., and Noble, E. P., An atypical neurocognitive profile in alcoholic fathers and their sons, *J. Stud. Alcohol*, 49, 240, 1988.
38. Berman, S. M., Whipple, S. C., Fitch, R. J., and Noble, E. P., P3 in boys as a predictor of adolescent substance use, *Alcohol*, 10, 69, 1993.
39. Noble, E. P., Berman, S. M., and Ozkaragoz, T. Z., Prolonged P300 latency in children with the D2 dopamine receptor A1 allele, *Am. J. Hum. Genet.*, 54, 658, 1994.
40. Blum, K., Braverman, E. R., Dinardo, M. J., Wood, R. C., and Sheridan, P. J., Prolonged P300 latency in a neuropsychiatric population with the D2 dopamine receptor A1 allele, *Pharmacogenetics*, 4, 313, 1994.
41. Cadoret, R. J., Froughton, E., O'Gorman, T., and Heywood, E., An adoption study of genetic and environmental factors in drug abuse, *Arch. Gen. Psychiatry*, 43, 1131, 1986.
42. Miller, N. S., Gold, M. S., Belkin, B. M., and Klahr, A. L., The diagnosis of alcohol and cannabis dependence in cocaine dependents and alcohol dependence in their families, *Br. J. Addict.*, 84, 1491, 1989.
43. Wallace, B. C., Crack cocaine smokers as adult children of alcoholics. The dysfunctional family link, *J. Substance Abuse Treat.*, 7, 89, 1990.
44. Schubert, D. S. P., Wolf, A. W., Patterson, M. B., Grande, T. P., and Pendleton, L., A statistical evaluation of the literature regarding the associations among alcoholism, drug abuse and antisocial personality disorder, *Int. J. Addict.*, 23, 797, 1988.
45. Caine, S. B. and Koob, G. F., Modulation of cocaine self-administration in the rat through D-3 dopamine receptors, *Science*, 260, 1814, 1993.
46. Noble, E. P., Blum, K., Khalsa, M. E., Ritchie, T., Montgomery, A., Wood, R.C., Fitch, J., Ozkaragoz, T., Sheridan, P.J., Anglin, M.D., Paredes, A., Treiman, J., and Sparks, R.S., Allelic association of the D2 dopamine receptor gene with cocaine dependence, *Drug Alcohol Depend.*, 33(3), 271, 1993.
47. Uhl, G., Blum, K., Noble, E. P., and Smith, S., Substance abuse vulnerability and D2 receptor genes, *Trends Neurosci.*, 16(3), 83, 1993.
48. Comings, D. E., Muhleman, D., Ahn, C., Gysin, R., and Flanagan, S. D., The dopamine D2 receptor gene: a genetic risk factor in substance abuse, *Drug Alcohol Depend.*, 34, 175, 1994.
49. Swan, G. E., Carmelli, D., Rosenman, R. H., Fabsitz, R. R., and Christian, J. C., Smoking and alcohol consumption in adult male twins: genetic heritability and shared environmental influence, *J. Substance Abuse*, 2, 39, 1990.
50. Comings, D. E., Ferry, L., Bradshaw-Robinson, S., Burchette, R., Chiu, C., and Muhleman, D., The dopamine D2 receptor (DRD2) gene: a genetic risk factor in smoking, *Pharmacogenetics*, 1996.
51. Noble, E. P., Jeor, S. T., Ritchie, T., Syndulko, K., Jeor, S. C., Fitch, R. J., Brunner, R. L., and Sparkes, R. S., D2 dopamine receptor gene and cigarette smoking: a reward gene?, *Med. Hypothesis*, 42, 257, 1994.
52. Bouchard, C., Genetics of obesity: an update on molecular markers, *Int. J. Obesity*, 19(3), S10, 1995.
53. Zhang, X., Proenca, R., Barone, M., Leopold, L., and Friedman, J. M., Positional cloning of the mouse obese gene and its human homologue, *Nature*, 372(6505), 425, 1994.
54. Turton, M. D., O'Shea, D., Gunn, I., Beak, S. A., Edwards, C. M. B., Meeran, K., Choi, S. J., Taylor, G. M., Heath, M. M., Lambert, P. D., Wilding, J. P. H., Smith, D. M., Gahel, M. A., Herbert, J., and Bloom, S. R., A role for glucagon-like peptide-1 in the central regulation of feeding, *Nature*, 379, 60, 1996.
55. Krahn, D., The relationship of eating disorders and substance abuse, *J. Stud. Alcohol*, 3, 239, 1991.
56. Newman, M. M. and Gold, M. S., Preliminary findings of patterns of substance abuse in eating, *Am. J Med. Genet., (Neuropsychiatr. Genet.)*, 18, 207, 1992.
57. Blum, K., Trachtenberg, M. C., and Cook, D. W., Neuronutrient effects on weight loss in carbohydrate bingers: an open clinical trial, *Curr. Ther. Res.*, 48(2), 217, 1990.
58. Syndul Ko, K., Bohlman, M. C., Noble, L. A., Zhang, Y., Sparkes, R. S., and Grandy, D. K., Allelic association of the human D2 dopamine receptor gene with obesity, *Int. J. Eating Disord.*, 15(3), 205, 1994.
59. Comings, D. E., Flanagan, S. D., Dietz, G., Muhleman, D., Knell, E., and Gysin, R., The dopamine D2 receptor (DRD2) as a major gene in obesity and height, *Biochem. Med. Metab. Biol.*, 50, 176, 1993.

60. Blum, K., Braverman, E. R., Wood, R. G., Gill, J., Li, C., Chen, T. J. H., Taub, M., Montgomery, A. R., Cull, J. G., and Sheridan, P. J., Increase prevalence of the *Taq I* A1 allele of the dopamine receptor gene (DRD2) in obesity with comorbid substance use disorder: preliminary findings, *Pharmacogenetics*, 1995 (in press).
61. Volberg, R. A. and Steadman, H. J., Refining prevalence estimates of pathological gambling, *Am. J. Psychiatry*, 145, 502, 1988.
62. Comings, D. E., Rosenthal, R. J., Leiseur, H. R., Rugle, L., Muhleman, D., Chiu, C., Dietz, F., and Gane, R., The molecular genetics of pathological gambling: the DRD2 gene, *Pharmacogenetics*, 1996 (in press).
63. Lopez, R., Hyperactivity in twins, *Can. Psychol. Assoc.*, 10, 421, 1965.
64. Willerman, L., Activity level and hyperactivity in twins, *Child Dev.*, 44, 288, 1973.
65. Comings, D. E., Comings, B. G., Muhleman, D., Deitz, G., Shahbahrami, B., Tast, D., Knell, E., Kocsis, P., Baumgarten, R., Kovacs, B. W., Levy, D. L., Smith, M., Kane, J. M., Lieberman, J. A., Klein, D. N., MacMurray, J., Tosk, J., Sverd, J., Gysin, R., and Flanagan, S., The dopamine D2 receptor locus as a modifying gene in neuropsychiatric disorders, *JAMA*, 266, 1793, 1991.
66. Comings, B. G. and Comings, D. E., A controlled study of Tourette syndrome. V. Depression and mania, *Am. J. Hum. Genet.*, 41, 804, 1987.
67. Comings, D. E., *Tourette Syndrome and Human Behavior*, Hope Press, Duarte, CA, 1990.
68. Knell, E. and Comings, D. E., Tourette syndrome and attention deficit hyperactivity disorder: evidence for a genetic relationship, *J. Clin. Psychiatry*, 54, 331, 1993.
69. Comings, D. E. and Comings, B. G., A controlled family history study of Tourette syndrome. I. Attention deficit hyperactivity disorder, learning disorders and dyslexia, *J. Clin. Psychiatry*, 1989.
70. Comings, D. E., Tourette syndrome: a hereditary neuropsychiatric spectrum disorder, *Ann. Clin. Psychiatry*, 6, 235, 1995.
71. Comings, D. E., Wu, H., Chiu, C., Ring, R. H., Deitz, G., and Muhleman, D., Polygenic inheritance of Tourette syndrome, stuttering, attention deficit hyperactivity, conduct and oppositional defiant disorder: the additive and subtractive effect of three dopaminergic genes — DRD2, DbetaH and DATA, *Am. J. Med. Genet. (Neuropsychiatr. Genet.)*, in press.
72. Cook, E. H., Stein, M. A., Drajowsi, M. D., Cox, W., Olkon, D. M., Kieffer, J. E., and Leventhal, B. L., Association of attention deficit disorder and the dopamine transporter gene, *Am. J. Hum. Genet.*, 56, 993, 1995.
73. Malison, R. T., McDougle, C. J., van Dyck, C. H., Scahill, L., Baldwin, R. M., Seibyle, J. P., Price, L. H., Leckman, J. F., and Innis R. B., [^{123}I]b-CIT SPECT imaging of striatal dopamine transporter binding in Tourette's disorder, *Am. J. Psychiatry*, 152, 1359, 1995.
74. Gittleman, R., Mannuzza, S., Shenker, R., and Bonagura, N., Hyperactive boys almost grown up. I. Psychiatric status, *Arch. Gen. Psychiatry*, 42, 937, 1985.
75. Comings, D. E., Muhleman, D., and Gysin, R., The dopamine D2 receptor (DRD2) gene in posttraumatic stress disorder: a study and replication, *Biol. Psychiatry*, in press.
76. Noble, E. P., Blum, K., Ritchie, T., Montgomery, A., and Sheridan, P. J., Allelic association of the D2 dopamine receptor gene with receptor binding characteristics in alcoholism, *Arch. Gen. Psychiatry*, 48(7), 648, 1991.
77. Ebstein, R. P., Novick, O., Umansky, R., Priel, B., Osher, Y., Blaine, D., Bennett, E., Nemanov, L., Katz, M., and Belmaker, R., Dopamine D4 receptor (D4DR) exon III polymorphism associated with the human personality trait of Novelty Seeking, *Nat. Genet.*, 12, 78, 1996.
78. Benjamain, J., Lin, L., Patterson, C., Greenberg, B. D., Murphy, D. L., and Hamer, D. H., Population and familial association between the D4 dopamine receptor gene and measures of novelty seeking, *Nat. Genet.*, 12, 81, 1996.
79. George, S. R., Cheng, R., Nguyen, T., Israel, Y., and O'Dowd, B. F., Polymorphisms of the D4 dopamine receptor alleles in chronic alcoholism, *Biochem. Biophys. Res. Commun.*, 196(1), 107, 1993.
81. Compton, P. A., Anglin, M. D., Khalsa-Denison, M. E., and Paredes, A., The D2 dopamine receptor gene, addiction and personality: clinical correlates in cocaine abusers, *Biol. Psychiatry*, 39, 302–304, 1996.
81. Harris, G. C. and Aston-Jones, G., Involvement of D2 dopamine receptors in the nucleus accumbens in the opiate withdrawal syndrome, *Nature*, 371(6493), 155, 1994.
82. Lawford, B. R., Young, R. M., Rowell, J., Qualichefski, J., Fletcher, B. H., Syndulko, K., Ritchie, T., and Noble, E. P., Bromocriptine in the treatment of alcoholics with the D2 dopamine receptor A1 allele, *Nat. Med.*, 1, 337, 1995.
83. Blum, K., Sheridan, P. J., Wood, R. G., Braverman, E. R., Chen, T. J. H., Cull, J. G., and Comings, D. E., The D2 dopamine receptor gene as a predictor of reward deficiency syndrome: Bayes theorem, *J. R. Soc. Med.*, in press.
84. Blum, K. and Kozlowski, G. P., Ethanol and neuromodulator interactions: a cascade model of reward, *Prog. Alcohol Res.*, 2, 131, 1990.

Section V

Substance Use Disorders

19 Polymorphisms of the D2 Dopamine Receptor Gene in Alcoholism, Cocaine and Nicotine Dependence, and Obesity

Ernest P. Noble

CONTENTS

A. Introduction 331
B. Initial Study of the DRD2 Gene in Alcoholism 332
C. Subsequent Studies of the DRD2 Gene in Alcoholism 333
D. The DRD2 Gene and Other Substance Use Disorders 336
    1. Cocaine Dependence 336
    2. Nicotine Dependence 338
    3. Obesity 340
E. Central Nervous System Expression of DRD2 Alleles 342
    1. Pharmacological Studies 342
    2. Neurophysiological Studies 343
    3. Neuropsychological Studies 343
F. The DRD2 Gene and Treatment of Alcoholics 344
G. Genetic Animal Models and the Dopaminergic System 345
H. The DRD2 Gene: A Reward Gene? 346
I. Conclusion 347
Acknowledgments 347
References 347

A. INTRODUCTION

It has been known since ancient times that alcoholism runs in families. However, it was not until the 19th century that investigators, for the first time, began systematic studies on alcoholics and their families. With very few exceptions, and regardless of country of origin, every family study of alcoholism revealed higher rates of alcoholism among the relatives of alcoholics when compared to the general population.[1]

Still, the question remains as to what factors in alcoholic families contribute to the development of alcoholism: genes or environment. This question began to be addressed directly about two decades ago when adoption and twin studies began to be applied to alcoholism. The results showed that virtually every adoption[2-5] and twin[6-10] study of alcoholics showed strong evidence for a hereditary component in alcoholism.

Like in alcoholism, hereditary factors have also been implicated in a variety of substance use problems. Moreover, recent studies on humans[9,11] and on animals[12] provide some evidence that hereditary factors are involved in the use and abuse of cocaine as well as other illicit drugs. Results from adoption, association, family, and twin studies indicate that the acquisition and maintenance of smoking are similarly influenced by heredity.[13] Of particular relevance are studies of twins, which show that concordance rates for smoking are consistently higher in monozygotic than dizygotic twins.[14-16] Finally, family, twin, and adoption studies[17-20] are also pointing to heredity as an important contributor to the development of obesity.

If alcoholism and other substance use disorders have a hereditary component, then they must have a molecular genetic basis. What might these genes be? Will they be specific for alcoholism? Or will they share a common diathesis with other substance use disorders? These questions and related issues are discussed in this chapter.

B. INITIAL STUDY OF THE DRD2 GENE IN ALCOHOLISM

The strong evidence accruing from twin, adoption, and other studies suggesting a hereditary component in alcoholism, and the advent of molecular genetic techniques for understanding hereditary disorders, prompted the author and co-workers to embark in 1988 on a search for a molecular genetic basis for alcoholism. However, before any studies were initiated, three major questions had to be addressed. (1) What genes should be considered? (2) What type of genetic analysis should be used? (3) What type of alcoholics should be studied?

Since a large number of previous studies had implicated the dopaminergic system in alcoholism as well as in a variety of other substance use disorders, the D2 dopamine receptor (DRD2) gene loomed as a strong candidate to explore. Moreover, previous research had also suggested the possible involvement of other biochemical systems in alcoholism. Therefore, we selected additional candidate genes for study, including those for alcohol dehydrogenase, protein kinase C, tryptophan hydroxylase, tyrosine hydroxylase, and monoamine oxidase.

Which type of genetic analysis to employ — linkage or association studies? In contrast to somatic disorders where linkage analysis had led to the identification of genes for Huntington's disease, cystic fibrosis, and many other inherited somatic diseases, this type of analysis has failed in identifying genes in complex behavioral disorders, including manic-depressive psychosis and schizophrenia. This is not surprising as psychiatric disorders are not simple Mendelian traits because a mode of transmission has not been demonstrated for them. Unlike Mendelian traits, genetic variance for behavior rarely accounts for more than half the phenotypic variance.[21] In part, contributing variables in the expression of behavioral disorders are environmental factors. Moreover, compounding these problems is the difficulty of determining which, if any, of the clinical types constitutes an etiologically homogeneous phenotype. It is for these reasons and others that we chose association studies, since a distinct advantage of this type of analysis is that no assumptions need to be made on the mode of inheritance, penetrance, associative mating, and age of onset of the disease.

The next issue that we addressed was which type of alcoholics should be considered for molecular genetic analysis. It is well recognized that alcoholism is not a homogeneous disorder, and the clinical literature is replete with studies showing the presence of different types of alcoholism. For example, Jellinek[22] has emphasized the distinction between alcoholics who had persistent alcohol-seeking behaviors (i.e., "inability to abstain entirely") and others who could abstain from alcohol for prolonged periods. Cloninger,[23] through empirical studies, has found that alcohol-seeking behaviors distinguish type 1 and type 2 alcoholism. Type 1, a less severe form of alcoholism, with necessary environmental provocative factors and hereditary background, is characterized by an **ability to abstain**. Type 2, a more severe form of alcoholism, with necessary hereditary background but irrespective of environmental provocative factors, is characterized by an **inability to abstain**. For our studies, we chose the more severe type of alcoholism. We reasoned that the overwhelming and unremitting alcohol-seeking behavior of the more severe alcoholic patients may have an underlying molecular genetic basis.

We were fortunate that brain samples from deceased subjects were being collected for us by the UCLA National Neurological Research Specimen Bank. For our association studies, 35 brains of severe alcoholics who had died from medical complications of alcoholism, and 35 brains of nonalcoholics who had died from other diseases were available for molecular genetic analysis. The findings of that study were published in 1990.[24] The results showed that of the nine candidate genes probed, the only one that associated with alcoholism was the DRD2, a gene that is located at the q22-q23 region of chromosome 11. Specifically, the *Taq I* A DRD2 A1 (minor) allele was found in 20% of the nonalcoholics and in 69% of the alcoholics. On the other hand, the absence of this allele (or the presence only of the A2 [major] allele) was found in 80% of the nonalcoholics and in 31% of the alcoholics. The prevalence of the A1 allele was significantly higher in alcoholics when compared to nonalcoholic controls ($p < .001$). Subsequent to this study, we have examined more than 40 other candidate genes, including adrenergic and serotonergic receptors, various subunits of the GABA receptor, D1 and D3 dopamine receptors, and others. None of the polymorphisms in these genes have so far been found to be associated with alcoholism.

C. SUBSEQUENT STUDIES OF THE DRD2 GENE IN ALCOHOLISM

After our first report appeared, numerous groups, both in the U.S. and abroad, have studied the role of the DRD2 gene in alcoholism. While many of these studies have found a significantly higher association of the DRD2 A1 allele with alcoholism, others have not. We believe that this variability may be due to at least two key factors: (1) the type of alcoholics (more severe or less severe) studied, and (2) the nature of the comparative controls.[25]

Table 19-1 shows the results of ten independent studies of Caucasians (non-Hispanics) wherein DRD2 allelic distribution in both heterogeneous alcoholics (more severe and less severe) and heterogeneous controls (assessed and unassessed for alcoholism) was ascertained within the same study. In the alcoholics, the prevalence of the A1 allele was 25.0 to 63.6%, with an average of 44.7% in the 566 alcoholics studied. In the controls, A1 allelic prevalence was 13.3 to 35.3%, with an average of 26.2% in the 608 controls examined. The prevalence of the A1 allele was significantly higher ($p < 10^{-10}$) in the heterogeneous alcoholics compared to the heterogeneous controls.

Table 19-2 presents a comparison of DRD2 allelic distribution in eight independent studies of Caucasians wherein severity of alcoholism was ascertained in each study. The differentiation of more severe and less severe alcoholics in these studies was determined using a variety of means. As Table 19-2 indicates, the odds ratio of the A1 allele was greater than 1 in all but one of the more severe than the less severe comparative groups. Furthermore, in the combined sample of 435 alcoholics, the 242 more severe alcoholics had 46.7% prevalence of the A1 allele compared to 32.1% prevalence of this allele in the 193 less severe alcoholics. The difference in A1 allelic prevalence between these two groups of alcoholics was significant ($p = 2.88 \times 10^{-3}$).

TABLE 19-1
Association of the A1 Allele of the DRD2 with Alcoholism in Ten Independent Studies of Caucasians

	Alcoholics[a]			Controls[b]			
	A1+	A1−	%A1+	A1+	A1−	%A1+	Odds Ratio
Blum et al.[24]	14	8	63.6	4	20	16.7	8.75
Bolos et al.[26]	15	25	37.5	38	89	29.9	1.41
Comings et al.[27]	44	60	42.3	24	84	22.2	2.75
Gelernter et al.[28]	19	25	43.2	24	44	35.3	1.39
Blum et al.[29]	42	47	47.2	6	25	19.4	3.72
Cook et al.[30]	5	15	25.0	6	14	30.0	0.78
Amadéo et al.[31]	21	28	42.9	7	36	16.3	3.86
Suarez et al.[32]	30	52	36.6	25	63	28.4	1.45
Noble et al.[25]	34	30	53.1	21	48	30.4	2.59
Neiswanger et al.[33]	29	23	55.8	4	26	13.3	8.20
Total sources (n = 1174)	253	313	44.7[c]	159	449	26.2[c]	2.28

Note: A1+ allele includes A1/A1 genotypes; A1− includes A2/A2 genotypes only.

[a] Includes both less severe and more severe alcoholics.
[b] Includes both nonalcoholics and subjects drawn from the general population (alcoholics not excluded).
[c] The prevalence of the A1 allele was significantly higher in the alcoholic than in the control group ($\chi^2 = 43.5$, df = 1, $p < 10^{-10}$).

TABLE 19-2
***Taq I* A DRD2 Allelic Distribution in Eight Independent Studies of More Severe and Less Severe Caucasian Alcoholics**

	More Severe Alcoholics			Less Severe Alcoholics			Odds Ratio
	A1+	A1−	%A1+	A1+	A1−	%A1+	of A1 Allele
Bolos et al.[26 a]	9	11	45.0	6	14	30.0	1.91
Parsian et al.[34 b]	6	4	60.0	7	15	31.8	3.21
Blum et al.[29 b]	33	19	63.5	15	29	34.0	3.36
Gelernter et al.[28 c]	12	11	52.2	7	13	35.0	2.03
Cook et al.[30 d]	4	11	26.7	1	4	20.0	1.45
Turner et al.[35 b]	4	18	18.2	5	20	20.0	0.89
Geijer et al.[36 e]	20	36	35.7	3	15	16.7	2.78
Geijer et al.[36 f]	6	4	60.0	3	6	33.3	3.00
Noble et al.[25 d]	19	15	55.9	15	15	50.0	1.27
Total sources (n = 435)	113	129	46.7[g]	62	131	32.1[g]	1.85

More severe and less severe alcoholics were differentiated by a variety of means:

[a] The Michigan Alcoholism Screening Test (MAST).[37]
[b] The presence or absence of medical complications of alcoholism.
[c] Alcohol consumption.
[d] Severity of Alcohol Dependence Questionnaire (SADQ).[38]
[e] DSM-III-R criteria (P2 group vs. P1 minus P2 group).
[f] Autopsy determination (P6 group vs. P5 minus P6 group).
[g] The prevalence of the A1 allele was significantly higher in the more severe than in the less severe alcoholic group ($\chi^2 = 8.88$, df = 1, $p = 2.88 \times 10^{-3}$).

Table 19-3 summarizes data on DRD2 allelic distribution in 11 Caucasian control groups which did or did not exclude alcoholics and/or drug abusers. A1 allelic prevalence was 16.5% in the 176 total controls which excluded alcoholics and/or drug abusers. The prevalence of this allele was 31.2% in the 353 controls in which alcoholics and/or drug abusers were not excluded. The difference in A1 allelic prevalence between these two control groups was significant ($p = 1.98 \times 10^{-4}$).

TABLE 19-3
Taq I A DRD2 A1 Allelic Distribution in 11 Caucasian Controls Which Did or Did Not Exclude Alcoholics and/or Drug Abusers

	A Controls (Alcoholics and/or Drug Abusers Not Excluded)				B Controls (Alcoholics and/or Drug Abusers Excluded)		
	A1+	A1−	%A1+		A1+	A1−	%A1+
Grandy et al.[39a]	16	27	37.2	Blum et al.[24d]	4	20	16.7
Bolos et al.[26a†]	21	41	33.9	Parsian et al.[34d]	3	22	12.0
Gelernter et al.[28a†]	24	44	35.3	Comings et al.[27d]	3	17	15.0
Comings et al.[27a]	21	67	23.9	Blum et al.[29d]	6	25	19.4
Smith et al.[40b]	6	14	30.0	Smith et al.[40e]	8	28	22.2
Amadéo et al.[31a]	5	18	21.7	Amadéo et al.[31d]	2	18	10.0
Noble et al.[25c]	17	32	34.7	Noble et al.[25f]	4	16	20.0
				Neiswanger et al.[33e]	4	26	13.3
Total sources (n = 559)	110	243	31.2[g]		34	172	16.5[g]

[a] Alcoholics and drug abusers not excluded.
[b] Alcohol and other drug abusers not excluded.
[c] Alcoholics excluded but not drug abusers or cigarette smokers.
[d] Alcoholics excluded.
[e] Alcohol and drug abusers excluded.
[f] Alcoholics and drug abusers and smokers excluded.
[g] The prevalence of the A1 allele was significantly higher in A Controls than B Controls ($\chi^2 = 13.9$, df = 1, $P = 1.98 \times 10^{-4}$).
[†] Excludes CEPH subjects included by Comings et al.[27]

When the data in Tables 19-2 and 19-3 are compared, the more severe alcoholics showed an almost 3-fold higher A1 allelic prevalence than controls wherein alcoholics and/or drug abusers were excluded ($\chi^2 = 44.6$, $p < 10^{-10}$). The more severe alcoholics had a significantly 1.5-fold higher A1 allelic prevalence than controls wherein alcoholics and/or drug abusers were not excluded ($\chi^2 = 14.1$, $p = 1.71 \times 10^{-4}$). Moreover, a significantly higher prevalence of the A1 allele was also found in less severe alcoholics than controls wherein alcoholics and/or drug abusers were excluded ($\chi^2 = 12.5$, $p = 4.15 \times 10^{-4}$). However, A1 allelic prevalence was virtually the same in the less severe alcoholics and in the controls wherein alcoholics and/or drug abusers were not excluded (32.1 vs. 31.2%, $\chi^2 = .018$, $p = .892$). Thus, the cumulative evidence suggests that significant or nonsignificant A1 allelic association with alcoholism may be obtained, depending on which types of alcoholics are compared to which types of controls.

Eight meta-analyses[41-48] have affirmed significant DRD2 A1 allelic association with alcoholism, whereas one has not.[49] However, it should be noted that this last meta-analysis excluded two studies of our own[24,25] and another by a French group[31] which showed significant DRD2 A1 allelic association with alcoholism. Although we could discern no reason for these curious exclusions, we have reanalyzed only the data that this last review chose to include. The results of that reanalysis[50] showed that unscreened controls (alcoholics not excluded) had a significantly higher frequency of the A1 allele than screened controls (alcoholics excluded). Moreover, alcoholics (both more severe and less severe) had a significantly higher frequency of the A1 allele than screened controls. This difference was even greater, and significantly more so, when more severe alcoholics were compared to screened controls.

Linkage analysis of the DRD2 gene in alcoholism has also been undertaken. Three linkage studies[26,33,34] of the DRD2 locus did not show evidence implicating this locus in alcoholism. As indicated earlier, the heterogeneous nature of alcoholism, alcoholism's yet unknown mode of inheritance, and the contribution of environmental factors may render difficult the interpretation of linkage results in this complex behavioral disorder. However, Gurling and colleagues employed sib-pair and extended sib-pair methods in large families multiply affected by alcoholism.[51,52] Utilizing both *Taq I* A DRD2 alleles and microsatellite repeat polymorphism,[53] these investigators found evidence in favor of a significant effect of the DRD2 locus in the liability to develop heavy drinking and alcoholism.

Additional studies further support a role of the DRD2 gene in alcoholism. Hauge et al.[53] have found a new polymorphism (*Taq I* B) located closer to the 5' regulatory region of this gene than the more distant 3' *Taq I* A polymorphism. Moreover, they showed that *Taq I* A and *Taq I* B polymorphisms were in linkage disequilibrium. This *Taq I* B DRD2 polymorphism also yields two alleles, a minor B1 and a major B2 allele. In a study of 115 Caucasian subjects, we found no significant difference in the prevalence of the B1 allele between nonalcoholics (n = 30) and less severe alcoholics (n = 36).[54] However, the prevalence of this allele was significantly higher in the more severe alcoholics (n = 49) when compared to either the nonalcoholics ($p = .008$) or the less severe alcoholics ($p = .005$).

We have also studied the relationship of other mutations in the DRD2 gene to alcohol consumption behaviors.[55] These included two new variants **within** the DRD2 gene (intron 6 and exon 7) which have been identified by Sommer and colleagues.[56,57] In an analysis of a large sample (n = 302) of alcoholics and nonalcoholics, the 1 haplotype of intron 6/exon 7 was found to be strongly and significantly associated not only with more severe alcoholism but also with the level of alcohol consumed.

In sum, the above findings show that mutations both **within** and **outside** the DRD2 gene are associated with alcoholism, particularly its more severe form.

D. THE DRD2 GENE AND OTHER SUBSTANCE USE DISORDERS

Neurochemical, neurophysiological, and neuropharmacological studies, in general, support the view that psychoactive substances of abuse exert their reinforcing properties in the dopaminergic system of the mesocorticolimbic pathway of the brain (for reviews, see References 58 and 59). Neurochemical studies have also shown a commonality of actions, through the dopaminergic system, in the reinforcing properties of these substances. Thus, alcohol,[60,61] cocaine,[62,63] nicotine,[64,65] and food,[66,67] when consumed, raise dopamine levels and affect dopamine metabolism in brain reward areas. The next question raised was whether DRD2 alleles associate with other substance use disorders.

1. Cocaine Dependence

We have recently assessed the role of the DRD2 gene in cocaine dependence.[68] Table 19-4 presents the genotypic distribution of *Taq I* A DRD2 alleles in Caucasian (non-Hispanic) cocaine-dependent subjects and controls. Of the 53 cocaine-dependent subjects, 27 (50.9%) had the A1 allele. In a sample of 100 non-substance-abusing controls, 16 (16.0%) carried the A1 allele. The difference in A1 allelic prevalence between these two groups was statistically

TABLE 19-4
Genotypic Distribution of *Taq I* A DRD2 Alleles in Caucasian (Non-Hispanic) Cocaine-Dependent Subjects and Controls

Group	N	% Genotype			%A1+	Significance (Yates)	Odds Ratio
		A1/A1	A1/A2	A2/A2			
A							
Cocaine-dependent subjects	53	5.7 (3/53)	45.3 (24/53)	49.1 (26/53)	50.9	—	—
Non-substance abusing controls[a]	100	0	16.0 (16/100)	84.0 (84/100)	16.0	$\chi^2 = 19.3$ df = 1; $p = 10^{-5}$	5.45
Population controls[b] (substance abusers not excluded)	265	3.4 (9/265)	27.5 (73/265)	69.1 (183/265)	30.9	$\chi^2 = 6.98$ df = 1; $p = 8 \times 10^3$	2.32
Non-substance abusing and general population controls[a,b] (substance abusers not excluded)	365	2.5 (9/365)	24.4 (89/365)	73.2 (267/365)	26.8	$\chi^2 = 11.6$ df = 1; $p = 6 \times 10^{-4}$	2.83
B							
Cocaine-dependent subjects without alcohol dependence	16	6.3 (1/16)	43.8 (7/16)	50.0 (8/16)	50.0	—	—
Cocaine-dependent subjects with alcohol dependence	37	5.4 (2/37)	45.9 (17/37)	48.6 (18/37)	51.4	$\chi^2 = 0.04$ df = 1; $p = 0.84$	0.95

[a] Obtained from References 24, 27, 29, 34.
[b] Obtained from References 26–29.

significant ($p = 10^{-5}$). In an independent sample of 265 population controls from which substance abusers were not excluded, 82 (30.9%) had the A1 allele. The difference between this group and the cocaine-dependent group was also statistically significant ($p = 8 \times 10^{-3}$). When the total sample of 365 control subjects was considered, 98 (26.8%) carried the A1 allele. The prevalence of the A1 allele in this total control group was, as expected, significantly different ($p = 6 \times 10^{-4}$) from the cocaine-dependent group.

Since a number of cocaine-dependent subjects in our sample were also alcohol dependent, A1 allelic presence was compared in subjects with and without comorbid alcohol dependence. The results showed that 19 (51.4%) of 37 cocaine-dependent subjects with alcohol dependence had the A1 allele, while 8 (50.0%) of 16 cocaine-dependent subjects without alcohol dependence carried the A1 allele. The difference between these two groups in the proportion of the presence and absence of the A1 allele was not significant. This suggests that the higher prevalence of the A1 allele in cocaine-dependent subjects, compared with controls, is not solely due to the contribution of comorbid alcohol dependence.

In this same study, we also ascertained the genotypic distribution of the *Taq I* B DRD2 alleles. Of the 52 cocaine-dependent subjects, 4 (7.7%) had the B1/B1, 16 (30.8%) had the B1/B2, and 32 (61.5%) had the B2/B2 genotypes. In the 53 non-substance-abusing controls, 0 (0%) had the B1/B1, 7 (13.2%) had the B1/B2, and 46 (86.8%) had the B2/B2 genotypes. The prevalence of the B1 allele in the cocaine-dependent subjects (38.5%) was significantly higher ($p = .006$) than in the non-substance-abusing controls (13.2%). Thus, like in alcoholism, both the minor *Taq I* A and *Taq I* B DRD2 alleles associate with cocaine dependence.

To determine whether phenotypic differences exist between cocaine-dependent subjects carrying the minor and the major DRD2 alleles, detailed behavioral and familial characteristics were obtained on these subjects. Logistic regression analysis identified potent routes of cocaine use (i.v., free base, and crack) and the interaction of early deviant behaviors (before regular cocaine use) and parental alcoholism (at least one parent alcoholic) as significant risk factors associated with the A1 allele. When each of these three risk factors were assigned a score of 1 and summed, scores ranging from 0 to 3 were obtained, depending on the number of risk factors counted for each subject. The results showed the A1 allele contributed to 16.7% in the 0-risk score group, 35.0% in the 1-risk score group, 66.7% in the 2-risks score group, and 87.5% in the 3-risks score group. When a verification of the count of risk factors was made using the Mantel-Haenzel test for linear association, increasing risk scores were positively and significantly ($p = .001$) related to A1 allelic classification. These findings suggest that DRD2 alleles differentiate cocaine-dependent phenotypes.

Additional studies by Uhl's group at the U.S. National Institute on Drug Abuse[40,69] and Comings' group[27,70] further support the significantly higher prevalence of the A1 (and B1) allele in cocaine dependence and polysubstance abuse (other than alcoholism) when compared to controls.

2. NICOTINE DEPENDENCE

As indicated earlier, nicotine, like alcohol and cocaine, increases brain dopamine levels and shares common reinforcement mechanisms with these drugs. It can then be hypothesized that nicotine addiction may also share a common molecular genetic diathesis with alcoholism and other drug addictions. To ascertain the role of the DRD2 gene in smoking behavior, *Taq I* A DRD2 alleles were determined in 354 Caucasian (non-Hispanic) smokers and nonsmokers.[71] Subjects were considered to be smokers if they had consumed 100 cigarettes or more in their lifetime. In the present sample, 57 were active smokers, 115 were ex-smokers, and 182 were nonsmokers.

The results of the study showed the following prevalence of the A1 allele: active smokers, 45.6%; ex-smokers, 40.0%; nonsmokers, 28.0%. Further analysis of the data showed that DRD2 allelic prevalence was significantly different among these three groups ($p = .018$). Specifically, the A1 allele occurred in a significantly larger proportion of active smokers compared to nonsmokers ($p = .021$). The prevalence of the A1 allele was also significantly higher in ex-smokers compared to nonsmokers ($p = .044$). Furthermore, ever smokers (ex- and active smokers) had a significantly higher prevalence of the A1 allele compared to nonsmokers ($p = .009$). Linear trend analysis of A1 allelic prevalence in the nonsmoker, ex-smoker, and active smoker groups, respectively, showed that as smoking severity increased so did the prevalence of the A1 allele ($p = .006$).

At a conference of the Tobacco-Related Disease Research Program of the University of California, Comings et al.[72] presented data on the role of the DRD2 gene in smoking. In 98 smokers attending a smoking cessation clinic, 47.0% carried the A1 allele. By contrast, the prevalence of the A1 allele in 600 nonalcoholic, nondrug-abusing (except for tobacco) Caucasians (non-Hispanics) was 25.8% ($p = .0002$). In a subset of 39 Caucasian controls screened to exclude alcoholism and drug abuse (including tobacco), 17.9% were carriers of the A1 allele ($p = .0017$). The similarity between the study by Comings et al.[72] and that of our own in the higher prevalence of the A1 allele in smokers compared to nonsmokers also supports a role of the DRD2 in smoking behavior.

Whereas our above study on smoking was carried out on generally younger healthy subjects drawn from the population at large, we have recently conducted a preliminary study on an older and medically ill group of nonsmokers and smokers.[73] The sample consisted of 41 nonsmokers (N), 69 ex-smokers (X), and 63 active smokers (A). *Taq I* A DRD2 alleles were determined, and their relationship to three behavioral characteristics were assessed: (1) Addictive Personality scores as determined by the Eysenck Personality Questionnaire-Revised,[74] (2) depression scores,[75] and (3) nicotine dependence scores.[76]

Table 19-5 gives the results of the Addictive Personality (AP) scores in the three groups studied. Data analysis showed no significant allele effect on the AP scores. However, a significant group effect, and allele by group interaction, was observed. In the next step, only the X and A groups were compared on the AP scores. No significant allele effect was found in the two groups. However, again, a significant group effect ($p = .002$) and a significant allele by group interaction ($p = .009$) was observed. Furthermore, comparisons were also made between current nonsmokers (N + X combined) and the A group. Here again, no significant allele effect was found. However, a strong group effect ($p = .0003$) and allele by group interaction ($p = .009$) was noted in these two groups.

TABLE 19-5
Addictive Personality Scores (Eysenck) for Nonsmoker (N), Ex-smoker (X), and Active Smoker (A) Groups

Allele	Smoking Characteristics			Total Groups
	N	X	A	
A1	10.3 ± 1.6 (10)	9.9 ± 0.9 (29)	14.9 ± 1.1 (23)	11.8 ± 0.7 (62)
A2	10.4 ± 0.8 (31)	11.2 ± 0.6 (40)	11.7 ± 0.8 (40)	11.1 ± 0.4 (111)
A1 + A2	10.4 ± 0.7 (41)	10.7 ± 0.5 (69)	12.9 ± 0.7 (63)	11.4 ± 0.4 (173)

Note: Two-factor ANOVA showed $p = .442$ for the effect of allele, (a), $p = .002$ for the three groups (b), and $p = .025$ for a × b interaction. Values in parentheses represent number of subjects.

Table 19-6 presents the results of the depression scores in the three groups studied. It should be noted that the higher the score the greater the depression. No significant allele effect was found, whereas a significant group effect was observed. However, in contrast to the AP scores, no significant allele by group interaction was found on the depression scores. In the next step, only the X and A groups were compared. No significant allele effect was found, whereas a significant group effect ($p = .0004$) was found. Moreover, in contrast to the AP scores, no significant allele by group interaction was observed in the two groups on the depression scores. Furthermore, comparison was also made between current nonsmokers (N + X combined) and the A group. Here again, no significant allele effect or allele by group interaction was found. However, again, a significant group effect ($p = .0001$) was found on the depression scores in these two groups.

TABLE 19-6
Depression Scores for the Nonsmoker (N), Ex-Smoker (X), and Active Smoker (A) Groups

Allele	Smoking Characteristics			Total Groups
	N	X	A	
A1+	5.6 ± 1.6 (10)	8.4 ± 1.4 (29)	13.1 ± 1.7 (23)	9.7 ± 1.0 (62)
A1-	6.6 ± 1.1 (31)	7.2 ± 1.0 (40)	10.3 ± 1.1 (40)	8.1 ± 0.6 (111)
A1+ + A1-	6.4 ± 0.9 (41)	7.7 ± 0.8 (69)	11.3 ± 1.0 (63)	8.7 ± 0.5 (173)

Note: Two-factor ANOVA showed $p = .388$ for the effect of the A1 allele (a), $p = .0004$ for the three groups (b), and $p = .446$ for a × b interaction.

The results of the nicotine dependence scores in ex-smokers and active smokers are shown in Table 19-7. It should be noted that the higher the score, the greater is the degree of nicotine dependence. No significant allele effect was found. However, a significant group effect was observed. Moreover, unlike the AP scores but like the depression scores, no significant allele by group interaction was found in these two groups on the nicotine dependence scores.

TABLE 19-7
Nicotine Dependence (Fagerström Test) for Ex-Smoker (X) and Active Smoker (A) Groups

Allele	Smoking Characteristics		Total Groups
	X	A	
A1$^+$	6.1 ± 0.5 (29)	7.2 ± 0.4 (23)	6.6 ± 0.3 (52)
A1$^-$	5.3 ± 0.4 (40)	7.1 ± 0.3 (40)	6.2 ± 0.3 (80)
A1$^+$ + A1$^-$	5.6 ± 0.3 (69)	7.2 ± 0.3 (63)	6.3 ± 0.2 (132)

Note: Two-factor ANOVA showed $p = .259$ for the effect of allele (a), $p = .0003$ for the two groups (b), and $p = .368$ for a × b interaction.

The results of this preliminary study support previous studies in the literature by showing (1) greater severity of depression in active smokers compared to ex-smokers and nonsmokers, and (2) greater degree of nicotine dependence in active smokers compared to ex-smokers. However, the above findings add a new dimension to smoking behavior when molecular genetic factors are taken into consideration. The results suggest a strong gene by group interaction in the three groups studied only on the Addictive Personality measure. Specifically, the data showed that subjects who currently engage in active smoking and carry the DRD2 A1 allele have the highest Addictive Personality scores. By the same token, the DRD2 gene does not appear to be implicated in depression or nicotine dependence, suggesting that environment and/or other genes may play a role in these behaviors in smokers. It is suggested that by differentially ascertaining how gene and environmental variables are involved in various aspects of smoking, a better understanding may be obtained about the complex set of behaviors associated with this drug habit.

3. OBESITY

Food, like a variety of reinforcing substances such as alcohol and other drugs of abuse, when consumed can produce euphoria or pleasure. In view of the involvement of the dopaminergic system in eating behaviors and central reward mechanisms, and because a hereditary basis underlies some forms of obesity, the next question we raised was whether the DRD2 gene is also implicated in obesity. To address this issue, a study of 73 patients who were significantly obese (average body mass index [BMI] = 35.1) was conducted.[77] *Taq I* DRD2 A1 allelic prevalence was 45.2% in these nonalcohol- and nondrug-abusing obese patients. This high A1 allelic prevalence is in the range observed in alcoholics, cocaine-dependent subjects, and smokers.

The relationship of DRD2 alleles to anthropomorphic and metabolic parameters was also ascertained. Neither BMI nor waist/hip ratio was significantly differentiated in these obese subjects carrying the A1$^+$ or A1$^-$ allele; although the A1 homozygotes showed the highest values. Furthermore, cardiovascular risk factors including blood pressure and lipids (cholesterol, HDL, LDL, and triglycerides) were not significantly differentiated in the A1$^+$ and the

A1⁻ allelic subjects. However, when familial and behavioral factors were examined, interesting differences were revealed in subjects carrying these two alleles.

Table 19-8 shows the relationship of DRD2 alleles to parental history of obesity, onset of obesity, and food preference. In obese subjects whose fathers and mothers were not obese, 31.0% carried the A1⁺ allele. A1⁺ allelic prevalence was 43.5% and 51.5% in subjects whose fathers and mothers, respectively, were obese. In subjects whose fathers and/or mothers were obese, 53.7% displayed the A1⁺ allele. The difference in A1⁺ allelic prevalence between this group and the group with negative parental history of obesity approached but did not achieve statistical significance ($p = .10$).

TABLE 19-8
Relationship of *Taq I* A DRD2 Alleles to Parental History, Onset of Obesity, and Food Preference

Background Characteristics	A1⁺	A1⁻	%A1⁺	Significance
Prenatal history of obesity[a]				
Neither fathers nor mothers obese	9	20	31.0	—
Fathers obese	10	13	43.4	$\chi^2 = 0.40, p = .53$
Mothers obese	17	16	51.5	$\chi^2 = 1.88, p = .17$
Fathers and/or mothers obese	22	19	53.7	$\chi^2 = 2.67, p = .10$
Onset of obesity[b]				
Child	5	15	25.0	—
Adolescent	5	8	36.5	$\chi^2 = 0.19, p = .66$
Adult	22	17	56.4	$\chi^2 = 4.07, p = .04$
Food preference				
Carbohydrates	18	10	64.3	—
Other[c]	4	15	21.1	$\chi^2 = 6.85, p = .009$

Note: A1⁺ allele includes A1/A1 and A1/A2 genotypes; A1⁻ allele includes A2/A2 genotype only.

[a] Comparison with neither fathers nor mothers obese.
[b] Comparison with child-onset obesity.
[c] Other includes proteins, fats, or food in general.

Table 19-8 also shows the relationship of DRD2 alleles to the age of onset of obesity. Subjects whose onset of obesity occurred when they were children, adolescents, and adults had the following progressive increase in A1 allelic prevalence, 25.0, 36.5, and 56.4%, respectively, with A1 allelic prevalence being significantly higher ($p = .04$) in adult-onset than in child-onset obesity. Moreover, when the relationship of age of obesity onset to A1 allelic prevalence was ascertained using a test for linear association, increasing age of onset was positively and significantly related to A1 allelic classification ($p = .02$).

The relationship of food preference of obese subjects to their DRD2 allelic distribution is further shown in Table 19-8. Comparison made of allelic prevalence between subjects who prefer carbohydrates and subjects who prefer other foods (fats, proteins, or food in general) showed that 64.3% of the carbohydrate preferrers carried the A1 allele, whereas 21.1% of the subjects who preferred other foods carried this allele ($p = .009$).

Next, a determination was made of the relationship of the three phenotypic factors shown in Table 19-8 to A1 allelic prevalence. Factor scores on each obese subject were obtained by assigning a score of 1 for the presence of each of the following: parental history of obesity (father and/or mother obese), onset of obesity (adolescent or adult), and food preference (carbohydrate preferrers). Thus, scores ranging from 0 to 3 were obtained depending on the number of these factors present in each subject. Because there were only two subjects in the

0-factor group, their allelic data were combined with the subjects in the 1-factor group. A1 allelic prevalence in these various factors score categories showed the following: the A1 allele contributed to 9.1% in the 0–1 factor group, 43.5% in the 2-factors group, and 84.6% in the 3-factors group. A significant difference in allelic prevalence was found among these three factor groups ($p = .001$). Furthermore, when the relationship of factor score to A1 allelic prevalence was determined using a linear association test, increasing factor score was positively and significantly related to A1 allelic classification ($p = .0002$).

These findings suggest that the DRD2 gene is implicated in a type of obesity that it is characterized by the presence of parental and late onset obesity and carbohydrate preference. That the DRD2 gene is involved in obesity also comes from another study carried out by Comings et al.[78] who found that haplotypes of intron 6/exon 7 associated with this disorder.

E. CENTRAL NERVOUS SYSTEM EXPRESSION OF DRD2 ALLELES

As the above studies suggest, the A1 allele is associated not only with alcoholism and other substance use problems, but it also shows unique behavioral phenotypes. The next question we then raised was whether there are CNS differences in the expression of the *Taq I* A DRD2 alleles.

1. Pharmacological Studies

One approach to answer this question would be to study the pharmacological characteristics of the D2 dopamine receptors in A1$^+$ and A1$^-$ allelic subjects. Unfortunately, receptor studies cannot be done on blood elements, since the DRD2 gene does not express itself in these peripheral cells. However, D2 dopamine receptors are expressed in the CNS, with the caudate nucleus being among brain regions having the highest concentration of these receptors. Fortunately, of the original 70 cerebral cortex tissues that we had probed earlier,[24] 66 samples of the caudate nucleus were available. Using [^{3}H]spiperone as the binding ligand, we studied[79] two important characteristics of the D2 dopamine receptor in these caudate nucleus samples: K_d (binding affinity) and B_{max} (number of binding sites). In the total subjects with the A1$^+$ allele (A1/A2 and A1/A1 genotypes), the B_{max} was found to be significantly reduced (by almost 30%) when compared with the B_{max} of the total subjects with the A1$^-$ allele (A2/A2 genotype only). However, no significant difference was found in the K_ds. Moreover, a significant progressively reduced B_{max} was observed in subjects with A2/A2, A1/A2, and A1/A1 genotypes. In this respect, it may be noteworthy that a recent study of Japanese patients[80] found that the average severity of alcoholism increased progressively in the order of A2/A2, A1/A2, and A1/A1 genotypes.

The lack of a difference in binding affinity between A1$^+$ and A1$^-$ allelic subjects, but reduced number of D2 dopamine receptors in A1$^+$ allelic subjects, suggest that the putative mutation(s) in the DRD2 gene may not involve a change in the structure but rather a change in the expression of D2 dopamine receptor numbers. To test the validity of this notion, we have recently examined[81] the expressed sequences of the DRD2 gene (exons 2 through 8) in 74 alcoholic (and 113 schizophrenic) patients and 34 controls. Using large-scale mutational analysis by denaturing gradient gel electrophoresis, no mutations were found in the DRD2 coding sequence that associated with alcoholism (or schizophrenia). Given the expected lack of structural mutation suggested from our binding studies, it is quite likely that there is a physiologically important alteration in DRD2 expression that is associated with the *Taq I* A RFLP (or a mutation in linkage disequilibrium with this RFLP) and with severe alcoholism.[53] This genetic defect, not yet identified by us, could be located in various genomic regions of this large 270-kb gene.[82] This could be located in exon 1 (a noncoding exon), introns, promoter, or other regulatory sequences. Research in our laboratory and in those of others are actively investigating these possibilities.

2. NEUROPHYSIOLOGICAL STUDIES

Another approach to study the differential expression of DRD2 A1+ and A1− alleles is to investigate the relationship of these alleles to relevant features of neurophysiological functioning. The rationale for undertaking such a study is, in part, based on evidence derived by others suggesting a hereditary component in the generation of the P300 (an event-related potential), and a growing number of studies implicating the dopaminergic system in the P300 or P3. We undertook such a study[83] using a sample of 98 healthy prepubescent Caucasian boys, mean age 12.5 years, and of above average intelligence. This sample consisted of three groups of children: (1) 32 sons of active (nonabstinent) alcoholic (SAA) fathers, (2) 36 sons of recovered alcoholic (SRA) fathers, and (3) 30 sons of social drinker (SSD) fathers. None of these boys had yet begun to consume alcohol, tobacco, or other psychoactive drugs, obviating the effects of these drugs on brain function. A significant difference ($p = .01$) was found in the frequency of the A1 allele among these three groups of boys. Specifically, the SAA group (i.e., sons of fathers with more severe alcoholism as revealed by background data) had the highest A1 allelic frequency (0.313). This was followed by the SRA (0.139) and the SSD (0.133) groups. Whereas a significant difference ($p = .02$) in A1 allelic frequency was found between the SAA and the SRA, and between the SAA and the SSD groups, the difference in A1 allelic frequency between the SRA and SSD groups was not significant. This evidence suggests that on a molecular genetic basis the SAA and the SRA are distinctly different groups. However, the SRA and the SSD, on the other hand, are not dissimilar.

Additionally, neurophysiological characteristics of these children were examined. Table 19-9 shows the relationship of target P300 amplitude and latency at P_z to *Taq I A* DRD2 alleles in sons of active alcoholic, recovered alcoholic, and social drinker fathers. Analysis of covariance (ANCOVA) of P300 amplitude showed that there were no significant main effect of allele (A1+,A1−) or group (SSA, SRA, SSD) and no interaction between allele and group. In contrast, allele × group ANCOVA of P300 latency showed a larger main effect of allele (A1 = 455 ± 12 ms; A2 = 412 ± 8 ms; $p = .004$), but no significant group effect or interaction between allele and group.

In conjunction with other studies in the literature on humans indicating a strong relationship of prolonged P300 latency to decreased dopaminergic function, and in view of our previous study showing a reduced number of D2 dopamine receptors in brains of A1+ allelic subjects, the evidence of prolonged P300 latency in A1+ allelic children suggests that they, too, may have fewer D2 dopamine receptors when compared to children carrying the A1− allele. However, this suggestion needs further confirmation by other studies, e.g., PET imaging techniques using radiolabeled D2 dopamine receptor ligands.

3. NEUROPSYCHOLOGICAL STUDIES

Since the above studies suggest reduced dopaminergic function in subjects with the A1+ allele we explored next,[84] in children of alcoholics and nonalcoholics, the influence of DRD2 alleles on visuospatial ability (i.e., how individuals perceive the relationship of objects in space). In this respect, there is a large body of evidence which indicates that chronic alcoholics are characterized by specific impairments of visuospatial functioning. These impairments extend to children of alcoholics, suggesting that their presence in alcoholics may be antecedent to their drinking behavior.

TABLE 19-9
Relationship of Target P300 Amplitude and Latency at P_z to *Taq I* A DRD2 Alleles in Sons of Active Alcoholic (SAA), Recovered Alcoholic (SRA), and Social Drinker (SSD) Fathers

	Amplitude (μV)		Latency (ms)	
	A1+	A1−	A1+	A1−
SAA[a]	26.3 ± 1.7	27.9 ± 1.8	437 ± 15	382 ± 17
SRA[b]	32.2 ± 2.3	26.4 ± 1.3	468 ± 21	422 ± 12
SSD[c]	31.0 ± 2.5	30.2 ± 1.5	460 ± 23	432 ± 14
Mean[d]	29.8 ± 1.2	28.2 ± 0.9	455 ± 12	412 ± 8
Significance	.28		.004	

Note: A1+ allele includes A1/A2 and A1/A2 genotypes; A1− allele includes A2/A2 genotype only. Values represent ANCOVA-adjusted mean ± SEM.

[a] N = 32; A1 = 17, A2 = 15.
[b] N = 36; A1 = 9, A2 = 27.
[c] N = 30; A1 = 8, A2 = 22.
[d] N = 98; A1 = 34, A2 = 64.

Evidence that the dopaminergic system may also be involved in visuospatial performance derives from studies of cognitive deficits in Parkinson's disease. These patients, who are characterized by dopamine deficiencies, have visuospatial problems that have been dissociated from generalized impairment as a result of depression and motoric disturbances. Additional evidence comes from animal research: lesions in the caudate nucleus of monkeys similarly produce deficits in visuospatial abilities.

We tested 182 alcohol and drug naive preadolescent sons of active alcoholic, recovered alcoholic, and nonalcoholic fathers using a visuospatial task (Benton's Judgment of Line Orientation Test) which makes minimal motor/verbal demands. Boys with the A1+ allele had significantly lower visuospatial scores than boys with the A1− allele ($p = .005$), with A1+ allele sons of active alcoholics showing the lowest scores of the six allele/clinical groups studied. These findings suggest that DRD2 alleles contribute to differential phenotypic expression of visuospatial performance.

In sum, the A1 allele has direct physiological effects, evidenced by decreased brain D2 dopamine receptor numbers, increased latency of the P300, and diminished visuospatial abilities. These findings suggest reduced CNS dopaminergic function in subjects carrying the DRD2 A1 allele.

F. THE DRD2 GENE AND TREATMENT OF ALCOHOLICS

If alcoholics with the A1 allele have an inherited deficit in their brain dopamine reward system, could a D2 dopamine receptor agonist such as bromocriptine have a more salutary effect on them than on alcoholics who carry only the major DRD2 allele (A2)? In collaboration with Australian colleagues at the University of Queensland, we conducted a study to answer this question.[85] In a double-blind bromocriptine (BRO)/placebo (PLA) study, 83 hospitalized alcoholics were studied over a 6-week trial. Besides ascertaining DRD2 alleles, three behavioral measures were assessed (craving, anxiety, and depression), and the patients' retention rate during the trial was also obtained. In the four groups studied (BRO A1+, BRO A1−, PLA A1+, PLA A1−), the greatest and most significant decreases in craving and anxiety occurred in the A1 allelic patients receiving bromocriptine (BRO A1+). However, no significant differences were found in decreased depression among the four groups. Moreover, the retention rate of the A1+ allelic alcoholics receiving bromocriptine during the 6-week trial was greater than the other three groups and significantly more so when compared to A1 allelic alcoholics receiving placebo (PLA A1).

The present study opens up the feasibility of a pharmacogenetic approach in the treatment of a certain type of alcoholics. Specifically, it suggests that alcoholics who carry the DRD2 A1 allele, having an inherited deficit of their dopaminergic system, are more amenable to treatment by a dopamine receptor agonist than those alcoholics who do not carry this allele. Additional studies are clearly needed to support or refute this notion.

G. GENETIC ANIMAL MODELS AND THE DOPAMINERGIC SYSTEM

Mardones in Chile and Ericksson in Finland were among the first investigators to develop, by selective breeding, rat lines that differed markedly in the amount of voluntary alcohol consumption. Since then, a variety of animal species have been developed for alcohol and other drug behaviors. These behaviors include drug traits such as preference, tolerance, sensitivity, hypothermia, and withdrawal seizures.[86] Whereas animal genetic models simulate only in part their more complex human condition, they do nevertheless offer, among other advantages, better control of experimental conditions and numerous stable genotypes for study. Consequently, they have been widely used in biochemical, pharmacological, and behavioral research.

The purpose of this section is not to review the burgeoning literature on genetic animal models; rather, it is to describe briefly those studies relevant to the role of the dopaminergic system, particularly the DRD2, in these animal models.

Dopamine (DA) concentrations in the nucleus accumbens (NAC) of alcohol-preferring (P) rats were found to be significantly lower when compared to alcohol nonpreferring (NP) rats.[87,88] More recently, two independent studies of alcohol-naive P and NP rats have shown the maximum number of binding sites (B_{max}) of the D2 dopamine receptor to be significantly reduced in the NAC and caudate nucleus of P compared to NP rats.[89,90] It is suggested that one possible interpretation of the reduced D2 dopamine receptor numbers in the P rats is that it may reflect a parallel reduction in dopaminergic neurotransmission in limbic areas of the brain. To compensate for this deficit, the P rats consume more alcohol in order to release enough DA to produce an adequate level of reward.

Whereas ethanol consumption involves several neurotransmitters in the brain, including norepinephrine, serotonin, and GABA, growing evidence derived from the administration of neurotransmitter receptor agonists and antagonists further supports an important role for the dopaminergic system in mediating the stimulating/reinforcing effects of ethanol. Decrease in alcohol intake of P and High-Alcohol Drinking (HAD) rats was found after the administration of the D2 dopamine receptor agonists bromocriptine[91] and quinpirole.[92] In another study,[93] the differential effects of naloxone (an opiate antagonist), bromocriptine, and methysergide (a 5-HT antagonist) on ethanol consumption was compared in P and unselected Wistar rats. In P rats, naloxone treatment resulted in a dose-dependent suppression in responding for both ethanol and water, but did not alter ethanol preference. This suggests to the authors of this last report[93] that the response decrements observed with naloxone is reflective of a more general depression in consummatory behavior. In contrast, bromocriptine produced a significant, dose-dependent shift in preference from ethanol toward water by inhibiting responding for ethanol while enhancing water consumption. In the Wistar rats, naloxone and bromocriptine treatments produced changes in ethanol preference patterns similar to but less than those observed in the P rats. However, no changes in ethanol preference and water or ethanol intake were observed with methysergide.

In contradistinction to the reduction in ethanol consumption by dopamine receptor agonists the administration of antagonists of this receptor, in general, show an opposite effect, although the data are less consistent. Thus, the microinjection of the D2 receptor antagonist sulpiride into the NAC of P rats caused a dose-dependent increase in alcohol intake.[94] Another D2 receptor antagonist, spiperone, at intermediate doses caused an increase in alcohol consumption in HAD rats.[92] Moreover, systemic injections of D2 receptor antagonists have also been found to increase intravenous cocaine self-administration.[95,96] However, the administration of other D2 receptor antagonists such as haloperidol and pimozide show either a reduction or no change in alcohol consumption.[97-99]

Recent developments in molecular genetics have provided new tools for exploring genes whose effects are not overwhelming, but which account for appreciable proportions of the variance. These quantitative trait loci (QTLs) offer attractive approaches to the identification of chromosomal loci with detectable effects on quantitatively distributed phenotypes, including pharmacological responses to alcohol and other drugs.[100]

Crabbe et al.[86] have recently summarized the extant literature on genomic locations of provisional QTLs affecting alcohol and other drug responses in genetic animal models. QTL analyses revealed that in a panel of recombinant inbred (RI) mouse strains, several genetic markers, including the DRD2 gene, were associated with responses to alcohol and other drugs. Specifically, Phillips et al.[101] compared ethanol consumption in BXD RI mice derived from an F2 cross of C57BL/6J (alcohol preferring) and DBA/2J (alcohol nonpreferring) inbred strains. The 6 BXD strains carrying the DRD2 allele of the C57BL/6J mice consumed over 2.5-fold more ethanol than the 11 BXD strains carrying the DRD2 allele of the DBA/2J mice. Moreover, tolerance to ethanol-induced hypothermia,[102] conditioned place preference,[103] and other drug-related behaviors including consumption of methamphetamine, methamphetamine-stimulated activity, and morphine-induced Straub Tail also map in the region of the DRD2 gene.[86]

Because of the great extent of linkage homology between human and mouse,[104] the involvement of the DRD2 gene using QTL analyses in alcohol and other drug-related behaviors adds a new dimension to supporting the role of this gene in human alcoholism and other drug-related disorders.

H. THE DRD2 GENE: A REWARD GENE?

The data presented above on humans suggest that the DRD2 is a **susceptibility** rather than a **necessary** gene[105] involved in severe alcoholism as well as in several other substance use disorders. What the precise biological mechanisms are that contribute to this susceptibility remain unknown. However, given the available body of evidence implicating the mesocorticolimbic dopaminergic pathways in drug reinforcement behaviors, and the observations of reduced dopaminergic function in brains of subjects with the A1 allele, we have hypothesized that the DRD2 may represent a reinforcement or reward gene.[71] Subjects with the A1 allele may compensate for the inherent deficiency of their dopaminergic system by the use of alcohol and other reinforcing substances, agents known to increase brain dopamine levels. The subsequent stimulation by dopamine of A1 allelic subjects' fewer D2 dopamine receptors may provide enhanced feelings of reward and pleasure and, with the continued and increasing use of these substances, the development of tolerance and dependence. It is important, however, to stress that this is only a hypothesized mechanism and only incremental future research can ascertain the validity of this notion.

In the available data, however, it should be noted that whereas about half the alcoholics (more severe) and other substance abusers carry the A1 allele, the other half do not. This suggests that in the latter half, environmental factors and/or genes other than the DRD2 are likely contributors to substance abuse vulnerability. Calculations based on data from twin studies indicate that genes may influence up to 60% of the vulnerability to severe substance use.[9] Computations based on the extant data on alcoholics and other substance abusers suggest that 27% of the risk for severe substance use can be attributed to DRD2 variants marked by the *Taq I* A1 allele, 33% to other genes and 40% to environmental factors.[44] Thus, the DRD2 variants could represent one of the most prominent single-gene determinants of susceptibility to severe substance use, but also the environment and other genes, when combined, still play the larger role.

I. CONCLUSION

There is now strong and growing evidence to implicate the DRD2 gene in alcoholism and other substance use disorders. It is hypothesized that the DRD2 is a gene that is involved in reward mechanisms and that one of its inherited forms (the A1 allele) increases susceptibility to alcohol and other substance use disorders. However, given the complexity of these disorders and their various consequences, including the development of tolerance, dependence, withdrawal seizures, cirrhosis, and other manifestations, it is doubtless that a number of other genes will be found in substance use disorders. At the same time, it is recognized that the environment also plays an important and, indeed, a critical role in inducing individuals to harmful use of these substances. The greater challenge that lies ahead is to understand how the interaction of hereditary and environmental factors either conspires to yield alcohol and other substance use problems, or acts to protect against the development of these problems.

ACKNOWLEDGMENTS

The author is grateful for the generous support by Mr. and Mrs. R. Brinkley Smithers and the Christopher D. Smithers Foundation, Inc. He thanks his many colleagues, including K. Blum, T. Ritchie, S. Berman, T. Ozkaragoz, R. Fitch, K. Syndulko, R. Noble, S. T. St. Jeor, R. Sparkes, P. Sheridan, D. Grandy, X. Zhang, and many others for their collaboration in various facets of the DRD2 studies. The excellent assistance of A. Jaeger in the preparation of this manuscript is also gratefully acknowledged.

REFERENCES

1. Cotton, N. S., The familial incidence of alcoholism: a review, *J. Stud. Alcohol*, 40, 89, 1979.
2. Goodwin, D. W., Biological factors in alcohol use and abuse: implications for recognizing and preventing alcohol problems in adolescence, *Int. Rev. Psychiatry*, 1, 41, 1989.
3. Goodwin, D. W., Schulsinger, F., Hermansen, L., Guze, S. B., and Winokur, G., Alcohol problems in adoptees raised apart from alcoholic biological parents, *Arch. Gen. Psychiatry*, 28, 238, 1973.
4. Cloninger, C. R., Bohman, M., and Sigvardsson, S., Inheritance of alcohol abuse: cross-fostering analysis of adopted men, *Arch. Gen. Psychiatry*, 38, 861, 1981.
5. Bohman, M., Sigvardsson, S., and Cloninger, C. R., Maternal inheritance of alcohol abuse: cross-fostering analysis of adopted women, *Arch. Gen. Psychiatry*, 38, 965, 1981.
6. Hrubec, Z. and Omenn, G. S., Evidence of genetic predisposition in alcoholic cirrhosis and psychosis: twin concordance for alcoholism and its biological end-points by zygosity among male veterans, *Alcoholism: Clin. Exp. Res.*, 5, 207, 1981.
7. Saunders, J. B. and Williams, R., Review. Genetics of alcoholism: is there an inherited susceptibility to alcohol-related problems?, *Alcohol Alcoholism*, 18, 189, 1983.
8. Pickens, R. W. and Svikis, D. S., Genetic influences in human substance abuse, *J. Addict. Dis.*, 10, 205, 1991.

9. Pickens, R. W., Svikis, D. S., McGue, M., Lykken, D. T., Heston, L. L., and Clayton, P. J., Heterogeneity in the inheritance of alcoholism, *Arch. Gen. Psychiatry*, 48, 19, 1991.
10. Kendler, K. S., Heath, A. C., Neale, M. C., Kessler, R. C., and Eaves, L. J., A population-based twin study of alcoholism in women, *J. Am. Med. Assoc.*, 268, 1877, 1992.
11. Cadoret, R. J., Troughton, E., O'Gorman, T. W., and Heywood, E., An adoption study of genetics and environmental factors in drug abuse, *Arch. Gen. Psychiatry*, 43, 1131, 1986.
12. George, F. R., Is there a common biological basis for reinforcement from alcohol and other drugs? *J. Addictive Dis.*, 10, 127, 1991.
13. Hughes, J. R., Genetics of smoking: a brief review, *Behav. Ther.*, 17, 335, 1986.
14. Mangan, G. L. and Golding, J. F., *The Psychopharmacology of Smoking*, Cambridge University Press, New York, 1984.
15. Hannah, M. C., Hopper, J. L., and Mathews, J. D., Twin concordance for a binary trait. Nested analysis of ever-smoking, ex-smoking traits and unnested analysis of a committed-smoking trait, *Am. J. Hum. Genet.*, 37, 153, 1985.
16. Heller, R. F., O'Connell, D. L., Roberts, D. C. K., Allen, J. R., Knapp, J. C., Steele, P. L., and Silove, D., Lifestyle factors in monozygotic and dizygotic twins, *Genet. Epidemiol.*, 5, 311, 1988.
17. Bray, G. A., The inheritance of corpulence, in *The Body Weight Regulatory System: Normal and Disturbed Mechanisms*, Cioffi, L. A., James, W. P. T., and Van Italie, T. B., Eds., Raven Press, New York, 61-64.
18. Stunkard, A. J., Sorensen, T. I. A., Hanis, C., Teasdale, T. W., Chakraborty, R., Schull, W. J., and Schulsinger, F., An adoption study of human obesity, *N. Engl. J. Med.*, 314, 193, 1986.
19. Zonta, L. A., Jayakar, S. D., Bosisio, M., Galante, A., and Pennetti, V., Genetic analysis of human obesity in an Italian sample, *Hum. Hered.*, 37, 129, 1987.
20. Sims, E. A. H., Destiny rides again as twins overeat, *N. Engl. J. Med.*, 322, 1522, 1990.
21. Plomin R., The role of inheritance in behavior, *Science*, 248, 183, 1990.
22. Jellinek, E.M., *The Disease Concept of Alcoholism*, Hill-House, New Haven, CT, 1960.
23. Cloninger, C. R., Neurogenetic adaptive mechanisms in alcoholism, *Science*, 236, 410, 1987.
24. Blum, K., Noble, E. P., Sheridan, P. J., Montgomery, A., Ritchie, T., Jagadeeswaran, P., Nogami, H., Briggs, A. H., and Cohn, J. B., Allelic association of human dopamine D2 receptor gene in alcoholism, *J. Am. Med. Assoc.*, 263, 2055, 1990.
25. Noble, E. P., Syndulko, K., Fitch, R. J., Ritchie, T., Bohlman, M. C., Guth, P., Sheridan, P. J., Montgomery, A., Heinzmann, C., Sparkes, R. S., and Blum, K., D2 dopamine receptor *Taq I* A alleles in medically ill alcoholic and nonalcoholic patients, *Alcohol Alcoholism*, 29, 729, 1994.
26. Bolos, A. M., Dean, M., Lucase-Derse, S., Ramsburg, M., Brown, G. L., and Goldman, D., Population and pedigree studies reveal a lack of association between the dopamine D2 receptor gene and alcoholism, *J. Am. Med. Assoc.*, 264, 3156, 1990.
27. Comings, D. E., Comings, B. G., Muhleman, D., Dietz, G., Shahbahrami, F., Tast, D., Knell, E., Kocsis, P., Baumgarten, R., Kovacs, B. W., Levy, D. L., Smith, M., Borison, R. L., Evans, D. D., Klein, D. N., MacMurray, J., Tosk, J. F., Sverd, J., Gysin, R., and Flanagan, S. D., The dopamine D2 receptor locus as a modifying gene in neuropsychiatric disorders, *J. Am. Med. Assoc.*, 266, 1793, 1991.
28. Gelernter, J., O'Malley, S., Risch, N., Kranzler, H. R., Krystal, J., Merikangas, K., Kennedy, J. F., and Kidd, K. K., No association between an allele at the D2 dopamine receptor gene (DRD2) and alcoholism, *J. Am. Med. Assoc.*, 266, 1801, 1991.
29. Blum, K., Noble, E. P., Sheridan, P. J., Finley, O., Montgomery, A., Ritchie, T., Ozkaragoz, T., Fitch, R. J., Sadlack, F., Sheffield, D., Dahlmann, T., Halbardier, S., and Nogami, H., Association of the A1 allele of the D2 dopamine receptor gene with severe alcoholism, *Alcohol*, 8, 409, 1991.
30. Cook, B. L., Wang, Z. W., Crowe, R. R., Hauser, R., and Freimer, M., Alcoholism and the D2 receptor gene, *Alcoholism: Clin. Exp. Res.*, 16, 806, 1992.
31. Amadéo, S., Abbar, M., Fourcade, M., Waksman, G., Leroux, M. G., Medec, A., Selin, M., Champiat, J.-C., Brethome, A., Leclair, Y., Castelnau, D., Venisse, J.-L., and Mallet, J., D2 dopamine receptor gene and alcoholism, *J. Psychiatr. Res.*, 27, 173, 1993.
32. Suarez, B. K., Parsian, A., Hampe, C. L., Todd, R. D., Reich, T., and Cloninger, C. R., Linkage disequilibria at the D2 dopamine receptor locus (DRD2) in alcoholics and controls, *Genomics*, 19, 12, 1994.
33. Neiswanger, K., Hill, S. Y., and Kaplan, B. B., Association and linkage studies of the *Taq I* A allele of the dopamine D2 receptor gene in samples of female and male alcoholics, *Am. J. Med. Genet. (Neuropsychiatr. Genet.)*, 60, 267, 1995.
34. Parsian, A., Todd, R. D., Devor, E. J., O'Malley, K. E., Suarez, B. K., Reich, T., and Cloninger, C. R., Alcoholism and alleles of the human D2 dopamine receptor locus. Studies of association and linkage, *Arch. Gen. Psychiatry*, 48, 655, 1991.
35. Turner, E., Ewing, J., Shilling, P., Smith, T. L., Irwin, M., Schuckit, M., and Kelsoe, J. R., Lack of association between an RFLP near the D2 dopmaine receptor gene and severe alcoholism, *Biol. Psychiatry*, 31, 285, 1992.

36. Geijer, T., Neiman, J., Rydberg, U., Gyllander, A., Jonsson, E., Sedvall, G., Valverius, P., and Terenius, L., Dopamine D2 receptor gene polymorphisms in Scandinavian chronic alcoholics, *Eur. Arch. Psychiatry Clin. Neurosci.*, 244, 26, 1994.
37. Selzer, M., The Michigan Alcoholism Screening Test: the quest for a new diagnostic-instrument, *Am. J. Psychiatry*, 127, 1653, 1976.
38. Stockwell, T. R., Hodgson, R. J., and Murphy, R., The severalty of alcohol dependence questionnaire: its use, reliability and validity, *Br. J. Addictions*, 78, 145, 1983.
39. Grandy, D. K., Litt, M., Allen, L., Bunzow, J.R., Marchionni, M., Makam, H., Reed, L., Magenis, R. E., and Civelli, 0., The human D2 dopamine receptor gene is located on chromosome 11 at q22-23 and identifies a *Taq I* RFLP, *Am. J. Hum. Genet.*, 45, 778, 1989.
40. Smith, S. S., O'Hara, B. F., Persico, A. M., Gorelick, D. A., Newlin, D. B., Vlahov, D., Solomon, L., Pickens, R., and Uhl, G. R., Genetic vulnerability to drug abuse: the dopamine D2 receptor *Taq* I B RFLP is more frequent in polysubstance abusers, *Arch. Gen. Psychiatry,* 49, 723, 1992.
41. Cloninger, C. R., D2 dopamine receptor gene is associated but not linked with alcoholism, *J. Am. Med. Assoc.*, 266, 1833, 1991.
42. Noble, E. P., The D2 dopamine receptor gene: a review of association studies in alcoholism, *Behav. Genet.*, 23, 119, 1993.
43. Pato, C. N., Macciardi, F., Pato, M. T., Verga, M., and Kennedy, H., Review of the putative association of dopamine D2 receptor and alcoholism: a meta-analysis, *Am. J. Med. Genet.*, 48, 78, 1993.
44. Uhl, G. R., Blum, K., Noble, E. P., and Smith, S., Substance abuse vulnerability and D2 receptor genes, *Trends Neurosci.*, 16, 83, 1993.
45. Cook, C. C. H. and Gurling, H. M. D., The D2 dopamine receptor gene and alcoholism: a genetic effect in the liability for alcoholism, *J. R. Soc. Med.*, 87, 400, 1994.
46. Gorwood, P., Ades, J., and Feingold, J., Are genes coding for dopamine receptors implicated in alcoholism?, *Eur. Psychiatry*, 9, 63, 1994.
47. Blum, K., Sheridan, P. J., Wood, R. C., Braverman, E R., Chen, I. J. H., and Comings, D. A., Dopamine D2 receptor gene variants: association and linkage studies in impulsive-addictive-compulsive behavior, *Pharmacogenetics*, 5, 121, 1995.
48. Neiswanger, K., Kaplan, B. B., and Hill, S. Y., What can the DRD2/alcoholism teach us about association studies in psychiatric genetics, *Am. J. Med. Genet. (Neuropsychiatr. Genet.)*, 60, 272, 1995.
49. Gelernter, J., Goldman, D., and Risch, N., The A1 allele of the D2 dopamine receptor gene and alcoholism, *J. Am. Med. Assoc.*, 269, 1673, 1993.
50. Noble, E. P. and Blum, K., Letters. Alcoholism and the D2 receptor gene, *J. Am. Med. Assoc.*, 270, 1547, 1993.
51. Cook, C. C. H., Brett, P., Curtis, D., Holmes, D., and Gurling, H. M. D., Linkage analysis confirms a genetic effect of the D2 dopamine receptor locus in heavy drinking and alcoholism, *Psychiatr. Genet.*, 3, 130, 1993.
52. Cook, C., Turner, A., Palsson, G., Petursson, H., and Gurling, H., Sib pair analysis of the role of the D2 dopamine (DRD2) gene in alcoholism: attempted replication employing a second sample of multiplex alcoholic families. Abstr. 1995 World Congr. Psychiatric Genetics, Vol. 5, S76, 1995.
53. Hauge, X. Y., Grandy, D. K., Eubanks, J. H., Evans, G. A., Civelli, 0., and Litt, M., Detection and characterization of additional DNA polymorphisms in the dopamine D2 receptor gene, *Genomics*, 10, 527, 1991.
54. Blum, K., Noble, E. P., Sheridan, P. J., Montgomery, A., Ritchie, T., Ozkaragoz, T., Fitch, R. J., Wood, R., Finley, O., and Sadlack, F., Genetic predisposition in alcoholism. Association of the D2 dopamine receptor *Taq I* B1 RFLP with severe alcoholics, *Alcohol*, 10, 59, 1993.
55. Zhang, X., Ritchie, T., Fitch, R. J., Sparkes, R. S., and Noble, E. P., Haplotypes of the D2 dopamine receptor gene in higher and lower alcohol consuming subjects, *Am. J. Hum. Genet.*, Suppl. 54, A168, 1994.
56. Sarkar, G., Kapelner, S., Grandy, D. K., Marchionni, M., Civelli, O., Sobell, J., Heston, L., and Sommer, S.S., Direct sequencing of the dopamine D2 receptor (DRD2) in schizophrenics reveals three polymorphisms but no structural change in the receptor, *Genomics*, 11, 8, 1991.
57. Sarkar, G. and Sommer, S.S., Haplotyping of double PCR amplification of specific alleles, *Biotechniques*, 10, 436, 1991.
58. Wise, R. A. and Rompre, P. P., Brain dopamine and reward, *Annu. Rev. Psychol.*, 40, 191, 1989.
59. Koob, S. F., Drugs of abuse. Anatomy, pharmacology and function of reward pathways, *Trends Pharmacol. Sci.*, 13, 177, 1992.
60. Fadda, R., Argiolas, A., Melis, M. R., Serra, G., and Gessa, G. I., Differential effect of acute and chronic ethanol on dopamine metabolism in frontal cortex, caudate nucleus and substantia nigra, *Life Sci.*, 27, 979, 1980.
61. Imperato, A. and Di Chiara, G., Preferential stimulation of dopamine release in the nucleus accumbens of freely moving rats by ethanol, *J. Pharmacol. Exp. Ther.*, 239, 219, 1986.
62. Boja, J. W. and Kuhar, M. J., [^{3}H]cocaine binding and inhibition of [^{3}H]dopamine uptake is similar in both rat striatum and nucleus accumbens, *Eur. J. Pharmacol.*, 173, 215, 1989.
63. Izenwasser, S., Werling, L. L., and Cox, B. M., Comparisons of the effects of cocaine and other inhibitors of dopamine uptake in rat striatum, nucleus accumbens, olfactory tubercle, and medial prefrontal cortex, *Brain Res.*, 520, 303, 1990.

64. Imperato, A., Mulas, A., and Di Chiara, G., Nicotine preferentially stimulates dopamine release in the limbic system of freely moving rats, *Eur. J. Pharmacol.*, 132, 337, 1986.
65. Brazell, M. P., Mitchell, S. N., Joseph, M. H., and Gray, J. A., Acute administration of nicotine increases the *in vivo* extracellular levels of dopamine, 3,4-dihydroxyphenylacetic acid and ascorbic acid preferentially in the nucleus accumbens of the rat. Comparison with caudate putamen, *Neuropsychopharmacology*, 29, 1177, 1990.
66. Heffner, T. G., Hartman, J. A., and Seiden, L. S., Feeding increases dopamine metabolism in the rat brain, *Science*, 208, 1168, 1980.
67. Hernandez L. and Hoebel G. B., Feeding and hypothalamic stimulation increases dopamine turnover in the accumbens, *Physiol. Behav.*, 44, 599, 1988.
68. Noble, E.P., Blum, K., Khalsa, M. E., Ritchie, T., Montgomery, A., Wood, R. C., Fitch, R. J., Ozkaragoz, T., Sheridan, P. J., Anglin, M. D., Paredes, A., Treiman, L. J., and Sparkes, R. S., Allelic association of the D2 dopamine receptor gene with cocaine dependence, *Drug Alcohol Depend.*, 33, 271, 1993.
69. Uhl, G., Persico, A. M., and Smith, S. S., Current excitement with D2 dopamine receptor gene alleles in substance abuse, *Arch. Gen. Psychiatry*, 49, 157, 1992.
70. Comings, D. E., Muhleman, D., Ahn, C., Gysin, R., and Flanagan, S. D., The dopamine D2 receptor gene: a genetic risk factor in substance abuse, *Drug Alcohol Depend.*, 34, 175, 1994.
71. Noble, E. P., St. Jeor, S. T., Ritchie, T., Syndulko, K., St. Jeor, S. C., Fitch, R. J., Brunner, R. L., and Sparkes, R. S., D2 dopamine receptor gene and cigarette smoking: a reward gene?, *Med. Hypotheses*, 42, 257, 1994.
72. Comings, D.E., Ferry, L., Bradshaw-Robinson, S., Burchette, R., Dino, M., Chiu, C., and Muhleman, D., Role of variants of the dopamine D2 receptor (DRD2) gene as genetic risk factors in smoking. Tobacco-Related Res. Dis. Prog. First Sci. Conf., San Francisco, December 2-3, p. 30, 1993.
73. Noble, E. P., Fitch, R. J., Syndulko, K., Ritchie, T., Zhang, X., and Sparkes, R. S., D2 dopamine receptor gene and behavioral characteristics in nicotine dependence, *Am. Soc. Hum. Genet.*, Suppl. 54, A168, 1994.
74. Eysenck, H. J. and Eysenck, S. B. G., *Eysenck Personality Questionnaire, Revised*, EdITS/Educational and Industrial Testing Service, San Diego, 1993.
75. Yesavage, J. A., Brink, T. L., Rose, T. L., Lum, O., Haug, V., Aden, M., and Von Otto, L., Development and validation of a geriatric depression screening scale: a preliminary report, *J. Psychiatr. Res.*, 17, 37, 1983.
76. Heatherton, T. F., Kozlowski, L. T., Frecker, R. C., and Fagerström, K.-0., The Fagerström test for nicotine dependence: a revision of the Fagerström Tolerance Questionnaire, *Br. J. Addiction*, 86, 1119, 1991.
77. Noble, E. P., Noble, R. E., Ritchie, T., Syndulko, K., Bohlman, M. C., Noble, L.A., Zhang, Y., Sparkes, R. S., and Grandy, D. K., D2 dopamine receptor gene and obesity, *Int. J. Eating Disord.*, 15, 205, 1994.
78. Comings, D. E., Flanagan, S. D., Dietz, G., Muhleman, D., Knell, E., and Gysin, R., The dopamine D2 receptor (DRD2) as a major gene in obesity and height, *Biochem. Med. Metab. Biol.*, 50, 176, 1993.
79. Noble, E. P., Blum, K., Ritchie, T., Montgomery, A., and Sheridan, P. J., Allelic association of the D2 dopamine receptor gene with receptor binding characteristics in alcoholism, *Arch. Gen. Psychiatry*, 48, 648, 1991.
80. Arinami, T., Itokawa, M., Komiyama, T., Mitsushio, H., Mori, H., Mifune, H., Hamaguchi, H., and Toru, M., Association between severity of alcoholism and the A1 allele of the dopamine D2 receptor gene Taq I A RFLP in Japanese, *Biol. Psychiatry*, 33, 108, 1993.
81. Gejman, P. V., Ram, A, Gelernter, J., Friedman, E., Cao, Q., Pickar, D., Blum, K., Noble, E. P., Kranzler, H. R., O'Malley, S., Hamer, D. H., Whitsitt, F., Rao, P., DeLii, L. E., Virkkunen, M., Linnoila, M., Goldman, D., and Gershon, E. S., No structural mutation in the dopamine D2 receptor gene in alcoholism or schizophrenia, *J. Am. Med. Assoc.*, 271, 204, 1994.
82. Eubanks, J. A., Djabali, M., Sellers, L., Grandy, D. K., Civelli, 0., McElligott, D. L., and Evans, G. A., Structure and linkage of the D2 dopamine receptor and neural cell adhesion molecular genes in human chromosome 11q23, *Genomics*, 14, 1010, 1992.
83. Noble, E. P., Berman, S. M., Ozkaragoz, T. Z., and Ritchie, T., Prolonged P300 latency in children with the D2 dopamine receptor A1 allele, *Am. J. Hum. Genet.*, 54, 658, 1994.
84. Berman, S. M. and Noble, E. P., Reduced visuospatial performance in children with the D2 dopamine receptor A1 allele, *Behav. Genet.*, 25, 45, 1995.
85. Lawford, B. R., Young, R. McD., Rowell, J. A., Qualichefski, J., Fletcher, B. H., Syndulko, K., Ritchie, T., and Noble, E. P., Bromocriptine in the treatment of alcoholics with the D2 dopamine receptor A1 allele, *Nat. Med.*, 1, 337, 1995.
86. Crabbe, J. C., Belknap, J. K., and Buck, K. J., Genetic animal models of alcohol and drug abuse, *Science*, 264, 1715, 1994.
87. Murphy, J. M., McBride, W. J., Lumeng, L., and Li, T.-K., Contents of monoamines in forebrain regions of alcohol-preferring (P) and nonpreferring (NP) lines of rats, *Pharmacol., Biochem. Behav.*, 26, 389, 1987.
88. Gongwer, M. A., Murphy, J. M., McBride, W. J., Lumeng, L., and Li, T.-K., Regional brain contents of serotonin, dopamine and their metabolites in the selectively bred high- and low-alcohol drinking lines of rats, *Alcohol*, 6 317, 1989.

89. Stefanini, E., Frau, M., Garau, M. G., Fadda, F., and Gessa, G. L., Alcohol-preferring rats have fewer dopamine D2 receptors in the limbic area, *Alcohol Alcoholism*, 27, 127, 1992.
90. McBride, W. J., Chernet, E., Dyr, W., Lumeng, L., and Li, T.-K., Densities of dopamine D2 receptors are reduced in CNS regions of alcohol-preferring P rats, *Alcohol*, 10, 387, 1993.
91. McBride, W. J., Murphy, J. M., Lumeng, L., and Li, T.-K., Serotonin, dopamine and GABA involvement in alcohol drinking of selectively bred rats, *Alcohol*, 7, 199, 1990.
92. Dyr, W., McBride, W. J., Lumeng, L., Li, T-K., and Murphy, J. M., Effects of D1 and D2 dopamine receptor agents on ethanol consumption in the High-Alcohol-Drinking (HAD) lines of rats, *Alcohol*, 10, 207, 1993.
93. Weiss, F., Mitchiner, M., Bloom, F. E., and Koob, G. F., Free-choice responding for ethanol vs. water in alcohol-preferring (P) and untreated Wistar rats is differentially modified by naloxone, bromocriptine and methysergide, *Psychopharmacology*, 101, 178, 1990.
94. Levy, A. D., Murphy, J. M., McBride, W. J., Lumeng, L., and Li, T.-K., Microinjection of sulpiride into the nucleus accumbens increases ethanol drinking in alcohol-preferring (P) rats, *Alcohol Alcoholism*, Suppl. 1, 417, 1990.
95. Britton, D. R., Curzon, P., MacKenzie, R. G., Kebabian, J. W., Williams, J. E. G., and Kerman, D., Evidence for involvement of both D1 and D2 receptors in maintaining cocaine self-administration, *Pharmacol., Biochem. Behav.*, 39, 799, 1991.
96. Corrigal, W. A. and Coen, K. M., Cocaine self-administration is increased by both D1 and D2 dopamine antagonists, *Pharmacol., Biochem. Behav.*, 39, 799, 1991.
97. Brown, Z. W., Gill, K., Abitbol, M., and Amit, Z., Lack of effect of dopamine receptor blockade in voluntary ethanol consumption in rats, *Behav. Neural Biol.*, 36, 291, 1982.
98. Pfeffer, A. O. and Samson, H. H., Effect of pimozide on home cage ethanol drinking in the rat. Dependence on drinking session length, *Drug Alcohol Depend.*, 17, 47, 1986.
99. Pfeffer, A. O. and Samson, H. H., Haloperidol and apomorphine effects on ethanol reinforcement in free-feeding rats, *Pharmacol., Biochem. Behav.*, 29, 343, 1988.
100. McClearn, G. E., Plomin, R., Gora-Maslak, G., and Crabbe, J. C., The gene chase in behavioral science, *Psychol. Sci.*, 2, 222, 1991.
101. Phillips, T. J., Crabbe, J. C., Metten, P., and Belknap, J. K., Localization of genes affecting alcohol drinking in mice, *Alcoholism: Clin. Exp. Res.*, 18, 931, 1994.
102. Crabbe, J. C., Belknap, J. K., Mitchell, S. R., and Crawshaw, L. I., Quantitative trait loci mapping of genes that influence the sensitivity tolerance to ethanol-induced hypothermia in BXD recombinant inbred mice, *J. Pharmacol. Exp. Ther.*, 209, 184, 1994.
103. Cunningham, C. L. and Malott, D., Ethanol-induced conditioned place preference in the BXD RI strains. Behavioral and QTL analyses, *Alcoholism: Clin. Exp. Res.*, 18, 451, 1994.
104. Copeland, N. G., Jenkins, N. A., Gilbert, D. J., Eppig, J. T., Maltais, L. J., Miller, J. C., Dietrich, W. F., Weaver, A., Lincoln, S. E., Steen, R. G., Stein, L. D., Nadeau, J. H., and Lander, E. S., A genetic linkage map of the mouse. Current applications and future prospects, *Science*, 262, 57, 1993.
105. Greenberg, D., Linkage analysis of "necessary" disease loci vs. "susceptibility" loci, *Am. J. Hum. Genet.*, 52, 135, 1993.

20 Polymorphisms of the D2 Dopamine Receptor Gene in Polysubstance Abusers

Antonio M. Persico and George R. Uhl

CONTENTS

A. Introduction 353
B. From "Genetic Factors" to Single Genes 354
    1. What Should We Expect From Single Genes in Polygenic Disorders? 354
    2. More Methodological Issues: Linkage and Association Analysis 354
C. Possible Associations Between Dopamine D2 Repector Gene Markers and Substance Abuse Vulnerability 356
    1. DRD2 Gene Structure, *Taq I* Polymorphisms, and Linkage Disequilibrium at the DRD2 Locus 359
    2. Other "Candidate Genes" 362
D. Conclusions 362
Acknowledgments 363
References 363

A. INTRODUCTION

Substance abuse vulnerability likely stems from complex interactions between genetic and environmental factors. Family, twin, and adoption studies all support the existence of genetic contributions, either directly conferring vulnerability to drug addiction or possibly predisposing to the development of clinical or personality disorders frequently associated with drug abuse (for review see Uhl et al., 1995). Siblings of drug abusers display significantly enhanced probabilities of abusing drugs, compared to the general population (Croughan, 1985; Mirin et al., 1991; Rousanville et al., 1991; Luthar et al., 1992; Luthar and Rousanville 1993). Monozygotic twins, who possess identical genomes, display much higher concordance rates for alcoholism and drug abuse than dizygotic twins, who share only 50% of their genes (Pickens et al., 1991; Tsuang et al., 1993). Adoption studies performed on children of biological parents with drug abuse or antisocial personality disorder reveal increased frequencies of drug addiction in the offspring, even when children are adopted into environments relatively free from these disorders (Cadoret et al., 1987; Cadoret et al., 1995). It is estimated from twin study data that 31% of vulnerability to drug abuse/dependence may be attributable to genetic influences in male twins identified on the basis of their alcoholic co-twins (Pickens et al., 1991). Although the mode of inheritance has not been conclusively proven, pedigree analyses display patterns compatible with polygenic transmission, modest penetrance, and high environmentally determined variability in expression (Pedersen 1984; Luthar et al., 1992).

Human genetic studies therefore suggest that genetic factors may exert a sizable impact on the etiology of substance abuse (Uhl et al., 1995). These results have prompted investigators to pursue single genes contributing to substance abuse liability. Their identification would significantly improve our understanding of the neurobiological bases of drug addiction and might aid treatment and prevention program design.

B. FROM "GENETIC FACTORS" TO SINGLE GENES

1. WHAT SHOULD WE EXPECT FROM SINGLE GENES IN POLYGENIC DISORDERS?

Attempts to identify a single gene contributing to drug addiction vulnerability require that the complexity of its likely genetic underpinnings be considered. Human genetic data available to date provide clear support for neither a single "drug addiction gene" hypothesis nor for simple Mendelian inheritance. Addiction-related "genetic factors" are more likely to consist of a combination of several genes, each present in the general population in distinct allelic versions. One or several possible combinations of alleles at these distinct loci could predispose to the development of drug addiction through the production of proteins altered in structure, function, or amount. Furthermore, only a few of these genes could be expected, when singly assessed, to exert detectable, albeit modest effects.

In this view, a single gene might have an impact on a specific feature of drug addiction and not necessarily on the overall presence/absence of a substance use disorder diagnosis. A specific gene might influence age of onset and exposure to drugs, possibly in connection with antisocial personality disorder, or presence and intensity of subjective and physiological drug-induced effects, characteristics and time course of tolerance and withdrawal, drug metabolism, and/or preferential use of specific substances. Therefore, potential contributions of single genes to the etiopathogenesis of a likely oligogenic disorder may then have to be sought in specific subgroups of patients, which should be sought within the framework of hypothesis-driven study designs. Their assessment may require use of specific markers, including neuroendocrine tests, subjective responses to pharmacological challenges, or receptor binding characteristics assessed postmortem or *in vivo* using PET scanning. Heterogeneity in amount of genetic susceptibility, in genes actually conferring such susceptibility at the individual level, in interactions between genetic and environmental factors, as well as the presence of phenocopy individuals who become addicted on a purely environmental basis, may dramatically reduce chances of detecting single-gene effects in polygenic disorders when defining "affected" and "unaffected" individuals purely on the basis of "presence" or "absence" of the disorder.

2. MORE METHODOLOGICAL ISSUES: LINKAGE AND ASSOCIATION ANALYSIS

The search for single genes can be typically carried out using two approaches, linkage and association analysis (Vogel and Motulsky, 1986). Both make use of polymorphic genetic markers, including Restriction-Fragment-Length-Polymorphisms (RFLPs), Variable-Number-Tandem-Repeats (VNTRs), simple sequence repeats, and Variations Affecting Protein Structure or Expression (VAPSEs). Except for VAPSEs, the vast majority of genetic markers most often have no functional meaning, but merely represent points of the genome where different individuals display different base sequences which can be identified using specific techniques. Genetic markers may either be anonymous, distributed throughout the genome, or located in close proximity of specific candidate genes, whose products are known to play relevant roles in addictive behaviors.

Classical pedigree linkage analysis searches for the cosegregation of genetic markers with a disorder in multigenerational families. Such cosegregation would be evidence that the marker is so close to the gene causing the disease that crossing-over during meiosis separates the two very rarely or never at all. Linkage analysis, which has proven extremely successful with monogenic disorders, displays serious limitations in sensitivity when applied to complex disorders such as drug addiction, characterized by polygenic transmission, frequent phenocopies and strong environmental influences on the expression of genetic predispositions (Gershon et al., 1989; Propping et al., 1993). Indeed, transgenerational changes in drug availability and in patterns of drug use make the drug addiction phenotype too unreliable to allow clearcut discrimination between "affected" and "unaffected" pedigree members. Although adoption studies suggest that alcoholism may share some genetic components with drug addiction (Cadoret et al., 1986; Cadoret et al., 1995), other studies clearly indicate that the overlap is far from complete. Rousanville and colleagues (1991), for example, found enhanced frequencies of alcoholics only in families of opiate addicts with comorbid alcoholism, and not in families of opiate addicts without drinking problems. Therefore, alcoholic parents and grandparents may not share the same affected phenotype present in young drug abusing probands, nor can we safely assume nonalcoholic relatives to be unaffected. Finally, linkage analysis requires specification of an exact mode of inheritance of a disease; despite some indications, conclusive evidence supporting a specific mode of inheritance in drug addiction is still lacking. Not surprisingly, no study applying classical linkage approaches to multiplex families with drug addiction has been published to date.

Association analysis, on the other hand, typically focuses upon candidate genes, whose protein products have been shown by neurobiological studies to be implicated in addictive behaviors. In the most common study design, a sample of unrelated affected individuals is contrasted with a sample of unrelated normal controls, matched for race and ethnicity. If distinct alleles of the candidate gene are present in different individuals, if one of these alleles provides significant contributions to substance abuse vulnerability, and if it can be recognized by the presence of a specific polymorphic marker, patients and controls will differ in marker distributions. A polymorphism specifically marking a predisposing allele will then display a positive association with the disease, appearing more frequently among affected than among unaffected individuals.

The process leading to recognition of the predisposing allele through the presence of a specific marker is called linkage disequilibrium. With the exception of VAPSEs, genetic markers do not usually bear any functional consequence per se, rather they merely represent polymorphic segments of the genome we can visualize using specific techniques. They may, however, be in linkage disequilibrium with another mutation exerting an impact on the disease phenotype. Linkage disequilibrium requires that the genetic marker and the pathogenic mutation be located within a relatively small distance so that marker and pathogenic mutation are rarely separated from each other during meiosis. They should thus cosegregate in the general population through many generations and not be separated by crossing over during meiotic recombination as frequently as anticipated for two equally distant DNA sites identified at random (for a more detailed discussion of linkage disequilibrium see Persico et al., 1993a).

Although association is more suitable than linkage analysis for studying polygenic disorders, it suffers from at least two major limitations. Linkage disequilibrium is not constant throughout the genome. Its extent should therefore be assessed for each candidate gene under scrutiny. Linkage disequilibrium can be measured by genotyping several polymorphic markers scattered throughout the locus of interest and analyzing their cosegregation rates in populations of interest (Persico et al., 1993a). Unfortunately, this preliminary assessment is not always available: for several genes of interest only single polymorphic sites have been described. In this case, lack of association may not necessarily exclude an etiological involvement of the gene. In the absence of strong linkage disequilibrium, in fact, studied markers might lie at even short distances from a relevant mutation, and yet display identical distributions in patients and controls. Single polymorphic markers, uninformative since in linkage equilibrium with pathogenetic mutations, may thus represent a relevant source of false-negative findings in association studies.

Secondly, both false-positive and false-negative results could stem from ethnic stratification. If patients and controls are not carefully matched for race and ethnicity, differences in genetic marker frequencies may merely reflect ethnic dishomogeneity between the two samples and have no relationship whatsoever with the disorder itself. Alternative approaches, such as haplotype relative risk (HRR)(Falk and Rubinstein, 1987; Knapp et al., 1993) and intrafamilial association (Swift et al., 1990; George and Elston, 1987; Hodge, 1993), have been developed to circumvent the risk of stratification. HRR compares the sample of affected individuals with a sample of ghost sibs, represented by the sets of two chromosomes each patient never inherited from his parents. Practically, this technique requires that each patient and both his parents be genotyped, providing information on the presence or absence of the genetic marker on the chromosomes that were not passed on to the affected offspring. Clinical characterization of the parents is thus irrelevant. HRR, however, suffers from problems frequently encountered working with drug addicts, namely, the unreliability of control frequencies in the presence of assortative mating, altered reproductive rates in families of affected individuals, and a limited power, which makes suitable sample-size adjustments or correct targeting of the appropriate clinical population perhaps even more critical than in classical case-control studies (Knapp et al., 1993).

These methodological issues, particularly those concerning association studies, each raise caution when interpreting results supporting allelic associations or rejecting this hypothesis. We can especially appreciate their relevance, when examining how they specifically apply to the association between dopamine D2 receptor (DRD2) gene polymorphisms and drug addiction.

C. POSSIBLE ASSOCIATIONS BETWEEN DOPAMINE D2 RECEPTOR GENE MARKERS AND SUBSTANCE ABUSE VULNERABILITY

Dopamine systems are prominently involved with the rewarding effects of addictive substances (Di Chiara and Imperato, 1988; Wise and Rompre, 1989; for review see Kuhar et al., 1991). Lesions or pharmacological blockade of mesolimbic dopaminergic terminals in the rat nucleus accumbens selectively block psychostimulant and ethanol, but not opiate self-administration (Pettit et al., 1984, Vaccarino et al., 1985; Pulvirenti et al., 1992; Rassnick et al., 1992; for review see Kuhar et al., 1991). Furthermore, cocaine and amphetamine act directly on dopaminergic systems (Ritz et al., 1987; Kuhar et al., 1991). Even addictive substances such as alcohol and nicotine, which do not directly alter dopamine function, can indirectly activate the mesolimbic/mesocortical dopaminergic system linked to reward (Di Chiara and Imperato, 1988). Genes involved in dopaminergic neurotransmission have therefore been recognized as reasonable candidate genes for genetic contributions to substance abuse.

The gene encoding the dopamine D2 receptor (DRD2) represents a candidate gene that displays population variants that may contribute to drug abuse liability and possibly modulate patterns of preferential substance use. Initial reports suggested that a specific variant of the

DRD2 gene, identified by the presence of the *Taq I* A1 RFLP marker, was much more frequent among alcoholics (24/35 = 69%) than in normal controls (7/35 = 20%) (Blum et al., 1990). Several groups subsequently explored this finding in alcoholics and in substance abusers using the *Taq I* A and other polymorphic markers identified at the DRD2 locus.

Five studies published to date compare frequencies of DRD2 gene polymorphic markers in U.S. drug abuser and control samples (Smith et al., 1992; O'Hara et al., 1993; Noble et al., 1993a; Comings et al., 1993; Gelernter et al., 1993a). *Taq I* A1 and B1 distributions, statistical comparisons between drug users and controls, and odds ratios for each study are summarized in Table 20-1. Merging results from separate studies is often the only procedure able to attain the statistical power required to identify modest genetic effects in complex disorders. Although caution is necessary when combining samples of individuals recruited according to different criteria and undergoing different degrees and types of assessment, nonetheless results from these studies display similar trends, with *Taq I* A1 and B1 markers more frequent in Caucasian drug abusers than in normal controls. Data from all five studies have therefore been combined (Table 20-1).

Meta-analyses of these studies appear to support a possible association between DRD2 genotypes and enhanced vulnerability to substance abuse, as shown by statistical significance of the differences between substance-abusing and control populations in both A1 and B1 frequencies (see Table 20-1). Polysubstance abusers display significantly increased A1 and B1 genotype frequencies, very similar to those recorded among severe alcoholics (for review see Uhl et al., 1992; Uhl et al., 1993; Persico et al., 1993a; Gelernter et al., 1993b; Noble et al., 1993b; Uhl et al., 1995). Conversely, lowest A1 and B1 frequencies are displayed by control groups selected for lack of relevant use of any addictive substance, through standardized interview-based methods (Smith et al., 1992; O'Hara et al., 1993; Noble et al., 1993a; Comings et al., 1993). Unassessed general population controls, likely including some individuals likely to manifest significant lifetime use of addictive drugs, display intermediate A1 and B1 frequencies (Noble et al., 1993a; Gelernter et al., 1993a).

In a polygenic disorder with large environmental influences such as substance abuse the effect of a single gene might conceivably modify specific clinical traits. Such an impact would be most readily detectable only in specific subsets of patients. The attempt to elucidate potential connections between DRD2 allelic status and specific clinical features of substance abuse has not thus far yielded univocal findings. Contrasting results have been produced regarding possible links between DRD2 polymorphisms and antisocial personality, commonly comorbid with substance abuse. Earlier onset of substance abuse and aggressive psychopathic behaviors were found associated with the DRD2 *Taq I* A1 marker by some (Commings et al., 1993; Noble et al., 1993a), but not by other groups (Smith et al., 1992; Smith et al., 1993). Moreover, initial evidence supported correlations of increased DRD2 A1 and B1 RFLP frequencies with severity of addictive processes, rather than with any one specific addiction alone (Blum et al., 1990; Smith et al., 1992; Uhl et al., 1993; Persico et al., 1993a). In fact, highest A1 and B1 DRD2 gene marker frequencies were initially recorded in polysubstance abusers with more severe lifetime histories of drug use, measured either as peak amount of daily drug intake (Smith et al., 1992) or as number of drugs abused (Commings et al., 1993). These observations, in conjunction with results obtained in other "compulsive diseases", have prompted the hypothesis that DRD2 alleles might be predictors of compulsivity and confer liability to a wide range of "impulsive-addictive-compulsive" behaviors (Blum et al., 1995a; Blum et al., 1995b).

Recent evidence, however, provides support for a potential role of DRD2 gene variants in modulating patterns of preferential drug use (Persico et al., 1996). We have selected polysubstance abusers reporting use of both psychostimulants and opiates, either concurrently or more often at different times in their life, and who had developed severe addiction to drugs in at least one of these two categories. On the basis of reported drug use patterns, and ***not*** of subjective drug liking, we have then defined a sample of "psychostimulant-preferring"

TABLE 20-1
Meta-Analyses of D2 Dopamine Receptor Gene *Taq I* A and B Polymorphisms in Caucasian Polysubstance Abusers and Controls

Ref.	*Taq I* A1					*Taq I* B1				
	Drug Abusers	Controls	Odds Ratio	χ^2 (2df)	P-value	Drug Abusers	Controls	Odds Ratio	χ^2 (2df)	P-value
Smith et al., 1992	40.4% (96/237)	28.1% (45/160)	1.74	5.86	0.015	32.1% (76/237)	21.9% (35/160)	1.69	4.43	0.035
O'Hara et al., 1993										
Noble et al., 1993a	50.9% (27/53)	16.0% (16/100)	5.44	19.24	0.00001	38.5% (20/52)	13.2% (7/53)	4.12	7.49	0.006
Comings et al., 1993	42.5% (45/106)	30.9% (25/81)	1.64	2.16	0.14					
Gelernter et al., 1993a	45.4% (49/108)	35.3% (24/68)	1.52	1.35	0.24					
Total	43.1% (217/504)	26.9% (110/409)	2.06	20.87	0.00003	33.2% (96/289)	19.7% (42/213)	2.03	10.36	0.006

Note: Data are expressed as percentage of *Taq I* A1 or B1 presence, either as homozygote or heterozygote form. Two-tail P-values and χ^2 tests for individual studies have been obtained using the CONTING program (Ott, 1991); odds ratios and χ^2 tests for the total sample are based on the Mantel-Haenszel statistic (Fleiss, 1981).

polysubstance users, with a history of daily, severe psychostimulant abuse and less severe peak lifetime use of opiate compounds; a sample of "opiate-preferring" addicts, who conversely reported prolonged daily use of opiates and sporadic use of psychostimulants; a "no preference" group, with equally severe use of both drugs; and a group of "heavy" polysubstance abusers with a lifetime history of severe, daily use of most or all drugs they had ever tried, including psychostimulants and opiates. Only polysubstance users with histories of heavy daily preferential psychostimulant use in our sample display significantly enhanced frequencies of the *Taq I* marker A1 (27/62 = 43.5% vs. 33/119 = 27.7% for controls; $\chi^2 = 3.92$; 1 df; one-tail P = 0.02), and B1 presence (20/62 = 32.3% vs. 23/119 = 19.3% for controls; $\chi^2 = 2.76$; 1 df; one-tail P = 0.04) (Persico et al., 1996). These results fit well with neurobiological data suggesting that molecular variants potentially implicated in mesolimbic/mesocortical dopaminergic function may more likely modulate psychostimulant, rather than opiate use. In fact, lesions or pharmacological blockade of mesolimbic dopaminergic terminals in the rat nucleus accumbens selectively block psychostimulant, but not opiate self-administration (Pettit et al., 1984; Vaccarino et al., 1985; Pulvirenti et al., 1992; for review see Kuhar et al., 1991). DRD2 genotype distributions in abusers with more prominent opiate use and in polysubstance abusers with no history of drug preference, including heavy polysubstance users, were similar to control genotypes. Our data are consistent with the hypothesis that DRD2 gene variants marked by these polymorphisms may work in concert with environmental factors to shape patterns of drug use, enhancing vulnerability to preferential psychostimulant abuse. Indeed, if these results were independently replicated, some of the sample-to sample-variability noted in the literature could be explained, based on differences in their clinical composition.

1. DRD2 GENE STRUCTURE, *TAQ I* POLYMORPHISMS, AND LINKAGE DISEQUILIBRIUM AT THE DRD2 LOCUS

In order to appreciate strengths and limitations of association studies using DRD2 gene markers in polysubstance abuse, some of the methodological issues previously raised must be applied to the specific context of DRD2 gene structure and linkage disequilibrium.

The D2 dopamine receptor (DRD2) gene comprises 8 exons, separated by 7 introns, and encodes a G-protein linked, 7-transmembrane domain receptor protein (Grandy et al., 1989a; Grandy et al., 1989b; Hauge et al., 1991). This locus also comprises polymorphic markers, produced by the restriction endonuclease *Taq I*, and defined Restriction Fragment Length Polymorphisms (RFLPs) (Botstein et al., 1980). As *Taq I* cuts genomic DNA whenever encountering its specific restriction sites, it generates DNA fragments that can be visualized following hybridization with radioactively-labeled DRD2 probes (i.e., Southern blotting). If enzymatic cleavage is prevented in one of these sites even by a single base gene variant, a different fragment pattern will be generated. The DRD2 *Taq I* A polymorphism thus stems from single-base mutations in the 3' flanking region of the DRD2 gene, affecting a restriction site located more than 10 kb from the eighth exon (Figure 20-1). These mutations yield at least four different restriction fragment patterns, termed A1 through A4 (Persico et al., 1993b). A1 and A2 are the most frequent patterns, accounting for more than 95% of *Taq I* A polymorphisms in the general population (Persico et al., 1993b; O'Hara et al., 1993). A second polymorphism, the *Taq I* B RFLP, lies close to the first intron/second exon junction (Hauge et al., 1991). Three B fragment patterns have been described, with B1 and B2 again accounting for the vast majority of individuals (Hauge et al., 1991; Persico et al., 1993b; O'Hara et al., 1993). These RFLPs have provided the data used in most studies so far. Another polymorphism located between the *Taq I* A and B sites is the *Taq I* C, initially termed "D" (Parsian et al., 1991). The restriction site generating the *Taq I* C polymorphism is located within the second intron, thus closer to *Taq I* B than to A, and generates two distinct fragment patterns, defined "C1" and "C2" (Parsian et al., 1991). Finally, several other DRD2 polymorphisms have been generated using other techniques, such as polymerase chain reaction (Bolos et al., 1991; Hauge et al., 1991; Sarkar et al., 1991; Flanagan et al., 1992), and will not be considered here.

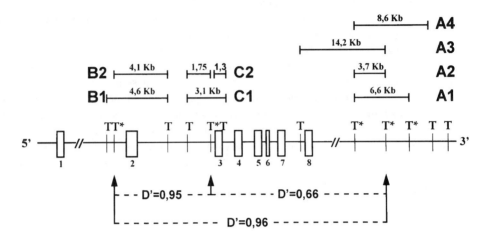

FIGURE 20-1 Diagram of the human D2 dopamine receptor gene locus with relative size and position of *Taq I* A, B, and C RFLPs.[21,24] Kb = kilobases; horizontal line = introns and flanking regions; open boxes = exons. Protein coding regions lie in exons 2 through 8. Intron 1 size (>100 to 250 kb) is not to scale (O. Civelli, J. Mallet, and R. Todd, personal communications). T* = polymorphic *Taq I* restriction site; T = nonpolymorphic *Taq I* restriction site. D' is a measure of linkage disequilibrium, ranging from 0 to 1. For example, D' = 0.96 corresponds to 96% of the maximum linkage disequilibrium possibly present between two genomic sites.

The potential association of *Taq I* A1 and B1 fragment patterns with drug addiction could possibly stem from two mechanisms. Pathogenetically relevant changes in DRD2 gene function could be produced by the same mutation that is responsible for differences in marker size. This event indeed appears quite unlikely. Contrary to VAPSE-generating mutations, RFLP-generating single-base changes are usually located in gene regions not encoding mRNA and are therefore unlikely to exert such direct impact. A more probable mechanism involves the existence of linkage disequilibrium between the mutation responsible for creating the *Taq I* A1 marker and another, currently unknown mutation influencing substance abuse behaviors. Linkage disequilibrium at the DRD2 locus would allow *Taq I* RFLP markers and the putative receptor-modifying mutation to cosegregate in the general population through hundreds of generations and not be separated by meiotic crossing-over as frequently, as anticipated for two equally distant DNA loci, should this process occur at random. An anonymous marker, such as A1, would then "mark" a specific allele of the DRD2 gene characterized by the presence of a mutation, predisposing to addictive behaviors possibly through alterations in receptor number or function. A 30% reduction in DRD2 receptor B_{max} found in the striata of individuals with at least one copy of the A1 marker, regardless of the presence of alcoholism (Noble et al., 1991), decreased DRD2 B_{max}/K_d ratios in alcohol-dependent subjects *in vivo* (Hietala et al., 1994), decreased DRD2 binding in brains of alcohol-preferring rodents (Stefanini et al., 1992; McBride et al., 1993), and reduced DRD2 sensitivity evidenced through blunted GH responses to apomorphine in chronic and abstinent alcoholics (Balldin et al., 1992, Balldin et al., 1993, Heinz et al., 1995) each could provide support for genetic control of variety in DRD2 receptor density. Quantitative trait loci (QTL) analyses supporting the existence of a gene in the proximity of the DRD2 locus providing genetic contributions to enhanced ethanol sensitivity (Crabbe et al., 1994; Phillips et al., 1994) and the lack of mutations in the exons of "A1-marked" DRD2 alleles (Geijman et al., 1994) all suggest that this putative mutation could well be located in 3' or 5' non-coding regions of the gene, possibly involved in transcriptional regulation and/or in mRNA stability, and ultimately affecting DRD2 receptor number. Alternatively, the mutation could occur in other genes located in close proximity of the *Taq I* A marker, and not in the DRD2 gene itself. Interestingly, relevant loci such as the N-CAM gene are located in the vicinity of these polymorphysms (Eubanks et al., 1992).

Haplotype analyses performed assessing several different DRD2 polymorphic markers in the same individuals, had revealed a surprising amount of linkage disequilibrium within the DRD2 locus, as shown in Figure 20-1. *Taq I* A1 and B1 RFLPs, located at roughly opposing ends of the gene, display more than 96% of the maximum possible disequilibrium (D_{max}), an impressive figure when compared with the 30 to 40% of D_{max} values anticipated if average recombination rates were manifest between the A and B sites (O'Hara et al., 1993; Uhl et al., 1993). Furthermore, the amount of linkage disequilibrium is not constant through the gene, but differs in distinct segments of the DRD2 locus (Figure 20-1). Finally, recombination rates are higher in some races than in others. Recent three-point analyses of A, B, and C loci indicate that overall disequilibrium is present in Caucasian-Americans, while it is much weaker among African-Americans (Persico, Uhl, Suarez, manuscript in preparation). Races also differ in haplotype frequencies: A1 and B1 tend to cosegregate with C2 in whites, while they most frequently cosegregate with C1 in blacks. These data suggest that the current complement of chromosomes with DRD2 polymorphic marker heterogeneity stems from complex population genetic phenomena. One plausible explanation could be that "primordial" chromosomal sequences were marked by A1 and B1 RFLPs at 3' and 5' ends of the DRD2 locus, respectively, prior to the time when decreased genetic flow between Africa and Europe originated the black and caucasian races recognized today (Persico et al., 1993a). Founder effects might then explain the higher frequency of A1/B1/C1 chromosomes in black than in white populations, where most of these chromosomes are marked by A1/B1/C2. Moreover, since the divergence of these races has occurred, recombination rates between the flanking segments of the gene marked by A and B and the region marked by C may have also been higher in blacks than in caucasians. Based on these phenomena, it is predictable that among African-Americans the *Taq I* RFLP and the putative receptor-modifying mutation should have been separated by crossing-over during meiotic recombination significantly more often than among Caucasian-Americans, possibly leading to weaker or no association between RFLP markers and disease phenotype. Indeed, no association between drug addiction and DRD2 alleles has been found in studies performed on African-American samples so far (O'Hara et al., 1993; Berrettini and Persico, manuscript in preparation).

The complex population genetic phenomena we have summarized clearly suggest caution in assuming that amounts of disequilibrium between polymorphic markers and putative pathogenetic mutations involved in addictive behaviors should be homogeneous in different racial or ethnic groups. Different amounts of disequilibrium represent one of the factors critically undermining generalizability of conclusions among ethnically distinct samples. Lack of association between DRD2 *Taq I* markers and addiction could occur in studies of populations displaying significantly less linkage disequilibrium between *Taq I* markers and the mutation relevant for addictive processes, compared with the Caucasian-American populations of mixed European heritage in which allelic association has been most frequently reported.

In addition to variability in linkage disequilibrium and haplotype frequencies, sample stratification represents another factor potentially capable of contributing to false-positive or false-negative errors in case-control association studies. Different racial and ethnic groups differ not only in linkage disequilibrium and haplotype frequencies, as we have seen, but also in A1 and B1 marker frequencies (Barr and Kidd, 1993; O'Hara et al., 1993; Gelernter et al., 1993b; Goldman et al., 1993). Unbalanced sampling of patients and controls from ethnic groups characterized by different A1 and B1 frequencies might produce false-positive or false-negative results, leading to over- or underestimation of DRD2 gene variant involvement in addictive processes. To reduce the chance of stratification bias, most studies have excluded subjects reporting Hispanic or American Indian origin, because allelic and genotype frequencies in these groups are known to differ significantly from those found in Caucasian Americans of largely European descent (Gelernter et al., 1993b; Barr and Kidd, 1993; Goldman et al., 1993). To date, no significant difference in DRD2 allelic distributions between distinct European ethnic groups has been described (Blum et al., 1995b). A preliminary assessment suggests that genotypic and allelic frequencies of Caucasian-Americans originating from different European countries may actually show more heterogeneity in human vesicular transporter gene polymorphisms, as previously described by Surratt and colleagues (1993), and found to display no association with substance abuse, than in DRD2 *Taq I* A and B markers (Persico and Uhl, manuscript in preparation). Furthermore, if stratification were the unique source of association between DRD2 markers and substance abuse, one would expect the same sample might yield similarly spurious associations with other dishomogeneously distributed polymorphisms. We have recently demonstrated that polymorphic markers for the dopamine transporter (DAT1) gene, while displaying significant interracial and interethnic differences in allelic distributions (Persico et al., 1996), show no association with polysubstance abuse (Persico et al., 1993c) or with psychostimulant preference in our sample (Persico et al., 1996). These results, together with an apparent lack of significant differences in DRD2 genotypic frequencies among distinct European ethnic groups or among Caucasian-Americans of distinct European origin, suggest that positive findings are less likely to stem from stratification and chance alone. Alternative approaches, such as Haplotype Relative Risk (HRR) (Falk and Rubinstein, 1987; Knapp et al., 1993) and intrafamilial association (Swift et al., 1990; George and Elston, 1987; Hodge, 1993), circumvent the risks of stratification and should also be employed in future studies.

2. OTHER "CANDIDATE GENES"

Even at maximal magnitude estimates, current attributable risk calculations suggest that the impact of DRD2 gene variants must be complemented by other genetic and environmental factors, responsible for the majority of variance in individual vulnerability to substance abuse (Uhl et al., 1993; Uhl et al., 1995). Polymorphic markers for other "candidate" genes potentially relevant to addictive behaviors, such as the dopamine transporter and the vesicular monoamine transporter, have not displayed significant associations with polysubstance abuse (Persico et al., 1993c; Persico and Uhl, manuscript in preparation). The limitations of these negative conclusions must be acknowledged, however, since the extent of linkage disequilibrium, making the *Taq I* A and B DRD2 gene markers so valuable in possibly identifying variants at regulatory or coding sequences, has not been assessed at these other loci. Among environmental factors, a recent study suggests that DRD2 allelic association with psychostimulant preferring-patterns of drug use can be accompanied by earlier onset of use of drugs "preferred" later in life (Persico et al., 1996). Interestingly, age of first use of any drugs, namely alcohol, tobacco or marijuana, displays no correlation with extent of later polysubstance abuse, while an early concomitant exposure to both psychostimulants and opiates appears to occur in most "heavy" polysubstance users. Conceivably, craving-inducing stimuli may more promptly precipitate use of heavy drugs which have "primed" genetically predisposed individuals at an earlier stage.

D. CONCLUSIONS

The association between DRD2 gene markers and substance abuse, supported by most studies published to date, continues to be worthy of exploration despite the caveats and limitations discussed here and elsewhere (Uhl et al., 1992; Uhl et al., 1993; Persico et al., 1993a; Goldman et al., 1993; Gelernter et al., 1993b; Uhl et al., 1995). The pathogenetic relevance of a putative DRD2 gene variant marked by the presence of A1 and B1 polymorphisms in addictive behaviors will require further support from unequivocal demonstrations of its pathophysiological correlates, such as potential differences in receptor binding characteristics *in vivo*, in psychostimulant-induced subjective effects, in event-related potentials, or in drug-induced neurohormonal responses. In fact, although decreased DRD2 B_{max}/K_d ratios have been recorded in alcohol-dependent subjects *in vivo* using PET (Hietala et al., 1994), whether they are present prior to alcohol abuse or only a consequence remains to be established. Blunted GH responses to apomorphine persist in abstinent alcoholics studied for 8 to 216 months following alcohol use cessation, strongly suggesting that reduced DRD2 function may be a permanent, possibly predisposing trait (Balldin et al., 1992; Balldin et al., 1993). Its association with early relapse in alcohol-dependent patients (Heinz et al., 1995) suggests a potential role for hyposensitive dopaminergic systems in severe alcoholism, which has in turn been found associated with DRD2 *Taq I* A1 and B1 RFLPs (Blum et al 1991; Blum et al., 1993). It will thus be of critical importance to directly assess the potential relationship between blunted GH responses to apomorphine and DRD2 genotypes. Using electrophysiological approaches, homozygote A1/A1 individuals may display prolonged P300 latency, but no difference between A1/A2 and A2/A2 individuals has been recorded in this sample (Blum et al., 1994). More work is therefore necessary to replicate, extend, and interpret these findings.

Pathogenetically relevant mutations will finally have to be identified in order for us to begin understanding the biological bases of the impact of the DRD2 gene on such complex behaviors as drug addiction and alcoholism. Only then will the dopamine D2 receptor gene have conclusively proven to be among the first individual genes whose role in a polygenic disorder with large environmental influences has been identified.

ACKNOWLEDGMENTS

The authors would like to thank Bruce O'Hara, Steven Smith, and Geoffrey Bird for their extensive contribution to the studies reviewed here. This work was supported by the intramural program of the National Institute on Drug Abuse.

REFERENCES

Balldin, J.I., Berggren, U.C., and Lindstedt, G., Neuroendocrine evidence for reduced dopamine receptor sensitivity in alcoholism, *Alc. Clin. Exp. Res.*, 16:71–74, 1992.

Balldin, J.I., Berggren, U.C., Lindstedt, G., and Sundkler, A., Further neuroendocrine evidence for reduced D2 dopamine receptor function in alcoholism, *Drug Alc. Depend.*, 32:159–162, 1993.

Barr, C.L. and Kidd, K.K., Population frequencies of the A1 allele at the dopamine D2 receptor locus, *Biol. Psychiatry*, 34:204–209, 1993.

Blum, K., Noble, E.P., Sheridan, P.J., Montgomery, A., Ritchie, T., Jagadeeswaran, P., Noble, E.P., Nogami, H., Briggs, A.H., and Cohn, J.B., Allelic association of human dopamine D2 receptor gene in alcoholism, *JAMA*, 263:2055–2060, 1990.

Blum, K., Noble, E.P., Sheridan, P.J., Finley, O., Montgomery, A., Ritchie, T., Ozkaragoz, T., Fitch, R.J., Sadlack, F., Sheffield, D., Dahlmann, T., Halbardier, S., and Nogami, H., Association of the A1 allele of the D2 dopamine receptor gene with severe alcoholism, *Alcohol*, 8:409–416, 1991.

Blum, K., Noble, E.P., Sheridan, P.J., Montgomery, A., Ritchie, T., Ozkaragoz, T., Fitch, R.J., Wood, R., Finley, O., and Sadlack, F. Genetic predisposition in alcoholism: association of the D2 dopamine receptor *Taq I* B1 RFLP with severe alcoholism, *Alcohol*, 10:59–67, 1993.

Blum, K., Braverman, E.R., Dinardo, M.J., Wood, R.C., and Sheridan, P.J., Prolonged P300 latency in a neuropsychiatric population with the D2 dopamine receptor A1 allele, *Pharmacogenetics*, 4:313–322, 1994.

Blum, K., Wood, R.C., Braverman, E.R., Chen, T.J.H., and Sheridan, P.J. The D2 dopamine receptor gene as a predictor of compulsive disease: Bayes' theorem, *Functional Neurol.,* 10:37–44, 1995a.

Blum, K., Sheridan, P.J., Wood, R.C., Braverman, E.R., Chen, T.J.H., and Comings, D.E. Dopamine D2 dopamine receptor gene variants: association and linkage studies in impulsive-addictive-compulsive behavior, *Pharmacogenetics,* 5:121–141, 1995b.

Bolos, A.M., Dean, M., Lucas-Derse, S., Ramsburg, M., Brown, G.L., and Goldman, D., Population and pedigree studies reveal a lack of association between the dopamine D2 receptor gene and alcoholism, *JAMA,* 264:3156–3160, 1990.

Botstein, D., White, R.L., Skolnick, M., and Davis, R.W., Construction of a genetic linkage map in man using Restriction Fragment Length Polymorphysms, *Am. J. Hum. Genet.,* 32:314–331, 1980.

Cadoret, R.J., O'Gorman, T., Troughton, E., and Heywood, E., An adoption study of genetic and environmental factors in drug abuse, *Arch. Gen. Psychiatry,* 43:1131–1136, 1986.

Cadoret, R.J., Yates, W.R., Troughton, E., Woodworth, G., and Stewart, M.A., Adoption study demonstrating two genetic pathways to drug abuse, *Arch. Gen. Psychiatry,* 52:42–52, 1995.

Commings, D.E., MacMurray, J., Johnson, J.P., Muhleman, D., Ask, M.N., Ahn, C., Gysin, R., and Flanagan, S., The dopamine D2 receptor gene: a genetic risk factor in substance abuse, *Drug Alc. Depend.,* 34:175–180, 1993.

Crabbe, J.C., Belknap, J.K., and Buck, K.J., Genetic animal models of alcohol and drug abuse, *Science,* 264:1715-1723, 1994.

Croughan, J.L., The contribution of family studies to understanding drug abuse, in, *Studying Drug Abuse, vol. 6,* Robbins L., Ed., Rutgers University Press, New Brunswick, Canada, pp 93–116, 1985.

Di Chiara G. and Imperato A. Drugs abused by humans preferentially increase synaptic dopamine concentrations in the mesolimbic system of freely moving rats, *Proc. Natl. Acad. Sci. U.S.A.,* 85:5274–5278, 1988.

Eubanks, J.H., Djabali, M., Selleri, L., Grandy, D.K., Civelli, O., McElligott, D.L., and Evans, G.A., Structure and linkage of the D2 dopamine receptor and neural cell adhesion molecule genes on human chromosome 11q23, *Genomics,* 14:1010–1018, 1992.

Falk, C.T. and Rubinstein, P., Haplotype relative risk: an easy reliable way to construct a proper control sample for risk calculations, *Ann. Hum. Genet.,* 51:227–233, 1987.

Flanagan, S.D., MacMurray, J., Comings, D., Johnson, J., Lopatin, G., and Gysin, R., Dopamine D2 receptor (DRD2) haplotype status and genetic risk for alcoholism and polysubstance abuse. Proc XVIIIth Congress of the Collegium Internationale Neuro-Psychopharmacologium, *Clin. Neuropharmacol.,* 15 (Suppl.1):97B, 1992.

Fleiss, J.L., *Statistical Methods for Rates and Proportions,* 2nd Ed, John Wiley and Sons, 1981.

Gabbay, F.H., Duncan, C.C., Bird, G., Uhl, G.R., and Mirsky, A.F., D2 dopamine receptor *Taq I* A genotypic differences in event-related brain potentials, *NIDA Research Monograph, Problems of Drug Dependence,* Proceedings of the 57th Annual Scientific Meeting, College on Problems in Drug Dependence, (in press).

Gejman, P.V., Ram, A., Gelernter, J., Friedman, E., Cao, Q., Pickar, D., Blum, K., Noble, E.P., Kranzler, H.R., O'Malley, S., Hamer, D.H., Whitsitt, F., Rao, P., DeLisi, L.E., Virkkunen, M., Linnoila, M., Goldman, D., and Gershon, E.S., No structural mutation in the dopamine D2 receptor gene in alcoholism or schizophrenia, *JAMA,* 271:204–208, 1994.

Gelernter, J., Kranzler, H., and Satel, S., No association between DRD2 alleles and cocaine abuse, *College on Problems of Drug Dependence,* Toronto, Canada, annual meeting, June 16, 1993a.

Gelernter, J., Goldman, D., and Risch, N. The A1 allele at the D2 dopamine receptor gene and alcoholism: a reappraisal, *JAMA,* 269:1673–1677, 1993b.

George, V.T. and Elston, R.C., Testing the association between polymorphic markers and quantitative traits in pedigrees, *Genet. Epidemiol.,* 4:193–201, 1987.

Gershon E.S., Martinez M., Goldin L., Gelernter J., and Silver J. Detection of marker associations with a dominant disease gene in genetically complex and heterogeneous diseases. *Am. J. Hum. Genet.,* 45:578–585, 1989.

Goldman, D., Brown, G.L., Albaugh, B., Robin, R., Goodson, S., Trunzo, M., Akhtar, L., Lucas-Derse, S., Long, J., Linnoila, M., and Dean, M. DRD2 dopamine receptor genotype, linkage disequilibrium, and alcoholism in American Indians and other populations, *Alc. Clin. Exp. Res.,* 17:199–204, 1993.

Grandy, D.K., Marchionni, M.A., Makam, H., Stofko, R.E., Alfano, M., Frothingham, L., Fischer, J., Burke-Howie, K.J., Bunzow, J.R., Server, A.C., and Civelli, O., Cloning of the cDNA and gene for a human D2 dopamine receptor, *Proc. Natl. Acad. Sci. U.S.A.,* 86:9762–9766, 1989a.

Grandy, D.K., Litt, M., Allen, L., Bunzow, J.R., Marchionni, M., Makam, H., Reed, L., Megenis, R.E., and Civelli, O., The human dopamine D2 receptor gene is located on chromosome 11 at q22-q23 and identifies a *Taq I* RFLP, *Am. J. Hum. Genet.,* 45:778–785, 1989b.

Hauge, X.Y., Grandy, D.K., Eubanks, J.H., Evans, G.A., Civelli, O., and Litt, M., Detection and characterization of additional DNA polymorphisms in the dopamine D2 receptor gene, *Genomics,* 10:527–530, 1991.

Heinz, A., Dettling, M., Kuhn, S., Dufeu, P., Graf, K.J., Kurten, I., Rommelspacher, H., and Schmidt, L.G., Blunted growth hormone response is associated with early relapse in alcohol-dependent patients, *Alc. Clin. Exp. Res.,* 19:62–65, 1995.

Hietala, J., West, C., Syvalahti, E., Nagren, K., Lehikoinen, P., Sonninen, P., and Ruotsalainen, U., Striatal D2 dopamine receptor binding characteristics in vivo in patients with alcohol dependence, *Psychopharmacology*, 116:285–290, 1994.

Hodge, S.E., Linkage analysis versus association analysis: distinguishing between two models that explain disease-marker associations, *Am. J. Hum. Genet.*, 53:367–384, 1993.

Knapp, M., Seuchter, S.A., and Baur, M.P., The Haplotype-Relative-Risk (HRR) method for analysis of association in nuclear families, *Am. J. Hum. Genet.*, 52:1085–1093, 1993.

Kuhar, M.J., Ritz, M.C., and Boja, J.W., The dopamine hypothesis of the reinforcing properties of cocaine, *Trends Neurosci.*, 14:299–302, 1991.

Luthar, S.S., Anton, S.F., Merikangas, K.R., and Rousanville, B.J., Vulnerability to substance abuse and psychopathology among siblings of opioid abusers, *J. Nerv. Ment. Dis.*, 180:153–161, 1992.

McBride, W.J., Chernet, E., Dyr, W., Lumeng, L., and Li, T.-K., Densities of dopamine D2 receptors are reduced in CNS regions of alcohol-preferring P rats, *Alcohol*, 10:387–390, 1993.

Mirin, S.M., Weiss, R.D., Griffin, M.L., and Michael, J.L., Psychopathology in drug abusers and their families, *Compr. Psychiatry*, 32:36–51, 1991.

Noble, E.P., Blum, K., Ritchie, T., Montgomery, A., and Sheridan, P.J., Allelic association of the D2 dopamine receptor gene with receptor-binding characteristics in alcoholism, *Arch. Gen. Psychiatry*, 48:648–654, 1991.

Noble, E.P., Blum, K., Khalsa, M.E., Ritchie, T., Montgomery, A., Wood, R.C., Fitch, R.J., Ozkaragoz, T., Sheridan, P.J., Anglin, M.D., Paredes, A., Treiman, L.J., and Sparkes, R.S., Allelic association of the D2 dopamine receptor gene with cocaine dependence, *Drug Alcohol Depend.*, 33:271–285, 1993a.

Noble, E.P., The D2 dopamine receptor gene: a review of association studies in alcoholism, *Behav. Genet.*, 23:119-129, 1993b.

O'Hara, B.F., Smith, S.S., Bird, G., Persico, A.M., Suarez, B., Cutting, G.R., and Uhl, G.R., Dopamine D2 receptor RFLPs, haplotypes and their association with substance use in black and caucasian research volunteers, *Hum. Hered.*, 43:209–218, 1993.

Ott, J., *Analysis of Human Genetic Linkage*, rev. ed., Baltimore, MD: Johns Hopkins University Press, 1991.

Parsian, A., Fisher, L., O'Malley, K.L., and Todd, R.D., A new *Taq I* RFLP within intron 2 of human dopamine D2 receptor gene (DRD2), *Nucleic Acids Res.*, 19:6977, 1991.

Pedersen, N.L., Multivariate analysis of familial and non-familial influences for commonality in drug use, *Drug Alcohol Depend.*, 14:67–74, 1984.

Persico, A.M., Smith, S.S., and Uhl, G.R., D2 receptor gene variants and substance abuse liability, *Seminars Neurosci.*, 5:377–382, 1993a.

Persico, A.M., O'Hara, B.F., Farmer, S., and Uhl, G.R., Dopamine D2 receptor gene *Taq I* "A" locus map including "A4" variant: relevance for alcoholism and drug abuse, *Drug Alcohol Depend.*, 31:229–234, 1993b.

Persico, A.M., Vandenbergh, D.J., Smith, S.S., and Uhl, G.R., Dopamine transporter gene markers are not associated with polysubstance abuse, *Biol. Psychiatry*, 34:265–267, 1993c.

Persico, A.M., Bird, G., Gabbay, F., and Uhl, G.R., The D2 dopamine receptor gene *Taq I* A1 and B1 RFLPs: enhanced frequencies in psychostimulant-preferring polysubstance abusers, *Biol. Psychiatry*, (in press).

Pettit, H.O., Ettenberg, A., Bloom, F.E., and Koob, G.F., Destruction of dopamine in the nucleus accumbens selectively attenuates cocaine but not heroin self-administration in rats, *Psychopharmacology*, 84:167–173, 1984.

Phillips, T.J., Crabbe, J.C., Metten, P., and Belknap, J.K., Localization of genes affecting alcohol drinking in mice, *Alcohol Clin. Exp. Res.*, 18:931–941, 1994.

Pickens, R.W., Svikis, D.S., McGue, M., Lykken, D.T., Heston, L.L., and Clayton, P.J., Heterogeneity in the inheritance of alcoholism: a study of male and female twins, *Arch. Gen. Psychiatry*, 48:19–28, 1991.

Propping, P., Nothen, M.M., Fimmers, R., and Baur, M.P., Linkage versus association studies in complex diseases, *Psychiat. Genet.*, 3:136, 1993.

Pulvirenti, L., Maldonado, R., and Koob, G.F., NMDA-receptors in the nucleus accumbens modulate intravenous cocaine but not heroin self-administration in the rat, *Brain Res.*, 594:327–330, 1992.

Rassnick, S., Pulvirenti, L., and Koob, G.F., Oral ethanol self-administration in rats is reduced by the administration of dopamine and glutamate receptor antagonists in the nucleus accumbens, *Psychopharmacology*, 109:92–98, 1992.

Ritz, M.C., Lamb, R.J., Goldberg, S.R., and Kuhar, M.J., Cocaine receptors on dopamine transporters are related to self-administration of cocaine, *Science*, 237:1219–1223, 1987.

Rousanville, B.J., Kosten, T.R., Weissman, M.M., Prusoff, B., Pauls, D.L., Anton, S.F., and Merikangas, K., Psychiatric disorders in relatives of probands with opiate addiction, *Arch. Gen. Psychiatry*, 48:33–42, 1991.

Sarkar, G. and Sommer, S.S., Haplotyping by double PCR amplification of specific alleles, *Biotechniques*, 10:436–440, 1991.

Smith, S.S., O'Hara, B.F., Persico, A.M., Gorelick, D.A., Newlin, D.B., Vlahov, D., Solomon, L., Pickens, R., and Uhl, G.R., Genetic vulnerability to drug abuse — The D2 dopamine receptor *Taq I* B1 restriction fragment length polymorphism appears more frequently in polysubstance abusers, *Arch. Gen. Psychiatry*, 49:723–727, 1992.

Smith, S.S., Newman, J.P., Evans, A., Pickens, R., Wydeven, J., Uhl, G.R., and Newlin, D.B., Comorbid psychopathy is not associated with increased D2 dopamine receptor *Taq I* A or B gene marker frequencies in incarcerated substance abusers, *Biol Psychiatry*, 33:845–848, 1993.

Stefanini, F., Frau, M., Garau, M.G., Garau, B., Fadda, F., and Gessa, G.L., Alcohol-preferring rats have fewer dopamine D2 receptors in the limbic system, *Alcohol Alcoholism*, 27:127–130, 1992.

Surratt, C.K., Persico, A.M., Yang, X.D., Edgar, S.R., Bird, G.S., Hawkins, A.L., Griffin, C.A., Li, X., Jabs, E.W., and Uhl, G.R., A human synaptic vesicle monoamine transporter cDNA predicts posttranslational modifications, reveals chromosome 10 gene location and identifies *Taq I* RFLPs, *FEBS Lett.*, 318:325–330, 1993.

Swift, M., Kupper, L.L., and Chase, C.L., Effective testing of gene-disease associations, *Am. J. Hum. Genet.*, 47:266–274, 1990.

Tsuang, M.T., Lyons, M.J., Eisen, S.E., Goldberg, J., and True, W.T., Heritability of initiation and continuation of drug use, *Psychiat. Genet.*, 3:141, 1993.

Uhl, G.R., Persico, A.M., and Smith, S.S., Current excitement with D2 receptor gene alleles in substance abuse, *Arch. Gen. Psychiatry*, 49:157–160, 1992.

Uhl, G.R., Blum, K., Noble, E., and Smith, S.S., Substance abuse vulnerability and D2 receptor genes. *Trends Neurosci.*, 16:83–88, 1993.

Uhl, G.R., Elmer, G., Labuda, M., and Pickens, R., Genetic influences in drug abuse, in *Psychopharmacology: The Fourth Generation of Progress*, Bloom, F.E. and Kupfer, F., Eds., Raven Press, New York, 1995.

Vaccarino, F.J., Bloom, F.E., and Koob, G.F., Blockage of nucleus accumbens opiate receptors attenuates intravenous heroin reward in the rat, *Psychopharmacology*, 85:37–42, 1985.

Vogel, F. and Motulsky, A.G., *Human Genetics — Problems and Approaches*, 2nd ed., Springer-Verlag, Berlin, Germany, 1986.

Wise, R.A. and Rompre, P.P., Brain dopamine and reward, *Annu. Rev. Psychol.*, 40: 191–225, 1989.

21 Association Vs. Linkage Analysis in Compulsive Disorders

Abbas Parsian, R. D. Todd, and Robert Cloninger

CONTENTS

A. Introduction 367
B. Molecular Genetic Studies 368
   1. Linkage Methods 368
   2. Relative Risk and Haplotype Relative Risk Association Methods 369
C. Special Problems With the Study of Alcoholism 371
D. Current Molecular Genetic Studies in Alcoholism 371
E. Future Directions 372
Acknowledgment 373
References 373

A. INTRODUCTION

Twin, adoption, and family studies have all demonstrated the importance of genetic factors in the development of alcoholism. Twin studies take advantage of the fact that monozygotic (MZ) twins are essentially genetically identical and dizygotic (DZ) twins are about 50% identical by descent on average (ignoring mutation, simple repeat expansion or contraction, etc.). A greater concordance among MZ than DZ twins is expected for a genetically determined trait. From these differences in concordance, the degree of heritability of the trait can be estimated. Early twin studies found that concordances for alcohol abuse, frequency and amount of drinking, psychosis, and liver cirrhosis were higher in MZ than in DZ twins.[1-3] In a recent twin study, Pickens et al.[4] used a Minnesota twin sample which included females. For the composite diagnosis of alcohol abuse and/or dependence, genetic factors appeared to have a modest influence on risk in both sexes. The degree of genetic influence, however, was distributed differently for abuse and dependence. For alcohol dependence, which reflects either pathologic use or impairment as well as evidence of tolerance and/or withdrawal, familial risk appeared to be due largely to the influence of genetic factors. The estimated heritability in Pickens et al.[4] study was 0.60 for male subjects and 0.42 for female subjects. Overall, the results of most twin studies support the existence of important genetic factors influencing predisposition to develop alcoholism.

Recent adoption studies support the concept that important genetic factors predispose to some subtypes of alcoholism. The early adoption studies used small samples and produced conflicting results.[5,6] These early findings are difficult to compare because of differences in the definition of the clinical characteristics of alcohol abuse in the parents and differences in the adoptive placements. In the 1980s, Cloninger and associates[7] undertook a large-scale (862 men and 913 women) adoption study of alcoholism in Sweden. Most of the subjects were separated from their biological relatives in the first months of life, and all had their final placement in the adoptive home before the age of three. Detailed information regarding characteristics of alcohol abuse permitted the definition of two subtypes of alcoholism. The biological fathers or mothers of type 1 alcoholics had an adult onset of alcohol abuse and no criminality. The biological fathers, but not mothers, of type 2 alcoholics had extensive treatment for alcohol abuse and serious criminality beginning in their adolescence or early adulthood. Type 2 alcoholism appeared to have a higher genetic liability than type 1. Alcohol abuse in the *adoptive* parents was not associated with an increased risk of alcohol abuse in the children they reared, so there was no evidence that alcoholism is familial because children imitate their rearing parents.

Family studies have suggested patterns of inheritance of a predisposition to alcoholism. Alcoholics who have positive family history of alcohol abuse have a poorer prognosis than alcoholics who have a negative family history.[8] In a large family study by Frances et al.,[9] alcoholics who had at least one first degree alcoholic relative did significantly worse at follow-up than did those who had no family history of alcoholism. In more recent years, with the development of computer programs which model transmission, it is possible to estimate genetic parameters for complex disorders. Gilligan et al.[10] performed segregation analysis with 286 alcoholic families collected through the Alcohol Research Center (ARC) at Washington University. The risk of alcoholism for different classes of relatives of alcoholics suggested that alcoholism has a complex mode of inheritance. No single gene Mendelian model with complete penetrance could explain its pattern of familial transmission. When families were distinguished according to whether the proband was a type 1 or type 2 alcoholic, there was significant evidence for the presence of a major gene effect for type 2 families.

B. MOLECULAR GENETIC STUDIES

1. Linkage Methods

There are two general methods of linkage analysis, parametric (LOD score method) and nonparametric (sib-pair method). Genetic linkage reflects the fact that two genes near one another on the same chromosome are not inherited independently in families. In the LOD score method, the primary assumption is that the traits being analyzed have a known (and usually simple) mode of inheritance.[11] Traditionally, a minimum LOD score of 3 or an odds ratio of 1000 to 1 is required to assert linkage. If the mechanism of disease transmission is complex, successful detection of linkage may be difficult. Unlike the Mendelian one-to-one correspondence between genotype and phenotype, the correspondence between phenotype and genotype may be weak. Disease status alone may not allow clear discrimination among genotypes at a susceptibility locus.[12] For non-Mendelian disorders, accurate specification of the mode of inheritance for LOD score analysis may be very difficult or impossible.

For complex disorders such as alcoholism, nonparametric methods such as affected-sib-pair (ASP) and affected-pedigree-member (APM) methods may be more appropriate. The ASP method was used in the early days of linkage analysis for Mendelian disorders.[13] ASP has been adopted for nonparametric linkage analysis and has been refined to include disease susceptibility loci closely linked to a marker locus.[14] The major advantages of ASP are that no genetic mechanism needs to be postulated and incomplete penetrance is not an issue. Since affected sibs can be assumed to have the same underlying cause for the disorder, the frequency of false positive cases (sometimes called sporadic cases or phenocopies) should be low. Suarez et al.[14] have reported that ASP method results in consistent estimates of the recombination fraction even if the disease susceptibility is due to an incompletely penetrant simple Mendelian locus.

The advantages and disadvantages of these two methods (LOD score and ASP) have been reviewed by Conneally.[15] The important advantages of the LOD score method are its relatively high power estimates of genetic distance (recombination fraction), and in case of simple genetic heterogeneity is the method of choice. The greatest disadvantage of this method (LOD score) is the requirement to specify the mode of transmission. In the ASP method, the advantages are that the mode of transmission is not specified and cases are relatively ease to ascertain. Some of the disadvantages of ASP are its relatively low power, lack of good' estimation of genetic distance, and the minimum information it provides on mode of inheritance. Therefore, the choice of which of these two methods to use is directly related to one's knowledge of the genetic structure of the disorder under study.

More recently, Weeks and Lange[16] have developed a new method which is a hybrid between LOD score and ASP methods called the affected-pedigree-member (APM) method. As the name of the method implies all combinations of affected members in families are used. This method, therefore, is usually considered as a nonparametric extension of the affected sib pair approach. The power and use of this approach in linkage analysis of complex disorders has not been extensively studied. Berritinni et al.[17] have used this approach to identify a candidate region for bipolar affective disorder on chromosome 18. Interestingly, LOD score analyses of the same data were not significant for a variety of transmission models.

2. RELATIVE RISK AND HAPLOTYPE RELATIVE RISK ASSOCIATION METHODS

When family material is not available or when the mode of genetic transmission of disorder is not well established, investigators frequently use association methods to test for the involvement of particular gene loci in a given disorder. Classical case control relative risk (RR) association studies match individuals with a given disorder with controls who either have no disorder or a different disorder. The frequency of alleles of polymorphic markers are then measured in the groups and differences in frequencies compared using contingency table analyses. Although this approach assumes no specific transmission model, the power to detect genetic effects does depend on a number of underlying factors such as mode of transmission, degree of heterogeneity in the population, etc. These factors are usually not known for the disorder under study (or this approach would not be used); hence, it is difficult to know *a priori* what sample sizes are necessary. However, since it is usually relatively easy to recruit large control or contrast groups for a given disorder, it is usually thought an association case control design is more economical. A more serious drawback to using the relative risk approach is the inability to test the assumption that the contrast groups differ only on the basis of factors directly related to the genetic etiology of the disorder. In particular, it assumes that any differences in allelic frequency for a given locus between the two groups are due to the disease rather than other factors such as ethnic or racial differences. As discussed by Kidd[18] and Crowe,[19] case control association studies are likely to generate false positive findings due to population stratification and statistical sampling problems.

As has been discussed by several authors, the finding of a statistically significant association of a given allele of a marker with the disease group indicates one of several situations. These include (1) the marker is etiologically associated with disease causation; (2) the marker locus alleles are in linkage disequilibrium with a second locus which is etiologically related to the disease; or (3) the case and control populations are stratified with respect to gene frequency on unknown bases unassociated with disease (such as admixture of two populations).[18–20] In the first case, the allele associated with the disease contains the mutation. In the second case, the allele associated with the disease is located near to the disease causing mutations, but is physically distinct. The third cause of association (population admixture) is probably responsible for conflicting reports for several disorders.[18]

Falk and Rubinstein[21] proposed an alternative relative risk approach in which the control group is constructed from the nontransmitted alleles of matings resulting in an affected case; that is, those parental alleles that were not transmitted to the identified case are used to construct an imaginary individual which is used as the control. In this fashion, each individual in the case group is exactly matched in the control group on the basis of parental gene frequency. This so-called haplotype relative risk (HRR) approach has become popular in that a variety of alternative analytical approaches have been proposed to test for significant differences between the transmitted (i.e., measured alleles) and nontransmitted (i.e., parental alleles which were not transmitted) groups. These include procedures developed by Terwilliger and Ott,[22] Spielman et al.,[23] Schaid and Sommer,[24] and Khoury.[25] All of these approaches use different test statistics for comparing transmitted and nontransmitted genotype data. The relative strengths and weaknesses of these different approaches have been reviewed by Schaid and Sommer[26] with respect to their relative power to detect deviations assuming different simple Mendelian transmission models (i.e., single locus models). Other factors being equal, for a given sample size, haplotype relative risk designs are less prone to both false positive and false negative findings than relative risk designs. However, this approach requires that the case and both parents are available for genotyping. Hence, for many disorders which have either late onset, significant mortality, or are associated with significant disruption of families, such case-parent trios will be difficult to recruit, for the addictive and compulsive disorders parents are frequently uncooperative, missing or dead.

Most of the proposed HRR statistics are forms of χ^2 analyses and assume that all of the cell sizes in the contingency tables are relatively large. Since most polymorphic genetic markers in use today have multiple alleles, many of which are relatively rare but may be associated with specific diseases, it is frequently difficult to satisfy this technical analytic assumption requirement. For markers with rare alleles, other statistical approaches such as sign analysis or exact tests are more appropriate.[27] Currently, these approaches are computationally intense and may be difficult to carry out.

Most haplotype relative risk test statistics require collapsing alleles into a dichotomy such as allele A vs. allele B through Z. The analysis of the data then involves a number of nonindependent tests. Rice and colleagues[28] have described an extension of the McNemar χ^2 statistics termed a generalized transmission disequilibrium test (GTDT). In the generalized transmission disequilibrium test, one first tests for significance of an overall χ^2 to avoid complications associated with multiple nonindependent test of alleles. A potential advantage of the Rice et al.[28] approach is that it can directly include other covariants such as the sex or affection status of one of the parents. However, this approach is also sensitive to the number of observations in given cells and hence requires a large sample size for markers with many alleles.

To incorporate information about the disease status of parents, Parsian et al.[27] suggest using a secondary test after detection of association in which it is determined whether or not the affected parent transmits the associated allele. One tests for departure from the expected 50% assuming a Poisson binomial distribution. The advantage of this approach, as a secondary test, is that it is a powerful, straightforward test of cosegregation. The disadvantage is that only matings in which the transmission of alleles from the affective parent can be unambiguously scored are informative.

It may be advantageous to combine RR and HRR association approaches as complementary tests of association. This is particularly true since it is more difficult to recruit cases with available parents than cases and controls. Thus, a reasonable compromise strategy may be to identify tentative associations in large populations using the classical case-control relative к approach and to confirm the involvement of specific alleles using a smaller haplotype relative risk sample to test for a specific allele. To reach statistical significance for a specific allele requires a much smaller group size than if all alleles are analyzed simultaneously. In this combined approach the haplotype relative risk sample is a secondary confirmatory analysis. In addition, looking at whether an affected parent transmits the suspected allele in the HRR trios can be further used as a tertiary test of significance. This three-tiered sequential approach has been applied with success to establishing an association between bipolar affective disorder and the tyrosine hydroxylase locus.[29]

C. SPECIAL PROBLEMS WITH THE STUDY OF ALCOHOLISM

Advances in using identity by descent and relative risk association approaches have greatly increased the power to conduct genetic studies of complex disorders such as alcoholism and the addictive syndromes. However, the behavioral features of these disorders give rise to several complications whose impact on genetic analyses is only partially understood. First, there is clear nonrandom mating among alcoholic individuals. Hence, a large number of alcoholics come from families in which either both parents were affected or there is a high frequency of disorder in both maternal and paternal relatives. If alcoholism has a very heterogeneous genetic basis then such assortative mating may lead to a high frequency of cases of different etiology in given families. This can impact both identity by descent and linkage approaches by misspecification of affected status. Marked genetic heterogeneity will decrease the power of RR and HRR association approaches but colineality of illness will not result in false negative or positive findings per se. As mentioned above, the high mortality associated with severe alcohol or drug abuse decreases the ability to find cases with living parents and intact families who have multiple affected relatives. In part, this problem can be overcome in family studies by ascertaining through two or more affected sibs. This increases the frequency of alcoholism in other relatives and may make up for losses of power due to mortality. This is the approach that has been adopted by the NIAAA Consortium on the Genetics of Alcoholism study (COGA). It is unknown, however, whether this strategy increases or decreases possible genetic heterogeneity within the study groups.

D. CURRENT MOLECULAR GENETIC STUDIES IN ALCOHOLISM

A brief review of the recent studies of candidate genes in alcoholism using molecular techniques will illuminate some of the points and problems mentioned above. Today, with advancements in biotechnology and recombinant DNA techniques, it is possible to use polymorphisms of anonymous DNA fragments and candidate genes in association and linkage studies. One topical example is the alleged association between the dopamine D2 receptor (DRD2) gene and alcoholism. Since the original report by Blum and Noble's groups in 1990,[30] there have been many reports of positive and negative associations of DRD2 with alcoholism and other disorders such as obesity, Tourette's syndrome, and attention deficit/hyperactivity disorder. Blum et al.[30] reported an association between alcoholism and the minor allele (A1) of a taq polymorphism in the 3' flanking region of the DRD2 gene. The sample consisted of 35 severe alcoholics who had reported treatment failures. Many had died from medical complications of their drinking. The control group consisted of 35 subjects. The A1 allele was present in 64% of 22 white alcoholics and only 17% of 24 white controls. However, this allele was present in 77% of 13 black alcoholics and 27% of 11 black controls.

Following this report, results of several positive and negative studies appeared. For example, Bolos et al.[31] studied 40 unrelated white alcoholics, 62 unrelated white patients who had cystic fibrosis, and 65 unrelated white individuals from CEPH reference families as control. The sex and age of this heterogeneous control panel was not described, and psychiatric interviews were not available to exclude alcoholism. The A1 allele was present in 38% of 40 alcoholics compared to 30% of 127 controls. Also, it is important to note that alcoholics with acutely active medical disorders or functional impairment from drug abuse were excluded. Given the differences in ascertainment and assessment procedures for both alcoholics and controls in this study, it is difficult to interpret the discrepancies with the Blum et al.[30] report.

Since 1990, we have tried to evaluate this association using both relative risk and linkage approaches. In our first study,[20] a group of probands from multiple incidence, alcoholic families and a group of unrelated, interviewed nonalcoholic controls were genotyped for the 3' *Taq I* polymorphism (*Taq I* A). The frequency of the A1 allele was 41% of 32 white alcoholics compared to 12% of 25 white controls. The association with the A1 allele was also significant when controls were compared with a subset of 10 alcoholics with severe medical problems (60 vs. 12%). Linkage analysis in 17 nuclear families found no evidence of cosegregation between A1 allele of DRD2 and alcoholism. The discrepancy between the two approaches could have been due to the small sample size. Subsequent studies were designed to control for potentially confounding factors. Three new RFLPs were developed within the human D2 gene and larger alcoholic and normal control samples were genotyped.[32] In this study there were no significant differences in the frequency of any polymorphism between alcoholics and normal controls. However, for marker phD2-244 (*Taq I* C) the alcoholic sample showed a significant departure from Hardy-Weinberg equilibrium.

E. FUTURE DIRECTIONS

Linkage analysis of complex disorders has many problems including poor power and dependency on the assumed transmission model. These shortcomings of the linkage approach make the use of genome wide and candidate gene identity by descent (IBD) and association studies more attractive. IBD analyses of diabetic sib pairs have recently proven useful in identifying multiple linkage markers in a total genome screen.[33] Eventually, it may be necessary to sequence candidate genes or transcriptional control regions of genes in affected and unaffected individuals to detect disease related sequence variations. This approach could be especially useful in the analysis of individual families.[34] The power of the candidate gene approach relies on the development of sequence and marker information for the gene of interest. In particular, after a positive allele association finding has been made and DNA sequence analysis of the proposed disease allele has been performed, it is necessary to know the normal sequence variation in the unaffected population. This is generally an unexplored area which will require significant efforts for each gene. In particular, for common disorders such as alcoholism, it may be critical how the unaffected population is defined. Semi-automated DNA sequencing, however, is ideally suited for the determination of sequence differences between disorders and for the determination of simple mutant frequencies within disorders. In combination with gene amplification approaches, such as asymmetric Polymerase Chain Reactions (PCR),[35] such sequencing can be done from genomic DNA without any cloning procedures. These approaches all rely on the repetitive application of standardized procedures and are easily adapted to automated procedures. If putative sequence differences are found in small samples, the sequencing of intronic DNA near any disease associated sequence difference should allow the development of PCR-based polymorphisms which can distinguish individuals carrying the mutant and normal genes. These could then be applied to larger population studies to identify individuals at risk for developing alcoholism. Finally, cDNAs coding for mRNAs containing disease associated sequence differences will be cloned. In combination with transfection studies, such clones should help identify functional differences between the normal and disease alleles. Such findings would help establish the pathophysiological mechanisms which lead to final disease expression. Such knowledge could lead to novel therapeutic interventions.

ACKNOWLEDGMENT

This work was supported in part by USPHS grants AA08028, AA09515, and MH 31302. We thank Nelly Mark for her help with manuscript preparation.

REFERENCES

1. Kaij, L., *Studies of the Etiology and Sequels of Abuse of Alcohol*, University of Lund Press, Sweden, 1960.
2. Partenan, J., Brunn, K., and Markanen, T., *Inheritance of Drinking Behavior*, Finnish Foundation on Alcohol Studies, Helsinki, 1966.
3. Hrubec, Z. and Omenn, G. S., Evidence of genetic predisposition to alcoholic cirrhosis and psychosis: twin concordance for alcoholism and its end points by zygosity among male veterans, *Alcohol. Clin. Exp. Res.*, 5, 207, 1981.
4. Pickens, R. W., Svikis, D. S., McGue, M., et al., Heterogeneity in the inheritance of alcoholism, *Arch. Gen. Psychiatry*, 48, 19, 1990.
5. Roe, A. and Burks, B., *Memoirs of Section on Alcohol Studies*, Yale University Press, New Haven, 1945.
6. Goodwin, D. W., Schulsinger, F., Hermansen, L., et al., Alcohol problems in adoptees raised apart from alcoholic biological parents, *Arch. Gen. Psychiatry*, 18, 238, 1973.
7. Cloninger, C. R., Bohman, M., and Sigvardson, S., Inheritance of alcohol abuse: cross-fostering analysis of adopted men, *Arch. Gen. Psychiatry*, 38, 861, 1981.
8. Penick, E. C., Read, M. R., Crowley, P. A., et al., Differentiation of alcoholics by family history, *J. Stud. Alc.* 39, 1944, 1978.
9. Frances, R. J., Bucky, S., and Alexopoulos, G. S., Outcome study of familial and nonfamilial alcoholism, *Am. J. Psychiatry*, 141, 1469, 1984.
10. Gilligan, J., O'Malley, S., Risch, N., et al., No association between an allele at the D2 dopamine receptor gene (DRD2) and alcoholism regardless of severity, *JAMA*, 266, 1801, 1991.
11. Ott, J., *Analysis of Human Genetic Linkage*, Revised Edition, The Johns Hopkins University Press, Baltimore, 1991.
12. Risch, N., Genetic linkage: interpreting LOD scores, *Science* 255, 803, 1992.
13. Penrose, L. S., The detection of autosomal linkage in data which consists of pairs of brothers and sisters of unspecified parentage, *Ann. Eugenics*, 6, 133, 1935.
14. Suarez, B. K., Rice, J. P., and Reich, T., The generalized sib-pair IBD distribution: its use in the detection of linkage, *Am. J. Hum. Genet.*, 42, 87, 1978.
15. Conneally, P. M., Criteria for optimal linkage studies in alcoholism, in *Genetic Aspects of Alcoholism*, Küanmaa, K., Tabakoff, B., and Saito, T., Eds., Finn Foundation on Alcohol Studies, Helsinki, 1989, 259.
16. Weeks, D. E. and Lange, K., The affected-pedigree-member method of linkage analysis. *Am. J. Hum. Genet.*, 42, 315, 1988.
17. Berritinni, W. H., Gerraro, T. N., Goldin, L. R., Weeks, D. E., Deterawadleigh, S., Nurnberger, J. I., and Gershon, E. S., Chromosome 18 DNA markers and manic-depressive illness: evidence for susceptibility gene, *Proc. Natl. Acad. Sci. U.S.A.*, 91, 5918, 1994.
18. Kidd, K. K., Associations of disease with genetic markers: deja vu all over again, *Am. J. Med. Genet.*, 48, 71, 1993.
19. Crowe, R., Candidate genes in psychiatry: an epidemiological perspective, *Am. J. Med. Genet.*, 48, 74, 1993.
20. Parsian, A., Todd, R. D., Devor, E. J., et al., Alcoholism and alleles of the human dopamine D2 receptor locus: studies of association and linkage, *Arch. Gen. Psychiatry*, 48, 654, 1991.
21. Falk, C. T. and Rubinstein, P., Haplotype relative risks: an easy reliable way to construct a proper control sample for risk calculating, *Ann. Hum. Genet.*, 51, 227, 1987.
22. Terwilliger, J. D. and Ott, J., A haplotype-based 'haplotype relative risk' approach to detecting allelic associations, *Hum. Hered.*, 42, 337, 1992.
23. Spielman, R. C., McGinnis, R. E., and Ewens, W. J., Transmission test for linkage disequilibrium: the insulin gene region and insulin-dependent diabetes mellitus (IDDM), *Am. J. Hum. Genet.*, 52, 506, 1993.

24. Schaid, D. J. and Sommer, S. S., Genotype relative risks: methods for design and analysis of candidate gene association studies, *Am. J. Hum. Genet.*, 53, 1114, 1993.
25. Khoury, M. J., Case parental control method in the search for disease susceptibility genes, *Am. J. Hum. Genet.*, 55, 414, 1994.
26. Schaid, D. J. and Sommer, S. S., Comparison of statistics for candidate-gene association studies using cases and parents, *Am. J. Hum. Genet.*, 55, 402, 1994.
27. Parsian, A., Chakraverty, S., and Todd, R. D., Possible association between the dopamine D3 receptor gene and bipolar affective disorder, *Am. J. Med. Genet.*, 54, 1, 1994.
28. Rice, J. P., Neuman, R. J., Hoshaw, S. L., Daw, E. W., and Gu, C., TDT tests with covariants and genome screens with MOD scores: their behavior on simulated data, *Genet. Epidemiol.*, 12, 659, 1996.
29. Todd, R. D., Parsian, A., Walczak, A., Hickok, J. M., Lobos, L. A., Simpson, S., and De Paulo, J. R., New evidence for association of bipolar affective disorder and the tyrosine hydroxylase locus, in preparation.
30. Blum, K., Noble, E. P., Sheridan, P. J., et al., Allelic association of human dopamine D2 receptor gene in alcoholism, *JAMA*, 263, 2055, 1990.
31. Bolos, A. M., Dean, M., Lucas-Derse, S., et al., Population and pedigree studies reveal a lack of association between the dopamine D2 receptor gene and alcoholism, *JAMA*, 264, 3156, 1990.
32. Suarez, B. K., Parsian, A., Hampe, C. L., Todd, R. D., Reich, T., and Cloninger, C. R., Linkage disequilibria at the D2 dopamine receptor locus (DRD2) in alcoholics and controls, *Genomics*, 19, 12, 1994.
33. Davies, J. L., Kawaguchi, Y., Bennett, S. T., Copeman, J. B., Cordell, H. J., Pritchard, L. E., Reed, P. W., Gough, S. C. L., Jenkins, S. C., Palmer, S. M., Balfour, K. M., Rowe, B. R., Farrall, M., Barnett, A. H., Bain, S. C., and Todd, J. A., A genome wide search for human type 1 diabetes susceptibility genes, *Nature (London)*, 371, 130, 1994.
34. Goate A. M., Chartier-Harlin, M.-C., Mullan. M. J., Brown, J., Crawford, F., Fidani, L., Giuffra, L., Haynes, A., Irving, N., James, L., Mant, R., Newton, P., Rooke, K., Roques, P., Talbot, C., Pericak-Vance, M., Roses, A., Williamson, R., Rossor, M., Owen, M., and Hardy, J., Segregation of a missense mutation in the amyloid precursor protein gene with familial Alzheimer's disease, *Nature (London)*, 349, 704, 1991.
35. McBride, L. J., Koepf, S. M., Gibbs, R. A., et al., Interfacing *in vitro* DNA amplification with DNA sequencing strategies, in *Polymerase Chain Reaction*, Erlich, H. A., Gibbs, R. A., and Kazazian, H. H., Eds., Cold Spring Harbor Laboratory Press, New York, 1989, 211.

22 Association of the A1 and B1 Alleles of the Dopamine D2 Receptor Gene in Severe Japanese Alcoholics

Tadao Arinami, Masanari Itokawa, Tokutaro Komiyama, Hiroshi Mitsushio, Hiroshi Mori, Hideo Mifune, and Michio Toru

CONTENTS

A. Introduction 375
B. Subjects and Method 376
    1. Subjects 376
    2. Typing of the D2/*Taq I* A and B RFLPs 377
C. Results 377
    1. D2/*Taq I* A and B RFLPs Allele Frequencies in the Japanese Population 377
    2. The Prevalence of the A1 and B1 Alleles and Severity of Alcoholism 377
    3. D2/*Taq I* A and B RFLPs Genotypes and Severity of Alcoholism 381
D. Discussion 382
References 385

A. INTRODUCTION

    Alcoholism is a heterogeneous entity that arises from a combination of biopsychosocial factors. Twin, adoption, high-risk, and family studies have generally supported some role for genetic factors in the etiology of alcoholism. A heritability analysis in a twin study conducted by Pickens et al. showed that genetic factors have a modest influence on an overall risk in both sexes, with the estimated heritabilities of 0.35 for men and 0.24 for women.[1] Complex segregation analysis suggested the presence of a major effect controlling liability to alcoholism.[2]
    There is evidence that alcohol stimulates brain reward systems in part through its action on central dopaminergic nervous systems.[3] Recently, an association between alcoholism and a restriction fragment length polymorphism (RFLP) (D2/*Taq I* A) recognized by the dopamine D2 receptor (DRD2) gene clone λhD2G1 was reported.[4] The polymorphic site of D2/*Taq I* A is thought to be located several kilobase 3' to the final exon of the DRD2 gene.[5] Southern blot hybridization of genomic DNA after *Taq I* digestion with this clone shows two common alleles of 6.6 kb (A1) and 3.7 kb (A2) bands. Blum et al. found that the allele A1 was present

in 69% of 35 deceased alcoholics, whereas it was present only in 20% of 35 controls. This difference was significant in both black and white subgroups, suggesting that a gene product or expression directly related to this RFLP allele or with linkage disequilibrium with this allele increases susceptibility to at least some type of alcoholism. While the allelic difference between nonalcoholics and alcoholics was confirmed in living subjects,[6] studies suggesting no significantly higher frequency of the subjects with the A1 allele for alcoholic groups compared with unclassified general controls were reported.[7-9] In a study reported from Munich, a decreased frequency of the A1 allele for the alcoholic group was found.[10]

Although the association of D2/*Taq I* A and alcoholism is still controversial, an association between severe alcoholism and the A1 allele of D2/*Taq I* A RFLP appears to be more likely.[9,11-14] Moreover, all data published so far show increased A1 allele frequencies in severe alcoholic subgroups compared with those in less severe alcoholic subgroups.[6,7,9]

Although the functional significance of the allelic difference is unknown, Noble et al. (1991) showed that in the subjects with the A1 allele, the B_{max} (number of binding sites) was significantly reduced compared with that of the subjects with the A2 allele. Moreover, the B_{max} decreased in the order A2/A2, A1/A2, and A1/A1 genotypes.[15]

More recently, Hauge et al. reported a new *Taq I* RFLP at the DRD2 locus (D2/*Taq I* B), which is located in the first intron close to the second exon.[16] Southern blot hybridization of genomic DNA after *Taq I* digestion with the clone λhD2G2 shows two common alleles of 4.6 kb (B1) and 4.1 kb (B2) bands. *Taq I* B RFLP was shown to be in strong linkage disequilibrium with *Taq I* A RFLP. Smith et al. reported that heavy polysubstance users displayed significantly higher *Taq I* A1 and *Taq I* B1 frequencies than control subjects, and *Taq I* B1 results were more robust compared with the *Taq I* A1 allele.[17] Blum et al. also reported that both *Taq I* A1 and B1 RFLP associate with severe alcoholism.[18]

Most of the studies on the association between D2/*Taq I* A RFLP and alcoholism were carried out in white populations in the U.S. Because large environmental influences on any gene's effect on alcoholism are suggested by heritability data, analyses in various racial/ethnic populations with various cultural backgrounds on drinking will contribute to a more complete evaluation of the association. We report our study on the association of *Taq I* A and B RFLPs with alcoholism carried out on 78 Japanese alcoholics and nonalcoholic controls.

B. SUBJECTS AND METHOD

1. Subjects

The subjects were 78 unrelated Japanese alcoholics, 74 men and 4 women, who were randomly selected among alcoholics receiving treatment at three hospitals in Japan. Their ages ranged from 25 to 79 with a mean (± SEM) age of 51 ± 1.0 years at the time of obtaining the specimens with their informed consent. All the subjects had been admitted to the hospitals for treatment of alcoholism. Diagnosis was made by two trained psychiatrists according to the *Diagnostic and Statistical Manual of Mental Disorders*, Revised Third Edition (*DSM-III-R;*) for alcohol dependence (code 303.90) and Criteria for the Diagnosis of Alcoholism by the Criteria Committee, National Council on Alcoholism (NCA).[19,20] The patients with a primary major affective disorder, schizophrenia, or psychoses other than those associated with alcohol use were excluded. Determination of parental alcoholism was based on the patient's report, which was corroborated by an interview with the patient's family members.

Severity of alcoholism was estimated by the number of items of diagnostic level 1 of the Major Criteria for the Diagnosis of Alcoholism by NCA. The mean (± SEM) number for all alcoholics was 4.9 ± 0.4 items and the median number was 5 items. For the classification by severity, "less severe" alcoholics were defined as those who scored fewer than 5 items and "more severe" alcoholics as those who scored 5 or more items. Clinical assessors were blinded to genetic data.

An unclassified control group for the study consisted of 100 unrelated healthy Japanese with a mean (± SEM) age of 44 ± 1.2 years. A gender-matched nonalcoholic control group consisted of 35 individuals with a mean (± SEM) age of 45 ± 1.5, who were recruited from among care staff members at a residential institution for the mentally retarded. They did not meet the criteria for alcoholism or other substance addiction including nicotine.

2. TYPING OF THE D2/*Taq I* A AND B RFLPs

Genomic DNAs were prepared from peripheral white blood cells using standard methods. Seven micrograms of DNA were digested with 21U of *Taq I* (Boelinger Manheim, Manheim) for 12 hr at 65°C, electrophoresed in 0.8% agarose gel in 1 × TBE (0.089-mol/L Tris-HCl, 0.089-mol/L boric acid, and 0.002-mol/L ethylenediaminetetraacetic acid disodium) and blotted onto a nylon membrane (Hybond N, Amersham, Buckinghanshire) by the technique of Southern. The *Taq I* blots were hybridized with a ^{32}P labeled 1.5-kb BamHI insert of clone λhD2G1 (*Taq I* A) or with a ^{32}P labeled whole insert of clone λhD2G2 (*Taq I* B) provided by Civelli (Vollum Institute of Biomedical Research, Oregon Health Sciences University, Portland). The DNA was labeled by the random primer method of Feinberg and Vogelstein to a specific activity of 10^9 cpm/mg. Hybridization was carried out at 42°C in 50% formamide, 5 × Denhardt's solution, 5 × SSPE, 0.5% sodium dodecylsulfate (SDS) and 200 mg/L of denatured sonicated salmon sperm DNA. Denatured human placenta DNA was also added to the hybridization solution at a concentration of 200 mg/L when using clone λhD2G2 as a probe. The filters were then washed at room temperature at 2 × SSC/0.5% SDS for 15 min and twice at 65°C at 0.2 × SSC/0.5% SDS for 15 min. Autoradiography was performed using Konica X-ray film with intensifying screens at −70°C.

C. RESULTS

1. D2/*Taq I* A AND B RFLPs ALLELE FREQUENCIES IN THE JAPANESE POPULATION

Allele frequencies were estimated by the gene counting method. The frequencies of the 6.6-kb allele (A1 allele) and the 3.7-kb allele (A2 allele) in the Japanese controls were 0.42 and 0.58, respectively, and the alleles were in Hardy-Weinberg equilibrium; ($\chi^2 = 1.21, df = 1$). The A1 allele frequency in the Japanese population was significantly higher than those in the white populations reported so far (Table 22-1). The frequencies of the 4.6-kb allele (B1 allele) and the 4.1-kb allele (B2 allele) in the Japanese controls were 0.31 and 0.69, respectively, and the alleles were in Hardy-Weinberg equilibrium ($\chi^2 = 1.58, df = 1$). The B1 allele frequency in the Japanese population was also significantly higher than that in the Caucasians (Table 22-1). Strong linkage disequilibrium between *Taq I* A and B RFLPs was suggested in the Japanese population as in Caucasians (Table 22-2).[16]

2. THE PREVALENCE OF THE A1 AND B1 ALLELES AND SEVERITY OF ALCOHOLISM

In this study, 21 (60%) of the 35 nonalcoholic controls and 54 (69%) of the 78 alcoholics carried the A1 allele (Table 22-3). The frequency of the subjects with the A1 allele was not significantly different between all the groups. Also, 14 (40%) of the 35 nonalcoholic controls

TABLE 22-1
D2/*Taq I* A and B Allele Frequencies in the Caucasian and the Japanese Populations

	Allele			
	A1 (6.6 kb)	B1 (4.6 kb)	χ^{2*}	p^*
Japanese population controls	84 (0.42)	62 (0.31)		
White population controls[5]	21 (0.24)		(8.0)	<.005
White population controls[7]	46 (0.18)		(31.2)	<.001
White population controls[8]	27 (0.20)		(17.9)	<.001
White population controls[16]		11 (0.16)	(6.1)	<.03

Note: * Comparison between Japanese controls and white controls using a two-tailed Yates χ^2 corrected for continuity.

TABLE 22-2
Observed and Expected Haplotypes for the *Taq I* A and B RFLPs in Unrelated Japanese Individuals

Loci		Haplotype	N (observed)	N (expected)
A	B			
1	1	1	45	18.2
1	2	2	18	44.8
2	1	3	0	26.8
2	2	4	93	66.2

and 40 (51%) of the 78 alcoholics carried the B1 allele (Table 22-4). The frequency of the subjects with the B1 allele was not significantly different between the nonalcoholic and alcoholic groups.

The average age of the subjects at the examination of the alcoholics with the A1 allele was 50.5 ± 1.3, which was almost the same as that of the alcoholics without the A1 allele, 50.7 ± 1.9. The average age of the subjects at the examination of the alcoholics with the B1 allele was 49.4 ± 1.3, which was almost the same as that of the alcoholics without the B1 allele, 51.8 ± 1.7. However, the A1 and B1 alleles were not equally distributed in respect to the age at the examination. The A1 and B1 alleles were significantly less frequently present in the alcoholics at the age of 60 or older (42 and 25%, respectively) than in those younger than the age of 60 (74 and 56%, respectively) ($p = 0.031$ and $p = 0.047$, respectively, by Fisher's Exact Test).

When the alcoholics were divided by the severity of alcoholism, 33 (77%) of the 43 more severe alcoholics carried the A1 allele, whereas 16 (59%) of the 27 less severe alcoholics carried the A1 allele (see Table 22-3). In the alcoholics under the age of 60, 29 (83%) out of the 35 more severe alcoholics had the A1 allele, whereas 15 (65%) out of the 23 less severe alcoholics had the A1 allele. The A1 allele was significantly more frequently present in the more severe alcoholics than in the nonalcoholic controls (21 out of 35; i.e., 60%) under the age of 60 ($p = 0.031$ by Fisher's Exact Test). In regard to *Taq I* B RFLP, 28 (65%) of the 43 more severe alcoholics carried the B1 allele, whereas 9 (33%) of the 27 less severe alcoholics carried the B1 allele (see Table 22-4). The B1 allele was significantly more frequently present

TABLE 22-3
Frequencies of Individuals with the A1 Allele of D2/Taq I A RFLP in the Controls, Total Alcoholics, and Subgroups of Alcoholics in the Present Study

	Total*	Total		Under the age of 60			The age of 60 or older		
		Subjects with the A1 allele	Homozygotes for the A1 allele	Total	Subjects with the A1 allele	Homozygotes for the A1 allele	Total	Subjects with the A1 allele	Homozygotes for the A1 allele
	N	N (%)	N (%)	N	N (%)	N (%)	N	N (%)	N (%)
Japanese nonalcoholics	35	21 (60)	2 (6)	35	21 (60)	2 (6)			
Japanese alcoholics	78	54 (69)	9 (12)	66	49 (74)	9 (14)	12	5 (42)	1 (8)
More severe ≥5 items of NCA	43	33 (77)	7 (16)	35	29 (83)	6 (17)	8	4 (50)	1 (13)
Less severe <5 items of NCA	27	16 (59)	0 (0)	23	15 (65)	0 (0)	4	1 (25)	0 (0)
Family history +**	33	24 (73)	4 (12)	29	23 (79)	8 (28)	4	1 (25)	0 (0)

Note: $p = .031$ for prevalence of the A1 allele in nonalcoholics *vs.* severe alcoholics under 60 years of age by Fisher's exact test; $p = .031$ for prevalence of the A1 allele in alcoholics under 60 years of age *vs.* those older by Fisher's exact test; *Eight alcoholics with no reliable data on each item of ICN and three alcoholics with no reliable data on the onset were excluded from the number; **Family history + includes subjects with one or more first degree family members diagnosed as alcoholics.

TABLE 22-4
Frequencies of Individuals with the B1 Allele of D2/Taq I B RFLP in the Controls, Total Alcoholics, and Subgroups of Alcoholics in the Present Study

	Total				Under the age of 60					The age of 60 or older				
	Total*	Subjects with the B1 allele		Homozygotes for the B1 allele		Total	Subjects with the B1 allele		Homozygotes for the B1 allele		Total	Subjects with the B1 allele		Homozygotes for the B1 allele	
	N	N	(%)	N	(%)	N	N	(%)	N	(%)	N	N	(%)	N	(%)
Japanese nonalcoholics	35	14	(40)	2	(6)	35	14	(40)	2	(6)					
Japanese alcoholics	78	40	(51)	6	(8)	66	37	(56)	5	(8)	12	3	(25)	1	(8)
More severe (≥5 items of NCA)	43	28	(65)	5	(12)	35	25	(71)	4	(11)	8	3	(38)	1	(13)
Less severe (<5 items of NCA)	27	9	(33)	0	(0)	23	9	(39)	0	(0)	4	0	(0)	0	(0)
Family history +**	33	21	(64)	2	(6)	49	20	(41)	2	(4)	10	1	(10)	0	(0)

Note: $p = .009$ for prevalence of the B1 allele in less severe alcoholics vs. severe alcoholics by Fisher's exact tests; $p = .047$ for prevalence of the B1 allele in nonalcoholics vs. severe alcoholics by Yates corrected χ^2 for continuity; $p = .049$ for prevalence of the B1 allele in alcoholics under 60 years of age vs. those older by Fisher's exact test; *Eight alcoholics with no reliable data on each item of ICN and three alcoholics with no reliable data on the onset were excluded from the number; **Family history + includes subjects with one or more first degree family members diagnosed as alcoholics.

in the more severe alcoholics than in the nonalcoholic controls (Yates corrected $\chi^2 = 3.94$, $df = 1$, $p = 0.047$) and than in the less severe alcoholics (Yates corrected $\chi^2 = 5.59$, $df = 1$, $p = 0.019$).

The numbers of the items satisfied by the alcoholics carrying the A1 allele, 5.3 ± 0.2, was significantly different compared with the numbers of the items satisfied by those alcoholics not carrying the A1 allele, 4.2 ± 0.3 ($p = 0.01$ by the Mann-Whitney U test). The number of the items satisfied by the alcoholics carrying the B1 allele, 5.5 ± 0.2, was significantly different compared with the numbers of the items satisfied by the alcoholics not carrying the B2 allele, 4.4 ± 0.2 ($p = 0.002$ by the Mann-Whitney U test).

Increasing severity of alcoholism scored by the numbers of the items satisfied by the alcoholics was significantly associated with an increase in the presence of the A1 and B1 allele ($p < .01$, $p < .005$, respectively by the test for a linear trend in proportion[21]) (Figure 22-1).

There was no specific item of NCA criteria that significantly frequent alcoholic subjects with the A1 or B1 allele satisfied compared with the frequency of the nonalcoholics with the A1 or B1 allele (by χ^2 analysis or Fisher's Exact Test; data not shown).

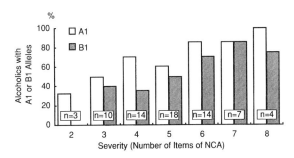

FIGURE 22-1 Severity of alcoholism and the presence of the A1 and B1 allele. The severity of alcoholism is scaled by the number of items of diagnostic level 1 of the Major Criteria for the Diagnosis of Alcoholism by NCA satisfied by the subjects.

3. D2/*Taq I* A AND B RFLPs GENOTYPES AND SEVERITY OF ALCOHOLISM

Two (6%) out of the 35 nonalcoholic controls and 9 (12%) out of the 78 alcoholics were homozygous for the A1 allele (Tables 22-3 and 22-5). All seven alcoholics homozygous for the A1 allele whose severity was evaluated were classified as more severe. The proportions of subjects with more severe alcoholism in the alcoholics with A1/A1, A1/A2, and A2/A2 were 100% (7 out of 7), 62% (26 out of 42), and 48% (10 out of 21), respectively. The proportion of alcoholics with more severe alcoholism in those with A1/A1 significantly increased as compared with that of those with A1/A2 ($p = 0.05$ by Fisher's Exact Test) and that of those with A2/A2 ($p = 0.016$ by Fisher's Exact Test). The average (± SEM) number of items of NCA satisfied by the alcoholics with A1/A1, A1/A2, and A2/A2 alleles were 6.0 ± 0.3, 5.1 ± 0.3, and 4.5 ± 0.3, respectively. There was a significant variability in the severity measured by the number of items among the three phenotypes (A1/A1, A1/A2, A2/A2) by a one way analysis of variance (ANOVA) ($F = 3.44$, $p = 0.03$). The numbers of the items satisfied by the alcoholics with the A1/A1 allele was significantly different compared with the numbers of the items satisfied by those alcoholics with the A2/A2 allele ($p = 0.005$ by the Mann-Whitney U test) (Table 22-5).

As for *Taq I* B RFLP, 2 (6%) out of the 35 nonalcoholic controls and six (8%) out of the 78 alcoholics were homozygous for the B1 allele (see Tables 22-4 and 22-5). All six alcoholics

TABLE 22-5
Proportion of Severe Alcoholics and Average Number of the Items of NCA According to *Taq I* A and B Genotype

Phenotype	Genotype		
	A1/A1	A1/A2	A2/A2
Proportions of the subjects with severe alcoholics	100% (7/7)	58% (21/36)	52% (13/25)
Satisfied numbers of items of NCA	6.0 ± 0.3	5.1 ± 0.3	4.5 ± 0.3

$p = 0.119*$ (A1/A1 vs A1/A2), $p = 0.041*$ (A1/A2 vs A2/A2), $p = 0.005*$ (A1/A1 vs A2/A2)

Phenotype	Genotype		
	B1/B1	B1/B2	B2/B2
Proportions of the subjects with severe alcoholics	100% (5/5)	72% (23/32)	45% (15/33)
Satisfied numbers of items of NCA	6.4 ± 0.2	5.3 ± 0.2	4.4 ± 0.2

$p = 0.087*$ (B1/B1 vs B1/B2), $p = 0.01*$ (B1/B2 vs B2/B2), $p = 0.004*$ (B1/B1 vs B2/B2)

* The Mann-Whitney U test.

homozygous for the B1 allele whose severity was evaluated were classified as more severe. The proportions of subjects with more severe alcoholism in the alcoholics with B1/B1, B1/B2, and B2/B2 were 100% (5 out of 5), 72% (23 out of 32), and 45% (15 out of 33), respectively. The proportion of alcoholics with more severe alcoholism in those with B1/B1 significantly increased as compared with that of those with B2/B2 ($p = 0.031$ by Fisher's Exact Test). The average (± SEM) number of items of NCA satisfied by the alcoholics with B1/B1, B1/B2, and B2/B2 alleles were 6.4 ± 0.2, 5.3 ± 0.2, and 4.4 ± 0.2, respectively. There was a significant variability in the severity measured by the number of items among the three phenotypes (B1/B1, B1/B2, B2/B2 by a one way analysis of variance (ANOVA) ($F = 6.80, p = 0.002$). The number of the items satisfied by the alcoholics with the B1/B1 allele was significantly different compared with the numbers of the items satisfied by the alcoholics with the B2/B2 genotype ($p = 0.004$, the Mann-Whitney U test) (see Table 22-5).

D. DISCUSSION

Racial differences in the allelic frequencies of the *Taq I* A RFLP have been implied because Blum et al. reported higher prevalences of the A1 allele in black alcoholics than in white alcoholics.[6] Our study showed a significantly higher frequency of the A1 allele of D2/*Taq I* A RFLP in the Japanese unclassified controls compared with those in the unclassified white populations in the U.S. Our study also showed a significantly higher frequency of the B1 allele in the Japanese than in whites, as expected from the fact that *Taq I* B RFLP is in linkage disequilibrium with *Taq I* A RFLP. Thus, the racial differences in the allele frequencies of the *Taq I* A and B RFLPs may be real. Therefore, the association studies related with these RFLPs should be carried out in populations in which race and ethnicity are well matched.

In our study, the prevalence of the A1 and B1 alleles was significantly different between two subgroups of alcoholics divided by the age of 60 at the time of the examination. The alcoholics at the age of 60 or older had significantly less frequency of the A1 allele (42%) and the B1 allele (25%) than those younger than the age of 60 (A1 allele, 74%; B1 allele, 56%). This difference seems to be caused by decreased frequencies of the A1 and B1 alleles in the elderly alcoholic group in our study. If this deviation did not result by chance nor sampling deviation, it might relate to a poorer outcome for the alcoholics with the A1 and/or B1 alleles compared with those without them, which was suggested by Parsian et al.[9] This hypothesis is consistent with the fact that the A1 allele was most frequently found in the deceased alcoholics among the white alcoholics studied so far (Table 22-6).[4]

TABLE 22-6
Frequency of Individuals with A1 and B1 Alleles of D2/*Taq I* A and B Polymorphisms in Controls (Nonalcoholics) and Alcoholics in the Eight Studies

Population	n	Presence of A1 allele n (percentage)	Presence of B1 allele n (%)	Ref.
Black nonalcoholics*	11	3 (27)		4
Black alcoholics*	13	10 (77)		4
White nonalcoholics*	24	4 (17)		4
White alcoholics*	22	14 (64)		4
White controls	127	38 (30)		7
White alcoholics	40	15 (38)		7
Less severe	20	6 (30)		7
More severe	20	9 (45)		7
White nonalcoholics	25	3 (12)		9
White alcoholics	32	13 (41)		9
With severe medical problems	10	6 (60)		9
White nonalcoholics	52	11 (21)		6
White alcoholics	96	48 (50)		6
More severe	52	33 (63)		6
Less severe	44	15 (34)		6
White controls	68	14 (20)		8
White alcoholics	44	10 (23)		8
Consumption ≤300 standard drinks in 2 months before the study	20	7 (35)		8
Consumption ≥300 standard drinks in 2 months before the study	23	12 (52)		8
White alcoholics	47	9 (19)		23
White nonalcoholics	41		6 (15)	18
White alcoholics	92		31 (34)	18
Less severe	40		7 (18)	18
Severe	52		24 (46)	18
Japanese nonalcoholics	35	21 (60)	14 (40)	
Japanese alcoholics	78	54 (69)	40 (51)	
More severe	43	33 (77)	9 (65)	
Less severe	27	16 (59)	28 (33)	

Note: *Deceased samples.

In our study, the frequency of the subjects with the A1 allele was not significantly different between the nonalcoholic controls and the total alcoholics. The difference was not also significant between the nonalcoholic controls and the more severe alcoholics. Because of the possibility that the prevalence of the A1 allele was biased in the elderly alcoholic subgroup in our study, a comparison was made between the nonalcoholic and alcoholic subjects who were younger than the age of 60. In this study, 60% of the nonalcoholics, 74% of the total alcoholics, 65% of those with less severe alcoholism, and 83% of those with more severe alcoholism had the A1 allele. The frequency of the individuals with the A1 allele was significantly higher in those with more severe alcoholism than in the nonalcoholic controls. The frequency of the individuals with the B1 allele was significantly higher in those with more severe alcoholism (65%) than in the nonalcoholic controls (40%) and than in those with less severe alcoholism (33%). We think that these data suggest the association between the A1 and B1 alleles and severe alcoholism in our Japanese population, as well as most of the populations studied in the U.S. so far (see Table 22-6).

High frequencies of the A1 and B1 alleles in the Japanese gave us the opportunity to explore the alcoholics homozygous for the A1 and B1 allele. The frequencies of the subjects homozygous for the A1 allele or the B1 allele in the alcoholic group were not different from those in the controls. However, all alcoholics homozygous for the A1 allele or the B1 allele in whom the severity was determined were classified as more severe. It was also shown that the average severity of alcoholism progressively increased in subjects with A2/A2, A1/A2, and A1/A1 allele, with subjects with A1/A1 being the most severe. This was also the case in *Taq I* B RFLP. These data support the hypothesis that the A1 and B1 alleles are associated with the severity of alcoholism, at least, when the severity was determined by scores on the NCA checklist.

We previously estimated the extent of the association of the A1 allele with severity of alcoholism, the relative contribution of the genetic variance associated with D2/*Taq I* A RFLP to the total phenotypic variance of the numbers of items of NCA satisfied by the Japanese alcoholics, assuming the severity to be scaled by the item numbers of NCA criteria.[22] The estimate of the total genotypic variance for the Japanese alcoholics studied (n = 70) is subdivided into a component ascribable to the average allelic effect (V_A^2) and the remainder due to genotypic differences not explained by additivity of the allelic effect (V_D^2) in the population at large. The total genetic variance associated with D2/*Taq I* A RFLP (V_G^2) was 0.23. The total phenotypic variance was 2.37. Therefore, the estimated relative contribution of the genetic variance associated with D2/*Taq I* A RFLP to the total phenotypic variance of severity of alcoholism was 9.5%. This relative contribution is very tentative, but considering that alcoholism is a highly heterogeneous entity and many factors are assumed to contribute to its development and progression, the relatively small role of D2/*Taq I* A RFLP in severity of alcoholism estimated here was thought to be reasonable.

On the other hand, the prevalences of the A1 and B1 alleles in the subjects with less severe alcoholism were close to those in the nonalcoholics. If the A1 and B1 alleles are associated with susceptibility to alcoholism, the A1 and B1 alleles are expected to be more frequent in both the less and more severe alcoholic subgroups than those in the nonalcoholic controls. In addition, the A1 and B1 alleles were not significantly frequent in the alcoholics with family history for alcoholism compared with those without family history. These data suggest that the A1 and B1 alleles do not have a major effect on susceptibility to alcoholism in most Japanese cases, being in line with the published data obtained from the white populations, which gave no evidence for the linkage between the A1 allele and alcoholism.[7,9]

The estimation of severity of alcoholism using item numbers of NCA criteria satisfied by the subjects might have highlighted the positive findings on the association, because of no specific items of NCA criteria found to be significantly associated with the A1 and B1 alleles in our study. This implies relatively weak, if any, associations of the A1 and B1 alleles with each item. Besides, our study suggests that there may be systems assessing severity of alcoholism with which the A1 and B1 alleles are not associated. For example, the average age of alcoholism onset in the alcoholics with the A1 or B1 alleles were not significantly different compared with that in the alcoholics without the A1 or B1 alleles in our study. If age of alcoholism onset was used as a scale for severity of alcoholism assuming early onset to be associated with greater severity, the conclusion that the A1 and B1 alleles have nothing to do with the severity determined by the age of alcoholism onset was drawn in our study.

Our study, as well as most of the studies carried out so far, was based on treatment samples, leaving the possibility that the association is not the case in alcoholics in the general population. Larger and randomly sampled studies will produce more rigorous results. However, we believe that the same tendency shown throughout the three different major racial groups in independent studies thus far is an indication of the association between severe alcoholism and D2/*Taq I* A and B RFLPs, even if these RFLPs do not exert direct influence to susceptibility of alcoholism.

Even if *Taq I* A and B RFLPs have a modest association with severe alcoholism and those with the A1 allele have the lower B_{max} of DRD2,[13] the mechanism of these associations is not known. It seems reasonable to assume that there might be a variant(s) in linkage disequilibrium with these alleles which alters the structure of the gene product or the expression of the gene.

In summary, the association of the A1 and B1 allele of D2/*Taq I* A and B RFLPs with severe alcoholism was suggested in the Japanese population. It is also suggested that D2/*Taq I* A and RFLPs are related to or have a linkage disequilibrium with a genetic factor that has a modest effect on severity of alcoholism.

REFERENCES

1. Pickens, R. W., Svikis, D. S., McGue, M., Lykken, D. T., Heston, L. L., and Clayton, P. J., Heterogeneity in the inheritance of alcoholism, *Arch. Gen. Psychiatry*, 48, 19, 1991.
2. Aston, C. and Hill, S. Y., Segregation analysis of alcoholism in families ascertained through a pair of male alcoholics, *Am. J. Hum. Genet.*, 46, 879, 1990.
3. Koob, G. F. and Bloom, F. E., Cellular and molecular mechanisms of drug dependence, *Science*, 242, 715, 1988.
4. Blum, K., Noble, E. P., Sheridan, P. J., Montgomery, A., Ritchie, T., Jagadeeswaran, P., Nogami, H., Briggs, A. H., and Cohn, J. B., Allelic association of human dopamine D2 receptor gene in alcoholism, *JAMA*, 263, 2055, 1990.
5. Grandy, D. K., Litt, M., Allen, L., Bunzow, J. R, Marchionni, M. A., Makam, H., Reed, L., Magenis, E., and Civelli, O., The human dopamine D2 receptor gene is located on chromosome 11 at q22-q23 and identifies a *Taq I* RFLP, *Am. J. Hum. Genet.*, 45, 778, 1989.
6. Blum, K., Noble, E. P., Sheridan, P. J., Finley, O., Montgomery, A., Ritchie, T, Ozkaragoz, T., Fitch, R. J., Sadlack, F., Sheffield, D., Dahlmann, T., Harbardier, S., and Nogami, H., Association of the A1 allele of the D2 dopamine receptor gene with severer alcoholism, *Alcohol*, 8, 409, 1991.
7. Bolos, A. M., Dean, M., Lucas-Derse, S., Ramsburg, M., Brown, G., and Goldman, D., Population and pedigree studies reveal a lack of association between the dopamine D2 receptor gene and alcoholism, *JAMA*, 264, 3156, 1990.
8. Gelernter, J., O'Malley, S., Risch, N., Kranzler, H. R., Krystal, J., Merikangas, K, Kennedy, J. L., and Kidd, K. K., No association between an allele at the D2 dopamine receptor gene (DRD2) and alcoholism, *JAMA*, 266, 1801, 1991.
9. Parsian, A., Todd, R. D., Devor, E. J., O'Malley, K. L., Suarez, B. K., Reich, T., and Cloninger, R., Alcoholism and alleles of the human D2 dopamine receptor locus, *Arch. Gen. Psychiatry*, 48, 655, 1991.
10. Schwab, S, Soyka, M., Niederecker, M., Ackenheil, M., Scherer, J., and Wildernauer, D. B., Alleleic association of human dopamine D2-receptor DNA polymorphism ruled out in 45 alcoholics, *Am. J. Hum. Genet.*, 49 (Suppl), 203, 1991.
11. Cloninger, C. R., D2 dopamine receptor gene is associated but not linked with alcoholism, *JAMA*, 266, 1833, 1991.
12. Conneally, P. M., Association between the D2 dopamine receptor gene and alcoholism, *Arch. Gen. Psychiatry*, 48, 757, 1991.
13. Noble, E. P. and Blum, K., The dopamine D2 receptor gene and alcoholism, *JAMA*, 265, 2667, 1991.
14. Uhl, G. R., Persico, A. M., and Smith, S. S., Current excitement with D2 dopamine receptor gene alleles in substance abuse, *Arch. Gen. Psychiatry*, 49, 157, 1992.

15. Noble, E. P., Blum, K., Ritchie, T., Montgomery, A., and Sheridan, P. J., Allelic association of the D2 dopamine receptor gene with receptor-binding characteristics in alcoholism, *Arch. Gen. Psychiatry,* 48, 648, 1991.
16. Hauge, X. Y., Grandy, D. K., Eubanks, J. H., Evans, G. A., Civelli, O., and Litt, M., Detection and characterization of additional DNA polymorphisms in the dopamine D2 receptor gene, *Genomics,* 10, 527, 1991.
17. Smith, S. S., O'Hara, B. F., Persico, A. M., Gorelick, D. A., Newlin, D. B., Vlahov, D., Solomon, L., Pickens, R., and Uhl, G. R., Genetic vulnerability to drug abuse. The D2 dopamine receptor *Taq I* B1 restriction fragment length polymorphism appears more frequently in polysubstance abusers, *Arch. Gen. Psychiatry,* 49, 723, 1992.
18. Blum, K., Noble, E. P., Sheridan, P. J., Montgomery, A., Ritchie, T., Ozkaragoz, T., Fitch, R. J., Wood, R., Finley, O., and Sandlack, F., Genetic predisposition in alcoholism: association of the D2 dopamine receptor *Taq I* B1 RFLP with severe alcoholics, *Alcohol,* 10, 59, 1993.
19. American Psychiatric Association, *Diagnostic and Statistical Manual of Mental Disorders,* 3rd ed. rev., Washington, DC: American Psychiatric Press, 173, 1987
20. The Criteria Committee, National Council on Alcoholism, Criteria for the Diagnosis of Alcoholism, *Am. J. Psychiatry,* 129, 127, 1972.
21. Snedecor, G. W. and Cochran, W. G., *Statistical Method,* The Iowa State University Press, Ames, 1980, 206.
22. Arinami, T., Itokawa, M., Komiyama, T., Mitsushio, H., Mori, H., Jifune, H., Hamaguchi, H., and Toru, M., Association between severity of alcoholism and the A1 allele of the dopamine D2 receptor gene *Taq I* A RFLP, *Biol. Psychiatry,* 33, 108, 1993 (in Japanese).
23. Turner, E., Ewing, J., Shilling, P., Smith, T. L., Irwin, M., Schuckit, M., and Kelsoe, J. R., Lack of association between an RFLP near the D2 dopamine receptor gene and severe alcoholism, *Biol. Psychiatry,* 31, 285, 1992.

23 Behavior Genetic Investigations of Cigarette Smoking and Related Issues in Twins

Gary E. Swan and Dorit Carmelli

CONTENTS

A. Overview 387
B. Background 388
    1. Neurochemical Actions of Nicotine 388
    2. Smoking Behavior and Reactions to Nicotine Are Genetically Determined: Previous Research 388
C. Study Populations 389
    1. The NAS-NRC Twin Registry 389
    2. The NHLBI Twin Study Subsample 390
D. Evidence for Genetic Influences on Smoking and Related Behaviors: Findings From the NAS-NRC Registry and the NHLBI Twin Study 391
    1. Smoking Has Significant Genetic Variance 391
    2. Different Components of Smoking Topography Have Significant Genetic Variance 392
    3. Smoking Shares a Genetic Commonality with Alcohol and Coffee Consumption 393
    4. Weight Gain Following Smoking Cessation Has a Genetic Component 401
E. Questions Remaining to Be Answered 402
    1. The Genetics of Nicotine Dependence 402
    2. The Study of Twins Discordant for Smoking and Nicotine Dependence 402
    3. Do Genetic Effects on Smoking and Nicotine Dependence Vary as a Function of Gender? 402
    4. Common or Specific Genetic Influences? 403
F. Summary 404
References 404

A. OVERVIEW

Smoking remains a large public health problem for the U.S., where it is estimated that the smoking of tobacco contributes to more than 415,000 premature deaths annually. Associated costs due to lost productivity and medical care are estimated to be $72 billion annually in the U.S. With approximately 50 million individuals currently smoking in the U.S., the smoking of tobacco remains of enormous concern to public health officials as well as those who manage and are responsible for the allocation of increasingly scarce health care resources.[1]

Over the past three decades, the various reports from the office of the U.S. Surgeon General have chronicled the progress being made in understanding smoking as a behavior, the health consequences of smoking, and the addictive nature of nicotine, the most psychoactive component of tobacco smoke. Smoking cessation programs ranging from simple counseling to relatively complex and multifaceted training programs have been developed and evaluated, with some showing good success in achieving end-of-treatment quit rates. A very vexing problem remains, however, in the form of relatively high relapse rates occurring for even the very best of programs. The relapse problem remains of concern to the public health community because it has been a difficult problem to solve. Even if solutions to the relapse problem are identified, it is not at all clear that those identified for today's smokers will be applicable to the smokers of tomorrow, primarily because of the shifting nature of smoker demography.

Our interest in the genetics of smoking began with the observation that family history for smoking was a significant predictor of relapse in both male and female ex-smokers.[2,3] This finding, from a study of unrelated individuals, did not allow for the study of the mode of transmission of the familial factors that predispose an individual to relapse.[3] Over the past 10 years, in a series of separate analyses we investigated the genetic contribution to smoking and related issues using the World War II Twin Registry of the National Academy of Sciences–National Research Council (NAS-NRC) and a smaller, more comprehensively studied subset of this registry, known as the National Heart, Lung, and Blood Institute (NHLBI) Twin Study. The NAS-NRC Twin Registry (described more completely later in this chapter) is the largest currently available U.S. twin registry and has proved to be ideal for the study of the genetics of smoking and related substance use.

The major objectives of the present review are to: (1) provide a brief overview of the neurochemistry of nicotine; (2) summarize evidence for the role of genetics from animal, human, and molecular investigations; (3) review the NAS-NRC Twin Registry and the NHLBI Twin Study subset, on which our work has been based; (4) review the findings from our studies; and (5) identify directions for future research.

B. BACKGROUND

1. NEUROCHEMICAL ACTIONS OF NICOTINE

The initial action of nicotine appears to be on the nicotinic cholinergic receptors, producing an increase in the rate of release and turnover of acetylcholine and catecholamines. Nicotine also acts on the neuropeptides, some of which are neuromodulators, such as cortisol, ACTH, and endogenous opioids (including beta-endorphin).[4] It has been hypothesized that nicotine enhances the experience of pleasure through its effects on the dopaminergic system, while its ability to reduce anxiety and tension is mediated through stimulation of β-endorphin, ACTH, and cortisol; nicotine's ability to enhance task performance may be due principally to its action on the cholinergic and noradrenergic pathways.[4] It is possible that gene variants play a role in receptor functioning in one or more of the systems on which nicotine exerts an effect. Current research is focused on the A1 allele of the DRD2 gene, which is partially responsible for the responsiveness of the dopaminergic system,[5] a central component to the pleasure-reward axis of addiction.[6]

2. SMOKING BEHAVIOR AND REACTIONS TO NICOTINE ARE GENETICALLY DETERMINED: PREVIOUS RESEARCH

Early studies on smoking in twins determined that monozygotic (MZ) twins are more similar than dizygotic (DZ) twins with respect to smoking behavior.[7,8] Estimates of heritability of smoking behavior derived from large cohorts of twins such as the Swedish and Finnish twin registries[9-11] support the conclusion of a significant contribution of genetic effects to smoking, accounting for 53% of the total variance (range 28 to 84%).[12] This finding can be interpreted as evidence either that the genotype exerts an influence on smoking behavior or that there is a particularly strong environmental pressure for conformity in smoking that is greater in MZ twins than in DZ twins.[13] Data from twins reared apart tend to support the conclusion that genes may be involved in smoking behaviors.[14,15]

With the recent recognition that nicotine is an addictive drug and that the treatment of nicotine addiction requires a knowledge of individual susceptibility to drug use,[16] it is apparent that the effect of genotype on various aspects of smoking behavior needs to be examined.[17] The possibility that there are important genetically determined differences in people's susceptibility to nicotine is supported by studies showing differences between strains of mice suggesting that genes mediate: (1) differences in sensitivity to nicotine between strains;[18] (2) the density in nicotine receptor binding sites;[19] (3) the ability to develop tolerance to nicotine;[20] (4) and the extent of nicotine self-administration.[21]

C. STUDY POPULATIONS

1. THE NAS-NRC TWIN REGISTRY

The data analyzed in several of our papers were from two epidemiologic surveys of the NAS-NRC Twin Registry, the first conducted in 1967 to 1969 and the second in 1983 to 1985. The methods used to construct this twin panel have been described elsewhere.[22] Briefly, multiple births of white males in the continental U.S. from 1917 to 1927 were identified by searching birth certificates. About 93% of all such births estimated from national statistics to have occurred during those years were located.[23] Among them, 15,924 pairs in which both twins had records in the Master Index File of the Department of Veterans Affairs were identified. These records yielded information on the physical examination at induction, hospital admissions and outpatient visits during military service, and home addresses for the twins. All members of the registry were screened for entry into the armed forces during World War II. Pairs in which one or both members had childhood diseases, such as diabetes or essential hypertension, were not included.

Apart from the power of the twin method to assess the roles of genetic and environmental factors in human behaviors, the NAS-NRC Twin Registry is noteworthy for several other important features. First, the cohort is sufficiently large to support powerful analyses; second, excellent longitudinal data on cardiovascular disease risk factors and health behaviors are available on these subjects from military induction (ages 17 to 24 years) and during middle age; and, third, the cohort, screened for health at induction, received maximum exposure for both the initiation and maintenance of smoking (free cigarettes were distributed as part of the rations given to members of the armed forces).

An initial questionnaire was mailed to 27,502 men in 1965 and 1966, and 20,946 (76%) replied. Zygosity was determined on the basis of self-reports of the degree of similarity between the two brothers. Later assessments of the accuracy of these reports based on fingerprints, physical characteristics, and blood typing suggested that approximately 95% of these determinations were correct.[24]

A second questionnaire was mailed between 1967 and 1969 in a collaborative study with investigators responsible for the twin registry of the Karolinska Institute in Stockholm.[25] The English-language questionnaire was essentially a translation of one used in Sweden and later adopted for the Finnish twin registry.[10,11] The objective of these surveys was to obtain information on histories of coronary and respiratory disease, tobacco and alcohol consumption, and related social and environmental factors.

The net response rate to this survey was 75%, excluding those not reached in the mailing and those found to have died. Respondents to the 1967 to 1969 epidemiologic survey were mailed a follow-up questionnaire during 1983 to 1985. Questions about tobacco and alcohol consumption were the same as those in the 1967 to 1969 questionnaire. Corrected for mortality, the response rate to the 1983 to 1985 epidemiologic survey was 66%.

Substance use in the NAS-NRC Twin Registry as well as substance use in the NAS-NRC Twin Registry and, for comparison purposes, that for the younger Finnish twin registry[26] are presented in Table 23-1. The extremely high prevalence of ever smoking (81%) in the NAS-NRC registry, compared with the Finnish registry (62%) and the general U.S. population, is noteworthy. The most probable explanation for this high prevalence is the subjects' participation in World War II, during which cigarettes were dispensed routinely, free of charge, to all servicemen.

TABLE 23-1
Substance Use in the NAS-NRC and Finnish Twin Registries

Substance	NAS-NRC Registry	Finnish Registry
Cigarettes		
Nonsmoker (%)	19.1	38.1
Current smoker (%)	58.7	42.5
Former smoker (%)	22.2	19.4
Alcohol (g/month)		
0–250	46.7	47.7
251–500	14.4	25.8
>500	38.9	26.5
Coffee (cups/day)		
0	11.0	6.1
1–5	66.1	51.0
>5	22.9	43.9

2. THE NHLBI TWIN STUDY SUBSAMPLE

Twins in this subsample were a subgroup of the NAS-NRC Twin Registry who lived in proximity to exam centers in California, Indiana, and Massachusetts. During the period 1969 to 1973, 1,046 male subjects, including 514 pairs of twins, participated in an extensive cardiovascular examination, laboratory tests, and zygosity assessment, as well as completed a health questionnaire.[27] Twin zygosity was determined by analysis of genetic variants at 22 chromosome locations, yielding a probability of less than .001 for identical markers at all loci in DZ twins. Of the pairs, 254 were determined to be MZ, and 260 were DZ. The analyses described below are of the smoking and alcohol data obtained from 360 complete pairs (176 MZ and 184 DZ) who participated in the second NHLBI Twin Study examination, conducted during 1979 to 1981. Of the surviving cohort, 81%, a total of 792 subjects, returned to participate in the second examination. The average age of these subjects was 59 years (range 52 to 66 years). Other characteristics of participants and nonparticipants in the repeated examinations of the NHLBI Twin Study are described elsewhere.[28]

D. EVIDENCE FOR GENETIC INFLUENCES ON SMOKING AND RELATED BEHAVIORS: FINDINGS FROM THE NAS-NRC REGISTRY AND THE NHLBI TWIN STUDY

In the work that we review below, the reader will see that we have applied a number of different analytic approaches to the study of twin similarity for smoking behavior. These approaches include the use of the "classic" analysis of variance approach to estimate broad heritability (i.e., percent of additive genetic variance of the total variance).[29] Before estimating broad heritability, we used multiple-regression techniques to remove variance shared between smoking and other substance use, such as coffee and alcohol consumption.[30] Other nonparametric techniques were used to estimate a rate ratio of the extent to which MZ concordances for smoking behaviors exceed those of DZ twins.[31] Most recently, we have used multivariate extensions of the biometrical twin path model[32-33] to study the genetic and environmental commonality of substance use.

1. SMOKING HAS SIGNIFICANT GENETIC VARIANCE

Smoking, alcohol use, and coffee consumption are consistently correlated across a wide variety of populations, with moderately strong associations between tobacco and alcohol consumption as well as between coffee drinking and cigarette smoking. Several models have been proposed to explain the clustering in the use of these substances. These include biobehavioral models in which the effects of one substance serve as cues for the use of others,[30] personality models in which an underlying psychological trait or set of traits (e.g., antisocial behavior, depression, or neuroticism) predispose an individual to polysubstance use,[13] and neural/genetic models in which the various substances are seen to act and interact on common neural pathways and receptors.[34-36] A recent investigation, in fact, has demonstrated a strong association between the *Taq I* B allele of the D2 dopamine receptor (DRD2) gene and heavy polysubstance use.[36]

Previous behavioral genetic studies estimated the heritability of smoking as if the behavior occurred independently of other appetitive behaviors, such as coffee and alcohol consumption. Reviews of this literature indicate the presence of significant heritability for both tobacco and alcohol consumption.[12,37] Other studies have shown the heritability for coffee drinking in male twins to range from 0.46 to 0.88.[10,37-39] From the perspective of behavioral genetics, however, the known phenotypic relationships among the various substance use behaviors present interesting problems for interpreting broad heritabilities derived for each behavior separately.

In the first analysis, we examined the heritability of cigarette smoking and alcohol consumption in twin pair participants in the second exam of the NHLBI Twin Study. Heritability estimates for smoking and alcohol use were calculated both before and after adjustment for shared variance between these behaviors and other characteristics, including coffee consumption. Before adjustment, heritability of both smoking and alcohol use was highly significant and accounted for 52 and 60% of the variance, respectively. After adjustment for covariates, the heritability of smoking remained at 52% while that for alcohol use decreased to 43%. The fact that these estimates remained significant after adjustment for covariates leads to increased confidence about the role of genetics in both smoking and alcohol consumption.[40]

In the second analysis,[38] we applied this approach to smoking, alcohol, and coffee consumption in twin pairs from the larger NAS-NRC registry. Before adjustment for covariance with other substances, heritability estimates for smoking, alcohol use, and coffee consumption were 52, 36, and 44%, respectively. After adjustment, these estimates were 42, 30, and 44%, respectively. These results reveal unadjusted heritability estimates for the use of all three substances to be consistent with previously published estimates.[12] Although the adjustment process resulted in heritability estimates that were somewhat lower, all estimates remained significant. The robustness of the genetic component supports the general conclusion that the use of each of the substances is, in part, genetically determined.

2. DIFFERENT COMPONENTS OF SMOKING TOPOGRAPHY HAVE SIGNIFICANT GENETIC VARIANCE

Smoking behavior has a complex topography and can vary in intensity (i.e., light or heavy), stage of development (i.e., initiation, maintenance, cessation, or relapse), and form of tobacco used (i.e., cigarettes, pipes, cigars, etc.). Previous genetic investigations have dealt only partially with the issue of genetic variance in the different manifestations of smoking.[17] Toward this end, we were the first to conduct a genetic analysis of several aspects of smoking behavior using the NAS-NRC Twin Registry.[31] Information on smoking history was available for 4,775 pairs of twins, who were first surveyed in 1967 to 1969, when they were 40 to 50 years old, and then resurveyed in 1983 to 1985, when they were 56 to 66. Of the subjects in this cohort, 80% had smoked at some time in their lives; 60% were smokers in 1967 to 1969; and 39% were smoking in 1983 to 1985. Twin pair similarities for smoking habits at baseline and at the second follow-up 16 years later (Table 23-2) were examined. The comparison of the concordance rates in MZ and DZ twins was used to assess the relative contribution of familial and genetic factors.

TABLE 23-2
Smoking Status on the 1967 to 1969 and 1983 to 1985 Questionnaires, According to Zygosity[a]

	1967–1969				1983–1985			
	Monozygotic		Dizygotic		Monozygotic		Dizygotic	
	(n)	(%)	(n)	(%)	(n)	(%)	(n)	(%)
Never smoked	913	(20)	863	(17)	718	(21)	655	(18)
Former smokers	873	(19)	984	(20)	690	(20)	819	(22)
Current cigarette smokers	2412	(52)	2638	(53)	918	(26)	999	(27)
Current cigar or pipe smokers	412	(9)	455	(9)	399	(11)	447	(12)
Quitters[b]	–	–	–	–	754	(22)	721	(20)
Total	4610	(100)	4940	(100)	3479	(100)	3641	(100)

[a] Percentages do not all sum to 100 because of rounding.
[b] Subjects who reported smoking in 1967 through 1969 but who were not regular smokers in 1983 through 1985.

In the 1967 to 1969 survey, the concordance rate for smoking was higher among MZ twins than among DZ twins for initiation (overall rate ratio, 1.38; 95% confidence interval, 1.25 to 1.54), for quitting prior to 1967 to 1969 (overall rate ratio, 1.59; 95% confidence interval, 1.35 to 1.85), for current cigarette smoking (overall rate ratio, 1.18; 95% confidence interval, 1.11 to 1.26), and for current cigar or pipe smoking (overall rate ratio, 1.60; 95% confidence interval, 1.22 to 2.06) (see Table 23-3). Evidence was also obtained for genetic influence on the intensity of smoking (see Table 23-4), with a higher concordance among

MZ twins than DZ twins for light smoking (1 to 10 cigarettes per day; overall rate ratio, 1.95; 95% confidence interval, 1.11 to 2.75) and heavy smoking (more than 30 cigarettes per day; overall rate ratio = 1.48, 95% confidence interval, 1.23 to 1.78), but not for moderate smoking (11 to 30 cigarettes per day; overall rate ratio = 1.10; 95% confidence interval, 0.98 to 1.22) (see Table 23-5). When changes in smoking over the follow-up were examined, the data suggested genetic influences on quitting smoking. MZ twins were more likely than DZ twins to be concordant for quitting smoking (overall rate ratio, 1.24; 95% confidence interval, 1.06 to 1.45) (see Table 23-6).

The finding that light and heavy smoking but not moderate smoking has a genetic basis is interesting. Shiffman[41,42] has described a group of smokers who have a long history of light smoking and who are apparently able to quit smoking with no or little difficulty ("chippers"), thus indicating an apparent resistance to the addictive nature of nicotine. He has speculated that the underlying biology of this type of smoker may well be determined by genetic factors, and our analysis lends support to this conclusion. Future work will need to determine whether chippers have gene variants that promote a super-functioning of the dopaminergic system, in contrast to heavy smokers, who may have a deficit in dopaminergic functioning.[5]

3. **SMOKING SHARES A GENETIC COMMONALITY WITH ALCOHOL AND COFFEE CONSUMPTION**

The first multivariate genetic analysis in our series of analyses focused on the full range of consumption of all three substances, as determined from a physician interview conducted for the NHLBI twin subsample. Multivariate structural equation modeling procedures[43] were used to estimate the genetic and environmental contributions to the variation in and covariation among tobacco use, alcohol consumption, and coffee intake. These modeling procedures follow directly from the univariate biometrical twin model.[32] In the typical univariate applications of the basic biometrical model, intraclass correlations of MZ twins are compared with those of DZ twins to estimate proportions of observed variation attributable to additive genetic (A) and environmental (E) factors, with the environmental determinants further partitioned into those that are shared by members of a twin pair, thus contributing to twin similarities (common environmental factors, C), and those that are nonshared or specific to individual twins, thus contributing to twin dissimilarities (nonshared environmental factors, E). Greater intraclass correlations for MZ than DZ twins are taken as evidence for genetic influences; situations in which MZ and DZ correlations are not statistically different from one another (and are nonzero) suggest common environmental influences on the trait, as do correlation patterns in which DZ resemblance is greater than half the MZ resemblance, since the excess similarity cannot be explained by differences in genetic resemblance. Estimates of nonshared environmental variance proportions are derived from the additive genetic and common environmental estimates by subtraction. It should be noted that a critical assumption of this model is that shared environmental influences on MZ twins do not differ from those on DZ twins.[44] Our earlier analysis of smoking, alcohol, and coffee use supports the tenability of this "equal environments" assumption in the NHLBI sample.[40]

Structural equation modeling procedures advance the basic biometrical model by minimizing some of the inherent disadvantages of correlation comparisons (including reduced statistical power, large standard errors of estimates, and the potential for variance proportion estimates to exceed 100%) and by providing a general and flexible framework for formal hypothesis testing. The multivariate approach used in these analyses estimates the same sources of variation as does the basic biometrical model (A, C, and E), extending the model to examine whether individual differences in tobacco use, alcohol consumption, and coffee intake are determined by the same genetic and/or environmental influences, or whether

TABLE 23-3
Pairwise Comparisons of Rates of Concordance, According to Smoking Status in 1967 to 1969[a]

	Never smoked		Former smokers		Current cigarette smokers		Current cigar or pipe smokers	
	Monozygotic (n = 297)	Dizygotic (n = 180)	Monozygotic (n = 165)	Dizygotic (n = 123)	Monozygotic (n = 890)	Dizygotic (n = 840)	Monozygotic (n = 61)	Dizygotic (n = 41)
Observed (%)	12.9	7.3	7.2	5.0	38.6	34.0	2.6	1.7
Expected (%)	3.9	3.0	3.6	4.0	27.4	28.5	0.8	0.8
Observed to expected ratio	3.29	2.39	2.00	1.26	1.41	1.19	3.27	2.05
95% Confidence interval	2.99–3.68	2.05–2.77	1.70–2.33	1.04–1.51	1.32–1.50	1.11–1.27	2.50–4.21	1.46–2.79
Overall rate ratio[b]	1.38		1.59		1.18		1.60	
95% Confidence interval	1.25–1.54		1.35–1.85		1.11–1.26		1.22–2.06	

[a] Numbers in parentheses are numbers of concordance pairs.
[b] The overall rate ratio is the ratio of observed to expected concordance in the MZ twins to the ratio of observed to expected concordance in the DZ twins.

TABLE 23-4
Distribution of Current Cigarette Smokers on the 1967 to 1969 and 1983 to 1985 Questionnaires, According to Amount Smoked[a]

	1967–1969				1983–1985			
	Monozygotic		Dizygotic		Monozygotic		Dizygotic	
	(n)	(%)	(n)	(%)	(n)	(%)	(n)	(%)
Light (1-10 cigarettes/day)	317	(13)	276	(10)	59	(6)	76	(8)
Medium (11-30 cigarettes/day)	1474	(61)	1636	(62)	517	(56)	550	(55)
Heavy (>30 cigarettes/day)	621	(26)	726	(28)	342	(37)	373	(37)
Total	2412	(100)	2638	(100)	918	(100)	999	(100)

[a] Percentages do not all sum to 100 because of rounding.

TABLE 23-5
Pairwise Comparisons of Rates of Concordance, According to Amount Smoked in 1967 to 1969[a]

	Light smokers (1–10 cigarettes per day)		Medium smokers (11–30 cigarettes per day)		Heavy smokers (>30 cigarettes per day)	
	Monozygotic (n = 33)	Dizygotic (n = 11)	Monozygotic (n = 379)	Dizygotic (n = 338)	Monozygotic (n = 116)	Dizygotic (n = 78)
Observed (%)	3.7	1.3	42.6	40.2	13.0	9.3
Expected (%)	1.2	0.8	38.4	39.9	7.0	7.7
Observed: expected ratio	3.16	1.62	1.11	1.01	1.78	1.20
95% Confidence interval	2.16-4.45	0.81-2.90	1.00-1.23	0.90-1.13	1.48-2.13	0.95-1.48
Overall rate ratio[b]	1.95		1.10		1.48	
95% Confidence interval	1.11-2.75		0.98-1.22		1.23-1.78	

[a] Numbers in parentheses are numbers of concordance pairs.
[b] The overall rate ratio is the ratio of observed to expected concordance in the MZ twins to the ratio of observed to expected concordance in the DZ twins.

TABLE 23-6
Comparisons of Rates of Concordance in Twins Who Quit Smoking During Follow-up[a]

Concordance	Monozygotic (n = 155)	Dizygotic (n = 117)
Observed rate (%)	16.2	14.5
Expected rate (%)	10.7	11.8
Observed to expected ratio	1.58	1.23
95% Confidence interval	1.29–1.77	1.02–1.47
Overall rate ratio[b]	1.24	
95% Confidence interval	1.06–1.45	

[a] Numbers in parentheses are numbers of concordant pairs.
[b] The overall rate ratio is the ratio of observed to expected concordance in the MZ twins to the ratio of observed to expected concordance in the DZ twins.

variation in each appetitive phenotype can be attributed to trait-specific genetic and/or environmental factors.

Our approach involved a model comparison series in which several alternative multivariate models were fitted to the data to determine the statistical significance of common- and/or trait-specific A, C, and E effects. The general multivariate model from which we initiate this series of significance tests is referred to as the "independent-pathway model."[45] In this model, a set of common genetic and environmental factors (A, C, and E, each with its own independent pathway to the phenotypes) are defined to account for all associations among the smoking, alcohol, and coffee measures. This model relates to "pleiotropic" genetic effects[32] in that the common genes contribute to individual differences in all of the phenotypes. The magnitude of effect may differ for each trait; for example, the common genes may have a greater influence on alcohol consumption than on coffee consumption. Each trait may be further influenced by trait-specific genetic and environmental effects. A reduced form of the independent-pathway model, known as the "common-pathway model,"[45] has also been described. This model defines a single latent variable having A, C, and E influences. The latent variable (labeled here as "polysubstance use") accounts for all phenotypic correlations such that common genetic and environmental effects are scalar multiples for each variable rather than having independent pathways. As in the independent-pathway model, there are trait-specific genetic and environmental effects on each observed measure.

The alternative models are fitted to the observed data using the LISREL program.[46] This program provides maximum-likelihood estimates of all model parameters and calculates a χ^2 goodness-of-fit measure under the assumption of multivariate normal distributions for all variables under study. For tests of statistical significance, parameters may be omitted from the model and the likelihood recomputed. The difference between nested χ^2 values is a likelihood-ratio value that is itself distributed as χ^2 with degrees of freedom (df) equal to the difference in the number of parameters estimated in the models. The Akaike Information Criterion (AIC = $\chi^2 - 2df$)[47] serves as an additional tool for model evaluation, with smaller values (larger negative values) indicating greater parsimony. We use the likelihood-ratio comparisons, goodness-of-fit indices, and parsimony criteria to derive a "best-fitting" model that adequately represents the observed correlations. Through this methodology, a common latent factor underlying twin similarity in the joint use of smoking, alcohol, and coffee was identified. The best-fitting, most parsimonious model is illustrated in Figure 23-1. This latent factor was explained entirely by additive genetic sources, suggesting that the observed associations among the use of these three substances are genetically mediated.[48]

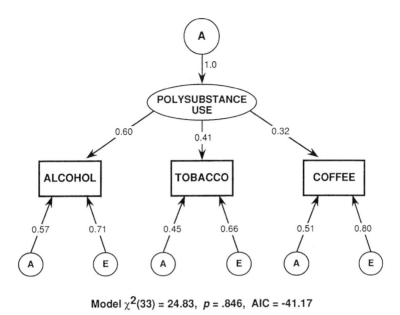

FIGURE 23-1 Path diagram of best-fitting model of tobacco, alcohol, and coffee use. Measured variables are contained in boxes; latent variables are shown in circles (A, additive genetic influences; E, environmental influences). *Polysubstance Use* is a latent variable representing the common cause of associations among the measured variables. Parameter estimates listed adjacent to causal pathways reflect the relative impact of genes and the environment on variation and covariation among the observed measures.

With the subsequent development of a constrained multivariate structural equation model postulating two pleiotropic sets of polygenes instead of one, we determined in the second analysis of this series, again over the entire range of consumption in the NAS-NRC sample of twins, the presence of two genetic factors: one for the joint use of tobacco and alcohol and the other for the joint use of tobacco and coffee (see Figure 23-2). The estimated correlation between these two genetic factors (r = 0.73) was significantly different from unity, thereby indicating some dissimilarity in the genetic etiologies of the two clusters of substance use.[49]

Because the factors determining joint heavy use may be different from those determining joint use over the entire range, we sought to determine in the third analysis in this series the best-fitting multivariate genetic model for the joint distribution of heavy alcohol, tobacco, and coffee consumption.[50] For the analysis of heavy polysubstance use, we defined heavy smoking as more than 30 cigarettes per day, heavy alcohol use as more than 67 drinks per month, and heavy coffee consumption as more than 5 cups per day. Prevalences and twin concordance rates for the three substances and their joint occurrence are presented in Table 23-7. About 20% of the NAS-NRC twins of either zygosity exceeded at least one of these thresholds, and 4% (81 MZ and 105 DZ twins) exceeded all three criteria for heavy polysubstance use.

The probandwise concordance rates shown in Table 23-7 reveal higher concordances for MZ twins than for DZ twins for combined heavy use, supporting the role of genes in the joint heavy use of these substances. The multivariate genetic analyses relied on a more efficient statistic, the tetrachoric correlation, based on the joint distribution of these variables. The calculation of a tetrachoric correlation assumes that a latent distribution consisting of the liability or vulnerability to heavy substance use underlies the division of twins into heavy and nonheavy users. The threshold on this liability is such that individuals with liability above the threshold will become heavy users while those with a liability below the threshold will be regular users. The tetrachoric correlation represents the correlation in twins for this

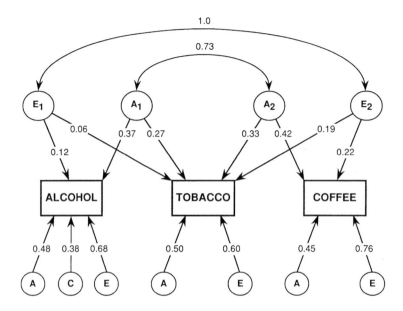

FIGURE 23-2 Path diagram of best-fitting model of tobacco, alcohol, and coffee consumption in the NAS-NRC Twin Registry. Measured variables are contained in boxes; latent variables are shown in circles (A, additive genetic influences; C, common environmental influences; E, nonshared environmental influences; A1, common additive genetic influence for joint use of alcohol and tobacco; A2, common additive genetic influences for joint use of tobacco and coffee; E_1, nonshared environmental influence for joint use of alcohol and tobacco; E_2, nonshared environmental influence for joint use of tobacco and coffee). Parameter estimates listed adjacent to causal pathways reflect the relative impact of genes and the environment on variation and covariation among the observed measures.

underlying liability. Since the model further assumes a multifactorial etiology involving multiple genetic and environmental risk factors of small to moderate effects, the liability distribution for heavy substance use is also assumed to be approximately normal.

Across the entire sample, the tetrachoric correlation between heavy alcohol and tobacco consumption was $0.29, p < 0.001$, while that between heavy coffee and tobacco consumption was $0.29, p < 0.001$. The correlation between heavy alcohol and coffee consumption was $-0.04, p < 0.01$. Tetrachoric correlations displayed by zygosity are presented in Table 23-8. Cross-twin correlations, shown in boldface, provide the information for estimation of genetic and environmental influences. Again, it can be seen that MZ twins are more similar than DZ twins for the liability of heavy use of each substance (diagonal elements) and their combined use (off-diagonal elements). In addition, MZ twins reveal greater correlations than DZ twins only for alcohol-tobacco and tobacco-coffee. For the joint use of alcohol and coffee, all correlations are close to zero.

The most parsimonious model fitting these data included genetic and nonshared environmental effects accounting for the joint heavy use of alcohol and tobacco and of coffee and tobacco, as well as respective trait-specific genetic influences. In addition, common environmental influences were found for heavy alcohol use. Standardized parameter estimates and summary statistics for this model are presented in Figure 23-3 and Table 23-9. Heavy polysubstance use shows no pleiotropic effects among all three substances. Rather, genetic and environmental effects are unique to the two clusterings in smoking and alcohol use and smoking and coffee use.

In our first multivariate genetic analysis of the joint use of tobacco, alcohol, and coffee over the entire range of consumption in the NHLBI Twin Study, we obtained evidence of a common genetic factor underlying the use of all three substances.[48] We interpreted the

TABLE 23-7
Prevalences and Concordance Rates for Heavy Use of Alcohol (ALC), Tobacco (TOB), and Coffee (COF) Among Male Monozygotic (MZ) and Dizygotic (DZ) Twins

	MZ Twins				DZ Twins			
	Both Affected (C)	One Affected (D)	Prevalence (2C + D)/2N	Concordance 2C/(2C + D)	Both Affected (C)	One Affected (D)	Prevalence (2C + D)/2N	Concordance 2C/(2C + D)
ALC	203	506	20.5	44.5	178	664	21.5	34.9
TOB	166	489	18.5	40.4	139	693	20.5	28.6
COF	238	551	23.1	46.4	226	775	25.8	36.8
ALC-TOB	29	208	6.0	21.8	19	285	6.8	11.8
ALC-COF	13	163	4.3	13.8	11	222	5.1	9.0
TOB-COF	40	231	7.0	25.7	23	330	7.9	12.2
ALC-TOB-COF	3	75	1.8	7.4	2	101	2.2	3.8

TABLE 23-8
Correlations for MZ (Below Diagonal) and DZ (Above Diagonal) Twins for Alcohol (ALC), Tobacco (TOB), and Coffee (COF)

	Twin 1			Twin 2		
	ALC	TOB	COF	ALC	TOB	COF
Twin 1						
Alcohol		0.27	−0.04	*0.35*	*0.11*	*−0.02*
Tobacco	0.31		0.33	*0.02*	*0.24*	*0.09*
Coffee	0.00	0.31		*0.05*	*0.12*	*0.28*
Twin 2						
Alcohol	*0.53*	*0.18*	*−0.04*		0.32	−0.01
Tobacco	*0.21*	*0.50*	*0.22*	0.28		0.24
Coffee	*0.02*	*0.24*	*0.51*	−0.05	0.32	

Note: Threshold criteria for heavy use: alcohol, > 67 drinks/month; smoking, > 30 cigarettes/day; coffee, > 5 cups of coffee/day.

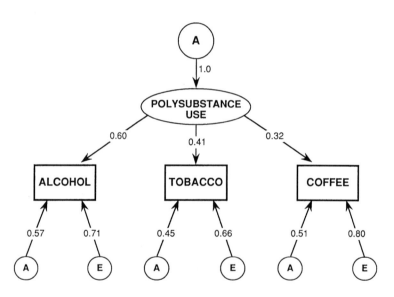

Model $\chi^2(33) = 24.83$, $p = .846$, AIC = −41.17

FIGURE 23-3 Parameter estimates from best-fitting, most parsimonious model of heavy use of alcohol, tobacco, and coffee in the NAS-NRC Twin Registry, $\chi^2_{22} = 14.13$; $p > 0.80$. Parameter estimates that were not significantly different from zero are omitted from the diagram.

underlying genetic component of variance as a "polysubstance use factor," possibly representing a common neurobiology of the joint use of all three substances through their effect on the mesolimbic/mesocortical dopaminergic system.[6,34] With the subsequent mathematical development of a two-factor model, we identified two *correlated* sets of polygenes underlying the observed concurrent use of tobacco and alcohol and of tobacco and coffee over the entire range of consumption in the larger NAS-NRC Twin Registry.[49] In the final analysis of heavy consumption, again in the NAS-NRC registry, the best-fitting account of the joint heavy use of the three substances identified two *independent* sets of polygenes underlying the combined

TABLE 23-9
Summary Statistics for Best Fitting Model of Heavy Use of Alcohol, Tobacco, and Coffee in the NAS-NRC Twin Registry, 1967 to 1969[a]

	Heritability (h^2)			Environmentality (c^2, s^2)			
	Total	Shared ALC/TOB	Shared TOB/COF	c^{2b}	s^2	Shared ALC/TOB	Shared TOB/COF
Alcohol	0.38	0.17(45%)	—	0.15	0.47	0.12(26%)	—
Tobacco	0.49	0.17(35%)	0.21(43%)	0.00	0.51	0.12(24%)	0.08(16%)
Coffee	0.51	—	0.21(41%)	0.00	0.49	—	0.08(16%)

[a] Model $\chi^2(22) = 14.13$, $p > 0.80$.
[b] All shared twin environmental influences are trait specific.

heavy use of tobacco and alcohol and of tobacco and coffee.[50] In our opinion, the study of heavy consumption provides the finest resolution, thus far, of the genetic variance underlying the joint use of these substances.

4. Weight Gain Following Smoking Cessation Has a Genetic Component

Until now in the sequence of presented analyses, we have focused on the genetics of smoking. We have also used data from the NAS-NRC Twin Registry to examine the extent to which smoking habits can operate as "environmental" factors to mediate another variable, weight gain, which also has genetic variance associated with it. In this respect, the next series of analyses represent the study of gene-environment interaction.

A majority of individuals who stop smoking subsequently gain weight.[50,51] There is growing concern that increasing numbers of people smoke because of the associated weight control effect, failure to quit for fear of weight gain, or a return to smoking after having quit because of the actual weight gain.[52] The precise mechanisms underlying postcessation weight gain are not well understood.[53,54] For these reasons, we sought to determine the extent to which genetic influences may be involved in weight gain or loss associated with changes in smoking behavior.

We have used data from the NAS-NRC Registry to determine the extent to which weight gain after the cessation of smoking has a demonstrable genetic component. In our first look at this issue,[55] cessation of smoking was associated with a gain in weight that was significantly greater than that in the twins who were nonsmokers or those who continued to smoke. The within-pair difference at baseline of 1.4 kg in the exsmoker-nonsmoker pairs disappeared by the time of the second survey, when both members of the pair were nonsmokers. Similarly, in the pairs in which one of the twins quit smoking while the other continued to smoke, cessation of smoking resulted in a significant within-pair weight difference of 3.8 kg. Weight gain averaged 4.6 kg in the twins who quit as compared with 1.4 kg in the twins who continued to smoke.

In a second approach to this issue, we used the twinning condition to examine twin pair similarity in weight change in the subset of 146 MZ co-twins and 111 DZ co-twins, both of whom quit smoking during a 14-year follow-up.[56] Among these, 56 MZ pairs and 34 DZ pairs reported a gain in weight of at least 2.3 kg, and 41 MZ pairs and 21 DZ pairs reported a loss in weight of more than 2.3 kg. The remaining 49 MZ and 56 DZ twin pairs were classified as discordant for weight gain or loss. The pairwise concordance rate for weight gain was 53% in MZ twin pairs and 38% in DZ pairs. Expected rates resulting from chance alone were 30 and 31%, respectively. The difference between the two observed rates for weight gain was statistically significant (mean difference in concordance rates = 0.15; 95% confidence interval, 0.03 to 0.26). Similarly, pairwise concordance for weight loss was 46% in MZ twin pairs and 27% in DZ twin pairs. Expected values from chance alone were 20 and 19%, respectively. The difference between the observed rates for weight loss was also statistically significant (mean difference = 0.19; 95% confidence interval, 0.08 to 0.30). The findings of these two studies suggest the presence of genetic influence on weight change following smoking cessation.

E. QUESTIONS REMAINING TO BE ANSWERED

1. THE GENETICS OF NICOTINE DEPENDENCE

In our work we have identified significant genetic contributions to various aspects of smoking behavior. However, because these studies did not employ relevant measures, the extent to which genetic variance exists in nicotine dependence per se remains unknown. There is a need to assess nicotine dependence using commonly accepted definitions in the context of a large twin registry. By doing so, it will be possible to advance to the next level of refinement for behavioral genetic studies of nicotine dependence, as called for by several leading investigators in the field.[12,17,57]

We are aware of no human genetic studies employing either the Fagerstrom Tolerance Questionnaire (FTQ)[58] or its newly revised version, the Fagerstrom Test for Nicotine Dependence (FTND).[59] A cost-effective genetic investigation of nicotine dependence would begin by using the FTND to identify MZ and DZ twin pairs concordant and discordant for nicotine dependence. Once identified, these twin pairs could be recruited for a variety of future genetic studies that rely on more refined indices of nicotine dependence, such as cotinine concentrations,[60,61] heart rate reactivity to a fixed dose of nicotine,[62] cardiovascular responsiveness to smoking cues,[63] or the use of a structured interview[64] to confirm the presence of nicotine dependence as defined by DSM-III-R criteria.[65]

2. THE STUDY OF TWINS DISCORDANT FOR SMOKING AND NICOTINE DEPENDENCE

Because MZ twins are matched by nature for genetic composition, the study of twin pairs discordant for nicotine dependence could provide us with a unique and powerful way to identify nonshared characteristics that may account for the discordance. A model to explain how genetically identical individuals come to be discordant for smoking was set forth by Pedersen and Floderus-Myrhed.[9] In this model, the probability that one or both members of a twin pair become smokers is a direct function of the environmental pressure to smoke. Discordance for smoking results when there is differential exposure to environmental pressure to smoke (i.e., one twin exceeds the threshold of liability while the other does not). Given the demonstrated genetic variance associated with smoking, the only way to examine environmental influences on smoking while controlling for genetic influences is to compare MZ twins discordant for smoking. We believe that this approach has been underutilized in the study of the environmental antecedents of smoking behavior.

3. DO GENETIC EFFECTS ON SMOKING AND NICOTINE DEPENDENCE VARY AS A FUNCTION OF GENDER?

There are gender differences in the use of tobacco. In their review, Grunberg, Winders, and Wewers[66] conclude that, whereas in the past men were more likely than women to have tried smoking, by 1980 this differential had reversed itself, with 83% of women and 72% of men having tried smoking a cigarette. However, when compared with male regular smokers, female regular smokers smoke fewer cigarettes per day, are less likely to smoke more than 25 cigarettes per day, are more likely to smoke fewer than 5 cigarettes per day, are less likely to inhale deeply, and take fewer, smaller, and shorter puffs when smoking.

Although it appears that nicotine intake is less in women than in men, women generally describe more withdrawal symptoms on cessation than do men[67] and may have lower rates than men of maintaining long-term abstinence from cigarettes.[66,68] Examination of our own data relevant to this topic reveals that, compared with men, women exhibit *less* dependence on some indices: they smoke fewer cigarettes per day: M = 24.5 vs. 27.9; $t(258) = -2.68$; $p < 0.01$; smoke cigarettes with less nicotine: M = .68 vs. .78 mg; $t(332) = -3.27$; $p < 0.01$; and have less cotinine in saliva: M = 304.4 vs. 363.9 ng/ml; $t(329) = -3.24$; $p < 0.01$. On the other hand, these same women respond to items on Russell's Reasons for Smoking Scale[69] in a way suggestive of **greater** dependence: higher scores on the Addiction, Sedative, and Stimulation subscales (all $p < 0.02$) as well as the composite Pharmacological scale ($p < 0.001$). Among the many potential psychosocial and biological explanations for these gender differences is the possibility that, compared with males, females are more sensitive to nicotine, metabolize nicotine more slowly, and are different with respect to how their bodies distribute and eliminate nicotine.[66]

Despite the evidence pointing to gender differences in tobacco use and nicotine dependence, most of the work concerning the genetics of smoking is based on surveys of male twins. The few studies that have been done using female twins provide support for a genetic effect on smoking in females similar to that found in males. One study of male and female twins of age 31 years or older found similar genetic effects on smoking persistence in males and females, but no genetic effect on smoking initiation in males, despite finding significant heritability for smoking initiation in females.[70] Another investigation of U.S. female twins[71] found significant genetic influences on smoking initiation and maintenance similar in magnitude to those found among Swedish and Finnish female twins.[72] In contrast to our findings in male twins,[31] however, Edwards et al.[71] did not find clear evidence in female twins for genetic variance in the initiation of smoking and amount smoked.

By virtue of their design, however, these studies did not determine the presence, if any, of differential genetic etiologies to smoking that may exist in males and females. Moreover, because they did not assess nicotine dependence directly, they do not provide any information with regard to either the genetic contribution to it or the possibility of gender mediation of the genetic effects. The analysis of covariation among male, female, and **opposite-sex** twin pairs is critical to the ability to investigate gender mediation of genetic effects.[43]

4. COMMON OR SPECIFIC GENETIC INFLUENCES?

Recent advances in both molecular and behavioral genetic studies point to a common factor underlying the use of several psychoactive substances, including nicotine and alcohol. Several recent molecular studies have found an increased prevalence in the A1 and B1 alleles of the dopamine receptor gene (DRD2) in smokers,[5,73] alcoholics,[74] and polysubstance users.[36] It is of great interest that addictive substances may share, entirely or in part, a common neurophysiological foundation through their effects on the dopaminergic system.[34–36]

Our behavioral genetic analyses in the NHLBI and NAS-NRC male twins have identified underlying genetic commonality in the joint use of tobacco, alcohol, and coffee over the entire range of consumption,[48,49] as well as in the heavy use of these substances (see results presented earlier in this chapter). Research from Kendler and associates in female twins points to common genetic effects on smoking and clinically defined depression[75] and on depression and alcoholism.[76] We believe that these studies imply that tobacco use, in all likelihood, does not occur in isolation of the genetically determined neurobiology underlying the use of other substances or of depression.

F. SUMMARY

Behavioral genetic studies of nicotine dependence in twin cohorts of the type reported here offer the promise of further refinement of the smoking phenotype. The identification of a candidate phenotype would then be useful in selecting probands and their families for further molecular studies. The study of heavy smokers who demonstrate nicotine dependence and their families offers the greatest promise for the identification of the specific gene variants responsible for tobacco use.

REFERENCES

1. Horgan, C., Marsden, M. E., and Larson, M. J., *Substance Abuse: The Nation's Number One Health Problem*, Robert Wood Johnson Foundation, Princeton, 1993, 8.
2. Swan, G. E. and Denk, C. E., Dynamic models for the maintenance of smoking cessation: event history analysis of late relapse, *J. Behav. Med.*, 10, 527, 1987.
3. Swan, G. E., Denk, C. E., Parker, S. D., Carmelli, D., Furze, C. T., and Rosenman, R. H., Risk factors for late relapse in male and female ex-smokers, *Addict. Behav.*, 13, 253, 1988.
4. Pomerleau, O. F. and Pomerleau, C. S., Neuroregulators and the reinforcement of smoking: towards a biobehavioral explanation, *Neurosci. Biobehav. Rev.*, 8, 503, 1984.
5. Noble, E. P., St. Jeor, S. T., Ritchie, T., Syndulko, K., St. Jeor, S. C., Fitch, R. J., Brunner, R. L., and Sparkes, R. S., DZ dopamine receptor gene and cigarette smoking: a reward gene? *Med. Hypoth.*, 42, 257, 1994.
6. Wise, R. A. and Bozarth, M. A., A psychomotor stimulant theory of addiction, *Psychol. Rev.*, 94, 469, 1987.
7. Cederlof, R., *The Twin Method in Epidemiological Studies on Chronic Disease*, Dissertation of the Academy of the University of Stockholm, 1966.
8. Cederlof, R., Friberg, L., and Lundman, T., The interactions of smoking, environment, and heredity and their implications for disease etiology. A report of epidemiological studies in the Swedish twin registries, *Acta Med. Scand.*, (Suppl.)612, 1977.
9. Pedersen, N. and Floderus-Myrhed, B., Twin analysis as a potential tool for examining psychosocial factors associated with and preceding smoking behaviors. *Acta Gene. Med. Gemellol.*, 33, 413, 1984.
10. Kaprio, J., Koskenvuo, M., and Sarna, S., Cigarette smoking, use of alcohol, and leisure time physical activity among same-sexed adult male twins, in Gedda, L., Parise, P., and Nance, W. E., Eds., *Twin Research 3: Part C, Epidemiological and Clinical Studies*, Alan R. Liss, New York, 1981, 37-46.
11. Kaprio, J., Koskenvuo, M., Langinvainio, H., Romanov, K., Sarna, S., and Rose, R. J., Genetic influences on use and abuse of alcohol: a study of 5638 adult Finnish twin brothers, *Alcohol. Clin. Exper. Res.*, 11, 349, 1987.
12. Hughes, J. R., Genetics of smoking: a brief review, *Behav. Ther.*, 17, 335, 1986.
13. Mangan, G. L. and Golding, J. F. *The Psychopharmacology of Smoking*, Cambridge University Press, New York, 1984.
14. Raaschou-Nielsen, E., Smoking habits in twins, *Dan. Med. Bull.*, 7, 82, 1960.
15. Shields, J., *Monozygotic Twins Brought Up Apart and Brought Up Together*, Oxford University Press, New York, 1962.
16. Department of Health and Human Services, The Health Consequences of Smoking: Nicotine Addiction: A Report of the Surgeon General, Government Printing Office, Washington, D.C., Publ. No. DHHS (CDC) 88-8406, 1988.
17. Kozlowski, L. T., Rehabilitating a genetic perspective in the study of tobacco and alcohol use, *Br. J. Addict.*, 86, 517, 1991.

18. Marks, M. J., Stitzel, J. A., and Collins, A. C., Genetic influences on nicotine responses, *Pharmacol., Biochem. Behav.*, 33, 667, 1989.
19. Marks, M. J., Romm, E., Campbell, S. M., and Collins, A. C., Variation of nicotinic binding sites among inbred strains, *Pharmacol., Biochem. Behav.*, 33, 679, 1989.
20. Collins, A. C. and Marks, M. J., Chronic nicotine exposure and brain nicotinic receptors--influence of genetic factors, *Prog. Brain Res.*, 79, 137, 1989.
21. Collins, A. C. and Marks, M. J., Progress towards the development of animal models of smoking-related behaviors, *J. Addict. Dis.*, 10, 109, 1991.
22. Jablon, S., Neel, J. V., Gershowitz, H., and Atkinson, G. F., The NAS-NRC twin panel: Methods of construction of the panel, zygosity diagnosis and proposed use. *Am. J. Hum. Genet.*, 19, 133, 1967.
23. Hrubec, Z. and Neel, J. V., The National Academy of Sciences-National Research Council Twin Registry: ten years of operation, in *Twin Research: Proceedings of the Second International Congress on Twin Studies, Part B. Biology and Epidemiology,* Nance, W. E., Ed., Alan R. Liss, New York, 1978, 153.
24. Cederlof, R., Friberg, L., Jonsson, E., and Kaij, L., Studies on similarity diagnosis in twins with the aid of mailed questionnaires, *Acta Genet. Med. Gemellol.*, 11, 338, 1961.
25. Cederlof, R., Epstein, F. H., Friberg, L., Hrubec, Z., and Radford, E. P., Twin Registries in the Study of Chronic Disease, *Acta Med. Scand.*, (Supp 1), 523, 1971.
26. Kaprio, J., Koskenvuo, M., Artimo, M., Sarna, S., and Rantasalo, I., Baseline characteristics of the Finnish Twin Registry, Section I: Materials, Methods, *Representativeness, and Results for Variables Special to Twin Studies*, Department of Public Health Science, Helsinki, Finland, M47, 1979.
27. Feinleib, M., Garrison, R. J., Fabsitz, R. R., Christian, J. C., Hrubec, Z., Borhani, N. O., Kannel, W. B., Rosenman, R. H., Schwartz, J. T., and Wagner, J. O., The NHLBI Twin Study: Methodology and summary of results, *Am. J. Epidemiol.*, 106, 284, 1977.
28. Fabsitz, R. R., Kalousdian, S., Carmelli, D., Robinette, D., and Christian, J. C., Characteristics of participants and nonparticipants in the NHLBI Twin Study — Part I: Sociodemographic and health status, *Acta Genet. Med. Gemellol. (Rome)*, 37, 217, 1988.
29. Christian, J.C., Testing twin means and estimating genetic variance: basic methodology for the analysis of quantitative twin data, *Acta Genet. Med. Gemellol.*, 28, 35, 1979.
30. Istvan, J. and Matarazzo, J. D., Tobacco, alcohol, and caffeine use: a review of their interrelationships, *Psychol. Bull.*, 95, 301, 1984.
31. Carmelli, D., Swan, G. E., Robinette, D., and Fabsitz R. R., Genetic influence on smoking: a study of male twins, *N. Engl. J. Med.*, 327, 829, 1992.
32. Falconer, D. S., *Introduction to Quantitative Genetics*, Longman Group, New York, 1990.
33. Neale, M. C. and Cardon, L. R., *Methodology for Genetic Studies of Twins and Families*, Kluwer, Boston, 1992.
34. Wise, R. A., The neurobiology of craving: implications for the understanding and treatment of addiction, *J. Abnorm. Psychol.*, 97, 118, 1988.
35. Collins, A. C., Interactions of ethanol and nicotine at the receptor level, *Recent Dev. Alcohol.*, 8, 221, 1990.
36. Smith, S. S., O'Hara, B. F., Persico, A. M., Gorelick, D. A., Newlin, D. B., Vlahov, D., Solomon, L., Pickens, R., and Uhl, G. R., Genetic vulnerability to drug abuse: The D2 dopamine receptor *Taq I* B1 restriction fragment length polymorphism appears more frequently in polysubstance users, *Arch. Gen. Psychiatry*, 49, 723, 1992.
37. Pedersen, N., Twin similarity for usage of common drugs, in *Twin Research 3: Part C, Epidemiological and Clinical Studies,* Gedda, L., Parise, P., and Nance, W. E., Eds., Alan R. Liss, New York, 1981, 53.
38. Carmelli, D., Swan, G. E., Robinette, D., and Fabsitz, R. R., Heritability of substance use in the NAS-NRC Twin Registry, *Acta Genet. Med. Gemellol. (Rome)*, 39, 91, 1990.
39. Partanen, J., Brunn, K., and Markkanen, T., *Inheritance of Drinking Behavior: A Study of Intelligence, Personality and Use of Alcohol of Adult Twins,* Finnish Foundation for Alcohol Studies, Oslo, 1966.
40. Swan, G. E., Carmelli, D., Rosenman, R. H., Fabsitz, R. R., and Christian, J. C., Smoking and alcohol consumption in adult male twins: genetic heritability and shared environmental influences, *J. Sub. Abuse*, 2, 39, 1990.
41. Shiffman, S., Tobacco "chippers": individual differences in tobacco dependence, *Psychopharmacology*, 97, 539, 1989.
42. Shiffman, S., Fischer, L. A., Zettler-Segal, M., and Benowitz, N. E., Nicotine exposure in non-dependent smokers, *Arch. Gen. Psychiatry*, 47, 333, 1990.
43. Neale, M. C., and Cardon L. R., *Methodology for Genetic Studies of Twins and Families*, Kluwer, Boston, 1992.
44. Plomin, R., The role of inheritance in behavior, *Science,* 248, 183, 1990.
45. Kendler, K. S., Heath, A. C., Martin, N. G., and Eaves, L. J., Symptoms of anxiety and symptoms of depression: same genes, different environments?, *Arch. Gen. Psychiatry*, 44, 451, 1987.
46. Jöreskog, K. G., and Sörbom, D., *LISREL 7: A Guide to the Program and Its Applications*, SPSS Inc., Chicago, 1989.
47. Akaike, H., Factor analysis and AIC, *Psychometrika,* 52, 317, 1987.
48. Swan, G. E., Cardon, L. R., and Carmelli, D., The consumption of tobacco, alcohol, and coffee in Caucasian male twins: a multivariate genetic analysis, *J. Sub. Abuse*, 8, 19, 1996.

49. Cardon, L. R., Swan, G. E., and Carmelli, D., A model of genetic and environmental effects for the joint use of alcohol, smoking, and caffeine, *Behav. Gene.*, in press.
50. Swan, G. E., Carmelli, D., and Larden, L. R., Heavy consumption of cigarettes, alcohol, and coffee in male twins, *J. Studies Alc.*, in press.
51. Klesges, R. C., Meyers, A. W., Klesges, L. M., and LaVasque, M. E., Smoking, body weight, and their effects on smoking behavior: a comprehensive review of the literature, *Psycholog. Bull.*, 106, 204, 1989.
52. U.S. Department of Health and Human Services, The Health Benefits of Smoking Cessation: A Report of the Surgeon General, DHHS Pub. No. CDC 90-8416, U.S. Government Printing Office, Washington, D.C., 1990.
53. Gritz, E. R., and St. Jeor, S. T., Implications with respect to intervention and prevention, in *Proc. National Working Conf. on Smoking and Body Weight, Health Psychology*, 11 (Suppl), 17, 1992.
54. Klesges, R. C. and Shumaker, S. A., Understanding the relations between smoking and body weight and their importance to smoking cessation and relapse, *Proc. National Working Conf. on Smoking and Body Weight, Health Psychology*, 11 (Suppl), 1, 1992.
55. Perkins, K. A., Weight gain following smoking cessation, *J. Consult. Clin. Psychol.*, in press.
56. Carmelli, D., Swan, G. E., and Robinette, D., Smoking cessation and weight gain in identical twins (Letter), *N. Engl. J. Medicine*, 325, 517, 1991.
57. Swan, G. E., and Carmelli, D., Characteristics associated with excessive weight gain following smoking cessation in men, *Am. J. Publ. Health*, 85, 73, 1995.
58. Benowitz, N. L., The genetics of drug dependence: tobacco addiction, *N. Engl. J. Med.*, 327, 881, 1992.
59. Fagerstrom, K. O., Measuring degrees of physical dependence to tobacco smoking with reference to individualization of treatment, *Addict. Behav.*, 3, 235, 1978.
60. Heatherton, T. F., Kozlowski, L. T., Frecker, R. C., and Fagerstrom, K. O., The Fagerstrom Test for Nicotine Dependence: a revision of the Fagerstrom Tolerance Questionnaire, *Br. J. Addict.*, 86, 1119, 1991.
61. Hughes, J. R., Defining the dependent smoker: validity and clinical utility, *Behav. Med. Abstr.*, 5, 202, 1984.
62. Fagerstrom, K. O., Towards better diagnoses and more individual treatment of tobacco dependence, *Br. J. Addict.*, 86, 543, 1991.
63. Pomerleau, O. F., Collins, A. C., Shiffman, S., and Pomerleau, C. S., Why some people smoke and others do not: new perspectives, *J. Consult. Clin. Psychol.*, 61, 723, 1993.
64. Niaura, R., Abrams, D., DeMuth, B., Pinto, R., and Monti, P., Responses to smoking-related stimuli and early relapse to smoking, *Addict. Behav.*, 14, 419, 1989.
65. Spitzer, R. L., Williams, J. B., and Gibbon, M., *Structured Clinical Interview for DSM-III-R*, New York State Psychiatric Institute, New York, 1987.
66. American Psychiatric Association, *Diagnostic and Statistical Manual of Mental Disorders (3rd rev. ed.)*, Washington, DC, 1987.
67. Grunberg, N. E., Winders, S. E., and Wewers, S. E., Gender differences in tobacco use, *Health Psychol.*, 10, 143, 1991.
68. Shiffman, S., The tobacco withdrawal syndrome, in *Cigarette Smoking as a Dependence Process*, Krasnegor, N. A., Ed., NIDA Research Monograph No. 23, U.S. Government Printing Office, Washington DC, 1979, 158.
69. Swan, G. E., Ward, M. M., Carmelli, D., and Jack L., Gender differences in the rate of relapse following smoking cessation, paper presented to the Annual Meeting of the American Psychological Association, San Francisco, 1991.
70. Russell, M. A. H., Peto, J., and Patel, U. A., The classification of smoking by factorial structure of motives, *J. Royal Stat. Soc.*, 137, 313, 1974.
71. Heath, A. C., and Martin, N. G., Genetic models for the natural history of smoking: Evidence for a genetic influence on smoking persistence, *Addict. Behav.*, 18, 19, 1993.
72. Edwards, K. L., Austin, M. A., and Jarvik, G. P., Evidence for genetic influence on smoking in adult women twins, *Clin. Genet.*, 47, 236, 1995.
73. Kaprio, J., Hammer, N., Koskenvuo, M., Floderus-Myrhed, B., Langinvainio, H., and Sarna, S., Cigarette smoking and alcohol use in Finland and Sweden: A cross-national twin study, *Int. J. Epidemiol.*, 11, 378, 1982.
74. Comings, D. E., Ferry, L., Bradshaw-Robinson, S., Burchette, R., Dino, M., Chiu, C., and Muhleman, D., Role of variants of the dopamine D2 receptor (DRD2) gene as genetic risk factors in smoking, paper presented at the First Annual Scientific Conference, Tobacco-Related Disease Research Program, University of California, San Francisco, December 2 and 3, 1993.
75. Noble, E. P., The D2 dopamine receptor gene: a review of association studies in alcoholism, *Behav. Genet.*, 23, 119, 1993.
76. Kendler, K. S., Neale, M. C., MacLean, C. J., Heath, A. C., Eaves, L. J., and Kessler, R. C., Smoking and major depression: a causal analysis, *Arch. Gen. Psychiatry*, 50, 36, 1993.
77. Kendler, K. S., Heath, A. C., Neale, M. C., Kessler, R. C., and Eaves, L. J., Alcoholism and major depression in women: a twin study of the causes of comorbidity, *Arch. Gen. Psychiatry*, 50, 690, 1993.

24 The Dopamine D2 Receptor Gene Locus in Reward Deficiency Syndrome: Meta-Analyses

Kenneth Blum, Peter J. Sheridan, Thomas C.H. Chen, Robert C. Wood, Eric R. Braverman, John G. Cull, and David E. Comings

CONTENTS

- A. Historical Perspective 408
- B. Discovery of the Association of DRD2 Gene Variants With Alcoholism 408
  - 1. The Controversy: Early View 409
  - 2. Issue of Severity 410
  - 3. Classical Linkage Studies: A Question of Phenotype and Power 411
  - 4. DRD2 Gene Expression: Association With Additional Variants 412
  - 5. Ethnicity 412
- C. DRD2 and Polysubstance Abuse 413
- D. DRD2 and Obesity 415
- E. Pathological Gambling 416
- F. Post-Traumatic Stress Disorder 417
- G. DRD2 Haplotypes and Defense Style 417
- H. Smoking 417
  - 1. Commentaries: Meta-Analyses 417
- I. Negative Studies: Statistical Problems 418
  - 1. The Gelernter Reappraisal 418
  - 2. Suarez Study 422
- J. Molecular Genetic Aspects 422
- K. Bayes' Theorem as a Predictive Model 424
- L. Conclusions 426
  - 1. Is There an Association Between the D2A1 Allele and All Forms of Alcoholism? 426
  - 2. Is there an Association Between the D2A1 Allele and Some Forms of Alcoholism? 426
  - 3. Are Some Disorders Other Than Alcoholism Significantly Associated With the D2A1 Allele? 427
- References 428

A. HISTORICAL PERSPECTIVE

The involvement of genetic factors in the susceptibility to develop alcoholism continues to be one of the most controversial areas of human behavioral genetics. The tendency for drinking patterns of children to resemble those of their parents has been recognized since antiquity as revealed, for instance, in the observations of Plato and Aristotle.[1] This observation constitutes the first potential evidence for familial alcoholism.

Numerous studies have appeared in the scientific literature, with a general consensus that alcoholism has a familial component.[1] In this regard, the work of Cotton,[2] Goodwin et al,[3] Schuckit et al.,[4] Cloninger,[5] and Bowman et al.,[6] and their associates evaluated the potential genetic contribution to alcoholism in twin studies, adoption studies, family linkage studies, in both male and females. Their work suggest that:

1. Alcoholism runs in families
2. If there is at least one alcoholic biological parent, there is a four times greater risk for males to become alcoholic relative to siblings of nonalcoholic parents, and a three times greater risk for females
3. Although still controversial, alcoholics have been classified into two basic types: ***Type I-milieu-limited-alcoholism*** — less severe and modest alcoholism occurring in both male and females where environmental factors play a major role with low genetic contribution and the onset of the disease occurs after 25 years of age; ***Type II- male-limited-alcoholism*** — the most severe alcoholism affecting mostly males where genetics play a major role and environment plays a less important role and the onset of the disease occurs prior to 25 years of age
4. The genetic contribution to female alcohol abuse is negligible-zero percent in one study whereas male alcohol abuse has an approximate 38% genetic contribution[7]
5. Severe alcoholism in females has a 60% genetic contribution,[8] which is similar to the genetic contribution of severe alcoholism in males.[7]

In terms of trying to link lesions in the brain with genetic susceptibility, numerous investigators have found that the brain electrical activity centered on the late positive complex concerned with event-related potentials (ERP), which includes the P300 wave, has been observed to be abnormal in male alcoholics and their sons.[9-11] Most recently, Berman et al.[12] found evidence that young adolescents of alcoholics having P300 deficits were significantly more vulnerable for substance abuse than adolescents of nonalcoholic parents without P300 deficits. The authors suggested that their work indicates that the P300 wave has utility as a vulnerability marker for substance abuse disorders and its use will depend on combining this marker potential with other measures (i.e., genetic testing).

To more specifically associate chromosomal mutations with familial alcoholism, a number of studies utilizing linkage analysis found evidence for potential association of chromosome 4, 6, and 11 with familial alcoholism.[13-15] Other molecular genetic studies utilizing restriction fragment length polymorphism (RFLPs) found an inverse correlation of the aldehyde dehydrogenase gene in various ethnic groups with regard to concordant rates of alcoholism.[16,17] While these findings are of interest, they do not provide evidence for a causal relationship between a specific genetic anomaly and vulnerability to alcoholism.

B. DISCOVERY OF THE ASSOCIATION OF DRD2 GENE VARIANTS WITH ALCOHOLISM

In order to perform scientifically sound genetic association studies in a complex disease such as alcoholism certain inclusion/exclusion criteria must be satisfied. Specifically, these include:

1. Assessment of alcohol, drugs, tobacco use and carbohydrate bingeing in controls
2. Characterization of the phenotype (i.e., severity and chronicity)

3. Power analysis and statistical appropriateness
4. Ethnicity

In terms of molecular genetic studies, the first identification of a putative gene candidate shown to associate with severe alcoholism comes from the work of Blum et al.[18]. These researchers found that a specific variant of the Dopamine D2 Receptor Gene (DRD2), the A1 allele, was associated with 69% of severe alcoholics compared to only 20% of nonalcoholics. This observation was maintained in blacks and whites. The study was immediately criticized by scientists at the National Institute of Alcohol Abuse and Alcoholism (NIAAA). In an accompanying editorial in the *Journal of the American Medical Association*, Enoch Gordis and associates at NIAAA criticized the use of DNA extracted from deceased individuals and argued that the frequency of alleles were surprisingly high in terms of a single gene contribution to this complex mental disorder.[19] In December 1990, scientists from the NIAAA reported that they could not replicate the original findings when they probed DNA from peripheral tissue (blood) in living alcoholic subjects.[20] Careful analysis of their work resulted in a letter to the journal from Noble and Blum which pointed out the weaknesses in this subsequent study.[21] In essence, the NIAAA scientists never screened for existing alcohol and drug abuse in their controls. Moreover, Noble and Blum pointed out that statistical analysis revealed a significant linear trend showing that the prevalence of D2A1 allele increases with severity of the disease. These basic weaknesses were further amplified by a similar letter in the same issue of the journal by Uhl's group at the National Institute of Drug Abuse (NIDA), pointing out that in a NIDA survey, a high rate of lifetime alcoholism, as much as 23%, has been found in the U.S. general population, calling for more rigorous control data.[22]

1. THE CONTROVERSY: EARLY VIEW

Following these two controversial papers, Parsian et al.[24] found association of the A1 allele of the DRD2 gene with severe alcoholics relative to well-characterized drug abuse and alcohol-free controls. This association significantly increased when the severity of alcoholism was taken into account. In fact, the prevalence of the A1 allele in severe alcoholics was quite similar to the percent prevalence found in the Blum et al.[18] study, approximately 60%. Additionally, it was reported that carriers of the D2A1 allele possessed significantly fewer dopamine D2 receptors than A2 carriers when alcoholism was excluded from the sample.[25] Moreover, a progressively reduced number of receptors was found in subjects with A2/A2, A1/A2, A1/A1 genotypes, respectively. The work of Comings and associates[26] found that the DRD2 receptor gene A1 allele variant was not only associated with alcoholism and polysubstance abuse, but also with other neuropsychiatric disorders including attention deficit/hyperactivity disorder (ADHD), Tourette's disorder, conduct disorder, autism, and post traumatic stress disorder. In the same issue of the *Journal of the American Medical Association*, Gelernter et al.[27] reported no association between the A1 allele of the DRD2 gene in poorly characterized nonsevere alcoholics compared to nonassessed controls. However, in a combined analyses of the then extant studies, Cloninger[28] showed D2A1 allele in alcoholics (n = 338) and the controls (n = 471) to be 45 and 27% respectively, a difference that was highly significant ($p < 10^{-7}$).

Moreover, in terms of association of the A1 allele of the DRD2 gene, a 1993 meta-analysis reveals strong positive results. In a total of 986 cases, an association of the A1 allele of the DRD2 gene with alcoholism in nine independent studies of Caucasians was observed.[23] The odds ratio for this association is 2.18 (Yates 2 corrected for continuity = 32.0; 95% confidence interval = 1.65–2.88; $p < 10^{-7}$).

2. Issue Of Severity

In order to more systematically characterize the prevalence of the D2A1 allele in alcoholics, Blum and associates reported[29] that the A1 allele was associated to a much greater degree with severe alcoholics than with less severe alcoholics. Specifically, the A1 allele is associated with 34% of less severe alcoholics compared to 63% of severe alcoholics. This difference was significant at $p = .007$. In fact a chi-square for linear trend showed that increasing the degree of alcoholism severity corresponds to a significant increase of D2A1 allele sample prevalence ($p < .001$). Moreover, to determine whether children at risk for developing alcoholism have a higher association of the D2A1 allele compared to a nonalcoholic sample, the presence or absence of this allele in children of alcoholics (COAs) was determined. The D2A1 allele in the Blum et al.[29] study associated with 55% of COAs compared to only 21% of known nondrug or -alcohol abusing controls. Following this study, Turner and coworkers[30] found a low D2A1 allele prevalence in a small number of alcoholics from the San Diego Alcoholism Treatment Unit. Although Turner and associates claimed to assess severity, they excluded from the study all individuals having a record of antisocial personality disorder, and severity was poorly determined. Moreover, the average age of onset of alcoholism was 35 years, suggesting a milieu-limited alcoholic type. It was further noted that Turner and associates relied exclusively on published population controls that included many subjects that had not been assessed for substance abuse.

Following these studies, George Uhl, Chief of the Molecular Neurobiology Branch, NIDA, organized on September 19, 1991, the first technical conference held in Baltimore, *D2 Receptor Alleles in Substance Abuse: Have We Identified a Relevant Gene*. The basic consensus of the attendees was that a meta-analysis on all extant studies revealed that despite the initial improbability of the finding and the cautions relating to ethnic variables, evidence from a number of different laboratories appeared to support an association between DRD2 alleles marked by *Taq I* RFLPs and substance abuse. According to Uhl, "....associating a DRD2 variant with substance abuse vulnerability would identify, for the first time , an individual gene which impacts a polygenetic disorder. These are the kinds of disorders that currently challenge the limits of molecular and behavioral genetics".[31]

Recently, it was found that the A1 allele of the DRD2 gene was once again positively associated with very ill, hospitalized severe alcoholics compared to a matched group of very ill, hospitalized less severe alcoholics and nonalcoholics.[32] The prevalence of the D2A1 allele in this study of 80 nonalcoholics and 73 alcoholic patients was 30.0 and 52.1%, respectively ($p = 0.0009$). In four subgroups of these patients, the prevalence of this allele was as follows:

LPSU (i.e., less severe polysubstance abuse) = 18.2%
MPSU (i.e., more severe polysubstance abuse) = 34.5%
LSA (less severe alcoholic) = 44.4%
MSA (i.e., more severe alcoholic) = 58.3%

Linear trend analysis showed that as the use of substance and severity of alcoholism increases, so does D2A1 prevalence ($p = 0.001$). Specific group comparisons showed D2A1 prevalence in MSA to be about 3-fold ($p = 0.007$) and 1.5-fold ($p = 0.04$) higher compared to LPSU and MPSU groups, respectively. In a combined analysis of independent studies, A1 prevalence in MSA was higher when compared to LSA ($p < 10^{-2}$), MPSU ($P < 10^{-4}$), and LPSU ($p < 10^{-8}$) groups. None of the medical or neuropsychiatric complications of alcoholism were associated with the D2A1 allele.

The severity of alcohol dependence in alcoholics and of substance use behaviors in controls are important variables in DRD2 association. The present review and converging lines of evidence suggest that DRD2 represents the most prominent single gene determinant of susceptibility to *severe* alcoholism identified to date. In this regard, a relevant meta-analysis[23] of the six independent published exant association studies of Caucasians and DRD2 and severe alcoholism revealed a strong positive association. In the four studies where alcoholics were excluded from the 148 controls, 23.0% subjects had the D2A1 allele. In the same studies of 115 severe alcoholics, 59.1% subjects had the D2A1 allele. The difference between these combined groups was significant (Yates χ^2 corrected for continuity = 34.1; odds ratio = 4.885; 95% confidence interval = 2.75–8.59; $p = < 10^{-8}$). In the two studies where alcoholics were not excluded from the general population sample of the 195 controls, 62 (31.8%) subjects had the D2A1 allele. In the same studies of 43 severe alcoholics, 21 (48.8%) subjects had the D2A1 allele. The difference between these combined groups did not attain statistical significance (Yates χ^2 corrected for continuity = 3.79; odds ratio = 2.05; 95% confidence interval = 1.00–4.21; $p = .052$). With regard to 501 total cases, a positive association of the A1 allele of DRD2 with severe alcoholism was obtained with an odds ratio of 3.32 (Yates χ^2 corrected for continuity = 36.1; 95% confidence interval = 2.20–5.01; $p < 10^{-8}$). However, other genes and environmental factors, when combined, still play the larger role.

3. **CLASSICAL LINKAGE STUDIES: A QUESTION OF PHENOTYPE AND POWER**

In a letter to the *Journal of the American Medical Association*, Eric Devor further confirmed the findings of Comings[33,34] in Tourette's disorder.[35] In this regard, Devor found an increasing prevalence of the D2A1 allele with increasing severity of Tourette's disorder. In contrast, other investigators utilizing linkage analysis excluded the DRD2 gene in Tourette's disorder.[36] Similarly, lack of association of the DRD2 gene in alcoholism utilizing linkage analysis was observed by a number of investigators.[20,24,37,38]

It is noteworthy that classical linkage analyses use family data to statistically link a trait to a genetic marker on a particular chromosome. Genetic markers at a single gene locus can be identified as linked with a disease if in different generations of families displaying a genetic familial disorder, one form of a genetic polymorphic marker at a gene locus is co-inherited with the disease. One gene marker form is thus present in family members with the disease, but not in those displaying normal phenotypes in the simplest case. Genetic linkage studies thus involve analysis of the ways in which the disease phenotype and each of a number of genetic genotypic markers cosegregate or appear together in different family members. The probable polygenic mode of inheritance of substance abuse vulnerabilities and the large environmental impact on substance abuse outcomes are both likely to dramatically weaken the power of the classical familial molecular genetic linkage approaches, in which tests are made to determine how genotypes at each of many genetic loci cosegregate with substance abuse phenotypes. If substance abuse in any individual could be caused by several different genes or by strictly environmental influences, then numerous "phenocopy" substance abusers will share the same clinical characteristics but differ in genotype.

The problems with the use of linkage analysis to study polygenic disorders has been examined by Propping et al.[39] Using computer simulation, they found that for disorders caused by four or more genes negative LOD scores could be obtained over loci shown by association studies to play a significant role in the disorder. Since alcoholism is very likely to be a polygenic and multifactorial disorder, it should come as no surprise that there are numerous reports of a positive association, but virtually no reports of a positive classical linkage between the DRD2 locus and alcoholism. Sib-pair-analysis (SPA) may be a more sensitive method of testing for linkage in genetically complex disorders since it is not necessary to make assumptions concerning unknown parameters such as penetrance and gene frequencies.[40] In this regard, Cook et al., utilizing affected-sib-pair analysis,[41] reported a genetic effect of the DRD2 locus on liability to develop heavy drinking ($p < 0.002$) and alcoholism ($p < 0.0002$). At the Third International Congress of Psychiatric Genetics, David Goldman (personal communication) pointed out that Cook's positive results may have come from a single large sibship and could be an artifact. However, most geneticists believe that in affected-sib-pair analysis, as long as the genotypes of the parents are known (assessed in the Cook study), data derived from single large sibships are equivalent to data derived from multiple sibships. By contrast, Neiswanger et al.[38] found a strong association between the D2A1 *Taq I* allele and alcoholism (gene frequency in alcoholism = 0.27; gene frequency in controls screened to exclude alcoholism or a family history of alcoholism = 0.07; $p = 0.01$), but found no evidence for linkage by unaffected-sib-pair analysis. A possible explanation for the discrepancy between these two studies may have to do with the type of sib-pair-analysis utilized. Therefore, it is possible that affected-sib-pair analyses could yield a positive result in the presence of association, but in the absence of classical linkage.

Furthermore, it has recently been recognized by both Elliot Gershon, Chief of the Neurogenetic Branch of the NIMH,[42] and Eric Lander[43] that when one is analyzing a complex, multicause disease, there is more power in an association analysis than a classical linkage analysis: "Applying classical RFLP linkage analysis to a complex disease is going to be much more difficult. Here researchers are looking for a gene that contributes only 10 to 20% of an individual's susceptibility, rather than 100%, as in the case of single-gene disorders. Such genes may be hard to detect by randomly scanning the genome for linkage between disease and a RFLP marker. In fact, recent fiascoes with linkage analysis of two complex disorders — manic-depression and schizophrenia — illustrate some of the hazards".[43]

4. DRD2 GENE EXPRESSION: ASSOCIATION WITH ADDITIONAL VARIANTS

A major question raised by some investigators focuses on the fact that the D2A1, A2 polymorphism is at a *Taq I* restriction site now known to be several thousand base pairs downstream from the protein coding portion of the D2 gene.[44] In this regard, the recent findings of a second *Taq I* polymorphism B1 allele between the first and second exon very near to the first coding exon,[45] provides further support for this gene, since this new polymorphism associates both with polysubstance abuse[46] and severe alcoholism.[46-48] Finally, as previously noted,[25] the presence of D2A1 allele is associated with a reduction in the number of DRD2 receptors in nonalcoholic controls. Furthermore, additional polymorphic loci of the DRD2 gene, specifically the $D2^{In6-Ex7}$ haplotype was assessed for an association with severe alcoholism and number of DRD2 receptors. In this regard, Flanagan and associates found that this polymorphism is also associated with not only severe alcoholism, polysubstance abuse, but with a decease in the number of DRD2 receptors as well.[49] Other unpublished work in the laboratories of Blum and Noble also found an association of the B1 allele and reduction of the number of DRD2 receptors independent of alcoholism. In addition, Parsian and associates[50] preliminarily reported an association of a polymorphic loci C1, recognizing the region spanning the entire 2 and 3 exons, with severe alcoholism to an even greater degree than the A1 allele (see chapter by Parsian).

5. ETHNICITY

Another criticism of the work centers around the suggestion of Barr and Kidd[51] that allelic differences in the Caucasian studies to date have not adequately taken into account stratification based on ethnicity and the lack of homogeneous sample, especially in Caucasians. While the studies of Finnish alcoholics,[52] American Indian alcoholics,[53] and German alcoholics[54] would on the surface support the view of Kidd and associates, a deeper scrutiny of their work reveals that the issue of appropriately categorizing subjects (i.e., assessment of severity) was not fully addressed. In all these studies, the investigators did not assess the severity of alcoholics or drug abusers that they genotyped and in the Schwab study controls were unscreened.[54] It is true that Goldman and associates found a very high frequency of the A1 allele in Cheyenne Indians and failed to observe any association with this allele of the DRD2 gene in subjects with alcoholism and drug abuse in both interviewed controls and noninterviewed population controls. This is not surprising, since it is well known that not only are native Americans at risk for alcohol and drug abuse,[55] but Levy and Kunitz[56] found a 60% abstention (i.e., "they took an oath not to use drugs") rate among native Americans as compared to 25% for the general population. This statistic alone would suggest a very strong bias where prevalence of the D2A1 allele in nonsubstance abusing controls in the Cheyenne population selected by these workers would be high. Therefore, any estimations involving genotyping this population with particular reference to substance abuse vulnerability must take into account this obvious bias in order to see actual differences between the various groups tested. In contrast, other studies subvert these findings and enhance further the association of the A1 allele in severe alcoholism. In this regard, Arinami and co-workers found a significantly higher frequency of the D2A1 allele in severe Japanese alcoholics than in less severe alcoholics. Moreover, the same group found that 100% of the alcoholics that had the A1/A1 genotype were in the severe category.[57] In addition, studies by Amadeo and associates found a positive association between the DRD2 gene polymorphism and French alcoholics. In fact these authors carefully assessed severely ill hospitalized alcoholics and further confirmed the association of the A1 allele of the DRD2 gene with alcoholics compared to nonalcoholic controls in this homogeneous population. The prevalence of the D2A1 allele was significantly higher in alcoholics (42.8%; N = 49) versus non-alcoholic controls (16.2%; N = 43). The frequency of the D2A1 allele was also significantly different in the total sample of the alcoholic population, .24 for the 49 alcoholics vs. .08 for the 43 nonalcoholic controls ($\chi^2 = 8.7$; $df = 1$; $p = .003$).[58]

Given that the A1 allele of the DRD2 varies significantly in frequency from one population to another, i.e., 0.09 in Yemenite Jews (known to have very low alcoholism rates) to 0.74 in Cheyenne American Indians (known to have high alcoholism rates),[51] it is possible to mask a real association between alcoholism and this allele or demonstrate an apparent association between the two, simply as a function of the population characteristics of the subject and control groups chosen for the study. This supports the need to test this association in sample groups where assessment of severity is characterized and population heterogeneity is minimized (see Figure 24-1).

C. DRD2 AND POLYSUBSTANCE ABUSE

Certainly, the DRD2 gene variants are not specific for alcoholism per se, but prevails in polysubstance abuse as well. Smith et al.[46] reported that similar to the original findings by Blum's group showing association of the *Taq I* A1 and B1 alleles of the DRD2 receptor gene with severe alcoholism,[47] heavy polysubstance users and subjects with DSM-III-R, psychoactive substance use diagnoses, displayed significantly higher *Taq I* A1 and *Taq I* B1 frequencies than control subjects. These authors suggested that these results are consistent with a role for a DRD2 gene variant marked by these restriction length fragment polymorphisms in enhanced substance abuse vulnerability.[46] In this regard, Comings et al. reported an association between A1 allele of the DRD2 gene and a number of neuropsychiatric disorders, including polysubstance abuse (see chapter by Comings). The prevalence of the D2A1 allele in patients with drug addiction compared to controls known not to be alcoholic yielded a significant difference, $p = 0.005$.[26]

In other work by Comings et al.,[59] 200 subjects on an addiction treatment unit were divided according to diagnosis. Of the 19 with alcohol abuse, only 21% carried the D2A1

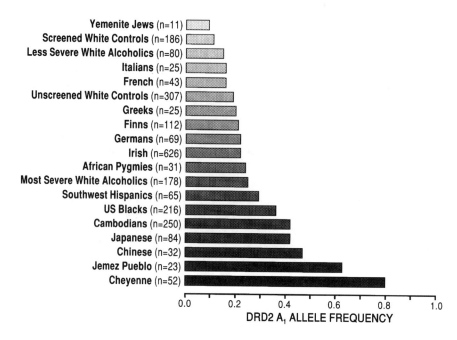

FIGURE 24-1 DRD2 *Taq 1* A1 allele frequency as a function of ethnicity derived from a number of independent investigations. The number in parenthesis denotes the proband size.

allele. Of 75 with alcohol dependence (severity not characterized), 32% carried the D2A1 allele. Neither of these was significantly different from controls, and the prevalence in those with alcohol abuse was lower than in the controls. By contrast, of 104 subjects with drug and alcohol dependence, 42.3% carried the D2A1 allele. This was significantly higher than in the 29% in 763 controls consisting of screened and unscreened subjects ($\chi^2 = 7.68$; $df = 1$; $p = .0056$). In this study logistic regression analysis revealed a highly significant association between multiple substance abuse, based on money spent on drugs, and the presence of the D2 A1 allele $p = 0.0003$, and age of onset of drug abuse ($p < .001$). D2A1 carriers exceeded D2A2A2 subjects for a history of being expelled from school for fighting ($p = 0.001$), and of those ever jailed for violent crimes, 53.1% carried the D2A1 allele versus 28.8% of those jailed for non-violent crimes ($p = .011$). This increased to 69.2% for those who were both jailed for violent crime and expelled from school. They concluded that possession of the D2 A1 allele is significantly associated with the severity of polysubstance abuse and some aggressive behaviors.

Further advancing this possibility, another study[25,60] found a strong association of the A1 allele and B1 allele in cocaine dependence. The prevalence of the A1 allele of the DRD2 gene in cocaine dependent subjects (n = 53) was 50.9% and was significantly different ($p < 10^{-4}$) from the 16.0% prevalence found in nonsubstance abusing controls (n = 100). In a larger control group (n = 365) consisting of nonsubstance abusers and subjects from the general population, 26.8% prevalence of the D2A1 allele was also significantly different from that of cocaine-dependent subjects ($p < 0.001$). The prevalence of the B1 allele of the DRD2 gene in these cocaine-dependent subjects (n = 52) and in a nonsubstance abusing group (n = 53) was 38.5 and 13.2%, respectively ($p < 0.01$). Logistic regression analysis in the cocaine-dependent subjects identified potent routes of cocaine use, early deviant behaviors, and parental alcoholism as significant risk factors associated with the A1 allele. The A1 allele contributed to 16.7% in the zero-risk score group, 35.0% in the one-risk score group, 66.7% in the two-risk group, and 87.5% in the three-risk score group. Risk score differences with allelic classification yielded a Pearson χ^2 of 10.9, $df = 3$ with a $p = 0.012$. Verification of the association of the number of risk factors with the allelic classification was made using a χ^2 test for linear trend.

Increasing risk scores are positively and significantly related to A1 classification with a χ^2 value of 10.5, $df = 1$, $p = 0.001$. The results indicate that the minor alleles A1 and B1 of the DRD2 gene are strongly associated with cocaine dependence in the samples studied. The polymorphic pattern of this dopamine receptor gene suggests that a gene conferring susceptibility to cocaine dependence is located on the q22-q23 region of chromosome 11.

In a recent review Uhl and associates[61] accessed the association between the DRD2 locus and substance abuse vulnerability and concluded that:

1. The *Taq I* A1 and B1 DRD2 RFLPs are interesting markers for events in significant portions of the DRD2 gene locus in Caucasians. *Taq I* A and B genotypes could reliably mark a structural or functional gene variant at the DRD2 locus that could be directly involved in altering behavior.
2. D2A1 and B1 markers appear more frequently in drug abusers than in control populations in each of four currently available studies.[46,59,60,62,63] Meta-analyses of these data suggest that differences between drug abuser and control populations are highly significant for both the four studies examining A1 and the two studies examining B1 frequencies.
3. The most severe abusers of addictive substances may manifest higher A1 and B1 DRD2 gene marker frequencies, while "control" comparison groups studied carefully to eliminate individuals with significant use of any addictive substance appear to display lower D2A1 and B1 frequencies than unscreened control populations.[61]
4. Current meta-analyses based on the three studies reported in full suggest that drug abusers may display an odds ratio of drug abuse likelihood for individuals possessing an D2A1 allele of 2.4- and 3.3-fold odds ratio for those having a B1 allele ($p < 0.001$) in both cases.

Moreover, one of us (Blum) along with James Halikas and associates in an unpublished study also found a twofold (37% vs. 19%)[64] higher prevalence of the DRD2 A1 allele in white cocaine addicts compared to screened controls ($p < 0.01$).

D. DRD2 AND OBESITY

The role of the DRD2 locus in obesity has been recently evaluated in both females and males.[65] In their study, the A1 allele of the DRD2 gene was present in 46% of obese subjects. A higher prevalence of the DRD2 A1 allele was observed in each of the non-Hispanic and Hispanic Caucasian and black groups examined when compared to their respective racial/ethnic controls. Furthermore, a more than twofold (46 vs. 19%) higher prevalence of the DRD2 A1 allele was found in these three obese groups when compared to population controls ($p < 10^{-3}$). The DRD2 A1 allele was not associated with a number of cardiovascular risk factors considered including: elevated levels of cholesterol, triglycerides, HDL, LDL, and blood pressure. However, a unique behavioral profile characterized by the presence of the following risk factors: parental history and later onset of obesity as well as carbohydrate preference was noted in obese subjects carrying the D2A1 allele. The cumulative number of these three risk factors was positively and significantly ($p < 10^{-4}$) related to D2A1 allelic prevalence. Specifically, the prevalence of the D2A1 allele in those probands carrying these three risk factors was observed in 11 out of 13 obese subjects or 85% (see chapter by Noble).

Blum et al.[66] genotyped 193 neuropsychiatrically ill patients with and without comorbid drug and alcohol/abuse/dependence and obesity for the prevalence of the A1 allele of the DRD2 gene. They found a significant linear trend ($\chi^2 = 40.4$, $df = 1$, $p < 0.00001$) where the percent prevalence of the D2A1 allele increased with increasing polysubstance abuse. Where the D2A1 allele was found in 44% of 40 obese subjects the D2A1 allele prevalence was found in as much as 91% of 11 obese subjects with comorbid polysubstance abuse. Of the obese subjects 53 having a mean body weight index (BMI) of 34.6 ± 8.8 were brain electrical activity mapped and compared with 15 controls with a BMI of 22.3 ± 3.0 ($p < .001$). The P300 amplitude was significantly different (two tailed; t = 3.24, $df = 16.2$, $p = 0.005$), whereas

P300 latency was not significant. Preliminarily, they found a significant decreased P300 amplitude correlated with parental polysubstance abuse ($p = 0.04$), and prolongation of P300 latency correlated with the three risk factors of parental substance abuse, chemical dependency, and carbohydrate bingeing ($p < 0.02$). Finally, in a small sample, the D2A1 allele was present in 25% of probands having zero risk compared to 66% in those obese subjects with any risk. This work represents the first electrophysiological data to implicate P300 abnormalities in a subtype of obesity and further confirms an association of the DRD2 gene and an electrophysiological marker previously indicated to have predictive value in vulnerability to addictive behaviors.[12]

Comings et al.[67] also suggested that genetic variants of the DRD2 gene may play a significant role in obesity. They rationalized that since dopaminergic agonists suppressed appetite and dopamine D2 receptor antagonists enhance it that allelic variance of the DRD2 locus may be associated with weight.[67] In this regard, they studied two DRD2 polymorphisms that could be haplotyped with PCR. Utilizing these haplotypes, Comings et al. found that haplotype 4 of the DRD2 gene was significantly associated with obese subjects ($p = 0.0003$) and suggested that this haplotype may play a major role in the regulation of obesity. They also found a significant association between haplotype 4 and height. This is consistent with the important role of dopamine D2 neurons in the regulation of growth hormone release, and provides further evidence that physiologically important genetic variants at the DRD2 locus exist.

Taken together these data showing an association of the minor allele of the DRD2 with obesity and with behavioral risk factors suggest that this gene, located on q22-q23 region of chromosome 11, confers susceptibility to subtypes of this disorder.

E. PATHOLOGICAL GAMBLING

Pathological or compulsive gambling has been termed "the pure addiction" since it is not associated with the intake of drugs.[68] Both the DSM-IIIR and DSM-IV criteria draw upon many similarities with psychoactive substance abuse including a preoccupation with gambling, development of tolerance with a need to gamble progressively larger amounts of money, withdrawal symptoms, and continued gambling despite the severe negative effects on family and occupation.[68] Because of its relevance to the question of the role of the DRD2 gene in addictive behaviors, Comings, along with R. Rosenthal, H. Lesieur and L. Rugle, initiated a study of pathological gamblers.[69] Of 171 non-Hispanic Caucasian subjects meeting the DSM-IV criteria for pathological gambling, 50.9% carried the D2A1 allele (χ^2 vs. nonsubstance abusing controls = 41.96; $df = 1$; $p < .00000001$). To evaluate severity, the subjects were asked 15 questions relating to severity and number of DSM-IV criteria met. Of those in the lower half of severity, 40.9% carried the D2A1 allele. Of those in the upper half, 63.8% carried the D2A1 allele. For males, the comparable figures were 42.9 and 67.4%, respectively. The Mantel-Haenszel χ^2 for a progressive trend (controls → less severe → more severe) for the total group and males only were 41.26 and 35.09 respectively, $p < .00000001$. Of the males without alcohol abuse or dependence, 49.1% carried the D2A1 allele. Of the males with alcohol abuse or dependence 76.2% carried the D2A1 allele (Mantel-Haenszel χ^2 = 33.4; $p < .00000001$). Despite this strong association of the prevalence of the D2A1 allele with alcoholism, substance abuse did not explain the association with pathological gambling. A logistic regression analysis showed that the presence or absence of the D2A1 allele showed a greater correlation with the severity of pathological gambling than any of 12 other variables including age of onset, sex, substance abuse, depression, and childhood conduct disorder. These studies indicate that the D2A1 allele plays a major role as a risk factor in a pure addictive behavior such as pathological gambling.

F. POST-TRAUMATIC STRESS DISORDER

Studies of mice show that one of the major effects of destruction of mesolimbic dopaminergic pathways is the development of a poor response to stress.[70] This led Comings and co-workers to examine the prevalence of the D2A1 allele in Vietnam veterans exposed to intense battle stress. An initial and a replication study gave the same results. For the total group of 56 non-Hispanic Caucasian veterans of those who developed PTSD, 59.5% carried the D2A1 allele while of those who did not develop PTSD only 5.3% carried the D2A1 allele ($\chi^2 = 15.2$; $df = 1$; $p = .0001$).

G. DRD2 HAPLOTYPES AND DEFENSE STYLE

The defenses of the ego represent one of the fundamental mechanisms of coping. The defense style questionnaire (DSQ) was developed;[71–72] Simerly, Swanson, and Gorski[74] to evaluate a various defense mechanisms. Comings et al.[75] examined DRD2 haplotypes in 57 subjects with substance abuse. For those carrying the 1 haplotype, there was a significant decrease in the mean score for mature defenses and a significant increase in the mean score for the immature defenses, compared to those not carrying the 1 haplotype.

H. SMOKING

Comings, along with L. Ferry and co-workers of the smoking cessation clinic at the Jerry L. Pettis V.A. Hospital, examined the D2A1 status of 312 non-Hispanic Caucasian smokers who were free of alcohol or drug abuse/dependence (except tobacco and who had made at least one prior unsuccessful attempt to stop smoking). Of these, 48.8% carried the D2A1 allele. When compared to 714 non-Hispanic nonalcoholics, nondrug abusing (except tobacco) controls, where 27.7% carried the D2A1; $\chi^2 = 51.4$; $p < 10^{-8}$.[76]

1. COMMENTARIES: META-ANALYSES

In reviewing the putative association of the DRD2 and alcoholism, a meta-analysis was conducted by Pato et al.[77] in which they showed that in eight published studies a statistically significant association between the A1 allele of DRD2 and alcoholism had an estimated risk ratio of 2.57 ($p < 0.0001$). In response to the review article by Pato et al.,[77] Kidd[78] highlighted two problems regarding whether or not a real association exists for the DRD2 gene variants and alcoholism. First, he suggested the use of gene frequency, where homozygotes were counted twice and heterozygotes once, was more appropriate than prevalence where homozygotes and heterozygotes are given equal weight.

According to Kidd, the meta-analysis by Pato et al.[77] used the approach that Blum et al.[18] originally used, classifying individuals by whether or not they had the DRD2 *Taq I* A1 allele. In contrast, the approach used by Gelernter et al.[27] was to calculate the frequency of the A1 allele. In this regard, there may be fundamental philosophical differences underlying the different approaches whereby the former approach, classifying by D2A1 positive or D2A1 negative, implies that the A1 allele itself is causative. However, Kidd fails to include other reports by investigators in this field utilizing the allele frequency method showing strong association with DRD2 variants with alcoholism and polysubstance abuse.[24,26,29,32,46–48,57–59,61,62] Moreover, O'Hara et al., utilizing multiple approaches including gene frequencies, genotypes, or by the presence vs. absence of the A1 allele, found no significant difference between these methods in terms of associating the A1 allele of the DRD2 receptor gene in heavy polysubstance abusers when compared to screened controls.[79] It is noteworthy that Pato et al. found an association between the A1 allele of DRD2 and alcoholism with an estimated odds ratio = 2.14, with a $\chi^2 = 25.40$, and with 1 df, whereby $p = 0.0001$. Recently, Comings[80] suggested that when association studies are limited to high quality candidate genes, such as the DRD2, instead of all genes, their power in psychiatric disorders is markedly enhanced. The finding of Pato et al.[77] at a $p = 0.0001$ is a statistically significant level in complex behaviors.[81]

Further confirming this meta-analysis, Uhl et al.[61] found that the A1 and B1 allele is a remarkably good predictor for substance abuse vulnerability. In their meta-analysis, the most severe abusers of addictive substances appear most likely to manifest higher A1 and B1 DRD2 marker frequencies. Conversely, when "control" comparison groups are studied more carefully to eliminate individuals with significant use of any addictive substance, D2A1 and B1 frequencies are at a lower level. Combined data drawn from all currently available sources suggest that DRD2 genotypes are associated with enhanced vulnerability to substance abuse.[81] The relative risk for severe substance abuse predicted from possession of an A1 RFLP marker in a Caucasian individual can be estimated to be 2.2 fold if 10% of the population abuse at least one addictive substance. While other genetic and environmental influences remain likely to determine the majority of variance in individual variability to substance abuse, computation based on the DRD2 data suggest that 27% of the risk for severe substance abuse can be attributed to DRD2 variance marked by the A1 genotype.[61]

According to Uhl et al.,[61] "as the strength of association increases, and as the means whereby D2 dopamine receptor gene variants influence substance abuse vulnerability are identified, we have increasing confidence that we have identified one of the first individual genes impacting a polygenic disorder with large environmental influences. These results can provide a window through which influences of a single gene on a pattern of complex behaviors can be glimpsed".

I. NEGATIVE STUDIES: STATISTICAL PROBLEMS

Since the original Blum et al. study,[18] a number of papers did not find an association of the *Taq I* A1 allele of the DRD2 gene and alcoholism.[19,27,30,37,52-54,82] A careful scrutiny of these studies reveals a number of statistical problems,[80] sampling errors in terms of characterization of severity of alcoholism,[19,27,30,37,52-54,82] unassessed controls for concomitant polysubstance abuse,[19,29,30] exclusion of medical complications (i.e., abnormal liver enzymes, liver cirrhosis, etc.),[19,27,54] as well as exclusion of comorbid personality disorders.[30] A detailed critique of these negative studies are the subject of other review(s).[23,77,81,83]

1. THE GELERNTER REAPPRAISAL

Most recently Gelernter et al.[84] reviewed a select number of studies and reported that in their view no association between the A1 allele of the DRD2 gene and alcoholism occurs. By their own admission, Gelernter et al.[84] utilized a number of heterogeneous studies which did not meet stringent and careful selection of both controls and subjects. Table 24-1 depicts most of the published DRD2 alcoholism association studies to date. While a number of factors may contribute to the variable results, the first positive association reported between the presence of the A1 allele and alcoholism[18] found a very high relative risk (i.e., 8.73). Due to the size of the relative risk between the two populations studied by Blum et al.,[18] this difference was detectable with a small number of subjects. It is important to consider the statistical power of this original study and realize that it would only allow detection of a relative risk equal to 4.5 or higher. This makes the study quite valid, and it has enough power to provide statistically significant results. In contrast, the studies where no association has been detected to date, the sample size is too small, and these studies may not have the power to provide statistically significant results unless the patients in these negative studies were selected with severity criteria comparable to the study by Blum. It is to be noted that a power analysis showed that a sample size of 80 subjects and 80 controls would be necessary in detecting a relative risk of 2.5.[18]

TABLE 24-1
Summary of Dopamine D2 Receptor Gene Variants in Impulsive-Addictive-Compulsive Behaviors

Substance Abuse	Allele	% Prevalence Disease	% Prevalence Controls	P Value ≤	Ref.
Alcoholism	D2A1	69	20	.001	1
Alcoholism (Less Severe)	D2A1	30	19	NS	2
Alcoholism (Less Severe)	D2B1	17	13	NS	3
Alcoholism[a]	$D2C_1$	57	33	.002	4
Severe Alcoholism	D2A1	47	17	.001	2
Severe Alcoholism	D2B1	47	13	.008	3
Severe Alcoholism	$D2^{In6-Ex7}$ Haplotype 1	39	16	.02	5
Cocaine Dependence	D2A1	51	18	.0001	6
Cocaine Dependence	D2B1	39	13	.01	6
Polysubstance Abuse	D2A1	44	20	.025	7
Polysubstance Abuse	D2B1	33	28	.001	7
Smoking[b]	D2A1	49	28	10^{-8}	8
Obesity Carbohydrate Bingeing	D2A1	46	19	.003	9
Obesity	D2A1	51	20	.0000197	10
Obesity[c]	$D2^{In6-Ex7}$ Haplotype 4	88	55	.0057	11
ADHD	D2A1	46	25	.0001	12
Pathological Gambling	D2A1	51	28	10^{-8}	13
Tourette's Disorder	D2A1	45	26	.0001	14

Note: [a] A1 allele denoted only with regard to homozygote genotype. Alcoholics (47/82); Controls (29/87), ($\chi^2 = 9.8$, $df = 1$, $p = .002$); [b] D. E. Comings et al. (personal communication); [c] Percentage of subjects in class 0 (normal weight-control) who carried the 4 haplotype was compared to class 4 (very high risk obese-diseased) who also carried the 4 haplotype; 1. Blum K, et al. *JAMA*, 1990;263(15):2055-2060; 2. Blum et al. *Alcohol* 1991;8(5):409–16); 3. Blum et al. *Alcohol* 1993;10(1):59–67; 4. Suarez et al. *Genomics* 1994;19:12–20; 5. Flanagan et al. Presented at American Psychopathological Association meeting, New York, March 5–7, 1992; 6. Noble et al. *Drug Alcohol Dep.* 1993;33(3):271–285; 7. OHara et al. *Hum. Hered.* 1993;43(4):209–218; 8. Comings et al. (personal communication); 9. Noble et al. *Int. J. Eating Dis.* (in press). 10. Blum et al. Presented at The American Society of Human Genetics Meeting, Montreal, Canada, October 28th; 11. Comings et al. *Biochem. Med. Metab. Biol.* 1993;50(2):176–185; 12. Comings et al. *JAMA* 1991;266(13):1793–1800; 13. Comings (personal communication); 14. Devor and Comings, *JAMA* 1992;267:651-652.

Several factors affect the comparability of the studies utilized by Gelernter and associates.[27] The studies by Blum's group,[18,29,47] Parsian's group,[24,50] and Amedeo's group,[57] as well as Noble's group,[32,60] used selection criteria that biased towards extreme severity of the disorder. It has been suggested by Pato et al. in a review of the putative association between the DRD2 and alcoholism[77] that divergence in study design may result in the inability to detect true association in a more general phenotype described by DSM-III3R alcohol dependence, but would be detected with severe alcoholism. It is also possible that as the level of severity decreases, the rate of phenocopies of nongenetic forms of alcoholism increases. This would make the detection of an association using a design focused on less severity much more difficult. In fact, the study by Blum et al. utilizing DNA derived from peripheral tissue[29] confirmed this notion by showing a lack of association of the A1 allele of the DRD2 gene in less severe alcoholics while showing a positive association with severe alcoholics. This work was further confirmed by others as previously pointed out in evaluating DRD2 gene variance in very ill, hospitalized alcoholics and nonalcoholics.[32]

In response to Gelernter's assertion that no association exists, we have reanalyzed the data they have presented and carefully addressed the issues pointed out by Gelernter et al.[84] and other analyses relevant to this association. We find the following:

1. In analyzing the data of Gelernter et al.[84], D2A1 frequency is 0.19 (n = 307) in unscreened controls vs. 0.11 (n = 186) in controls screened for alcoholism. A significant difference is found between these two controls ($\chi^2 = 5.17$; $p = 0.023$). It is important that association studies of alcoholics should use controls carefully screened for alcoholism for comparison.
2. When the data on alcoholics severe and not severe are compared with the screened controls presented by Gelernter et al.,[84] D2A1 frequency is 0.19 in the alcoholics compared to 0.11 (n = 186) in screened controls. The difference between these two groups is significant ($\chi^2 = 5.94$; $p = 0.015$).
3. When severe alcoholics are compared to screened controls, excluding the Blum et al. first study[18] as Gelernter et al.[84] chose to do, A1 frequency in the former group is 0.23 (n = 143) vs. 0.11 (n = 151) in the latter group. The difference was significant ($\chi^2 = 6.46$; $p = 0.011$). If the first Blum et al.[18] study of severe alcoholics had been included, D2A1 frequency would be .25 (n = 178) in the severe alcoholics compared to 0.11 (n = 186) A1 frequency in screened controls. The difference between these two groups is highly significant ($\chi^2 = 11.3$; $p = 0.0008$).
4. Utilizing the same four studies and pooling the data similar to Gelernter's review[84] similarly reveals a positive association between the A1 allele of the DRD2 gene and severity when compared against controls ($\chi^2 = 20.32$; $df = 1$; $p = 0.0001$). Since differences have been observed between screened and unscreened controls, we decided to also pool the data in the manner of Gelernter and analyze allelic frequency in alcoholics compared to their screened controls and found a positive association of the D2A1 allele ($\chi^2 = 8.4$; $df = 1$; $p = .004$).

From these calculations, at least two factors emerge as important in D2A1 allelic association with alcoholism: (1) The type of alcoholics (i.e., severe and not severe) and (2) the type of controls (i.e., assessed and unassessed) chosen for the study. Reanalysis of the data presented by Gelernter's group[84] show that significant differences exist in A1 frequency between unassessed and assessed controls and between severe alcoholics and assessed controls.

The rationale used by Gelernter et al.[84] for excluding the first Blum et al.[18] study from analysis is not clear in spite of being different from other less rigid studies. These latter authors carefully compared D2A1 frequency in severe alcoholics with assessed nonalcoholics. It is also curious why Gelernter et al.[84] chose not to include in their analyses two other relevant studies which showed significantly higher frequencies of the D2A1 allele in hospitalized alcoholics compared to controls. These studies, one on a French population[57] and the other on a U.S. sample,[32] were presented at the ISBRA Satellite conference in Bordeaux, France, June 1992, with published abstracts available to all invitees to that meeting and now have been subsequently published.

Without clear justification, the authors excluded the two Blum et al. studies[18,29] from their analysis which meet the stringent criteria pointed out above. In reanalysis, utilizing six independent Caucasian studies[18,24,26,29,32,58] on alcoholism only, which meet three out of four inclusion/exclusion criteria, an opposite conclusion is observed whereby there is a positive association ($\chi^2 = 6.71$; $df = 1$; $p = .009$), when proper meta-analysis techniques are utilized.

It is noteworthy that whereas there was heterogeneity among Gelernter's studies for both alcoholics and controls in contrast, there was no heterogeneity in the six studies utilized in our meta-analysis ($\chi^2 = 4.98$; $df = 5$; $p = .419$ for alcoholics; $\chi^2 = 1.83$; $df = 5$, $p = .872$ for controls). Figure 24-2 explains the heterogeneity in the studies reviewed by Gelernter,[84] who fails to characterize either alcoholics or controls in terms of the inclusion/exclusion criteria. For our analysis, we carefully rated each individual study according to a scoring of 0 to 4, whereby 1 equaled only one inclusion/exclusion criteria satisfied and 4 equaled all criteria satisfied. While it is easy to agree on criteria scores assigned to power analysis, scores assigned for inclusion/exclusion of ethnicity, assessment of controls for substance abuse, and characterization of the experimental subjects (i.e., severity) carries a less rigid and greater degree of uncertainty, thereby making it difficult to reach significant agreement among investigators. For example, in terms of ethnicity, we assumed in our assessment inclusion was based strictly on race rather than ethnic origin, especially in the Caucasian population. We would like to seriously caution interpretation of our statistical approach in an attempt to explain heterogeneity, nevertheless, as depicted in Figure 24-2, a Pearson's Correlation Coefficient revealed that in alcoholics a direct and significant positive correlation was observed whereby $r = 0.808$ ($p = .0014$), whereas in controls an indirect and significant negative correlation was also observed whereby $r = -0.762$ ($p = .004$). These results explain some of the existing heterogeneity in Gelernter's[84] selected studies.

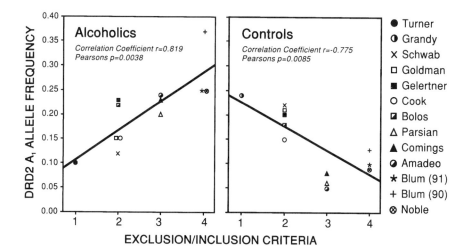

FIGURE 24-2 DRD2 *Taq 1* A1 allele frequency as a function of exclusion/inclusion criteria in both alcoholics and controls. The data is derived from: Blum, K. et. al. *JAMA*, 263: 2055, 1990; Blum, K. et. al. *Alcohol*, 8: 409, 1991; Bolos, A. et. al. *JAMA*, 264: 3156, 1990; Gelernter, J. et. al. *JAMA*, 266: 1801, 1991; Comings, D. et. al. *JAMA*, 266: 1833, 1991; Parsian, A. *Arch. Gen. Psychiatry* 48: 655, 1991; Cook, B. et. al. *Alcoholism: Clinical and Experimental Research* 16:806, 1992; Schwab, H., *Am. J. Hum. Genet*. 49: 203A, 1094, 1991; Goldman, D. et. al. *Acta Psychiat. Scan*. 86: 351, 1992; Grandy, D. et. al. *Proc. Natl. Acad. Sci. USA*, 86: 9762, 1989; Noble, EP, et. al. *Alcohol Alcohol.* in press; Amadeo, S. et. al. *J. Psychiat. Res*. 27: 173, 1993; Turner, M. et. al. *Biol. Psychiat.* 31: 285, 1992.

Therefore, it is our contention that reanalysis of the Gelernter data reveals the following positive findings:

1. An association of the DRD2 gene with alcoholism vulnerability in the general population
2. A positive association of the A1 allele and severe alcoholism
3. An association of the D2A1 allele with alcoholism utilizing six heterogeneous studies
4. Increasing frequency of the A1 allele in alcoholics when stringent inclusion/exclusion criteria are met

5. Decreasing frequency of the A1 allele in controls when stringent inclusion/exclusion criteria are met.

2. SUAREZ STUDY

Most recently, a paper by Suarez and associates[37] examined five different polymorphisms at different locations along the DRD2 gene for linkage disequilibrium and for association with alcoholism. No significant differences for RFLP frequency between alcoholics and controls was found. The prevalence of the D2A1 allele was 36.5% for the 82 alcoholics and 28% for the 88 controls screened to exclude psychiatric disease including alcohol and drug abuse. They concluded their results did not support the involvement of the DRD2 region in the etiology of alcoholism. In reviewing this paper, unlike their earlier investigation where the D2A1 allele associated with severe alcoholics, these authors failed to adequately classify their alcoholic probands with regard to chronicity and severity of alcoholism. Therefore, the subjects utilized in this study represent a rather diffuse and unclassified group, which includes less severe alcoholics, moderate alcoholics, and severe alcoholics. The lack of characterization in our opinion could significantly impact on attempts to perform association studies with alcoholism, whereby the phenotype is quite complex. However, the most interesting finding of the Suarez et al.[37] paper relates to an apparent increase in the number of D2C1 allelic homozygotic genotypes that correlate with alcoholic probands relative to control. Specifically, Suarez et al.[37] found that two copies of the D2C1 allele of the DRD2 gene was significantly more prevalent in alcoholics compared to nonalcoholics, 57 and 33%, respectively ($\chi^2 = 9.8$; $df = 1$; $p = .002$).

Furthermore, as originally stated in the Blum et al.[18] study, the A1 allele of the DRD2 gene is probably not specific for alcoholism but is associated with one virulent form and may be associated with multiple forms of drug abuse and compulsive behaviors and conduct disorders.[26,59,60] In this regard, as previously noted by Uhl et al.[61] evaluating all extant data they reported a positive association of both the A1 and B1 alleles of the DRD2 gene with substance abuse. Moreover, as previously stated, Arinami et al.[47] also found an association of the B1 allele in severe Japanese alcoholics compared to less severe alcoholics. In our view the following represents the most logical stance with regard to the negative findings:

1. Alcoholism is a complex, multifactorial, polygenic disorder with a significant environmental component.
2. Some forms of alcoholism (severe) are more highly associated with the D2A1 allele than others (less severe or abusers).
3. We still do not have a clear idea of what characterizes those with an increased prevalence of the D2A1 allele versus those with a normal or even less than normal prevalence of the D2A1 allele.
4. Studies where alcoholics were excluded if they had other comorbid disorders in themselves or their families tended to have a lower prevalence of the D2A1 allele than those without such exclusions.
5. Given the above, it is not surprising that some studies are negative and some are positive.
6. If all the above are true, it seems pointless to endlessly criticize those studies that were negative. They are probably absolutely correct for the group of alcoholics and controls they examined.

J. MOLECULAR GENETIC ASPECTS

The ultimate question to answer is "Where is the mutation (in and around the gene) that causes the observable reductions in DRD2 receptors in variant carriers, and what other gene defects are also potentially involved?"

In this regard, Ram and a number of associates[85], examined expressed sequences of the DRD2 gene (exons 2 through 8) in a sizeable series of unrelated patients (113 patients with schizophrenia and 74 with alcoholism and controls). The researchers performed large scale mutational analysis by denaturing gradient gel electrophoresis (DGGE). DGGE is a powerful method for the detection of mutations in DNA.[86,87] They found no evidence of mutations in the coding sequence of the DRD2 gene associated with either alcoholism or schizophrenia. However, since no change occurred with the association constant (K_d) or the ability of the D2 Receptor to bind dopamine,[25] it follows that no structural changes within the exons of A1 carriers should occur. Conceivable explanations for the lack of any detectable mutation associated with disease may be reduced to three: (1) lack of a DRD2 coding region mutation may exclude a role for the coding region of this gene in the disorders studied, which is further supported by the lack of change in the binding constant (K_d) of the DRD2 receptor in A1 allele probands; (2) the method of DGGE failed to detect an actual DRD2 disease mutation in a coding sequence; or (3) there is a physiologically important alteration in DRD2 associated with polymorphic loci of the DRD2 gene, but the mutation is not located in the coding region of the DRD2 gene.

In terms of the third explanation, examples of gene regulation involving untranslated regions in and around the coding region of gene have been well documented. In this regard, a number of 3' flanking sequences have been identified that play a role in mRNA stability.[88-92] An interesting example which may have relevance to the DRD2 gene is the finding by Koeller et al.[93] that the turnover of the full length transferrin receptor mRNA is regulated by iron, and this regulation is mediated by the 3' untranslated region of the transcript. Moreover, there have been reports of enhancers modulating gene transcription within the first intron of the growth hormone gene.[94,95] Furthermore, there is evidence that intronic mutations can lead to defective proteins which could result in disease. Naylor et al. found an unusual cluster of mutations involving regions of intron 22 which lead to defective joining of exons 22 and 23 in the mRNA and causes hemophilia A in 50% of severely affected probands. This explains why previous analyses of the putative promotor, exons, and most exon/intron boundaries of the Factor VIII gene failed to detect any anomalous structural defects in one half of the severely affected patients.[96] Moreover, others have shown that the DNA comprising the 22nd intron of factor VIII actually consists of two structural genes that are expressed independent of Factor VIII and may be involved in the defective joining of exons 22 and 23 of the mRNA for Factor VIII causing hemophilia A in a number of cases. In an analogous condition, Eubanks et al.[97] found hypomethylated region of the DRD2 gene which strongly suggest the presence of structural genes. Therefore, until genomic analyses of the 5' promotor region, introns, including intron 1 (250kb), and the untranslated 3' region have been completed, no definitive statements should be made concerning the failure to find structural abnormalities in the exons of the DRD2 gene and the association of polymorphic loci on the DRD2 gene with substance vulnerability.[80]

The experience with HRAS polymorphisms and the risk of cancer may provide some important lessons for the D2A1 story. In 1985, Krontiris et al.[98] reported the presence of rare alleles at the HRAS locus three times more often in cancer patients than in controls without cancer. Following this, of a series of 23 studies, 9 claimed a significant positive association with cancer and 14 did not.[99] However, a meta-analysis of those 14 alone showed a significant positive association. All studies combined, excluding the first report, gave an odds ratio of only 1.9, considerably lower than that for severe alcoholism. Despite this modest odds ratio, the combined results suggested that 1 of 11 cancers of the breast, colon, and bladder may be attributed to this genetic locus.[99] It is of considerable interest that the polymorphism involved was an unstable minisatellite located 3' of the coding sequence of the gene. Studies showed that the minisatellite binds at least four members of the rel/NF-kB family of transcriptional regulatory factors.[100] The minisatellite was capable of activating and repressing transcription, and allele specific effects were observed.[101] While the number of genes that possess such regulatory minisatellites 5' and 3' to the coding sequences is unknown, a similar mechanism could account for the genetic variations in the expression of the DRD2 locus. Ironically, the 3' D2A1 *Taq I* polymorphism could be closer to the mutation that affects the expression of the DRD2 locus than variants within the gene itself.

Therefore, until genomic analyses of the 5' promoter region, introns, including intron 1 (250 kb), the untranslated 3' region, and extensive segments 5' and 3' of the gene, have been completed, no definitive statements regarding important segments of the DRD2 gene should be made.

K. BAYES' THEOREM AS A PREDICTIVE MODEL

Bayes' Theorem is standard in the field of medicine to predict the likelihood that a particular event (defect) such as possessing the *Taq I* A1 allele of the DRD2 gene will result in an another event (disease) such as having abnormal drug and alcohol seeking behavior. Table 24-2 below illustrates the predictive value of impulsive-addictive-compulsive behavioral prevalence in carriers of the A1 allele of the DRD2 gene utilizing Bayes' Theorem.

According to Bayes' rule, the ***predictive value positive*** (PV$^+$) of a screening test is the probability that a person has disease given that the test is positive:

$$Pr(\text{disease/test}^+)$$

The ***predictive value negative*** (PV$^-$) of a screening test is the probability that a person does not have disease given that the test is negative:

$$Pr(\text{no disease/test}^-)$$

In this mathematical model, the ***sensitivity*** of a symptom (or set of symptoms or screening test) is the probability that the symptom is present given that the person has disease. The ***specificity*** of a symptom (or set of symptoms or screening test) is the probability that the symptom is not present given that a person does not have disease. A ***false negative*** is defined as a person who tests out negative but who is actually positive. A ***false positive*** is defined as a person who tests out as positive but who is actually negative.

In order to calculate Bayes' theorem we utilized the following formula:

$$\text{predictive value} = \frac{(\text{prevalence})(\text{sensitivity})}{(\text{prevalence})(\text{sensitivity}) + (1 - \text{prevalence})(1 - \text{specificity})}$$

To calculate the ***specificity***, we utilized very well-characterized accessed controls screened for alcohol, drug, and tobacco use (in some samples). Moreover, to calculate the ***sensitivity*** of alcoholism, cocaine dependence, polysubstance abuse, overeating, and attention deficit/hyperactivity disorder (ADHD), we utilized proband genotyping from studies where the probands were characterized for chronicity or severity of the disease (see Table 24-1).

The predictive value of a positive test result is defined as the percentage of positive results that are true positives when the test is applied to a population containing both healthy and diseased subjects. Common sense would suggest that the predictive value of a test is dependent on the positivity of the test in disease and its negativity in health.[102] With this in mind, interpretation of these data suggest that a negative result from the *Taq* A1 allelic genotype is not predictive, since we found that PV$^-$ = 0.548 or 54.8%. It is noteworthy that the low negative predictive value is due to the lack of specificity since control assessment for all exant studies have not utilized rigid inclusion/exclusion criterion and is quite heterogeneous.

TABLE 24-2
The Dopamine D2 Receptor Gene as a Predictor of Compulsive Disease Utilizing Bayes' Theorem

Risk Behavior[1]	Bayes' Theorem (F)	Predictive Value (%)
Alcoholism (Severe)[2]	0.1433	14.3
Cocaine Dependence (Severe)[3]	0.1235	12.3
Polysubstance Abuse[4]	0.1218	12.8
Chemical Dependency[5]	0.2836	28.3
Overeating (Severe)[6]	0.1860	18.6
Ingestive Behavior[7]	0.3500	35.0
ADHD[8]	0.1602	16.0
Smoking[9]	0.415	41.5
Pathological Gambling[10]	0.046	4.6
Tourette's Disorder[11]	0.055	5.5
Total Impulsive-Addictive-Compulsive Behavior[12]	0.744	74.4

Note: To calculate Bayes' theorem we utilize the following formula:

$$\text{predictive value} = \frac{(\text{prevalence})(\text{sensitivity})}{(\text{prevalence})(\text{sensitivity}) + (1 - \text{prevalence})(1 - \text{specificity})}$$

Assumptions: For a gross calculation we utilized the following:
- Alcoholism - prevalence - 0.055, sensitivity = 0.5632, specificity = 0.804
- Cocaine - prevalence = 0.0285, sensitivity = 0.8750, specificity = 0.804
- Polysubstance Abuse - prevalence = 0.06, sensitivity = 0.4260, specificity = 0.804
- Food - prevalence = 0.04, sensitivity = 0.8460, specificity = 0.804
- ADHD - prevalence = 0.075, sensitivity = 0.4615, specificity = 0.804
- Smoking - prevalence = 0.25, sensitivity = 0.419
- Pathological Gambling - prevalence = 0.0185, sensitivity = 0.0508
- Tourette's disorder - prevalence = 0.025, sensitivity = 0.448
- Impulsive-Addictive-Compulsive Behaviors - prevalence = 0.552, sensitivity 0.462

To calculate prevalence the following assumptions were made:
- Severe alcoholics in the U.S. population constitute an estimated 11 million out of 200 million adults.
- Severe cocaine addicts in the U.S. population constitute an estimated 5.7 million out of 200 million adults.
- Morbidly obese food addicts in the U.S. population constitute 8 million out of 200 million adults.
- Polysubstance abusers in the U.S. population constitute an estimated 14.9 million out of 200 million adults.
- Attention deficit/hyperactivity disorder (ADHD) in the U.S. population constitute an estimated 7.5% of 45,250,000 school age children between the age of 5 and 17 or 3.39 million out of 249 million people.
- The prevalence of smoking in the U.S. population has been estimated at 25% or 62 million.
- The prevalence of pathological gambling in the U.S. population is approximately 1.8%.
- The prevalence of severe Tourette's disorder in the U.S. population is 1 out of 40 o4 2.5%.
- The prevalence of total Impulsive-Addictive-Compulsive Behavior is 55.2%. This estimate may vary as a function of overlap among these spectrum disorders.

To calculate the specificity, we utilized very well-characterized accessed controls screened for alcohol, drug and tobacco use (in some samples). Studies utilized include Blum et al., 1990; Blum et al., 1991; Blum et al., 1992; Parsian et al., 1991; Noble et al., 1993; Amadeo et al., 1993; Comings et al., 1991; and Smith et al., 1992.
- To calculate the sensitivity of severe alcoholism we utilized proband genotyping from the following studies where the proband was characterized for chronicity or severity of the disease: Blum et al., 1990; Blum et al., 1991; Bolos et al., 1990; Parsian et al., 1991; Gelernter et al., 1991; Noble et al., 1993; Amadeo et al., 1993.
- To calculate the sensitivity of severe cocaine dependence we utilized Noble et al., 1993 (probands with 3 risk factors).
- To calculate the sensitivity to polysubstance abuse we utilized a number of studies (Smith et al., 1992; Noble et al., 1993; Comings et al., 1991; and Gelernter et al. (unpublished), reviewed by Uhl et al., 1993.
- Chemical dependency is the combination of alcoholism, cocaine dependence and polysubstance abuser probands.
- To calculate the sensitivity of severe overeating we utilized Noble et al.[36] (probands with 3 risk factors).
- Ingestive behavior is the combination of alcoholism, cocaine dependence, polysubstance abuse and overeating probands.
- To calculate the sensitivity of ADHD we utilized Comings et al., 1991.
- To calculate the sensitivity of smokers, we utilized Comings et al., 1994.
- To calculate the sensitivity of pathological gambling we utilized Comings et al., 1994.
- To calculate the sensitivity of Tourette's disorder we utilized Comings et al., 1991.
- To calculate the sensitivity of Impulsive-Addictive-Compulsive Behavior we utilized a composite of all available data.

We believe that as we begin to exclude a number of related impulsive-addictive-compulsive behaviors in controls and more accurately define the prevalence of "reward" behaviors, the negative predictive value will significantly increase. However, a positive result predicts that the proband would have a 74% risk for impulsive-addictive-compulsive disease.[103]

Finally, an analysis of the pooled data utilizing the above referenced studies (see Table 24-2) related to CD resulted in an odds ratio of 3.62 (95% confidence limits [2.59 to 5.07]) with a p value of 0.00001 indicating a similarly strong positive correlation of the DRD2 gene variant and this disease (Yates $\chi^2 = 64.9$; $df = 1$; $p < 1\chi\ 10^{-5}$). We consider this Bayes' Theorem analysis preliminary calling for cautious interpretation of these results. Nevertheless, according to our analysis, DRD2 gene screening test would have valuable predictive value.

L. CONCLUSIONS

Where Do We Go From Here? The studies to date suggest it is now possible to answer the following series of questions.

1. IS THERE AN ASSOCIATION BETWEEN THE D2A1 ALLELE AND ALL FORMS OF ALCOHOLISM?

The answer to this is clearly no. There are too many well designed and well executed studies that have reported a lack of association between the D2A1 allele and alcoholism to support the claim that all forms of alcoholism are associated with the D2A1 allele.

2. IS THERE AN ASSOCIATION BETWEEN THE D2A1 ALLELE AND SOME FORMS OF ALCOHOLISM?

The answer to this seems to be just as positive as the answer to the previous question was negative. As reviewed above and elsewhere,[23] severe alcoholism is more likely to be associated with the D2A1 allele than milder forms, especially alcohol abuse. However, the definition of severe alcoholism warrants further clarification. Future studies might productively compare personality and other characteristics of D2A1 positive versus D2A1 negative forms of alcoholism. The data to date suggest that when alcoholics with any comorbid disorders are excluded, the association with DRD2 variants are decreased or eliminated. In this regard, it is of interest that in the studies of DiChiara and Imperato[104] ranked alcohol lower than most other addicting substances in its ability to enhance dopaminergic activity in the nucleus accumbens.

It seems reasonably clear that alcoholism associated with comorbid disorders such as drug dependence, smoking, conduct disorder, pathological gambling, obesity, and PTSD is associated with significant increases in the prevalence of the D2A1 allele. Among these disorders, individuals with comorbid severe alcoholism tend to show a higher prevalence of the D2A1 allele than those with lower degrees of alcoholism. However, when those without alcoholism are analyzed separately the prevalence of the D2A1 allele is still significantly greater than in controls for both groups.

3. ARE SOME DISORDERS OTHER THAN ALCOHOLISM SIGNIFICANTLY ASSOCIATED WITH THE D2A1 ALLELE?

The answer to this also appears to be yes (see Figure 24-3). Examples include ADHD, Tourette's disorder, pathological gambling, polysubstance dependence, smoking, PTSD, obesity, conduct disorder, and in Japanese schizophrenics.[105] We believe this finding to be supportive when you take into consideration the lack of DRD2 binding differences between D2A1 and B1 carriers compared to A2 and B2 carriers in spite of the significant reduction in the number of DRD2 receptors in the former group.

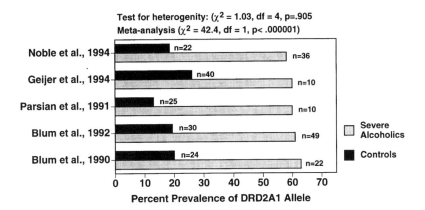

FIGURE 24-3 Percent prevalence of the DRD2 A1 allele in severe alcoholics and controls.

Since the original paper on the D2A1 allele involved alcoholism,[18] it is not surprising that the majority of subsequent papers have been concerned with this common and important disorder. However, from the point of view of understanding the role of the DRD2 locus in psychiatric disorders, it may turn out that conditions with more severe degrees of impulsive, compulsive, addictive behaviors, and certain personality traits, may provide a far more fertile ground for investigation than alcoholism per se (Figure 24-3).

In this regard, the A1 allele of the DRD2 gene has been shown to associate with a well characterized neurophysiological parameter in children of alcoholics. Specifically, it has been reported that the D2A1 allele significantly correlated with delayed latency of the P300 wave, a known phenotypic predictor of drug and alcohol seeking behavior,[12] in alcohol-naive sons of active alcoholic fathers[9,107] and in a neuropsychiatric population.[66] Moreover, the DRD2 gene appears to be a major gene in these disorders while the larger role seems to be played by a combination of other genes and environmental factors such as nicotine abuse.[76,108]

Because of its importance in reward pathways, defects in dopamine metabolism have been one of the prominent theories for the biological component of alcoholism. However, defects in serotonin metabolism have also been widely proposed.[108–115] The examination of genes affecting serotonin metabolism have lagged far behind that for dopamine.

Most recently, Wiesbeck et al.[116] found neuroendocrinological support to the assumption that a reduced D2 receptor function in alcohol dependent men is not only a state marker of residual heavy drinking, but also a genetically determined trait marker. We anticipate that one or more such genes will make additional significant contributions to the biological basis of alcoholism and impulsive-addictive-compulsive diseases or RDS.

REFERENCES

1. Blum, K. (in collaboration with J. Payne), *Alcohol and the Addictive Brain*, The Free Press, New York, NY, 1991.
2. Cotton, N. S., The familial incidence of alcoholism, *J. Stud. Alc.*, 40, 89, 1979.
3. Goodwin, D. W., Schulsinger, F., Hermansen, L., Guz, S. B., and Winkur, G., Alcohol problems in adoptees raised apart from alcoholic biological parents, *Arch. Gen. Psychiatry*, 28, 238, 1973.
4. Schuckit, M. A., Goodwin, D. W., and Winokur, G., A study of alcoholism in half-siblings, *Amer. J. Psychiatry*, 128,1132, 1972.
5. Cloninger, C. R., Genetic and environmental factors in the development of alcoholism, *J. Psychiat. Treat. Eval.*, 5, 487, 1983.
6. Bohman, M., Sigvardsson, S., and Cloninger, C. R., Maternal inheritance of alcohol abuse: cross-fostering analysis of adopted women, *Arch. Gen. Psychiatry*, 38, 965, 1981.
7. Pickens, R. W., Svikis, D. S., McGue, M., Lykken, D. T., Heaton, L. L., and Clayton, P. J., Heterogeneity in the inheritance of alcoholism, *Arch. Gen. Psychiatry*, 48, 19, 1991.
8. Kendler, K. S., Heath, A. C., Neale, M. C., Kessler, R. C., and Eaves, L. T., A population-based twin study of alcoholism in women, *JAMA*, 268, 1882, 1992.
9. Whipple, S. C., Parker, E. S., and Noble, E. P., A typical neurocognitive profile in alcoholic fathers and their sons, *J. Stud. Alcohol*, 49, 240, 1988.
10. Elmasian, R., Neville, H., Woods, D., Schuckit, M., and Bloom, F. E., Event-related brain potentials are different in individuals at high and low risk for developing alcoholism, *Proc. Natl. Acad. Sci. U.S.A.*, 79, 7900, 1982.
11. Begleiter, H., Porjesz, B., Rawlings, R., and Eckardt, M., Auditory recovery function and P3 in boys at risk for alcoholism, *Alcohol*, 4, 315, 1987.
12. Berman, S. M., Whipple, S. C., Fitch, R. J., and Noble, E. P., P3 in young boys as a predictor of adolescent substance use, *Alcohol*, 10, 69, 1993.
13. Hill, S. Y., Armstrong, J., Steinhauer, S. R., Baughman, T., and Zubin, J., Static ataxia as a psychobiological marker for alcoholism, *Alcoholism (NY)*, 11, 345, 1987).
14. Shigeta, Y., Ishii, H., Takagi, S., et al., HLA antigens as immunogenetic markers of alcoholism and alcoholic lever disease, *Pharmacol. Biochem. Behav.*, 13(1), 89, 1980.
15. Hill, S. Y., Goodwin, D. W., Cadoret, R., Osterland, C. K., and Doner, S. M., Association and linkage between alcoholism and eleven serological markers, *J. Stud. Alcohol*, 36, 981, 1975.
16. Agarwal, D. P., Harada, S., Goedde, and H. W., Radical differences in biological sensitivity to ethanol, *Alcoholism (NY)*, 5, 12, 1981.
17. Yoshida, A., Huang, I-Y., and Ikawa, M., Molecular abnormality of an inactive aldehyde dehydrogenase variant commonly found in Orientals, *Proc. Natl. Acad. Sci. U.S.A.*, 81, 258, 1984.
18. Blum, K., Noble, E. P., Sheridan, P. J., Montgomery, A., Ritchie, T., Jagadeeswaran, P., Nogami, H., Briggs, A. H., and Cohn, J. B., Allelic association of human dopamine D2 receptor gene in alcoholism, *JAMA*, 263, 2055, 1990.
19. Gordis, E., Tabakoff, B., Goldman, D., and Berg, K., Finding the gene(s) for alcoholism, *JAMA*, 263, 2094, 1990.
20. Bolos, A. M., Dean, M., Lucas-Derse, S., Ramsburg, M., Brown, G. L., and Goldman, D. Population and pedigree studies reveal a lack of association between the dopamine D2 receptor gene and alcoholism, *JAMA*, 264, 3156, 1990.
21. Noble, E. P. and Blum, K., The dopamine D2 receptor gene and alcoholism, *JAMA*, 265, 2667, 1991.
22. Smith, S. S., Gorelick, D. A., O'Hara, B. F., and Uhl, G. R., The dopamine D2 receptor gene and alcoholism, *JAMA*, 265, 2667, 1991.
23. Noble, E. P., The D2 dopamine receptor gene: a review of association studies in alcoholism, *Behav. Genet.*, 23, 119, 1993.
24. Parsian, A., Todd, R. D., Devor, E. J., O'Malley, K. L., Suarez, B. K., Reich, T., and Cloninger, C. R., Alcoholism and alleles of the Human D2 dopamine receptor locus, *Arch. Gen. Psychiatry*, 48, 655, 1991.
25. Noble, E. P., Blum, K., Khalsa, M. E., Ritchie, T., Montgomery, A., Wood, R. C., Fitch, R. J., Ozkaragoz, T., Sheridan, P. J., Anglin, M. D., Paredes, A., Treiman, L. J., and Sparkes, R. J., Allelic association of the D2 dopamine receptor gene with cocaine dependence, *Drug Alc. Dep.*, 33, 271, 1993.
25. Noble, E. P., Blum, K., Ritchie, T., Montgomery, A., and Sheridan, P. J., Allelic association of the D2 dopamine receptor gene with receptor-binding characteristics in alcoholism, *Arch. Gen. Psychiatry*, 48, 648, 1991.
26. Comings, D. E., Comings, B., Muhleman, D., Dietz, G., Shahbahrami, B., Tast, D., Knell, E., Kocsis, P., Baumgarten, R., Kovacs, B. W., Levy, D. L., Smith, M., Borison, R. L., Evans, D., Klein, D. N., MacMurray, J., Tosk, J. M., Sverd, J,. Gysin, R., and Flanagan, S. D., The Dopamine D2 receptor locus as a modifying gene in neuropsychiatric disorders, *JAMA*, 266, 1793, 1991.

27. Gelernter, J., O'Malley, S., Risch, N., Kranzier, H. R., Krystal, J., Merikangas, K., Kennedy, J. L., and Kidd, K. K., No association between an allele at the D2 dopamine receptor gene (DRD2) and alcoholism, *JAMA*, 266, 1801, 1991.
28. Cloninger, C. R. D2 dopamine receptor gene is associated but not linked with alcoholism, *JAMA*, 266, 1833, 1991.
29. Blum, K., Noble, E. P., Sheridan, P. J., Finley, O., Montgomery, A., Ritchie, T., Ozkaragoz, T., Fitch, R. J., Sadlack, F., Sheffield, D., Dahlmann, T., Halbardier, S., and Nogami, H., Association of the A1 Allele of the D2 dopamine receptor gene with severe alcoholism, *Alcohol*, 8, 409, 1991.
30. Turner, E., Ewing, J., Shilling, P., Smith, T. L., Irwin, M., Schuckit, M., and Kelsoe, J. R., Lack of association between an RFLP near the D2 dopamine receptor gene and severe alcoholism, *Biol. Psychiatry*, 31, 285, 1992.
31. Uhl, G., Persico, A. L., and Smith, S. S., Current excitement with D2 receptor gene alleles in substance abuse, *Arch. Gen. Psychiatry*, 49, 157, 1992.
32. Noble, E. P., Syndulko, K., Fitch, R. J., Ritchie, T., Bohlman, M. C., Guth, P., Sheridan, P. J., Montgomery, A., Heinzmann, C., Sparkes, R. S., and Blum, K., D2 receptor *Taq I* A alleles in alcoholic and nonalcoholic patients, *Alcohol Alcohol.*, 1994, in press.
33. Comings, D. E., The D2 dopamine receptor and Tourette's disorder, *JAMA*, 267, 652, 1992.
34. Comings, D. E., Muhleman, D., Dietz, G., Dino, M., LeGro, R., and Gade, R., Association between Tourette's disorder and homozygosity at the dopamine D3 receptor gene, *Lancet*, 341, 906, 1993.
35. Devor, E. J., The D2 dopamine receptor and Tourette's disorder, *JAMA*, 267, 651, 1992.
36. Nöthen, M. M., Hebebrand, J., Knapp, M., Hebebrand, K., Camps, A., von Gontard, A., Wettke-Schäfer, R., Cichon, S., Poustica, F., Schmidt, M., Lehmkuhl, G., Remschmidt, H., and Proping, P., Association analysis of the dopamine D2 receptor gene in Tourette's disorder using the haplotype relative risk method, *Am. J. Med. Genet. (Neuropsychiatr. Genet.)*, 54, 249, 1994.
37. Suarez, B. K., Parsian, A., Hampe, C. L., Todd, R. D., Reich, T., and Cloninger, C. R., Linkage disequilibria at the D2 dopamine receptor locus (DRD2) in alcoholics and controls, *Genomics*, 19, 12, 1994.
38. Neiswanger, K., Hill, S. Y., and Kaplan, B. B., What can the DRD2/alcoholism story teach us about association studies in psychiatric genetics, *Am. J Med. Genet. (Neuropsychiatr. Genet.)*, 60, 272, 1995.
39. Propping, P., Nöthen, M. M., Fimmers, R., and Baur, M. P., Linkage versus association studies in complex diseases, *Psychiat. Genet.*, 3, 136 (Abst.), 1993.
40. Sandkuyl, L. A., Analysis of affected sib pairs using information from extended families, in *Multipoint Maping and Linkage Based Upon Affected Pedigree Members*, Elston, R. C., et al., Eds., Alan R. Liss, New York, 1989.
41. Cook, C. C. H., Brett, P., Curtis, D., Holmes, D., and Gurling, H. M. D., Linkage analysis confirms a genetic effect at the D2 dopamine receptor locus in heavy drinking and alcoholism, *Psychiatr. Genet.*, 3, 130, 1993.
42. Gershon, E. S. and Rieder, R. O., Major disorders of mind and brain, *Sci. Am.*, 267, 126, 1992.
43. Marks, J., Dissecting the complex diseases, *Science*, 247, 1540, 1990.
44. Karp, R. W., D2 or not D2. *Alcoholism: Clinical and Experimental Research*, 16, 786, 1992.
45. Hauge, X. Y., Grandy, D. K., Eubanks, J. H., Evans, G. A., Civelli, O., and Litt, M., Detection and characterization of additional DNA polymorphisms in the dopamine D2 receptor gene, *Genomics*, 10, 527, 1991.
46. Smith, S. S., O'Hara, B. F., Persico, A. M., Gorelick, D. A., Newlin, D. B., Vlahov, D., Solomon, L., Pickens, R., and Uhl, G. R., The D2 dopamine receptor *Taq I* B1 restriction fragment length polymorphism appears more frequently in polysubstance abusers, *Arch. Gen. Psychiatry*, 49, 723, 1992.
47. Blum, K., Noble, E. P., Sheridan, P. J., Montgomery, A., Ritchie, T., Ozkaragoz, T., Fitch, R. J., Wood, R., Finley, O., and Sadlack, F., Genetic predisposition in alcoholism: association of the D2 dopamine receptor *Taq I* B1 RFLP with severe alcoholism, *Alcohol*, 10, 59, 1993.
48. Arinami, T., Itokawa, M., Komiyama, T., Mitsushio, H., Mori, H., Mifune, H., and Toru, M., Association of the A1 and B1 alleles of the dopamine D2 receptor gene in severe Japanese alcoholics, in *Handbook of Psychneurogenetics*, Blum, K. and Noble, E.P. Eds., CRC Press, Boca Raton, Fl, in press.
49. Flanagan, S. D., Noble, E. P., Blum, K., MacMurray, J., Comings, D., Ritchie, T., Sheridan, P. J., Lopatin, G., and Gysin, R., Evidence for a third physiologically distinct allele at the dopamine D2 receptor locus (DRD2), presented at the American Psychopathological Meeting, New York, March 5–7, 1992.
50. Parsian, A., Todd, R. D., O'Malley, K. L., Suarez, B. K., and Cloninger, C. R., Association and linkage studies of new human dopamine D2 receptor polymorphisms (RFLPS) in alcoholism, *Clin. Neuropharm.*, 15(1), Pt B, 1992.
51. Barr, C. L. and Kidd, K. K., Population frequencies of the A1 allele at the dopamine D2 receptor locus, *Biol. Psychiatry*, 34, 204, 1993.
52. Goldman, D., Dean, M., Brown, G. L., Bolos, A. M., Tokola, R., Virkkunen, M., and Linnoila, M., D2 dopamine receptor genotype and cerebrospinal fluid homovanillic acid, 5-hydroxyindoleacetic acid and 3-methoxy-4-hydroxyphenylglycol in alcoholics in Finland and the United States, *Acta Psychiatr. Scand.*, 86, 351, 1992.

53. Goldman, D., Brown, G. L., Albaugh, B., Robin, R., Goodson, S., Trunzo, M., Akhtar, L., Lucas-Derse, S., Long, J., Linnoila, M., and Dean, M., DRD2 dopamine receptor genotype, linkage disequilibrium, and alcoholism in American indians and other populations, *Alcohol. Clin. Exper. Res.*, 17, 199, 1993.
54. Schwab, S., Soyka, M., Niederecker, M., Ackenheil, M., Scherer, J., and Wildenauer, D. B., Allelic association of human D2 receptor DNA polymorphism ruled out in 45 alcoholics, *Am J. Hum. Genet.*, 49, 203 (Abstract), 1991.
55. Lamarine, R. J., Alcohol abuse among native Americans, *J. Comm. Health*, 13, 143, 1988.
56. Levy and Kunitz, *In Indian Drinking: Navajo Practices and Anglo-American Theories*, John Wiley & Sons, New York, 1974.
57. Arinami, T., Itokawa, M., Komiyama, T., Mitsushio, H., Mori, H., Mifune, H., Hamaguchi, H., and Toru, M., Association between severity of alcoholism and the A1 Allele of the dopamine D2 receptor gene *Taq I* a RFLP in Japanese, *Biol. Psychiatry*, 33, 108, 1993.
58. Amadeo, S., Abbar, M., Fourcade, M. L., Waksman, G., Leroux, M. G., Madec, A., Selin, M., Champiat, J-Claude, Brethome, A., Lucclaire, Y., Castelnau, D., Venisse, J-L., and Mallet, J. D2 dopamine receptor gene and alcoholism, *J. Psychiat. Res.*, 27, 173, 1993.
59. Comings, D. E., MacMurray, J., Johnson, J. P., Muhleman, D., Ask, M. N., Ahn, C., Gysin, R., and Flanagan, S. D. The dopamine D2 receptor gene: a genetic risk factor in polysubstance abuse, *Drug Alcohol Depend.*, 1994.
60. Blum K. and Halakas, J., personal communication.
61. Uhl, G. R., Molecular and genetic studies of the targets of acute drug action, substrates for inter individual differences in vulnerability to substance abuse, and candidate mechanisms for addiction, in *NIDA Research Monographs*, Chiarello, E., Ed., in press.
62. Uhl, G., Blum, K., Noble, E., and Smith, S., Substance abuse vulnerability and D2 receptor genes, *Trends Neurosci.*, 16, 83, 1993.
63. Gelernter, J., Kranzler, H., and Satel, S., No association between DRD2 dopamine receptor alleles and cocaine abuse, presented at the Fifty-fifth Annual Scientific Meeting, College on Problems of Drug Dependence, Poster, San Francisco, California, June 16, 1993.
64. Noble, E. P., Polyomorphisms of the D2 dopamine receptor in alcoholism, cocaine and nicotine dependence, and obesity, in *Handbook of Psychiatric Genetics*, K. Blum and E. P. Noble, Eds., CRC Press, Boca Raton, FL, 1996, 323.
65. Noble, E. P., Noble, R. E., Ritchie, T., Syndulko, K., Bohlman, M. C., Noble, L. A., Zhang, Y., Sparkes, R. S., and Grandy, D. K., D2 dopamine receipt gene and obesity, *Int. J. Eating Dis.*, 15, 205, 1994.
66. Blum, K., Braverman, E. R., Dinardo, N. J., Wood, R. C., and Sheridan, P. J., Prolonged P300 latency in a neuropsychiatric population with the D2 dopamine receptor A1 allele, *Pharmacogenetics*, 4, 313, 1994.
67. Comings, D. E., Flanagan, S. D., Dietz, G., Muhleman, D., Knell, E., and Gysin, R., The dopamine D2 receptor (DRD2) as a major gene in obesity and height, *Biochem. Med. Metab. Biol.*, 50, 176, 1993.
68. Rosenthal, R. J., Pathological gambling, *Psychiatr. Ann.*, 22, 72, 1992.
69. Comings, D. E., Rosenthal, R. J., Leiseur, H. R., Rugle, L., Muhleman, D., Chin, C., Dietz, F., and Gawe, R., The molecular genetics of pathological gambling: the DRD2 gene, *Pharmacogenetics*, in press.
70. Comings, D. E., Muhleman, D., Ahn, C., Gysin, R., and Flanagan, S. D., The dopamine D2 receptor (DRD2) gene in posttraumatic stress disorder: a study and replication, *Biol. Psychiatry*, in press.
71. Andrews, G., Pollock, C., and Stewart, G., The determination of defense style by questionnaire, *Arch. Gen. Psychiatry*, 46, 455, 1989.
72. Bond, M., Defense style questionnaire, in *Emperical Studies of Ego Mechanisms of Defense*, Vaillant, G.E., Ed., American Psychiatric Press, Washington, D.C., 1986.
73. Bond, M., Gardner, S. T., Christian, J., and Sigal, J. J. Empirical study of self-rated defense styles, *Arch. Gen. Psychiatry*, 40, 333, 1983.
74. Simerly, R. B., Swanson, L. W., and Gorski, R. A., Reversal of the sexually dimorphic distribution of serotonin-immunoreactive fibers in the medial preoptic nucleus by treatment with perinatal androgen, *Brain Res.*, 340, 91, 1985.
75. Comings, D. E., MacMurray, J., Johnson, P., Dietz, G., and Muhleman, D., Dopamine D2 receptor gene (DRD2) haplotypes and the defense style questionnaire in substance abuse, Tourette's disorder and controls, *Biol. Psychiatry*, in press.
76. Comings, E. D., Ferry, L., Bradshaw, S., Robinson, R., Burchette, C., Chin, C., and Muhleman, D., The dopamine D2 receptor (DRD2) gene: a genetic risk factor in smoking, *Pharmacogenetics*, in press.
77. Pato, C. N., Macciardi, F., Pato, M. T., Verga, M., and Kennedy, J. L., Review of the putative association of dopamine D2 receptor and alcoholism: a meta analysis, *Am. J. Med. Gen. (Neuropsychiatr. Genet.)*, 48, 78, 1993.
78. Kidd, K. K., Associations of disease with genetic markers: Deja vu all over again, *Am. J. Med. Gen. (Neuropsychiatr. Genet.)*, 48, 71, 1993.

79. O'Hara, B. F., Smith, S. S., Bird, G., Persico, A., Suarez, B., Cutting, G. R., and Uhl, G. R., Dopamine D2 receptor RFLPs, Haplotypes and their association with substance use in black and caucasian research volunteers, *Hum. Hered.*, 43, 209, 1993.
80. Comings, D. E., The role of the D2 dopamine receptor gene variants in neuropsychiatric disorders, in *Handbook of Psychoneurogenetics*, Blum, K., and Noble, E. P., Eds., 1995, CRC Press, Boca Raton, Florida, in press.
81. Crowe, R. R., Candidate genes in psychiatry: an epidemiological perspective, *Am. J. Med. Gen. (Neuropsychiatr. Genet.)*, 48, 74, 1993.
82. Geijer, T., Neiman, J. Rydberg, A., Jowsson, E. Sedvall, G., Valverius, P., and Terenius, L., Dopamine D2 receptor gene polymorphisms in Scandinavian chronic alcoholics, *Clin. Neurosci.*, 244, 26, 1994.
83. Blum, K., Sheridan, P. J., Wood, R. C., Braverman, E. R., Chen, T. J. H., and Comings, D. E., Dopamine D2 receptor gene variants: association and linkage studies in impulsive-addictive-compulsive behavior, *Pharmacogenetics*, 5, 121, 1995.
84. Gelernter, J., Goldman, D., and Risch, N., The A1 allele at the D2 dopamine receptor gene and alcoholism: a reappraisal, *JAMA*, 269, 1673, 1993.
85. Ram, A., Gejman, P. V., Gelernter, J., et al., No structural mutation in the dopamine D2 receptor gene in alcoholism or schizophrenia: Analysis using denaturing gradient gel electrophoresis, *JAMA*, 271, 204.
86. Myers, R. M., Fischer, S. G., Lerman, L. S., and Maniatis, T., Nearly all single base substitutions in DNA fragments joined to a GC-clamp can be detected by denaturing gradient gel electrophoresis, *Nucleic Acids Res*, 13, 3131, 1985.
87. Sheffield, V. C., Cox, D. R., Lerman, L. S., and Myers, R. M., Attachment of a 40-base-pair G+ C-rich sequence (GC-clamp) to genomic DNA fragments by polymerase chain reaction results in improved detection of single-base changes, *Proc. Natl. Acad. Sci. U.S.A.*, 86, 232, 1989.
88. Beck, I., Ramirez, S., Weinmann, R., and Carok, J., Enhancere element at the 3' flanking region controls transcription response to hypoxia in human erythropoietin gene, *J. Biol. Chem.*, 266, 15563, 1991.
89. Berstein, P. and Ross, J., Poly (A), poly (A) binding protein and the regulation of mRNA stability, *TIBS*, 14, 373, 1989.
90. Rastinejad, F. and Blau, H. M., Genetic complementation reveals a novel regulatory role for 3' untranslated regions in growth and differentiation, *Cell*, 72, 903, 1993.
91. Lake, R. A., Wotton, D., and Owen, M. J., A 3' transcriptional enhancer regulates tissue-specific expression of the CD2 gene, *EMBO J.*, 9, 3129, 1990.
92. White, J. W., Sobnosky, M., Rogers, B. L., Walker, W. H., and Suanders, G. F., Nucleotide sequence of a transcriptional enhancer located 2.2 Kb 3' of a human placental lactogen-encoding gene, *Gene*, 84, 521, 1989.
93. Koeller, D. M., Horowitz, J. A., Casey, J. L., Klausner, R. D., and Harford, J. B. Translation and the stability of mRNAs encoding the transferrin receptor and c-fos, *Proc. Natl Acad. Sci. U.S.A.*, 88, 7778, 1991.
94. Moore, D. D., Marks, A. R., Buckley, D. I., Kapler, G., Payvar, F., and Goodman, H. M., The first intron of the human growth hormone gene contains a binding site for glucocorticoid receptor, *Proc. Natl. Acad Sci. U.S.A.*, 82:, 699, 1985.
95. Slater, E. P., Rabenau, O., Karin, M., Baxter, J., and Beato, M., Glucocorticoid receptor binding and activation of a heterologous promotor by dexamethasone by the first intron of the human growth hormone gene, *Mol. Cell. Biol.*, 5, 2984, 1985.
96. Naylor, J. A., Green, P. M., Rizza, C. R., and Giannelli, U. K., Factor VIII gene explains all cases of haemophilia A, *Lancet,* 340, 1066, 1992.
97. Eubanks, J. H., Djabali, M., Selleri, L., Grandy, D. K., Civelli, O., McElligott, D. L., and Evans, G. A., Structure and linkage of the D2 dopamine receptor and neural cell adhesion molecule genes on human chromosome 11q23, *Genomics*, 14, 1010, 1992.
98. Krontiris, T. G., Devlin, B., Karp, D. D., Robert, N. J., and Risch, N., An association between the risk of cancer and mutations in the Hras1 minisatellite locus, *N. Eng. J. Med.*, 329, 517, 1993.
99. Krontiris, T. G., DiMartino, N. A., Colb, M., and Parkinson, D. R., Unique allelic restriction fragments of the human Ha-ras locus in leukocyte and tumor DNAs of cancer patients, *Nature*, 313, 369, 1985.
100. Trepicchio, W. L. and Krontiris, T. G., Members of the rel/NF-kB family of transcriptional regulatory factors bind the HRAS1 minisatellite DNA sequence, *Nucleic Acids Res.*, 21, 977, 1922.
101. Green, M. and Krontiris, T. G., Allelic variations of reporter gene activation by the HRAS1 minisatellite, *Genomics*, 17, 429, 1993.
102. Galen, R. S. and Gambino, R., *Beyond Normality: The Predictive Value And The Efficiency Of Medical Diagnosis*, Wyley Biomedical Publications, New York, 1975.
103. Blum, K., Sheridan, P. J., Wood, R. C., Braverman, E. R., Chen, T. J. H., Cull, J. G., and Comings, D. E., The D2 dopamine receptor gene as a predictor of reward deficiency syndrome: Bayes theorem, *J. Royal Soc. Med.*, in press.
104. DiChiara, G. and Imperato, A., Drugs abused by humans preferentially increase synaptic dopamine concentrations in the mesolimbic system of freely moving rats, *Proc. Natl. Acad. Sci. U.S.A.*, 85, 5274, 1988.

105. Arinami, T., Itokawa, M., Enguci, H., Tagaya, H., Yano, S., Shimizu, H., Hamaguchi, H., and Toru, M., Association of dopamine D2 receptor molecular variant with schizophrenia, *Lancet*, 343, 703, 1994.
106. Comings, D. W., Wu, S., Chiu, C., Ring, R. H., Gake, R., Ahn, C., MacMurray, J. P., Dietz, G., and Muhleman, D., Polygenic inheritance of Tourette's disorder, stuttering, attention deficit/hyperactivity, conduct and oppositional defiant disorder: the additive and sustractive effect of three dopaminergic genes in DRD2, DbetaH and DAT1, *Am. J. Med. Genet. (Neuropsychiatr. Genet.)*, in press.
107. Noble, E. P., Berman, S. M., Ozkaragoz, T. Z., and Ritchie, T., Prolonged P300 latency in children with the D2 dopamine receptor A1 allele, *Am. J. Hum. Genet.*, in press.
108. Ballenger, J. C., Goodwin, F. K., Major, L. F., and Brown, G. L., Alcohol and central serotonin metabolism in man, *Arch. Gen. Psychiatry*, 36, 224, 1979.
109. Banki, C., Factors influencing monamine metabolites and tryptophan in patients with alcohol dependence, *J. Neural. Transm.*, 50, 98, 1981.
110. Blum, K., Calhoun, W., Merritt, J. H., and Wallace, J. E., Synergy of ethanol and alcohol-like metabolites: tryptophol and 3,4-dihydroxyphenyl-ethanol, *Pharmacology*, 9, 294, 1973.
111. LeMoal, M. and Simon, H., Mesocorticolimbic dopaminergic network: Functional and regulatory roles, *Physiol. Rev.*, 71, 155, 1991.
112. Myers, R. D. and Veale, W. L., Alcohol preference in the rat: reduction following depletion of bran serotonin, *Science*, 160, 1469, 1968.
113. Roy, A., DeJong, J., Lamparski, D., George, T., and Linnoila, M., Depression among alcoholics - relationship to clinical and cerebrospinal fluid variables, *Arch. Gen. Psychiatry*, 48, 428, 1991.
114. Tollefson, G. D., Anxiety and alcoholism - a serotonin link, *Br. J. Psychiatry*, 159, 34, 1991.
115. Virkkunen, M. and Linnoila, M., Serotonin in early onset, male alcoholics with violent behaviour, *Ann. Med.*, 22, 327, 1991.
116. Wiesbeck, G. A., Mauerer, C., Thome, J., Jakob, F., and Boening, J., Alcohol dependence, family history, and D2 dopamine receptor function as neuroendocrinologically assessed with apomorphine, *Drug Alcohol Dep.*, 40, 49, 1995.

Section VI

From Animal Research to Society: Genetic Impact on Behavior

25 Mapping Quantitative Trait Loci for Behavioral Traits in the Mouse

*John K. Belknap, Christopher Dubay,
John C. Crabbe, and Kari J. Buck*

CONTENTS

A. Introduction 435
    1. Crosses Widely Used in the Mouse 436
    2. Importance of Microsatellite Marker Loci 436
    3. Statistical Models 436
B. Relevance to Human QTL Mapping 437
C. QTL Mapping in the Mouse 438
    1. The Question of Statistical Significance for QTL Mapping 438
    2. One-Step Mapping Strategies 439
    3. Two- or Multistep Mapping Strategies 442
    4. Statistical Programs for QTL Mapping 445
D. Identification of Genes Underlying a QTL 445
    1. Narrowing the Region of a QTL: Fine Mapping 445
    2. Identification of Genes in a QTL Region 446
    3. Evaluation of Detected Genes 446
    4. Demonstration of Gene Action 447
    5. Congenic Strains as a Tool for QTL Characterization 447
E. Informatics 448
F. Conclusion 449
References 450

A. INTRODUCTION

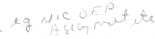

A quantitative trait locus (QTL) is a site on a chromosome containing alleles (i.e., genes) that influence a continuously distributed or quantitative trait (i.e., phenotype). Quantitative traits are also referred to as complex traits due to their polygenic and polyenvironmental determination. This is in contrast to qualitative (single locus) traits, which are generally much easier to map. To map a QTL, its influence on the trait must be detected amid considerable "noise" from other QTLs and nongenetic sources. This previously daunting task has been made quite feasible recently through the implementation of technologies to identify genetic variation (i.e., polymorphisms) at marker loci throughout the genome, and the development of statistical methods to detect and map QTLs from marker and trait data.[1–4] The identification of those chromosomal regions where marker allelic and trait variation significantly co-vary, thus implicating the presence of QTLs, is now a straight-forward (although large-scale) enterprise. The power of this approach was first demonstrated in plants[5] and later in rodents.[6–9]

We shall refer to a QTL in the singular and QTLs as the plural. The singular is used when a QTL is detected statistically as an apparent single entity. However, it must be kept in mind that more than one locus could be the basis for any QTL mapped using the methods commonly in use. This is because the mapping resolution attainable is not sufficient to resolve closely-linked loci in many cases, as noted below.

1. CROSSES WIDELY USED IN THE MOUSE

For QTL mapping, a major advantage of laboratory species over *Homo sapiens* is the ability to make crosses that are maximally informative concerning the presence of QTLs. These crosses generally involve two inbred strains (i.e., progenitors), which insures that only two alleles exist per locus in equal frequencies, and the parental source of each allele can be unambiguously determined at any locus. Since inbred strains can be regarded as genotypic constants, crosses between them can be replicated in any laboratory. The most common crosses involve (1) crossing two inbred strains (the progenitors) to obtain an F_1, and intercrossing them to produce an F_2 population or (2) crossing an F_1 to one of the two progenitor inbred strains (a backcross or BC). A third type of "cross" is to start with an F_2 population and carry out maximal inbreeding in multiple, parallel, brother by sister matings until a new set of inbred strains is created, known as recombinant inbred (RI) strains.[10–12] In all three crosses, all genotypes are exposed to the same laboratory environment. This minimizes problems posed by genotype by environment correlations or interactions which trouble many human studies. Because inbred strains are central to these crosses, the mouse, with its bounty of genetically well-characterized inbred strains, has been the major vertebrate species used for gene mapping except for humans.[13]

2. IMPORTANCE OF MICROSATELLITE MARKER LOCI

Most recent QTL mapping efforts have primarily utilized microsatellite markers, also known as simple sequence length polymorphisms (SSLPs), which are highly polymorphic, naturally occurring variations in the number of repetitive base pair sequences.[13] These are readily genotyped by (polymerase chain reaction PCR; see Jagadeeswaran, Chapter 4) amplification using oligodeoxynucleotide primer pairs specific to each marker, followed by resolution of PCR products (alleles) on standard agarose or denaturing polyacrylamide gels.[3,4] This development has greatly facilitated the genotyping of large numbers of individuals in a BC or F_2 population to determine which alleles each animal possesses at many marker loci. In the mouse, there are presently over 8000 PCR-based microsatellite markers available, each of known chromosomal location, that can be used in a genome-wide search. This strategy is well exemplified by a number of recent reports in mice reviewed below.

3. STATISTICAL MODELS

For a qualitative (single locus) trait, a one-to-one relationship between genotype and phenotype is common; for example, humans possessing the allele causing Huntington's disease almost invariably display the disease phenotype. The relationship between genotype and phenotype for a quantitative trait is only probabilistic in that those possessing a given allele at a QTL have a probability of an increased (or decreased) value of a trait in concert with other QTLs and nongenetic determinants. Analyzing these relationships requires an unusually heavy reliance on statistical methods and stochastic models. In recent years, several models have been developed specifically as tools for QTL mapping in laboratory species.[1,14-20] These have increased the power to detect and to map QTLs while minimizing the risk of errors.

B. RELEVANCE TO HUMAN QTL MAPPING

The mouse and human species share a number of genetic similarities, as do most mammals. While the genetic (linkage) map length is about 3300 cM in humans and 1500 cM in mice, the physical length of the genome appears to be similar — about 3 billion base pairs (bp) in both species. The number of estimated genes is also similar at about 100,000.[13] The difference in linkage map length is largely due to the higher rate of recombination (crossing-over) per generation seen in humans, thus expanding the linkage map.

The question of the human relevance of mouse QTL data depends on the degree of homology that exists between mice and humans at primarily three levels — gene homology, linkage homology, and trait homology. Molecular cloning has shown that almost all genes in mice have homologues in humans and vice versa.[13] Gene homology exists when similarity in base pair sequence can be shown between the two species, which often implies some similarity in function for the homologous gene. Most identified homologous genes code for enzymes, receptors, or structural components that are essential for cellular function. In the CNS, many are involved in neurotransmission and signal transduction, where similarities in function often exist between the two species.

Linkage homology exists when loci close together (closely linked) on a chromosome in the mouse are also located close together in the human genome. This implies that sizeable segments of chromosomes have remained relatively intact (syntenic conservation) since the common ancestor to both species existed some 70 million years ago.[13] Over 1000 homologous genes have been mapped in both species, allowing the identification of over 100 chromosomal regions in the mouse genome showing linkage homology with portions of the human genome, each averaging about 9 cM in length. It has been estimated that 80% of the mouse genome shows linkage homology with portions of the human genome.[21] This greatly increases the probability that a QTL mapping result in the mouse will immediately suggest a map site in the human genome and vice versa. For example, the μ-opioid receptor gene (*OPRM*) was recently mapped in the mouse to proximal chromosome 10.[22] Based on linkage homology, the human μ-opioid receptor gene (*OPRM*) should map to human chromosome 6q24-25. This is indeed where human *OPRM* is located. Since QTL mapping is much easier in the mouse, this genetically well studied laboratory species is likely to become an important tool in mapping human QTLs. This is a major reason why sequencing the mouse genome is an important objective of the Human Genome Project.[23] The identification of genes underlying QTLs should also be much easier in the mouse, thus providing a valuable testing ground for candidate genes prior to testing in human populations.

Trait homology implies that a trait measured in two species will have some similarity in its determinants and function. For many cellular and physiological traits, trait homology can be demonstrated, e.g., regulation of blood pressure and gonadal hormone output, and homologous genes can be identified that appear to contribute to the trait homology. For behavioral traits, those showing prominent biological determinants, such as consummatory behavior, sexual behavior, stress reactions, seizure activity, drug withdrawal syndromes, and some aspects of drug reward, appear to show sufficient trait homology to make mouse models useful as a testing ground for genetic hypotheses relevant to human behavior. However, the lack of knowledge concerning the determinants and function of many behavioral traits in either species makes trait homology difficult to assess. Thus, we are often forced to rely on mouse behaviors where the degree of trait homology is not well known.

C. QTL MAPPING IN THE MOUSE

During the past decade, the majority of mapped mouse genes have been localized using several *in vitro* mapping techniques such as somatic cell hybrids, *in situ* hybridization, and interspecific backcrosses. The strength of all three is seen when mapping genes that have been cloned, so that appropriate probes can be used *in vitro* based on cloned sequences.[13,21] However, for most behavioral traits, we rarely know what genes are involved, let alone their cloned sequence. Thus, at the present state of our knowledge, the mapping methods most suited to behavioral traits measured *in vivo* rely mostly on linkage analysis using the crosses described above.

Virtually all proven QTL mapping strategies for polygenically determined (quantitative) traits have four common features.[1,9,16,24] First, one or more populations (e.g., F_2, BC, RI set) are tested for the trait of interest and genotyped at many marker loci distributed throughout the genome. (*Note:* Genotyping has already been done for most RI sets, but must be done for each F_2 or BC mouse since each possesses a unique genotype.) Second, the population is divided into genotypic classes at each marker locus, i.e., the two homozygote and heterozygote classes in an F_2 population, or the two homozygote classes in an RI set. Alternatively, the population can be divided into two phenotypic classes based on the highest and lowest scoring animals for a trait. Third, a statistical test (e.g., correlation, regression, χ^2, t, F, LOD) is used to determine if the genotypic classes differ with respect to the phenotype, or if the phenotypic classes differ with respect to genotype (or allele) frequency. Fourth, if the statistical test is significant (at least $p < .0001$ or LOD 3.3), it is concluded that a QTL exists in the same chromosomal region as the marker, i.e., the marker and QTL are linked. Many sophisticated statistical variants of this basic approach have been developed, for example, interval mapping,[1] and multiple regression.[17,18]

1. THE QUESTION OF STATISTICAL SIGNIFICANCE FOR QTL MAPPING

The technology to carry out a genome-wide scan is a powerful capability, but it comes at a rather high statistical price — greatly inflated Type I errors (false positives) arising from the large number of markers required.[1,2] The standard remedy for this problem is to use more stringent α levels (reduce acceptable Type I error risk). A general guideline for which there is some agreement is to use an α value for single markers (α_S), or single points on a chromosome, that yields $p < .05$ protection against ***even one*** Type I error occurring anywhere in the genome, i.e., $\alpha_G = .05$.[1] A genome-wide $\alpha_G = .05$ implies that there will be a 95% probability of no Type I errors in the entire marker set for a given trait. Thus, the conventional $p < .05$ significance level is still operative, but it applies to ***all*** markers used in a genome-wide search (α_G), and ***not*** to individual markers (or single points) examined singly (α_S).

The estimated α_S for individual markers that results in $\alpha_G = .05$ has been the subject of differences in opinion in the past. However, the recently recommended α_S values for genome-wide searches reported by Lander and Schork[2] and Lander and Kruglyak[25] have set a new standard for which there is some consensus. Their recommendations are shown graphically in Figure 25-1. They set $\alpha_S = .0001$ as the significance threshold in an F_2 or BC, which approximates a LOD score of 3.3 for the additive effects of a QTL ($df = 1$). To obtain the same $\alpha_G = .05$ value in RI sets, they recommend $p < .00002$, or LOD 3.9 for single markers or points.

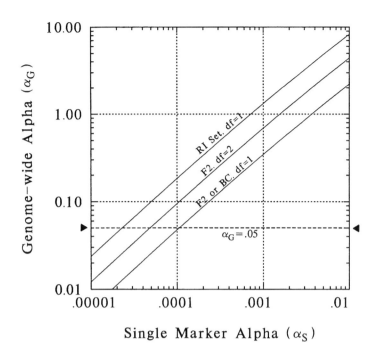

FIGURE 25-1 The number of false positive QTLs expected by chance (α_G) in a full genome search for a single trait as a function of α_S, the preset alpha for single markers. Values for a recombinant inbred (RI) set or an F_2 or backcross population are shown and were calculated from the expression given by Lander and Schork[2] and Lander and Kruglyak[25] as follows: $\alpha_G = [C + 2\rho G\chi^2] \alpha_S$. The quantity $[C + 2\rho G\chi^2]$ is a Bonferroni correction (k), where C is the number of chromosomes (20 in the mouse), G is the length of the genome expressed in Morgans (15 in the mouse), ρ reflects the relative cross-over density (equals 1.0 for an F_2 or BC and 4.0 for RI sets), and χ^2 is the chi-square value corresponding to α_S, and has either one degree of freedom ($df = 1$) for additive QTL effects or two degrees of freedom ($df = 2$) for additive and dominance effects estimated separately (F_2 only). The dotted line bounded by triangles denotes the $\alpha_G = .05$ threshold for significance recommended by these authors. It is assumed that the number of distinct genotypes (hence, the number of recombinations) and number of markers are both potentially very large ($\rightarrow \infty$). Since the limited number of genotypes available in an RI set do not meet this assumption, Belknap et al.[37] have suggested that for the largest existing mouse RI sets (e.g., BXD, AXB/BXA, LSXSS, AKXD), the function should be approximately the same as that shown for an F_2 with $df = 2$ (e.g., when $\alpha_G = .05$; $\alpha_S = .00005$). LOD scores can be estimated from p values[25] using the expression LOD = $\chi^2_{\alpha s}/4.6$. For example, $p = .001$ and $p = .0001$ (two-tailed) are approximately equivalent to LOD 2.4 and 3.3, respectively, when $df = 1$, and LOD 3.0 and 4.0 when $df = 2$.

2. ONE-STEP MAPPING STRATEGIES

Two general mapping strategies have been used in mice to effectively detect and map QTLs. The first strategy uses a single large F_2 or BC population to screen the entire genome in the search for QTLs. The first genome-wide searches in mammals involved blood pressure in rats[6,7] and seizures in mice.[8] We shall refer to this as a one-step approach, since mapping is carried out in a single population independently of other populations. In contrast, the two- or multistep approaches use more than one population in a sequential fashion, where the search in later steps is contingent on earlier results. The one-step approach has been used successfully for a number of behavioral traits or traits with behavioral implications. Three such studies exemplifying this approach are reviewed below.

Rise et al.[8] used 67 markers to screen most of the mouse genome for QTLs influencing seizures thought to model those seen in human complex partial epilepsy. They used the EL inbred strain known to be highly susceptible to these seizures and crossed them to each of two nonsusceptible inbred strains, DBA/2 and ABP/Le, to form two BC populations. Using the MAPMAKER/QTL program,[15] they found a QTL on chromosome 9 (~45 cM, LOD > 4.4) with a major influence on this trait in both BCs and a second QTL on chromosome 2 (~75 cM) that emerged primarily in only one BC. These QTLs were named *El1* and *El2*, respectively. Additional crosses carried out by Frankel et al.[26] led to the detection and mapping of four additional QTLs, *El3* through *El6*, on chromosomes 10, 9, 14, and 11, respectively, in one or more of these crosses. Most of the QTLs were highly cross-dependent, suggesting extensive interactions with the genetic background. To control for these background effects, congenic strains for three of the QTLs were developed[27] and are reviewed in the section on congenic strains. *El5* maps to the same region of chromosome 14 as the serotonin-2 receptor locus, *Htr2a*, and *El1* is in the same region as the serotonin-1B receptor locus, *Htr1b*. Since drugs that affect serotonin neurotransmission are also known to influence this type of seizure, the two serotonin receptor genes are plausible candidate genes for two of the QTLs.[26]

The second study involved voluntary morphine drinking in a large ($N = 606$) F_2 population derived from the C57BL/6J (B6) and DBA/2J (D2) inbred strains (i.e., B6D2F_2).[28] Saccharin was used to partially mask the usually avoided bitter taste of morphine. The mice were presented with a choice between morphine/saccharin and quinine/saccharin (quinine approximates the bitter taste of morphine). The progenitor strains were known to differ markedly for this behavior,[29–31] which insured that this trait would be highly heritable in the F_2 population. The extreme ends of the F_2 trait distribution (93 mice) were genotyped for 157 microsatellite markers distributed throughout the genome. Each marker and the interval between markers were examined to determine whether the allele frequencies differed between the high and low ends of the trait distribution using the MAPMAKER/QTL program. They found three QTLs with LOD scores >3.0 for both additive and dominance QTL effects ($df = 2$), which collectively accounted for an estimated 85% of the genetic variance. These QTLs were mapped to chromosomes 1 (~90 cM), 6 (~60 cM), and 10 (~5 cM).

The chromosome 10 QTL was the largest detected and mapped to the same proximal region of chromosome 10 as does the μ-opioid receptor locus (*OPRM*) recently mapped by Kozak et al.[22] Since μ-opioid receptors are known to mediate most effects of morphine, the possibility is raised that the QTL and *OPRM* may be one and the same. Support for this possibility comes from recent mapping studies for other morphine traits. A QTL that influences morphine analgesia and morphine hypothermia also maps to this same chromosomal region with LOD scores of about 4.0 ($df = 1$).[32] Overall, these findings provide a good example where a plausible candidate gene emerged (i.e., *OPRM*) from QTL mapping studies suitable for further testing in higher resolution and functional studies. Several other examples are noted below.

The third one-step study is that of Flint et al.[33] of a series of behavioral measures related to emotionality (anxiety or fearfulness) in 879 F_2 mice. They were tested for open field activity and defecation, open arm entries in the plus maze, and Y-maze activity. The highest and lowest 10% of the trait distribution for the first three measures were genotyped at 84 microsatellite loci distributed throughout the genome in a search for evidence of co-variation between a marker and each trait. The F_2 was derived from a cross of two fully inbred strains who were descendants of lines that had been selectively bred for either high or low open field activity for many generations starting from a C57BL/6 × BALB/c cross.[34] This strategy guaranteed that the progenitors to the F_2 would differ markedly in open field activity and also in a known genetically correlated trait, open field defecation.[34,35] It also insured that these traits would be maximally heritable in the F_2. Using the MAPMAKER/QTL program,[15] they found six promising QTLs for open field activity with LOD ≥ 3.7 ($df = 2, p < .0002$). The largest was on chromosome 1 (~100 cM), accounting for 9% of the trait variance, or about one-third of the genetic variance. Three of the six QTLs were also associated with open field defecation and plus maze open entries in a manner suggesting a common genetically mediated emotionality trait across the behavioral measures. These three QTLs were on chromosomes 1 (~100 cM), 12 (~20 cM), and 15 (~45 cM).

The one-step mapping studies described above nicely illustrate two important attributes of this approach. First, QTLs of very small effect can be detected and mapped if a sufficiently large F_2 population is used, and each mouse is genotyped for a sufficient number of markers to adequately cover the genome. The latter two studies reviewed above were capable of detecting individual QTLs accounting for 5 to 7% of the trait variance with a 90% probability (power = 0.9) at significance levels recommended by Lander and Schork.[2] Since QTLs of this effect size are probably fairly common for many traits, it is clear that large sample sizes are needed to detect them. As a guideline for future studies, the F_2 or BC sample sizes needed to attain statistical significance are shown in Figure 25-2 as a function of the QTL effect size to be detected when power = 0.9. Second, selective genotyping, where only the extreme scoring animals in both tails of the trait distribution are genotyped, can be used to markedly reduce genotyping costs. This incurs some loss of power to detect QTLs compared to genotyping all phenotyped animals, but the loss is usually relatively small.[1]

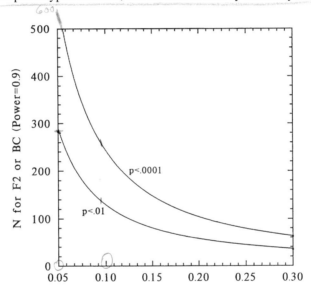

FIGURE 25-2 The estimated number of F_2 or BC mice needed to attain either α_s = .01 or .0001 (two-tailed) for single markers as a function of the QTL effect size expressed as a proportion of the phenotypic variance, or s^2_{QTL}/s^2_P for a single QTL. It is assumed that the statistical power (1-β), or the probability of correctly detecting a QTL, is set at 0.9, and that only the additive effects of a QTL are being detected (df = 1). The required N is given by the following expression from Soller et al.[109] and Lander and Botstein:[1] N = $(Z_\alpha + Z_\beta)^2(s^2_{RES}/s^2_{QTL})$, where Z_α and Z_β are the normal variates for the desired α_s (e.g., .0001, two-tailed, Z_α = 3.89) and β (0.1, one-tailed, Z_β = 1.28), s^2_{QTL} is the phenotypic variance accounted for (explained by) an individual QTL and s^2_{RES} is the residual (unexplained) phenotypic variance due to other QTLs and environmental effects. The total phenotypic variance, s^2_P, is equal to $s^2_{QTL} + s^2_{RES}$ for each QTL. Since s^2_{QTL} in a BC is about half that in a comparable F_2 for the same QTL, the needed BC sample sizes will be about double that of an F_2.[1] The plotted values presume that all phenotyped animals are also genotyped. When selective genotyping is used to reduce genotyping costs (genotyping only the extreme ends of the trait distribution), a somewhat larger sample size will need to be phenotyped to maintain the same power. For example, when 15, 23, or 30% of the phenotyped population is genotyped (the extreme scorers at both ends of the trait distribution), the plotted values of N must be increased by about 80, 50, and 33%, respectively (Figure 5 in Lander and Botstein[1]). When both additive and dominance QTL effects are to be detected (F_2 only), the above estimates of N should be increased by an additional 15% for p < .0001.

3. Two- or Multistep Mapping Strategies

The most common form of two- or multistep QTL mapping is to use recombinant inbred (RI) strains as an initial preliminary screen of the genome to identify provisional QTLs (Step 1), which are then specifically tested in other populations, usually an F_2 derived from the same progenitors (Step 2). RI strains have been tested for a variety of behaviorally-relevant traits, as shown in Table 25-1. The same criterion for significance is required as in the one-step approach, but combining the two (or more) steps can be carried out, e.g., adding LOD scores, or combining p values using R.A. Fisher's method,[36] to obtain an overall LOD or p value.[37]

RI strains are the fully inbred descendants of an F_2 cross between two inbred strains (the progenitors), and they have proven to be valuable tools in mapping qualitative (single locus) traits.[10] More recently, they have come into widespread use as a preliminary genome screen for QTLs.[9,39-40] The largest existing mouse RI sets are much preferred for this purpose; this includes the BXD set with 26 strains,[12] the AKXD set with 25 strains,[12] the AXB/BXA set with 31 strains,[41,42] and the LSXSS set with 27 strains.[43] More recently, recombinant congenic strains have come into use.[44,45] They are similar to RI strains except that several generations of backcrossing were carried out prior to the application of brother by sister inbreeding.[46]

In the first step, several mice of each RI strain are tested for the phenotype of interest, and the RI strain means are taken as an estimate of the phenotypic value associated with each genotype (strain). The variation in strain means for a trait (which estimates the additive genetic variance) is correlated with allelic variation at each marker locus. A significantly nonzero correlation suggests the presence of a QTL in the region of the correlated marker.[9,38-40] This initial step usually requires no genotyping, since most RI sets have been genotyped for hundreds of markers throughout the genome as a result of the cumulative effort of many researchers over many years.[47,48] This accumulation of allelic information for each genotype (i.e., strain) was possible because of the nature of inbred strains: they are highly replicable, stable over time, and widely available genotypes.

Since the number of available mouse RI strains is generally too small to map all but the largest QTLs unequivocally, the RI QTL results (Step 1) must be followed by at least a second step — confirmation testing in an independent test.[37,39,49-51] Usually an F_2 intercross between the two RI progenitor strains is used for Step 2. Each F_2 animal must be genotyped, but only for chromosomal regions which Step 1 results indicate contain QTLs. We shall refer to this as a two- or multistep approach — the domain of the Step 2 search is determined by Step 1 results.[37,40] This limited genome search in the F_2 typically involves 5 to 15% of the genome for provisional QTLs attaining $p < .01$ in the RI QTL analysis, and about twice this extent when $p < .05$ is used. This represents a considerable savings in genotyping effort compared to the full genome search required in a one-step approach.

The limited number of genotypes (i.e., strains) in existing mouse RI sets limits the power to detect QTLs of small effect,[37] but this concern is somewhat mitigated by testing several mice per strain, which is equivalent to making multiple measurements on each genotype, which allows a more accurate assessment of the phenotype associated with each genotype. This is possible because each genotype is readily replicable by breeding more mice of a given strain. In an F_2, in contrast, each genotype is represented by only a single mouse that cannot be replicated. Thus, the accuracy of predicting phenotype from genotype, which is important for efficient QTL mapping, is potentially much greater in the RIs.[1]

An important advantage of RI strains is the ready accumulation of phenotypic, QTL, and marker locus information across many traits, years, and laboratories on essentially the same genotypes.[10] This greatly facilitates the detection and assessment of genetic correlations, which index the degree to which two traits share common genetic influences,[52] and the QTL basis for the genetic correlations.[40] These advantages are not readily obtainable with F_2 populations since each genotype (individual mouse) can usually only be behaviorally tested

TABLE 25-1
Published Recombinant Inbred Studies of Behaviorally Related Traits in the Larger RI Sets (N Strains > 20), or in the Smaller Sets When Confirmation Testing Data Are Also Available

Phenotype (trait)	RI set[a]	Confirmation (Step 2)	Ref.
Locomotor activity	BXD	None	84, 85
	LSXSS	None	86
	AXB/BXA	None	87
Alcohol-induced activity/sensitization	CXB	Congenics, BC[88]	88
	BXD	None	84, 85
	LSXSS	None	86
Scopolamine-induced activity	CXB	Congenics[89]	89
Morphine-induced activity	BXD	F_2[32]	31
Cocaine-induced activity and tolerance	BXD	None	90, 91
Alcohol acceptance: single bottle under thirst	BXD	Std. inbreds[93]	92
	BXD	F_1[95]	94
Alcohol preference: two-bottle choice, no thirst	BXD	Selection[56]	96
	BXD	F_1[95]	94
Morphine/saccharin preference	BXD	F_2[28]	30, 31
Saccharin preference	BXD	BC[97]	97
	BXD	None	98
Quinine preference	BXD	None	30
Multiple tastant preferences (e.g., glycine, sucrose)	BXD	None	97, 99
Sucrose octaacetate (SOA) preference	SWXL	Congenics, BC[76,77]	76, 77
	SWXJ		
Alcohol conditioned place preference	BXD	None	84
Alcohol conditioned taste aversion	BXD	None	100
Audiogenic seizures	BXD	BC, congenics[57,101]	57
	B10.D2	None	44
Cocaine-induced seizures	BXD	None	102
Alcohol withdrawal seizures	BXD	F_2, Selection[54,103]	53
	BXD	None	91
Nitrous oxide withdrawal seizures	BXD	None	53
High pressure seizures	BXD	None	104
Light entrees in light-dark apparatus (with and without diazepam)	AXB/BXA	None	87
Restraint stress effects on activity	BXD	None	105
Spatial learning (Morris water maze)	BXD	None	106
Alcohol-induced loss of righting reflex	LSXSS	F_2[20,49]	43, 20
	BXD	None	93
Propofol-induced loss of righting reflex	LSXSS	Transgenics[62]	62
Multiple anesthetic-induced loss of righting reflex	LSXSS	None	59
Alcohol-induced ataxia/tolerance	BXD	None	107
Brain weight	BXD	Std. Inbreds[108]	108
Haloperidol catalepsy	BXD	F_2[110]	110

Note: The remaining studies for the smaller RI sets are reviewed by Broadhurst,[79] Shuster[80–82] and Neiderhiser et al.[83] Provisional QTL analyses for many of the earlier BXD studies (before 1991) can be found in Plomin et al.[9] and Gora-Maslak et al.[38] While RI QTL analyses often provide the important first step toward QTL mapping, they are likely to contain false positive errors. On average, a genome-wide search with 1500 markers at $p < .01$ in the BXD RI set will generate almost four false positive QTLs[37] or about half of the provisional QTLs identified in a typical genome search. [a]The CXB set, comprised originally of 7 strains, was derived from a BALB/c (C) × C57BL/6 (B6) cross.[10] The BXD set (26 strains) was derived from a C57BL/6 × DBA/2 cross.[12] The SWXL (6 strains) set was derived from an SWR × C57L cross.[12] The AXB/BXA set (31 strains) was derived from crosses between the A/J and C57BL/6 strains.[41] The LSXSS set (27 strains) was derived from the SS and LS strains known to differ markedly in alcohol loss of the righting reflex.[43] B10.D2 are recombinant congenic strains (36 strains) derived from C57BL/10 and DBA/2 strains and subjected to several generations of backcrossing to the D2 strain before brother by sister inbreeding was applied.[46]

once for a trait of interest (or a few times if the measurements per mouse are not confounding). This greatly limits multiple trait comparisons for the same genotypes.

A major concern in the use of a preliminary RI screen is Type II (false negative) errors — failing to detect and map important QTLs. For this reason, it has been customary to use relatively relaxed α_S values of .01 or .05 to identify provisional QTLs, which reduces the risk of false negatives compared to more stringent α_S values. This is feasible as long as adequate protection against Type I errors (false positives) comes from Step 2. When the number of genotypes is limited, as in Step 1 when RI strains are used, there is a trade-off between false positive and false negative error risks that must be considered.[37,39]

Three examples of the two- or multistep approach are described below. In our laboratories, the BXD RI set was studied for acute ethanol withdrawal severity as indexed by withdrawal convulsions,[53] which was followed by an F_2 and short-term selection studies to test the BXD QTL results.[54] For the latter, selective breeding (selection) was carried out for alcohol withdrawal severity in both the high and low directions starting from an F_2 population derived from the same progenitors as the BXD set, i.e., B6D2F_2. The phenotype diverged rapidly in only two generations of selection. The presence of QTLs was detected by the divergence in allele frequencies at a marker in the two oppositely selected lines significantly exceeding that expected from random (i.e., genetic) drift.[55,56] Pooling all three experiments (BXD, F_2, and selection), three statistically significant QTLs were found with LOD scores >4 on chromosomes 1 (~90 cM), 4 (~40 cM), and 11 (~20 cM). The latter QTL maps to the same region as several GABA$_A$ receptor subunit genes; most are attractive candidate gene(s). It is also of interest that the chromosome 4 QTL is in the same region (*Tyrp* or *b* locus) as *Asp2*, a locus known to influence audiogenic seizures in BXD mice.[57]

Phillips et al.[30] carried out a BXD study of two bottle choice alcohol preference drinking, and Belknap et al.[56] carried out a short-term selection study to test the BXD QTL results. After combining LOD scores from both experiments, no QTLs have as yet reached Lander and Schork significance levels, but four probable QTLs (LOD > 2.6; $p < .0005$) were found on chromosomes 2 (~45 cM), 3 (~70 cM), 9 (~30 cM), and 15 (~40 cM). The chromosome 3 presumed QTL (LOD 3.4) is in the same region as the alcohol dehydrogenase (ADH) cluster of genes (*Adh1*, *Adh2*, ~72 cM), which code for the enzyme that catalyzes the rate-limiting step in the metabolism of alcohol. Since ADH variants appear to be associated with the abuse of alcohol in some human populations,[58] *Adh1-2* is an intriguing candidate gene(s) for this presumed QTL.

The LSXSS RI set was used to identify provisional QTLs affecting loss of the righting reflex (LORR), an index of anesthesic sensitivity, to several general depressant drugs.[34,59] An F_2 study of over 1000 mice derived from closely similar progenitor strains as the LSXSS RI set was carried out for alcohol LORR.[20,49] Four QTLs were found that attained statistical significance on chromosomes 1 (~45 cM), 2 (~85 cM), 11 (~50 cM), and 15 (~50 cM). Interestingly, the chromosome 2 QTL maps to the same region as the neurotensin receptor gene (*Ntsr*). Considerable evidence exists for a modulatory role for neurotensin in several effects of alcohol.[60,61] For propofol, an anesthetic agent, one QTL was found in the region of the *Tyr* (or *c*) locus on chromosome 7 that accounted for 80% of the genetic variance for LORR in the LSXSS RI set.[62] Sensitivity to propofol LORR appears to be determined largely by a single locus, in marked contrast to alcohol LORR, which shows multilocus inheritance.

Finally, it should be noted that not all traits are equally suited to QTL analysis. The traits most likely to be successfully mapped are those showing (1) the highest heritabilities (the proportion of the total variation due to genetic sources), (2) the highest test-retest or split-half reliability coefficients, and (3) by relatively large effect QTLs.[37,39] Traits not meeting this description can be successfully mapped, but unusually large sample sizes may be required.

4. STATISTICAL PROGRAMS FOR QTL MAPPING

The MAPMAKER and MAPMAKER/QTL programs developed by Dr. Eric Lander's group have become the standard for QTL mapping in BC or F_2 populations. First, the MAPMAKER/EXP program is used to construct the primary linkage map for several microsatellite markers flanking each QTL, and the MAPMAKER/QTL program is used to assess the presence of a QTL within this framework.[1,15,63] The latter program uses linear regression of phenotype on gene dosage (number of alleles from one progenitor), but adds several features not seen in conventional linear statistics. The most important of these are: (1) both additive and dominance effects ($df = 2$) of a QTL are assessed, (2) interval mapping using maximum likelihood estimation is used, conferring greater power to detect QTLs that may be located between markers, (3) a genotyping error check routine is built-in,[14] and (4) correction for missing genotyping data, which is critical when the middle of the trait distribution is "missing" due to selective genotyping.[1,63]

A second valuable program is Map Manager or Map Manager QTL.[48,64] This program focuses primarily on RI sets for mapping purposes, both for qualitative and (very recently) quantitative traits. A valuable feature is the extensive marker data bases for the existing mouse RI sets that are integral to the program.

D. IDENTIFICATION OF GENES UNDERLYING A QTL

Knowing the approximate map location of a QTL is often the first important step toward identifying the specific gene underlying a QTL. This is because of the large number of mapped genes of obvious neurochemical import in the mouse, making it likely that QTL map sites emerging from genome searches will immediately suggest plausible candidate genes previously mapped to the same region. A good example cited above is the mapping of a QTL influencing morphine-induced analgesia, hypothermia, and morphine drinking to the same proximal chromosome 10 region as the μ-opioid receptor gene (*OPRM*). This candidate gene can then be tested by looking for variants (polymorphisms) between the two progenitor strains in or around *OPRM*, and by functional studies to determine if the progenitor strains differ in μ-opioid receptor binding or function. What happens if no candidate gene emerges in the chromosomal region of a known QTL? One answer is simply to wait until a candidate gene is mapped in the appropriate region, but it is impossible to know how long this might take. A long range strategy when there are no candidate genes is positional cloning based on map position.

Positional cloning usually requires three basic steps: first, the region of the QTL is narrowed as much as possible; then, genes within the most narrowed region are identified and cloned. Finally, the cloned genes are evaluated for their involvement with the trait and analyzed for the presence of genetic variants which segregate with the trait. Each of these steps is briefly discussed below.

1. NARROWING THE REGION OF A QTL: FINE MAPPING

If a QTL is initially identified using a full genome scan of 1500 mice using an unlimited number of well-distributed microsatellite markers, the mapping resolution, expressed as the 95% confidence interval, is estimated to be ~15 cM for QTLs accounting for 3% of the phenotypic variance in an F_2.[65] This QTL effect size is probably fairly typical for many traits. (*Note:* The confidence interval would be smaller for larger effect QTLs and larger for smaller effect QTLs.) The situation is similar with RI mapping data.[66] A chromosomal region of 15 cM would involve about 30 million base pairs (Mb) in size (1 cM equals on average ~2 Mb in the mouse).[13] With an estimated 100,000 genes in the mouse genome, we can expect roughly 1000 genes to be found in an 15 cM region or about 65 genes per cM.

It is clear that a region of interest must be narrowed much further before starting to look for genes. Progress toward this goal can be achieved by higher resolution linkage analysis, which requires a large number of recombinant genotypes in the region of interest and a high density of markers to saturate the region.[65] Congenic strains can also be used to increase mapping resolution, as noted below. To achieve even finer genetic mapping, one could use the most closely linked markers to screen a large insert genomic library (e.g., YAC or BAC) to identify clones spanning the region.[13] By subcloning and screening this region for additional informative polymorphic markers, the region containing a QTL can be further narrowed.

2. IDENTIFICATION OF GENES IN A QTL REGION

To identify genes in the narrowed region, molecular biology techniques can be applied to cloned regions of DNA which span the QTL region. If the QTL has been isolated to a 0.1 cM region, ~200,000 base pairs or 200 kb would need to be sequenced in both progenitor strains to find all the genes and genetic variants in the region. This objective can be facilitated by using techniques to identify cloned subregions which are more likely to contain genes. One such technique is to use subclones from the region to screen whole genome or tissue-specific cDNA expression libraries to identify transcribed regions. Another is to find CpG-islands which represent regions rich in the hypomethylated sequence CpG and which have been shown to be often associated with coding regions of DNA.[66] Yet another technique, which has been used with some success, is exon-trapping.[67] A special vector is used to subclone genomic DNA from the region of interest. If the subcloned region contains splice sites for a gene, which are found at exon-intron boundaries, it will combine with splice sites present in the vector and create a unique spliced product which can be detected by PCR after the vector is transfected into a host transcription system.

3. EVALUATION OF DETECTED GENES

Once genes are detected, those regions can be sequenced and the structure of the genes present elucidated. Some valuable short cuts have recently been developed to facilitate this task, in addition to the use of faster and larger-scale sequencing technologies (see Kolakowski, Chapter 7). After a partial sequence of a gene in a region of interest is determined, it can be used to search databases of expressed sequences and known genes that have been previously identified at a variety of sites, as noted in the Informatics section. Homologies with known genes and expressed sequences can be used to provide clues to the gene's function, which can aid in evaluation of the gene as a candidate for the QTL. Additionally, these homologies can tell you when the gene you have "found" has been previously cloned and sequenced, which allows the anchoring of the known gene in the QTL region.

Once a gene has been located, it must be analyzed for differences between the two progenitor strains within and around the gene. Any differences found can be used in two ways, first as markers which can be tested to see if they are closer to the QTL than previously identified markers, and second as hypothesized mutations which may be the genetic basis for the QTL.

Several methods have been developed to help find variations. Single bp differences can be detected by Single Strand Conformation Polymorphisms (SSCP) in a short region of DNA.[13,68] Comparative restriction digests can also be used to detect single bp variations

(RFLPs).[66] A promising variation is one present in the coding region of the gene, which would cause an alteration of the amino acid sequence of the peptide product. However, mutations at a variety of other sites, including splice junctions, the promoter, and enhancer/repressor regions, which may be located many kilobases from the gene itself, could also play potential roles. This is an important caveat, as the QTL may be caused by variations not necessarily within the coding region of a gene.

4. DEMONSTRATION OF GENE ACTION

Once a plausible candidate variation has been found that is consistently associated with the observed phenotypes, the final task is to prove the variation actually is the genetic basis for the QTL. Creating a transgenic animal with either zero (i.e., knockout), one, two (i.e., homozygous), or more (i.e., duplicated) copies of the gene can provide evidence that a particular gene influences a trait, as can mutations of any type affecting the function of known genes.[69–71] The phenotypes of these animals can then be used to infer the effect of the gene on the trait of interest. A problem that must be considered is that compensation for the effects of the mutation during early development, possibly involving many other loci, may influence trait expression. Therefore, the difference in a trait between a transgenic and a normal control mouse may involve the effects of loci other than the gene specifically targeted. For this reason, much research effort has been directed at the development of conditional transgenic methods that produce knockout or overexpression only regionally or only in adults.[69] A direct demonstration that genetic differences in a behavioral trait involve a specific gene can be tested using gene (i.e., allele) substitution techniques using transgenic or congenic approaches.

Based on the complexity of each of the three steps discussed above, it can be concluded that the positional cloning strategy is a difficult and time-intensive task in even the most advanced laboratories. However, technological innovations and the expanding knowledge base of mapped and cloned genes are constantly decreasing the time needed for going from a regional association (QTL) to a demonstrated DNA sequence variation underlying a QTL.

5. CONGENIC STRAINS AS A TOOL FOR QTL CHARACTERIZATION

The development of congenic strains involves the transfer of a chromosomal segment from one strain onto the genetic background of another (background) strain. This is accomplished by repeated backcrossing to the background inbred strain and selection of the desired allele at a marker or markers at each backcross generation.[10,72–74] After 10 backcross generations, the congenic and background strains can be expected to be about 98% genetically identical except for the transferred (introgressed) chromosomal region of about 20 cM.[74] The primary advantage of the congenics is that the influence of an individual QTL on any trait can be tested using the congenic vs. background strain comparison at any level from the molecular to the behavioral. Any differences found would strongly implicate a QTL in the introgressed chromosomal region as the cause of the differences. If there are several congenic strains for a given QTL, their differing sites of recombination can aid in attaining higher resolution mapping of the QTL with respect to neighboring markers. The near elimination of "genetic noise" due to unlinked loci should greatly aid the search for candidate genes associated with each QTL and for studies of differential gene expression such as representational difference analysis.[75] Ultimately, congenic strains can greatly facilitate positional cloning of a QTL.

Whitney and co-workers[76,77] have created congenic strains where alleles that confer taste sensitivity to a bitter tastant, sucrose octaacetate (SOA) have been transferred to an inbred background of a nontaster strain. Based on congenic, RI, BC, and outbred data, a single locus, named *Soa*, appears to largely determine SOA sensitivity. *Soa* was mapped to chromosome 6 near the *Prp* locus. Dudek and coworkers[72,73] have developed congenic strains where alleles predisposing to high alcohol-induced activity have been transferred to a genetic background of a nonactivated strain. Unlike the usual congenic development strategy of selection for a marker or flanking markers for a QTL at each backcross generation, they used the phenotype as the basis for selection, which can lead to the transfer of more than one QTL. A similar approach has been used to isolate loci influencing male-to-male aggression.[78] These loci have not as yet been mapped.[73,78]

Frankel et al.[27] have developed congenic strains for the epilepsy seizure QTLs *El1*, *El2* and *El3* isolated on an nonseizure inbred strain background. The detection and mapping of these QTLs in various BC and F_2 crosses was reviewed in an earlier section. *El2* isolated in a congenic strain had the expected effect on seizures, but *El1* was much weaker than expected from the BC QTL data, and *El3* had no detectable effect. The background genotype appears to have a marked influence on the expression of some QTLs influencing this trait. There is not enough data for other traits to know how often background effects can be expected.

E. INFORMATICS

The advent of electronic mail (e-mail) and other global methods of rapid electronic communication via the Internet and other networks has revolutionized how scientific research is carried out in many fields. From collaborations between distant researchers to communication between members of a laboratory informatics tools such as e-mail have had significant effects. Bioinformatics tools encompass basic informatics tools and specialized programs and databases for the analysis, manipulation, and storage of biological information.

Table 25-2 presents a list of Internet resource references of interest in QTL mapping and characterization. The examples cited use what is currently the most popular method for accessing information on the Internet: the World Wide Web (WWW). WWW is a network of information servers, which are accessed by remote users with client programs called "Web Browsers". Web Browsers allow a user to enter a Universal Resource Locator (URL) as an address to network resources which are then accessed via the Internet and viewed using the client browser. The most common documents accessed are coded in Hypertext Mark-Up Language (HTML) and contain hypertext regions which when selected cause the browser to load additional resources to which they pertain.

To give an appreciation of the potential utility of the above mentioned bioinformatics tools, we will take an closer look at two of them, physiology knowledge bases and DNA/Protein motif searches. Most genome scans begin with regions of candidate genes. Starting with a knowledge of the biology of the trait of interest, any major enzyme, structural protein, signaling peptide, or other component of the metabolic pathways or gene products associated with the trait becomes a potential candidate. Our growing knowledge of biology and its relation to genes can be documented using informatics tools. Currently an on-line knowledge base of the entire *E. Coli* genome and metabolism has been made available and work in the future will be directed at higher organisms including humans (http://www.mcs.anl.gov/home/compbio/PUMA/Production/ puma_graphics.html).

The question "What do we do with the DNA sequence once we have it?" applies to large genome sequencing projects as well as positional cloning projects. Databases and programs have been developed to search DNA sequences for coding regions, (http://avalon.epm.ornl.gov/Grail-bin/EmptyGrailForm), compare and align DNA sequences (http://genome.eerie.fr/fasta/ fastan-query.html), find regions of high conservation in proteins (http://www.blocks.fhcrc.org/), and predict the final protein's secondary structure (http://www.cmpharm.ucsf.edu/~nomi/nnpredict.html). With these and other tools, a

TABLE 25-2
As Examples of the Importance of Bioinformatics to the Mapping of Complex Traits, a List of Tools that Might be Employed in a Genome Scan and Their Current URLs are Provided

- A physiological knowledge base for identifying candidate genes.
  (http://www.ai.sri.com/ecocyc/ecocyc.html)
- A database of homologies between species for identifying syntenic regions.
  (http://www.hgmp.mrc.ac.uk/Comparative/home.html)
- A database of markers for identifying SSLPs in the mouse genome for typing.
  (http://www-genome.wi.mit.edu/genome_data/mouse/mouse_index.html)
- A database of PCR conditions and allele sizes for those markers.
  (http://www2.resgen.com/cgi-bin/sql/map2.pl).
- Software for performing linkage analysis.
  (http://www-genome.wi.mit.edu/ftp/distribution/software/).
- A database of human genes, to find genes in a syntenic region.
  (http://gdbwww.gdb.org/)
- An internet newsgroup to keep up on sequencing strategies and techniques.
  (news:bionet.molbio.methds-reagnts).
- A database of expressed sequences for homology searching.
  (http://www.ncbi.nlm.nih.gov/dbEST/index.html).
- Check your sequence for protein coding potential.
  (http://avalon.epm.ornl.gov/Grail-bin/EmptyGrailForm).
- Compare your sequence to known sequences using BLAST, etc.
  (http://gc.bcm.tmc.edu:8088/bio/bio_home.html).
- Submit your new gene sequence to GenBank.
  (http://www3.ncbi.nlm.nih.gov/BankIt/index.html).
- Read about latest successful genome scans in Science.
  (http://science-mag.aaas.org/science).
- Search mouse genome information
  (htt://www.informatics.jax.org/)

Note: To facilitate exploring these references, this text is available with hyperlinks to the mentioned resources at http://www.ohsu.edu/~dubayc/crcinformatics.html. Since this information is subject to change, an updated version can be obtained at this address. Other mouse data bases can also be found in Reference 13.

researcher can begin to answer the preceding question with a rapidity and depth unobtainable just a few years ago.

F. CONCLUSION

Since the first mouse genes were mapped about 80 years ago, the number of mapped genes increased by only 1 or 2 per year in the first two thirds of this century. The advent of recombinant molecular biology has led to a meteoric increase in mapped genes to over 5000, largely due to the development of techniques to detect polymorphisms, which can serve as markers and used to detect linkage with newly discovered genes. Most of these genes can be described as single locus traits determined *in vitro*, such as RFLPs, based on already cloned genes.[21] However, the same marker technologies have opened the door to the "last frontier" of gene mapping, the detection and mapping of loci influencing quantitative traits (QTLs) *in vivo*. Given that quantitative traits are far more numerous than qualitative (single locus) traits when studying living organisms,[24] the traits of most interest to behavioral scientists are now amenable to gene mapping, the all important first step toward identifying the genes involved and their modes of action. This promises to be a very lively and exciting field of research in the years to come.

REFERENCES

1. Lander, E. S. and Botstein, D., Mapping mendelian factors underlying quantitative traits using RFLP linkage maps, *Genetics*, 121:185-199, 1989.
2. Lander, E. S. and Schork, N. J., Genetic dissection of complex traits, *Science*, 265:2037-2048, 1994.
3. Dietrich, W., Katz, H., Lincoln, S. E., Shin, H-S., Friedman, J., Dracopoli, N. C. and Lander, E. S., A genetic map of the mouse suitable for typing intraspecific crosses, *Genetics*, 131:423-447, 1992.
4. Dietrich, W. F., Miller, J. C., Steen, R. G., Merchant, M., Damron, D., Nahf, R., Gross, A., Joyce, D. C., Wessel, M., Dredge, R. D., et al., A genetic map of the mouse with 4,006 simple sequence length polymorphisms, *Nat. Genet.*, 7:220-245, 1994.
5. Paterson, A. H., Lander, E. S., Hewitt, J. D., Peterson, S., Lincoln, S. E., and Tanksley, S. D., Resolution of quantitative traits into Mendelian factors using a complete linkage map of restriction fragment length polymorphisms, *Nature (London)*, 335: 721-726, 1988.
6. Jacob, H. J., Lindpaintner, K., Lincoln, S. E., Kusumi, K., Bunker, R. K., Mao, Y.-P., Ganten, D., Dzau, V. J., and Lander, E. S., Genetic mapping of a gene causing hypertension in the stroke-prone spontaneously hypertensive rat, *Cell*, 67: 213-224, 1991.
7. Hilbert, P, Lindpaintner, K., Beckmann, J. S., Serikawa, T., Soubrier, F., Dubay, C., Cartwright, P., De Gouyon, B., Julier, C., Takahasi, S., et al., Chromosomal mapping of two genetic loci associated with blood-pressure regulation in hereditary hypertensive rats, *Nature*, 353: 521-529, 1991.
8. Rise, M. L., Frankel, W. N., Coffin, J. M., and Seyfried, T. N., Genes for epilepsy mapped in the mouse, *Science*, 253:669-673, 1991.
9. Plomin, R., McClearn, G. E. and Gora-Maslak, G., Use of recombinant inbred strains to detect quantitative trait loci associated with behavior, *Behav. Genet.*, 21:99-116, 1991.
10. Bailey, D. W., Recombinant inbred strains and bilineal congenic strains, in *The Mouse in Biomedical Research*, Vol I., Foster, H.L., Small, J.D. and Fox, J.G., Eds., Academic Press, NY, pp. 223-239, 1981.
11. Taylor, B. A., Recombinant inbred strains: Use in gene mapping, in *Origins of Inbred Mice*, H.C. Morse, Ed., Academic Press, NY, pp 423-438, 1978.
12. Taylor, B. A., Recombinant inbred strains, in *Genetic Variants and Strains of the Laboratory Mouse*, 3rd Ed., M.F. Lyon and A.G. Searle, Eds., Oxford University Press, Oxford, 1995a.
13. Silver, L. M., *Mouse Genetics: Concepts and Applications*, Oxford Press, Oxford, UK, 1995.
14. Lincoln, S. E. and Lander, E. S., Systematic detection of errors in genetic linkage data, *Genomics*, 14:604-610, 1992.
15. Lincoln, S. E., Daly, M., and Lander, E. S., Mapping genes controlling quantitative traits with MAPMAKER/QTL 1.1. Whitehead Inst. Tech. Rep. (manual and tutorial), 2nd ed., 1993.
16. Haley, C. S. and Knott, S. A., A simple regression method for mapping quantitative trait loci in line crosses using flanking markers, *Heredity*, 69:315-324, 1992.
17. Zeng, Z.-B., Precision mapping of quantitative trait loci, *Genetics*, 136:1457-1468, 1994.
18. Jansen, R. C., Interval mapping of multiple quantitative trait loci, *Genetics*, 135:205-211, 1993.
19. Jiang, C. and Zeng, Z-B., Multiple trait analysis of genetic mapping for quantitative trait loci, *Genetics*, 140:1111-1127, 1995.
20. Markel, P. D., Fulker, D. W., Bennett, B., Corley, R. P., De Fries, J. C., Erwin, V. G., and Johnson, T. E., Quantitative trait loci for ethanol sensitivity in the LSXSS recombinant inbred strains: interval mapping, *Behav. Genet.*, in press.
21. Copeland, N. G., Jenkins, N. A., Gilbert, D. J., Eppig, J. T., Maltais, L. J., Miller, J. C., Dietrich, W. F., Weaver, A., Lincoln, S. E., Steen, R. G., Stein, L. D., Nadeau, J. H., and Lander, E. S., A genetic linkage map of the mouse: current applications and future prospects, *Science*, 262:57-66, 1993.
22. Kozak, C.A., Filie, J., Adamson, M.C., Chen, Y., and Yu, L., Murine chromosomal location of the mu and kappa opioid receptor genes, *Genomics*, 21:659-661, 1994.
23. Collins, F. and Galas, D., A new five-year plan for the U.S. human genome project, *Science*, 262:43-46, 1993.
24. Tanksley, S. D., Mapping polygenes, *Annu. Rev. Genet.*, 27:205-233, 1993.
25. Lander, E. S. and Kruglyak, L., Genetic dissection of complex traits: guidelines for interpreting and reporting linkage results, *Nat. Genet.*, 11:241-247, 1995.
26. Frankel, W. N., Valenzuela, A., Lutz, C. M., Johnson, E. W., Dietrich, W. F., and Coffin, J. M., New seizure frequency QTL and the complex genetics of epilepsy in EL mice, *Mammal. Genome*, 6:830-838, 1995a.
27. Frankel, W. N., Johnson, E. W., and Lutz, C. M., Congenic strains reveal effects of the epilepsy quantitative trait locus, *El2*, separate from other *El* loci, *Mammal. Genome*, 6:839-843, 1995b.

28. Berrettini, W. H, Ferraro, T. N., Alexander, R. C., Buchberg, A. M., and Vogel, W. H. Quantitative trait loci mapping of three loci controlling morphine preference using inbred mouse strains. *Nature Genetics* 7:54-58, 1994.
29. Belknap, J. K., Physical dependence induced by the voluntary consumption of morphine in inbred mice, *Pharmacol. Biochem. Behav.*, 35: 311-315, 1990.
30. Phillips, T. J., Belknap, J. K., and Crabbe, J. C., Use of voluntary morphine consumption in recombinant inbred strains to assess vulnerability to drug abuse at the genetic level, *J. Addict. Dis.*, 10:73-88, 1991.
31. Belknap, J. K. and Crabbe, J. C., Chromosome mapping of gene loci affecting morphine and amphetamine responses in BXD recombinant inbred mice, in *The Neurobiology of Alcohol and Drug Addiction*, Kalivas, P. and Samson, H., Eds., Annals of the N.Y. Academy of Sciences 654:311-323, 1992.
32. Belknap, J .K., Mogil, J. S., Helms, M. L., Richards, S. P, O'Toole, L. A., Bergeson, S. E., and Buck, K. J., Localization to chromosome 10 of a locus influencing morphine-induced analgesia in crosses derived from C57BL/6 and DBA/2 mice, *Life Sci. (Pharmacol. Lett.)*, 57: PL117-PL124, 1995.
33. Flint, J., Corley, R., DeFries, J. C., Fulker, D. W., Gray, J. A., Miller, S., and Collins, A. C., A simple genetic basis for a complex psychological trait in laboratory mice, *Science*, 269:1432-1435, 1995.
34. DeFries, J. C. and Hegmann, J. P. (1970), Genetic analysis of open-field behavior, in *Contributions to Behavior-Genetic Analysis: The Mouse as a Prototype*, Lindzey, G. and Thiessen, D., Eds., Appleton-Century-Crofts, NY, pp. 23-56.
35. Blizard, D. A. and Bailey, D. W., Genetic correlation between open-field activity and defecation: Analysis with CXB recombinant inbred strains, *Behav. Genet.*, 9:349-357, 1994.
36. Sokal, R. R. and Rohlf, F. J., *Biometry*, Freeman, San Francisco, 1981.
37. Belknap, J. K., Mitchell, S. R., O'Toole, L. A., Helms, M. L., and Crabbe, J. C., Type I and Type II error rates for quantitative trait loci (QTL) mapping studies using recombinant inbred mouse strains, *Behav. Genet.*, 26:149-160, 1996.
38. Gora-Maslak, G., McClearn, G. E., Crabbe, J. C., Phillips, T. J., Belknap, J. K., and Plomin, R., Use of recombinant inbred strains to identify quantitative trait loci in pharmacogenetic research, *Psychopharmacology*, 104:413-424, 1991.
39. Plomin, R. and McClearn, G. E., Quantitative trait loci analysis and alcohol-related behaviors, *Behav. Genet.*, 23:197-212, 1993.
40. Crabbe, J. C., Belknap, J. K., and Buck, K. J., Genetic animal models of alcohol and drug abuse, *Science*, 264:1715-1723, 1994.
41. Marshall, J. D., Mu, J.-L., Cheah, Y.-C., Nesbitt, M. N., Frankel, W. N., and Paigen, B., The AXB and BXA set of recombinant inbred mouse strains, *Mammal. Genome*, 3:669-680, 1992.
42. Taylor, B. A., Nonindependence of AXB RI strains, *Mouse Genome*, 93:1030, 1985b
43. DeFries, J. C., Wilson, J. R., Erwin, V. G., and Petersen, D. R., LSXSS recombinant inbred strains of mice: initial characterization, *Alc. Clin. Exp. Res.*, 13:196-200, 1989.
44. Martin, B., Marchaland, C., Phillips, J., Chapouthier, G., Spach, C., and Motta, R., Recombinant congenic strains of mice from B10.D2 and DBA/2: their contribution to behavior genetic research and application to audiogenic seizures, *Behav. Genet.*, 22:685-701, 1992.
45. Groot, P. C., Moen, C. J. A., Dietrich, W., Stoye, J. P., Lander, E. S., and DeMant, P., The recombinant congenic strains for analysis of multigenic traits: genetic composition, *FASEB J.*, 6:2826-2835, 1992.
46. Demant, P. and Hart, A., Recombinant congenic strains — a new tool for analyzing genetic traits determined by more than one gene, *Immunogenetics*, 24:416-422, 1986.
47. Silver, L. M., Nadeau, J. H., and Goodfellow, P. N., Encyclopedia of the Mouse Genome IV, *Mammal. Genome*, 6:S1-S295 (Special issue), 1995.
48. Manly, K. E. and Cudmore, R., Map Manager: *A Program for Genetic Mapping* (v. 2.6.5). Roswell Park Cancer Inst., Buffalo, NY, 1995.
49. Johnson, T. E., DeFries, J. C., and Markel, P., Mapping quantitative trait loci for behavioral traits in the mouse, *Behav. Genet.*, 22:635-653, 1992.
50. Belknap, J. K., Empirical estimates of Bonferroni corrections for use in chromosome mapping studies with the BXD recombinant inbred strains, *Behav. Genet.*, 22:677-684, 1992.
51. Neumann, P. E., Inference in linkage analysis of multifactorial traits using recombinant inbred strains of mice, *Behav. Genet.*, 22:665-676, 1992.
52. Crabbe, J. C., Phillips, T. J., Kosobud, A., and Belknap, J. K. Estimation of genetic correlation: interpretation of experiments using selectively bred and inbred animals, *Alc. Clin. Exp. Res.*, 14: 141-151, 1990.
53. Belknap, J. K., Metten, P. A., Helms, M. L., O'Toole, L. A., Angeli-Gade, S., Crabbe, J. C., and Phillips, T. J., Quantitative Trait Loci (QTL) applications to substances of abuse: physical dependence studies with nitrous oxide and ethanol, *Behav. Genet.*, 23:211-220, 1993.
54. Buck, K. J., Metten, P., Belknap, J. K., and Crabbe, J. C., Loci influencing genetic predispositions to acute alcohol dependence and withdrawal mapped to murine chromosomes 1, 4 and 11, submitted.
55. Keightley, P. D. and Bulfield, G., Detection of quantitative trait loci from frequency changes of marker alleles under selection, *Genet. Res.*, 62:195-203, 1993.

56. Belknap, J. K. Richards, S. P., O'Toole, L. A. Helms, M. L., and Phillips, T. J., Short-term selective breeding as a tool for QTL mapping: Alcohol preference drinking in mice, *Behav. Genet.*, in press.
57. Neumann, P. E. and Seyfried, T. N., Mapping of two genes that influence susceptibility to audiogenic seizures in crosses of C57BL/6J and DBA/2J mice, *Behav. Genet.*, 20:307-323, 1990.
58. Thomasson, H. R., Crabb, D. W., Edenberg, H. J., and Li, T.-K., Alcohol and aldehyde dehydrogenase polymorphisms and alcoholism, *Behav. Genet.*, 23:131-136, 1993.
59. Christensen, S. C., Johnson, T. E., Markel, P. D., Clark, V., Fulker, D. W., Corley, R. P., Collins, A. C., and Wehner, J. M., Quantitative trait locus analyses of sleep times induced by sedative-hypnotics in LSXSS recombinant inbred strains of mice, *Alc. Clin. Exp. Res.*, 20:543-550, 1996.
60. Erwin, V. G. and Jones, B. C., Genetic correlations among ethanol-related behaviors and neurotensin receptors in Long Sleep (LS) and Short Sleep (SS) recombinant inbred strains of mice, *Behav. Genet.*, 23:191-196, 1993.
61. Erwin, V. G., Jones, B. C., and Myers, R. Effects of acute and chronic ethanol administration on neurotensinergic systems, *NY Acad Sci.*, 739:185-196, 1994.
62. Simpson, V. J., Markel, P. D., Fulker, D. W., Corley, R., DeFries, J. C., and Johnson, T. E., Identification of a gene in mice specifying general anesthetic sensitivity, submitted.
63. Lander, E. S., Green, P., Abrahamson, J., Barlow, A., Daly, M. J., Lincoln, S. E., and Newburg, L., MAPMAKER: An interactive computer package for constructing primary genetic linkage maps of experimental and natural populations, *Genomics*, 1: 174-181, 1987.
64. Manly, K. E., A Macintosh program for storage and analysis of experimental genetic mapping data, *Mammal. Genome*, 4:303-313, 1993.
65. Darvasi, A. and Soller, M., Advanced intercross lines, an experimental population for fine genetic mapping, *Genetics*, 141:1199-1207, 1995.
66. Silver, J., Confidence limits for estimates of gene linkage based on analysis of recombinant inbred strains, *J. Hered.*, 76:436-440, 1985.
66. Lewin, B., *Genes V.*, Oxford Press, Oxford, UK, 1994.
67. Buckler, A. J., Chang D. D., Graw, S. L., Brook, J. D., Harber, D. A., Sharp, P. A., and Housman, D. E., Exon amplification: a strategy to isolate mammalian genes based on RNA splicing, *Proc. Natl. Acad. Sci. U.S.A.*, 88:4005-4009, 1991.
68. Orita, M., Iwahana, H., Kanazawa H., Hayashi, K., and Sekiya, T., Detection of polymorphisms of human DNA by gel electrophoresis as single-strand conformation polymorphisms, *Proc. Natl. Acad. Sci. U.S.A.*, 86:2766-2770, 1989.
69. Herrup, K., Transgenic and ES cell chimeric mice as tools for the study of the nervous system, *Discuss. Neurosci.*, 10:1-64, 1995.1
70. Wehner, J. M. and Bowers, B., Use of transgenics, null mutants, and antisense approaches to study ethanol's actions, *Alc. Clin. Exp. Res.*, 19:811-820, 1995.
71. Smithies, O. and Kim H. S., Targeted gene duplication and disruption for analyzing quantitative genetic traits in mice, *Proc. Natl. Acad. Sci. U.S.A.*, 91:3612-3615, 1994.
72. Dudek, B. C. and Underwood, K., Selective breeding, congenic strains, and other classical genetic approaches to the analysis of alcohol-related polygenic pleiotropisms, *Behav. Genet.*, 23:179-190, 1993.
73. Dudek, B. C. and Tritto, T., Classical and neoclassical approaches to the genetic analysis of alcohol-related phenotypes, *Alc. Clin. Exp. Res.*, 19:802-810, 1995.
74. Flaherty, L., Congenic strains, in *The Mouse In Biomedical Research. Volume I: History, Genetics, and Wild Mice*, Foster, H.L., Small, J.D., and Fox, J.G., Eds., Academic Press, NY, pp. 215-222, 1981.
75. Lisitsyn, N. A., Segre, J. A., Kusumi, K., Lisitsyn, N. M., Nadeau, J. H., Frankel, W. N., Wigler, M. H., and Lander, E. S., Direct isolation of polymorphic markers linked to a trait by genetically directed representational difference analysis, *Nat. Genet.*, 6:57-63, 1994.
76. Capeless, C. G., Whitney, G., and Azen, E. A., Chromosome mapping of *Soa*, a gene influencing gustatory sensitivity to sucrose octaacetate in mice, *Behav. Genet.*, 22:655-664, 1992.
77. Whitney, G. and Harder, D. B., Genetics of bitter perception in mice, *Physiol. Behav.*, 56:1141-1147, 1994.
78. Schneider-Stock, R. and Epplen, J. T., Congenic AB mice: A novel means for studying the (molecular) genetics of aggression, *Behav. Genet.*, 25:475-482, 1995.
79. Broadhurst, P., *Drugs and the Inheritance of Behavior*, Plenum, NY, 1978.
80. Shuster, L., Genetic determinants of responses to drugs of abuse: an evaluation of research strategies. NIDA Research Monograph No. 54, pp 50-69, USGPO, Washington, D.C., 1984.
81. Shuster, L., Pharmacogenetics of drugs of abuse, *Ann. N.Y. Acad. Sci. U.S.A.*, 562:56-73, 1989.
82. Shuster, L., Genetic markers of drug abuse in mouse models, in *Genetic and Biological Markers in Drug Abuse and Alcoholism*, Braude, M.C. and Chao, H.M., Eds., NIDA Research Monograph 66, GSGPO, Washington, D.C., pp. 71-85, 1986.
83. Neiderhiser, J. M., Plomin, R., and McClearn, G. E., The use of CXB recombinant inbred mice to detect quantitative trait loci in behavior, *Physiol. Behav.*, 52:429-439, 1992.
84. Cunningham, C. L., Localization of genes influencing ethanol-induced conditioned place preference and locomotor activity in BXD recombinant inbred mice, *Psychopharmacology*, 1995.

85. Phillips, T. J., Huson, M., Gwiazdon, C., Burkhart-Kasch, S., and Shen, E. H., Effects of acute and repeated ethanol exposures on the locomotor activity of BXD recombinant inbred mice, *Alc. Clin. Exp. Res.*, 19:269-278, 1995.
86. Erwin, V. G., Jones, B. C., and Radcliffe, R. A., Further characterization of the LSXSS recombinant inbred strains of mice: activating and hypothermic effects of ethanol, *Alc. Clin. Exp. Res.*, 14:200-204, 1990.
87. Mathis, C., Neumann, P. E., Gershenfeld, H., Paul, S. M., and Crawley, J., Genetic analysis of anxiety-related behaviors and responses to benzodiazepine-related drugs in AXB and BXA recombinant inbred mouse strains, *Psychopharmacology*, in press.
88. Oliverio, A. and Eleftheriou, B. E., Motor activity and alcohol: a genetic investigation in the mouse, *Physiol. Behav.*, 16:577-581, 1976.
89. Oliverio, A., Eleftheriou, B. E., and Bailey, D. W., Exploratory activity; genetic analysis of its modification by scopolamine and amphetamine, *Physiol. Behav.*, 10:893-901, 1973.
90. Tolliver, B. K., Belknap, J. K., Woods, W. E., and Carney, J. M., Genetic analysis of sensitization and tolerance to cocaine, *J. Pharmacol. Exp. Ther.*, 270:1230-1238, 1994.
91. Miner, L. L. and Marley, R. J., Chromosomal mapping of the psychomotor stimulant effects of cocaine in BXD recombinant inbred mice, *Psychopharmacology*, 122:209-214, 1995a.
92. Crabbe, J. C., Kosobud, A., Young, E. R., and Janowsky, J. S., Polygenic and single-gene determination of responses to ethanol in BXD/Ty recombinant inbred mouse strains, *Neurobeh. Toxicol. Teratol.*, 5:181-187, 1983.
93. Goldman, D., Lister, R. G., and Crabbe, J. C., Mapping of a putative genetic locus determining ethanol intake in the mouse, *Brain Res.*, 420:220-226, 1987.
94. Rodriguez, L. A., Plomin, R., Blizard, D. A., Jones, B. C., and McClearn, G. E., Alcohol acceptance, preference and sensitivity in mice. II. Quantitative trait loci mapping analysis using BXD recombinant inbred strains, *Alc. Clin. Exp. Res.*, 19:367-373, 1995.
95. Plomin, R., Rodriguez, L. A., Blizard, D. A., Jones, B. C., and McClearn, G. E., Alcohol acceptance, preference and sensitivity in mice. III. Using F_1 crosses between BXD recombinant inbred strains to replicate BXD-nominated quantitative trait loci (QTL), submitted.
96. Phillips, T. J., Crabbe, J. C., Metten, P., and Belknap, J. K., Localization of genes affecting alcohol drinking in mice, *Alc. Clin. Exp. Res.*, 18:931-941, 1994.
97. Lush, I., The genetics of tasting in mice. VI. Saccharin, acesulfame, dulcin and sucrose, *Genet. Res.*, 53:95-99, 1989.
98. Belknap, J. K., Crabbe, J. C., Plomin, R., McClearn, G. E., Sampson, K. E., O'Toole, L. A., and Gora-Maslak, G., Single locus control of saccharin intake in BXD/Ty recombinant inbred mice: some methodological implications for RI strain analysis, *Behav. Genet.*, 22: 81-100, 1992.
99. Lush, I. and Holland, G., The genetics of tasting in mice. V. Glycine and cycloheximide, *Genet. Res.*, 52:207-212, 1988.
100. Risinger, F. O. and Cunningham, C. L., Identification of genetic markers associated with sensitivity to ethanol-induced conditioned taste aversion, *Alc. Clin. Exp. Res.*, 18:451, 1994.
101. Neumann, P. E. and Collins, R. L., Genetic dissection of susceptibility to audiogenic seizures in inbred mice, *Proc. Natl. Acad. Sci. U.S.A.*, 88:5408-5412, 1991.
102. Miner, L. L. and Marley, R. J., Chromosomal mapping of loci influencing sensitivity to cocaine-induced seizures in BXD recombinant inbred strains of mice, *Psychopharmacology*, 117:62-66, 1995.
103. Buck, K. J., Strategies for mapping and identifying quantitative trait loci specifying behavioral responses to alcohol, *Alc. Clin. Exp. Res.*, 19:795-801, 1995.
104. McCall, R. D. and Frierson, D., Evidence that two loci predominantly determine the difference in susceptibility to the high pressure neurological syndrome type I seizure in mice, *Genetics*, 99:285-307, 1981.
105. Tarricone, B. J., Hingtgen, J. N., Belknap, J. K., Mitchell, S. R., and Nurnberger, J. I., Jr., Quantitative trait loci associated with the behavioral response of BXD recombinant inbred mice to restraint stress, *Behav. Genet.*, 25:489-496, 1995.
106. Wehner, J. M., Sleight, S., and Upchurch, M., Hippocampal protein kinase C activity is reduced in poor spatial learners, *Brain Res.*, 523:181-187, 1990.
107. Gallaher, E. J., Jones, G. E., Belknap, J. K., and Crabbe, J. C., Identification of genetic markers for initial sensitivity and rapid tolerance to ethanol-induced ataxia using quantitative trait locus analysis in BXD recombinant inbred mice, *J. Pharmacol. Exp. Ther.*, in press.
108. Belknap, J. K., Phillips, T. J., and O'Toole, L. A., Quantitative trait loci (QTL) associated with brain weight in the BXD/Ty recombinant inbred mouse strains, *Brain Res. Bull.*, 29:337-344, 1992.
109. Soller, M., Brody, T., and Genizi, A., On the power of experimental designs for the detection of linkage between marker loci and quantitative loci in crosses between inbred lines, *Theoret. Appl. Genetics*, 47:35-39, 1976.
110. Kanes, S., Dains, K., Cipp, L., Gatley, J., Hitzemann, B., Rasmussen, E., Sanderson, S., Silverman, M., and Hitzeman, R., Mapping the genes for haloperidol-induced catalepsy, *J. Pharmacol. Exp. Ther.*, 277:1016-1025, 1996.

26 Drug Discrimination: A Tool to Unravel the Genetic Determinants of Alcohol Preference and Aversion

Giancarlo Colombo, Roberta Agabio, Carla Lobina, Roberta Reali, Fabio Fadda, and Gian Luigi Gessa

CONTENTS

A. Drug Discrimination Procedure and Discriminative Stimulus Effects 455
B. The Mixed Discriminative Stimulus Effects of Ethanol 456
C. Selectively Bred Ethanol-Preferring Rats 459
D. Ethanol Discrimination in Ethanol-Preferring Rats 461
References 462

A. DRUG DISCRIMINATION PROCEDURE AND DISCRIMINATIVE STIMULUS EFFECTS

It is widely accepted that the subjective effects or interoceptive cues of addictive drugs constitute a critical factor in the initiation and maintenance processes of drug abuse in a large number of individuals.[1,2] Thus, advances in understanding the neurobiological substrates of the internal stimulus effects produced by addictive drugs may yield a valuable clarification on a major feature of drug addiction and lead to a rational design of effective pharmacotherapies.

A large body of experimental evidence over the last 30 years has indicated the suitability and sensitivity of the drug discrimination procedure in investigating the discriminative stimulus effects of psychoactive drugs in laboratory animals[3–6] and their highly predictive validity as the animal correlate of human subjective effects.[3,7]

In drug discrimination procedures, laboratory animals are trained to recognize the interoceptive cues of a given dose of a certain drug (training drug) and associate these with a specific behavior. Namely, animals are trained to behave in a particular manner (e.g., pressing a lever in a two-lever operant procedure or running an arm of a T-maze) every time they detect that drug state (drug condition) and to behave differentially (e.g., pressing the second lever or running the opposite arm) when they do not perceive that particular drug state (i.e., control or nondrug condition). Thus, the response behavior is the means by which the laboratory animal "self-reports" its feelings after being drugged. Administration of a different drug (testing drug) in place of the training drug, yielding the training drug-associated behavior, is indicative of the similarity of the discriminative stimulus effects of the testing drug to those of the training drug. Alternatively, combination of both testing and training drugs, resulting in the selection of the nondrug-associated behavior, accounts for an attenuated perception of the discriminative stimulus effects of the training drug.

Food-reinforcement and shock-avoidance are commonly used to motivate responding. A variety of methodological protocols have been used to assess the discriminative stimulus effects of psychotropic drugs. A detailed technical description of these paradigms is beyond the scope of the present paper (this issue has been reviewed in References 3, 8, and 9). Briefly, at present virtually all laboratories employ two-lever operant chambers (Skinner boxes), in which animals are trained to select one of two levers depending upon they have been treated with the training drug or the drug vehicle. However, Figure 26-1 concisely describes the T-maze, food-reinforced drug discrimination paradigm used in the authors' laboratory. In this procedure, rats are trained to approach the far end of one of the two arms of a T-maze for food-reinforcement when drugged and the opposite arm when not drugged.[10]

B. THE MIXED DISCRIMINATIVE STIMULUS EFFECTS OF ETHANOL

Following an early report by Conger,[11] the suitability of ethanol to develop discriminative control has been shown in a number of animal species, including monkeys, pigeons, gerbils, rats and mice.[4,5,12]

In alcohol research, the drug discrimination paradigm has been widely used to investigate (1) the brain pathways that transmit information on the discriminative stimulus effects of ethanol, by assessing the degree of similarity of its discriminative stimulus effects with those of drugs with known action on the CNS and (2) the pharmacological agents that may attenuate or block the discriminative stimulus effects of ethanol.

In order to define the pharmacological profile of the perceptual effects of ethanol, a number of drugs belonging to different pharmacological classes have been tested for substitution, potentiation, or blockade of the discriminative stimulus effects of ethanol.[4,5,12]

Accumulating experimental evidence indicates that the $GABA_A$ and NMDA receptor complexes, as well as the $5-HT_1$ subtype of the serotonin receptor, are involved in the mediation of the discriminative stimulus effects of ethanol. A number of ligands at different sites of the $GABA_A$ receptor complex, which increase chloride flux through the $GABA_A$ receptor-associated channel, substitute for the interoceptive cues of ethanol. Indeed, barbiturates,[13-29] benzodiazepines,[14,22,24,28,30-37] and endogenous neurosteroids[26,29] have been reported to elicit ethanol-like discriminative stimulus effects. Moreover, the benzodiazepine inverse agonist Ro 15-4513 attenuates ethanol discrimination,[38,39] although this result has been not confirmed in other studies.[23,40-42] Recent electrophysiological and biochemical evidence has shown that ethanol possesses antagonistic properties on the NMDA-mediated conductance of calcium ions.[43] Consistently, NMDA antagonists have been reported to substitute for the discriminative stimulus effects of ethanol.[10,28,29,35,44-48] Finally, the $5-HT_1$ agonists, trifluoromethylphenylpiperazine (TFMPP), CGS 12066B, and RU24969, fully substitute for the ethanol cue.[49-51]

Recently, Grant has proposed that the discriminative stimulus effects of ethanol can be viewed as a mixture of multiple components (i.e., cues), where each component is the effect of ethanol on a specific receptor system.[27,45,50,52] Accordingly, the effects of ethanol on the

Drug Discrimination

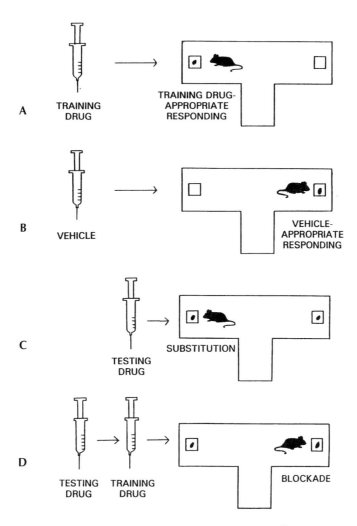

FIGURE 26-1 T-maze food-reinforced drug discrimination technique.[10] The apparatus is made of black plexiglas and consists of a central stem (start point) and two opposite runways (right and left arm). A sliding door divides the start point from the arms. A recessed food cup is placed at the far end (goal area) of each arm and sunflower seeds are used as reinforcements to motivate responding of food-deprived rats. Panels A and B show an example of training sessions under the drug- and the vehicle-condition, respectively. Each daily session consists of 10 consecutive trials (runs). Rats are trained to run the drug-appropriate arm (the left arm for half the rats and the right arm for the other half) after the administration of the training drug (panel A) and the vehicle-appropriate arm (the opposite arm) after the administration of drug vehicle (panel B). Only the selection of the condition-appropriate arm is food-reinforced. Sessions under drug- and vehicle-condition are daily alternated. Five consecutive correct training sessions define the criterion for acquisition of the discrimination. Once rats are trained to criterion, substitution (panel C) and blockade (panel D) tests are performed. In substitution (or generalization) tests, the degree of similarity between the discriminative stimulus effects of the training drug and those of different doses of the testing drug is assessed. Average selection of the drug-appropriate arm higher than 80% after administration of the testing drug is indicative of complete substitution for the discriminative stimulus effects of the training drug. Blockade tests are aimed at evaluating the ability of the testing drug to interfere in the perception of the discriminative stimulus effects of the training drug. Average selection of the vehicle-appropriate arm lower than 20% after administration of the testing drug indicates complete blockade of the discriminative stimulus effects of the training drug. In test sessions, the choice of each arm is food-reinforced.

GABA$_A$ and NMDA receptor complexes, as well as those on the 5-HT$_1$ receptor subtype, result in qualitatively different cues and constitute different components of the mixture.[27,45,50,52] Interestingly, the rules regulating the discrimination of a drug with multiple effects, such as ethanol, closely match the discrimination characteristics of drug mixtures.[45,53] Drug discrimination studies using mixtures composed of drugs acting at different receptor systems have shown that each component of the mixture can substitute for the mixture, indicating that it can be perceived separately and that the mixture is not producing a novel, homogeneous stimulus.[54-57] As reviewed above, a number of drugs acting at the GABA$_A$[13-37] and NMDA[10,28,29,35,44-48] receptor complexes and at the 5-HT$_1$[49-51] receptor subtype have been reported to substitute fully for ethanol. This evidence indicates (1) the presence of a GABA$_A$, NMDA, and 5-HT contribution in the perception of the interoceptive cues of ethanol and (2) the distinctiveness of each component. However, rats trained to discriminate any of the drugs which substitute for ethanol commonly select the nondrug-associate response when tested with ethanol. In detail, ethanol does not consistently substitute for benzodiazepines,[22,23,31,32,34,36,58-60] barbiturates,[13,17,18,22,58,61-66] uncompetitive NMDA antagonists,[45,67,68] and 5-HT$_1$ agonists.[69] These data indicate the presence of asymmetrical or one-way generalization with ethanol. The lack of symmetrical generalization might be due to the inability of the compound interoceptive stimuli (the ethanol cue) to substitute for a single component. Recently, Barry III[12] has elegantly represented the lack of cross-substitution with a model of distinctive perception of an auditory stimulus. In rats trained to discriminate a five pitch sound (the ethanol cue) from a generalized noise (12 pitches, nondrug condition), one pitch sound (i.e. a benzodiazepine or a NMDA antagonist) is more similar to the five pitch sound (ethanol) than to the twelve pitch one (vehicle). In contrast, in animals trained to discriminate a one pitch sound from the generalized noise, five pitches are more similar to the generalized noise than to the one pitch sound.

A second major aspect of Grant's hypothesis on the mixed discriminative stimulus effects of ethanol is the role of the training dose.[27,45,50,52] Previous studies have shown that the discriminative stimulus effects of a drug mixture vary as a function of the ratio between the components.[54-57] As the dose of a single component is increased, its relative contribution to the discriminative stimulus effects of the mixture is raised in prominence, whereas the salience of the other component(s) proportionally decreases. A similar situation occurs with the compound discriminative stimulus effects of ethanol, since the relative contributions of the GABA$_A$-, NMDA-, and serotonin-mediated neurotransmission to the interoceptive cues of ethanol vary with the training dose of ethanol. Indeed, the stimulus effects elicited by pentobarbital are similar to those of low (1.0 g/kg, i.g.) and moderate (1.5 g/kg, i.g.) doses of ethanol, but not to those of a higher (2.0 g/kg, i.g.) dose.[27] Administration of the 5-HT$_1$ agonist, TFMPP results in complete substitution only for 1.5 g/kg ethanol i.g. and not for lower and higher doses.[50] Finally, the NMDA uncompetitive antagonists dizocilpine and phencyclidine substitute more potently for the high dose (2.0 g/kg, i.g.) ethanol than for the lower ones.[45] These data indicate that (1) GABA$_A$-mediated effects of ethanol appear to be a more salient component of the discriminative stimulus effects of low compared with high doses of ethanol; (2) activation of serotonin neurotransmission via 5-HT$_1$ receptor subtype results in a prominent cue of the discriminative stimulus effects of a narrow range of ethanol intermediate doses; and (3) the blockade of NMDA-mediated neurotransmission is a salient component of the high doses of ethanol and also overshadows the contribution of the other components. These results suggest that the discriminative stimulus effects of low and high doses of ethanol vary quantitatively (different intensity of the net cue) and qualitatively (different proportion of the component cues).

C. SELECTIVELY BRED ETHANOL-PREFERRING RATS

Laboratory animals selectively bred for high and low ethanol preference provide a powerful tool in alcohol research. Since the selection procedure has been designed in order to obtain a differential segregation only at those genes involved in the regulation of ethanol preference, ethanol-preferring and -nonpreferring animals may yield clarification of the genetically transmitted determinants predisposed to ethanol preference and dependence. These animals constitute an appropriate animal model to address the study on the neurophysiological mechanisms that make ethanol pleasurable and compelling in some individuals, leading them to excessive consumption and aversive and noxious in others, determining complete avoidance of alcoholic beverages.

Over the last 30 years, a number of pairs of rat lines have been successfully selected for their difference in voluntary ethanol intake, namely, UChA/UChB,[70] AA/ANA,[71] P/NP,[72] HAD/LAD,[73] and sP/sNP.[74] Table 26-1 briefly reviews these rat strains, reporting the line denomination, foundation stock, method of selection, generation reached, and location of the colony.

These pairs of rat lines have all been selectively bred under similar criteria. Starting from heterogeneous base populations of unselected rats, ethanol-preferring and -nonpreferring rats have been selected by systematic mating, repeated over many generations, of the highest drinking rats in the ethanol-preferring lines and of the lowest drinking rats in the ethanol-nonpreferring lines. After 10 to 20 generations of this bidirectional selection, a high degree of separation between the two lines has been observed.[75] Rat strains are usually constituted by at least 15 to 20 breeding families in order to ensure a sufficiently large individual genetic variability.

Ethanol preference is evaluated under the two-bottle, free choice regimen when the rats are 2 to 3 months old. Namely, rats are individually housed and offered two bottles, containing

TABLE 26-1
Rat Strains Selectively Bred for Divergent Ethanol Preference

Denomination	Foundation stock	Method of selection	Generation reached (year)*	Location	Ref.
Low/high ethanol consuming (UChA/UChB)	Wistar	Inbreeding	50 (1983)	University of Chile, Santiago, Chile	70
ALKO Alcohol/NonAlcohol (AA/ANA)	Wistar	Outbreeding	62 (1993)	ALKO Finnish Monopole, Helsinki, Finland	71
Preferring/NonPreferring (P/NP)	Wistar	Outbreeding	39 (1994)	Indiana University, Indianapolis, IN, USA	72
High/low alcohol drinking (HAD/LAD)	N/Nih	Outbreeding	15 (1993)	Indiana University, Indianapolis, IN, USA	73
Sardinian alcohol-preferring/-Nonpreferring (sP/sNP)	Wistar	Outbreeding	32 (1994)	University of Cagliari, Cagliari, Italy	74

Note: * Determined on the basis of recent papers. Recently, selective breeding of a new line pair of ethanol-preferring and -non preferring rats (named Warsaw High Preferring, WHP, and Warsaw Low Preferring, WLP), derived from a Wistar stock has been reported.[135]

ethanol solution (usually at the 10% v/v concentration) and tap water, respectively. Ethanol and water intakes are monitored every day and the bottle position is daily alternated to avoid development of position preference. Food is available *ad lib*. Rats belonging to the ethanol-preferring lines daily consume more than 5 g/kg ethanol, while rats of the ethanol-avoiding lines drink less than 1 g/kg ethanol per day.[70–74] Daily ethanol intake in ethanol-preferring rats usually occurs in three or four separate binges during the nocturnal phase of the light/dark cycle and in pharmacologically relevant amounts at each binge, as revealed by intoxicating blood ethanol concentrations.[76–78]

Ethanol-preferring and -nonpreferring rats have been shown to differ in a number of genetically controlled neurochemical and behavioral traits, likely associated to the development of ethanol preference and dependence.

For instance, a positive relationship between ethanol preference and development of tolerance to some ethanol effects has been demonstrated in P/NP,[79–81] AA/ANA,[82] and sP/sNP[74] line pairs. The more rapid acquisition of tolerance in ethanol-preferring rats might primarily involve the aversive effects of ethanol, leading to the unmasking of its reinforcing properties and, consequently, to higher ethanol intakes. Moreover, ethanol-preferring P,[79,83] AA,[84] and HAD[25,85] rats showed a higher sensitivity to the ethanol-induced locomotor stimulation compared to ethanol-nonpreferring NP, ANA, and LAD rats, respectively, suggesting that this trait might be a behavioral correlate of the stronger reinforcing properties of ethanol in these rat strains. Finally, ethanol-preferring P[86] and sP[87] rats showed inherent higher levels of anxiety than ethanol-nonpreferring NP and sNP, and furthermore voluntary ethanol consumption ameliorated anxiety in sP rats.[87] These results strengthen the hypothesis that anxiety may constitute a critical factor in the etiology of ethanol preference and that the anxiolytic effects of ethanol are part of its reinforcing properties. However, inconsistent results have been reported on the anxiety levels of ethanol-preferring AA and ethanol-nonpreferring ANA rats, perhaps due to the different methodological approaches employed.[88,89]

Ethanol-preferring P[90,91] and HAD[92] rats, compared with -nonpreferring NP and LAD ones, have been reported to possess lower levels of dopamine, a candidate neurotransmitter in the mediation of the reinforcing properties of ethanol,[93] and of its metabolites, 3,4-dihydroxyphenylacetic acid (DOPAC) and homovanillic acid (HVA), in the nucleus accumbens. Likewise, fewer dopamine fibers have been found in the medial mesolimbic region of ethanol-preferring P rats in comparison to ethanol-avoiding NP rats.[94] Dopamine metabolism in discrete brain areas was more sensitive to ethanol-induced stimulation in sP than in sNP rats.[95,96] Voluntary ethanol consumption resulted in a significant increase of dopamine release from nigrostriatal and mesolimbic neurons in both ethanol-preferring P[97] and sP[95] rats. Finally, lower density of the dopamine D2 receptor has been found in AA,[98,99] P,[100] and sP[101] lines, compared to the ethanol-nonpreferring ANA, NP, and sNP rat strains, respectively. A large number of observations suggest a negative correlation between ethanol preference and functionality of serotonergic neurotransmission.[102] Accordingly, lower levels of serotonin have been detected in the cortex, striatum, hippocampus and hypothalamus of P[91,103,104] and HAD[92] rats, with respect to NP and LAD rats. Moreover, a reduced number of serotonergic fibers[104–106] and differences in density of the 5-HT_{1A},[107,108] 5-HT_{1B},[109] and 5-HT_2[110] subtypes of the serotonin receptor have been monitored in selected brain regions of P rats, compared with NP rats. In complete contrast to these findings, concentrations of serotonin in cortex, striatum, hippocampus, and hypothalamus were higher in ethanol-preferring AA than in -nonpreferring ANA rats,[111] and no difference in the number of 5-HT_1 and 5-HT_2 binding sites was reported between AA and ANA rats.[112]

In spite of the bulk of studies conducted with ethanol-preferring and -avoiding rats, a number of issues still need to be clarified. For instance, a review of the current literature reveals the lack of complete consistency on the candidate neurochemical and behavioral traits predisposing to ethanol preference among the different rat strains selected for divergent ethanol preference. It cannot be ruled out that some misleading false-positive, or spurious genetic differences between the two rat lines unrelated to ethanol preference, might have occurred. However, ethanol preference is a phenotypic trait determined by multiple genetic determinants.[113,114] Thus, the contribution of a genetic factor predisposing to ethanol preference might be different among rat lines (i.e., being more prominent in one rat line than in others), leading to the concept of different forms of excessive ethanol intake.[75]

D. ETHANOL DISCRIMINATION IN ETHANOL-PREFERRING RATS

The high amount of daily ethanol intake,[70–74] the typical pattern of consumption under the two-bottle free choice regimen,[76–78] and the performance of operant behavior (i.e., lever pressing) to orally,[103,115–121] intragastrically,[122,123] and intracranially,[124] self-administer ethanol clearly indicate that ethanol-preferring rats drink ethanol in preference over water for its reinforcing effects. In contrast, the virtually complete rejection of ethanol solution by the ethanol-nonpreferring rats is consistent with the hypothesis that ethanol-induced aversive effects overshadow its reinforcing properties in these rat lines. Thus, the divergent ethanol intake between ethanol-preferring and -nonpreferring rats might be due to a different ratio between the reinforcing and aversive properties of ethanol.

Both the reinforcing and aversive properties of ethanol are part of its perceivable and discriminable effects. The different sensitivity to the reinforcing and aversive properties of ethanol between ethanol-preferring and -nonpreferring rats may underlie a different perception of the discriminative stimulus effects of ethanol. In this light, the drug discrimination method, applied to rats selectively bred for differential ethanol preference, constitutes a promising tool for unraveling the neurochemical mechanisms regulating ethanol preference and aversion.

A limited number of studies to date have investigated the possible existence of genetic differences in the perception of the interoceptive cues of ethanol between ethanol-preferring and -nonpreferring rats. To our knowledge, the first study on this issue was conducted by York,[125] using ethanol-preferring AA and -nonpreferring ANA rats trained to discriminate 1.0 g/kg ethanol i.p. from saline in a T-maze, shock-motivated drug discrimination procedure. Interestingly, ANA rats required approximately half the number of training sessions than AA rats to learn to discriminate ethanol from saline and maintained a better level of performance throughout the entire experimental period. It has been shown that the number of training sessions needed to achieve the discrimination criterion is negatively related to the intensity of the drug cue.[126] The more salient the drug stimulus, the higher its degree of discriminability and, therefore, the shorter the training procedure required. Thus, in the study by York,[125] the stimulus effects of 1.0 g/kg ethanol i.p. at 20-min postinjection were more easily perceivable and discriminable in the ethanol-avoiding ANA compared to the ethanol-accepting AA rats. The higher degree of discriminability of ethanol in ethanol-nonpreferring ANA rats than in ethanol-preferring AA rats might account for the salient features of its aversive properties in the former rat strain selected for ethanol avoidance. In this regard, York found that acetaldehyde partially substituted for the ethanol cue in ANA but not in AA rats,[125] suggesting that the acetaldehyde-like effects of ethanol might contribute, at least in part, to its aversive effects in ethanol-nonpreferring rats, leading to its avoidance in the free choice paradigm. An interesting confirmation on the greater strength of the ethanol cue in ethanol-nonpreferring rats comes from a recent study that compared the ability of ethanol to establish stimulus control in three different rat strains, namely, Sprague-Dawley, N/Nih, and Fawn-Hooded, employing a two-lever, food-motivated drug discrimination assay.[127] Fawn-Hooded rats, which possess genetic abnormalities in serotonergic functioning[128,129] and show levels of voluntary ethanol consumption as high as those of selectively bred, ethanol-preferring

P rats,[130,131] required a longer discrimination training and a higher training dose than unselected ethanol-avoiding[132] N/Nih and Sprague-Dawley rats.

In contrast, no differences in the time needed to acquire the discrimination and in the ED_{50} values of ethanol were observed between ethanol-preferring HAD and -nonpreferring LAD rats trained to discriminate 0.5 g/kg ethanol i.p. with a 2-min presession interval[85] or 0.75 g/kg ethanol i.p. with either 2- or 30-min time onset[25] from saline in a two-lever, food-reinforced drug discrimination paradigm. However, the training doses of ethanol and the pretreatment times employed in these two studies should have resulted, at the time of testing, in lower blood ethanol levels compared to those expected in the experiments with AA and ANA rats.[125] Thus, differences in magnitude of the discriminative stimulus effects of ethanol between ethanol-preferring and -nonpreferring rats might occur only at blood ethanol concentrations corresponding to moderate to high doses of ethanol. On the other hand, a more recent study with ethanol-preferring P and -nonpreferring NP rats trained to discriminate 1.0 g/kg ethanol i.p. at 10-min postinjection failed to provide information on the time needed to reach the discrimination criterion.[133] The slight difference in the ED_{50} values of ethanol reported between the two rat strains might be indicative of both quantitative and qualitative differences in the perceived effects of ethanol. A subsequent study by Meehan et al.[134] reported no difference in the ability to acquire the ethanol (0.6 g/kg i.p., 10-min pretreatment) discrimination between ethanol-preferring HAD and -nonpreferring LAD rats. It has been hypothesized that the prolonged training might have obfuscated possible differences in ethanol sensitivity.[134]

The potential usefulness of drug discrimination procedures in assessing the pharmacological profile of the interoceptive cues of ethanol in ethanol-preferring and -nonpreferring rats has not yet been properly exploited. In a recent paper by Krimmer,[25] pentobarbital completely substituted for ethanol in both ethanol-preferring HAD and -nonpreferring LAD rats trained to discriminate 0.75 g/kg ethanol i.p. at either 2- or 30-min postinjection. However, HAD rats of the 30-min postinjection group showed a greater sensitivity to the ethanol-like effects of pentobarbital in comparison to the other groups, as revealed by a lower ED_{50} value for pentobarbital, which might be indicative of a more prominent contribution of the GABAergic component to the ethanol cue[27] in ethanol-preferring rats compared with -nonpreferring ones.

Genetic differences in the degree of substitution of the discriminative stimulus effects of nicotine for those of ethanol has recently been found in ethanol-preferring P and -nonpreferring NP rats trained to discriminate 1.0 g/kg ethanol i.p. from water in a two-lever, food-motivated operant task.[133] Nicotine administration resulted in partial substitution for ethanol in P but not in NP rats, indicating that the ethanol stimulus effects of P rats comprise a nicotine-like cue.

More recently, Meehan and colleagues have described a dose–dependent substitution for 0.6 g/kg ethanol i.p. with MDMA (3,4-methylenedioxymethamphetamine; "ecstasy") in ethanol-preferring HAD but not in ethanol-nonpreferring LAD rats,[134] suggestive of a greater relevance of serotonergic activation in the discriminative stimulus effects of ethanol in HAD than LAD rats.

Further investigations on the possible differences in the pharmacological profile of the internal cues of ethanol, i.e. the multi-component frame proposed by Grant,[27,45,50,52] between ethanol-preferring and -nonpreferring rats would likely advance our understanding on the neurobiological substrates of ethanol preference and aversion.

REFERENCES

1. Colpaert, F. C., Drug discrimination: Methods of manipulation, measurement, and analysis, in *Methods of Assessing the Reinforcing Properties of Abused Drugs*, Bozarth, M. A., Ed., Springer-Verlag, New York, 1987, 291.

2. Koob, G. F., Rassnick, S., Heinrichs, S. and Weiss, F., Alcohol, the reward system and dependence, in *Toward a Molecular Basis of Alcohol Use and Abuse*, Jansson, B., Jörnvall, H., Rydberg, U., Terenius, L., and Vallee, B. L., Eds., Birkäuser, Basel, 1994, 103.
3. Overton, D. A., Applications and limitations of the drug discrimination method for the study of drug abuse, in *Methods of Assessing the Reinforcing Properties of Abused Drugs*, Bozarth, M. A., Ed., Springer-Verlag, New York, 1987, 291.
4. Samele, C., Shine, P. J. and Stolerman, I. P., A bibliography of drug discrimination research, 1989–1991, *Behav. Pharmacol.*, 3, 171, 1992.
5. Stolerman, I. P., Samele, C., Kamien J. B., Mariathasan, E. A. and Hague, D. S. A bibliography of drug discrimination research, 1992–1994, *Behav. Pharmacol.*, 6, 643, 1995.
6. NIDA Research Monograph 116, *Drug Discrimination: Applications to Drug Abuse Research*, Glennon, R. A., Järbe, T. U. C., and Frankenheim, J., Eds., U.S. DHHS, Rockville, 1991.
7. Kamien, J. B., Bickel, W. K., Hughes, J. R., Higgins, S. T., and Smith, B. J., Drug discrimination by humans compared to nonhumans: current status and future directions, *Psychopharmacology*, 111, 259, 1993.
8. Barry III, H., Classification of drugs according to their discriminable effects in rats, *Fed. Proc.*, 33, 1814, 1974.
9. Goudie, A. J. and Leathley, M. J., Drug-discrimination assays, in *Behavioural Neuroscience: A Practical Approach, Volume II*, Sahgal, A., Ed., IRL Press, Oxford, 1993, 145.
10. Colombo, G., Agabio, R., Balaklievskaia, N., Lobina, C., Reali, R., Fadda, F., and Gessa, G. L., T-maze and food-reinforcement: an inexpensive drug discrimination procedure, *J. Neurosci. Methods*, in press.
11. Conger, J. J., The effects of alcohol on conflict behavior in the albino rat, *Quatl. J. Stud. Alcohol*, 12, 1, 1951.
12. Barry III, H., Distinctive discriminative effects of ethanol, in *Drug Discrimination: Applications to Drug Abuse Research, NIDA Research Monograph, Vol. 116*, Glennon, R. A., Järbe, T. U. C. and Frankenheim, J., Eds., U.S. DHHS, Rockville, 1991, 131.
13. Overton, D. A., State-dependent learning produced by depressant and atropine-like drugs, *Psychopharmacologia (Berlin)*, 10, 6, 1966.
14. Kubena, R. K. and Barry III, H., Generalization by rats of alcohol and atropine stimulus characteristics to other drugs, *Psychopharmacologia (Berlin)*, 15, 196, 1969.
15. Bueno, O. F. A., Carlini, E. A., Finkelfarb, E., and Suzuki, J. S., Δ^9-tetrahydrocannabinol, ethanol, and amphetamine as discriminative stimuli-generalization tests with other drugs, *Psychopharmacologia (Berlin)*, 46, 235, 1976.
16. Järbe, T. U. C., Alcohol-discrimination in gerbils: Interactions with bemegride, DH-524, amphetamine, and Δ^9-THC, *Arch. Int. Pharmacodyn.*, 227, 118, 1977.
17. Overton, D. A., Comparison of ethanol, pentobarbital, and phenobarbital using drug vs. drug discrimination training, *Psychopharmacology*, 53, 195, 1977.
18. York, J. L., A comparison of the discriminative stimulus effects of ethanol, barbital, and pentobarbital in rats, *Psychopharmacology*, 60, 19, 1978.
19. Schechter, M. D., Extended schedule transfer of ethanol discrimination, *Pharmacol. Biochem. Behav.*, 14, 23, 1981.
20. York, J. L. and Bush, R., Studies on the discriminative stimulus properties of ethanol in squirrel monkeys, *Psychopharmacology*, 77, 212, 1982.
21. Schechter, M. D., Ethanol and pentobarbital have different behavioral effects in the rat, *Prog. Neuro-Psychopharmacol. Biol. Psychiatr.*, 8, 271, 1984.
22. De Vry, J. and Slanger, J. L., Effects of training dose on discrimination and cross-generalization of chlordiazepoxide, pentobarbital and ethanol in the rat, *Psychopharmacology*, 88, 341, 1986.
23. Emmett-Oglesby, M. W., Tolerance to the discriminative stimulus effects of ethanol, *Behav. Pharmacol.*, 1, 497, 1990.
24. Krimmer, E. C., HAS and LAS rats respond differentially to behavioral effects of ethanol, pentobarbital, chlorpromazine and chlordiazepoxide, *Pharmacol. Biochem. Behav.*, 39, 5, 1991.
25. Krimmer, E. C., Biphasic effects of ethanol tested with drug discrimination in HAD and LAD rats, *Pharmacol. Biochem. Behav.*, 43, 1233, 1992.
26. Ator, N. A., Grant, K. A., Purdy, R. H., Paul, S. M., and Griffiths, R. R., Drug discrimination analysis of endogenous neuroactive steroids in rats, *Eur. J. Pharmacol.*, 24, 237, 1993.
27. Grant, K. A. and Colombo, G., Pharmacological analysis of the mixed discriminative stimulus effects of ethanol, in *Advances in Biomedical Alcohol Research*, Taberner, P. V. and Badawy, A. A., Eds., Pergamon Press, Oxford, 1993, 445.
28. Shelton, K. L. and Balster, R., Ethanol discrimination in rats: Substitution with GABA agonists and NMDA antagonists, *Behav. Pharmacol.*, 5, 441, 1994.
29. Azarov, A. V., Purdy, R. H. and Grant, K. A., The discriminative stimulus effects of ethanol in cynomolgus monkeys (*Macaca Fascicularis*), *Alc. Clin. Exp. Res.*, 19(Suppl.), 63A, 1995.
30. Schechter, M. D., Behavioral evidence for different mechanisms of action for ethanol and anxiolytics, *Prog. Neuro-Psychopharmacol. Biol. Psychiat.*, 6, 129, 1982.

31. Järbe, T. U. C. and McMillan, D. E., Interaction of the discriminative stimulus properties of diazepam and ethanol in pigeons, *Pharmacol. Biochem. Behav.*, 18, 73, 1983.
32. Schechter, M. D. and Lovano, D. M., Ethanol-chlordiazepoxide interactions in the rat, *Pharmacol. Biochem. Behav.*, 23, 927, 1985.
33. Hiltunen, A. J. and Järbe, T. U. C., Discrimination of Ro 11-6896, chlordiazepoxide and ethanol in gerbils: generalization and antagonism tests, *Psychopharmacology*, 89, 284, 1986.
34. Rees, D. C., Knisely, J. S., Breen, T. J., and Balster, R. L., Toluene, halothane, 1,1,1-trichloroethane and oxazepam produce ethanol-like discriminative stimulus effects in mice, *J. Pharmacol. Exp. Ther.*, 243, 931, 1987.
35. Sanger, D. J., Substitution by NMDA antagonists and other drugs in rats trained to discriminate ethanol, *Behav. Pharmacol.*, 4, 523, 1993.
36. Lytle, D. A., Egilmez, Y., Rocha, B. A., and Emmett-Oglesby, M. W., Discrimination of ethanol and of diazepam: Differential cross-tolerance, *Behav. Pharmacol.*, 5, 451, 1994.
37. Sanger, D. J. and Zivkovic, B., Discriminative stimulus effects of alpidem, a new imidazopyridine anxiolytic, *Psychopharmacology*, 113, 395, 1994.
38. Rees, D. C. and Balster, R. L., Attenuation of the discriminative stimulus properties of ethanol and oxazepam, but not of pentobarbital, by Ro 15-4513 in mice, *J. Pharmacol. Exp. Ther.*, 244, 592, 1988.
39. Gatto, G. J., Patterson, D. J. and Grant, K. A., Antagonism by Ro 15-4513 on the discriminative stimulus effects of ethanol, *Alc. Clin. Exp. Res.*, 19(Suppl.), 63A, 1995.
40. Hiltunen, A. J. and Järbe, T. U. C., Ro 15-4513 does not antagonize the discriminative stimulus- or rate-depressant effects of ethanol in rats, *Alcohol*, 5, 203, 1988.
41. Hiltunen, A. J. and Järbe, T. U. C., Discriminative stimulus properties of ethanol: effects of cumulative dosing and Ro 15-4513, *Behav. Pharmacol.*, 1, 133, 1989.
42. Middaugh, L. D., Bao, K., Becker, H. C., and Daniel, S. S., Effects of Ro 15-4513 on ethanol discrimination in C57BL/6 mice, *Pharmacol. Biochem. Behav.*, 38, 763, 1991.
43. Gonzales, R. A., NMDA receptors excite alcohol research, *Trends Pharmacol. Sci.*, 11, 137, 1990.
44. Grant, K. A., Knisely, J. S., Tabakoff, B., Barrett, J. E., and Balster, R. L., Ethanol-like discriminative stimulus effects of non-competitive n-methyl-d-aspartate antagonists, *Behav. Pharmacol.*, 2, 87, 1991.
45. Grant, K. A. and Colombo, G., Discriminative stimulus effects of ethanol: effect of training dose on the substitution of N-methyl-D-aspartate antagonists, *J. Pharmacol. Exp. Ther.*, 264, 1241, 1993.
46. Schechter, M. D., Meehan, S. M., Gordon, T. L., and McBurney, D. M., The NMDA receptor antagonist MK-801 produces ethanol-like discrimination in the rat, *Alcohol*, 10, 197, 1993.
47. Emmett-Oglesby, M. W., Tolerance and cross-tolerance to ethanol discriminative stimulus effects, *Behav. Pharmacol.*, 5, 19, 1994.
48. Koek, W., Kleven, M. S., and Colpaert, F. C. Effects of the NMDA antagonist, dizocilpine, in various drug discriminations: Characterization of intermediate levels of drug lever selection, *Behav. Pharmacol.*, 6, 590, 1995.
49. Signs, S. A. and Schechter, M. D., The role of dopamine and serotonin receptors in the mediation of the ethanol interoceptive cue, *Pharmacol. Biochem. Behav.*, 30, 55, 1988.
50. Grant, K.A. and Colombo, G., Substitution of the 5-HT_1 agonist trifluoromethylphenylpiperazine (TFMPP) for the discriminative stimulus effects of ethanol: effect of training dose, *Psychopharmacology*, 113, 26, 1993.
51. Grant K. A., Colombo, G., and Gatto, G. J., Characterization of 5-HT_1 receptor agonists and the mediation of ethanol mimetic discriminative stimulus effects, *Alc. Clin. Exp. Res.*, 17, 497, 1993.
52. Grant, K. A., The multiple discriminative stimulus effects of ethanol, *Behav. Pharmacol.*, 5, 9, 1994.
53. Stolerman, I. P., Mariathasan, E. A., and White, J.-A. W., Implications of mixture research for discrimination of single drugs, *Behav. Pharmacol.*, 5, 18, 1994.
54. Garcha, H. S. and Stolerman, I. P., Discrimination of a drug mixture in rats: role of training dose and specificity, *Behav. Pharmacol.*, 1, 25, 1989.
55. Stolerman, I. P., Rauch, R. J., and Norris, E. A., Discriminative stimulus effects of a nicotine-midazolam mixture in rats, *Psychopharmacology*, 93, 256, 1987.
56. Stolerman, I. P. and Mariathasan, E. A., Discrimination of an amphetamine-pentobarbitone mixture by rats in an AND-OR paradigm, *Psychopharmacology*, 102, 557, 1990.
57. Mariathasan, E. A., Garcha, H. S., and Stolerman, I. P., Discriminative stimulus effects of amphetamine and pentobarbitone separately and as mixtures in rats, *Behav. Pharmacol.*, 2, 405, 1991.
58. Krimmer, E. C. and Barry H., III, Relationship among pentobarbital, chlordiazepoxide and alcohol tested by their discriminative effects, *Fed. Proc.*, 39, 402, 1980.
59. Haug, T., Neuropharmacological specificity of the diazepam stimulus complex: Effects of agonists and antagonists, *Eur. J. Pharmacol.*, 93, 221, 1983.
60. Shannon, H. E. and Herling, S., Discriminative stimulus effects of diazepam in rats: evidence for maximal effect, *J. Pharmacol. Exp. Ther.*, 227, 160, 1983.
61. Krimmer, E. C. and Barry, H., III, Discriminative pentobarbital stimulus in rats immediately after intravenous administration, *Eur. J. Pharmacol.*, 38, 321, 1976.

62. Barry III, H. and Krimmer, E. C., Similarities and differences in discriminative stimulus effects of chlordiazepoxide, pentobarbital, ethanol, and other sedatives, in *Stimulus Properties of Drugs: Ten Years of Progress*, Colpaert, F. C. and Rosecrans, J. A., Eds., Elsevier, Amsterdam, 1978, 31.
63. Herling, S., Valentino, R. J., and Winger, G. D., Discriminative stimulus effects of pentobarbital in pigeons, *Psychopharmacology*, 71, 21, 1980.
64. Krimmer, E. C., Barry H., III, and Alvin, J. D., Discriminative, disinhibitory, and depressant effects of several anticonvulsants, *Psychopharmacology*, 78, 28, 1982.
65. Kline, F. S. and Young, A. M., Differential modification of pentobarbital stimulus control by d-amphetamine and ethanol, *Pharmacol. Biochem. Behav.*, 24, 1305, 1986.
66. Massey, B. W. and Woolverton, W. L., Discriminative stimulus effects of combinations of pentobarbital and ethanol in rhesus monkeys, *Drug Alcohol Dep.*, 35, 37, 1994.
67. Balster, R. L., Grech, D. M., and Bobelis, D. J., Drug discrimination analysis of ethanol as an N-methyl-D-aspartate receptor antagonist, *Eur. J. Pharmacol.*, 222, 39, 1992.
68. Butelman, E. R., Baron, S. P., and Woods, J.H., Ethanol effects in pigeons trained to discriminate MK-801, PCP or CGS-17955, *Behav. Pharmacol.*, 4, 57, 1993.
69. Schechter, M. D., Use of TFMPP stimulus properties as a model 5-HT$_{1B}$ receptor activation, *Pharmacol. Biochem. Behav.*, 31, 53, 1988.
70. Mardones, J. and Segovia-Requelme, N., Thirty-two years of selection of rats by ethanol preference: UChA and UChB strains, *Neurobehav. Toxicol. Teratol.*, 5, 171, 1983.
71. Kiianmaa, K., Stenius, K. and Sinclair, J. D., Determinants of alcohol preference in the AA and ANA rat lines selected for differential ethanol intake, in *Advances in Biomedical Alcohol Research*, Kalant, H., Khanna, J. M. and Israel, Y., Eds:, Pergamon Press, Oxford, 1991, 151.
72. Li, T. -K., Lumeng, L., McBride, W. J., and Murphy, J. M., Rodent lines selected for factors affecting alcohol consumption, in *Advances in Biomedical Alcohol Research*, Lindros, K. O., Ylikahri, R. and Kiianmaa, K., Eds., Pergamon Press, Oxford, 1987, 91.
73. Li, T. -K., Lumeng, L., Doolittle, P. L., McBride, W. J., Murphy, J. M., Froehlich, J. C., and Morzorati, S., Behavioral and neurochemical associations of alcohol-seeking behavior, in *Biomedical and Social Aspects of Alcohol and Alcoholism*, Kuriyama, K., Takada, A. and Ishii, H., Eds., Excerpta Medica, Amsterdam, 1988, 435.
74. Gessa, G. L., Colombo, G., and Fadda, F., Rat lines genetically selected for differences in voluntary ethanol consumption, in *Current Practices and Future Developments in the Pharmacotherapy of Mental Disorders*, Meltzer, H. Y. and Nerozzi, D., Eds., Elsevier, Amsterdam, 1991, 193.
75. Sinclair, J. D., Lê, A. D., and Kiianmaa, K., The AA and ANA rat lines, selected for differences in voluntary alcohol consumption, *Experientia*, 45, 798, 1989.
76. Waller, M. B., McBride, W. J., Lumeng, L., and Li, T. -K. Induction of dependence on ethanol by free-choice drinking in alcohol-preferring rats, *Pharmacol. Biochem. Behav.*, 16, 501, 1982.
77. Aalto, J., Circadian drinking rhythms and blood alcohol levels in two rat lines developed for their alcohol consumption, *Alcohol*, 3, 73, 1986.
78. Agabio, R., Cortis, G., Fadda, F., Gessa, G. L., Lobina, C., Reali, R., and Colombo, G., Circadian drinking patterns of Sardinian alcohol preferring rats, *Alcohol Alcohol.*, in press.
79. Waller, M. B., McBride, W. J., Lumeng, L., and Li, T. -K., Initial sensitivity and acute tolerance to ethanol in the P and NP lines of rats, *Pharmacol. Biochem. Behav.*, 19, 683, 1983.
80. Gatto, G. J., Murphy, J. M., Waller, M. B., McBride, W. J., Lumeng, L., and Li, T. -K., Persistence of tolerance to a single dose of ethanol in the selectively-bred alcohol-preferring P rat, *Pharmacol. Biochem. Behav.*, 28, 105, 1987.
81. Murphy, J. M., Gatto, G. J., McBride, W. J., Lumeng, L., and Li, T. -K., Persistence of tolerance in the P line of alcohol-preferring rats does not require performance while intoxicated, *Alcohol*, 7, 367, 1990.
82. Lê, A. D. and Kiianmaa, K., Characteristics of ethanol tolerance in alcohol drinking (AA) and alcohol avoiding (ANA) rats, *Psychopharmacology*, 94, 479, 1988.
83. Waller, M. B., Murphy, J. M., McBride, W. J., Lumeng, L., and Li, T. -K., Effect of low dose ethanol on spontaneous motor activity in alcohol-preferring and -nonpreferring lines of rats, *Pharmacol. Biochem. Behav.*, 24, 617, 1986.
84. Päivärinta, P. and Korpi, E. R., Voluntary ethanol drinking increases locomotor activity in alcohol-preferring AA rats, *Pharmacol. Biochem. Behav.*, 44, 127, 1993.
85. Krimmer, E. C. and Schechter, M. D., HAD and LAD rats respond differently to stimulating effect but not discriminative effects of ethanol, *Alcohol*, 9, 71, 1991.
86. Stewart, R. B., Gatto, G. J., Lumeng, L., Li, T. -K., and Murphy, J. M., Comparison of alcohol-preferring (P) and nonpreferring (NP) rats on tests of anxiety and for the anxiolytic effects of ethanol, *Alcohol*, 10, 1, 1993.
87. Colombo, G., Agabio, R., Lobina, C., Reali, R., Zocchi, A., Fadda, F., and Gessa, G. L., Sardinian alcohol-preferring rats: A genetic animal model of anxiety, *Physiol. Behav.*, 57, 1181, 1995.

88. Tuominen, K., Hilakivi, L. A., Päivärinta, P., and Korpi, E. R., Behavior of alcohol-preferring AA and alcohol-avoiding ANA rat lines in tests of anxiety and aggression, *Alcohol*, 7, 349, 1990.
89. de Beun, R., Schneider, R., Lohmann, A., Kuhl, E., and De Vry, J. Relationship between emotional behavior and subsequent alcohol preference in alcohol preferring AA rats. *Alcohol Alcohol.*, 30, 542, 1995.
90. Murphy, J. M., McBride, W. J., Lumeng, L., and Li, T. -K., Regional brain levels of monoamines in alcohol-preferring and -nonpreferring lines of rats, *Pharmacol. Biochem. Behav.*, 16, 145, 1982.
91. Murphy, J. M., McBride, W. J., Lumeng, L., and Li, T. -K., Contents of monoamines in forebrain regions of alcohol-preferring (P) and -nonpreferring (NP) lines of rats, *Pharmacol. Biochem. Behav.*, 26, 389, 1987.
92. Gongwer, M. A., Murphy, J. M., McBride, W. J., Lumeng, L., and Li, T. -K., Regional brain content of serotonin, dopamine and their metabolites in the selectively bred high- and low-alcohol drinking lines of rats, *Alcohol*, 6, 317, 1989.
93. Samson, H. H., The function of brain dopamine in ethanol reinforcement, in *Alcohol and Neurobiology: Receptors, Membranes and Channels*, Watson, R. R., Ed., CRC Press, Boca Raton, 1992, 91.
94. Zhou, F. C., Zhang, J. K., Lumeng, L., and Li, T. -K., Mesolimbic dopamine system in alcohol-preferring rats, *Alcohol*, 12, 403, 1995.
95. Fadda, F., Mosca, E., Colombo, G., and Gessa, G. L., Effects of spontaneous ingestion of ethanol on brain dopamine metabolism, *Life Sci.*, 44, 281, 1989.
96. Fadda, F., Mosca, E., Colombo, G., and Gessa, G. L., Alcohol-preferring rats: Genetic sensitivity to alcohol-induced stimulation of dopamine metabolism, *Physiol. Behav.*, 47, 727, 1990.
97. Weiss, F., Lorang, M. T., Bloom, F. E., and Koob, G. F., Oral alcohol self-administration stimulates dopamine release in the rat nucleus accumbens: Genetic and motivational determinants, *J. Pharmacol Exp. Ther.*, 267, 250, 1993.
98. Korpi, E. R., Sinclair, J. D., and Malminen, O., Dopamine D2 receptor binding in striatal membranes of rat lines selected for differences in alcohol-related behaviors, *Pharmacol. Toxicol.*, 61, 94, 1987.
99. Syvälahti, E. K. G., Pohjalainen, T., Korpi, E. R., Pälvimäki, E. -P., Ovaska, T., Kuoppamäki, M., and Hietala, J., Dopamine D2 receptor gene expression in rat lines selected for differences in voluntary alcohol consumption, *Alc. Clin. Exp. Res.*, 18, 1029, 1994.
100. McBride, W. J., Chernet, E., Dyr, W., Lumeng, L., and Li, T. -K., Densities of dopamine D2 receptors are reduced in CNS regions of alcohol-preferring P rats, *Alcohol*, 10, 387, 1993.
101. Stefanini, E., Frau, M., Garau, M. G., Garau, B., Fadda, F., and Gessa, G. L., Alcohol-preferring rats have fewer dopamine D2 receptors in the limbic system, *Alcohol Alcohol.*, 27, 127, 1992.
102. Sellers, E. M., Higgins, G. A., and Sobell, M. B., 5-HT and alcohol abuse, *Trends Pharmacol. Sci.*, 13, 69, 1992.
103. Penn, P. E., McBride, W. J., Lumeng, L., Gaff, T. M., and Li, T. -K., Neurochemical and operant behavioral studies of a strain of alcohol-preferring rats, *Pharmacol. Biochem. Behav.*, 8, 475, 1978.
104. Zhou, F. C., Bledsoe, S., Lumeng, L., and Li, T. -K., Immunostained serotonergic fibers are decreased in selected brain regions of alcohol-preferring rats, *Alcohol*, 8, 425, 1991.
105. Zhou, F. C., Bledsoe, S., Lumeng, L., and Li, T. -K., Reduced serotonergic immunoreactive fibers in the forebrain of alcohol-preferring rats, *Alc. Clin. Exp. Res.*, 18, 571, 1994.
106. Zhou, F. C., Pu, C. F., Murphy, J., Lumeng, L., and Li, T. -K., Serotonergic neurons in the alcohol preferring rats, *Alcohol*, 11, 397, 1994.
107. Wong, D. T., Reid, L. R., Li, T.-K., and Lumeng, L., Greater abundance of serotonin$_{1A}$ receptor in some brain areas of alcohol-preferring (P) rats compared to nonpreferring (NP) rats, *Pharmacol. Biochem. Behav.*, 46, 173, 1993.
108. McBride, W. J., Guan, X.-M., Chernet, E., Lumeng, L., and Li, T.-K., Regional serotonin$_{1A}$ receptors in the CNS of alcohol-preferring and -nonpreferring rats, *Pharmacol. Biochem. Behav.*, 49, 7, 1994.
109. McBride, W. J., Chernet, E., Lumeng, L., and Li, T.-K., Densities of serotonin-1B receptors in the CNS of P and NP rats, *Alc. Clin. Exp. Res.*, 15, 315, 1991.
110. McBride, W. J., Chernet, E., Rabold, J. A., Lumeng, L., and Li, T. -K., Serotonin-2 receptors in the CNS of alcohol-preferring and -nonpreferring rats, *Pharmacol. Biochem. Behav.*, 46, 631, 1993.
111. Korpi, E. R., Sinclair, J. D., Kaheinen, P., Viitamaa, T., Hellevuo, K., and Kiianmaa, K., Brain regional and adrenal monoamine concentrations and behavioral responses to stress in alcohol-preferring AA and alcohol-avoiding ANA rats, *Alcohol*, 5, 417, 1988.
112. Korpi, E. R., Päivärinta, P., Abi-Dargham, A., Honkanen, A., Laruelle, M., Tuominen, K., and Hilakivi, L. A., Binding of serotonergic ligands to brain membranes of alcohol-preferring AA and alcohol-avoiding ANA rats, *Alcohol*, 9, 369, 1992.
113. Deitrich, R. A. and McClearn, G. E., Neurobiological and genetic aspects of the etiology of alcoholism, *Fed. Proc.*, 40, 2051, 1981.
114. Schuckit, M. A., Li, T. -K., Cloninger, C. R., and Dietrich, R. A., Genetics of alcoholism, *Alc. Clin. Exp. Res.*, 9, 475, 1985.
115. Ritz, M. C., George, F. R., De Friebre, C. M., and Meisch, R. A., Genetic differences in the establishment of ethanol as a reinforcer, *Pharmacol. Biochem. Behav.*, 24, 1089, 1986.

116. Hyytiä, P. and Sinclair, J. D., Demonstration of lever pressing for oral ethanol by rats with no prior training or ethanol experience, *Alcohol*, 6, 161, 1989.
117. Murphy, J. M., Gatto, G. J., McBride, W. J., Lumeng, L., and Li, T. -K., Operant responding for oral ethanol in the alcohol-preferring P and alcohol-nonpreferring NP lines rats, *Alcohol*, 6, 127, 1989.
118. Hyytiä, P. and Sinclair, J. D., Differential reinforcement and diurnal rhythms of lever pressing for ethanol in AA and Wistar rats, *Alc. Clin. Exp. Res.*, 14, 375, 1990.
119. Schwarz-Stevens, K., Samson, H. H., Tolliver, G. A., Lumeng, L., and Li, T.-K., The effects of ethanol initiation procedures on ethanol reinforced behavior in the alcohol-preferring rat, *Alc. Clin. Exp. Res.*, 15, 277, 1991.
120. Dyr, W., Cox, R., McBride, W. J., Lumeng, L., Li, T.-K., and Murphy, J. M., Operant responding for ethanol in high HAD and low LAD alcohol-drinking rats, *Alc. Clin..Exp. Res.*, 17, 479, 1993.
121. Files, F. J., Andrews, C. M., Lewis, R. S., and Samson, H. H., Effects of manipulating food availability on ethanol self-administration by P rats in a continuous access situation, *Alc. Clin. Exp. Res.*, 17, 586, 1993.
122. Waller, M. B., McBride, W. J., Gatto, G. J., Lumeng, L., and Li, T. -K., Intragastric self-infusion of ethanol by ethanol-preferring and -nonpreferring lines of rats, *Science*, 225, 78, 1984.
123. Murphy, J. M., Waller, M. B., Gatto, G. J., McBride, W. J., Lumeng, L., and Li, T.-K., Effects of fluoxetine on the intragastric self-administration of ethanol in the alcohol preferring P line of rats, *Alcohol*, 5, 283, 1988.
124. McBride, W. J., Murphy, J. M., Gatto, G. J., Levy, A. D., Yoshimoto, K., Lumeng, L., and Li, T -K., CNS mechanisms of alcohol self-administration, in *Advances in Biomedical Alcohol Research*, Taberner, P. V. and Badawy, A. A., Eds., Pergamon Press, Oxford, 1993, 463.
125. York, J. L., The ethanol stimulus in rats with differing ethanol preferences, *Psychopharmacology*, 74, 339, 1981.
126. Overton, D. A., Leonard, W. R., and Merkle, D. A., Methods for measuring the strength of discriminable drug effects, *Neurosci. Biobehav. Rev.*, 10, 251, 1986.
127. Schechter, M. D. and Meehan, S. M., Ethanol discrimination in Fawn-Hooded rats is compromised when compared to other strains, *Alcohol*, 10, 77, 1993.
128. Dumbrill-Ross, A. and Tang, S. W., Absence of high affinity [^{3}H]-imipramine binding in platelets and cerebral cortex of Fawn-Hooded rats, *Eur. J. Pharmacol.*, 72, 137, 1981.
129. Arora, R. C., Tong, C., Jackson, H., Stoff, D., and Meltzer, Y. H., Serotonin uptake in imipramine binding in blood platelets and brain of Fawn-Hooded and Sprague-Dowley rats, *Life Sci.*, 33, 437, 1983.
130. Rezvani, A. H., Overstreet, D. H., and Janowsky, D. S., Genetic serotonin deficiency and alcohol preference in the fawn hooded rat, *Alcohol Alcohol.*, 25, 573, 1990.
131. Overstreet, D. H., Rezvani, A. H., and Janowsky, D. S., Genetic animal models of sepression and ethanol preference provide support for cholinergic and serotonergic involvement in depression and alcoholism, *Biol. Psychiat.*, 31, 919, 1992.
132. Khanna, J. M., Kalant, H., Shah, G., and Sharma, H., Comparison of sensitivity and alcohol consumption in four strains of rats, *Alcohol*, 7, 429, 1990.
133. Gordon, T. L., Meehan, S. M., and Schechter, M. D., P and NP rats respond differently to the discriminative stimulus effects of nicotine, *Pharmacol. Biochem. Behav.*, 45, 305, 1993.
134. Meehan, S. M., Gordon, T. L., and Schechter, M. D., MDMA (Ecstasy) substitutes for ethanol discriminative cue in HAD but not LAD rats, *Alcohol*, 12, 569, 1995.
135. Bisaga, A. and Kostowski, W., Selective breeding of rats differing in voluntary ethanol consumption, *Pol. J. Pharmacol.*, 45, 431, 1993.

27 Ethical Issues in Genetic Screening and Testing, Gene Therapy, and Scientific Conduct

Lisa S. Parker and Elizabeth Gettig

CONTENTS

A. Introduction 469
B. Preventive Ethics 471
    1. Preventive Ethics in Genetic Research 472
C. Gene Therapy 474
D. Screening for Genetic Diseases 475
E. Transplants 476
F. Psychiatric Disorders 476
G. Conclusion 477
References 477

A. INTRODUCTION

    Bioethics as an interdisciplinary field involving clinicians, lawyers, philosophers, theologians, and other humanists was born in the early 1970s amid technological advances in medicine and growing respect for persons in society. The era was marked by the end of the Tuskegee syphilis study, the first wide-spread use of hemodialysis and mechanical ventilation, abortion reform, and the first human heart transplant. Technological capabilities clashed with individuals' values. In short, bioethics was born of conflict.[26]

    Respect for individuals' rights of self-determination came into conflict with some social values and with the medical profession's previously largely unchallenged paternalistic concern for patient welfare, as the medical profession and individual professionals — not patients — defined that well-being. In 1970, for example, Paul Ramsey published his patient-centered medical ethics treatise, *The Patient as Person*.[31] The field of bioethics emerged in the wake of landmark legal cases, such as Karen Quinlan's parents' bid to remove her from her respirator[30] or the paralyzed Mr. Canterbury's suit claiming that he had not been fully informed of the risks of his surgery.[7] Bioethics evolved to provide a legal and ethical framework within which to resolve conflicts between physician and patient as well as between social consensus and individual values. The values of the individual patient came to trump the traditional values of medicine, and the privacy both of individuals and of the physician–patient relationship erected a boundary against the intrusion of society's interests.

Historically, the physician–patient relationship has been the primary focus of bioethics, but it is clear that the crisis of funding health care is emerging as the fundamental challenge of the 1990s.[9,21,35] Social policies and institutional contexts are now considered in association with or occasionally instead of the physician or health care provider and patient relationship.[20]

Developments in theoretical ethics, specifically feminist ethics, support this new focus. Feminist philosophers suggest that in order to provide just answers to ethical questions, ethics must pay increased attention to their social context and political dynamics, the balance of power, and the history of oppression.[14,15,33]

So it was in this intellectual and social climate that the Human Genome Project (HGP) was initiated in 1990 to support and to coordinate efforts of the National Institutes of Health (NIH) and the Department of Energy (DOE) to produce a complete genetic map of all human genes. The Ethical, Legal, and Social Implications (ELSI) Program of the HGP was charged with anticipating the social consequences of the acquisition of this knowledge and developing policies to guide its use. With 5% of the genome budget supporting ELSI activities, the ELSI Program is both the first federally supported extramural research initiative in ethical issues and the largest source of public funds for bioethics.[44]

However the bioethical issues of the HGP are not unique. The topics of informed consent, justice, gender justice, privacy, confidentiality, discrimination, genetic discrimination, health care needs, or private health insurance vs. a national health service are familiar ones. Even those challenges to the premises of a private health insurance system presented by genetic screening are, for example, also presented by other predictive medical tests, e.g., cholesterol screening for hypertension.

Advances in genetics may, however, place ethical concerns on a grander scale because everyone is at an increased risk for developing some disease. New genetic technologies may cause ethical concerns to arise at a different stage of life or of decision making (e.g., prior to conception or at a presymptomatic stage of a disease).

If the conflict between paternalism and autonomy is seen to have been played out in the context of the doctor–patient relationship since the 1970s, the genetic counselor–consultand relationship of the 1980s and 1990s seems to reflect the resolution of that conflict. Prior to the 1970s, a priestly model accurately described the typical, paternalistic doctor–patient relationship. According to this model, decision making is taken from the patient and placed in the hands of the expert professional who is charged with benefiting the patient; in the extreme, the physician's "moral authority so dominates the patient that the patient's freedom and dignity are extinguished".[39] In contrast, the physician–patient relationship model which is currently advocated is a contractual model: "The basic norms of freedom, dignity, truth-telling, promise-keeping, and justice are essential to a contractual relationship. The promise is trust and confidence even though it is recognized that there is not a full mutuality of interests... With the contractual relationship there is a sharing in which the physician recognizes that the patient must maintain freedom of control over his own life and destiny when significant choices are to be made".[39]

The relationship between professional genetic counselors and their consultands reflects this shared decision making process which guarantees to consultands the authority to make choices reflecting their own values. The Code of Ethics of the National Society of Genetic Counselors (NSGC) states that genetic counselors strive to: "respect their clients' beliefs, cultural traditions, inclinations, circumstances, and feelings ... and refer clients to other competent professionals when they are unable to support the clients".[25] Thus, the consultand-centered, autonomy-oriented conception of the genetic counseling relationship reflects the outcome of at least two decades of bioethical discussions of patients' rights, of the therapeutic advantage of involving patients in their own care, and of value pluralism.

The nonpaternalistic, nondirective process of genetic counseling also embodies aspects of the doctrine of informed consent, the most prominent bioethical and legal doctrine to emerge in the early years of bioethics. Informed consent is the process whereby competent patients or research subjects are informed of the risks and benefits of proposed therapeutic or research protocols (i.e., "disclosure"), are asked to weigh these risks and benefits in light of their own values and desires, and are asked to give their informed, voluntary consent to undertake the therapy or participate in the research.[2] Health care professionals and researchers are obligated to disclose the information in a manner so that a reasonable layperson can understand it and to answer the specific questions which the individual client or research subject may raise. In so far as the professional or researcher becomes aware of a particular client's or subject's desire to have additional information disclosed, the professional or researcher incurs an obligation to make reasonable attempts to satisfy that desire (i.e., "dialogue"). The doctrine of informed consent has two justifications — first and most fundamentally, respect for persons and their autonomy, and second, protection of individuals' welfare by requiring their consent as a prerequisite to incur the risks of research or treatment.[5]

The fundamental role of genetic counselors is to provide information to enable consultands to make free and informed reproductive and health care decisions. The NSGC Code of Ethics states that counselors "strive to enable their clients to make informed independent decisions, free of coercion, by providing or illuminating the necessary facts and clarifying alternatives and anticipated consequences".[25] Supplying information in an understandable manner, answering consultands' questions, helping consultands develop the understanding necessary to make their own decisions, and supporting those choices are the primary tasks of genetic counselors. Whereas these tasks, which comprise the disclosure and dialogue stages of the process of informed consent, are just one facet of other health care providers' jobs, they constitute, in broad outline, the primary tasks of genetic counselors.

Thus, in an important sense, the first two decades of bioethics not only provided background for current ethical consideration of issues arising from genetic research and the management of genetic disease, but actually laid the foundation for the process of modern genetic counseling. As the HGP progresses and genetic services become a more integral part of health care, ethical analysis of issues concerning these rapid advances in genetic technology and knowledge will continue to reflect this individual-oriented bioethical tradition.

Growing attention to more socially-oriented bioethical concerns, such as allocation of health care resources, however, also coincides with and will be influenced by advances in genetics. Allocation issues, for example, are no longer primarily questions of micro-allocation or triage, but instead focus on macro-allocation concerns, such as how to provide a decent minimum of health care to all of society's members, what constitutes a decent minimum, and whether certain types of health care should be available at all ("should organ transplantation research or gene therapy trials receive funding?").

B. PREVENTIVE ETHICS

Although bioethics was born in an atmosphere of conflict and was initially concerned with the resolution of ethical conflicts, it is gradually beginning to address the social and institutional factors which may create or exacerbate ethical problems. In this way bioethics may be said to parallel preventive medicine.[10,27] The practice of "preventive ethics," including its anticipatory stance and its attention to social and institutional factors, mirrors the practice of preventive medicine.[12]

Waiting until a conflict arises makes resolving ethical quandaries more difficult, because by then medical and institutional factors may limit options or opposing parties may have become deeply entrenched and personally identified with their (conflicting) positions. The inadvertent disclosure of confidential information to family members about their risk for having or transmitting a genetic condition can result in long term dysfunction in family dynamics which might be avoided by establishing and observing preventive ethics policies for information management. Therefore, even successfully resolved crises incur high human costs in terms of time and emotion expended in their resolution.

Furthermore, the crisis–resolution approach measures success in terms of whether a settlement to the particular crisis can be found and thus too readily accepts patterns of recurring ethical problems. In its early years, bioethics neglected the underlying causes of ethical conflicts, such as routine aspects of health care or social and institutional structures which have exacerbated or even directly caused ethical conflicts in the provision of health care.[4] The traditional approach defines the scope of bioethics in terms of discrete problems; thus, it necessarily fails to attend adequately to the ethical aspects of health care in which no specific "problem" has been identified. Outside of genetic counseling, for example, even though the process of informed consent is important in defining the ethical character of every provider–patient interaction, the disclosure and dialogue inherent in informed consent are often ignored until the physician and patient disagree about the proper treatment.

In contrast, a preventive approach to bioethical issues can help overcome these limitations of early crisis-oriented bioethics by placing a greater emphasis on preventing the development of ethical conflicts.[28] A preventive approach seeks to detect potential ethical conflicts at stages where "symptoms" of the conflicts are not yet present or are relatively mild. By identifying the predictable patterns of "pathophysiology"[3] and "ethical risk factors" shared by common ethical problems (e.g., institutional structures or differences in cultural or religious views), preventive ethics facilitates the development of mechanisms to avert serious conflicts or to reach ethically defensible plans more readily, thereby minimizing unnecessary personal anguish and social conflict.

Because preventive ethics correctly recognizes that the absence of ethical conflict is an inadequate measure of the ethical provision of health care or conduct of research, preventive ethics not only seeks to avoid conflicts, but also strives to create and preserve relationships of trust and understanding between health care providers and recipients and between researchers and the public.[12] It seeks to maximize opportunities for the exercise of autonomy and the provision of quality patient or consultand-centered care.

According to a preventive bioethics approach, alternative social policies should be judged not merely according to whether they will prevent open ethical conflicts, but also according to their capacity to promote ethical health care and the opportunity for society's members to pursue life plans reflecting their own values. In expanding the focus of bioethics from decisions in problematic cases to a general concern with both the routine aspects of health care and the social and institutional factors which affect health care, preventive ethics more fully integrates ethical considerations into health care and research.

By identifying recurrent problems, a preventive ethics approach enables researchers and health care providers to develop ethics "protocols" for addressing them, particularly in advance of individual occurrences. Thus, the preventive approach (1) may avoid some individual hardship (or at least permit individuals to anticipate and prepare for future hardships); (2) may, by identifying the problems, invite their innovative solution; and (3) may prompt changes in existing structures and policies, if they are themselves contributing to the problems.

1. PREVENTIVE ETHICS IN GENETIC RESEARCH

In genetic research, practicing preventive ethics in the presymptomatic testing of individuals at risk for Huntington's disease (HD) has lead to a tentative code of conduct for genetic researchers.[17] The code evolved from research on samples from families with genetic disease and from the development of new molecular tests. The proposed code of conduct intends to protect both the subject and researchers. Harper[17] admits that most problems encountered in genetic testing are a result of not paying adequate attention to the ethics of gene testing and therapy. HD protocols have been examined by review committees often, unfortunately, with more attention given to the risks of the sampling procedure (dangers and discomfort of venipuncture) rather than the social, psychological, and economic consequences of the test results, i.e., the detection of a genetic defect.

The proposed code also addresses the conflicts of interest between the patient's needs and the physician's or researcher's interests. Financial ties with industry, through research, personal investment in commercial ventures, or consulting fees, appear to be greater in genetics than in other fields of medicine due to the technology-driven nature of genetic research. Norman Fost[13] has written, "sometimes it is difficult to distinguish a conflict of interest from a congruence of interest. The scientist's desire of fame and fortune may drive him or her to the extra effort that results in a discovery that benefits others. The physician's desire for income may stimulate him or her to work long hours and provide beneficial services to others. But there is also evidence that self-interest can adversely affect clinical judgement, whether it be for suggesting elective surgery or for ordering expensive diagnostic tests".[13]

Disclosure statements have become commonplace to minimize the possible effects of conflicts of interest, and some groups, notably a multicenter clinical trial of treatment after coronary-artery bypass-graft surgery, have moved toward prohibiting ties with industry when such ties are not necessary for the practice of medicine or the advancement of science.[19]

The code of conduct proposed by Harper[17] also points to some of the difficulties that will be faced as genetic technologies developed in the research context are applied in the clinical diagnostic or therapeutic context. The code states:

1. Family members "at risk" for a genetic disorder should not be sampled unless strictly necessary for the research, especially in late onset or variable disorders. This statement applies particularly to children. Proposals should clearly justify the testing of unaffected subjects and should include a clear plan stating what will be done in the event a genotypic abnormality is detected.
2. When consent is given for sampling by an unaffected person, to assist a family member in determining his/her risk status, it should be made clear that the risk status of the unaffected person will not be disclosed and the result of the test should not be expected nor will be sent to his/her doctor nor placed in his/her medical record unless specifically requested.
3. If the sample is to be stored and used for future tests, new consent should be obtained if the implications for the person at risk resulting from the new research are likely to be considerably different — for example, if direct mutations analysis, rather than a general linkage analysis, is possible.
4. If the possibility of identifying defects in people at risk is foreseeable or inevitable, then such samples should be coded or made anonymous for the purpose of these tests unless the person concerned has specifically requested that relevant information should be disclosed and has received information that allows him\her to fully understand the implications of such disclosure.
5. If a person at risk, who gave a research sample later requests presymptomatic testing or other genetic services, a new sample should be taken and the request handled in the same way as it would be for any other person electing presymptomatic testing.
6. When a test may show a specific genetic defect in people affected by a disorder not previously known to be genetic, the possible genetic implications (as well as psychosocial implications) should be made clear and new consent obtained if samples previously obtained are being restudied.
7. Ethics committees should pay at least as much attention to the consequences of a sample being taken as to the risks attached to the sampling procedure.

The presymptomatic HD testing programs have attempted to create and preserve trust and understanding between researchers and test providers. Presymptomatic testing is a multistep process involving numerous visits to testing centers. The HD protocols prescribe review of the subject's family history, neurologic examination, psychiatric examination, review of medical charts of extended family members for confirmation of diagnostic information, psychological testing, pretest counseling, and disclosure of results. Follow-up both clinically and for research purposes is a standard feature of presymptomatic testing protocols.[11]

The HD model sometimes limited the subject's right to privacy because of the need for extensive review of family medical data and the need for samples for linkage analysis (prior to the recent discovery of the HD gene). The protocol was born from the traditional pre-1970s model of the physician-patient relationship. It is therefore criticized on paternalistic grounds. The protocols were neither publicly reviewed nor discussed. As individuals have "graduated" from the testing program, the protocols are being revisited. Suggestions and recommendations from participants are being sought in order to evaluate and possibly to modify the protocols. Moreover, the recent discovery of the gene responsible for HD has pushed the scientific community to re-evaluate the protocols because extended family review is no longer necessary.

The HD model represents the first testing program which enables a person to choose to know with a high degree of certainty that he or she will die of a fatal, inherited, and presently untreatable disease. The psychiatric and social consequences of having such knowledge were anticipated and prompted the rigid protocol structure to preserve the most basic of ethical tenets — to do no harm. Experience with the HD protocols has shown that explaining genetic risks is a complex subject and that understanding comes slowly.[24]

The counseling steps of the HD protocols may be included in future genetic testing models. Testing without giving information, counseling, and support must be agreed upon in order to be unacceptable. Concern about stigmatization and discrimination in employment, insurance, and personal relationships should provoke society to monitor and regulate the availability and use of genetic testing to assure that abuse or coercion does not occur.[18] A preventive ethics approach allows for better planning and more open discussion of these ethical concerns.

C. GENE THERAPY

The creation of the NIH's Recombinant DNA Advisory Committee (RAC) represents an attempt to anticipate and address ethical concerns pertaining to gene therapy. The RAC was responsible for what some consider one of the most important milestones in the history of medicine, the approval of a human gene transfer study and human gene therapy protocols.[40] The gene therapy protocols currently involve only somatic cell gene therapy. Somatic cell gene therapy refers to the insertion of new DNA into a particular tissue (such as bone marrow) of an affected individual. The reproductive system is not targeted, so the new DNA material serves the individual only and is not transmitted to progeny.

A preventive ethics approach is evident in the public review process of RAC. By serving to inform the public of perhaps the most controversial advances in genetics and permitting public comment on the use of gene therapy technology, the RAC provides a mechanism to minimize public concern and social conflict. The guidelines of the RAC evolved over a decade.

The RAC is not, however, without its critics. The RAC is a committee of the NIH, which, in turn, is the primary funding agent for biomedical research. The initial protocols were submitted by NIH scientists. The RAC has acknowledged the conflict in simultaneously promoting and regulating a single field of research.

In addition, because RAC review provides a safeguard against employment of potentially high risk gene therapy in the absence of safety and efficacy data, the 1992 decision to exempt one therapeutic protocol on a compassionate plea basis raises concern.[19,36,37] By responding to the crisis of the moment and not fully addressing the precedent-setting ramifications of its departure from its peer review protocol, the NIH's departure from its preventive ethics stance invited criticism concerning the susceptibility of the NIH's peer review process to political pressure and constituted a potentially serious breach of public trust.[8,23]

Still, the RAC again embraced the notion of preventive ethics by introducing for public debate the concept of germ line gene therapy. Germ line gene therapy means that the new DNA introduced into an individual may be passed to future generations. In 1990 Francis Collins, now director of the Human Genome Project, published the following statement, "germ line gene therapy ... is an approach that carries such risks of unknown damage to future generations that virtually all geneticists and lay organizations have concluded that it would not be appropriate to attempt it in humans".[16]

The very next year LeRoy Walters stated, "the time has now come to begin a formal public process for the ethical assessment of germ line genetic intervention".[40] Organizations such as the Council for Responsible Genetics oppose the use of germ line gene modification in humans.[29] RAC chairman Nelson Wivel, writing as a citizen and not in his official capacity, and Walters state, "it would, in our view, be a useful investment of time and energy to continue and in fact intensify the public discussion of germ line gene modification for disease prevention, even though the application of this new technology to humans is not likely to be proposed in the near future".[43] The debate continues regarding gene therapy and its application to human subjects.

D. SCREENING FOR GENETIC DISEASES

A national policy has yet to be developed governing population based screening of genetic diseases. The debate over population based screening for the gene for Cystic Fibrosis (CF) has begun. Earlier screening programs, particularly screening for the gene for Sickle Cell Anemia in the African American population, failed to clearly establish program goals, failed to distinguish promotion of patient autonomy from the motivation of the public health community, (i.e., distinguishing reproductive choices from the public health concern to decrease the incidence of the disease or the gene in the population), and led to discrimination by employers, including the military, as well as the loss of insurance. The development of the screening program for gene for Tay-Sachs disease in the Jewish community benefited from the sickle cell experience and has resulted in successful population based screening with high community acceptance and minimal adverse effects.[22]

Before population based screening for genetic conditions occurs, programs should consider five points.[41] First, screening programs should clearly state their purpose and goal. Second, peer-reviewed pilot studies are necessary to demonstrate that the stated goals of the program can be achieved at a reasonable cost and with few adverse effects. This proposed safeguard may come under pressure because the potential of screening for literally hundreds of genetic traits or susceptibilities creates pressure to begin screening prior to either adequate review of pilot studies or public debate. Third, the target population must be educated about the disease or condition in question and receive counseling about the risks and benefits of screening. Fourth, the traditional standards of informed consent must be observed. Screening should remain voluntary. Individuals must be able to exercise their "right not to know". In addition, although Fost does not specifically address this concern, universal access to testing must be assured by public health agencies if genetic technologies are not to exacerbate existing social inequalities. Fifth, confidentiality of the individual must be maintained.

A national policy for genetic screening should have procedural mechanisms in place at both the State and Federal levels to prevent harm to the individual being screened.[42] Discrimination as a consequence of genetic testing has been documented.[6] The health and life insurance industries use genetic test results to deny coverage, and the presence of pre-existing condition clauses in many policies have led to "job lock" for families or the loss of coverage for either the individual with a genetic disease or the carrier of a gene for a disease. Stigmatization in the form of loss of services or entitlement has also been reported.

E. TRANSPLANTS

The use of fetal tissue transplants for Parkinson's disease (PD) has prompted considerable debate by the public and within scientific communities. In 1987 the NIH submitted a request to the Assistant Secretary of Health seeking approval for fetal tissue transplantation. In May of 1988, however, a moratorium on federal funding was declared on fetal tissue transplant research. Although at least one center voluntarily discontinued its research in response to the moratorium, two centers, at Yale and at the University of Colorado, elected to use private funds to continue research efforts with fetal tissue transplants.[1] The central objection to the use of fetal tissue is political rather than scientific. Because the tissue is obtained from aborted fetuses, it is the source of the tissue rather than its use that creates conflict.

Political responses and both public and scientific debate about fetal tissue transplants for PD patients clearly illustrates the shift from the physician and patient-based bioethic to one influenced by social and political interests. Indeed, the 1993 lifting of the ban on fetal tissue research by the Clinton administration was both responsive to public debate and guarantees that the debate will continue, while permitting research protocols to be judged on scientific merit rather than political precepts.

F. PSYCHIATRIC DISORDERS

The genes responsible for schizophrenia, bipolar disorders, and Alzheimer's disease (AD) have yet to be clearly elucidated though familial predispositions have been identified. The etiology of schizophrenia and affective disorders is unknown. These conditions are probably heterogeneous resulting from both biologic and environmental components. The cause of AD is also unknown, but may have several genetic etiologies.

Family studies involving the affective disorders have confirmed clear genetic factors including: increased risks for early onset probands vs. late onset probands; an increased risk for unipolar depression in women and a slightly increased risk for bipolar depression in women; an increased risk for women who have first degree relatives with a bipolar disorder for developing bipolar disease while no such association has been noted for unipolar conditions; relatives of bipolar probands have a higher risk of affective disorders (primarily unipolar) than relatives of unipolar probands; affected relatives of unipolar probands usually have unipolar depression; and an increased risk (50 to 75%) is present when both parents have bipolar disorders.[32]

Hereditary risks are also present in schizophrenia. Empiric risks are dependant upon the relationship to the affected individual. Second degree relatives have the lowest risk, about 2 to 3% while an individual with an affected identical twin has a 40 to 60% chance of developing the condition. The individual with one affected parent has about a 10 to 15% risk for schizophrenia.

In the late 1980s, the genetic material responsible for schizophrenia was mapped to chromosome 5 and bipolar disorders to chromosome 11. This work could not be replicated, and the initial findings were discovered to have been reversed in the original samples. The data further polarized the debate of the role of genetics in psychiatric conditions. The arguments of nurture vs. nature resurfaced in the mental health community.

AD fared better with pathophysiological characterization of β-amyloid containing placques and eventual mutation identification of early onset AD on chromosome 21 and another early onset gene on chromosome 14. A late onset gene has been mapped to chromosome 19.

Alzheimer's disease demonstrates that genetic studies can be applied to psychiatric conditions despite the confounding factors of genetic heterogeneity, ascertainment of late onset conditions, and variable ages of onset.

Psychiatric disorders are complex and will probably result in the identification of complicated rather than straightforward modes of inheritance and uncertainty in defining inherited psychiatric conditions.[38]

The preventive ethics model is again well suited to these complex conditions. Psychiatric genetic research poses two specific issues that other genetic conditions do not, namely, the subject's competence to participate in research and determining the legal and ethical acceptability of substituted judgement for subjects not competent to consent.[34]

Legal requirements of competence must be met, and if a research participant is not competent to consent, provisions could be obtained from a legally authorized representative approved by the local internal review board (IRB) in accordance with prevailing state regulations. Harper's guidelines previously reviewed do not address the competency issue. Psychiatric conditions also may involve the circumstance where clinical information is communicated to a third party for the subject's safety. Consent documents might include a section allowing the subject to designate a physician or another individual to receive such information.

The protection of the rights of the individual are not unique to psychiatric disorders or genetic conditions but pose significant issues in the context of genetic research and discovery of the genes responsible for psychiatric disorders.

G. CONCLUSION

The preventive ethics paradigm provides a model for considering clinical and scientific conduct which accommodates more than the factors immediately apparent in a particular circumstance. By anticipating ethical concerns, by seeking comment from the relevant parties, and by examining background social factors and institutional structures, preventive ethics anticipates the effects of policies and practices on people of different social, economic, and educational backgrounds. By adopting an anticipatory rather than a reactive stance, the preventive ethics model encourages the development of policies governing genetic research and the provision of genetic services which build upon the experience of health care providers and researchers in nongenetic contexts. Finally, in seeking to anticipate and minimize ethical conflict and to explore possible ethical solutions to problems before actual conflicts develop, preventive ethics seeks to provide individuals with the opportunity to make use of genetic and other medical technologies in the pursuit of their life plans in accordance with their sets of values.

REFERENCES

1. Annas, G.J., *Standard of Care*, New York: Oxford University Press, 1993, 181-186, 154-159, 164-166.
2. Appelbaum, P.S., Lidz, C.W., and Meisel, A., *Informed Consent: Legal Theory and Clinical Practice*, New York: Oxford University Press, 1987.
3. Appelbaum, P.S. and Roth, L.H., Patients who refuse treatment in medical hospitals, *J. Am. Med. Assoc.*, 1983, 250:1296-1301.
4. Barnard, D., Unsung questions of medical ethics, *Soc. Sci. Med.*, 1985, 21:243-249.
5. Beauchamp, T.L. and Childress, J.F., *Principles of Biomedical Ethics*, 3rd ed. New York: Oxford University Press, 1989.
6. Billings, P. et al., Discrimination as a consequence of genetic testing, *Am. J. Human Genet.*, 1992, 51(4):899-901.
7. Canterbury vs. Spence, 464 F.2d 772, 1972.
8. Emmitt, R.J., Tardy compassion (Letter), *Lancet*, 1993, 341:1157-1158.
9. Epstein, A.N., Changes in the delivery of care under comprehensive health care reform, *N. Engl. J. Med.*, 1993, 329:1672-1676.
10. Fisher, M., Ed., *U.S. Preventive Services Task Force. Guide to Clinical Preventive Services: An Assessment of the Effectiveness of 169 Interventions*, Baltimore: Williams & Wilkins, 1989.
11. Folstein, S., *Huntington Disease*, Baltimore: The Johns Hopkins University Press, 1989, 177-187.

12. Forrow, L., Arnold, R.M., and Parker, L.S., Preventive ethics: expanding the horizons of clinical ethics, *J. Clin. Ethics*, in press.
13. Fost, N., Genetic diagnosis and treatment, *AJDC*, 1993, 146:1190-1195.
14. Friedman, M., Care and context in moral reasoning, in *Women and Moral Theory*, Kittay, E.F. and Meyers, D.T., Eds., New Jersey: Rowman and Littlefield, 1987.
15. Frye, M., *The Politics of Realty: Essays in Feminist Theory*, California: Crossing Press, 1983.
16. Gelehrter, T. and Collins, F., *Principles of Medical Genetics*, Baltimore: Williams and Wilkins, 1990, 289-297.
17. Harper, P., Research samples from families with genetic diseases: a proposed code of conduct, *Br. Med. J.*, 1993, 306:1391-1394.
18. Harper, P., Clinical consequences of isolating the gene for Huntington's disease, *Br. Med. J.*, 1993, 307:397-398.
19. Healy, B., Remarks for the RAC committee meeting of January 14,1993, regarding compassionate use exemption, *Human Gene Ther.*, 1993, 4:196-7.
20. Jennings, B., Bioethics and democracy, *Centennial Rev.*, 1990, 35(2):207-225.
21. *JAMA*, Special Issue: Caring for the Uninsured and Underinsured, 1993, 265:2437-2624.
22. Kaback, M.M. and Zeiger, R.S., The John Kennedy Institute Tay Sachs Program: practical and ethical issues in an adult genetic screening program, in *Ethical Issues in Genetic Counseling and the Use of Genetic Knowledge*, Condliffe, P., Callanhan, D., and Hilton, B., Eds., New York: Plenum Press, 1972.
23. *Lancet*, Hasty Compassion (Editorial). 1993, 341:663
24. Murray, T., Ethical issues in human genome research, *FASEB*, 1991, 5:55-60.
25. National Society of Genetic Counselors, *Code of Ethics*, Wallingford, PA, 1992.
26. Parker, L., Bioethics for Human Geneticists: Models for Reasoning and Methods for Teaching, *Am. J. Hum. Gen.*, 1994, 54:137-47 [portions of this chapter have appeared previously in this article].
27. Payer, L., *Medicine and Culture*, New York: Henry Holt and Company, 1988.
28. Pincoffs, E., Quandary ethics, *Mind*, 1971;80:552-571.
29. Position Paper on Human Germ Line Manipulation Presented by Council for Responsible Genetics, Human Genetic Committee Fall, 1992, *Human Gene Ther.*, 1993, 4:35-37.
30. *In the Matter of Karen Quinlan*, 70 NJ 10, 1976.
31. Ramsey, P., *The Patient as Person*, New Haven: Yale University Press, 1970.
32. Robinson, A. and Linden, M., *Clinical Genetic Handbook*, 2nd ed., Boston: Blackwell Scientific Publications, 1993, 465-469.
33. Sherwin, S.S., *No Longer Patient: Feminist Ethics and Health Care*, Philadelphia: Temple University Press, 1992.
34. Shore, D., Berg, K., Wynne, O., and Folstein, M.F., Legal and ethical issues in psychiatric genetic research, *Am. J. Med. Gen.*, 1993, 48(1): 17-21.
35. Starr, P., The framework of health care reform, *N. Engl. J. Med.*, 1993, 329:1666-1672.
36. Thompson, L., Harkin seeks compassionate use of unproven treatments, *Science*, 1992, 258:1728.
37. Thompson, L., Healy approves an unproven treatment, *Science*, 1993, 259:172.
38. Tsuang, M.T. and Faraone, S.V., [Editorial] Neuropsychiatric genetics: a new specialty section of the American Journal of Medical Genetics, *Am. J. Med. Gen.*, 1993, 48(1):1-3.
39. Veatch, R.M., Models for ethical medicine in a revolutionary age, *Hastings Center Rep.*, 1972, 2(3): 5-7.
40. Walters, L., Human Gene Therapy: Ethics and Public Policy, *Human Gene Ther.*, 1991, 2:115-122.
41. Wilfond, B.S. and Fost, N., The introduction of cystic fibrosis carrier screening into clinical practice: policy considerations, *Milbank Quarterly*, 1992, 70(4):629-659.
42. Wilfond, B.S. and Nolan, K., National policy development for the clinical application of genetic diagnostic technologies. Lessons from cystic fibrosis, *JAMA*, 1993, 270(24):2948-2954.
43. Wivel, N. and Walters, L., Germ line gene modification and disease prevention: some medical and ethical perspectives, *Science*, 1993, 262:533-538.
44. Yesley, M.S., Bibliography: Ethical, Legal, and Social Implications of the Human Genome Project, U.S. Department of Energy, Washington, D.C., 1992.

Index

INDEX

A

A1 allele. *See Taq I* A1 allele
A2 allele. *See Taq I* A2 allele
Abeorphine, 102
Aβ protein
   Alzheimer's disease and, 221–224
   amino acid sequence of, 221
   apoE as molecular charperone for, 229
   normal production of, 224
Ability to abstain, 333
ABI sequencing system, 52
Absorption, 277
Acetylcholine, 122, 132, 315, 388
ACHC, 174, 185, 186, 388
   in GABA transport counterflow, 188–189, 192
   pharmacology, 187
Achievement, 277, 284
AD. *See* Alzheimer's disease
Addictive disorders, 238
Addictive Personality scores, 338, 339
Additive effect of dopamine genes, 249–250
Adenylyl cyclase, 78, 84, 96
   coupling with serotonin receptors, 116, 123, 132
   effects of adrenergic receptors on, 150, 152
   in inbred mouse strains, 100
   inhibition by serotonin receptors, 119
   regulation, by dopamine, 103–104
ADHD. *See* Attention-deficit/hyperactivity disorder
Adjustment, 277, 284
Admixture, population, 13, 14
Adoption studies, 202
   alcoholism, 314, 367–368
   mood disorders, 262
   obesity, 318
   path analysis of, 275, 278–279
   of personality, 283, 285–287
   substance abuse, 317, 353
α-Adrenergic receptor agonists, 150
β-Adrenergic receptor agonists, 150, 153
α-Adrenergic receptor antagonists, 149, 150
β-Adrenergic receptor antagonists, 122, 149, 150
β-Adrenergic receptor kinase(βARK), 153–154
Adrenergic receptors
   alcoholism, lack of association with, 333
   of the brain, 147–157
   cellular and physiological functions, 155–157
   molecular biology of, 149, 151
   pharmacologic classification, 148–149, 150
   regulation of, 152–154
α-Adrenergic receptors, 148–148
   distribution, 154–155
β-Adrenergic receptors, 104, 120, 148–149, 153–154
   distribution, 155
Adult-onset diseases, 10
Affected individuals, 11
Affected-pedigree-member (APM), 3, 368, 369
Affected samples, 39
Affected-sib-pair (ASP), 3, 368–369
Affective disorders. *See* Mood disorders
Affiliation, 277, 284
Affinity capture, 68, 71
Affinity chromatography, 83
African Americans, frequency of *Taq I* A1 allele, 383
African pygmies, *Taq I* A1 allele in, 414
AGG. *See* Aggressive behavior
Aggression, 120, 122, 277
Aggressive Behavior (AGG), 202
   developmental stability of, 204
   genetics of, 209–210, 211
Agreeableness, 276, 277, 282
   analyses for SATSA and MISTRA, 286, 287
   Loehlin's modeling, 283
   Nichols meta-analysis, 283, 284
β-Alanine, 174
Alcohol
   mixed discriminative stimulus effects of ethanol, 456–458
   preference and aversion, 455–462
   use, commonality with coffee drinking and smoking, 393, 396–401
Alcohol dehydrogenase, 314, 332
Alcoholism, 28, 236, 311, 332. *See also* Ethanol; Serotonin receptors
   absence of linkage in families, 44
   ADHD and, 298
   association with *Taq I* A1 allele, 103, 363, 371–372, 375–385
   association with *Taq I* B1 allele, 336, 375–385
   co-aggregation of disorders with, study design and, 39
   DRD2 gene and, 29–31, 245, 313, 314, 315–317, 332–336, 363, 371–372, 408–413
   gender and, 408
   Gts genes and, 241
   number of alcoholics in U.S., 323
   parental, 338, 343
   pharmacological treatment, 323, 344–345
   P300 wave, 316–317, 343, 344, 408, 426
   severity of, and *Taq I* alleles, 375–385, 410–411
   studies. *See* Adoption studies; Family studies; Twin studies
   types of, 333, 368, 408
Alcohol-preferring rats, 345
Alcohol-seeking behaviors, 333
Aldehyde dehydrogenase, 314, 408
Algorithms to identify coding sequences, 51
Alkaline phosphatase, 180, 182, 183
Alzheimer's disease (AD), 3, 20, 27–28. *See also* Familial Alzheimer's disease
   false positives in linkage analysis, 8–9
   molecular genetics of, 219–230
   preventive ethics, 476, 477
   receptor mutations and, 90, 91
   serotonin receptors and, 125, 132

Amacrine cells, 164
Amalgam, 83
Amber codon, 177, 181
Amber suppression, 177–179, 181
American Indians, 13, 362, 412, 413, 414
Amino acids. *See* Proteins
4-Aminobutyrate. *See* GABA
γ-Aminobutyric acid. *See* GABA
6-Aminocaproate (6-ACA), 186, 187
^{3}H-*para*-Aminoclonidine, 155
cis-4-Aminocrotonic acid (CACA), 189–190, 191
trans-4-Aminocrotonic acid (TACA), 189–190, 191
3-Aminopropanoic acid (3-APA), 186, 187, 190, 192
5-Aminovaleric acid (5-AVA), 184, 185, 186, 187, 190, 192
Amish, bipolar disorder in, 264–267
Amphetamine, 300, 306
Amyloid deposits, 221, 222
Amyloid precursor protein gene (APP), 219, 221–224
β-Amyloid protein. *See* Aβ protein
Angiotensin-converting enzyme (ACE), 27
Anhedonia, 261
Antibodies, in identification of coding sequences, 50, 182
Antibody epitopes, 179
Anticipation, 262
Anticonvulsants, 172
Antidepressants, serotonin receptors and, 118, 119–120, 125, 126, 129, 130
Antiepileptic drugs, 172
Antisocial behavior, 210, 288–289, 317
Anxiety disorders, 39, 90, 238
 childhood, 201
 comorbidity with ADHD, 302
 gender bias in diagnosis, 204
 serotonin receptors and, 119, 121, 126
Anxiolytics
 GABA-mimetics as, 172
 serotonin receptors and, 118, 119, 120, 130
Anxious Depressed syndrome, 208
AP (attention problem), 202, 209–210
APM (affected-pedigree-member), 3, 368, 369
APOE gene, 27, 220, 227–230
Apomorphine, 248, 363
Appetite disturbance, 261
Applied Biosystems, Inc. sequencing system (ABI), 52
βARK, 153–154
Arousal, 156
β-Arrestins, 154, 155
Ascertainment bias, 241, 243, 307
ASP (affected-sib-pair), 3, 368–369
Aspartate, 174, 176
Association analysis, 2, 9, 10, 236, 238, 355
 for ADHD, 300–301
 disadvantages, 28–29
 family-based, 301
 with genes "excluded" by linkage studies, 243–250
 in identification of candidate loci, 20
 independence from mechanism of inheritance, 238
 in meta-analysis, 420
 mutations in receptors and, 90, 92
 population-based, 37, 40–41, 42
 population stratification and, 42–44
 prior probability and, 42, 43
 recent history of, 25–32
 statistical significance and, 15
 vs. linkage analysis, 7–21, 40–41, 243–250, 251, 255, 367–373
Associations
 causes of, 11–13
 indication of gene modifiers by, 40–41
 persistence of, 14
Assortative mating, 45, 288, 371
Assumptions, 10
 genetic homogeneity, 26
 LOD scores, 15
 path analysis, 279–280
 that mutations are in the exons, 236
 twin studies, 279–280
Atenolol, 148
Attentional dysfunction, 298, 299
Attention-deficit disorder, number afflicted in U.S., 323
Attention-deficit hyperactivity disorder (ADHD), 201, 202, 237
 comorbidity of, 299, 302
 diagnostic criteria, 298–300, 307
  DSM-III-R, 301, 302, 303
  DSM-IV, 204, 207, 208
 dopamine transporter and, 299, 301–307
 DRD2 and, 244–245, 320
 genetics of, 209–210, 304
 Gts genes and, 241
 HRAS and, 249
 longitudinal studies, 204
 monoamine oxidase and, 250
 nongenetic etiological factors, 298–299
 polygenic convergence and, 251
 serotonin 1A receptor gene and, 250
Attention disorder, 320–321
Autism, 236, 249
Autoimmune disease, 9, 20
Autoradiography
 adrenergic receptors, 154, 155
 dopamine receptor mRNA, 101
 serotonin receptors, 128
Autoreceptors, 99, 100, 102, 156
 serotonin, 122, 123
 somatodendritic, 118, 119, 120
Autosomal dominant model
 inappropriate use of the, 236–238
 incomplete penetrance model, 239, 240
 three modes of inheritance, 238–240
Azapirone, 118, 119
Azide, 188

B

Backcross, 436
Barbiturates, 172, 456, 458
Batten disease, 17
Battle eagerness, 246
Bayes' Theorem, 324, 424–426
Bead-capture method, 54–55
Behavioral tendencies, 39

Index

behavior genetics *vs.* socialization theory, 290
genetic analysis of kinship data, 280–289
genetics of personality, 273–290
Benzazepine compounds, 96
Benzodiazepine, 90, 172, 456, 458
Betaxolol, 148
Big Five, The, 275, 276, 277
modeling of extended kinship data, 281–283
Swedish Adoption Twin Study of Aging analysis, 285–287
Big Nine, The, 275, 277
Big Three, The, 274, 277, 281, 282
Bilineal pedigrees, 237
Binding affinity, 342
Binding sites, number of, 321, 322, 345
Bioethics, history of, 469, 470, 471. *See also* Ethics
Biogenic amine, 170, 174
Bioinformatics, 448–449
Biological plausibility, 203
Bipolar cells, 164
Bipolar disorder (BP), 2–3
linkage analysis of, 264–267
reproduction of results, 8
symptom descriptions and epidemiology, 261–264
Blockade tests, 457, 458
Blood-brain barrier, 172, 173
Blood group, 1, 26
Blood pressure, 26
"Blunt off," 183
Body mass index (BMI), 340, 415
BP. *See* Bipolar disorder
Brain
adrenergic receptors, distribution, 154–155
cDNA sequencing in the, 51–53, 66
dopamine receptors, distribution, 97–99, 100–101
electrical activities in the, 316–317
genetic expression in the, 50
pleasure mechanisms, 312, 319. *See also* Reward deficiency syndrome
recombinant serotonin receptors, distribution, 133, 134, 135
serotonin receptors, distribution, 118–119, 122, 123, 125, 127, 128, 130, 132
Bromocriptine, 96, 102, 323, 344–345
"Brunner's syndrome," 28
*Bst*XI recognition sequence, 58
Bupropion, 299, 306
Burst firing, 172
Buspirone, 118, 119
BXD set, 439, 442, 443, 444

C

CACA, 189–190, 191
CAD (coronary artery disease), 8, 11, 20
Caenorhabditis elegans, 226
CAIGES, 55
Calcium channels, 130
Calf thymus, 55
California Psychological Inventory (CPI), 274, 276
Cambodians, *Taq I* A1 allele in, 414
cAMP, 96
in adrenergic receptor regulation, 152, 153, 156
in GABA transport regulation, 172
Cancer, 423
Candidate genes, 20, 50–51
ADHD, 301
Alzheimer's disease, 27
polymorphisms near, 28–29
Carbachol, 91–92
Carbohydrates, 165, 319, 341, 342
Carbon, GABA as source of, 175
Cardiomyopathy, hypertrophic, 27
Cardiovascular disease, 8, 11, 20
Case-control design, 11, 13, 19
Cassette mutagenesis, 183
Catalepsy, 120
Catalytic mechanisms, 168, 190
Catecholamines, 141, 388. *See also* Norepinephrine; Norepinephrine system
Caucasians, U.S.
frequency of *Taq I* A1 allele, 44, 378, 383, 414
frequency of *Taq I* B1 allele, 383
CBCL (Child Behavior Checklist), 205, 207
C57BL/6J inbred mouse strain, 100–102
cDNA
clone of amyloid precursor protein, 221
hydropathy analysis of, 179
sequencing of, in the brain, 51–53, 66
cDNA amplification for the identification of genomic expressed sequences (CAIGES), 55
Central nervous system, expression of DRD2 alleles, 342–344
CEPH reference families, 372
Chain elongation, 56
Channel activity in GABA transport, 167–168, 190, 192
Chemical potential, 190
χ^2 analyses, 370
Child Behavior Checklist (CBCL), 205, 207
Childhood psychopathology, phenotypes for molecular genetic studies, 201–213
Child-rearing behavior, 290
Children of alcoholics (COA), 410, 427
Chimeras, genomic-cDNA, 56
Chinese, *Taq I* A1 allele in, 414
Chlorethylclonidine, 149
Chloride and GABA transport, 164, 165, 166, 167, 168–170, 172, 456
Cholesterol production, 319
Cholinergic agonists, 90
Cholinesterase inhibitors, 172
Chromosome 1, familial Alzheimer's disease, 226
Chromosome 2, number of genes in, 52
Chromosome 4, Kit proto-oncogene region YACs, 72
Chromosome 5
α-receptor gene, 14
schizophrenia and, 8
Chromosome 6
MHC region, 72
Quantitative Trait Locus on, 212
Chromosome 7, 54
Chromosome 11
alcoholism and, 41–42, 315
cocaine addiction and, 317

lipidemia and, 8
Novelty Seeking and, 289
Chromosome 13, number of genes in, 52
Chromosome 14, Alzheimer's disease and, 9, 27, 219, 225–226
Chromosome 19
    Alzheimer's disease and, 27, 219, 227–230
    exon trapping and coding sequences of, 58
    number of genes in, 52
Chromosome 20, 151
Chromosome 21
    Alzheimer's disease and, 9, 219
    amyloid precursor protein gene and, 221–224
    expressed region of, 72, 73
    mapping, 59
    trisomy 21, 219
Cigarette smoking. See Smoking
Citalopram, 120
Classical likelihood method. See LOD score
Clomiphene, 248
Clonidine, 148, 150, 156, 157
Cloning
    from cDNA libraries, 51–53
    in direct selection method, 54–55
    familial Alzheimer's disease genes, 225–226
    of GABA transporters, 167, 173–175, 191
    of *gabP*, 175
    of GAT-1, 170–171
    molecular, of dopamine receptors, 96
    of neurotransmitter receptors, 149
    of opioid binding cell adhesion molecules, 83–84
    of opioid receptors, 78–79, 103, 104
    positional, 2, 17, 50, 445
    recombinant serotonin receptors, 133, 134, 135
    of serotonin receptor genes, 120–121, 126, 127, 132
Clozapine, 96
CNBr-cleavage, 174
COA (children of alcoholics), 410, 427
Co-aggregation of disorders within the family, 39
Cocaine, 32, 100, 130, 313, 314
    DRD2 gene and, 336–338
    genetic predisposition to addiction, 317–318
    number of addicts in U.S., 323
    self-administration, 105, 346
Coding sequences
    direct selection of, by hybridization, 53–57
    expressed tags from brain cDNA, 51–53
    as percent of human genome, 49
    sequence accuracy in ESTs, 52
    traditional method of identification, 50–51
Coffee drinking
    commonality with alcohol use and smoking, 393, 396–401
    heritability in twins, 391, 392
COGA (Consortium on the Genetics of Alcoholism study), 371
Cognitive ability, 261
Comorbidity, 243, 299, 302
Complementation, 175
Complex diseases, 2, 20
Complex traits, 8, 435. See also Quantitative trait loci
Compulsive gambling. See Gambling

Compulsivity, 39
Concordance, 367
Conduct disorder (CD), 201, 202
    comorbidity with ADHD, 299, 302, 306
    DRD2 and, 244–245
    DSM-IV criteria for, 207
    genetics of, 209–210
    Gts genes and, 241
    monoamine oxidase and, 250
    serotonin 1A receptor gene and, 250
    twin study correlation, 284, 285
Conformation, 168
Congenic strains, 447–448
Conscientiousness, 276, 277, 289
    analyses for SATSA and MISTRA, 286, 287
    Loehlin's modeling, 283
    Nichols meta-analysis, 283
Consensus genetic control elements, 51
Consortium on the Genetics of Alcoholism study (COGA), 371
Constitutive activity, 91, 92
Control, as personality trait, 277
Control chromosomes, 11
Control samples
    in alcoholism and *Taq I* A1 allele studies, 335, 376–377
    exclusion/inclusion criteria for, 421
    inadequate examination of, 242
    reliability of "normal" diagnosis, 30
    screened *vs.* unscreened, 29–30, 40, 45
    in study design, 37, 38
    "supernormal," 40, 45
"Controls without, cases without, cases with" strategy, 242, 243
Coronary artery disease (CAD), 8, 11, 20
Cortisol, 388
COS-7 cells, 57, 66, 132, 133
Cosegregation, 203, 298, 354
Cosmid vectors, 65, 67
Cost considerations, 9, 10, 20–21
Cot1 DNA, 54
Cot-1 quencher, 70
Cotransport with GABA, 165, 166, 168
Council for Responsible Genetics, 475
Counterflow in GABA transport, 186–190, 192
Counterscreening, 55
CpG islands, 51, 66
CPI (California Psychological Inventory), 274, 276
Crime, 244, 288–289
cRNA, exon trapping and, 57–58
Cross hybridization, 51, 72
Crossovers, 238, 355
Cyanopindolol, 121
Cystic fibrosis, 17, 19, 51, 475

D

Data sets, 9–10
DBA/2J inbred mouse strain, 100, 101
DB (Delinquent behavior), 204
Defense style questionnaire, 417
Deglycosylation, 179

Index

Delinquent behavior (DB), 204
Dementia, 27, 224
Denaturation of DNA, 68
Denaturing gradient gel electrophoresis (DGGE), 423
Deoxythymidine, 65
Deoxyuracil, 65
Department of Veterans Affairs, 389
Dependability, 277, 284
Depression, 90, 237, 261
  childhood, 201
  Gts genes and, 241
  monoamine oxidase and, 250
  serotonin receptors and, 119, 121, 126
Depression scores, 338, 339
Desensitization, 103, 104, 105
  of adrenergic receptors, 153
  of serotonin receptors, 119, 125, 126
Developmental disorders, comorbidity with ADHD, 302
Developmental sensitivity, 203–204
Developmental stages, 204
  gender differences in, 205
  gene expression and, 50
Dextroamphetamine, 299
5,7-DHT, 118
Diabetes, 13, 25, 300
Diacylglycerol, 151
Diagnoses, 203
  accuracy of, 206
  of ADHD, 298–300, 301, 302, 307
  categorical approach *vs.* quantitative differentiation, 207
  considerations with association studies, 242
  gender bias in, 204
  polygene model and, 255–256
  prenatal, 256
  reliability of, 30
Diagnostic and Statistical Manual of Mental Disorders. *See* DSM-III-R; DSM-IV
Dialogue, 471, 472
Dibenzyline, 114
^{3}H-Dihydroalprenolol, 154
3,4-Dihydroxyphenylacetic acid (DOPAC), 460
5,7-Dihydroxytryptamine (5,7-DHT), 118
Dimensional Assessment of Personality Pathology - Basic Questionnaire (DAPP-BQ), 284
Dimers, dopamine receptor, 104
Direct selection of coding sequences, 31
  matrix-based methods, 53–55
  solution-based methods, 55–57
Disclosure, 471, 472, 473
Discrimination
  drug, 455–462
  social and employment, 474, 475
Disease-associated haplotypes, 19
Disease mapping, 1, 2, 17–18. *See also* Genetic mapping
Disease models, 16, 236–238
Disequilibrium. *See* Linkage disequilibrium
Dissociated cell preparation, 164
Distractibility, 301
DNA amplification, 54–55

DNA analysis
  identification of microsatellite repeat markers, 63–73, 436
  isolation of coding sequences from human genome, 49–59
cDNA libraries, 50
  dT-primed, 67, 73
  "equalized," 53
  expressed sequence tags from, 51–53
  input for selection, 68–69
  microsatellite-enriched, 68
  random sequencing of, 66
  screening, 67, 372–373
  strategy to increase CA repeats, 65, 70–71
cDNA selection, 69, 72–73
Doctor-patient relationship, 469–470
DOPA, 100
DOPAC, 460
Dopamine
  in polygenic spectrum model, 252
  role of muscarinic receptor genes in release, 89–93
Dopamine-β-hydroxylase, 170, 249, 321
D2 Dopamine receptor agonists in substance abuse treatment, 323, 324, 344–345, 346
Dopamine receptor antagonists, 102
Dopamine receptor dimers, 104
Dopamine receptor genes, 103, 106, 249–250
Dopamine receptors. *See also individual dopamine receptor types*
  biochemical classification of, 96, 97
  desensitization/resensitization and, 103–104, 105
  distribution, 97–99, 104, 106
  internalization, 104, 105
  pharmacological classification, 96, 97
  role in drug addiction, 95–106
  structural classification of, 96–97, 98
  in vitro regulation, 103–106
D1 Dopamine receptors, 95, 96
  distribution of, 97–98, 104, 106
  lack of association with alcoholism, 333
  proposed topology, 98
D1-like Dopamine receptors, 95, 96
D2 Dopamine receptors, 95, 96, 243–245
  alcoholism, 29–31, 41–45, 103, 244, 315–317, 332–336, 371–372, 408–413
  attention disorder, 244–245, 320
  central nervous system expression of, 342–344
  density of binding sites, 321, 322, 342, 345, 376
  distribution, 97–99, 104, 106
  gambling, 247, 416
  impulse disorders, 244–245
  obesity, 31, 247, 319, 340–342, 415–416
  polymorphism in, 31
  polysubstance abuse, 353–363, 413–415
  the "real" mutation, 253
  reward deficiency syndrome, 312–324
    mechanism, 321–323
    meta-analysis, 407–427
  smoking, 247, 338–340, 417
  structure, 97, 359–360, 423–424
  substance abuse, 103, 244, 317, 336–338, 356–362
  Tourette's disorder, 31, 244–245

D2-like Dopamine receptors, 95, 96
D3 Dopamine receptors, 95, 96
   alcoholism, lack of association with, 333
   association with schizophrenia, 248
   distribution, 99
   excess homozygosity and, 31–32
   origin of, 97
   polymorphism in, 95, 96, 247–248
   the "real" mutation, 253
   substance abuse, 317–318
   Tourette's disorder, 248
D4 Dopamine receptors, 95, 96
   alcoholism, 103, 323
   distribution, 99
   nicotine dependence, 323
   Novelty Seeking, 212, 289, 322
   origin of, 97
   proposed topology, 98
   the "real" mutation, 253
D5 Dopamine receptors, 95, 96
   distribution, 99
D1-D2 Dopamine receptor synergism, 101
Dopaminergic system, 100, 170, 336, 345–346
Dopamine transporter, 176, 249
   ADHD and, 299, 301–307
   GABA and, 170
Dopamine transporter gene, 322
DOR. *See* δ-Opioid receptor
Downregulation
   of adrenergic receptors, 152, 153, 154
   of dopamine receptors, 104, 105
   of proteins, 85
   of serotonin receptors, 119, 125–126
DRD2. *See* D2 Dopamine receptors
DRD3. *See* D3 Dopamine receptors
DRD4. *See* D4 Dopamine receptors
DRD5. *See* D5 Dopamine receptors
D2 Receptor Alleles in Substance Abuse: Have We Identified a Relevant Gene conference, 410
Drug addiction. *See* Substance abuse
Drug discrimination, 455–462
Drug reward. *See* Reward, drug; Reward deficiency syndrome
DSM-III-R
   ADHD, 301, 302, 303
   aggressive behavior, 211
   alcohol dependence, 376
   depressive symptoms, 208
DSM-IV, 39, 202
   aggressive behavior, 211
   in combination with empirically based approaches, 208–209, 211
   empirically based taxonomy and, 206–208
   relating to multi-informant approach, 206
Dyskinesia, 125
Dyslexia, 236, 237

E

Early-onset familial Alzheimer's disease, 225–226, 230
Eating disorders, 318–319. *See also* Obesity
*Eco*RI, 54, 68, 70, 72

"Ecstasy," 462
Effector systems, second messenger, 151–152
Electrochemical gradient, 166, 167
Electrogenic mechanisms, 165
Electrophoresis, 31, 67
Electrophysiological markers for alcoholism, 316–317
Emergenesis, 282
Emotionality, 440
Empirically-based taxonomy, 206–209, 211
Endogenous opioids, 314
β-Endorphin, 77, 84, 388
Energy coupling, 179
Energy-transduction, 168
Enkephalins, 77, 78, 312, 313
Environmental factors
   ADHD, 298–299
   aggressive behavior, 211
   love style, 288
   mood disorders, 263
   risk of severe substance abuse, 347
   role of, 8
   socialization theory, 290
   socioeconomic status, 280
Epidemiological Catchment Area survey, 40
Epidemiology, 38, 220
Epilepsy, 172, 440
Epistasis, 15
Epitope insertions, 179, 180, 181, 183
ERCC1 DNA excision repair gene, 58
ERP (event-related potentials), 408
Erythropoietin gene, 54
Escherichia coli
   C-terminal reporter enzymes, 180
   epitope insertions, 180
   GABA transporter. *See* GabP; *gabP*
   inherent advantages as a model system, 175–184
   metabolism of GABA, 165
   natural selection of point mutants, 175–177
   PacI-mediated gene fusion, 181–184
   site-directed amber suppression in, 177–179, 181
   topology by transposition, 179–180
EST. *See* Expressed sequence tags
Ethanol. *See also* Alcoholism; Alcohol
   drug abuse research on, 100–102
   drug discrimination, 461–462
   mixed discriminative stimulus effects of, 456–458
   selectively bred rats preferring, 459–461
Ethical issues, 469–477
Ethics committees, 473
Ethnicity, 26, 28, 41
   alcoholism and, 43–44, 251–252, 371, 382, 412–413
   Alzheimer's disease and, 219, 226
   false associations due to, 252–253, 356, 361–362
   subpopulations, 12, 28
   Taq I polymorphism and, 361–362
Etiologic heterogeneity, 202–203
Event-related potentials (ERP), 408
Evolution
   evolutionary-conserved subclones as probes, 51
   of GABA transporters, 175, 176
   of G-protein-coupled receptors, 115
   metabolism of GABA, 165

Index

of psychological mechanisms, 275
significance of receptor subtypes, 148
Excitatory amino acid carriers, 174
Exocytosis, 170
Exon amplification, 53, 58–59, 66
Exons, 49, 66
assumption that mutations are in the, 236
in GAT-1, 171
identification of individual, 66
in serotonin receptors, 126
Exon trapping, 57–58, 66–67
Expressed sequence tags (EST), 59
from brain cDNA, 51–53
encoding of familial Alzheimer's disease, 226
Expression systems, 167
Extroversion, 274, 276, 277
analyses for SATSA and MISTRA, 286, 287
The Big Five, 276
Loehlin's modeling, 283
MZT and DZT correlations, 281, 282, 283, 284
Nichols meta-analysis, 283
Eye color, 26
Eysenck Personality Questionnaire (EPQ), 274, 338, 339
Eysenck Public Opinion inventory, 287

F

FAD. See Familial Alzheimer's disease
Fagerström Test for Nicotine Dependence, 340, 402
Fagerström Tolerance Questionnaire, 402
Falk-Rubinstein study, 252, 253
False negatives (Type II errors), 9, 424, 444
diagnostic accuracy and, 206
in exon amplification, 59
False positives (Type I errors), 3, 14, 424, 438, 444
in association studies, 300–301
in CAIGES, 55
diagnostic accuracy and, 206, 213
in direct selection of coding sequences, 54
in exon amplification, 58
in linkage analysis, 8–9, 10, 11, 17, 19
Familial Alzheimer's disease (FAD), 219, 220. See also Alzheimer's disease
and amyloid precursor protein gene mutations, 222–224
early-onset loci, 225–226, 230
late-onset locus, 226, 230
Familial association, 203. See also Family studies
Familial Mediterranean fever, 17
Family studies, 7, 9, 202
ADHD, 204, 298, 301–307
alcoholism, 368
Alzheimer's disease, 220
finding pedigrees for linkage studies, 237
limitations to pedigree approach, 9–10
mood disorders, 262, 263–264
preventive ethics and, 476–477
random selection for, 39
sib-pair approach, 212
substance abuse, 317
Tourette's disorder, 31
Fasciclin II, 83

Fat transport, 319
Fetal tissue transplants, ethical issues, 476
Finnish Twin Registry, 389–390
Finns, alcoholism in, 412, 414
5-HT receptors. See individual serotonin receptor classifications; Serotonin receptors
5-ht receptors. See Serotonin receptors, recombinant
Flashbacks, 246
Folk concepts of psychiatry, 274
Food-reinforcement, 456, 457
Founder effects, 13
Fragile X, 262
Free-energy, 168, 169
Freidrich's ataxia, 17
French, *Taq I* A1 allele in the, 414
Frequency
of disease, 20
of genes in a subpopulation, 10, 12, 28, 32
of genotype in population, study design and, 40
Functional GI symptoms, 241

G

GABA, 102, 125, 131
alcoholism and, 314, 456, 458
biochemical specialization, 165
clinical relevance of, 172–173
historical perspective, 164–165
pharmacology, 187
prokaryotic, 192
release of, 170
reward and, 312, 313
uptake of, 164–165, 166–167
GABA carrier, 168
GABA metabolism, 165
GABA-mimetic compounds, 172, 184, 191
GABA receptor agonists, 189, 190
GABA receptors
bacterial models for, 163–192
lack of association with alcoholism, 333
TACA and CACA activity, 189
GABA transport
bacterial models for, 163–192
counterflow in, 186–190
GabP-mediated, 186–188
inhibition of, 172, 185–188
pharmacology of inhibitors, 187, 192
summary of potencies and counterflow, 189
kinetics and ion-dependence, 165–167, 169
rate, 168–170
regulation of, 171–172
stoichiometry of, 165–167
GABA transporters
cloning of, 173–175
mutagenesis of
natural selection, 175–177
site-directed amber suppression, 177–179, 181
GabP, 165, 176, 180, 190
-mediated counterflow of GABA, 186–188
model for structure of, 192
pharmacology, 184–186
substrate specificity, 186–190

gabP, 166, 175, 177–178
β-Galactosidase, 180, 182, 183
Gambling, 247, 319
   number of compulsive gamblers in U.S., 323
   Taq I A1 allele and, 319, 416
Ganglion cells, 164
Gastrointestinal symptoms, 241
GAT-1, 166, 167
   cloning of, 170–171, 174
   pharmacological profile, 173
   rat, 174
   relationship to *gabP*, 175
   topology, 171
Gating, 192
Gelernter reappraisal, 418–422
Gel filtration, 83
GenBank database, 51, 226
Gender
   alcoholism and, 408
   nicotine dependence and, 402–403
Gender sensitivity, 204–205
Gene expression
   from brain cDNA, 51–53
   of DRD2 gene, association with additional variants, 412
   of GABA transporters, 173
   overexpression, 175, 181, 191
   of proteins, 49–50, 52
Gene flow, 13
Gene frequency, 10
Gene fusion, 180, 181–184
Gene homology, 437
GeneId, 51
Gene mapping, 10, 11, 19, 20. *See also* Candidate genes; Quantitative trait loci
   Alzheimer's disease, 27–28
   candidate loci, 20
   historical overview, 1–2
   Huntington's disease, 17, 18
   in identification of disease markers, 59, 64
   identification of transcribed sequences, 63–73
   sizes of regions around gene/disease, 17
   statistical significance and, 15–16
GeneModeler, 51
Gene Parser, 51
Generalized resistance to thyroid hormone (GRTH), 301
Generalized transmission disequilibrium test (GTDT), 370
Gene therapy
   germ line, 475
   preventive ethics, 474–475
   somatic, 474
Genetic analysis, components, 9–11
Genetic counseling, 470–471
Genetic drift, 13
Genetic homogeneity
   assumption of, 26
   Hardy-Weinberg squares law and, 32
   in study design, 38–39
Genetic loading
   clinical genetics, 240–241
   molecular genetics, 241–242

Genetic mapping
   chromosome 14 and Alzheimer's disease locus, 225
   insert-resistance maps, 180
   for personality and intelligence, 289
   quantitative trait loci mapping, 435–450
Genetic markers, 1–2, 354–355. *See also* Restriction fragment length polymorphisms
   apparent association to disease, 13
   frequency of, and LOD score, 15
   phenotype and locus of, 11
   polymorphisms near *vs.* within, 28–29
   spacing of, 19, 20, 40, 65
   unlinked, 16–17, 18
   YAC-specific, 71–72
Genetic nosologic approach, 202–206
Genetic predisposition, 262, 307, 353
   to alcoholism, 314, 316
   to cocaine, 317–318
Genetic research
   ethical issues, 472–474, 477
   financial issues, 473
Genetic screening, 475
Genetic susceptibility to ethanol abuse, 102
Genetic testing, 93, 323
Genetic variance, nonadditive, 282–283
Genotyping methods, 303–304, 441
Gepirone, 118, 119
Germans
   alcoholism in, 412, 414
   familial Alzheimer's disease gene in, 219, 226
Germ line gene therapy, 475
Glial cells, 172, 174
Glucagon-like peptide (GLP-1), 318
Glutamate
   GABA transport and, 170, 174, 176
   modulation by 5-HT, 122
Glutamate receptors, 104
Glycerol kinase gene, 55
Glycosylation, 98, 170–171, 174
Gordis, Enoch, 409
G-protein-coupled receptors, 95, 96, 97
   activation of, 104, 105
   development of subtypes, 147, 149
   evolution of, 115
   schematic diagram, 115
   second messenger effector systems and, 151–152, 153, 154
   structural features of, 147–148
Gq family, 151
GRAIL, 51, 66, 72
Greeks, *Taq I* A1 allele in, 414
Green Fluorescent Protein, 183
G6PD gene, 17, 55
GTDT (generalized transmission disequilibrium test), 370
Gts genes, 239, 241
Guvacine, 173, 185, 190, 191, 192

H

Hallucinogens, 124, 125, 126
   LSD, 114–115, 124, 125, 128
Haloperidol, 102, 114, 346

Index

Haplotype relative risk (HRR), 28
  association studies, 300, 301, 305, 370–371
  elimination of false association due to ethnicity, 32, 252–253, 356
Haplotypic background, 18, 19, 417
Hardy-Weinberg equilibrium, 13, 372
Hardy-Weinberg squares law, 32
Harm aviodance, 39
^{125}I-HEAT (BE2254), 154
α-Helices, 166, 171
Hematochromatosis, 17
Hereditary cerebral hemorrhage with amyloidosis of the Dutch type (HCHWA- D), 222, 224
Heritability, 203, 279, 288
Heroin, 77
Heterocycles
  conformationally constrained, 184–185
  constrained nonplanar, 185–186
  constrained planar, 186
  planar, 185
Heterogeneity
  etiologic, 202–203
  phenotypic, 209–210, 263
Heteroreceptors, 122
Hexahistidine, 183
High-throughput screening, 89, 92
Hind III, 72
Hippocampus, cDNA library of the, 51
Hispanics, 362, 414
Historical background
  GABA, 164–165
  lysergic acid diethylamide, 114–115
  molecular genetics of Alzheimer's disease, 220
  psychiatric genetics, 1–4
HLA associations, 2, 9, 26
Homogeneity. *See* Genetic homogeneity
Homology
  among transporters, 176, 179
  gene, 437
  linkage, 437
  trait, 437
Homosexuality, 236
Homovanillic acid, 460
Homozygosity, excess, 31–32
Horizontal cells, 164, 170
HRAS gene, 249, 253–254, 423
HRR. *See* Haplotype relative risk
HTF islands, 64
Human genome
  isolation of coding sequences from, 49–59
  mapping. *See* Gene mapping; Human Genome Project
  number of base pairs, 49
  number of genes expressed in the brain, 50
Human Genome Project, 64, 470, 471, 475
Human immunodeficiency virus, 57
Human-rodent somatic cell lines, 52
Huntington's disease (HD), 1, 17, 18, 237
  anticipation, 262
  code of conduct for researchers, 472–474
  distribution of serotonin receptors, 123
  exon amplification and, 59

Hybridization
  cross-species, 51
  direct selection of coding sequences by, 53–57
  polymerase chain reaction-based method, 67
  selection for YAC-specific CA repeats, 71–72
  subtractive strategy, 85
  in vitro methods, 67
Hybridization screening, of λ phage library, 174
Hybridization selection method, 68–70
Hydropathy analyses, 79, 85, 174, 179
Hydropathy plots, 166
cis-4-Hydroxy-nipecotic acid, 173
5-Hydroxy-tryptamine. *See* Serotonin
Hyperactivity, 300, 301, 303
Hyperreactivity, 246
Hypersomnia, 261
Hypomania, 261
Hypomethylated CpG islands, 51, 66

I

ICI-118,551, 148
Identical by descent (IBD), 3
Identical by state (IBS), 3
Ig heavy-chain, 17
IMAGE consortium, 53
Imidazoline receptors, 157
Imprinting, 255
Impulsivity, 39, 120, 122, 300, 301
Inability to abstain, 333
Inbred laboratory animal strains, 100–103, 436, 442–444, 459–461
Independent pathway model, 396
Individualism, 277, 284
Informatics, 448–449
Informed consent, 470, 471, 472, 473, 477
Inheritance. *See* Mode of inheritance
Inheritance models, 238–240
Inhibitory postsynaptic potential (IPSP), 164
Injury of neuronal response proteins, 229–230
Insert-resistance maps, 180
In situ hybridization, 101, 154, 155
Insomnia, 261
Institute of Genomic Research, 52
Insulin, 318
Intellectance, 277, 284
Intelligence, 280, 288, 289, 302
Internalization, 104, 105
Internal review boards, 477
Internet, bioinformatics on, 448
Interspecific extrapolation, 26
Introns, 49, 58, 150
  allowing for splicing, 149
  in GAT-1, 171
In vitro hybridization methods, 67, 85
In vitro translation, 174
In vivo transcription methods, 57–58, 66–67
Ipsapirone, 118, 119
IPSP (inhibitory postsynaptic potential), 164
IQ (intelligence quotient), 280, 288, 289
Iron oxide-coated linkers, 55
Isoguvacine, 190, 191

Isomerization, 167
Isonipecotic acid, 190
Isoproterenol, 148, 150
Isotope exchange, 188
Italians, *Taq I* A1 allele in, 414

J

Japanese, *Taq I* alleles in alcoholics, 375–385, 413, 414
Joint transmission, 10, 11
Jung, Carl, 274

K

Karolinska Institute, 389
Ketanserin, 124, 127, 128
Ketocyclazocine, 78
Kidneys, cDNA from, 54
Kinship data, 280–289
Kit region, 72
"Knock-out," 170, 178
Kojic amine, 189, 190, 191
KOR. *See* κ-Opioid receptor

L

lac genes, 180, 181
lac permease, 179
Lambda vectors, 67, 70
Late-onset diseases, 10, 226, 230
LDL receptors, 229
"Leak current," 169
Learning disabilities, 209, 249, 303
 ADHD and, 298
 monoamine oxidase and, 250
Lectin chromatography, 83
Leucine zipper, 121, 126
Libido, 261
Ligand-gated ion channel, 116, 130
Ligand recognition, 165, 176, 184, 190
Linkage analysis, 2, 10–11, 368, 411, 438
 bipolar affective disorder, 2
 for DRD2 gene in alcoholism, 336
 familial Alzheimer's disease, 225
 in identification of coding sequences, 50
 limitations of, 238–240
 manic-depressive disorder, 264–267
 methodology, 3
 mutations in receptors and, 90–91, 92
 posterior probability of linkage, 14–15, 19
 prior probability of linkage, 14, 19
 schizophrenia, 2–3
 in search of "drug addiction" gene, 354–355
 trade-offs, 37
 vs. association, 7–21, 40–41, 243–250, 367–373
 vs. polygene inheritance model, 236–238, 251, 254–255, 411
Linkage disequilibrium, 10, 11, 238, 300
 assessment in family-based association, 305, 355–356
 causes of, 11–13
 decay of, 14
 at the DRD2 locus, 360–361, 375–376, 377
 extent of, 16, 17–18
 in human populations, 16–19
 persistence of, 14
 population stratification and, 12
 statistical significance and, 14–16, 369
Linkage disequilibrium coefficient, 12, 14
Linkage homology, 437
LIPED, 1
Lipid bilayer, 190
Lipidemia, 8
Liposomes, 165
LISREL program, 396
Liver, cDNA library, 55
LOD score, 3, 10, 368
 advantages and disadvantages, 369
 bipolar disorder locus, 264, 266
 inherent assumptions of, 15
 phenotype definition and, 3
 QTL mapping, 438
 statistical significance of, 14–16
Loehlin's modeling of kinship data, 281–283
Love styles, 288
LSXSS set, 439, 442, 443, 444
Lysergic acid diethylamide (LSD)
 historical background, 114–115
 serotonin receptors and, 124, 125, 128

M

MAG (myelin-associated glycoprotein), 83
Major-gene, 15
Mania, 250, 261
Manic-depressive disorder, 236, 237, 261–267
MAO (monoamine oxidase), 28, 250, 332
MAPMAKER, 440, 445
Map Manager, 445
Markers. *See* Genetic markers
Master Index File, 389
MAST (Michigan Alcohol Screening Test), 29, 324
Maternal transmission, 267
Mating. *See also* Population stratification
 assortative, 45, 288, 371
 random, 12, 13, 14, 32
Matrix-based methods of direct selection, 53–55
Maximum likelihood, 9, 10, 11
Mbo I, 68, 70
MBTI (Myers-Briggs Type Indicator), 274
MDMA, 462
Meiosis, 10
Membrane proteins, 180, 182, 190
Membrane vesicles, 166
Mendelian genetics, 2, 38
Menke's disease, 59
Mental disorders
 late-onset, 226, 230
 rare, 10
 social stigma and nonparticipation in studies, 9–10
Mental retardation, 250
Mesocorticolimbic dopaminergic pathways, 346
Mesulergine, 122, 128
Meta-analyses, 335, 358, 407–427

Methiothepin, 118
3,4-Methylenedioxymethamphetamine, 462
Methylphenidate, 299, 306
Methysergide, 345, 346
Mianserin, 121
Michigan Alcohol Screening Test (MAST), 29, 324
Microiontophoresis, 164
Microsatellite markers, identification of, 63–73, 436
Migraine headaches, 122, 123, 126, 127, 128–129, 131
Migration
    Hardy-Weinberg squares law and, 32
    of a population, 13
Minnesota Multiphasic Personality Inventory, 274
Minnesota Study of Twins Reared Apart (MISTRA), 274, 282, 288
    Big Five analyses, 285–287
Missing data, 10
Modeling
    disease, 16, 236–238
    Loehrin's, 281–283
    misspecification of model, 15–16
    mode of inheritance, 238–240
Mode of inheritance, 8, 9, 10
    autosomal dominance model, 238–240
    bipolar disorder, 264
    independence from mechanism in association studies, 238
    linkage analysis vs. polygene inheritance, 236–238
    single gene vs. polygene, 251
    Tourette's disorder, 238–240
Modifier genes, 40–41
Modulation of synaptic responses, 156
Molecular recognition, 184
Monoamine oxidase (MAO), 28, 250, 332
Monoamine oxidase (MAO) inhibitors, 120
Monogenic disorders, 20
Mood disorders, 238
    bipolar disorder, 3, 8, 261–267
    gene localization of, 2–3
    HLA locus and, 2
    linkage analysis of, 255–261
    unipolar, 255–257
MOR. See μ-Opioid receptor
Morphine, 77, 78, 100
    dopaminergic pathway and, 314
    and serotonin activity, 114
    voluntary drinking of, in rats, 440
Mortality, 10, 387
Mouse
    crosses widely used in the, 436
    recombinant inbred strains used, 442–444
MPQ. See Multidimensional Personality Questionnaire
mRNA
    adrenergic receptors and, 152, 154, 155
    in cloning of GABA transporters, 173
    of dopamine receptor, 100, 101
    noncoding sequences transcribed into, 49
    recombinant serotonin receptors, 133, 134, 135
    of serotonin receptors, 117, 120, 121, 124, 126, 127, 128, 131
Multidimensional Personality Questionnaire (MPQ), 274, 275, 276, 277, 288

Multifactorial diseases (complex), 2, 20
Multi-informant sensitivity, 205–206
Muscarinic receptor gene, 89–93
M1 Muscarinic receptors, 104
M2 Muscarinic receptors, 104
M5 Muscarinic receptors, 89, 91, 92
Muscimol, 185, 186
    pharmacology, 187, 191
    transport, 188
Muscular dystrophy, Duchenne, 51
Mutagenesis
    cassette, 183
    natural selection of point mutants, 175–177
    of serotonin receptor genes, 121, 126
    site-directed, 171, 177–179
Mutations, 13
    in exons vs. introns, 251
    independent, 20
    linkage disequilibrium, 238
    in neurotransmitter receptor genes, 89
    in nonexonic DNA, 254
Myelin-associated glycoprotein (MAG), 83
Myers-Briggs Type Indicator (MBTI), 274
Myotonic dystrophy, 17, 237, 262

N

Naloxone, 345
National Academy of Sciences, 388
    twin study, 391, 392, 397, 400, 401, 403
National Council on Alcoholism (NCA), 376
National Heart, Lung, and Blood Institute (NHLBI) Twin Study, 388, 390, 393, 398, 403
National Institute of Alcohol Abuse and Alcoholism (NIAAA), 409
National Research Council, 388
National Society of Genetic Counselors, Code of Ethics, 470, 471
Native Americans, 13, 362, 412, 413, 414
Nature vs. nurture, 290
Nausea, 131
N-CAM, 83
Negative results, 16
NEO-PI-R facet scales, 275, 276
Neural cell adhesion modecule (N-CAM), 83
Neuraminidase, 170
Neuroamines, 313
Neurofibromatosis, 59, 237
Neuroglian, 83
Neuroleptics in serotonin activity, 114, 120, 124, 125
Neuronal injury-response proteins, 229–230
Neurophysiological studies, 342
Neuropsychological studies, 343–344
Neuroticism, 274, 276, 277
    analyses for SATSA and MISTRA, 286, 287
    Loehlin's modeling, 283
    MZT and DZT correlations, 281, 282
    Nichols meta-analysis, 283
Neurotransmission, noradrenergic, 155
Neurotransmitter receptors, 89, 147–157
Neurotransmitters, 147. See also Dopamine receptors; Muscarinic receptor gene; Serotonin receptors

inhibitory, 164. *See also* GABA
  in reward mechanism, 312–313
Neurotransmitter transporter superfamily (NTT), 174
NG108-15 cells, 78, 84, 85
Nichol's meta-analysis, 283–284
Nicotine, 313, 318, 338–340. *See also* Smoking
  discriminative stimulus effects of, 462
  neurochemical actions of, 388
  number of addicts in U.S., 323
Nicotine dependence scores, 338, 340
Nigrostriatal/mesolimbic dopaminergic activity, 100
Nipecotic acid, 170, 173, 184–185, 186
  in GABA transport counterflow, 188–189, 190
Nitrogen, 165, 175
NMDA receptor, 456, 458
NO-711, 173
Nonadditive genetic variance, 282–283
Noncoding sequences, 49, 58
Nonparametric approach, 3, 9, 16, 40
Nonpaternity, 9
Nonpenetrance, 239
Nonvesicular release, 170
Noradrenalin, 102
Norepinephrine (NE)
  effect on adrenergic receptors, 155–156
  in reward mechanism, 313
  Tourette's disorder, 321
Norepinephrine system, 313
Northern blot analysis, 57, 134
Nosologic approach, 202–206
Novelty Seeking, 212, 289, 322
NTT superfamily, 174
Nucleotide repeating sequences. *See also* Coding sequences
  dinucleotide, 1, 2, 64, 254
  direct sequencing, 31
  identification of, in cloned genomic DNA, 63–73
  methodology to block, 54
  microsatellite markers, 63–73, 436
  in polygenes, 254
  tetranucleotide, 2, 64
  trinucleotide, 64, 254, 262
Null hypothesis, 16

O

OBCAM, 83–84
Obesity, 311
  DRD2 alleles and, 31, 247, 340–342
  drug dependence and, 319
  number overweight in U.S., 323
  onset of, 341
  parental history of, 341
  Taq I A1 allele, 319, 415–416
Obsessive-compulsive disorder, 236, 237, 244
Ochre stop codons, 181
8-OH-DPAT, 118, 121, 122, 127
Olds, James, 312
Oligonucleotide linkers, 54
One-step mapping, 439–441
Open-chain amino acids, 184
Openness, 276, 277, 282
  analyses for SATSA and MISTRA, 286, 287

Loehlin's modeling, 283
Nichols meta-analysis, 283
Opioid binding cell adhesion molecules (OBCAM), 83–84
Opioid peptides, 312, 313
Opioid receptor-associated proteins, 83–85
δ-Opioid receptor (DOR), 78–80, 85
δ-Opioid receptor (DOR) gene, 82, 84
κ-Opioid receptor (KOR), 78, 80, 81, 85
κ-Opioid receptor (KOR) gene, 82, 84
μ-Opioid receptor (MOR), 78–80, 85
μ-Opioid receptor (MOR) gene, 81–82, 84
Opioid receptors, 77–78
  cloning of, 78–79
  distribution of, 79–81
Oppositional defiant disorder (ODD), 201, 203
  comorbidity with ADHD, 299, 302, 306
  dopamine β hydroxylase and, 249
  DSM-IV criteria for, 210
  genetics of, 210
  Gts genes and, 241
  serotonin 1A receptor gene and, 250
Orphan receptors, 78
Orphan receptor X (X-OR), 81, 83–85
Orphan transporters, 174
Overexpression, 175, 181, 191

P

PacI-mediated gene fusion, 181–184
Palmitoylation, 97, 104, 105
Panic disorder, 129, 236, 237
Parametric approach, 3. *See also* LOD score
Paranoia, 26
Parental Bonding Instrument, 290
Parental input for diagnosis of child, 205
Parenting, 290
Parkinson's disease, 123, 476
Partial reactions, 167
Path analysis, 275, 278–280
Patient as Person, The, 469
PCR. *See* Polymerase chain reaction; Polymorphic chain reaction
PDI (Personnel Decisions Incorporated), 275
Pedigrees, 9–10, 16, 237, 238
Pemoline, 299
Penetrance functions, 10, 237
  incomplete, 239, 240
  pedigrees mimicking reduced, 238
Pentylenetetrazol, 172
Personality, 311
  behavior genetic analyses of kinship data, 280–289
  behavior genetics *vs.* socialization theory, 290
  crime, 288–289
  disorders, 284
  genetics of, 273–290
  IQ and, 280
  love styles, 288
  mapping genes for, 289
  psychometric approaches, 274–275
  quantitative genetic methods, 275, 278–280
  role of intelligence, 289
  social attitudes, 287–288

Personnel Decisions Incorporated group (PDI), 275
PET studies, 210
p53 gene, 58
Phage vectors, 65, 174, 182
Pharmacological profile of GAT-1, 173
Pharmacological studies, dopamine receptors and ethanol, 102
Pharmacologic classification
  adrenergic receptors, 148–149, 150
  dopamine receptors, 96, 97
  serotonin receptors, 118, 121, 122–123, 124–125, 127, 128, 129–130, 131–132
Phenocopies, 263
Phenotypes
  in cocaine-dependent subjects, 337
  in diagnosis, DSM and empirically-based, 208–209
  of drug addiction, unreliability of, 355
  empirically-based approach to taxonomy, 206–208
  genetic *vs.* environmental factors, 8, 411
  identification of, genetic nosologic approach, 202–206
  LOD score and, 3
  multiple genes influencing, 26
  new molecular and statistical genetic methods, 211–212
  relationship to allele at marker locus, 11
  study design and, 37, 45
  taxonomy for molecular genetic studies, 201–213
  variance of, specific gene *vs.* polygene, 255
Phenotypic analysis, 89–93
Phenotypic testing, 91–92
Phenylephrine, 148, 150
Phenylketonuria, 19
Phenylpiperazine, 118
phoA, 180
Phobias, 237
Phosphate, 175
Phosphoinositides, 125
Phosphoinositol, 151
Phospholipase A, 152
Phospholipase C
  activation of, 151, 152
  serotonin and, 116, 125, 127, 128
Phospholipase D, 152
Phosphorylation, 97, 98, 104, 105
  in GABA transporter cloning, 174
  GABA transport regulation and, 171–172
  in second messenger effector systems, 151, 152–153, 154
Photoreceptors, 164
Phylogenetic divergence, 175
Physician-patient relationship, 469–470
Pimozide, 346
Pindolol, 118, 120, 155
Pirenperone, 128
pKK3535 quencher, 70
pL1a quencher, 70
Plaques, 221
Plasmids, genomic DNA from, 54, 56
Plausibility, biological, 203
Pleasure centers in the brain, 312, 319
Poisson binomial distribution, 370
Polygenic cloud, 251

Polygenic convergence, 251
Polygenic diseases (complex), 2, 20
Polygenic inheritance
  impact of single genes in, 354
  in psychiatric disorders, 235–256
  tobacco, alcohol, and coffee consumption, 400
Polygenic model, 239, 240, 243
  of alcoholism, 422
  of behavioral disorders, 250–251
Polymerase chain reaction (PCR)
  -based hybridization selection method, 67
  cloning of opioid receptors, 78
  gabP synthesis by, 181
  in genotyping methods, 303–304
  identification of serotonin receptors, 123–124
  of *PacI* sites, 183
Polymorphic chain reaction (PCR), 2
  in cDNA sequences, 52
  in DNA amplification, 54, 55–57, 91, 372
Polymorphisms
  amyloid precursor protein gene, 222, 224
  of APOE gene, 227
  in association studies, 32, 90, 92
  of dopamine receptors, 97
  of DRD2 allele, 31, 103, 331–347
    in polysubstance abuse, 353–363
  of DRD3 allele, 31
  near the candidate gene, 28–29
  protein, 1, 27
  repeat, in polytenes, 254
  sequence, 90
  of serotonin receptors, 121, 124
  simple sequence length, 436
  single strand conformation, 121, 446
  Taq I, 359–362
  VNTR, 301
Polysubstance abuse, 337, 353–363, 391, 396–397, 413–415
Polytenes, 236
Population admixture, 13, 14
Population genetics, 38
Population stratification, 12–13, 28, 42–44, 300, 301, 369
Positional cloning, 2, 17, 50, 445
Posttraumatic stress disorder (PTSD)
  DRD2 and, 244, 245–247
  Taq I A1 allele and, 321, 417
Potassium channels, 119, 125, 130, 132
Potassium ions, GABA transport and, 170
Potency, 277, 284
Power. *See* Statistics, power
Prazosin, 148, 149, 150, 151, 156
Predictive value, 424
Prenatal diagnoses, 256
Prenatal effects upon personality, 280
Presenilin, 219, 225
Presymptomatic testing, 472, 473, 474
Presynaptic terminal, 157
Preventive ethics, 471–474
pRibH7 quencher, 70
pRibH15 quencher, 70
Primary neurons, 172

Primers, 56, 66, 71
Primer sequences, 56, 57
Prior probability, 14, 19, 42, 43
Privacy, right to, 474
Probes, 51, 64
Prokaryotic site model, 184, 192
Propranolol, 148, 150
PROP tasting, 28
n-Propyl-nor-apomorphine, 323
Protease, 182
Protein kinase A, 132, 151, 153, 171–172
Protein kinase C
    alcoholism, 332
    in GABA transporter cloning, 174
    GABA transport regulation and, 171–172
Proteins
    coding sequences and expression of, 49–50
    neuronal injury-response, 229–230
    number expressed by the brain, 52
    open-chain amino acids, 184
    opioid receptor-associated, 83–85
Proteolysis, 179
^{3}H-Prozosin, 154
pR 5.8 quencher, 70
pR 7.3 quencher, 70
Pseudogenes, 97, 124
pSPL1 splicing vector, 57, 58
pSPL3 splicing vector, 58
Psychoticism, 274, 276, 277, 281, 282, 289
pTAG4, 67
P300 wave, 316–317, 343, 344, 408, 427
PTSD. See Posttraumatic stress disorder
Purification, 165, 183

Q

QTL. See Quantitative trait loci
Qualitative trait, 11
Quantitative approach to diagnoses, 207
Quantitative trait loci (QTL), 26, 202, 212, 346, 360, 435–437
    behavioral traits in the mouse, 435–450
    identification of genes underlying, 445–448
    one-step strategies, 439–441
    relevance to human mapping, 437
    two- or multi-step strategies, 442–444
Quenching agents, 54, 69, 70
Quinpirole, 96, 101, 345

R

Race, 26, 28, 238. See also Ethnicity; Subpopulations
RACE (rapid amplification of cDNA ends), 67
RAC (Recombinant DNA Advisory Committee), 474, 475
Radioligand autoradiography, 98
Radioligand binding assays, 90, 96, 114
    adrenergic receptors, 154, 155
    serotonin receptors and, 123, 124, 132
Random mating, 12, 13, 14, 32
Random sample, 12, 39
Random sequencing of cDNAs, 66

Rapid amplification of cDNA ends (RACE), 67
Rare diseases, 10
Rats, inbred strains of, 459–461
Rauwolscine, 148, 150
Reading disability, 26
Reading frames, 51
Receptor genes
    dopamine, 103, 106
    muscarinic, 89–93
    opioid, 77–86
    phenotypic analysis of, 89–93
    relating to phenotypes, 212
    serotonin, 120–121, 123–124, 126–127, 129
Receptor internalization, 103
Receptor isoforms, 97
Receptor Selection and Amplification Technology (R-SAT), 91
Receptor subtypes, 90
Recombinant DNA Advisory Committee (RAC), 474, 475
Recombinant inbred strains, laboratory animals, 436, 442–444, 459–461
Recombinant receptors. See Cloning, recombinant serotonin receptors; Serotonin receptors, recombinant
Recombination fraction, 14, 369
Reconstitution, 165
Region-specific transcript identification approaches, 66–67
Regression analysis, 242–243
Relative-pair methods, 9
Relative risk (RR) association studies, 369, 370
Reliability of diagnosis, 30
Religiousness, 288
Repeatability of results, 16
Reporter enzymes, 181, 182
    alkaline phosphatase, 180, 182, 183
    β-galactosidase, 180, 182, 183
    C-terminal, 180
Reporter genes, 182, 183
Representativeness, 37, 38–39, 45
Reproduction. See Mating
Research design. See Study design
Response behavior of study animals, 455–456
Restriction enzymes, 56
Restriction fragment length polymorphisms (RFLPs), 1, 63, 354
    alcoholism and, 29, 31, 315, 316, 342, 363, 372, 375
    of dopamine receptor genes, 103, 361
Retina, 164–165
Reverse genetics. See Positional cloning
Reverse transcription and PCR amplification. See RT-PCR
Reward
    cascade theory of, 313–314, 323
    drug, 105, 244, 345, 346–347
Reward deficiency syndrome, 324
    DRD2 receptor in, meta-analysis, 407–427
    neurobiological and genetic aspects, 311–324
    pharmacological treatment, 323, 344–345
RFLPs. See Restriction fragment length polymorphisms
Rhodopsin, 154

Ribosome, 177
Risperidone, 128
RNA
   from GAT-1 clone, 174
   nuclear into mature cytoplasmic, 57–58
   yeast, 54
R-SAT, 91
RT-PCR, 57–58, 67
   in cloning of serotonin receptors, 132
   in detection of opioid receptors expression, 79
Rugged individualism, 277, 284
Russell's Reasons for Smoking Scale, 403

S

Salmon sperm DNA, 55
Sample stratification, 13, 38
Sampling techniques, 12, 37, 39. *See also* Control samples
SATSA (Swedish Adoption Twin Study of Aging), 283, 285–287
Schizophrenia, 25, 236
   DRD2 and, 29, 342
   DRD3 and, 31–32, 248
   gene localization of, 2–3
   HLA locus and, 2
   HRAS and, 249
   muscarinic receptors and, 89, 90, 92
   preventive ethics, 476, 477
   reproduction of results, 8
   serotonin receptors and, 125, 126
Scottish, bipolar disorder in the, 267
Screening
   of control samples, 29–30, 40, 45
   of cDNA libraries, 67
   high-throughput, 89, 92
SDS-PAGE, 184
Secondary structure, 180
Second messenger effector systems, 151–152
α-Secretase pathway, 224
Sedatives, 172
Segregation analyses, 38, 237, 264, 368
Seizures, 172, 440
Selection, 13, 191
Selective genotyping, 441
Selective serotonin reuptake inhibitors (SSRI), 120
Self-administration of drug, 105, 346, 440
Semidominant-semirecessive inheritance model, 239, 240
Sensitivity
   of analysis, 11, 19
   developmental, 203–204
   gender, 204–205
   multi-informant, 205–206
Sequence alignments, 176
Sequence analysis, 372
   identification of receptor mutations, 91, 93
   between species, 254
Sequence-based approaches to gene mapping, 66
Sequence polymorphisms, 90
Sequences. *See* Coding sequences; Nucleotide repeating sequences

Sequence tagged sites (STS), 64
Sequestration of adrenergic receptors, 152, 153, 154
Serotonergic receptors, lack of association with alcoholism, 333
Serotonin, 90, 114. *See also individual serotonin receptor classifications;* Serotonin receptors
   in ethanol-drinking behavior, 100, 102–103, 456–458, 460
   in polygenic spectrum model, 252
   reuptake of, 120
   reward and, 312, 313
   transporter, 168
Serotonin receptor genes, 120–121, 123–124, 126–127, 129
Serotonin receptors
   classification of subtypes, 115–117
   desensitization, 119, 125, 126
   downregulation, 119, 125–126
   human locus, 117
   mRNA distribution, 117
   protein distribution, 117
   recombinant, 115, 117, 133–135
   role in psychiatry, 113–135
Serotonin transporter, 176
Serotonin-1A receptor (5-HT$_{1A}$)
   clinical correlates, 119–120
   distribution and function in brain, 118–119
   molecular biology and genetics, 120–121
   pharmacology, 118
   Tourette's disorder and, 250
Serotonin-1A receptor (5-HT$_{1A}$) antagonists, 118, 120
Serotonin-1B receptor (5-HT$_{1B}$), 104
   distribution and function in brain, 122
   pharmacology, 121
Serotonin-1B receptor (5-HT$_{1B}$) antagonists, 121
Serotonin-1D receptor (5-HT$_{1D}$)
   clinical correlates, 123
   distribution and function in brain, 123
   molecular biology and genetics, 123–124
   pharmacology, 122–123
Serotonin-1D receptor (5-HT$_{1D}$) antagonists, 122–123
Serotonin-2A receptor (5-HT$_{2A}$)
   clinical correlates, 125–126
   distribution and function in brain, 125
   molecular biology and genetics, 126–127
   pharmacology, 124–125
Serotonin-2A receptor (5-HT$_{2A}$) antagonists, 124–125, 126
Serotonin-2B receptor (5-HT$_{2B}$)
   distribution and function in brain, 127
   molecular biology, 127–128
   pharmacology, 127
Serotonin-2B receptor (5-HT$_{2B}$) antagonists, 127
Serotonin-2C receptor (5-HT$_{2C}$)
   clinical correlates, 128–129
   distribution and function in brain, 128
   molecular biology and genetics, 129
   pharmacology, 128
Serotonin-2C receptor (5-HT$_{2C}$) antagonists, 128, 129
Serotonin-3 receptor (5-HT$_3$)
   clinical correlates, 130–131
   distribution and function in brain, 130

molecular biology, 131
   pharmacology, 129–130
Serotonin-3 receptor (5-HT$_3$) antagonists, 130, 131
Serotonin-4 receptor (5-HT$_4$)
   distribution and function in brain, 132
   molecular biology, 132–133
   pharmacology, 131–132
Serotonin-4 receptor (5-HT$_4$) antagonists, 131–132
7q22, 59
Severity of Alcohol Dependence Questionnaire, 334
Sexual arousal, 312
Sexual disorders
   Gts genes and, 241
   serotonin 1A receptor gene and, 250
Shock-avoidance, 456
Short-circuit currents, 168
"Shotgun subcloning," 56
Shuttle vector, 57
Sialic acid, 170
Sib-pair studies, 3, 212, 336, 368, 411
Sickle cell anemia, 475
Signal-to-noise ratio, 38, 67, 156, 435
Significance, statistical, 14–16, 265, 438–439
Simple sequence length polymorphisms (SSLP), 436
Single strand conformation polymorphisms (SSCP), 121, 446
Site-directed amber suppression, 177–179, 181
(+)-SKF-38393, 102
SKF-81297, 102
SKF-89976A, 167
Skinner boxes, 456
Sleep dysfunction, 261
Smoking, 247, 318, 338–340
   annual number of premature deaths in U.S., 387
   commonality with alcohol and coffee consumption, 393, 396–401
   gender and nicotine dependence, 402–403
   number of smokers in U.S., 387
   Taq I A1 allele, 417
   twin studies, 387–404
      cigarette *vs.* cigar or pipe, 392, 394
      discordance with nicotine dependence, 402
      weight gain following cessation of, 401–402
Social attitudes, 287–288, 474
Socialization theory, 290
Socioeconomic status, 280, 288
Sodium
   in GABA release, 170
   in GABA transport, 165, 166, 167, 168–170
Solution-based methods of direct selection, 55–57
Somatic cell gene therapy, 474
Somatodendritic autoreceptors, 118, 119, 120, 157
Southern blot analysis, 55, 71
Specificity, 203, 424
Spectrum disorders, 243
Spinal muscular atrophy, 17
Spinocerebellar ataxias, 262
Spiperone, 96, 118, 121, 124, 346
Spiroperidol, 114
SSCP (single strand conformation polymorphisms), 121, 446

SSRI (selective serotonin reuptake inhibitors), 120
State independence, 203
Statistics, 212
   χ^2 analyses, 370
   guidelines for accepting linkages, 265
   Poisson binomial distribution, 370
   power, 11, 16, 18, 418–419
      increase in, for association studies, 241–242
      sample stratification and, 38–39
   problems in negative studies, 418–422
   for quantitative trait loci, 436–437, 445
   regression analysis, 242–243
   statistical significance, 14–16, 438–439
   tests, 14–16
Stigmatization, 474, 475
Stoichiometric coefficients, 166
Stoichiometry of GABA transport, 165–167
Stop codons, 177, 183
   amber, 177, 181
   ochre, 181
Stratification. *See* Population stratification
Streptavidin-coated magnetic beads, 54, 55
Stress, 156
Stressor, 246
Stress reaction, 277
Structure-function relationships, 177
STS (sequence tagged sites), 64
Study design, 8, 19–21
   association studies and, 25–27
   power, 16
   repeatability, 16
   in search of perfect, 38–41
   statistical significance level, 15
Stuttering, 250
Suarez study, 422
Subpopulations, 12, 28
Substance abuse, 31, 237. *See also* Alcoholism; Smoking
   age of initial exposure, 362
   D2 dopamine receptors, 103, 244, 317–318
      polymorphism in, 331–347
   D3 dopamine receptors, 32, 317
   dopamine receptors and, 95, 99–103
   obesity and, 319
   polygenic convergence and, 251
   polysubstance abuse, 353–363
   posttraumatic stress disorder and, 246
   reward deficiency syndrome, 312–319
   Taq I alleles and, 42, 317–318, 336–338
   treatment, 323–324, 344–345
Substrate recognition, 179
Succinic semialdehyde, 165
Suicidal ideation, 261
Sulfate, 176
Sulpiride, 346
"Supernormal" controls, 40, 45
Suppressor strains, 177
Susceptibility, genetic, 102
Swedish Adoption Twin Study of Aging (SATSA), 283, 285–287
Synapse, 164
Synaptic vesicles, 170

Index

T

TACA, 189–190, 191
Tandospirone, 118, 119
Taq I A1 allele, 300. *See also* D2 Dopamine receptors
    alcoholism, 41–45, 315–316, 333–336, 371–372, 375–385, 409–413
    central nervous system expression of, 342–344
    in children of alcoholics, 410, 427
    ethnicity of, 412–415. *See also individual ethnic groups*
    frequency in African American population, 383
    frequency in Japanese population, 377, 378, 379, 383
    frequency in U.S. Caucasians, 44, 378, 383
    gambling, 319, 416
    and lower density of D2 receptors, 321–322, 346
    obesity, 319, 340–342, 415–416
    polysubstance abuse, 413–415
    posttraumatic stress disorder, 321, 417
    reward deficiency syndrome, 312, 315–316, 317–318
    smoking, 317–318, 338–340, 403, 417
    substance abuse, 103, 318, 336–338, 357–361, 362, 363
    Tourette's disorder, 320–321, 411
    typing of, 377
Taq I A2 allele, 321, 324
    alcoholism and, 315, 316, 317, 376
Taq I alleles, 56, 57, 243. *See also Taq I* A1 allele; *Taq I* A2 allele; *Taq I* B1 allele
Taq I B1 allele
    alcoholism, 336, 375–385, 412
    frequency in Caucasian U.S. population, 383
    frequency in Japanese population, 377, 378, 380, 383
    smoking, 403
    substance abuse, 357, 359, 360, 362, 363, 413–415
    typing of, 377
Taq I C allele
    alcoholism, 412
    substance abuse, 359–360
Target cells for opioid receptors, 78
tat intron, 57, 58
Tay-Sachs disease, 475
TdT (transmission/disequilibrium test), 300, 301, 305
Teacher Report Form (TRF), 205, 207
Tertiary structure, 180
TFMPP, 456, 458
Thermodynamic gradient, 168, 170
Thiomuscimol, 189, 190, 191
THIP, 185, 186, 187
Third International Congress of Psychiatric Genetics, 411
THPO, 173, 174, 185, 186, 187
Thymus, 55
Tissue-specific gene expression, 50, 52
TM. *See* Transmembrane domains
T-maze, 455, 456, 457
Topological analysis, 191
    of *E. coli* GabP, 179
    of human GAT-1, 171
    membrane proteins of *E. coli,* 182
    by transposition, 179–184

Tourette's disorder (TS), 29, 31, 236, 237, 320
    comorbidity with other disorders, 243
    DRD3 and, 31, 32, 248
    genetic loading, 240–241
    HRAS and, 249
    modes of inheritance of, 238–240
    muscarinic receptors and, 89, 90, 92
    polygenic convergence and, 251
    serotonin 1A receptor gene and, 250
    Taq I A1 allele and, 320–321, 411
Trait homology, 437
Transcribed sequences
    background, 65
    identification of, 63–73
    region-specific transcript identification approaches, 66–67
    sequence-based approaches, 66
Transcription Factors Database, 82
Transcription methods
    identification of 3' ends, 66–67
    in vivo, 66–67
Transfection, 167, 373
Transferrin, 314
Transgenic mice, 122, 129
Transition states, 168, 169, 170
Transmembrane domains (TM), 79, 97, 116, 121, 124
Transmembrane isomerization, 167, 168
Transmembrane pH, 166
Transmission/disequilibrium test (TdT), 300, 301, 305
Transplants, 476
Transport inhibition. *See* GABA transport, inhibition of
Transposition, topology by, 179–184
Transposons, 180, 182
Trazadone, 128
TRF (Teacher Report Form), 205, 207
Tricyclic antidepressants, 120
Trifluoromethylphenylpiperazine (TFMPP), 456, 458
Tris-EDTA extraction method, 303–304
Trisomy 21, 219
Tryptophan 2,3-dioxygenase, 248
TS. *See* Tourette's disorder
Tsuang guidelines, 203, 206
Tumor-suppressor genes, 59
Tunicamycin, 170
Turnover number, 168
25-kDa protein, 84–85
Twin Registry, NAS-NRC, 388, 389–390
Twin studies, 7, 202, 367
    aggressive behavior, 211
    alcohol use and abuse, 353, 367, 391
    Alzheimer's disease, 220
    assumptions of, 279–280
    attention problem, 210
    Big Five analyses for MISTRA, 285–287
    coffee drinking, 391, 392
    correlations for Big Three, 281
    correlations for DAPP-BQ, 285
    gender sensitivity and, 204
    Loehlin's modeling, 281–283
    mood disorders, 262–263
    Nichols meta-analysis, 283–284

path analysis, 275, 278–279
role of intelligence, 289
smoking, 318, 387–404
social attitudes, 287–288
using empirically-based taxonomy, 212
Type I errors. *See* False positives (Type I errors)
Type II errors. *See* False negatives (Type II errors)
Tyrosine hydroxylase, 332

U

UK-14, 304, 148, 150
Uncoupling, 152, 167–168
Unilineal pedigrees, 237
Unipolar disorder (UP), 255–257
Unlinked loci, 14, 16
Unrelated individuals, 9, 10, 11, 26
Upregulation of dopamine receptors, 104

V

Valproate, 172
VAPSE (variations affecting protein structure or expression), 354, 360
Variable-number-tandem-repeats (VNTR), 32, 301, 354
Variables, tied *vs.* untied, 38–39
Variations affecting protein structure or expression (VAPSE), 354, 360
Vector-specific primers, 56, 66
Vectrex-avidin matrix column, 68, 70
Vesicular release, 170
Vigilance, 156, 246
γ-Vinyl GABA, 172
Violent crime, 244
Visuospatial ability, 343–344
VNTR (variable-number-tandem-repeats), 32, 301, 354
Voltage-clamp, 167

W

WB-4101, 149
Wechsler Preschool and Primary Scale of Intelligence (WPPSI), 302
Weight (body mass index), 349, 415
Weight gain following cessation of smoking, 401–402
Weight of samples, 38
Wild-type receptor, 91, 92
Wilms' tumor, 51
Wilson Patterson Conservativism scale, 287
Wilson's disease, 17
WPPSI (Wechsler Preschool and Primary Scale of Intelligence), 302

X

Xenopus, 166, 167, 172, 173–174
X-OR (orphan receptor X), 81, 83, 84, 85
Xp11, 59
Xp21, 59
Xq13.3, 59
Xq26, 266, 267
Xq28, 59, 265, 267

Y

YAC contig, 65, 72
YAC DNA, 54, 58, 69–70, 72
YAC-specific markers, 71–72
YAC vectors, 67
Yeast DNA, 54, 71
Yemenite Jews, *Taq I* A1 allele in, 413, 414
Yohimbine, 148, 150, 156
Youth Self-Report Form (YSR), 205, 207

Z

Zinc finger protein, 85